Principles and Practice of

Mathematics

Project Director

Solomon Garfunkel, *COMAP, Inc.*

Coordinating Editor

Walter Meyer, *Adelphi University*

AUTHORS

David C. Arney, *USMA, West Point*

Robert Bumcrot, *Hofstra University*

Paul Campbell, *Beloit College*

Joseph Gallian, *University of Minnesota, Duluth*

Frank Giordano, *USMA, West Point*

Rochelle Wilson Meyer, *Nassau County Community College*

Michael Olinick, *Middlebury College*

Alan Tucker, *SUNY at Stony Brook*

CONTRIBUTORS

Sheldon Gordon, *Suffolk County Community College*

Zaven Karian, *Denison University*

Principles and Practice of Mathematics

COMAP

Springer

Textbooks in Mathematical Sciences

Series Editors:

Thomas F. Banchoff, *Brown University*
Keith Devlin, *St. Mary's College*
Gaston Gonnet, *ETH Zentrum, Zürich*
Jerrold Marsden, *California Institute of Technology*
Stan Wagon, *Macalester College*

PRINCIPLES AND PRACTICE OF MATHEMATICS is produced by the Consortium for Mathematics and its Applications (COMAP) through a grant from the Division of Undergraduate Education of the National Science Foundation, grant number USE9155853, to COMAP, Inc.

PROJECT ADVISORS:

Saul Gass, *University of Maryland*
Andrew Gleason, *Harvard University*
Joseph Malkevitch, *York College, City University of New York*
David Moore, *Purdue University*
Henry Pollak, *Columbia Teachers College*
Paul Sally, *University of Chicago*
Laurie Snell, *Dartmouth College*
Gail Young, *Columbia Teachers College*

Library of Congress Cataloging-in-Publication Data

Principles and practice of mathematics / COMAP, Inc.
     p. cm. — (Textbooks in mathematical sciences)
   Includes bibliographical references (p. –) and index.
   ISBN 0-387-94612-8 (hardcover : alk. paper)
   1. Mathematics. I. Consortium for Mathematics and its Applications (U.S.)
   II. Series.
QA39.2.P698   1996
510—dc20   96-1763
                                         CIP

Printed on acid-free paper.

Production managed by Steven Pisano; manufacturing supervised by Rhea Talbert.
Photocomposed by Impressions Book and Journal Services, Inc., Madison, WI
Printed and bound by Edwards Brothers, Inc., Ann Arbor, MI
Printed in the United States of America.

9 8 7 6 5 4 3 2 1

ISBN 0-387-94612-8 Springer-Verlag New York Berlin Heidelberg   SPIN 10490859

PREFACE

Principles and Practice is a revolutionary text which we wrote in order to change the undergraduate mathematics curriculum. Revolutions are started for two reasons: to overturn the status quo, and to realize a vision of the future. Why do we at COMAP think now is the time to incite change through the preparation of this text which, after all, proposes a radically new introductory course for undergraduate mathematics?

The central importance of mathematics in our technologically complex world is undeniable, and the possibilities for new applications are almost endless. But at the undergraduate level, little of this excitement is being conveyed to our students. Currently, attention is being focused on reforming calculus, the traditional gateway course into the undergraduate curriculum. No one is questioning the importance and beauty of continuous mathematics. However, reformed or not, calculus is one branch (and a highly technical one) of a very rich subject. We know the breadth and richness of our subject; how, then, do we expect the students who are starting their study to gain these insights?

In seeking answers to this question, we identified a model which has been in place throughout college science curricula. Every science department offers an introductory course that focuses on developing basic principles and concepts, and at the same time introduces students to the range of the subject—chemistry, physics, biology, etc. The "101–102" sequence usually serves as a prerequisite for further courses. Our proposed new start or gateway into the college mathematics curriculum is only a revolutionary idea for our discipline; other disciplines have had such courses in place for years.

Our project started more than five years ago, with funding from the Division of Undergraduate Education of the National Science Foundation. We asked ourselves a simple question: in designing the first undergraduate course for math and science majors, what should such a course look like? The contents of *Principles and Practice of Mathematics* represents our most considered answer to this provocative question. The course content stresses the breadth of mathematics, discrete and continuous, probabilistic as well as deterministic, algorithmic and conceptual. We emphasize applications that are both real and immediate. And the text includes topics from modern mathematics that are currently homeless in the undergraduate curriculum.

We should stress that the level of mathematics included here is not trivial. The audience for this text should not be confused with that of terminal courses such as surveys of mathematics for liberal arts students, or finite mathematics. Throughout the writing, class-testing, and revising, we have aimed for a level of presentation equivalent to a conventional first-year calculus course. The typical student for this text will have completed the standard prerequisites for studying calculus.

Even for those students who take just a year of college-level mathematics, we will have achieved something of considerable value. For the year's worth of attention and effort they give to mathematics, they will gain a wider understanding of what mathematics is all about, including some of its most modern ideas and applications. Currently, if a student drops out of mathematics after a year of calculus, he or she has no idea of how mathematics provides a conceptual base for computer science, has only a limited concept of abstraction, and, perhaps most damning as we approach the twenty-first century, has seen little or no mathematics more modern than the eighteenth century. Such a student is unaware of subjects such as graph theory, linear programming, and combinatorial optimization, subjects which are taught to beginning students in other disciplines and which appear from time to time in newspapers and popular science magazines.

The adoption of this text, we realize, might mean redesigning the curriculum—and that will not happen overnight. Because of the variety in kinds of schools and programs of study, we expect many trial sections and experimental courses to be offered. While the book is intended for use over two semesters, it is organized so that chapters and sections can be covered selectively and adapted easily for one-term courses.

This book is a team effort in which authors looked over each other's shoulders during numerous team meetings. However, primary writing responsibilities were: David Arney and Frank Giordano, Chapter 1; Robert Bumcrot, Chapter 2; Alan Tucker, Chapter 3; Rochelle Wilson Meyer, Chapters 4, 6, 7;

Paul Campbell, Chapter 5; Michael Olinick, Chapter 8; Joseph Gallian, Chapter 9. The editor wishes to thank each of these authors for being splendid team players as well as talented expositors.

A revolution is the work of many hands, and a project of this magnitude could not be completed without extensive assistance. In particular we would like to thank the National Science Foundation for its steadfast support under the DUE program.

We were fortunate to have a highly talented group of project advisors: Saul Gass, Andrew Gleason, Zaven Karian, Joseph Malkevitch, David Moore, Henry Pollak, Paul Sally, Laurie Snell, Marcia Sward, and Gail Young.

In addition, special thanks go to: John Burns, for careful reading of early drafts, Sheldon Gordon who contributed much to the early work on Chapter 1, Zaven Karian for technology advice, Harald Ness who wrote many of the problems in Chapter 8, and Yves Nievergelt for contributing spotlights.

We also wish to thank all the participants at the West Point workshops of 1994 and 1995, including especially P. Baker, M. Gallit, A. Lebow, C. Lindsey, J. Orlett, B. Reid, P. Rose, S. Seltzer, F. Serio, M. Vanisko; field testers of the early drafts; those who have read and criticized early drafts, including R. Bradley, J. Buonocristiani, M. Fegan, M. Grady, R. Griego, D. Knee, P. Lindstrom, F. Meyer, T. Walsh, W. Williams, J. Wynn.

Last, but not least, we would like to thank our editors, Jerry Lyons, Liesl Gibson, and Teresa Shields for their commitment to this book.

CONTENTS

4 COMBINATORICS

5 GRAPHS AND ALGORITHMS

6 ANALYSIS OF ALGORITHMS

7 LOGIC AND THE DESIGN OF "INTELLIGENT" DEVICES

8 CHANCE

9 MODERN ALGEBRA

Principles and Practice of

Mathematics

CHAPTER

1

CHANGE

SECTION 1.1 *Introduction*

Change is all around us. Wherever we look, things are changing. We see it in such varied phenomena as

➤ the path of a perfectly thrown pass in football or the regal motion of the planets around the sun,

➤ the growth of a human being from infancy through old age or the growth of the entire human population on Earth,

➤ the temperature change of a cold potato put into a hot oven or the global warming patterns that threaten our very existence,

➤ the growth of money deposited in a bank ac-count or the growth pattern of our balance of trade deficit.

All of these changing phenomena can be investigated mathematically. Many of them can be described very effectively by mathematics, and in such cases we can accurately predict what will happen. For instance, the orbits of the planets are ellipses; knowing that precise mathematical relationship allows scientists to calculate the trajectories of man-made satellites so that they reach the desired target. Similarly, the path of the football is a parabola; the quarterback, like the rocket scientist, wants his "satellite" to arrive at the correct spot at the precise instant that the receiver passes through that point.

Population growth patterns tend to be exponential in nature; if current population trends continue unchanged, the population of Mexico will surpass that of the United States somewhere around the year 2051.

In the present chapter, we will develop some extraordinarily powerful, yet simple, mathematical tools that will enable us to study change in a wide variety of areas. To do so, we must consider some basic ideas. First and foremost, the very fact that a quantity *changes* means that the quantity varies with respect to some other quantity. That is, the quantity of interest to us, say position or temperature or population, depends on some other quantity, say time. Consequently, the quantity of interest will always be a *function* of time or some other *independent variable.* For example, if we roll a bowling ball with a forward velocity of 30 ft/sec, after t seconds the distance (d) is approximately expressed as a function of t by $d = 30t$. An example of a function where the independent variable is not time is the circumference of a circle (C), which is a function of the radius (r) expressed by the formula $C = 2\pi r$.

DEFINITIONS

A *function* is a rule or procedure for producing output values from input values of the independent variable.

The set of the possible values of the independent variable is known as the *domain* of the function.

The possible values for the dependent variable are known as the *range* of the function.

Because the independent variable represents an actual quantity, in the real world, such as time, it naturally is limited in terms of the values we can intelligently use. This set of the possible values of the independent variable is known as the *domain* of the function. Similarly, because the quantity of interest, the *dependent variable,* represents an actual quantity, it is also limited in terms of the values it can assume. These possible values for the dependent variable are known as the *range* of the function.

For instance, if we are considering the population of North America over time, then the domain might be limited to the interval from $-15,000$ (that is, 15,000 B.C., approximately when anthropologists believe the first settlers crossed the land bridge between Siberia and Alaska) to A.D. 2100 (it is extremely difficult to extrapolate very far into the future with any hope of accuracy). The range for population values would then extend from a minimum of zero to a maximum of potentially half a billion (approximately double the present population). Alternatively, if we are interested in the population of the United States, then the domain is limited from 1776 to 2100, say, and the range would be from 2.75 million to about half a billion.

Let's consider how we might represent functions. In previous courses you were probably led to believe that all functions are expressed as an explicit formula of the form $y = f(x)$. While this is true of many situations, we often have to deal with cases where no such formula is known. For example, in a daily lottery, the winning number is a function of the day—in the sense that for each day, there is a definite winning number associated with it—but we have no formula for the lottery to help us get rich.

The study of population presents interesting issues in how we represent functions. Typically, we begin with a table of values, which is a way of presenting a function. For example, the population of the United States in millions from 1780 to 1990 is presented in Table 1. Here the domain is the 22 years that end in 0: 1780, 1790, . . . , 1990.

These population numbers are all approxima-

TABLE 1 The U.S. Population.			
Year	Population (millions)	Year	Population (millions)
1780	2.78	1890	62.95
1790	3.93	1900	75.99
1800	5.31	1910	91.97
1810	7.24	1920	105.71
1820	9.64	1930	122.77
1830	12.87	1940	131.67
1840	17.07	1950	150.70
1850	23.19	1960	179.32
1860	31.44	1970	203.30
1870	39.82	1980	226.55
1880	50.16	1990	248.71

qualitative feel for the trends present in the change we are observing.

Notice in Figure 1 that the graph is reasonably smooth, showing an upward trend. We will see that we can capture the trend with the following formula (or function), which approximately predicts the U.S. population, $p(x)$, in year x:

$$p(x) = (203{,}211{,}926)2^{0.0216(x - 1970)}. \tag{1}$$

For example, for the year $x = 1980$, the formula predicts

$$\begin{aligned} p(1980) &= (203{,}211{,}926)2^{0.0216(1980 - 1970)} \\ &= 236{,}032{,}426. \end{aligned}$$

tions. No census is completely accurate, and the number for 1780 was not obtained from a systematic census anyhow (the first official U.S. census occurred in 1790).

We might choose to represent such a function via a graph, as shown in Figure 1, which rises from 0 to the population in 1990, about 250 million. Notice that although the graph is not as precise as the table of values appears to be, it quickly gives us a

which we can compare to the 226.55 million actually recorded.

We have seen that there are different ways to express a function: as a formula, as a graph, and as a table of observed data. All of these arise in the real world, and we must be able to interpret the behavior of the quantity they represent in each instance.

Equation (1) can clearly have any value of time plugged into it, not just 1780, 1790, 1980, etc. This gives us a way to estimate the population for intermediate times, like 1982, 1943.78, and so on. If we wish to do this, we are saying, in effect, that the domain of the function now consists of all the infinitely many numerical values between 1780 and 1990, i.e., the interval [1780, 1990]. We could even go out on a limb and use the formula to try to predict the population of 2000, by declaring the domain to be [1780, 2000]. If we were to draw the graph of the function with the interval [1780, 2000] as its domain, we get an unbroken curve like that of Figure 2 instead of a series of dots.

We have now seen a number of examples of functions and domains, so let's summarize the important points.

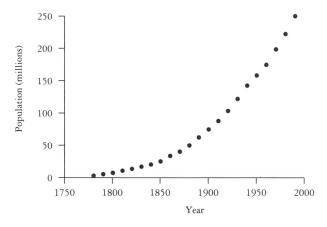

FIGURE 1 U.S. population.

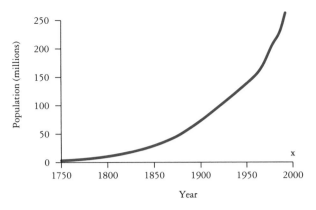

FIGURE 2 A continuous curve for the U.S. population.

TABLE 2 Important Characteristics of a Function.

When is it increasing?
When is it decreasing?
What is its maximum value?
What is its minimum value?
When is the rate of increase increasing?
When is the rate of increase decreasing?
When is the rate of decrease decreasing?
When is the rate of decrease increasing?
When is it increasing or decreasing most rapidly?
What are its roots? (When is its value equal to 0?)
Is it periodic? (Do the function values form a repeating pattern?)

One quantity y, often called the dependent variable, is said to be a function of another quantity x, often called the independent variable, if there is a certain domain of x-values where each x-value has a single, definite y-value associated with it. A set from which these associated y-values come is called the range. The domain and range can be finite or infinite sets. Not all functions can be represented by a formula, although most of the ones of interest to us will be. Tables and graphs are other ways in which functions can be represented.

In particular, when we speak of the behavior of a function or a quantity, there are several specific aspects that are typically important to know, as shown in Table 2.

Difference Equation Models and Their Solutions

In this chapter we will learn about a particular kind of mathematical model called a *difference equation model*. We will learn how to build such a model, find what it means to solve such a model, and see how solutions can be found. The solutions will be functions, represented as formulas, graphs, tables, or sequences of numbers. Finally, we will learn how to analyze our solutions to determine the characteristics outlined in Table 2. Let's preview what is in store for us.

Discrete and Continuous Change

In many cases, the behavior we are observing changes abruptly at instants of time separated by periods when no change occurs (Figure 3). For example, the amount owed on a mortgage or car loan changes when interest is charged and a payment is made. The value of a stock portfolio changes when a dividend is declared or the market value of a share changes at

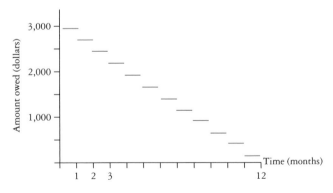

FIGURE 3 Amount of a loan still outstanding at various times.

Population, Misery, and Vice: Is There a Connection?

Sometimes it's fun to be part of a crowd—for example, at this ticker-tape parade on New York's Broadway. But what if there were crowds everywhere, all the time? The most famous western thinker to warn of the perils of too many people was Thomas Malthus, a

country parson with a taste for mathematics. Malthus noticed that many people in 18th century England lived amidst misery and vice. He tried to explain this with a mathematical argument that today we can give in the language of difference equations. Malthus claimed that the food supply grows by a constant rate, so the population that can be sustained does also, as shown by the "sustainable population" curve $S(t)$ in the accompanying figure. When not choked off by excessive death rates due to starvation, disease, and war-

fare, population grows according to a graph like $P(t)$ in the figure. The figure shows a scenario where the population is small in the beginning—many more could be sustained by the available food. However, population grows faster than food supply, and eventu-

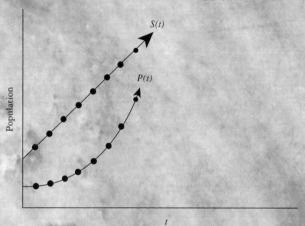

ally, when $P(t)$ approaches $S(t)$, people can barely get enough to stay alive. At this point, deaths due to what Malthus called misery and vice (starvation, infanticide, murder, etc.) rise to stop the population curve from following its exponentially increasing sweep.

People argue about Malthus' ideas to this day. Do you think there are parts of the world where population pressures are causing hardships? Are there places where the population is not a problem? Is the U.S. birthrate too high or too low?

the end of a day. In other cases, the process is taking place continuously. For example, if a warm soda is placed in the refrigerator, the temperature of the soda changes continuously. Likewise, the exact location in space of a planet, space capsule, or weather satellite changes continuously with time. We will first learn how to model and analyze behavior that is changing at discrete instants before modeling and analyzing continuous change.

Models of Discrete Change

A *mathematical model* is a mathematical construct used to capture or approximate the relationship between variables being observed. From one point in time to the next, we might observe a population level, the amount of money in a bank account, or the number of people infected with a contagious disease. For example, suppose you place $1000 in a bank account that pays 2% interest each month. Assuming you do not add or remove any money and a_n represents the amount of money after n months, the following sequence of numbers represents the value of the account:

$$\{a_n\} = \{1000, 1020, 1040.40, 1061.208, \ldots\},$$

$$n = 0, 1, 2, 3, \ldots,$$

where a_0 represents the initial $1000 deposit. Since $a_{n+1} = a_n + 0.02a_n$, the relationship between successive numbers in the sequence can be captured by the following function (model or formula):

$$a_{n+1} = 1.02a_n. \tag{2}$$

Notice that the function in Eq. (2) gives the bank a rule, method, or procedure for calculating the value of the account given the previous value but does not give a specific formula for the value of the account after n months. In the example at hand, Eq. (2) is our model. The model captures the change being observed. We will see that the solution to our model is the function

$$a_n = (1.02)^n(1000), \tag{3}$$

which allows us to compute the value of the account at the end of month n. For example, after 20 months, we would have

$$a_{20} = (1.02)^{20}(1000) = 1485.95,$$

or about $1486. Thus, Eq. (3) provides a solution to the model represented by Eq. (2). Alternatively, we could have built a table of values or a graph to represent our solution function.

D E F I N I T I O N S

A *difference equation* is an equation which shows how any term of a sequence is computed from one or more previous terms.

The *initial condition* is the first term of the sequence.

An equation like Eq. (2), which shows how any term of a sequence is computed from one or more previous terms, is called a *difference equation.* The first term of the sequence, a_0, is called the *initial condition.*

In this chapter we will build various types of difference equation models and see that they lead to remarkably different solution functions. Some solution functions will predict no change, others will predict change that increases or decreases in a linear fashion, and still others will predict change that occurs nonlinearly, perhaps following a polynomial or exponential pattern. Some solution functions will exhibit periodicity, represented perhaps by a trigonometric function. Finally, in Section 1.7 we will see that certain types of *nonlinear* models lead to solutions that have no pattern whatsoever—completely defying our ability to make predictions! We call such behavior *chaotic*. We will see that behaviors modeled by these nonlinear functions are quite sensitive to initial conditions and that the outcomes of the model plotted against the initial conditions often exhibit interesting patterns, which we call *fractal patterns*.

In the process of building and analyzing models, you will learn some simplifying assumptions of classical "laws" of nature underlying biology, ecology, economics, engineering, mechanics, physics, and other areas of human interest. You will learn to look for patterns and conjecture or hypothesize functions that capture the observed patterns. You will learn to fit functions you have selected to the observed data. You will also learn to conjecture or hypothesize solution functions based on the patterns you have observed. Finally, you will see how formal proofs allow us to characterize entire classes of models regardless of the particular application under study—demonstrating the immense power of mathematics to help understand and predict the world in which we live. Your ability to understand and predict change will increase dramatically—exponentially, we hope. So let's get started!

SECTION 1.2 *Sequences and Differences*

The concept of a sequence is central to much of our study of the mathematics of change. Informally, a *sequence* is any ordered set (finite or infinite) of real numbers. That is, there is a first number a_1, a second number a_2, a third number a_3, and so forth. Sometimes we start the subscripts at 0—then the initial term is denoted a_0. These numbers are called the *terms* or *elements* of the sequence. For example, when we use Section 1.1's table of values on the growth of the U.S. population, we have a sequence of values for the population.

We denote the nth term of a sequence by a_n, where the subscript n gives order to the sequence by its place in the ordered set of positive integers.

Therefore, an entire infinite sequence can be denoted by $A = \{a_1, a_2, \ldots, a_n, \ldots\}$.

As examples of sequences, we have

$$A = \{2, 4, 6, 8, 10, \ldots\},$$
$$B = \{2, 4, 8, 16, 32, \ldots\},$$
$$C = \{\tfrac{1}{2}, \tfrac{2}{3}, \tfrac{3}{4}, \tfrac{4}{5}, \ldots\}.$$

In each of these examples, it is possible to determine the patterns from which we can supply meaning to the infinite number of terms represented by the dots (. . .). In particular, this is done by obtaining a formula for the *general term,* or *nth term,* for each sequence. As a useful shorthand notation for the times when we know a formula for the general term, we express the sequence A in terms of the general term, $A = \{a_n\}$. Thus, for each value of $n = 1, 2, 3, \ldots$, we have the following formulas for general terms of the sequences given above:

$$A = \{a_n\} = \{2n\},$$

$$B = \{b_n\} = \{2^n\},$$

$$C = \{c_n\} = \left\{\frac{n}{(n + 1)}\right\}.$$

Here is a more formal definition of *sequence:*

DEFINITION

A *sequence* is a function whose domain is the set of all positive integers (or sometimes, for convenience, all nonnegative integers) and whose range is in the set of all real numbers.

In other words, for each positive integer (from the domain) $n = 1, 2, 3, \ldots$, there corresponds a real number (in the range) a_n as the nth term in the sequence. That is, a_n is a function of n. This is illustrated by the general terms found for the previous examples. For instance, in sequence A, the first term is $a_1 = 2(1) = 2$, the second term is $a_2 = 2(2) = 4$, the tenth term is $a_{10} = 2(10) = 20$, and so forth.

EXAMPLE 1 Graph of a Sequence If a sequence is defined by the general term $a_n = 3n - 5$, then the first six terms of the sequence are $\{-2, 1, 4, 7, 10, 13\}$. Of course, there is an infinite number of terms following these six terms. Moreover, the terms of a sequence can be graphed by plotting points formed by the pair of values (n, a_n). Therefore, these six terms can be used to form the following six points: $(1, -2)$, $(2, 1)$,

$(3, 4)$, $(4, 7)$, $(5, 10)$, and $(6, 13)$. Then these points can be plotted on coordinate axes, n and a_n, as shown in Figure 1. ▲

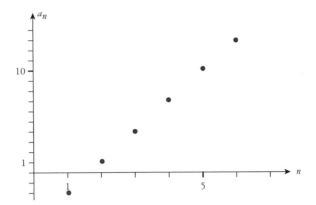

FIGURE 1 Graph of the sequence $\{3n - 5\}$ for $n = 1, 2, \ldots, 6$.

The graph in Figure 1 enables us to visualize how the function changes over the interval of values that are plotted. This visualization is important in analyzing trends and understanding the properties of the sequence. Economists and financial planners often use graphs of sequences, like the prices of stocks and bonds over time, to predict future trends by visualization of past performance data.

Finding the General Term

As we saw earlier, it is often important that we be able to determine the pattern or formula for the general term of a sequence. Knowing the formula helps us to understand how the sequence changes and the trends in its behavior, such as maximum and minimum values and location of intervals where the sequence increases or decreases. We will examine sequence A above to analyze one approach that can be used to find such patterns. In this example we will assume that the term following $a_5 = 10$ should be

12, since each element is 2 more than the preceding one. Alternatively, we can express this fact by noting that

$$a_2 - a_1 = 4 - 2 = 2,$$

$$a_3 - a_2 = 6 - 4 = 2,$$

$$a_4 - a_3 = 8 - 6 = 2,$$

and so on. Thus, if the general term is a_n, then the difference between it and the following term, a_{n+1}, is given by

$$a_{n+1} - a_n = 2, \tag{1}$$

for any value of n. Thus, the difference between any two successive terms of this sequence will equal 2.

In formulating Eq. (1) and asserting that it continues to hold for all positive integer values of n, we are going out on a limb. To illustrate this, let's take a simple example where we have just three terms of that sequence known to us, 2, 4, 6, and we want to predict the fourth term. These three terms are fit not only by the equation $a_{n+1} - a_n = 2$, but also by

$$a_{n+1} = a_n^2 - 5a_n + 10 \tag{2}$$

(try substituting $n = 1, 2, 3$ to verify these patterns). Using Eq. (1), we predict the next term is $a_4 = 8$, but using Eq. (2), we get $a_4 = 16$. There is no simple answer to what the correct equation is. Either could be correct, or even a different pattern or equation may be correct. Mathematicians usually leave finding the right pattern to the scientists observing the behavior. The job of mathematicians is mostly to draw conclusions about the equations that fit the patterns determined by the scientists.

E X A M P L E   2   Special Sequence   Consider the sequence $\{1, 3, 6, 10, 15, 21, \ldots\}$. If we look at the difference between successive terms, we find that

$$a_2 - a_1 = 2, \qquad a_3 - a_2 = 3, \qquad a_4 - a_3 = 4,$$

$$a_5 - a_4 = 5, \qquad a_6 - a_5 = 6,$$

and so the pattern, $a_{n+1} - a_n = n + 1$, seems like a good guess. The next term, a_7, will be 7 more than a_6, so that $a_7 = a_6 + 7 = 28$; further, $a_8 = a_7 + 8 = 36$, and so on. So, in general,

$$a_{n+1} = a_n + n + 1.$$

Therefore, for this pattern, each term a_n represents the sum of the first n positive integers. It turns out that it is possible to express the general term of this sequence with the formula

$$a_n = \tfrac{1}{2}n(n + 1). \tag{3}$$

Later in this chapter we will show how to obtain Eq. (3). ▲

Differences of Sequences

The approach of using differences of terms does not help for all problems involving sequences. However, for many sequences, the use of differences can be very profitable and, moreover, the differences themselves prove extremely valuable in many other contexts. In order to study differences in detail, it is convenient to introduce a special symbol to denote the difference between two successive terms of a sequence.

DEFINITION

The *difference operator* Δ (called delta) is defined, for any sequence $A = \{a_1, a_2, \ldots\}$, as follows:

$$\Delta a_1 = a_2 - a_1, \qquad \Delta a_2 = a_3 - a_2,$$
$$\Delta a_3 = a_4 - a_3,$$

and, in general,

$$\Delta a_n = a_{n+1} - a_n,$$

for any value of n. Therefore, applying this operator Δ forms a new sequence ΔA from the original sequence A.

This difference operator represents the main topic of study in this chapter: the change in the sequence.

A useful application of the difference operator Δ arises when we apply it to a sequence of terms arising from a linear function.

E X A M P L E 3 **Sequence of Differences**

Suppose we have the sequence $\{a_n\} = \{3n - 5\}$ and consider the sequence for values of $n = 1, 2, 3, \ldots$ (Table 3). We organize these values along with the corresponding sequence of values for the function as a table and consider the difference of successive terms. Notice that the first six terms in the sequence are the same as those in Example 1 and the differences Δa_n are all constant. In particular, their constant value, 3, is precisely equal to the slope of the graph of the original linear function $a_n = 3n - 5$ as shown in the graph of the function in Figure 2. ▲

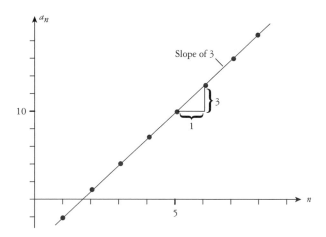

FIGURE 2 Graph of $\{3n - 5\}$, showing the constant slope of 3.

This is not an accident, as Theorem 1 shows.

Theorem 1. If c and b are constants and $a_n = cn + b$, where this holds for $n = 1, 2, 3, \ldots$, then

1. the differences of the sequence $\{a_n\}$ are constant for all n, and

2. when we plot a_n against n, the points fall on a straight line.

Proof of Theorem 1.

1. $\Delta a_n = a_{n+1} - a_n = [c(n + 1) + b] - [cn + b] = c$, a constant independent of n.

2. Successive points on the line are (n, a_n) and

TABLE 3 A Table of Sequence $a_n = 3n - 5$.

n	a_n	Δa_n
1	-2	3
2	1	3
3	4	3
4	7	3
5	10	3
6	13	3
7	16	3
8	19	

$(n + 1, a_{n+1})$. The slope m of the segment connecting these points is obtained in the usual way as the difference in y-coordinates (rise) divided by the difference in x-coordinates (run). Therefore, the slope, m, is

$$m = \frac{(a_{n+1} - a_n)}{(n + 1 - n)}$$
$$= [c(n + 1) + b] - [cn + b]$$
$$= c,$$

a constant, which doesn't depend on n.

▲ ▲ ▲

Since much of our interest in this chapter is to obtain formulas for a sequence, we are also quite interested to know if the converse of Theorem 1 holds. If we know that the differences of the sequence $\{a_n\}$ are constant for all n, can we conclude that there are constants c and b so that $a_n = cn + b$ for all n? The answer is yes, as we see in Theorem 2.

Theorem 2. If $\Delta a_n = c$, where c is a constant independent of n, then there is a linear function for a_n (i.e., there exists a constant b so that $a_n = cn + b$.)

To see how to prove this theorem, try taking a specific sequence with a constant difference, for example 4, 7, 10, 13, . . . , and try working out a formula for a_n. If you can do that, you have the main idea. In any case, we will see a proof of this theorem in a later section.

▲ ▲ ▲

To see how to apply the previous theorem, suppose we had started with the values in Table 3 or its graph but did not know the function from which the data came. We could reconstruct this function in the following manner. The fact that the differences are all 3 shows that we want a formula of the form $a_n = 3n + b$. To find b, use the fact that if we substitute 1 for n, we should get -2, since $a_1 = -2$. Thus,

$$-2 = 3(1) + b, \quad \text{so } b = -5.$$

Therefore, our function is

$$a_n = 3n - 5.$$

EXAMPLE 4 Car Trip Suppose that while your family is on a car trip, you record the mileage readings from the odometer every hour. Let $A = \{a_n\} = \{22{,}322; 22{,}352; 22{,}401; 22{,}456; 22{,}479; 22{,}511\}$ be the first six terms of the sequence of these odometer readings. The graph of these data is shown in Figure 4. In this case, the a_1 term actually represents the initial odometer reading. With this notation, a_2 is the odometer reading after the first hour of the trip. Then, using the formula for the difference operator,

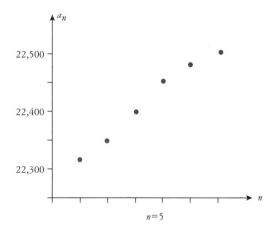

FIGURE 4 Odometer data for the car trip example.

Global Warming: Is It Happening?

This plot of atmospheric temperatures shows that we have had an overall warming trend over the period from 1970 until the 1990s. If the atmosphere continues to warm up, the polar ice caps will melt, raising the ocean levels so that cities on the sea coasts will be swallowed by the ocean. This part of the argument is not in doubt. But will the current warming continue long enough for this to happen? If you simply extend the trend line of the temperature curve, warming will continue without end. However, extending trend lines (a process called *extrapolation*) is not very scientific. After all, it is not just baseballs that follow the rule "What goes up must come down." Indeed, other portions of the graph show that upward trends have been followed by downward ones. Is the current upward trend one that will continue for a long time, or will it reverse itself soon? To

answer this question, it is necessary to understand the causes of temperature change in the atmosphere and to

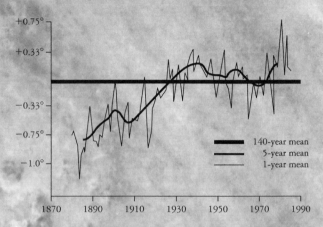

be able to express next year's temperature, T_{t+1}, with some kind of recurrence relation of this general form:

$$T_{t+1} = f(T_t, \text{environmental factors, output of the sun, etc.})$$

Here, f stands for some formula, not yet known, that deals with T_t and the other factors of the kind indicated. This is work for mathematically inclined atmospheric scientists.

$\Delta A = \{\Delta a_n\} = \{30, 49, 55, 23, 32\}$. Notice that for six values of a_n, only five values of Δa_n can be computed. For this sequence, the Δa_n terms represent the distance your family traveled during the nth hour of the trip. These values of Δa_n

could also represent the average speed or average velocity during every hour of the trip. Using this technique, the average speed or average velocities of an object can be determined directly from the distance data. ▲

EXAMPLE 5 Special Sequence Revisited

We consider the sequence for the sum of the first n integers presented in Example 2, $A = \{1, 3, 6, 10, 15, 21, \ldots\}$. If we apply the difference operator to the elements of this sequence, we obtain the new sequence of differences

$$\Delta A = \{2, 3, 4, 5, 6, \ldots\}.$$

Since this itself is just a sequence of numbers, we can apply the difference operator to this new sequence. Thus, for the first term,

$$\Delta(\Delta a_1) = \Delta a_2 - \Delta a_1 \text{ (since } \Delta a_2 \text{ is the}$$
$$\text{term after } \Delta a_1, \text{ in the sequence } \Delta A)$$
$$= 3 - 2 = 1.$$

Further,

$$\Delta(\Delta a_2) = \Delta a_3 - \Delta a_2 = 4 - 3 = 1,$$

and so forth. The above notation is cumbersome and can be expressed more efficiently by writing

$$\Delta(\Delta a_n) = \Delta^2 a_n.$$

This sequence is called the *second difference.* Therefore, in this example, we find that

$$\Delta^2 a_1 = \Delta^2 a_2 = \Delta^2 a_3 = \ldots = 1,$$

or equivalently,

$$\Delta^2 A = \{1, 1, 1, \ldots\}. \ \blacktriangle$$

Behavior of Sequences

What does it mean for all the second differences to be constant, as in Example 5? In particular, it might indicate something about the function from which the data values originally came. Let's consider the interpretation of the first and second differences. The first difference tells us the difference in successive values for a sequence. When the first difference is large, the sequence is growing rapidly. When the first difference is small but positive, the sequence is growing more slowly. When the first difference is negative, the sequence is decreasing. When the first difference is constant, the sequence is changing at a constant rate. In total, the first difference measures "change" of the sequence.

Now suppose that the second difference is a positive constant, as in Example 5, and the first difference is positive. This tells us that the *rate* of growth is growing; that is, the sequence is growing ever faster. Consequently, we see that the original sequence cannot be defined by a linear function, because linear functions grow at a constant rate and have a second difference equal to 0. (Verify this for the sequence in Example 1.) Thus, when the second differences are positive, the sequence has to grow faster than a linear function. While there are many possible candidates for such a function, the two most prominent are exponential functions and polynomials. Let's consider an example of each to see how the second difference behaves.

EXAMPLE 6 Table of Differences— Exponential Function

Construct a table of differences for the exponential function

$$\{a_n\} = 3^n,$$

for $n = 1, 2, \ldots, 7$.

We have produced this in Table 4. Notice that seven terms of a_n produce six terms of Δa_n and five terms of $\Delta^2 a_n$. From this we see that, for this exponential function, neither the first nor the second differences are constant. By the way, you will notice that each number of the third column

TABLE 4 A Table of Differences, Exponential Function.			
n	a_n	Δa_n	$\Delta^2 a_n$
1	3	6	12
2	9	18	36
3	27	54	108
4	81	/007193/	324
5	243	486	972
6	729	1458	
7	2187		

is two times the number in the second column. Likewise, the fourth column is twice the third. This is no accident, and it can be readily proved. See Exercise 12 for a hint. In any case, the second differences are not constant. ▲

E X A M P L E 7 Table of Differences— Quadratic Function Construct a table of differences for the quadratic polynomial sequence

$$\{a_n\} = \{n^2 - 3n + 5\},$$

for $n = 1, 2, \ldots, 6$.

We have produced this in Table 5. Notice that for this quadratic polynomial, the second differences are all constant while the first differences follow a linear pattern. ▲

TABLE 5 A Table of Differences, Quadration.			
n	a_n	Δa_n	$\Delta^2 a_n$
1	3	0	2
2	3	2	2
3	5	4	2
4	9	6	2
5	15	8	
6	23		

The result in Example 7 illustrates the general case, so we summarize the result in the following theorem.

Theorem 3. If a sequence $\{a_n\}$ is defined by a quadratic polynomial, then the sequence has the property that the second differences are constant, $\Delta^2 a_n = c$.

Once again, the converse of this theorem is also true:

Theorem 4. If a sequence of values $\{a_n\} = \{a_1, a_2, \ldots\}$ has the property that $\Delta^2 a_n = c$, a constant, for all values of n, then the data values follow a quadratic pattern and the formula for the general term is a quadratic polynomial.

Try proving Theorem 3 in the same style used to prove the first part of Theorem 1. We'll see a proof of Theorem 4 in a later section.

E X A M P L E 8 Special Sequence, Revisited Again We again consider the sequence from Examples 2 and 5 in which $A = \{1, 3, 6, 10, 15, 21, \ldots\}$, so that

$$\Delta A = \{2, 3, 4, 5, 6, \ldots\}$$

and

$$\Delta^2 A = \{1, 1, 1, 1, \ldots\}.$$

Therefore, from Theorem 4, we conclude that sequence A has a general term in the form of a quadratic polynomial. ▲

E X A M P L E 9 Special Sequence, Revisited Once More Find the equation for the quadratic function that assumes the values $\{1, 3, 6, 10, 15, 21, \ldots\}$ at $n = 1, 2, 3, 4, 5, 6, \ldots$. Since we know that the terms of the sequence satisfy some quadratic equation, we can assume that it is in the general form $q(n) = An^2 + Bn + C$, where A, B, and C are three unknown coefficients that we must determine. To find them, we require three conditions, and we can choose any three of the given values to use for these conditions. Substituting $n = 1, 2,$ and 3 should give the first three terms of the sequence, so

$$A(1^2) + B(1) + C = 1,$$

$$A(2^2) + B(2) + C = 3,$$

$$A(3^2) + B(3) + C = 6,$$

or equivalently, these three equations can be written as

$$A + B + C = 1,$$

$$4A + 2B + C = 3,$$

$$9A + 3B + C = 6.$$

This system of three simultaneous equations can be solved in several ways. One way is to subtract the first equation from both the second and third equations to get two equations in the two remaining unknowns A and B. Then the two equations can be solved by a similar technique. You will learn several new techniques to solve systems of simultaneous equations in a later chapter. In this case, you can easily verify that $A = 1/2$, $B = 1/2$, and $C = 0$ solve all three equations. As a result, the desired quadratic polynomial is

$$q(n) = An^2 + Bn + C = \tfrac{1}{2}n^2 + \tfrac{1}{2}n,$$

or equivalently,

$$q(n) = \tfrac{1}{2}n(n + 1).$$

This agrees with the known result given by Eq. (3) in Example 2. Remember from Example 2 that the nth term of this sequence represents the sum of the first n positive integers. For example,

$$a_4 = 1 + 2 + 3 + 4 = 10 = \tfrac{1}{2}(4)(5).$$

So with this formula, we can quickly determine the sum of the first 100 positive integers as

$$a_{100} = \tfrac{1}{2}(100)(101) = 5050. \; \blacktriangle$$

Formula for the Second Difference

In general, the second difference, $\Delta^2 a_n$, can also be expressed in terms of a simple formula. Starting with $\Delta a_n = a_{n+1} - a_n$, for any n, we have

$$\Delta^2 a_n = \Delta(a_{n+1} - a_n)$$
$$= (a_{n+2} - a_{n+1}) - (a_{n+1} - a_n)$$
$$= a_{n+2} - 2a_{n+1} + a_n.$$

You will see this formula again later in this chapter when we model the acceleration of a falling object. Since the first differences represent average velocities, as shown in Example 4, the second differences of a sequence of distance traveled at uniform time intervals represent *average acceleration.* In general, second differences measure the "change of the change" of a sequence.

Finally, we apply the difference operator to a sequence with a known general term a_n. With this technique, the formula for the general term is used to determine $\Delta^2 a_n$ directly, without having to operate on specific terms in the sequence and then finding the formula for the general term of $\Delta^2 a_n$.

EXAMPLE 10 Differences in Symbolic Form Consider the sequence $A = \{1, 4, 9, 16, 25, \ldots\}$, in which the general term is $a_n = n^2$. Therefore,

$$\Delta a_n = a_{n+1} - a_n$$
$$= (n+1)^2 - n^2$$
$$= (n^2 + 2n + 1) - n^2$$
$$= 2n + 1.$$

Hence, $\Delta A = \{2n + 1\} = \{3, 5, 7, 9, 11, \ldots\}$. Similarly,

$$\Delta^2 a_n = a_{n+2} - 2a_{n+1} + a_n$$
$$= (n+2)^2 - 2(n+1)^2 + n^2$$
$$= 2.$$

Therefore, $\Delta^2 A = 2$. This is a constant, as was expected for a sequence described by a quadratic general term like n^2. ▲

Exercises for Section 1.2

1. Write out the first six terms of the following sequences, given the general term as defined in the following formulas:

(a) $a_n = 3n$
(b) $a_n = 5n + 2$
(c) $a_n = (1/2)^n$
(d) $a_n = n^2 - 4$
(e) $a_n = n^3 + n$
(f) $a_n = (n^2 + 2)/(n^2 + 1)$
(g) $a_n = 2^n/3^n$
(h) $a_n = n^2/(2^n)$

2. Apply the difference operator Δ to each of the terms (first six terms) found for the sequences in Exercise 1.

3. Graph the first six terms of each of the sequences in Exercise 1.

For Exercises 4–9, determine an expression for the general term in each sequence, and use it to predict the next two terms.

4. $\{3, 5, 7, 9, 11, \ldots\}$

5. $\{2, 5, 8, 11, 14, 17, \ldots\}$

6. $\{192, 96, 48, 24, 12, \ldots\}$

7. $\{2, 5, 10, 17, 26, 37, 50, 65, 82, \ldots\}$

8. $\{1/3, 2/4, 3/5, 4/6, \ldots\}$

9. $\{2/5, 4/25, 8/125, \ldots\}$

For Exercises 10 and 11, use the pattern of the differences to find the next term in the sequence.

10. $\{2, 5, 11, 21, 36, 57, \ldots\}$

11. $\{1, 5, 15, 35, 70, 126, \ldots\}$

12. Show, using the difference operator and the algebra of exponents, that if $a_n = 3^n$, then $\Delta a_n = 2a_n$.

13. Use the formula in Example 9 to find the value of the sum of the first 10,000 positive integers.

14. Prove Theorem 3.

For Exercises 15–18, apply the difference operator Δ to the formula for the general term a_n of each sequence to find the formula for the general term of Δa_n.

15. $a_n = 7n$

16. $a_n = 10n^2 + 3n$

17. $a_n = n^4$

18. $a_n = 3n$

19. The stands in one section of a stadium are arranged so that each row has two more seats than the row in front of it. The first row of the stands has 15 seats.
   (a) If s_n represents the number of seats in row n, write a formula for s_{n+1} in terms of s_n.
   (b) How many seats are there in row 10?
   (c) How many total seats are there in this section if there are 20 rows?

Find the second difference for each sequence in Exercises 20–23.

20. $\{3, 6, 9, 12, 15, \ldots\}$

21. $\{2, 4, 8, 16, 32, \ldots\}$

22. $\{4, 7, 12, 19, 28, 39, \ldots\}$

23. $\{-2, -4, 0, 16, 50, 108, \ldots\}$

24. Explain how differencing a sequence (finding the first and second differences) can help analyze its behavior.

Properties and Applications of Difference Tables

The use of the difference operator Δ to construct tables enables us to develop methods to analyze the change or trends of sequences and determine properties of the graphs of sequences. We will establish procedures to find where the graphs of sequences increase, decrease, reach relative maxima and minima, are concave up and concave down, and possess inflection points. While most of these words may be familiar to you, we need to give more precise mathematical definitions to each of these concepts as they relate to a general sequence a_n, the related sequences Δa_n and $\Delta^2 a_n$, and the difference table for a_n.

Before we construct the formal definitions, let's look at a graph of a sequence in order to visualize these properties. Increasing, decreasing, maximum, and minimum are familiar concepts, and these properties could be determined easily from the graph of the sequence shown in Figure 1. A place where a graph is "bent upward" is called *concave up* in this region. Similarly, a place where a graph is "bent downward" is called *concave down.* An *inflection point* is a point where a graph changes concavity. The graph in Figure 1 is labeled to show its concavity and location of inflection points.

Now we define these terms:

D E F I N I T I O N S

Sequence $A = \{a_n\}$ is *increasing* at the kth term if $a_k < a_{k+1}$ (or, in operator notation, $\Delta a_k > 0$).

Sequence A is *decreasing* at the kth term if $a_k > a_{k+1}$ (or $\Delta a_k < 0$).

Sequence A reaches a *relative maximum* at index k if $a_k > a_{k+1}$ and $a_k \geq a_{k-1}$ (or, in operator notation, $\Delta a_{k-1} \geq 0$ and $\Delta a_k < 0$).

Sequence A reaches a *relative minimum* at index k if $a_k < a_{k+1}$ and $a_k \leq a_{k-1}$ (or $\Delta a_{k-1} \leq 0$ and $\Delta a_k > 0$).

Sequence A is *concave up* at k if $\Delta a_k > \Delta a_{k-1}$ (or, using the second difference operator notation, $\Delta^2 a_{k-1} > 0$).

Sequence A is *concave down* at k if $\Delta a_k < \Delta a_{k-1}$ (or $\Delta^2 a_{k-1} < 0$).

Notice that it is the second difference at $k - 1$ that determines the concavity at k. Another perspective on the determination of concavity is that the sequence is concave up when the first differences are increasing and concave down when the first differences are decreasing.

D E F I N I T I O N

Sequence A has an *inflection point* at k if $\Delta^2 a_k$ has a different sign than $\Delta^2 a_{k-1}$.

Difference Table

The difference operator can be used to build a table of values to help analyze sequences and determine

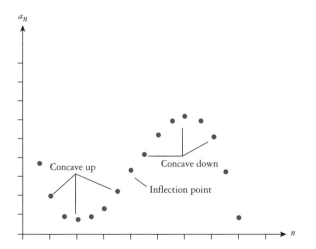

FIGURE 1 Graph of sequence showing properties of concavity and inflection points.

their properties. Given a sequence A and a set of values of the index n for that sequence, a *difference table* includes columns of values for n, a_n, Δa_n, and $\Delta^2 a_n$ (and higher differences, as needed).

E X A M P L E   1   Properties of $\{n^2 - 4n + 3\}$
Construct a difference table for the first seven sequence values of $a_n = n^2 - 4n + 3$, and use the table to determine where the sequence increases, decreases, reaches a relative maximum and minimum, is concave up and concave down, and has an inflection point.

The difference table is built by directly entering the column for n and using the general formula $n^2 - 4n + 3$ to determine and enter the column for a_n. Then the differences are taken between the terms of a_n and Δa_n, respectively, to fill in the columns for Δa_n and $\Delta^2 a_n$. The seven terms of the sequence produce six terms for Δa_n and five terms for $\Delta^2 a_n$ (Table 1).

We use the values from Table 1 to determine the properties of the sequence. The sequence decreases for $n = 1$, since $\Delta a_1 < 0$. The sequence increases for $n = 2, 3, 4, 5$, and 6, since $\Delta a_n > 0$ for these values of n. A relative minimum is reached at a_2, since $\Delta a_1 < 0$ and $\Delta a_2 > 0$. There is no relative maximum for these seven terms. The sequence is concave up for all of its possible values ($n = 2, 3, 4, 5$, and 6) since $\Delta^2 a_n > 0$.

These properties can be verified by looking on the graph for a_n given in Figure 2. ▲

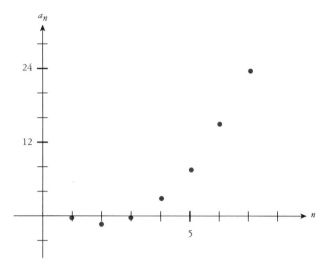

FIGURE 2 Graph of sequence $\{n^2 - 4n + 3\}$.

E X A M P L E   2   Rainfall Data   Table 2 presents data that were recorded for the monthly rainfall for a city in 1993.

TABLE 1

n	a_n	Δa_n	$\Delta^2 a_n$
1	0	-1	2
2	-1	1	2
3	0	3	2
4	3	5	2
5	8	7	2
6	15	9	
7	24		

TABLE 2 Monthly Rainfall Data.

Month	Number of month (n)	Rainfall in inches (r_n)
January	1	2.46
February	2	1.93
March	3	2.92
April	4	4.79
May	5	1.28
June	6	2.56
July	7	2.60
August	8	2.37
September	9	2.77
October	10	2.91
November	11	1.13
December	12	1.13

Just by looking at the data, it is difficult to determine the trends of the sequence. The graph of this data is shown in Figure 3 in order to visualize the obvious properties and patterns.

Next, the difference table (Table 3) is produced with extra columns to indicate where the sequence is increasing or decreasing (based on the sign of Δr_n) and its concavity (based on the sign of $\Delta^2 r_n$). Recall that geometrically Δr_n represents the slope of the straight line between the points (n, r_n) and $(n + 1, r_{n+1})$ on the graph of r_n.

In Table 3, the "change" entry for $n = 11$ is "neither," because neither of the definitions for increasing and decreasing is satisfied when Δr_n is 0.

The relative maxima and minima are determined by locations of changes in the growth trends. The relative minima occur at $n = 2, 5,$ and 8 because the sequence changes from decreasing to increasing at these indices. The relative maxima occur at $n = 4, 7,$ and 10 because of the change from increasing to decreasing. The inflection points occur at $n = 3, 4, 5, 7, 8,$ and 10 because of the change of the sign of $\Delta^2 r_n$ at these indices. ▲

n	r_n	Δr_n	Change	$\Delta^2 r_n$	Concavity
1	2.46	-0.53	Decreasing	1.52	
2	1.93	0.99	Increasing	0.88	Up
3	2.92	1.87	Increasing	-5.38	Up
4	4.79	-3.51	Decreasing	4.79	Down
5	1.28	-1.28	Increasing	-1.24	Up
6	2.56	0.04	Increasing	-0.27	Down
7	2.60	-0.23	Decreasing	0.63	Down
8	2.37	0.40	Increasing	-0.26	Up
9	2.77	0.14	Increasing	-1.92	Down
10	2.91	-1.78	Decreasing	1.78	Down
11	1.13	0.00	Neither		Up
12	1.13				

TABLE 3 The Difference Table.

Higher-Order Differences

The application of the difference operator need not end at the second difference. Differences of the second differences $\Delta^2 a_n$ can be calculated to find the third differences $\Delta^3 a_n$, and so on. These *higher-order differences* are useful in determining if the sequence is defined by a polynomial and the degree of the polynomial. Remember from Section 1.2 that the second differences of a linear function were 0. Theorem 5 in Section 1.2, states that the second differences of a quadratic polynomial are constant.

Since constant second differences produce zeros for the third differences, a quadratic produces zero entries in the $\Delta^3 a_n$ column of a difference table. Similarly, a sequence defined by an nth-degree polynomial has zero entries in the $\Delta^{n+1} a_n$ column of its difference table. The converse of this principle is also true: If a sequence has zero entries in the $\Delta^{n+1} a_n$ column of its difference table, then there is an nth-degree polynomial that generates the sequence.

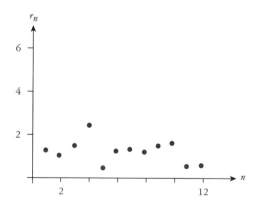

FIGURE 3 Graph of rainfall data from Example 2.

EXAMPLE 3 Higher-Degree Polynomial

Determine the degree of the polynomial that defines the general term for the following terms of a sequence: $b_n = \{-2.9, -8.4, -12.9, -10.4, 7.5, 51.6, 135.1, 273.6\}$. We build the difference table (Table 4) and continue to add columns of higher-order differences until an entire column becomes zero.

From this difference table, we see that the fifth differences are all zero. Therefore, we know that the general term of b_n is a fourth-degree polynomial. We could use any five data points in this sequence to determine the five coefficients for the fourth-degree polynomial in the general formula for the sequence. To do this, follow the method of Example 9 in Section 1.2—the system of equations is merely larger here. Of course, the formula we find is only valid for the known terms of the sequence. If there are other terms that we don't know, we have no way of knowing whether the formula fits them. Furthermore, the formula could be one of many. Suppose that the fourth-degree polynomial that fits the b_n is $p(n)$. Other functions can be constructed to fit the b_n. For example, if we take $q(n) = p(n) + K(n - 1)(n - 2)(n - 3) \cdots (n - 8)$ for any constant K, then $q(n)$ produces a sequence that also fits the b_n for $n = 1, 2, \ldots, 8$. Notice that for $n = 1, 2, \ldots, 8$, $q(n) = p(n)$. ▲

Limit Values

Sometimes we want to determine the long-term trends or behavior of a sequence. In other words, what happens to a_n as n becomes large? Some of the possibilities are as follows: a_n approaches (or gets very near) a constant value; a_n keeps growing without any bound; a_n decreases without bound; or a_n oscillates and never approaches any one value.

It may happen that for very large values of n, a_n is close to some number that we will designate as L. If increasing n causes a_n to get even closer to L or to equal L, we call L the *limit value* of a_n and write

$$\lim_{n \to \infty} a_n = L.$$

If a_n has a limit value, then we also say that a_n converges to L. To make this concept more precise, we will describe this circumstance in mathematical terms:

DEFINITION

A sequence a_n converges to a *limit value L* if, whenever we pick any arbitrary number ϵ (epsilon) greater than 0, there is an N so that we can make the difference $|a_n - L|$ smaller than ϵ by choosing $n > N$.

If a_n converges to L, then we say a_n becomes arbitrarily close to L as $n \to \infty$, as shown in Figure 4.

In the definition we make use of the absolute value function $|x|$. Recall that this function strips a number of its negative sign if it has one and leaves nonnegative numbers unaltered:

$$|-3| = 3 \quad \text{and} \quad |3| = 3.$$

TABLE 4 Difference Table for Example 3.

n	b_n	Δb_n	$\Delta^2 b_n$	$\Delta^3 b_n$	$\Delta^4 b_n$	$\Delta^5 b_n$
1	-2.9	-5.5	1.0	6.0	2.4	0
2	-8.4	-4.5	7.0	8.4	2.4	0
3	-12.9	2.5	15.4	10.8	2.4	0
4	-10.4	17.9	26.2	13.2	2.4	
5	7.5	44.1	39.4	15.6		
6	51.6	83.5	55.0			
7	135.1	138.5				
8	273.6					

SPOTLIGHT

Canny Canning

Each year many millions of tons of aluminum are used to create containers for beverages. In order to cut down on the expense of all this metal, it is helpful to know what the ideal shape is for a container that needs to hold 12 ounces of liquid. At first you might suppose that it makes no difference. However, if you made a container that was extremely skinny, it would have to be very tall to hold 12 ounces, and the total metal used would be greater than if you had used the conventional shape. The radius (r) you choose determines the height (h) needed to hold 12 ounces and determines the amount of metal as well. The following formulas give the height that is needed and the amount of aluminum for a soda can (r and h are measured in inches):

$$h = \frac{17.89}{\pi r^2}$$

$$a = 0.02\pi \left[\frac{4r^2 + 35.78}{r} \right]$$

where 0.02 stands for the thickness of the sides and bottom. (You can derive this yourself based on formulas for the circular cylinders, but you need to know that the top of the can is three times the thickness of the bottom and sides. This additional thickness prevents you from tearing off the entire top when you pop the can open.)

With this formula for a, you can use the methods of this section to determine the value of r that gives the minimum amount of metal. You will find that you get values for r and h that are very close to those found on actual soda cans. Beverage companies have done these mathematical calculations themselves to find the best shape.

A handy way to express distance between two values a and b on the number line is $|a - b|$ or $|b - a|$. For example, if we want the distance from 3 to 4: $|4 - 3| = |3 - 4| = 1$. This also works when one or more of the values is negative. The distance from -5 to -7 is $|-5 - (-7)| = |-5 + 7| = 2$. The distance from -5 to 7 is $|-5 - 7| =$ $|-12| = 12$. Returning to our definition of limit, $|a_n - L|$ is simply a symbolic way of referring to the distance from a_n to L on the number line.

EXAMPLE 4 Limit of 1/n If $a_n = 1/n$, then we see that as n grows larger, a_n becomes smaller yet stays positive. We see that a_n becomes

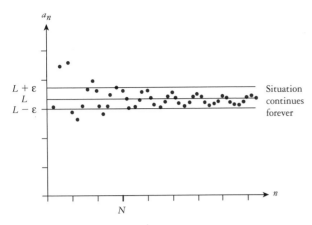

FIGURE 4 The geometry of convergence of a_n to the limit value L. A similar picture can be produced for any ϵ.

closer and closer to zero as n increases, but it will never reach zero. No matter how small a number ϵ is given, there is a value N so that $1/N < \epsilon$ and, therefore, $1/n < \epsilon$ for all $n > N$. We write

$$\lim_{n \to \infty} \left(\frac{1}{n}\right) = 0. \blacktriangle$$

A sequence can reach its limit value, but it does not have to, as illustrated in Example 4. A real-life example where the limit is reached involves a ball being dropped and then bouncing up and down until it stops. If we take the height at times $t = 1, 2, 3, \ldots$, eventually the heights will always be 0.

EXAMPLE 5 Limit of n^2 If $b_n = n^2$, then we see that as n grows larger, b_n continues to grow larger, never getting arbitrarily close to any number. Since b_n grows without bound, we say that b_n *diverges* to ∞. We write

$$\lim_{n \to \infty} (n^2) = \infty. \blacktriangle$$

EXAMPLE 6 Limit of $(-1)^n$ If $c_n = (-1)^n$, then we see that as n grows larger, c_n oscillates between 1 and -1. When n is an even integer, $c_n = 1$; when n is odd, $c_n = -1$. In this case we say c_n *nonconverges* since it has no limit value and does not diverge to $+\infty$ or $-\infty$. $\blacktriangle$

Exercises for Section 1.3

1. Build the difference table including the first and second differences for each of the following sequences (given terms only), and use it to find where the sequence is increasing or decreasing and its concavity:

   (a) $\{2, 5, 6, 4, 3, 2, -3\}$
   (b) $\{-2, 2, 6, 11, 14, 22\}$
   (c) $\{1, 1, 2, 3, 5, 8, 13, 21\}$
   (d) $\{1.3, 2.5, -3.7, -5.7, -8.9, -10.2\}$
   (e) $\{22, 31, 31, 31, 22, 22, 31\}$
   (f) $\{0, -1, 0, 1, 0, -1, 0, 1, 0, -1\}$

2. Determine the relative minima, relative maxima, and inflection points for the sequences in Exercise 1.

3. Build a difference table for the given terms of the following sequences, and use it to determine the degree of the polynomial that defines the general term of the sequence:

   (a) $\{3.4, 3.6, 4.6, 6.4, 9, 12.4, 16.6, \ldots\}$
   (b) $\{1, -1, -1, 1, 5, 11, \ldots\}$
   (c) $\{-7.9, -13.2, -21.3, -31.6, -43.5, -56.4, -69.7, \ldots\}$

(d) $\{-24.1,\ -17.2,\ 1.7,\ 38.6,\ 99.5,\ 190.4,$
    $317.3,\ldots\}$

4. Find the following limits (if the sequence converges):
   (a) $\lim_{n\to\infty}(1/n^2)$
   (b) $\lim_{n\to\infty}1/(1+n)$
   (c) $\lim_{n\to\infty}(2n+3)$
   (d) $\lim_{n\to\infty}(n-1)/(n+1)$

5. The following data set represents the number of wins per month for a hypothetical major league baseball team.

Month	Number of month in season (n)	Number of wins during the month (w_n)
April	1	10
May	2	7
June	3	12
July	4	18
August	5	21
September	6	28

Graph these data, and label the relative extrema (minima and maxima), concavity, and inflection points.

6. Suppose we have the following set of observations of the temperature, in Celsius, for the first 13 days of March: $(1, 1)$, $(2, 5)$, $(3, 7)$, $(4, 6)$, $(5, 2)$, $(6, 1)$, $(7, 3)$, $(8, 10)$, $(9, 12)$, $(10, 8)$, $(11, 0)$, $(12, -5)$, $(13, -2)$. Graph the given values of the sequence, and construct the difference table including the first and second differences. What are the relative maxima and minima for this data set? When is it increasing or decreasing? When is it concave up or concave down?

7. The soda can in our Spotlight in this section is a circular cylinder plus two circles for the top and bottom.
   (a) If we let h denote the height and r the radius, the volume of the cylinder portion of the can is $v = \pi r^2 h$. In order for the can to hold the desired amount of soda, this volume must be 17.89 cubic inches. (This is how the first formula in the Spotlight is derived.) From this, derive an equation for h in terms of r.

(b) The surface area of the cylinder (not including the top and bottom) is $2\pi r h$. The amount of aluminum is

$$a = 0.02(\text{surface area of the cylinder})$$
$$+ 0.06(\text{area of top circle})$$
$$+ 0.02(\text{area of bottom circle}).$$

Using this, part (a), and the surface area formula, derive the second equation of the Spotlight.
   (c) Plot the graph of a versus r, and find the r-value where the amount of aluminum is a minimum.

Find a sequence (list at least six terms) with the following properties:

8. Increasing and concave down for the first five terms.

9. Decreasing and concave down for the first five terms.

10. Increasing for the first three terms and decreasing for the next two terms.

11. Containing a relative maximum at the third term and a relative minimum at the fifth term.

12. Having an inflection point at the third term.

13. Increasing and concave up for the first five terms.

14. Can a sequence be increasing, concave up, and have a relative maximum at the same term? Explain your answer.

15. Why must a sequence defined by a sixth-degree polynomial have a sequence of zeros for its seventh difference?

16. Provide three examples of limits that do not exist. For each case, explain why the limit does not exist.

17. Provide an example of a limit that nonconverges.

18. Provide examples of limits that converge to the following values:
    (a) 0   (b) 10   (c) -5

SECTION 1.4 Modeling Change with Difference Equations

In this section we will *build* difference equation models to describe change in behavior that we observe. In subsequent sections, we will *solve* these difference equations and *analyze* how good our resulting mathematical explanations and predictions are. The solution techniques that we employ in subsequent sections take advantage of certain characteristics that the various models enjoy. Consequently, after building the models, we will *classify* the models based on mathematical structures and the solution techniques that apply.

When we observe change, we are often interested in understanding why change occurs the way it does; perhaps to analyze the effects of different conditions or perhaps to predict what will happen in the future. Often a mathematical model can help us better understand a behavior while allowing us to experiment mathematically with different conditions affecting the behavior. For our purposes we will consider a mathematical model to be a mathematical construct designed to study a particular real-world system or behavior. The model allows us to use mathematical operations to reach mathematical conclusions about the model, as illustrated in Figure 1.

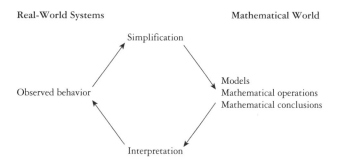

FIGURE 1 A difference equation modeling diagram.

Exact Models of Discrete Change

Some behavior can be described exactly with mathematical models. For example, suppose we deposit $10,000 in a bank that pays 7% interest compounded annually. If n is the number of years and a_n the amount of money in the bank after n years, the difference equation

$$\Delta a_n = a_{n+1} - a_n = 0.07 a_n$$

models the change each period. The solution of the model,

$$a_k = (1.07)^k 10,000,$$

which we will derive later, explicitly represents the amount of money accrued in the account after k years.

Discrete Versus Continuous Change

For the purposes of constructing models, an important distinction regarding change is that some change takes place at *discrete* time instants, such as the depositing of interest in an account. In other cases the change is taking place *continuously,* such as the change in temperature of a cold can of soda on a warm day. Difference equations represent change at discrete time instants. We will see the relationship between discrete change and continuous change (for which calculus was developed) at the end of the chapter. However, we often approximate continuous change with difference equations by collecting data at discrete points in time. We will illustrate this in this section with models of falling objects and the warming of a cold soda.

Approximating Change

Few models are exact representations of the real world. More generally, mathematical models *approximate* real-world behavior. That is, some *simplification* is required in order to represent a real-world behavior with a mathematical construct. For example, suppose you wanted to represent the spotted owl population in a habitat for the purpose of predicting the effect of changes in environmental policy.

The spotted owl population depends on many variables, including the following:

Variables Affecting Spotted Owl Growth

➤ birth rate,
➤ death rate,
➤ availability of resources,
➤ competition for resources,
➤ predators,
➤ natural disasters,

and many other factors. Obviously, we cannot hope to capture precisely every detail. We will eventually build models considering each of the variables boxed above, but first let's construct a simple model. To learn model construction, it is useful to consider the following steps:

Constructing Difference Equation Models

STEP 1:   Identify the problem.

STEP 2:   Make assumptions on which variables to include and the relationships among the variables.

STEP 3:   Find the function or functions that satisfy the model.

STEP 4:   Verify the model.

In this section we will concentrate on the first two steps only. In subsequent sections we will solve the models that we formulate in this section and verify them against observed data. For example, to predict the owl population, we may initially want to consider only the birth and death rates and neglect the effect of other variables. Now what relationships do we consider among the variables? A powerful simplifying relationship is that of *proportionality*. Let's define and illustrate the concept before modeling the owl population.

D E F I N I T I O N

Two varying quantities p and q are said to be *proportional* (to one another) if one varying quantity is always a constant multiple of the other, that is

$$p = kq,$$

for some constant k.

Note from the definition that the graph of p versus q must be a straight line through the origin. This geometric understanding or visualization allows us to collect and plot data in order to test the reasonableness of the proportionality assumption and to estimate k. Let's consider an example.

E X A M P L E 1   Driver Reaction Distance
During a panic stop, the driver of a car must react to the emergency, apply the brakes, and bring the car to a stop. Let's call the distance traveled before the brakes are applied the *reaction distance*

and the remaining distance the *braking distance*. In the exercise set, you will consider mathematical models for the braking distance. Here we consider a model for the reaction distance. The U.S. Bureau of Public Roads collected data for reaction and braking distances. In Figure 2, the data for reaction distance are presented. We plot the reaction distance versus speed and observe the graph.

Since the graph approximates a straight line passing through the origin when projected, we can use a proportionality model. To estimate r, the constant of proportionality, we "eyeball" a straight line through the data and estimate the slope of the line, which is the constant of proportionality. (You will learn more sophisticated techniques for finding a "best" straight line later in your careers, but for the purposes at hand, the "eyeball" method is sufficient.) Using the first and last points, the slope appears to be about 66/60 = 1.1, giving the model

$$d_r = 1.1v,$$

where d_r is the reaction distance in feet and v is the velocity in mph. ▲

EXAMPLE 2 Growth of a Yeast Culture

Consider the data collected from an experiment (Figure 3) measuring the growth of a yeast culture. The associated graph tests the assumption that the change in population is proportional to the current size of the population. That is,

$$\Delta p_n = (p_{n+1} - p_n) \propto p_n,$$

where p_n represents the size of the biomass after n hours and the symbol "$\propto$" means "is proportional to." As an example of why such an equation is reasonable, suppose it takes four time periods for a yeast cell to divide in two. Then, since their divisions are probably not happening at the same time, in any given time period, one-fourth of the yeast will have divided, adding $0.25\, p_n$ yeast to the population. In favorable circumstances yeast cells don't die, so the net change is an addition of

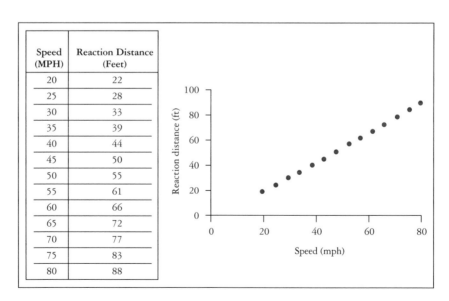

Speed (MPH)	Reaction Distance (Feet)
20	22
25	28
30	33
35	39
40	44
45	50
50	55
55	61
60	66
65	72
70	77
75	83
80	88

FIGURE 2 Driver reaction distance.

Time in Hours	Observed Yeast Biomass	Change in Biomass
	p_n	$p_{n+1} - p_n$
0	9.6	8.7
1	18.3	10.7
2	29.0	18.2
3	47.2	23.9
4	71.1	48.0
5	119.1	55.5
6	174.6	82.7
7	257.3	

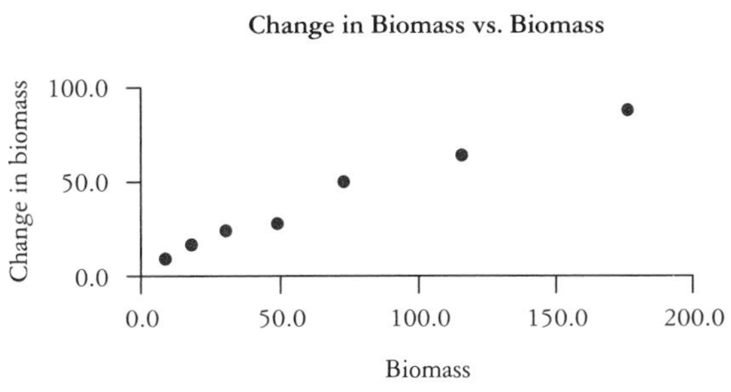

Figure 3 Growth of a yeast culture versus observed yeast biomass [data from Pearl (1927)].

$0.25\ p_n$. We can describe this as $p_{n+1} - p_n = 0.25\ p_n$. Of course, the four time periods was just an illustration. If we don't know the exact number, we just represent the rate as k, giving

$$p_{n+1} - p_n = kp_n.$$

Note that the graph of the data does not lie precisely on a straight line nor does it pass exactly through the origin. However, the data can be approximated with a straight line passing through the origin. Placing a ruler over the data to approximate a straight line, we can estimate the slope of the line to be about 0.6. Estimating $k = 0.6$ for the slope of the line, one might use proportionality to hypothesize the model

$$\Delta p_n = p_{n+1} - p_n = 0.6p_n. \ \blacktriangle$$

Modeling Births, Deaths, and Resources

If both births and deaths during a period are proportional to the population, then the change in population itself should be proportional to the population, as was illustrated in Example 2. Note that the model predicts a population that increases forever. However, certain resources (food, for instance) can support only a limited population level. As this maximum level is approached, growth should slow. Let's see what happens to the yeast culture, which is grown in a restricted area, as time increases beyond the eight observations included in Figure 3.

We note from column 3 in Figure 4 that the change in population per hour becomes small as resources become constrained. From the graph of the populations versus time, the population appears to be approaching a limiting or carrying capacity. Suppose from our graph that we estimate that 665 is the carrying capacity. (Actually our graph doesn't allow us to tell by looking that the correct number is 665 and not 664 or 666, for example. More advanced techniques can be used to obtain 665.) As p_n approaches 665, the change should slow considerably. Consider the following model:

$$\Delta p_n = p_{n+1} - p_n = kp_n(665 - p_n).$$

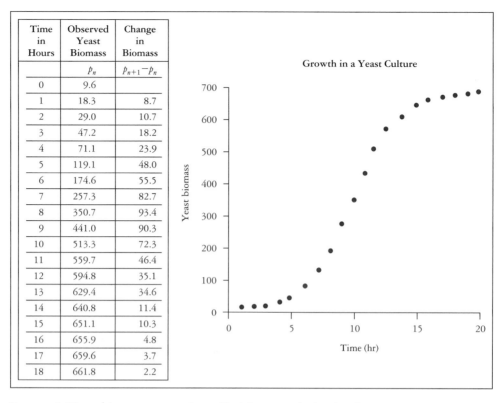

Time in Hours	Observed Yeast Biomass	Change in Biomass
	p_n	$p_{n+1} - p_n$
0	9.6	
1	18.3	8.7
2	29.0	10.7
3	47.2	18.2
4	71.1	23.9
5	119.1	48.0
6	174.6	55.5
7	257.3	82.7
8	350.7	93.4
9	441.0	90.3
10	513.3	72.3
11	559.7	46.4
12	594.8	35.1
13	629.4	34.6
14	640.8	11.4
15	651.1	10.3
16	655.9	4.8
17	659.6	3.7
18	661.8	2.2

FIGURE 4 Yeast biomass approaches a limiting population level.

Intuitively, the factor $(665 - p_n)$ will approach 0 as p_n approaches 665, causing Δp_n to become small. Let's check the hypothesized model against some data (Figure 5).

To test the model and find k, let's plot $(p_{n+1} - p_n)$ versus $p_n(665 - p_n)$ to see whether these quantities are proportional.

If we accept the proportionality argument, we estimate k as the slope of the line approximating the data to be about 0.00082, giving the model

$$p_{n+1} - p_n = 0.00082 p_n(665 - p_n).$$

Solving the Model and Verifying Results

Using algebra to solve for p_{n+1} gives

$$p_{n+1} = p_n + 0.00082 p_n(665 - p_n).$$

Note that the right-hand side is a quadratic equation in p_n. Such equations are classified as *nonlinear* and in general cannot be solved for analytic solutions. That is, we cannot find a formula expressing p_n in terms of n. However, if we are given that p_0 is 9.6, we can substitute to compute p_1. That is,

$$\begin{aligned} p_1 &= p_0 + 0.00082 p_0(665 - p_0) \\ &= 9.6 + 0.00082(9.6)(665 - 9.6) \\ &= 14.76. \end{aligned}$$

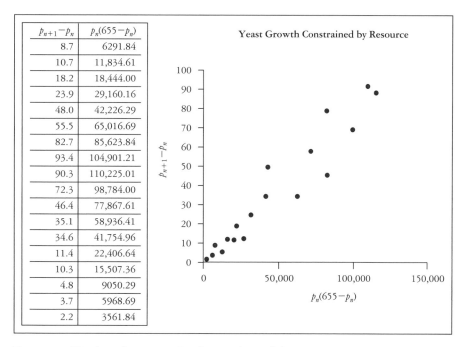

$p_{n+1}-p_n$	$p_n(655-p_n)$
8.7	6291.84
10.7	11,834.61
18.2	18,444.00
23.9	29,160.16
48.0	42,226.29
55.5	65,016.69
82.7	85,623.84
93.4	104,901.21
90.3	110,225.01
72.3	98,784.00
46.4	77,867.61
35.1	58,936.41
34.6	41,754.96
11.4	22,406.64
10.3	15,507.36
4.8	9050.29
3.7	5968.69
2.2	3561.84

FIGURE 5 Testing the constrained growth model.

DEFINITIONS

Iteration is a procedure in which repetition of a sequence of operations yields results successively closer to a desired result.

A *numerical solution* is a table of values obtained by iterating a difference equation beginning with an initial value or values.

In a similar manner, we can use $p_1 = 14.76$ to compute $p_2 = 28.00$. *Iterating,* we can compute a table of values, or a *numerical solution* to the model. The numerical solution, or *model predictions,* are presented in Figure 6. The predictions and observations are plotted on the same graph versus time. Note that the model captures the trend of the observed data very well. This verifies that our model is reasonable.

We will now build several models that we will solve in subsequent sections.

EXAMPLE 3 The Spotted Owl Population
We will build four models of the spotted owl population, each incorporating different assumptions. We will begin with a very simple model— one assuming an abundance of resources. The second model will add the assumption that resources are restricted. In the third and fourth models, we assume the existence of another species in the ecosystem. In the third model, we assume the two species compete against one another for the scarce resources. In the fourth model, we assume that the two species have a predator–prey relationship.

a. Unconstrained Growth Consider only births and deaths. Assume that during each period, births are a percentage of the current population, bp_n, for b a positive constant. Similarly, assume that during each period, a percentage of the current populations dies, say dp_n, for d a positive constant. Neglecting all other variables, the

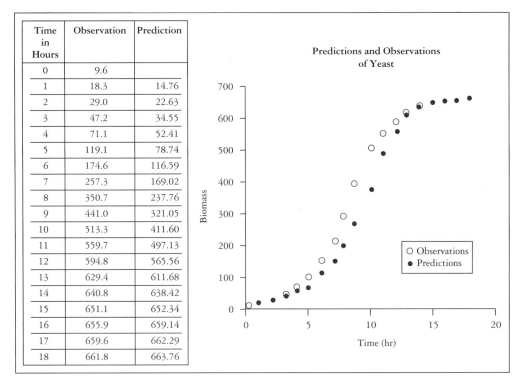

Time in Hours	Observation	Prediction
0	9.6	
1	18.3	14.76
2	29.0	22.63
3	47.2	34.55
4	71.1	52.41
5	119.1	78.74
6	174.6	116.59
7	257.3	169.02
8	350.7	237.76
9	441.0	321.05
10	513.3	411.60
11	559.7	497.13
12	594.8	565.56
13	629.4	611.68
14	640.8	638.42
15	651.1	652.34
16	655.9	659.14
17	659.6	662.29
18	661.8	663.76

FIGURE 6 Plotting the model predictions and the observations.

change in population is the births minus the deaths, or

$$\Delta p_n = p_{n+1} - p_n = bp_n - dp_n = kp_n,$$

where $k = b - d$ represents the growth constant.

b. Constrained Growth Suppose the habitat can support an owl population of size M, where M represents the *carrying capacity* of the environment. That is, if there are more than M spotted owls, the growth rate should be negative. Also, the growth rate should slow as p nears M. One such model is

$$\Delta p_n = p_{n+1} - p_n = kp_n(M - p_n),$$

for k a positive constant.

c. Competing Species Now suppose a second species is in the habitat. Let's represent the population of the competing species after n periods as c_n. Suppose in the absence of a second species, either species exhibits unconstrained growth, that is

$$\Delta c_n = c_{n+1} - c_n = k_1 c_n,$$

$$\Delta p_n = p_{n+1} - p_n = k_2 p_n,$$

where k_1 and k_2 represent the constant growth rates when the other species is not present. The effect of the presence of the second species is to diminish the growth rate of the first species, and vice versa. While there are many ways to model the mutually detrimental interaction of the two species, consider the following model, where the decrease in the growth rate is proportional to the size of the competing species:

$$\Delta c_n = c_{n+1} - c_n = (k_1 - k_3 p_n)c_n = k_1 c_n - k_3 c_n p_n,$$

$$\Delta p_n = p_{n+1} - p_n = (k_2 - k_4 c_n)p_n = k_2 p_n - k_4 c_n p_n.$$

Positive constants k_3 and k_4 represent the relative intensity of the competitive interaction. We call this a *system of difference equations* since there is more than one equation. Do you see why the constants k_3 and k_4 should be positive?

d. Predator–Prey Species   Let's now assume that the spotted owl's primary food source is a single prey, say rabbits. We will represent the size of the rabbit population after n periods as r_n. In the absence of the predatory spotted owl, the rabbit population prospers. We can model the detrimental effect on the rabbit growth rate of the spotted owl population in a manner similar to the way we modeled detrimental competition (by decreasing the growth rates):

$$\Delta r_n = r_{n+1} - r_n = (k_1 - k_2 p_n)r_n = k_1 r_n - k_2 p_n r_n,$$

where k_1 and k_2 are positive constants. On the other hand, if the single food source of the spotted owl is the rabbit, then without the rabbits the spotted owl population would diminish to zero, let's say at a rate proportional to p_n or $-k_3 p_n$ for k_3 a positive constant. The presence of the rabbits increases the growth rate. One such model is

$$\Delta p_n = p_{n+1} - p_n = (-k_3 + k_4 r_n)p_n$$
$$= -k_3 p_n + k_4 p_n r_n.$$

Summarizing our *predator–prey model*, we have

$$\Delta r_n = r_{n+1} - r_n = (k_1 - k_2 p_n)r_n$$
$$= k_1 r_n - k_2 p_n r_n,$$

$$\Delta p_n = p_{n+1} - p_n = (-k_3 + k_4 r_n)p_n$$
$$= -k_3 p_n + k_4 p_n r_n.$$

Note the similarities between the above model and the competing species model of part c. ▲

E X A M P L E 4   The Spread of a Contagious Disease   Suppose there are 400 students in a college dormitory and that one or more students has a severe case of the flu. Let i_n represent the number of infected students after n time periods. Assume that some interaction between those infected and those not infected is required to pass on the disease. If all are susceptible to the disease, $(400 - i_n)$ represents those susceptible but not yet infected after n time periods. If those infected remain contagious, we may model the change in infecteds as proportional to the product of those infected by those susceptible but not yet infected, or

$$\Delta i_n = i_{n+1} - i_n = k i_n (400 - i_n).$$

There are many refinements to this model. For example, we can consider that a portion of the population is not susceptible to the disease, that the contagious period is limited, that infected students are removed from the dorm, and so forth. The more sophisticated models treat the infected and susceptible populations separately, as we did in the predator–prey model in Example 3. ▲

E X A M P L E 5   Warming of a Cold Object   Now we consider a physical example where the behavior is taking place continuously. Suppose a cold can of soda is taken from a refrigerator and placed in a warm classroom and we measure the temperature periodically. Suppose the temperature of the soda is initially 40°F and the room temperature is 72°F. Temperature is a measure of energy per unit volume. Since the volume of soda is small relative to the volume of the room, we would expect the room temperature to remain constant. Further, we may want to assume that the entire can of soda is the same temperature, neglecting variation within. We might expect the change in temperature per time period to be greater when the difference in temperatures between the soda and the room is large and the change in temperature per unit time to be small when the difference in temperatures is small. Letting t_n represent the temperature of the soda after n time periods and k a positive constant of proportionality, we propose

$$\Delta t_n = t_{n+1} - t_n = k(72 - t_n).$$

It turns out that this is a very good model if k is properly chosen. Again, many refinements are possible. While we have assumed k to be constant, it depends on the shape and conductivity properties of the container, the time period of the measurements, and so forth. The temperature of the environment may not be constant in many examples, and often it is necessary to recognize that the temperature is not uniform throughout the object. The temperature of the object may vary in one dimension, as in a thin wire, or in two dimensions, such as a flat plate, or in three dimensions, as in a space capsule reentering the atmosphere. ▲

E X A M P L E 6 From Galileo to Newton to Galileo Galileo observed an object rolling down a smooth inclined plane. Noting the cumulative distance traveled after 1 sec, 2 sec, etc., he then computed the differences of the cumulative distances traveled, or the average velocities during the periods. Knowing the velocities, he computed the differences of the velocities, or average accelerations, and observed that these "second differences" of distance traveled were constant. Galileo studied balls rolling down an inclined plane as an indirect way of understanding what happens when a body falls through the air, for example as if it were dropped from the Leaning Tower of Pisa. As far as we can tell, Galileo never actually conducted experiments in which balls dropped through the air, probably because getting the time and distance measurements was much harder than getting them for balls rolling down ramps. However, let's suppose he had dropped balls from the Leaning Tower of Pisa. What number would he get? Modern experiments show that the numbers would be as shown in Figure 7. Consider the difference table of hypothetical data in Figure 7 for motion under gravity without resistive forces.

The first two columns in the table in Figure 7 are the observed data. There is no proportionality between these columns (the graph at the right is not well fit with a straight line), so we calculate the differences of the distances Δd_n and enter these in the third column. Notice that $\Delta d_n = d_{n+1} - d_n$ is the average velocity between times n and $n + 1$ so we label it v_n. This column also has no proportionality relation to either of the first two columns. We take another difference (column 4) and find that

$$\Delta v_n = v_{n+1} - v_n = 32.$$

However, we want a difference equation involving the distances. To get this, we note that, by definition of velocity,

n seconds	d_n feet	$\Delta d_n = v_n$ ft./sec	Δv_n ft./sec/sec
0	0	16	32
1	16	48	32
2	64	80	32
3	144	112	32
4	256	144	32
5	400	176	32
6	576	208	
7	784		

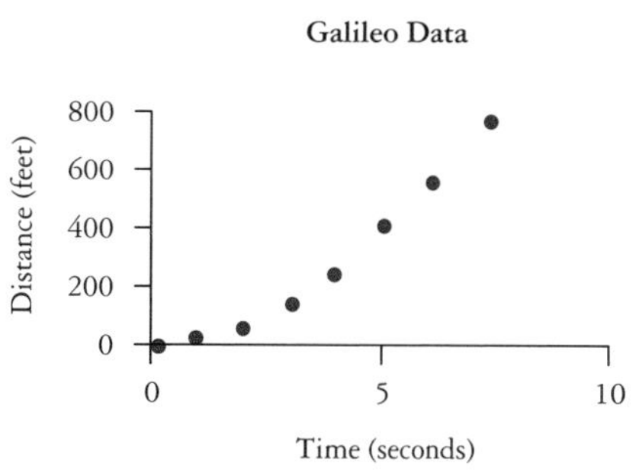

FIGURE 7 Galileo's data for falling bodies.

$$v_{n+1} = d_{n+2} - d_{n+1}, \quad \text{and}$$

$$v_n = d_{n+1} - d_n.$$

Substituting, we have

$$d_{n+2} - d_{n+1} - (d_{n+1} - d_n) = 32, \quad \text{or}$$

$$d_{n+2} - 2d_{n+1} + d_n = 32.$$

However, the motion is taking place continuously and not in discrete time intervals. What if we take time intervals smaller than one second? Newton observed that the ratio of the change in velocity per unit time approached a limiting value as the time interval became arbitrarily small, and he invented the calculus to address "continuous" change using such limits. This was a big advance over the mathematics available to Galileo. ▲

Galileo to Newton and Back

In all of the difference equations we have discussed, we have chosen a specific time unit as the basis for the equation. For example, Galileo's law, $\Delta v_n = 32$, tells us how the velocity after n seconds compares with the velocity after $n + 1$ seconds. Two criticisms can be made of this approach. First, why should one particular time unit be chosen over another? Why not let n tick off minutes or microseconds instead of seconds? Another criticism is that the equation is ignoring what happens during the in-between times.

Although it is far from obvious, neither of these criticisms is serious in relation to Galileo's law. Thus, one might say that difference equations were fine for expressing Galileo's thoughts. However, when Newton took up the matter of gravitation, he found it necessary to deal with the in-between times and with "instantaneous changes" rather than changes taking place in some particular time unit. Continuous models of this sort required the invention of calculus (mostly due to Newton and Leibniz), a subject so profound and useful that entire courses are given over to it. With calculus, it was possible to deal with bodies whose fall is retarded by the friction of air molecules rubbing on the body or colliding with it—a problem Galileo could not deal with, even though he was certainly aware of its importance.

Trying to draw useful conclusions from these calculus models based on instantaneous changes often proved difficult. Currently, the best way over these

Galileo's "Law" of Falling Bodies

The painting shows Galileo measuring how objects gain speed as they roll down to the bottom of a ramp. Based on experiments such as these, Galileo concluded

that all objects will reach the bottom at the same time if they start at the same time and place and if other things are equal. This is related to what is sometimes called *Galileo's law:* Objects gain speed at the same rate, independent of their weight or physical shape. Of course, Galileo's law is not a law, but more like a simple model—it's a rough description that is approximately correct in some circumstances.

The Cub Scouts compete each year in a Pine Car Derby to see which of their homemade cars will roll down a ramp the fastest. They intuitively understand that "other things" are not equal among the cars: Friction in the wheels and air drag play a significant role

in determining how the car gains speed. If we held the Pine Car Derby on the moon, Galileo's simple model would be more nearly correct because air drag would be the same—namely, zero—for all cars. And if we replaced the cars with greased pigs on a waterslide on the moon, Galileo's simple model would be better yet and all the pigs would arrive at the bottom of the ramp at the same time. But then, what would there be for the Cub Scouts to build?

difficulties is to find a difference equation model to approximate the calculus model and to draw conclusions from it using ideas from this chapter. Thus, we need both the sophistication of Newton's mathematics and the simplicity of Galileo's in order to make progress.

We have gotten but a glimpse of the power of difference equations and the idea of proportionality to model change in the world about us. In the next section we will classify the difference equations by type and begin solving them. Thus, we are back to Galileo in the sense that the mathematical formulation would have been completely understood in his day.

1. In the table below, y represents the distance in feet that automobiles traveled after applying the brakes during a panic stop; x represents the speed in mph at which the automobiles were traveling at the time the brakes were applied.

x	20	25	30	35	40	45	50	55	60	65	70
y	20	28	41	53	72	93	118	149	182	221	266

The data were collected by the U.S. Bureau of Public Roads and represent average values for the drivers tested. Plot the data—is it "smooth"? Are there observations that do not follow the general trend? Plot the distance traveled versus the square of the speed. Is it reasonable to say y is proportional to x^2? If so, estimate the constant of proportionality from your plot.

2. In the table below, x represents the girth of a pine tree measured in inches at shoulder height; y represents the board feet of lumber finally obtained.

x	17	19	20	23	25	28	32	38	39	41
y	19	25	32	57	71	113	123	252	259	294

Plot the data. Is it smooth? Are there unusual observations? Formulate and test the following two models: that usable board feet is proportional to (a) the square of the girth and (b) the cube of the girth. Which is better?

3. The following data represent the weight w in ounces of New York black bass for various lengths l in inches.

l	12.5	12.63	12.63	14.13	14.5	14.5	17.25	17.75
w	17	16	17	23	26	27	41	49

Test the model that weight is proportional to the cube of the length.

4. These data show the resistive force in a spring versus the distance stretched beyond the equilibrium length.

d (distance)	0.1	0.2	0.3	0.4
f (force)	14.1	29.8	40.7	58.5

Is the resistive force proportional to the distance stretched? If so, estimate the constant of proportionality from your plot.

5. Digoxin is used in the treatment of heart disease. In the table below, y represents the amount of digoxin in the bloodstream, and t represents the time in days after taking a single dose. The initial dosage is 0.5 mg.

t	0	1	2	3	4	5	6	7	8
y	0.500	0.345	0.238	0.164	0.113	0.078	0.054	0.037	0.026

(a) Formulate a model using a difference equation that asserts that the change in amount of digoxin per day is proportional to the amount of digoxin present.

(b) Now assume that after the "initial dose" of 0.5 mg, each day a "maintenance dose" of 0.1 mg is taken. Formulate the difference equation model.

(c) Using the starting values indicated, build a table of values or numerical solution for Exercises 5a and 5b for 15 days.

6. A certain drug is effective in treating a disease if the concentration remains above 100 mg/l. The initial concentration is 640 mg/l. It is known from laboratory experiments that, in each hour, the drug decays at the rate of 20% of the amount present at the start of the hour.

(a) Formulate a model representing the concentration each hour.

(b) Build a table of values to determine when the concentration reaches 100 mg/l.

7. A humanoid skull is discovered near the remains of an ancient campfire. Archaeologists are convinced that the skull is the same age as the original campfire. It is

determined from laboratory testing that only 1% of the original amount of carbon-14 remains in the burned wood taken from the campfire. It is known that carbon-14 decays at a rate proportional to the amount remaining. Carbon-14 decays 50% in 5700 years. Formulate a model for "carbon-14 dating." How old is the humanoid skull?

8. Place a cold can of soda in a room. Measure the temperature of the room and periodically measure the temperature of the soda. Formulate a model to predict the change in the temperature of the soda. Estimate any constants of proportionality from your data. What are some of the sources of error in your model?

SECTION 1.5 *Solving Linear Homogeneous Difference Equations*

In the previous section we built models by noting that change often leads to a difference equation model. Furthermore, in many cases we assumed that the change or difference was proportional to some quantity. When we estimated and substituted the constant of proportionality from a set of observed data, a difference equation with constant coefficients resulted. For example, if each month 7% of the amount of a pollutant p_n were removed from a contaminated lake, the change or difference each month would be

$$\Delta p_n = p_{n+1} - p_n = -0.07 p_n,$$

which can be rewritten with the highest term in the sequence isolated on the left,

$$p_{n+1} = p_n - 0.07 p_n,$$

which is the form typically used for studying difference equations.

In this section we will define what we mean by a *solution to a difference equation,* discuss what types of difference equations have analytical (formula) solutions by classifying the equations by their mathematical structure, and find analytical solutions to two important classes of difference equations whose solutions predict polynomial or exponential change.

Solutions to Difference Equations

What do we mean by a *solution* to a difference equation? Is it a table of values, a graph, or a formula or function? Actually all three are forms of solutions, each having its particular advantages. Let's begin by formalizing the notion of a numerical solution.

As stated earlier, a *numerical solution* is a table of values obtained by iterating a difference equation beginning with an initial value or values. For example, money accruing in a bank account at 7% interest is modeled by the difference equation

$$a_{n+1} = a_n + 0.07 a_n,$$

where a_n represents the amount in the bank after n months. If the initial value a_0 is \$1000, then Figure 1 shows a numerical solution for the first 10 months. It is obtained by substituting $a_0 = 1000$ and computing a_1. Successive terms for the sequence are obtained iteratively. To assist in analyzing the behavior, we have included a graph of the sequence for the first 11 terms.

DEFINITION

An *analytical solution* to a difference equation is a function that results in an identity when substituted in the difference equation and satisfies any initial conditions given.

Month	Principal	Interest
n	a_n	
0	$1000.000	$70.0000
1	1070.000	74.9000
2	1144.900	80.1430
3	1225.043	85.7530
4	1310.796	91.7557
5	1402.552	98.1786
6	1500.730	105.0510
7	1605.781	112.4050
8	1718.186	120.2730
9	1838.459	128.6920
10	1967.151	137.7010

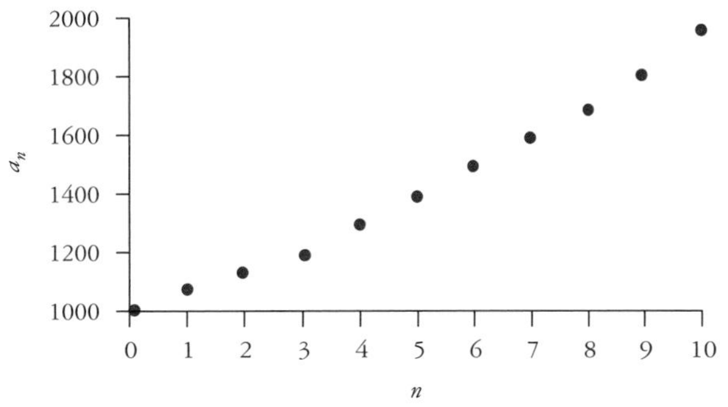

FIGURE 1 Interest in a bank account.

In the preceding example, given the difference equation

$$a_{n+1} = a_n + 0.07a_n,$$

if the function $a_k = (1.07)^k c$ for any constant c is substituted, an identity results:

$$a_{k+1} = (1.07)^{k+1} c$$
$$= (1.07)^k c + 0.07(1.07)^k c,$$
$$(1.07)^{k+1} c = (1.07)^{k+1} c.$$

Note that no initial condition was needed. If we knew that a_0 was 1000, then we could determine c by substituting 1000 for a_k for $k = 0$:

$$a_0 = 1000 = (1.07)^0 c,$$
$$c = 1000.$$

DEFINITION

A *general solution* to a difference equation is a function where substituting individual values gives all the individual solutions corresponding to the different values of the initial condition.

We see that the constant c allows us to solve for the initial value, and vice versa. We call $a_k = (1.07)^k c$ a *general solution* to the difference equation because substituting individual values of c gives all the individual solutions corresponding to the different values of the initial condition a_0.

Let's compare numerical and analytical solutions before discussing graphical solutions. Note that all that was needed to find a numerical solution to our banking model was a difference equation and an initial value. This is a powerful property of numerical solutions—the difference equation does not need to possess special properties for you to find numerical solutions. Merely begin with the initial value or values and *iterate* or substitute values and evaluate. On the other hand, since we did not have a general formula for the kth term, each term had to be computed from the previous term. Predicting long-term behavior may be difficult from a numerical solution, as we shall see in Section 1.8 when we study nonlinear difference equations.

An analytical solution provides a function from which we can directly compute any particular term in the sequence. For example, knowing that $a_k = 1000(1.07)^k$ in the previous example allows us to compute the amount after 95 months, $a_{95} = (1000)(1.07)^{95}$, by pushing 18 buttons on a calculator with a log function and inverse log function. How many buttons do you have to push to calculate a_{95} by iteration?

Another advantage of analytic solutions is that when we find one, we usually get the general solution at the same time. By contrast, a solution found by iteration pertains only to one initial condition. Unfortunately, we can find analytical solutions for only a relatively small number of difference equations that possess certain properties, which we classify ahead.

Graphing a numerical or analytical solution provides visual intuition of the behavior under study, as we can see by referring to the graph in Figure 1. More importantly, we will see in Section 1.8 that we will be able to construct a solution to a difference equation using a graphical procedure. The method is called *cobwebbing* and provides great insight into the long-term behavior of difference equations that are otherwise difficult to analyze.

Let's now begin our quest to find analytical solutions to difference equations. The simplest class that possesses analytical solutions are difference equations that are linear, have constant coefficients, are homogeneous, and are first-order. What do these terms mean?

Classifying Difference Equations

In ordinary algebra, we call $y = mx + b$ a *linear equation*. Certain difference equations look a little like this, namely those of the form

$$a_{n+1} = ra_n + b,$$

and are examples of *linear difference equations*. Here, r

and b could be constants or functions of n, but in this course we stick to the case where they are constants. This is a linear difference equation with *constant coefficients*. There are other kinds of linear difference equations as well.

DEFINITION

If the terms of the difference equation that involve sequence variables (i.e., which contain a's) involve no products of sequence variables, no powers of sequence variables, nor functions of sequence variables such as exponential, logarithmic, or trigonometric functions, then we call the difference equation *linear*. Otherwise the difference equation is *nonlinear*. Note that the restrictions apply only to the terms involving the sequence variable and not to any added terms which don't involve the sequence variables.

The following difference equations are linear and have constant coefficients:

$$a_{n+2} = 3a_n + n^2, \qquad (1)$$

$$a_{n+1} = 5a_n, \qquad (2)$$

$$a_{n+2} - 3a_{n+1} + 4a_n = 6, \qquad (3)$$

while the following are nonlinear:

$$a_{n+1} = (a_n)^2, \qquad (4)$$

$$a_{n+2} = (a_{n+1})(a_n). \qquad (5)$$

The definition of linearity also applies to systems of difference equations such as the competing species and predator–prey models of Section 1.4. Those particular systems were nonlinear because of the existence of the terms $c_n p_n$ and $p_n r_n$.

DEFINITION

A linear difference equation is called *homogeneous* if the only terms that appear are ones involving a sequence variable.

Equation (1) is not homogeneous, because of the n^2 term. Equation (2) is homogeneous. Equation (3) is not homogeneous, because of the 6. We are not concerned with classifying (4) and (5) since they are not linear. The following equations are homogeneous:

$$a_{n+2} = 3a_n, \qquad (6)$$

$$a_{n+2} - 3a_{n+1} + 4a_n = 0. \qquad (7)$$

A system is called *homogeneous* if each of its equations is homogeneous.

If one neglects the terms in a nonhomogeneous linear equation that do not contain a sequence variable, a homogeneous equation results. In Eq. (1), if we remove the term n^2, we get Eq. (6), which is called the *associated homogenous equation*. In Eq. (3), if we replace the 6 with 0, we get the associated homogeneous equation (7).

If one were to take the lowest subscript appearing in a difference equation and subtract it from the highest, the resulting number is called the *order of the difference equation*. For example,

$$a_{n+2} = 3a_{n+1} + 2a_n$$

is a *second-order equation* because $(n + 2) - n = 2$, while

$$a_{n-1} = a_{n+3} - a_{n-3}$$

is a *sixth-order equation* because $(n + 3) - (n - 3) = 6$.

We will mainly be concerned with first-order equations, but there is one important lesson about higher-order difference equations; it concerns how many initial values are needed to generate a numerical solution. This point is illustrated in Example 1.

EXAMPLE 1 Galileo's Problem   In Example 4 of Section 1.4, we derived a model for the distance d that an object would fall after n seconds, assuming no resistance:

$$d_{n+2} - 2d_{n+1} + d_n = 32,$$

$$d_{n+2} = 32 + 2d_{n+1} - d_n.$$

Note that the equation is second-order. Furthermore, note that in order to build a numerical solution, one must have two starting values. If one observed that $d_0 = 0$ and $d_1 = 16$, one could then compute the sequence representing the cumulative distance traveled,

$$d_n = \{0, 16, 64, 144, 256, 400, \ldots\}.$$

In general, a difference equation of order n requires n initial conditions to build a numerical solution. We note before moving on that in the case at hand, if one only knew $d_0 = 0$ and had no knowledge of the solution function, one could not enumerate the other terms in the sequence. ▲

For a system of difference equations, the order is defined to be the larger of the two orders of the individual equations. The following system is first-order, linear, and homogeneous:

$$a_{n+1} = 3a_n - b_n,$$

$$b_{n+1} = 2a_n - b_n,$$

while the following system is second-order, nonlinear (products of terms in the sequence), and nonhomogeneous (n^2 term):

$$a_{n+2} = a_{n+1} - 3a_n b_n - n^2,$$

$$b_{n+2} = b_{n+1} - 2a_n b_n.$$

Finding Analytic Solutions by Conjecture

To find an analytical solution to a difference equation, we need to find a function that when substituted in the difference equation yields an identity and that coincides with any initial values that are specified. The *method of conjecture* is a powerful mathematical technique that enables us to hypothesize a solution, substitute, and either accept or reject our hypothesis. In essence, we prove or disprove our conjecture. How does one find a function to test by conjecture?

Mathematical science is often called the science of patterns. We will now illustrate several methods for observing a pattern which enable us to conjecture a solution to accept or reject. First we examine difference equations that lead to polynomial growth.

In Section 1.2 we noted from a difference table that linear functions have first differences that are constants. Thus, one might expect that the solution to first-order difference equations such as

$$\Delta a_n = a_{n+1} - a_n = 5,$$
$$a_{n+1} = a_n + 5 \tag{8}$$

have the form

$$a_k = mk + b. \tag{9}$$

Substituting in Eq. (8) for a_k and $a_{k+1} = m(k+1) + b$, we have

$$m(k+1) + b - (mk + b) = 5,$$

yielding $m = 5$ and b arbitrary, or $a_k = 5k + b$. This is the general solution of Eq. (8). The arbitrary constant b can be determined if an initial value a_0 is given, as we see in the next example. If we allow the constant 5 to be any constant, we can prove Theorem 4, in Section 1.2, similarly.

EXAMPLE 2   A Linear Solution Function
Find the analytic solution to the following difference equation:

$$a_{n+1} = a_n + 6,$$

$$a_0 = 3.$$

The first difference is a constant, and the solution is given by Eq. (9) with $m = 6$. Thus, $a_k = 6k + b$. Substituting $k = 0$ allows us to evaluate the arbitrary constant,

$$3 = 6(0) + b,$$

$$b = 3,$$

$$a_k = 6k + 3. \ \blacktriangle$$

Had we not observed from the difference table the pattern that linear functions produce first differences that are constant, we could have constructed a numerical solution to Example 2 to observe that the graph of the solution was a series of points falling on a line. You may want to verify this for Example 2.

Similarly, from difference tables, we noted that quadratic functions have second differences that are constant. Thus, we would conjecture that the solution to Galileo's problem (from Section 1.4)

$$\Delta^2 d_n = d_{n+2} - 2d_{n+1} + d_n = 32$$

How Long Have These Bodies Been Dead?

The murder victim at the left has been dead for a few hours. But how long exactly? Knowing the time of death can help find the killer. Based on current body temperature, the elapsed time since death can be estimated from the equation

$$T_{n+1} = T_n - 0.0081(T_n - 70), \qquad (1)$$

where T_n is the temperature in Fahrenheit after n hours have elapsed and where the 70 signifies a sur-

rounding temperature of 70°F. T_0 would be normal body temperature of 98.6°F.

The "Iceman" was discovered in the Alps in 1991 and has attracted intense interest from scientists. He is clearly very old, but how old? The body long ago lost all body heat, so the method based on Eq. (1) is not useful. But this body has been "cooling off" in another

way. When living things die, the carbon-14 atoms in their body start to disappear (they turn into nitrogen by radioactive decay), somewhat like body heat disappears, except that it takes hundreds of years for the reduction to become measurable. The equation describing this is

$$f_{n+1} = 0.8860 f_n, \qquad (2)$$

where f_n is the fraction of the original amount of carbon-14 left after n thousands of years.

Scientists have used Eq. (2) to find out that the Iceman died about 5000 years ago. His ax, his bow and arrows, and the tattoos on his body are giving scientists a priceless glimpse into the world of the Copper Age.

has the form

$$d_k = ak^2 + bk + c.$$

This is an example of a *quadratic change.*

From Example 1, we expect two arbitrary constants will be involved in a general solution since the equation is second-order. In the exercise set you are asked to show that the conjecture is true and to solve for the constants a, b, c. You will see that two arbitrary constants result and can be evaluated knowing that $d_0 = 0$ and $d_1 = 16$.

We have seen that equations with constant first differences or second differences lead to linear and quadratic functions, respectively. Similarly, we expect equations with constant higher-order differences to lead to higher-degree polynomials (and they do!). Now let us consider equations leading to exponential growth.

Suppose we have a sequence $\{a_n\}$, which has the property that each term is some constant r times the preceding one. That is,

$$a_2 = ra_1,$$

$$a_3 = ra_2,$$

and, in general,

$$a_{n+1} = ra_n$$

for any value of n. Further, suppose that the initial term in this sequence is a_0. It follows that

$$a_1 = ra_0,$$

$$a_2 = ra_1 = r(ra_0) = r^2a_0,$$

$$a_3 = ra_2 = r(r^2a_0) = r^3a_0,$$

and so forth. This leads to a formula for the general term,

$$a_k = r^k a_0,$$

$$k = 0, 1, 2, 3, \ldots.$$

We note that whenever the constant $r > 1$, the values for a_n get successively larger. On the other hand, if $0 \le r < 1$, then the values become successively smaller. Such problems are called *exponential growth* and *decay problems,* respectively, and they arise in many different situations. We now consider a series of examples that use these ideas.

EXAMPLE 3 An Investment Problem Suppose $1000 is deposited in a bank at 6% interest, compounded annually. After one year, the initial balance $b_0 = \$1000$ has earned 6% of $1000 = 0.06(1000) = \$60$ in interest, so that there is balance of

$$b_1 = 1000 + 0.06(1000) = 1.06(1000)$$

in the account. Letting b_n represent the balance after n years, we have the model

$$b_{n+1} = b_n + 0.06b_n = 1.06b_n,$$

which displays exponential growth with a growth rate $r = 1.06$. Therefore, the balance after k years is given by

$$b_k = 1.06^k(1000). \quad \blacktriangle$$

EXAMPLE 4 Population Growth The population of Latin America is currently increasing at a rate of about 2.2% a year. That is, the population at the end of any given year is 2.2% greater than at the beginning, or

$$p_{n+1} = p_n + 0.022p_n = 1.022p_n.$$

Assuming the current Latin American population to be 450 million, find

a) the population in 25 years,

b) how long it takes for the population to double,

c) how long it takes for the population to reach 1 billion.

For a growth rate $r = 1.022$ and an initial population of 450 million, the population in year k, p_k, is given by

$$p_k = (1.022)^k(450{,}000{,}000),$$

which for $k = 25$ gives a population of 775,327,000. The population will have doubled when

$$p_n = 2p_0 = (1.022)^n p_0,$$

or when $1.022^n = 2$.

If we take logs of both sides of this last equation, we find that

$$\log(1.022)^n = n\log(1.022) = \log(2), \quad \text{so that}$$

$$n = \frac{(\log(2))}{(\log(1.022))} = 31.85.$$

Therefore, if the current growth rate continues, the population of Latin America will double in about 32 years.

Finally, the population of Latin America will reach one billion when

$$p_n = (1.022)^n(450{,}000{,}000) = 1{,}000{,}000{,}000.$$

Solving this last equation for n,

$$(1.022)^n = 2.222,$$

$$\log(1.022)^n = n\log(1.022) = \log(2.222),$$

$$n = \frac{(\log(2.222))}{(\log(1.022))} = 36.69,$$

or about 37 years. ▲

E X A M P L E 5 Sewage Treatment A sewage treatment plant processes raw sewage to produce usable fertilizer and clean water by removing all other contaminants. The process is such that each hour, 12% of remaining contaminants in a processing tank are removed. What percentage of the contaminants would remain after one day? How long would it take to lower the amount of contaminants by half? How long until the level of contaminants is down to 10% of the original level?

Let the initial amount of sewage contaminants be s_0, and let the amount remaining after n hours be s_n. We then build the model

$$s_{n+1} = s_n - 0.12s_n = 0.88s_n.$$

The solution to the difference equation is given by the sequence

$$s_k = (0.88)^k s_0.$$

After one day (24 hours), when $k = 24$, we obtain

$$s_{24} = (0.88)^{24}s_0 = 0.0465s_0,$$

so that the level of contaminants has been reduced by over 95%. To find when half the original amount of contaminant remains,

$$s_n = 0.5s_0 = (0.88)^k s_0.$$

We solve for k:

$$(0.88)^k = 0.5,$$

$$k = \frac{(\log(0.5))}{(\log(0.88))} = 5.42,$$

or 5.42 hours.

Similarly, to reduce the level of contaminants by 90%, we find

$$(0.88)^k s_0 = 0.1 s_0,$$

$$k = \frac{(\log(0.1))}{(\log(0.88))} = 18.0,$$

or 18 hours. ▲

Comparing Polynomial and Exponential Change

Many functions we deal with in applications grow toward ∞ as n gets larger. Often, we need a more specific idea of how swift the progress toward ∞ is. For example, in a later chapter, you will study algorithms and become aware of the concern with the ever-increasing time it takes to compute solutions to problems when the required number of arithmetic operations grows. For another example, the Spotlight in Section 1.1 deals with the question of whether population tends to outstrip the growth in food. Let us now consider an example in detail in which the two growth rates are compared.

EXAMPLE 6 Wedding Gifts Suppose your parents have offered you a choice of two investment plans for your wedding gift. In plan A, your parents will place $7000 in a certificate of deposit redeemable at the time of your marriage which pays 12% annual interest compounded

monthly (1% per month). For plan B your parents will provide $7000 plus an additional $100 for each month until you marry, but no interest. Letting a_n and b_n represent the amount that would exist after n months for plan A and plan B, respectively, we have the following models:

$$a_{n+1} = 1.01 a_n,$$

$$a_0 = 7000,$$

$$b_{n+1} = b_n + 100,$$

$$b_0 = 7000.$$

The solutions to the models are

$$a_k = (1.01)^k 7000,$$

$$b_k = 100k + 7000.$$

You quickly realize that plan A returns $70 the first month while plan B returns $100 above the $7000. Since you do not currently have any wedding plans, you compute a few more points. If you marry after graduation, about 48 months from now,

$$a_{48} = (1.01)^{48} 7000 = 11,285.83,$$

$$b_{48} = 100(48) + 7000 = 11,800.$$

Plan B is still better, but not by much. If you start your career before marrying, you may marry in 10 years or so:

$$a_{120} = (1.01)^{120} 7000 = 23,102.60,$$

$$b_{120} = 100(120) + 7000 = 19,000.$$

Plan A is now better by more than $4000. To emphasize the difference between the growth patterns, you compute the status of the two plans in 50 years:

$$a_{600} = (1.01)^{600}7000 = 2{,}741{,}083.78,$$

$$b_{600} = 100(600) + 7000 = 67{,}000.$$

The two plans now differ by a factor of over 40! Notice that the base of the exponential is only slightly greater than 1. Even so, the exponential eventually dwarfs the polynomial, in this case a linear model. Growth rates are important in many applications, and we will return to this fascinating subject later in the course. ▲

Summary

In this section we have found solutions to all first-order linear difference equations with constant co-efficients if the equations were homogeneous. Additionally, we discussed solutions to difference equations where the nth differences were constants (nonhomogeneous equations of the form $\Delta^k a_n = c$, for c any constant). We saw that the solutions led to behavior that changes in an exponential or polynomial manner. The method of conjecture allowed us to find solutions to entire classes of problems for all initial conditions, while iteration allowed us to build a numerical solution for a particular problem given specific initial conditions. In the next section we consider additional first-order linear equations with constant coefficients that are nonhomogeneous.

Exercises for Section 1.5

Find the solutions to the following difference equations for any n:

1. $a_{n+1} = 3a_n$, $a_0 = 1$.

2. $a_{n+1} = 3a_n$, $a_0 = 5$.

3. $a_{n+1} = 5a_n$, $a_0 = 10$.

4. $a_{n+1} = a_n$, $a_0 = 64$.

5. $x_{n+1} = (3/4)x_n$, $x_0 = 64$.

6. $x_{n+1} = x_n$, $x_0 = 6$.

7. $x_{n+1} = x_n$, $x_0 = 200$.

8. Find the balance in a bank account in which $500 was deposited at 5% annually compounded interest for 10 years.

9. Find the balance after one year if $500 is deposited at 4% interest compounded quarterly. What is the balance after 10 years?

10. Assume world population increases at 3% per year. Find the population in 25 years and the time needed for the population to double. Do the same for a 4% annual growth rate.

11. Suppose world food production increases at a rate of 1% a year. How much of an increase will occur in 10 years?

12. A bug population increases at the rate of 30% per hour. If there are initially one million bugs, how many are there after four hours? After a full day? How long does it take until the population has increased a hundredfold?

13. Suppose the dose for a drug is 640 mg. If the amount of drug dissipates at a rate of 20% of the amount of drug present each hour, how long will it take to bring the drug level down to under 100 mg? To under 10% of the original level?

14. *Galileo's Problem:* By substituting $d_k = ak^2 + bk + c$ in the difference equation

$$d_{n+2} - 2d_{n+1} + d_n = 32,$$

show that

$$d_k = ak^2 + c,$$

for a and c arbitrary constants. If $d_0 = 0$, and $d_1 = 16$, show that

$$d_k = 16k^2.$$

15. *Method of Conjecture:* Show by substitution that

$$a_k = r^k a_0$$

is a solution to the linear, first-order, homogeneous difference equation with constant coefficient r:

$$a_{n+1} = ra_n.$$

SECTION 1.6 *Nonhomogeneous Linear Difference Equations*

You receive your credit card bill. You note that the minimum payment is $20 and the interest rate is 12% per year billed 1% per month. You have frequented the bookstore buying some really interesting math books, and so you have run up a not-so-arbitrarily-small bill. Because your hard-earned summer funds have dried up, you decide to pay the minimum amount. You wonder how long it will take you to pay the bill if each month you pay the minimum amount. You have given your credit card to your younger brother for safekeeping, so you are sure there will be no new charges. Letting a_n equal the amount due at the beginning of month n results in the model

$$a_{n+1} = a_n + 0.01a_n - 20. \tag{1}$$

You note that Eq. (1) is linear with constant coefficients but is *nonhomogeneous*. You have solved nonhomogeneous equations before when the first or second differences were constant. But that is not the case here. You would like to find a solution for all possible initial values. Remembering that you can build a numerical solution for particular initial conditions, you decide to explore a bit with your trusty computer. Noting that the total amount due is $400, you produce Figure 1.

You note that in month 23, the amount turns negative, which means you have finally paid the bill (and about $60 interest to pay off the $400!). While you were on the computer, your roommate handed you your mail—no letters from home, but a new credit card bill and a note from your brother. The total amount due is $2500. The note from your brother states he was in a bind and charged his room and board to your account with a promise to "catch up" later. He also suggested contacting home if you needed help since his parental credit line was exceeded—some brother! You decide to deal with him later, and then you change a cell in your spreadsheet knowing you can afford only the $20 monthly payment (see Figure 2).

You look at the increasing interest and amount owed and realize for the first time why the credit card company considered you such a fine customer and raised your credit line as you continued to pay the minimum amount. After reaching for the antacid, your curiosity is piqued. What if the amount owed were $2000 and you continued to pay $20 per month? (See Figure 3).

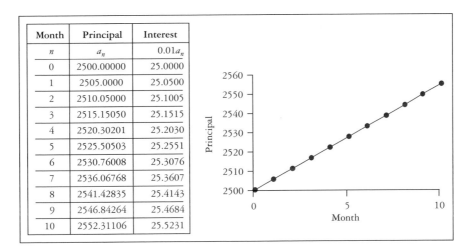

Month	Principal	Interest
n	a_n	$0.01a_n$
0	400.0000	4.00000
1	384.0000	3.84000
2	367.8400	3.67840
3	351.5184	3.51518
4	335.0336	3.35034
5	318.3839	3.18384
6	301.5678	3.01568
7	284.5834	2.84583
8	267.4293	2.67429
9	250.1036	2.50104

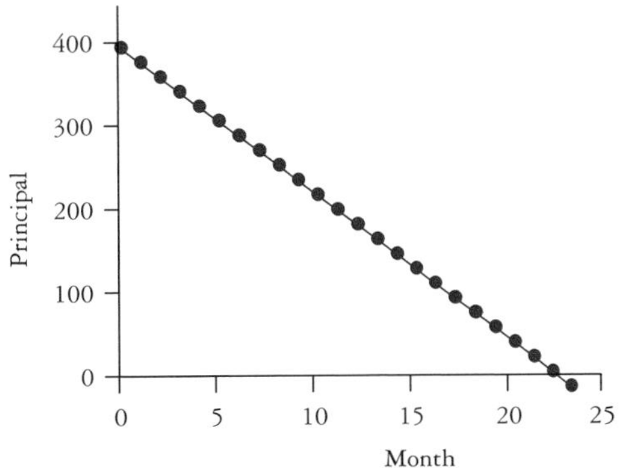

FIGURE 1 Initial balance of $400.

Month	Principal	Interest
n	a_n	$0.01a_n$
0	2500.00000	25.0000
1	2505.0000	25.0500
2	2510.05000	25.1005
3	2515.15050	25.1515
4	2520.30201	25.2030
5	2525.50503	25.2551
6	2530.76008	25.3076
7	2536.06768	25.3607
8	2541.42835	25.4143
9	2546.84264	25.4684
10	2552.31106	25.5231

FIGURE 2 Initial balance of $2500.

Very interesting! No wonder the credit card company is happy to have customers pay the "minimum amount"! You overlay the three graphs you have produced (Figure 4).

Let's analyze Figure 4. Whether you pay off the bill or not seems to be sensitive to the initial condition. The models are the same—only the initial conditions vary. Yet the three graphs are remarkably different and predict wildly distinct conclusions! You note that if the interest equals exactly the minimum payment of $20, the total amount owed remains constant. If the amount owed were greater than $2000, then the interest exceeds your payment and each month the interest increases. If the amount owed were less than $2000, each month you could not only pay the interest but a portion of the bill.

Month	Principal	Interest
n	a_n	$0.01a_n$
0	2000	20
1	2000	20
2	2000	20
3	2000	20
4	2000	20
5	2000	20
6	2000	20
7	2000	20
8	2000	20
9	2000	20
10	2000	20

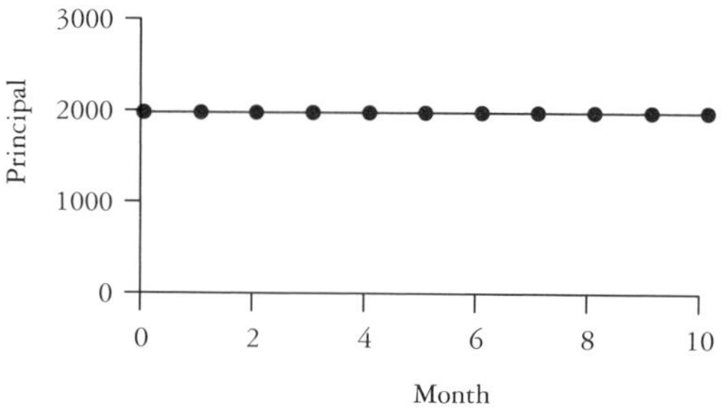

FIGURE 3 Initial balance of $2000.

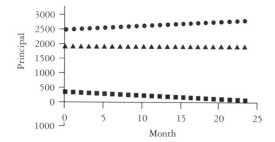

FIGURE 4 An unstable equilibrium.

In addition, each month the interest would decrease, allowing you to pay off a greater portion of the bill. So the shapes of the graphs in Figure 4 seem logical. (Note that while the graphs for the initial values of $400 and $2500 appear to be lines in Figure 4 because of the scale, they are really nonlinear, as can be seen by examining the appropriate table.)

But what kind of function can represent all three graphs and allow the *constant solution* $a_k = 2000$? The constant solution $a_k = 2000$ is an *equilibrium solution* in the sense that if you start there, you remain there forever. Furthermore, it is *unstable* in the sense that if you start close, you do not remain close. In

fact, you diverge from the equilibrium value. Try $a_k = 2000.01$ and $a_k = 1999.99$.

You are anxious to find the general solution so you can analyze alternative payment schemes before writing home for help. Noting the nonlinear nature of two of the solutions in Figure 4 and remembering that polynomials solved the class of difference equations when the differences were constant, you conjecture a quadratic solution function of the form $a_k = bk^2 + ck + d$, for constants b, c, and d.

Equation:

$$a_{n+1} = 1.01a_n - 20$$

Conjecture:

$$a_k = bk^2 + ck + d$$
$$a_{k+1} = b(k + 1)^2 + c(k + 1) + d$$

Substitution:

$$b(k^2 + 2k + 1) + ck + c + d$$
$$= 1.01(bk^2 + ck + d) - 20$$

$$bk^2 + (2b + c)k + b + c + d$$
$$= 1.01bk^2 + 1.01ck + 1.01d - 20$$

Equating coefficients of like powers of k:

$$b = 1.01b \rightarrow b = 0$$

$$2b + c = 1.01c \rightarrow c = 0$$

$$b + c + d = 1.01d - 20 \rightarrow 0.01d = 20 \rightarrow d$$
$$= 2000$$

Solution:

$$b = 0, \quad c = 0, \quad d = 2000,$$

or

$$a_k = 2000.$$

Very interesting—your conjecture produced the constant equilibrium solution but did not pick up the nonlinear solutions. You need another conjecture—a quadratic will not suffice. However, you've become so anxious about the bill that your creative conjecturing brain cells have closed down; all this finance has caused you to work up quite a sweat.

You go to the refrigerator and remove the last cold soda, noting that the refrigerator temperature is exactly 40°F. You place a few cans of soda in the refrigerator since it was really your roommate's soda in the first place. You realize your roommate is due back in a few minutes and you have already taken a sip from that last cold can. A few minutes—will that be enough to get the cans you have added "reasonably close" to 40°F? Thanking the stars that you know so much mathematics, you hypothesize that the *change* in temperature each minute would be proportional to the difference between the temperature of the refrigerator (which you assume remains at 40°F) and the temperature of the soda, which initially is at room temperature, 72°F. Letting t_n rep-

resent the temperature of the can after n minutes, you have the following model:

$$t_{n+1} = t_n - r(t_n - 40),$$

$$t_0 = 72,$$

for r a constant of proportionality. But what constant? Aha! You grab your math notebook and refer to Exercise 8 in Section 1.4—there was a purpose in working that mindless problem! After referring to the answers in the back of the book, you write the following difference equation:

$$t_{n+1} = t_n - 0.0081(t_n - 40) = 0.9919t_n + 0.324,$$

$$t_0 = 72,$$

where -0.0081 is the constant of proportionality from Exercise 8 of Section 1.4. You note that the difference equation is again nonhomogeneous—my math professor said there would be days like this! Back to the computer to find a numerical solution by iteration (see Figure 5).

You note that the temperature of the soda approaches 40°F but never quite gets there, the difference becoming *arbitrarily small.* You hope you do not have to arbitrate with your roommate over the difference. What if the initial temperature of the soda had been 35°F? (See Figure 6.)

Or 40°F? (See Figure 7.)

You overlay the three graphs (Figure 8).

Do the graphs in Figure 8 make sense for the problems at hand? What kind of function can allow those three solutions? You start to think about what is likely to happen. The refrigerator is kept at a temperature of 40°F, so that initially there is a large difference between the temperature of the soda and that of the refrigerator. Eventually, the temperature of the soda will be fairly close to the temperature of

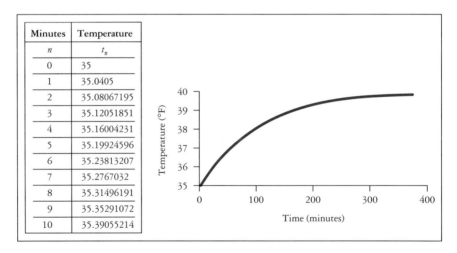

Minutes	Temperature
n	t_n
0	72
1	71.7408
2	71.4836995
3	71.2286816
4	70.9757292
5	70.7248258
6	70.4759547
7	70.2290995
8	69.9842438
9	69.7413714
10	69.5004663
11	69.2615125
12	69.0244943

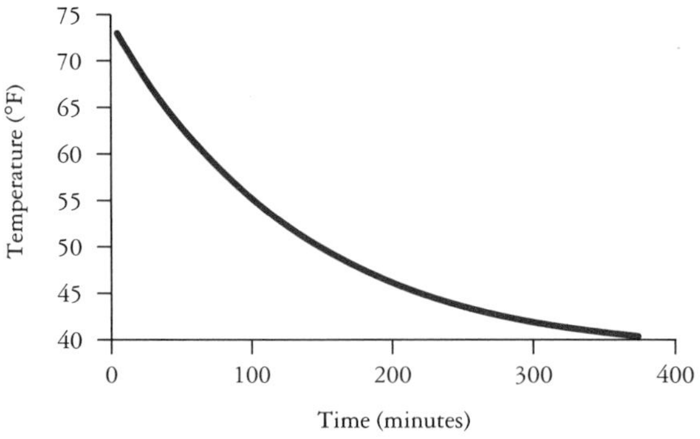

FIGURE 5 The cooling of a warm soda.

Minutes	Temperature
n	t_n
0	35
1	35.0405
2	35.08067195
3	35.12051851
4	35.16004231
5	35.19924596
6	35.23813207
7	35.2767032
8	35.31496191
9	35.35291072
10	35.39055214

FIGURE 6 An initial temperature of 35°F.

Minutes	Temperature
n	t_n
0	40
1	40
2	40
3	40
4	40
5	40
6	40
7	40
8	40
9	40
10	40

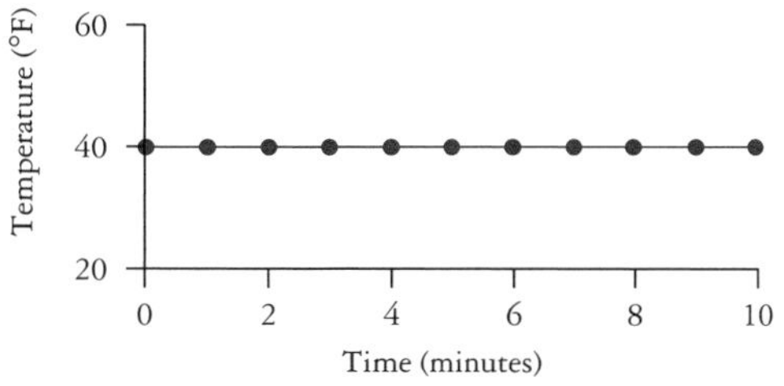

FIGURE 7 An initial temperature of 40°F.

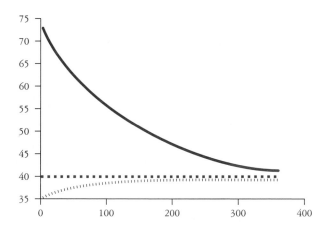

FIGURE 8 A stable equilibrium at 40°F.

the refrigerator. What kind of functional behavior best describes the way the temperature of the soda decreases? Is it linear with a negative slope? Is it decreasing and concave down? Is it decreasing and concave up? After a little thought, you decide that it is much harder to drop the last few degrees than the first few. You expect that the behavior pattern for the temperature of the soda will drop faster at the beginning and less so later on. That is, the pat-

tern is one where the temperature drops in a concave-up fashion. Just what the graph shows! You analyze the other two graphs in a similar manner.

From the graph you again note that there is a constant solution, $t_n = 40$, an equilibrium value for the problem. This time, however, it appears to be *stable*, in the sense that if you start close, you remain close apparently forever. You try initial values of 39.99 and 40.01, and note that not only do you remain close to the equilibrium value, but you come closer and closer as time goes on. The stability of a difference equation is of paramount concern in mathematics and its applications. For example, engineers desire that the mathematical model representing vibrations in a bridge or automotive suspension system be stable—oscillations caused by external forces should die away quickly. But how do practitioners know if the system is stable, like the cooling of soda, or unstable, like your personal finances?

Having seen the numerical solution to several problems with various initial conditions, you are anxious to solve the entire class of difference equations of the form

$$a_{n+1} = ra_n + b \qquad (2)$$

for *any* initial condition. In other words, you want a formula that is a general solution. Twice you have observed that there is a constant solution if the initial condition is equal to the equilibrium value. A difference equation such as Eq. (2) has the value a as an equilibrium value if, whenever $a_n = a$ for some particular n, it follows that $a_{n+1} = a$ as well.

To find the equilibrium value for Eq. (2), you substitute $a = a_{n+1} = a_n$:

$$a = ra + b.$$

Solving for a, you obtain the desired formula,

$$a = \frac{b}{(1-r)}, \quad \text{for } r \neq 1. \qquad (3)$$

You are anxious to continue the search for the solution to Eq. (2) but you cannot help checking your work. For the credit card model, $r = 1.01$ and $b = -20$, which upon substitution into Eq. (3) yields $a = 2000$, in agreement with your numerical work. Similarly, for the soda model, $r = 0.9919$, $b = 0.324$, which upon substitution gives $a = 40$, again as expected. So far so good. As a check on your work, you substitute the formula for the equilibrium value in Eq. (2) to verify that it is indeed a solution:

Equation:

$$a_{n+1} = ra_n + b$$

Solution:

$$a_k = a_{k+1} = \frac{b}{(1-r)}$$

Substitution:

$$\frac{b}{(1-r)} = r\frac{b}{(1-r)} + b$$

Identity:

$$\frac{b}{(1-r)} = \frac{b}{(1-r)}.$$

Interesting, but you realize that the constant solution will not generate the nonlinear solutions that you obtained numerically. Again you look at the graphs. You have tried a polynomial. You look at Eq. (2) and note that the associated homogeneous equation is

$$a_{n+1} = ra_n, \qquad (4)$$

which has the solution $a_k = r^k c$ for c, a constant dependent on the initial condition. We call this an exponential solution, since the variable k is an exponent of the solution function. Placing the associated homogeneous equation next to the class of equations you are trying to solve,

Nonhomogeneous equation: $a_{n+1} - ra_n = b$, (5)

Associated homogeneous equation: $a_{n+1} - ra_n = 0$, (6)

you realize that the exponential satisfies Eq. (6) and the constant equilibrium solution satisfies Eq. (5). On a hunch, you add the two solutions together to form a conjecture:

Equation:

$$a_{n+1} = ra_n + b$$

Conjecture:

$$a_k = r^k c + \frac{b}{(1 - r)}$$

$$a_{k+1} = r^{k+1} c + \frac{b}{(1 - r)}$$

Substitution:

$$r^{k+1} c + \frac{b}{(1 - r)} = r \frac{r^k c + b}{(1 - r)} + b$$

Identity:

$$r^{k+1} c + \frac{b}{(1 - r)} = r^{k+1} c + \frac{b}{(1 - r)}.$$

Eureka! Finally, you have solved the following class of equations for all constants r (except $r = 1$; do you see why this is an exception?) and b and for all possible initial conditions:

Class of difference equation: $\quad a_{n+1} = r a_n + b, \quad$ (7)

General analytic solution: $\quad a_k = r^k c + \dfrac{b}{(1 - r)},$

(8)

where $b/(1 - r) = a$ is the equilibrium value. Now you are ready to explain why some equilibria are stable and some are unstable. Rewriting the analytic solution by abbreviating the equilibrium value a, you get

$$a_k = r^k c + a.$$

You realize that a will be approached only if the term $r^k c$ goes to 0. This happens if $|r| < 1$. Clearly, if $|r| > 1$, then $r^k c$ will eventually become large compared to a, making a unstable. You decide to investigate later the case for $r = 1$, which you have been avoiding.

Anxious to apply your knowledge, you decide to answer several questions that have been bugging you. First, you hope to pay off your debt in 12 monthly installments with some subsidy from home, given your brother's antics. Knowing your parents will not "bite" without some evidence, you form the appropriate model and solve it.

EXAMPLE 1 The Credit Card Given an initial debt of \$2500, what monthly payment will be required to pay off the debt in 12 installments assuming no new purchases and 1% monthly interest? Letting a_n be the amount owed *after* making the nth payment, you want to find a monthly payment p such that $a_{12} = 0$. Forming the model,

$$a_{n+1} = a_n + 0.01 a_n - p,$$

$$a_{n+1} = 1.01 a_n - p,$$

$$a_0 = 2500,$$

$$a_{12} = 0.$$

You note that $r = 1.01$ and that the equilibrium value is $a = 100p$. Thus, the solution is

$$a_k = (1.01)^k c + 100p.$$

To find the constants c and p, you substitute $k = 0$ and $k = 12$:

$$a_0 = 2500 = c + 100p,$$

$$a_{12} = 1.01^{12} c + 100p.$$

Solving the equations simultaneously, you find that $c = -19{,}712.20$ and $p = 222.13$, giving the solution for the given conditions:

$$a_k = 1.01^k(-19{,}712.20) - 222.13.$$

Strange Attractors

The "connect-the-dots" diagram on the right shows how the two-dimensional dynamical system of equations

$$x_{n+1} = \left(\frac{1}{3}\right) x_n - \left(\frac{\sqrt{3}}{3}\right) y_n,$$

$$y_{n+1} = \left(\frac{\sqrt{3}}{3}\right) x_n - \left(\frac{1}{3}\right) y_n$$

moves toward an equilibrium point E over time, starting at $x_0 = 0$, $y_0 = 2$. Notice that this is a different kind of plot from the others in this section: There is no axis representing n. Just recently, mathematicians have turned their attention to processes like those in the diagram at the right, where the system meanders around instead of tending toward equilibrium. (In this figure, the

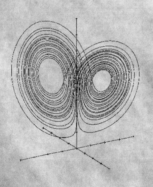

dots to be connected are close enough that they have the appearance of a smooth curve.) However, the meandering is not random. There seems to be a basic underlying shape around which the system wanders. This shape is called a *strange attractor*. The figure on the left is derived from a weather model. The "patterned wandering" that it shows seems to conform to what we know

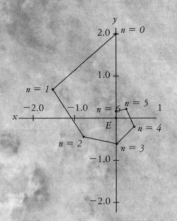

about the weather: It is constantly changing (no equilibrium), but its overall pattern of change is predictable even though its day-to-day variations are unpredictable. Some scientists think that the study of strange attractors will help us understand dynamical systems like the weather.

Thus, you can pay the bill in 12 months with monthly installments of $222.13, and your total interest will be about $165.56 (because $(12)(222.13) - 2500 = 165.56$). ▲

You write a note to Dad, sending him Example 1 and asking him to pay the 12 monthly payments

of $222.13. You throw in a few spreadsheets and your analysis to impress him. You add your formula for the general case, which he can use to solve his annuity investments and those mortgage options with which he has been wrestling. Then you remember he has been investigating new car loans for that new family car he thinks he can afford when you graduate. The loan options seem mindboggling—three-, four-, and five-year payoff periods and each with a different interest rate. You realize the model should have the same form as the equations you just solved. After analyzing each situation by using your solution for the general case to the particular problems, you decide that your consulting services deserve the $400 you owe the credit card company. You decide to go for the whole bundle. You want to formalize your finding on first-order linear difference equations of the form

$$a_{n+1} = ra_n + b$$

by stating and proving a theorem. But first, what about the case $r = 1$ that you have been neglecting? (The equilibrium value $b/(1 - r)$ is not defined for $r = 1$.) However, if $r = 1$, then the difference equation becomes

$$a_{n+1} = a_n + b,$$

which means the first difference is the constant b, a difference equation you have already solved. You state your theorem.

Theorem 1. The solution to the first-order linear difference equation $a_{n+1} = ra_n + b$ is

$$a_k = r^k c + \frac{b}{(1 - r)}, \quad \text{if } r \neq 1, \quad \text{and}$$

$$a_k = bk + c, \quad \text{if } r = 1.$$

You include your proof by conjecture, remembering that Dad likes you to back up your statements. Then, you remember your roommate—how long will that hot soda you placed in the refrigerator take to get to 42°F?

EXAMPLE 2 Hot Soda Given that a soda at 72°F was placed in a refrigerator that you expect to remain at 40°F, how long will it take for the soda to reach a temperature of 42°F, assuming a constant of proportionality of 0.0081? Letting t_n be the temperature after n minutes, you restate the model you formulated earlier:

$$t_{n+1} = t_n - 0.0081(t_n - 40),$$

$$t_{n+1} = 0.9919t_n + 0.324,$$

$$t_0 = 72.$$

You solve for the equilibrium value a:

$$t_{n+1} = t_n = a = \frac{0.324}{(1 - 0.9919)} = 40,$$

giving the solution for any k and any initial condition

$$t_k = 0.9919^k c + 40.$$

To find c, you substitute the initial temperature:

$$t_0 = 72 = c + 40,$$

$$c = 32.$$

This gives the solution to your problem:

$$t_k = (0.9919)^k 32 + 40.$$

To find k for $t_k = 42$, you substitute and solve:

$$42 = 0.9919^k(32) + 40,$$

$$2 = 0.9919^k(32),$$

$$\ln(\tfrac{1}{16}) = k \ln(0.9919),$$

$$k = 340.$$

The solution of 340 minutes is more than 5 hours! Knowing only 20 minutes have passed, you quickly compute the temperature for $k = 20$:

$$t_{20} = 0.9919^{20}(32) + 40 = 67.$$

Not nearly cold enough—too close to the room temperature. Just then, your roommate enters and goes straight to the refrigerator. Remembering the modeling process, you assume the existence of some ice, say hello, and go to the movies. ▲

At the movies you see *Jurassic Park.* The hero is a mathematician! Someone who develops models to explain behavior and predict results *before* mindless experimentation! Nonlinear difference equations, unstable equilibria, chaotic behavior, fractal patterns, sensitivity to initial conditions. . . . You re-member your math professor promised to introduce you to these fascinating subjects in Section 1.8 if you handled the linear difference equations well. This linear information, especially the nonhomogeneous equations, has been interesting and practical, but that nonlinear world seems downright fascinating. Your interest piqued, you absolutely cannot wait to get to the next math lesson, but you realize you bet-ter do the assigned exercises if you have a hope of grasping what comes next. Before doing so, you de-cide to finish your note to Dad. You do some research to see if your results are publishable and perhaps even get a *grant* from Dad, who would be so impressed. You look up "cooling" in the Index, *Newton's law of cooling,* Isaac Newton, calculus, differential equa-tions, similar format and same terminology as dif-ference equations but time is allowed to vary contin-uously, similar solution techniques. Isaac Newton— as a kid you always appreciated the fig cookies he invented—but calculus, some genius!

Exercises for Section 1.6

1. For each of the exercises below, find the equilibrium value (solution) if one exists. Find the solution to the difference equation and discuss the long-term behavior of the solution for various initial or starting values. Is the equilibrium stable or unstable? Why?

   (a) $a_{n+1} = 0.5a_n + 30$
   (b) $a_{n+1} = 3a_n + 40$
   (c) $a_{n+1} = a_n + 20$
   (d) $a_{n+1} = 2a_n - 20$
   (e) $a_{n+1} = 0.8a_n - 30$
   (f) $a_{n+1} = a_n - 20$

2. *Your Graduation Car:* You are considering the pur-chase of one of three cars:

   (a) Sporty—it's "you," but it costs $20,000
   (b) Basic—a new car with no "frills" that costs $10,000
   (c) Nerdo—it's "not you," a used car at $6000

You visit the bank. Based upon your pending "recent college graduate" status, the bank waives a down pay-ment and offers you three possible loan options:

Option 1: a five-year loan at 1% interest per month

Option 2: a four-year loan at 0.67% interest per month

Option 3: a three-year loan at 0.5% interest per month

You must finance the entire purchase price of the car you

decide to buy. You are concerned about the monthly payments but do not want to reduce the monthly payment at "any cost." To help you make your decision, for each of the cars and loan options, compute the monthly payment and total interest paid for the loan. Should you go "sporty," or buy the used car your math professor has been driving?

3. *Paying off the Mortgage:* You are advising your parents on whether or not to pay off their mortgage. They financed $120,000 for 20 years and have made 72 monthly payments (6 years). The monthly interest rate is 0.75%. They have enough cash through an inheritance to pay off the mortgage, but the bank tells them to use the money to make an investment instead.

   (a) What are their monthly payments?

   (b) How much do they owe after 72 payments?

   (c) (Before-tax analysis) Neglecting any "tax break," at what monthly interest rate would they need to invest the inheritance in order to "break even"—that is, investing the inheritance instead of paying off the mortgage? Assume the inheritance equals exactly the amount they currently owe on the mortgage.

   (d) (After-tax analysis) Assume your parents are in the 30% income tax bracket. Their situation is such that any interest paid on the loan will reduce their "gross income," the base on which their taxes are computed. To assist them in making a decision, compute the amount of interest they will pay each of the remaining years.

   (e) What is your advice now on paying off the mortgage instead of investing the money? What rate of return on the investment do they need to break even after taxes?

4. *Annuities:* You are advising your grandparents on investment options. They would like to supplement their monthly income for the next 10 years until other investments mature. They plan to place $50,000 in a savings account that pays 0.5% interest each month. They have asked you how much they should withdraw each month in order to reduce the amount remaining in the account to 0 after 10 years, or 120 months.

5. *Financing an Education:* You plan to invest part of your paycheck after graduation to finance your children's education. You want to have enough in the account to draw $1000 a month every month for 8 years beginning 20 years from now. The account pays 0.5% interest each month.

   (a) How much will you need 20 years from now in order to accomplish the financial objective? Assume you stop investing when your first child begins college—a safe assumption!

   (b) How much must you deposit each month over the next 20 years?

SECTION 1.7 *First-Order Nonlinear Difference Equations*

In Example 2, in Section 1.4, we investigated the change in the population of a yeast culture that was living in a restricted area with a finite amount of resources to support the population. We saw in the model and its solution that as the population grew, the resources became constrained, and the growth slowed. If p_n is the population size at time period n,

k is the growth rate, and b is the carrying capacity of the environment, then the difference equation that models the population change is

$$p_{n+1} = p_n + kp_n(b - p_n). \tag{1}$$

This model is called the *logistic equation.* Since the expanded form

$$p_{n+1} = p_n + kp_n b - kp_n^2$$

contains the *nonlinear* term p_n^2, Eq. (1) is a *nonlinear difference equation.*

E X A M P L E 1 Constrained Population If we have a population under constrained growth with $k = 0.0001$, $b = 10,000$, and $p_0 = 50$, use iteration and a graph to characterize the change of the population over the first 50 time periods. Substituting the given values of k and b into Eq. (1) produces the difference equation

$$p_{n+1} = 2p_n - 0.0001p_n^2. \qquad (2)$$

Iteration starting with $p_0 = 50$ produces the sequence {50, 99.75, 198.5, 393, 770, 1482, 2744, 4735, 7228, . . .}. The graph of the population over the 50 time periods is shown in Figure 1. ▲

10,000 is an equilibrium value for the equation. The population value of 0 is also an equilibrium value for this equation. Equilibrium values are found by substituting a common variable (we will use "q" for the equilibrium value) into the equation for both p_n and p_{n+1} and solving for q in this equilibrium equation. For this population model, the equilibrium equation is $q = 2q - 0.0001q^2$ and the solutions are $q = 0$ and $q = 10,000$.

E X A M P L E 2 New Approach to Equilibrium Iterate and graph the difference equation from Example 1, $p_{n+1} = 2p_n - 0.0001p_n^2$ for 15 time periods starting with $p_0 = 12,000$. This initial value provides a population greater than the carrying capacity. The graph of this sequence is shown in Figure 2. ▲

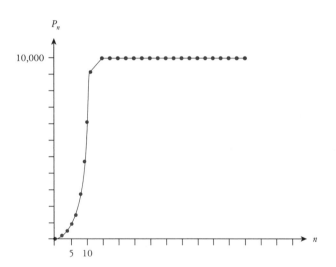

FIGURE 1 Connected graph of the population model Eq. (2) starting with $p_0 = 50$.

Equilibrium Value

In this problem, the population grows steadily, approaching the carrying capacity of 10,000. If the population reaches 10,000, it stays there; therefore,

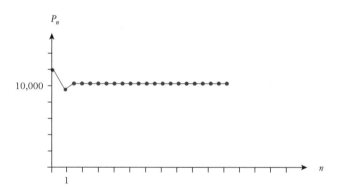

FIGURE 2 Connected graph of the population model Eq. (2) starting with $p_0 = 12,000$.

The graph in Figure 2 also shows convergence to the equilibrium value of 10,000. The iterations start above 10,000; dip below 10,000 and eventually increase, tending toward 10,000. These two experiments suggest that 10,000 is a *stable equilibrium* value for Eq. (1), since further iteration of population values close to 10,000 become even closer to that equilibrium value. On the other hand, we could conduct experiments to show that the equilibrium value of 0 is *unstable* and that the population will move farther away from this equilibrium value (unless the population is exactly 0, in which case it will remain at 0). It must be noted that experiments involving a finite number of iterations or even a finite number of experiments can never guarantee stability. One never knows what will happen in the next iteration or experiment.

Stability is related to the limit concept introduced in Section 1.3 and therefore involves an infinite process. If the difference equation starts with an initial condition near its equilibrium value, then for a stable equilibrium the limit value of the sequence produced by the difference equation is the equilibrium value. These terms are defined as follows:

DEFINITION

Suppose that the difference equation defining a_n has an equilibrium value q. Then q is a *stable equilibrium* if there is a number $\epsilon > 0$ such that when $|a_0 - q| < \epsilon$, then $\lim_{n \to \infty} a_n = q$. If the equilibrium value is not stable, then it is *unstable*.

No Analytic Solution

We have seen how helpful numerical and graphical solutions produced by iteration are to conduct analysis of solution behavior. Iteration and graphing are always possible for reasonably small values of n, but often they are not possible for extremely large values of n. In the previous two sections, we saw how useful the analytic solution is to understand the long-term behavior of a solution or to determine values of the solution for large values of the index n. Therefore, we desire general, analytic solutions to nonlinear difference equations, like those we determined in the

previous two sections for linear equations. Unfortunately, there is no general theory for solving nonlinear equations, and only in special cases is it possible to find an analytic solution for such equations. Therefore, for now, we have to be satisfied with iteration and graphing to try to conduct analysis of the long-term behavior for nonlinear equations. In another course you will see how calculus can help with this analysis to determine the stability of equilibrium values and other properties of the long-term behavior of solutions to nonlinear difference equations.

EXAMPLE 3 Experiment on Changing Behavior   Conduct an experiment on the difference equation

$$a_{n+1} = (1 + r)(a_n - a_n^2), \qquad (3)$$

with $a_0 = 0.2$ for four different values of parameter r, $r = 1.9, 2.4, 2.55$, and 2.7. Iterate for the first 40 values, and produce a linearly connected graph for each of the four values of r. A linearly connected graph contains straight line segments to connect the plotted points of the sequence. Determine if it is possible to use the graphs produced to predict future values of a_n for values of n much larger than 40.

For $r = 1.9$, the graph of the first 40 iterations of Eq. (3) is given in Figure 3. The graph suggests that the solution of Eq. (3) is converging to a stable equilibrium value of about 0.65. If this convergence continues in the form of a limit as $n \to \infty$, future values of a_n can be predicted to be at or extremely near the equilibrium value.

For $r = 2.4$, the graph of the first 40 iterations of Eq. (3) is shown in Figure 4. The graph shows that after a period of adjustment, the solution of Eq. (3) oscillates between two values of

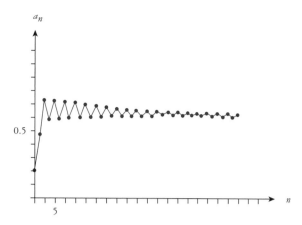

FIGURE 3 Graph of the iteration solution of Eq. (3) for $r = 1.9$.

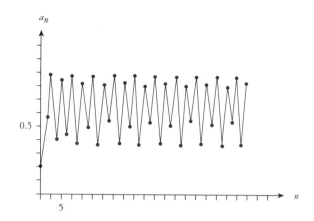

FIGURE 5 Graph of the iteration solution of Eq. (3) for $r = 2.55$.

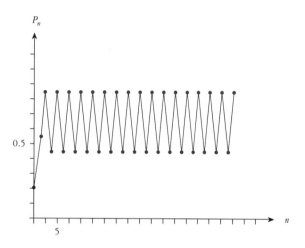

FIGURE 4 Graph of the iteration solution of Eq. (3) for $r = 2.4$.

approximately 0.42 and 0.82. This kind of oscillation is called a *two-cycle*. Once this pattern is established in the solution, it is easy to predict future values of a_n.

For $r = 2.55$, the graph of the first 40 iterations of Eq. (3) is shown in Figure 5. This graph shows a similar pattern as that for $r = 2.4$, except the oscillations repeat over four values. This

kind of patterned oscillation is called a *four-cycle*. Once again, future values of a_n can be predicted quite easily because of this repeating pattern. The act of splitting from a single, stable equilibrium to a two-cycle and from a two-cycle to a four-cycle is called *bifurcation*.

Some interesting questions arise about the overall behavior of solutions to Eq. (3). For what value of r does the solution change from approaching a stable equilibrium to a two-cycle? For what value of r does the two-cycle change to a four-cycle? Is there a three-cycle or other types of behavior between the two-cycle and the four-cycle?

Finally, for $r = 2.7$, the graph of the first 40 iterations of Eq. (3) is given in Figure 6. You need to look carefully at this graph. It does not contain any of the predictable patterns of the previous graphs of solutions. Absent of any pattern or repetition, the general long-term behavior and specific values of a_n for large n are impossible to predict from the graph. Of course, we only calculated 40 values of a_n. There may be repetition over a larger interval of values of n. After per-

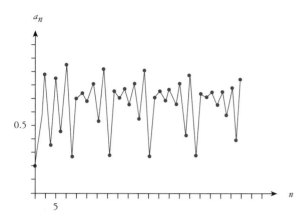

FIGURE 6 Graph of the iteration solution of Eq. (3) for $r = 2.7$.

tive dependence on r and a_0 is an aspect of what has been called mathematical *chaos*. Although "chaos" has been in the mathematical literature for some time, it was poorly understood; hardly anyone knew about it or thought it worth studying. In recent decades, largely because of the computer, our familiarity with chaos has increased. It has turned out to be surprisingly widespread, and our knowledge of chaos is growing greatly. Many mathematicians and scientists are working in this field to better understand chaos and the strange behavior often found in the solutions of nonlinear difference equations. ▲

forming this fourth calculation, we could ask more questions. Did the solutions continue to bifurcate? What values of r produce an eight-cycle? Sixteen-cycle? Can we be sure that $r = 2.7$ does not produce a cycle greater than 40, like a 64-cycle? These questions reveal some of the problems of this type of experimentation: We never really know how far to calculate and look for patterns.

DEFINITION

Chaos is the inherent unpredictability in the behavior of a natural system. More specifically, a difference equation is called *chaotic* if its solutions depend sensitively on the initial conditions.

The solutions of Eq. (3) not only show a sensitive dependence on r, but they can also exhibit a *sensitive dependence on initial conditions,* since a small shift in a_0 causes a big alteration in the long-term behavior and, therefore, unpredictability. Experiments showing the sensitive dependence on a_0 are contained in the exercises for this section. Sensi-

Cobweb Graphs

Besides plotting the solution sequence, there is another graphical method that enables direct visualization of iteration and equilibrium values. This graphing technique can be constructed with just the plot of a function determined from the difference equation and a straightedge or ruler. The resulting graph is called a *cobweb graph,* or just *cobweb.* We will show an example first and then outline the general algorithm for producing cobwebs.

EXAMPLE 4 Cobweb Construction Produce the cobweb graph for the linear, nonhomogeneous difference equation

$$c_{n+1} = -0.5c_n + 4, \quad \text{with } c_0 = -4. \qquad (4)$$

Convert the difference equation into an equation of continuous variables by replacing c_{n+1} with y and c_n with x. This produces the equation

$$y = -0.5x + 4.$$

Then, plot the two curves (lines in this case), $y =$

Chaos

In its early history, mathematics was mostly applied to slowly and smoothly varying phenomena. Our forefathers gave less thought to natural phenomena that gave rise to data points like those in the plot below. But much of life is like that plot—full of sudden (and therefore surprising) changes. Recently, difference equations related to simple natural phenomena have been discovered which give data points that career

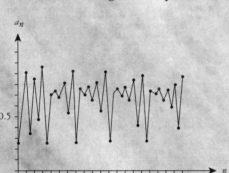

sharply and suddenly all over the lot. For example, the data in the figure on the right are generated by the following equation, which describes the growth of simple organisms like yeast in a test tube:

$$a_{n+1} = 3.7(a_n - a_n^2).$$

In this equation a_n is a measure (on a scale of zero to one) of how close the number of yeast comes, at a time period n, to the maximum limit for the test tube.

How could such a simple-appearing equation, involving only ordinary algebra, give such a chaotic pattern of population levels? And if this can happen with this equation, is it possible that many things we think we understand can "get out of hand" and give unpredictable results? Many mathematicians are coming to feel that chaos often lurks within things we think are under control.

The scene above, from the movie *Jurassic Park,* shows a dinosaur being hatched after being cloned from fossil DNA. In this film, it is a mathematician (the sexy guy in the sunglasses) who warns that the attempt to clone dinosaurs and control their reproductive behavior would fail. He intuitively understood, from having studied much simpler situations like the equation above, that there was a real chance of things "getting out of hand" even when they appear simple.

x and $y = -0.5x + 4$, on the same set of coordinate axes. Start the graphical iteration along the x-axis at $x = c_0 = -4$. Draw a vertical line from the initial value $x = c_0 = -4$ to the graph of $y = -0.5x + 4$. The point of intersection will have height $y = c_1 = -0.5c_0 + 4$. (This produces c_1.) From there, draw a horizontal line to the graph of $y = x$. Now we are at the point (c_1, c_1). We draw the vertical line through this point; it has the equation $x = c_1$. Thus, it crosses the graph of $y = -0.5x + 4$ at a point of height c_2. Continue to alternate the drawing of horizontal lines (to $y = x$) and vertical lines (to $y = -0.5x + 4$), always starting from where you left off. We can perform this graphical iteration as long as the size of the plots allows. The cobweb for $c_{n+1} = -0.5c_n + 4$ starting with $c_0 = -4$ is presented in Figure 7.

Notice that the cobweb in Figure 7 tends toward the equilibrium value of $q = 2.66$. This is the kind of behavior we expect when the equilibrium value is stable. ▲

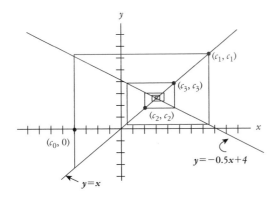

FIGURE 7 Cobweb graph for Eq. (4).

Producing Cobwebs

The general algorithm to produce a cobweb graph for a difference equation is as follows:

1. Plot the graph of the difference equation and that of $y = x$ on a set of coordinate axes.

2. Draw a vertical line from the initial value on the x-axis to the graph of the difference equation.

3. Draw a horizontal line from the point on the graph of the difference equation to the line $y = x$.

4. Draw a vertical line from the point on the line $y = x$ to the graph of the difference equation.

5. Continue steps 3 and 4 until you have achieved the insight you want about the process.

The most important insight we want from a cobweb concerns how the iterated values vary in relation to the equilibrium. The equilibrium can be located where $y = x$ crosses the graph of the difference equation. To see this, consider Example 4 and suppose (q, q) is the point where $y = x$ crosses $y = -0.5x + 4$. Since y stands for c_{n+1} and x stands for c_n, at the crossing point we have

$$c_{n+1} = -0.5c_n + 4,$$

$$c_{n+1} = c_n = q.$$

However, this is what it means for q to be an equilibrium value for the difference equation. Summarizing, the x- (or y-) coordinate of any crossing of $y = x$ with the graph of the difference equation is an equilibrium of the difference equation.

EXAMPLE 5 Cobwebs for Experiment
Draw the cobweb graphs for the difference equation Eq. (3) from Example 3

$$a_{n+1} = (1 + r)(a_n - a_n^2),$$

with $a_0 = 0.2$, for these four different values of r:

$$r = 1.9, 2.4, 2.55, 2.7.$$

The cobweb graph for Eq. (3) showing the cobweb structure of an inward spiral for the stable equilibrium produced when $r = 1.9$ is shown in Figure 8.

The cobweb graph for Eq. (3) showing the cobweb structure for a two-cycle when $r = 2.4$ is shown in Figure 9. Notice that the boxlike shape is retraced over and over again.

The cobweb graph for Eq. (3) showing the cobweb structure for a four-cycle when $r = 2.55$ is shown in Figure 10. In this case, two boxlike structures are retraced repeatedly.

The cobweb graph for Eq. (3) showing the complete lack of a pattern or structure produced by chaos when $r = 2.7$ is shown in Figure 11. If the graphical iteration for the cobweb were continued, it could completely fill in the boxlike region around the curve in the first quadrant. ▲

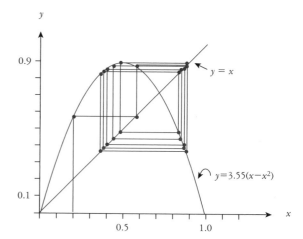

FIGURE 10 Graph of the cobweb graph of Eq. (3) for $r = 2.55$.

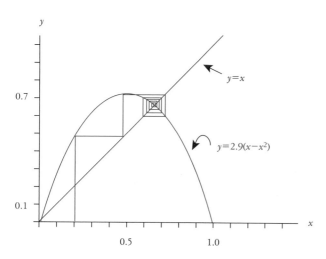

FIGURE 8 Graph of the cobweb graph of Eq. (3) for $r = 1.9$.

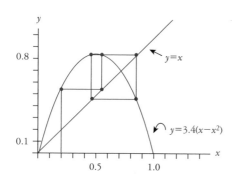

FIGURE 9 Graph of the cobweb graph of Eq. (3) for $r = 2.4$.

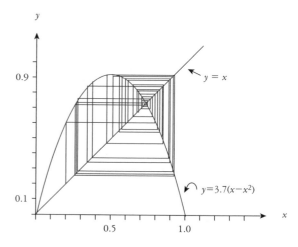

FIGURE 11 Graph of the cobweb graph of Eq. (3) for $r = 2.7$.

EXAMPLE 6 Spread of a Contagious Disease Recall Example 2 of Section 1.4, where the spread of contagious disease was modeled by a difference equation. The general scenario involved a total of 400 students in a dormitory, with i_n students infected with the flu. If all the students are susceptible and the possible interaction of infected and susceptible students is proportional (with constant k) to the product of infected and susceptible students, a model is $i_{n+1} = i_n + ki_n(400 - i_n)$.

If 2 students return from a weekend trip having just contracted a strain of flu that is contagious for two weeks, and the rate of spreading per day of this strain of flu is characterized by $k = 0.002$, how many of the 400 students will contract the flu over the next two weeks?

Substitution of $k = 0.002$ produces the difference equation

$$i_{n+1} = i_n + 0.002i_n(400 - i_n), \quad \text{with } i_0 = 2. \quad (5)$$

Equation (5) is iterated 14 times to produce the population of infected students for each day of the two-week period. These values produce the graph of flu-infected students given in Figure 12.

As seen on this graph, by the 14th day, all 400 students have contracted the flu. Of course, at that time the first two students with the flu are no longer contagious. A related equation can be derived for the recovery model.

We should point out some of the limitations in our model for infected students. In reality, i_n can only be a nonnegative integer, but the values produced by iteration of our model are not all integers. Also, convergence to the equilibrium of the model actually takes infinitely many steps, not 14. However, by $n = 14$, $a_{14} = 399.6$, which for this application is practically 400. The truth is that for this application we have a "rounded-off" version of this model in the back of our mind. We could change the model so that after each step of the iteration, we round off i_n to its nearest integer. The effect of this modification to the model is investigated in an exercise at the end of this section.

The cobweb graph for iteration of Eq. (5) provides visualization of the solution's approach to the equilibrium value of 400. This cobweb graph is given in Figure 13. ▲

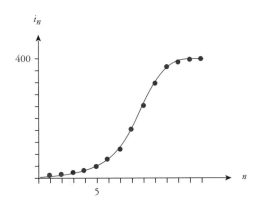

FIGURE 12 Connected graph of the iterated solution of Eq. (5).

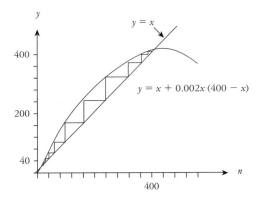

FIGURE 13 Cobweb of Eq. (5), showing the convergence to the equilibrium of $i_n = 400$.

1. Iterate the given difference equation for 10 values and graph these values:
   (a) $a_{n+1} = 2.5a_n - 0.25a_n^2, \quad a_0 = 9$
   (b) $b_{n+1} = b_n - 0.2b_n^3, \quad b_0 = 0.9$
   (c) $c_{n+1} = c_n^3 - 3c_n^2 - 4c_n, \quad c_0 = 3$
   (d) $d_{n+1} = (d_n - 1)/d_n, \quad d_0 = 3$

2. Conduct further experiments on the difference equation in Example 3, $a_{n+1} = (1 + r)(a_n - a_n^2)$, with $a_0 = 0.2$.
   (a) Through experimentation, estimate to two decimal places of accuracy the value of r where the bifurcation takes place from a two-cycle to a four-cycle.
   (b) Through experimentation, find a value of r that produces an eight-cycle.

3. Find all the equilibrium values for the following nonlinear difference equations:
   (a) $a_{n+1} = a_n^2 - 5a_n + 6$
   (b) $b_{n+1} = b_n^3 - 5b_n^2 + 4b_n$
   (c) $c_{n+1} = c_n^2 - 2$

4. Draw a cobweb for the first four iterations of the following difference equations:
   (a) $c_{n+1} = 1.5c_n - 2, \quad c_0 = 2$
   (b) $d_{n+1} = d_n - d_n^2 + 3, \quad d_0 = 2$

5. Explain why the cobweb algorithm works. What is the significance of the line $y = x$?

6. Can the equilibrium value be a complex (or imaginary) number? What is the significance of this condition? Investigate the equilibrium for the difference equation $a_{n+1} = a_n^2 + 2$ using a cobweb graph.

7. Consider a scenario of a rumor spreading through a company with 1000 employees all working in the same building. Assume that spreading the rumor is similar to spreading a contagious disease in that the number of people hearing the rumor each day is proportional to the product of the number who have heard the rumor and the number who have not heard the rumor.

   (a) Write a difference equation model for the number of employees who have heard the rumor.
   (b) If the constant of proportionality is 0.0005 and initially just one person in the company has heard the rumor, how long does it take for half the company to hear the rumor? How long does it take for everyone in the company to hear the rumor?
   (c) The constant of proportionality may depend on several factors such as the information in the rumor and how busy the company is when the rumor is being spread. If the constant of proportionality is doubled from 0.0005 to 0.001, what is the effect on the length of time it takes for the rumor to spread to all the company's employees?

8. Conduct an experiment to show the sensitive dependence of the nonlinear difference equation $a_{n+1} = (1 + r)(a_n - a_n^2)$ with $r = 2.7$ on its initial value a_0. Try to determine the effect on a_{35} of changing a_0 slightly from an original value of 0.2. At a minimum, try the following values: $a_0 = 0.21, 0.19, 0.205, 0.195$. What have you learned from studying the effects of changing a_0? Based on your experimental values, can you accurately predict a_{35} for $a_0 = 0.2025$ and $a_0 = 0.199$?

9. Modify the model of Example 6 to include rounding off the number of infected students at each step (day) of the iteration to the nearest integer. This new model can be described by

$$i_{n+1} = \text{ROUND-OFF}[i_n + 0.002i_n(400 - i_n)].$$

Compare the values for the first 14 days using both the round-off and nonround-off models.

10. Construct a difference equation model for the recovery of the students infected by the flu in Example 6. Assume it takes 14 days to recover from this strain of flu. Be sure to define your variables. (*Hint:* Your equation may have to involve both infected students and recovered students.)

SECTION 1.8 *Systems of Difference Equations*

Most of our discussion thus far has been restricted to a single difference equation containing only one dependent variable. Many problems involve more than one dependent quantity with more than one relationship existing between these quantities. Some of these problems can be modeled by a *system* of two or more difference equations. To ensure that such models are solvable, we need the same number of equations as dependent variables.

E X A M P L E   1   Rental Cars Consider a small rental car firm that operates on an island containing two cities, A and B. The firm has only two agencies, one in each of the two cities. Each day, 10% of the available cars of the agency in city A are dropped off at city B by the customers. Also, each day, 12% of the available cars from city B end up in city A. If we let a_n represent the number of rental cars available at city A on day n, and b_n represent the number of rental cars available at city B on day n, then the following system of two difference equations can be used to model this scenario:

$$a_{n+1} = 0.9a_n + 0.12b_n, \qquad (1a)$$

$$b_{n+1} = 0.1a_n + 0.88b_n. \qquad (1b)$$

The major concerns of the agency are to maintain the required minimum numbers of cars at both locations and to reduce the cost of shuffling cars between the two locations. Initially, there are 120 rental cars in city A and 150 rental cars in city B. Setting $a_0 = 120$ and $b_0 = 150$, we iterate Eqs. (1a) and (1b) to find the number of cars in both cities for the future days. Using Eq. (1a), we calculate

$$
\begin{aligned}
a_1 &= 0.9a_0 + 0.12b_0 \\
&= 0.9(120) + 0.12(150) \\
&= 108 + 18 = 126.
\end{aligned}
$$

Using Eq. (1b), we calculate

$$
\begin{aligned}
b_1 &= 0.1a_0 + 0.88b_0 \\
&= 0.1(120) + 0.88(150) \\
&= 12 + 132 = 144.
\end{aligned}
$$

Once we have found both a_1 and b_1, we can continue the iteration to calculate a_2 and b_2 as follows:

$$
\begin{aligned}
a_2 &= 0.9a_1 + 0.12b_1 \\
&= 0.9(126) + 0.12(144) = 130.68,
\end{aligned}
$$

$$
\begin{aligned}
b_2 &= 0.1a_1 + 0.88b_1 \\
&= 0.1(126) + 0.88(144) = 139.32.
\end{aligned}
$$

In this problem, we must realize that decimal parts of cars are not realistic and that the percentages given were estimates for planning. At some point in our calculations and analysis, we need to round our approximate values for a_n and b_n to whole numbers to represent the numbers of cars at the two cities. We can continue the iteration in this manner indefinitely. However, in order to obtain a_{n+1} or b_{n+1}, we must have already iterated both equations to know a_n and b_n. In other words, both equations must be iterated, even if we are interested in values for only one of the dependent variables.

The model given by Eqs. (1a) and (1b) can be classified as a 2×2 *linear system* of first-order, homogeneous difference equations. The "2×2" represents the number of equations and number of dependent variables, called *components* of the system. The system is linear since both equations are linear. Similarly, both equations are first-order and homogenous. One way to graph the solution to a system of difference equations is to overlay the graphs of all the components of the system on the same coordinate axes. For this example we can find the first seven values for a_n and b_n using Eq. (1) by continued iteration and then plot the first seven values of both sequences. The resulting graph of this system using linearly connected graphs for both components (a_n and b_n) is shown in Figure 1.

We can determine from the iteration of the model for rental car movement on the island for the first week that $a_7 = 142$ and $b_7 = 128$. In order to prevent the agency in city A from holding too large a share of the company's cars, the company currently has 22 cars moved each week from city A to city B. Therefore, the initial conditions are reinitialized every seven days. Since this shuffling of cars is expensive, we can use analysis of the model to help the company explore other options in their schedule to move cars between the cities.

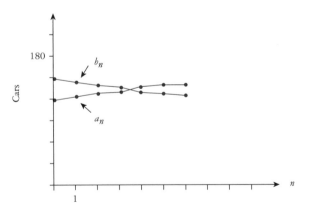

FIGURE 1 Graph of the solution (components a_n and b_n) for Eq. (1).

We can iterate further to see the two-week ($n = 14$) and one-month ($n = 30$) information produced by continued use of the model without reshuffling cars. The results are

$$a_{14} = 146, \qquad b_{14} = 124, \qquad a_{30} = 147, \quad \text{and}$$
$$b_{30} = 123.$$

Since this system seems to approach acceptable limit values for a_n and b_n of 147 and 123, respectively, there seems to be no need to reshuffle cars and the company can eliminate the reshuffling and save the costs associated with that operation. The numbers of cars at the two locations is an equilibrium that is acceptable to the company's operation.

Once again, we see the utility of iteration and graphing. However, it is still desirable to find a general method to determine the analytic solution to systems of linear difference equations. A general theory that enables analytic solutions to be computed does exist, but that theory uses results from the subject of matrix algebra, which we present in a later chapter. Therefore, we will not develop the method to calculate analytic solutions now. Be sure to learn about the powerful methods of matrix algebra when the opportunity presents itself.

To show how useful analytic solutions can be, we give the analytic solution to the system in Eqs. (1a) and (1b) with the given initial conditions, $a_0 = 120$ and $b_0 = 150$:

$$a_n \approx -27.3(0.78)^n + 147.3, \qquad (2a)$$

$$b_n \approx 27.3(0.78)^n + 122.7. \qquad (2b)$$

The "approximately equal" sign $\approx$ is used because the numbers in the functions have been rounded off to one or two decimal points. We can verify that these two functions Eqs. (2a) and (2b) are solutions to Eqs. (1a) and (1b) by showing equality when the functions are directly substituted into the equations.

We can use the analytic solution, functions Eqs. (2a) and (2b) to understand the long-term behavior of the model and to determine directly values of the components for large values of n. For instance, $a_{100} \approx -27.2(0.78)^{100} + 147.3 \approx 147.3$. Without the formula for the analytic solution given in Eq. (2a), we would have to perform 100 iterations of Eqs. (1a) and (1b) to find a_{100}. Such calculations would be tedious and would definitely require the use of a computer or calculator. From Eqs. (2a) and (2b), we also see explicitly what happens to a_n and b_n as n grows very large. Since the exponential part of the solution $(0.78)^n$ becomes smaller and smaller as n increases, eventually becoming very close to 0, this part of Eqs. (2a) and (2b) can be ignored when analyzing long-term behavior. Therefore, as we expected from the results of iteration, we know now that both a_n and b_n approach the constant part of Eq. (2a) and Eq. (2b), 147.3 and 122.7, respectively. These values are the equilibrium values of the system. Analytic solutions with the base of exponential terms less than 1 in absolute value eventually decay and approach 0. This makes this equilibrium value stable. This is one of the key factors in determining long-term behavior for linear systems. ▲

Now, let's look at a 2 × 2 system of nonlinear difference equations. Recall the scenario involving competing species from Section 1.4. The presence of a competing species diminishes the growth rate of the populations of both species. When the decrease

of the growth rates is proportional to the product of the populations of the two species, the following nonlinear system is used to model the change in the populations c_n and p_n at time period n:

$$c_{n+1} = c_n + k_1 c_n - k_3 c_n p_n, \tag{3a}$$

$$p_{n+1} = p_n + k_2 p_n - k_4 c_n p_n. \tag{3b}$$

The values of k_1, k_2, k_3, and k_4 are positive constants that represent the growth rates and rates of competition between the two species.

Once again, the techniques of iteration and graphing are directly available to analyze and solve nonlinear systems of difference equations, just as we did in Example 1 for a system of linear equations. However, because the equations are nonlinear, there is no general method to determine analytic solutions for nonlinear systems.

EXAMPLE 2 Competing Species If $k_1 = 0.005$, $k_2 = 0.01$, $k_3 = 0.000005$, $k_4 = 0.000012$, $c_0 = 6000$, and $p_0 = 5000$, iterate Eq. (3a) to find the populations after the first two time periods. The calculations are as follows:

$$c_1 = c_0 + k_1 c_0 - k_3 c_0 p_0$$
$$= 6000 + 0.005(6000) - 0.000005(6000)(5000)$$
$$= 5880,$$

$$p_1 = p_0 + k_2 p_0 - k_4 c_0 p_0$$
$$= 5000 + 0.01(5000) - 0.000012(6000)(5000)$$
$$= 4690,$$

$$c_2 = c_1 + k_2 c_1 - k_3 c_1 p_1$$
$$= 6000 + 0.005(5880) - 0.000005(5880)(4690)$$
$$= 5771.5,$$

Mathematics at Work: Tom Black

Tom Black is a meteorologist who works in the Development Division of the National Meteorological Center (NMC) in Maryland. That division develops computer models that originate both inside and outside of NMC to forecast the weather

Tom Black

from 12 hours to 10 days into the future. Tom Black talked with Frank Giordano about the mathematics used in meteorology:

Frank: How important is mathematics to your work?

Tom: Mathematics is critical in the field of meteorology. It is an essential part of both undergraduate and graduate studies in the field and is used extensively by professional meteorologists involved in research and development.

Frank: What sort of mathematics do you use?

Tom: Difference equations are the mathematical soul of most short-range forecasting. They provide the information needed in our final forecasts. The fundamental equations that describe the continuously changing state of the atmosphere are actually a system of partial differential equations called the Navier–Stokes equations. But we cannot use these equations directly.

Frank: Why not?

Tom: The Navier–Stokes equations do not have an analytical solution, so we use difference equations based on them. The difference equations do not have an analytical solution either, but they do give us approximate numerical solutions. First an initial state of the atmosphere is measured at each point of a three-dimensional grid; then the rates of change or "tendencies" of the primary variables (e.g., temperature, water vapor, wind components) are determined from the difference equations. These "tendencies" are used to update the variables at one time step into the future. With this new atmospheric state, the process then repeats, leading to subsequent updating of the variables into the future.

Frank: You mention a system of equations. How many equations are there?

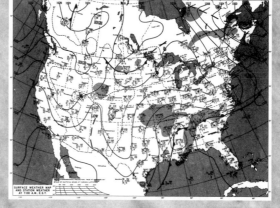

Tom: The most fundamental system of equations that describes the atmosphere is made up of six "primitive equations." They describe the temperature, the x and y components of the wind, and three basic physical and/or chemical relationships in the atmosphere. However, while these relations are indeed the bottom line, our forecast models must solve many dozens if not hundreds of ancillary equations for other unknowns. For example, the model I am using also predicts variables like cloud liquid water, turbulent kinetic energy, and soil temperatures. Knowing these and many other quantities ultimately leads to improved forecasts arising from the primitive equations themselves. In this particular model I am referring to, it takes about 30,000 lines of computer code to solve all of the equations.

$$p_2 = p_1 + k_2 p_1 - k_4 c_1 p_1$$
$$= 4690 + 0.01(4690) - 0.000012(5880)(4690)$$
$$= 4406.0.$$

A helpful and powerful computational tool, like a computer or programmable calculator, can continue the iteration and possibly produce a graph. The graph of both populations (c_n and p_n) determined by Eqs. (3a) and (3b) for the first 20 time periods is shown in Figure 2.

Calculations for iteration can become even more tedious and complicated as the number of equations and dependent variables increase. If the system is linear, the power of linear algebra to determine an analytic solution (this is covered in Chapter 3) helps analysis considerably. However,

as the size of any type of system grows, the only recourse is to perform tremendous numbers of iterations. This need for more large-scale iteration capacity is one of the primary reasons we continue to develop larger, faster, and special kinds of supercomputers. Many of the calculations performed on the large-scale supercomputers involve iteration of large systems. ▲

Complexity

We conclude this section with an example that gives a taste of the growth of computational complexity as the size of the system increases. We tackle a problem that produces and requires iteration of a 3×3 system of first-order, nonhomogeneous linear difference equations.

EXAMPLE 3 Car Rentals Revisited The Island Rental Car Company has opened an agency at a third location on the island. If c_n represents the number of rental cars at this new location at day n, then the following 3×3 system is used to model the change in location of the company's cars:

$$a_{n+1} = 0.80 a_n + 0.08 b_n + 0.19 c_n - 2, \quad (4a)$$

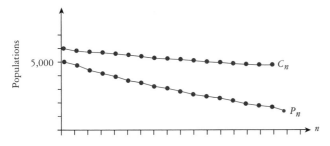

FIGURE 2 Graph of the iterative solution for components c_n and p_n determined by Eq. (3).

I shot an arrow into the air . . . it fell to earth I know not where.

What is the trajectory of an object propelled into the air at an angle? In the Middle Ages, armies wanted to

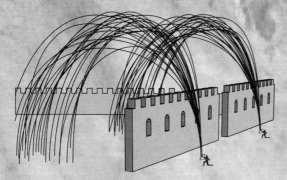

catapult balls of fire into cities. They needed to know trajectories so they could aim their projectiles. When gunpowder and cannons were developed during the Renaissance, the question became even more important. Early theories about trajectories were inaccurate, as illustrated by this 1551 drawing in which the object is imagined to fall back to Earth vertically for the last part of its trip. Leonardo da Vinci knew that the shape was a parabola, which has no straight segments. He illustrated this with the sketch at the left in which he shows how a bastion can be defended by bombard fire. His intuition was eventually confirmed by deductions from Galileo's law of falling bodies. This law implies that the x- and y-coordinates of an object shot

from the origin at velocity v, aimed at an angle θ to the positive x-axis, can be approximated by the fol-

lowing system of equations:

$$x_{n+1} = (v \cos \theta)(n + 1),$$
$$y_{n+1} = y_n + v \sin \theta - 32n.$$

In these equations, n denotes the number of seconds that have elapsed since the shot and (x_n, y_n) is the position of the object after n seconds.

You can see that the shape of such a trajectory looks like a parabola by picking specific values for v and θ generating some points for $n = 0, 1, 2, \ldots$, and plotting. You can also proceed to prove that the shape of the trajectory is a parabola. Can you do this proof?

$$b_{n+1} = 0.09a_n + 0.78b_n = 0.06c_n - 1, \quad (4b)$$

$$c_{n+1} = 0.11a_n + 0.14b_n + 0.75c_n - 1. \quad (4c)$$

The last term of each equation——-2, -1, and -1, respectively—represents rental cars that leave the island on a ferry boat. The presence of these constant terms makes the linear system nonhomogeneous, as described in Section 1.5. The initial 270 cars are located so that $a_0 = 100$, $b_0 = 90$, and $c_0 = 80$. To find the values of the 3 components over the first week (7 days) requires iteration of 21 equations (7 each of Eqs. (4a), (4b), and (4c)). The graph of the values of the three components (a_n, b_n, and c_n) over the first week is shown in Figure 3.

It is interesting to see how many computations it takes to keep track of car movement for a year using this model. If we count three multiplications and four additions (or subtractions) in each equation, then each iteration of one equation uses seven arithmetic operations. There are 3

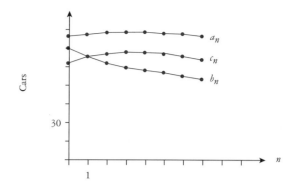

FIGURE 3 Graph of three components of Eq. (4).

equations and 365 iterations of the system (one for each day of the year). This produces a total of 7665 operations. Problems this size definitely require a computer or calculator. Some large-scale problems require millions or billions of operations to perform the iterations necessary to fully analyze and solve the problem. You will encounter more about computational complexity later in this text. ▲

Exercises for Section 1.8

1. Iterate the following 2×2 systems of difference equations to find the first four values of the components:
   (a) $a_{n+1} = 2a_n - b_n, \quad a_0 = 2$
   $b_{n+1} = 2a_n - 2b_n, \quad b_0 = 2$
   (b) $c_{n+1} = -c_n + 0.3d_n, \quad c_0 = 0$
   $d_{n+1} = 0.5c_n - 0.2d_n, \quad d_0 = 5$
   (c) $e_{n+1} = 0.5e_n + 0.4f_n, \quad e_0 = 100$
   $f_{n+1} = 0.9e_n, \quad f_0 = 30$
   (d) $g_{n+1} = -6g_n + 6h_n, \quad g_0 = 0.15$
   $h_{n+1} = -5g_n + 6h_n, \quad h_0 = 0.13$

2. Graph (using linearly connected graphs) the iteration solutions found in Exercise 1, using the same set of axes for both the components.

3. Verify by substitution that the solutions given in a–d satisfy (solve) the systems of difference equations provided in Exercise 1a–d, respectively.
   (a) $a_n = (1 - \sqrt{2}/2)(-\sqrt{2})^n + (1 + \sqrt{2}/2)(\sqrt{2})^n$
   $b_n = (1 - \sqrt{2}/2)(2 + \sqrt{2})(-\sqrt{2})^n +$
   $\qquad (1 + \sqrt{2}/2)(2 - \sqrt{2})(\sqrt{2})^n$

(b) $c_n = (-0.24\sqrt{31})(-0.6 - \sqrt{31}/10)^n$
$\quad + (0.24\sqrt{31})(-0.6 + \sqrt{31}/10)^n$

$\quad d_n = (-0.24\sqrt{31})(4/3 - \sqrt{31}/3)$
$\quad \cdot (-0.6 - \sqrt{31}/10)^n + (0.24\sqrt{31})$
$\quad \cdot (4/3 + \sqrt{31}/3)(-0.6 + \sqrt{31}/10)^n$

(c) $e_n = 20(-0.4)^n + 80(0.9)^n$
$\quad f_n = 20(-10/4)(-0.4)^n + 80(0.9)^n$

(d) $g_n = (0.075 + \sqrt{6}/100)(-\sqrt{6})^n$
$\quad + (0.075 - \sqrt{6}/100)(\sqrt{6})^n$

$\quad h_n = (0.075 + \sqrt{6}/100)(1 - \sqrt{6}/6)(-\sqrt{6})^n$
$\quad + (0.075 - \sqrt{6}/100)(1 + \sqrt{6}/6)(\sqrt{6})^n$

4. Which of the solutions given in Exercise 3 converge to a stable equilibrium value? (*Hint:* Study the size of the base of the exponential terms.)

5. Calculate the first two iterations of the following 3×3 nonlinear systems of nonhomogeneous difference equations:

(a) $a_{n+1} = a_n^2 + 0.5b_n c_n + 1, \quad a_0 = 1$
$\quad b_{n+1} = 0.25b_n^2 + 0.2c_n^3 - 2, \quad b_0 = 0.5$
$\quad c_{n+1} = 0.1b_n - n, \quad c_0 = 1.5$

(b) $d_{n+1} = nd_n + 1, \quad d_0 = 0.5$
$\quad e_{n+1} = d_n e_n, \quad e_0 = 0.5$
$\quad f_{n+1} = f_n^2 - 2e_n, \quad f_0 = 1$

6. Recall the predator–prey model from Section 1.4, where the spotted owl was preying on rabbits. Supposing that o_n represents the population of the predator owls at time period n, r_n represents the population of the rabbits, and specific values for the rates of change have been determined for the model, the following 2×2 system of nonlinear difference equations models the change in the owl and rabbit populations:

$$o_{n+1} = 0.97o_n + 0.0013o_n r_n,$$

$$r_{n+1} = 1.05r_n - 0.002o_n r_n.$$

(a) Conduct a numerical experiment by determining the direction of the change in the two populations over the first four time periods given

the following four sets of initial conditions:

first set: $o_0 = 5, r_0 = 150$;
second set: $o_0 = 100, r_0 = 100$;
third set: $o_0 = 150, r_0 = 5$;
fourth set: $o_0 = 15, r_0 = 15$.

What can you conclude about the changes in the populations based on the results of this experiment?

(b) Conduct another experiment using the initial conditions $o_0 = 15$ and $r_0 = 15$. Perform 200 iterations to see if any patterns develop. It may be helpful to plot the results of the iteration on two coordinate axes representing the two populations. What can you conclude about the changes in the populations based on the results of this experiment?

7. A decontamination process of water in a holding tank is conducted by iterating a process five times. Three separate containers are used. To start the process, all the contaminated water is in the holding tank designated as container 1. The other two containers hold uncontaminated water. In the first step, 30% of the contaminated water mixture is drawn out of the contaminated holding tank (container 1), placed into container 2, and container 1 is filled back to capacity with uncontaminated water. The process in container 2 involves chemical elimination of 10% of the contaminant (without reducing volume of the mixture), and 30% of the resulting mixture is removed from container 2 and placed in container 3. In the third container, 20% of the contaminant is chemically eliminated, and 35% of the resulting mixture is dumped back into the water supply from container 3.

(a) Write a system of difference equations to model the amount of contamination in each of the three tanks and the water supply.

(b) Given that the amount of contamination in the initial water in container 1 is 350 grams, iterate the system five times to determine the amounts of the contaminant in each of the containers. What is the total amount of contaminant left in the three tanks?

(c) Iterate the same system 10 times to determine the amounts of the contaminant in each of the containers. What is the total amount of contaminant left in the three tanks? Does continuation of this process help the decontamination process?

(d) How much contaminant is returned to the water supply if the process is continued indefinitely?

8. (a) Using the equation for y_n in the Spotlight on trajectories, show that $\Delta^2 y_n$ is constant, namely, -32.

(b) Using Theorem 4, in Section 1.2, on the results of Exercise 8a allows you to deduce that $y_n =$

$an^2 + bn + c$ for some constants a, b, and c. Find out what the constants are by substituting $n = 0$, $n = 1$, $n = 2$, and comparing to the information given in the Spotlight.

(c) Continuing with Exercise 8b, solve for n from the equation for x_{n+1} and substitute into the equation found in Exercise 8b. Now you have a quadratic equation involving x_n and y_n which holds for all values of n. Can you sketch the graph of this equation? (Where are its intercepts? Is it pointing up or down?)

9. Explain why a system of difference equations containing k variables needs k equations. Does it make sense to have more equations than variables?

SECTION 1.9 *Sums and Series*

Sometimes the terms of a sequence need to be summed to form a new sequence. For instance, if the original sequence gives the amount of rainfall in the city per month, then successively adding the terms of the original sequence produces the cumulative rainfall through the months of the year. Similarly, a sequence of the numbers of seats in the rows of stands of a stadium starting from the bottom is summed to produce the cumulative seat total for all rows up to and including the given row number. The new sequences formed in this manner are called sequences of *partial sums*.

EXAMPLE 1 Money in the Bank You open a savings account and over the first 10 months you deposit and earn interest in the following amounts (in dollars): {25, 75, 30, 40, 27, 100, 43, 75, 80, 125}. Then the following partial sums represent a sequence of the cumulative amounts in the account (in dollars) for the first 10 months: {25, 100, 130, 170, 197, 297, 340, 415, 495, 620}. ▲

We denote the *sequence of partial sums* of a sequence a_n as the new sequence s_n and write

$$s_n = \sum_{k=1}^{n} a_k = a_1 + a_2 + \cdots + a_n,$$

with the lower index ($k = 1$) indicating where to start the sum, and the upper index ($k = n$) indicating when to stop summing. In algorithm form, this partial summation is as follows:

Algorithm for Producing the Partial Sums s_k

Input: upper index n and formula or values for a_k

Output: sequence of partial sums of $a_k(s_k)$ for $k = 1$ to n

$s[0] \leftarrow 0$
For k from 1 to n,
$s[k] \leftarrow s[k-1] + a[k]$,
end for.

The Σ symbol (called *sigma*) represents the summation operator just like the Δ (*delta*) symbol represented the difference operator in Sections 1.2 and 1.3. Later, we will also consider the infinite sum or series

$$\sum_{k=1}^{\infty} a_k = a_1 + a_2 + a_3 + \cdots + a_n + \cdots.$$

For example,

$$\sum_{k=1}^{\infty} \frac{1}{k} = 1 + \frac{1}{2} + \frac{1}{3} + \frac{1}{4} + \cdots. \qquad (1)$$

EXAMPLE 2 Special Sum We write the sequence of partial sums for the sequence {1, 1/2, 1/3, 1/4, . . .} as

$$s_n = \sum_{k=1}^{n} \frac{1}{k}.$$

Then the sequence s_n starts like this:

$$s_1 = 1,$$
$$s_2 = 1 + \tfrac{1}{2} = \tfrac{3}{2},$$
$$s_3 = 1 + \tfrac{1}{2} + \tfrac{1}{3} = \tfrac{11}{6},$$

and so on. ▲

Determining Area Using Sums

One useful application of partial sums is the determination of area. If we graph a sequence by filling in a constant value over the interval $n - 1$ to n for

a_n, then instead of the graph of a_n being a set of points, the graph of the sequence a_n looks like stair steps.

EXAMPLE 3 Graph of 1/n Graph $a_n = 1/n$ for $n = 1, 2, \ldots, 8$ using piecewise-constant values to the left of n to represent the sequence as a function from 0 to 8. This graph is shown in Figure 1.

In order to compute the area between the horizontal axis and the graph of a sequence plotted in this manner, we simply add the area of the rectangles whose tops are the stair steps (see Figure 2). ▲

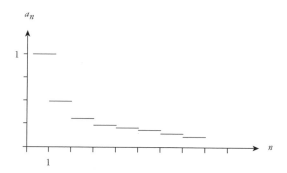

FIGURE 1 Graph of 1/n using piecewise-constant values.

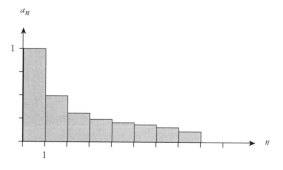

FIGURE 2 Shaded region under the graph of 1/n.

EXAMPLE 4 Area Under 1/*n* Find the area of the shaded region on the graph of the sequence $a_n = 1/n$ in Figure 2.

The rectangles are $1/n$ units high and 1 unit wide; the total area of these eight rectangles is the partial sum

$$\sum_{n=1}^{8} \left(\frac{1}{n}\right)(1) = 1 + \frac{1}{2} + \frac{1}{3} + \frac{1}{4} + \frac{1}{5}$$
$$+ \frac{1}{6} + \frac{1}{7} + \frac{1}{8} = \frac{761}{280}. \ \blacktriangle$$

EXAMPLE 5 Area of Paper Strips If you need to cut strips of paper 1 inch wide with lengths set at increments of 1 inch, starting at 1 inch long and ending at 10 inches long, what is the total amount of paper needed? The area of the paper is simply the sum of the area of 10 rectangles, or

$$\sum_{k=1}^{10} (k)(1).$$

In Section 1.2, Example 2, we determined the formula for the sum of the first n integers to be $(n)(n + 1)/2$; therefore, these 10 strips of paper use $(10)(11)/2 = 55$ square inches of paper. $\blacktriangle$

EXAMPLE 6 Car Trip While your family is on a car trip, you record your family's average speed (in miles per hour) during each hour. This reading is easy to determine since it is just the distance traveled during each hour. The sequence of average speeds recorded for your trip is $v_n = \{30, 49, 55, 23, 32\}$. Using the summation operator, the distance traveled in k hours is

$$d_k = \sum_{n=1}^{k} v_n.$$

Therefore, the new sequence of cumulative distances traveled by your family over k hours is $d_k = \{30, 79, 134, 157, 189\}$. These distances can be converted to the odometer readings of the car by knowing the initial odometer reading. If the odometer read 22,322 at the beginning of the trip, then the sequence of odometer readings at each hour m was $o_m = \{22,322; 22,352; 22,401; 22,456; 22,479; 22,511\}$. Here, o_m represents the reading after $m - 1$ hours. $\blacktriangle$

Notice that Example 6 involves the same data and a similar scenario as Example 4 in Section 1.2, except that the sequence given in one example must be determined in the other, and vice versa. In Example 4, Section 1.2, the odometer readings were recorded and the average speeds were determined by the difference operator. In the current example the average speeds were recorded and the odometer readings were determined using the summation operator. In this sense, the difference operator and the summation operator are reverse operations of one another. One operator undoes what the other does. We generalize this relationship in the following theorem, the *sum of a difference theorem*.

Theorem 1. If $\Delta a_n = a_{n+1} - a_n$, then

$$\sum_{n=1}^{k-1} \Delta a_n = \Delta a_1 + \Delta a_2 + \cdots + \Delta a_{k-1}$$
$$= a_k - a_1.$$

Proof. Substituting for Δa_n with differences $(a_{n+1} - a_n)$ gives

$$\sum_{n=1}^{k-1} \Delta a_n = (a_2 - a_1) + (a_3 - a_2) + (a_4 - a_3)$$
$$+ (a_{k-1} - a_{k-2}) + (a_k - a_{k-1}).$$

In this sum, the a_2 cancels with the $-a_2$ and the a_3 cancels with the $-a_3$, etc. After all the cancellations are completed, only a_k and $-a_1$ remain. Therefore,

$$\sum_{n=1}^{k-1} \Delta a_n = a_k - a_1.$$

▲ ▲ ▲

Geometrically, this theorem means that the area between the stair-step graph of Δa_n and the horizontal axis over the interval 0 to k is equal to the total change in the sequence a_n from the first term to the kth term ($a_k - a_1$). In order for this to hold for all sequences, we need to discuss what we mean by area when Δa_n is negative. For this case, the areas of the rectangles that are below the axis are subtracted from the areas of the rectangles above the axis. In other words, these areas are given a sign of $+$ or $-$ depending on whether they are above or below the horizontal axis. Therefore, if there is more area of Δa_n below the axis, the result of the summation would be a negative number and $a_k < a_1$. This "signed area" concept is depicted in Figure 3 of Example 7. In this example the sum of the first six terms is negative. This indicates that there is more area below the axis (darker shading) than above the axis (lighter shading) in Figure 3.

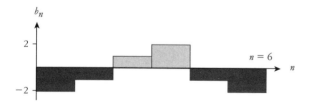

FIGURE 3 $\sum_{n=1}^{6} b_n$ represented graphically by areas above and below the axis.

EXAMPLE 7 **Area Above and Below the Axis** Given the sequence $\{b_n\} = \{-2, -1, 1, 2, -1, -2, \ldots\}$, we can represent the sum of the first six terms $\sum_{n=1}^{6} b_n = -2 - 1 + 1 + 2 - 1 - 2 = -3$ on the graph in Figure 3.

The sum $\sum_{n=1}^{6} b_n$ is represented graphically as the area above the axis (lighter shading, positive area $= 3$) minus the area below the axis (darker shading, negative area $= 6$). Therefore, $\sum_{n=1}^{6} b_n = 3 - 6 = -3$. ▲

Using the result of Theorem 1, we can find a_n whenever we are given the sequence Δa_n and the first term a_1.

EXAMPLE 8 **Determining a_n from Δa_n** Find a_n given $\Delta a_n = \{-1, 1, 3, 5, 7, 9\}$. Using Theorem 1,

$$a_k = a_1 + \sum_{n=1}^{k-1} \Delta a_n.$$

Therefore,

$$a_2 = a_1 + \Delta a_1 = a_1 - 1,$$

$$a_3 = a_1 + \Delta a_1 + \Delta a_2 = a_1 - 1 + 1 = a_1.$$

Similarly, the other terms of a_n can be calculated in terms of a_1 to obtain $a_n = \{a_1, a_1 - 1, a_1, a_1 + 3, a_1 + 8, a_1 + 15, a_1 + 24\}$. Therefore, we need to know a_1 to determine the values of a_n explicitly. If $a_1 = 0$, then $a_n = \{0, -1, 0, 3, 8, 15, 24\}$. Notice how this example compares with Example 1, in Section 1.3, where the same sequences were used but the unknown sequence Δa_n was determined from a_n. ▲

EXAMPLE 9 Oscillating Sequences

Given the following sequence of velocities $\Delta d_n = \{1, 0, -1, -1, 0, 1, 1, 0, -1, -1, 0, 1, \ldots\}$, find d_n if $d_1 = 0$ and graph the two sequences Δd_n and d_n. Notice how Δd_n oscillates in a predictable pattern. Using Theorem 1, one can perform the following summation calculations:

$$d_2 = d_1 + \Delta d_1 = 1,$$

$$d_3 = d_1 + \Delta d_1 + \Delta d_2 = 1,$$

$$d_4 = d_1 + \Delta d_1 + \Delta d_2 + \Delta d_3 = 0, \text{ etc.}$$

These calculations can be continued to produce the sequence $d_n = \{0, 1, 1, 0, -1, -1, 0, 1, 1, 0, -1, -1, 0, \ldots\}$. The graphs of these two oscillating functions with the points representing the sequence connected by straight lines are provided in Figure 4.

The two sequences and their graphs are very similar. The sequence for d_n is just shifted two terms to the right of Δd_n. Sometimes, as in this case, the relationship between Δd_n and d_n is easy to discern. ▲

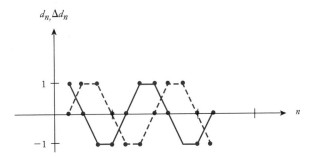

FIGURE 4 Graphs of d_n and Δd_n.

Difference of a Sum

In Theorem 1, the order of the operations is important. The first difference operation of the sequence is performed first, and the summation operation is performed second. We now investigate what happens when that order is reversed. What is

$$\Delta \left(\sum_{k=1}^{n} a_k \right)_{n=1,2,3,\ldots} ?$$

As we let $n = 1, 2, 3, \ldots$ in the summation, the sequence $\Delta \{a_1, a_1 + a_2, a_1 + a_2 + a_3, a_1 + a_2 + a_3 + a_4, \ldots\}$ is produced. Performing the difference operation produces

$$\Delta \left(\sum_{k=1}^{n} a_k \right)_{n=1,2,3,\ldots} ? = \{(a_1 + a_2) - (a_1), (a_1 + a_2 + a_3) - (a_1 + a_2), (a_1 + a_2 + a_3 + a_4) - (a_1 + a_2 + a_3), \ldots\}$$

$$= \{a_2, a_3, a_4, \ldots, a_n\}.$$

Notice that the a_1 term is lost through the performance of the two operations. However, we see once again that the two operations Δ and Σ reverse one another.

Series

What if we add all the terms of an infinite sequence? For example, if $a_r = 1/r$, then we get Eq. (1). For another example, if $a_k = 1/k^2$, then

$$\sum_{k=1}^{\infty} \frac{1}{k^2} = 1 + \frac{1}{4} + \frac{1}{9} + \cdots. \qquad (2)$$

Such an infinite sum is called an *infinite series,* or just a *series.*

On first glance, it might seem that such a sum is either meaningless or perhaps ∞. Remarkably, sometimes a series adds to a finite number. We define the sum of an infinite series through the partial sums

$$s_n = \sum_{k=1}^{n} a_k.$$

Fractal Shapes

The Koch snowflake at the right illustrates the concept that infinitely many numbers can be added up without the sum approaching infinity. This curve, an

example of a *fractal,* is produced in stages. The zero stage is an equilateral triangle whose area is 1. In the next stage, we add three smaller equilateral triangles along the middle thirds of the sides. Each of these little triangles has $^1/_3$ the linear dimension of the previous triangle, thus $^1/_9$ the area. So the area added in stage 1 is $^1/_3$, making the total area $1 + ^1/_3$, or $^4/_3$. Yet, it seems visually apparent that the area of the Koch snowflake won't get too big. In fact, it looks as if at no stage of the construction will the snowflake push outside of a square drawn around the second stage. However, what about the perimeter? We can get a good numerical fix on the area and perimeter of the snowflake with a little mathematical reasoning.

We need to focus on the number of line segments at each stage, since at the next stage, each segment is replaced by four segments and one triangle is "born" on each old segment. At each new stage, there are 4 times as many segments as before, but each is $^1/_3$ the length of the segment at the previous stage. Thus the perimeter is $^4/_3$ as large as in the previous stage. This implies that the perimeter gets larger and larger, exceeding any number we can name.

If we tabulate how many segments make up the boundary, we get the sequence in the second line of the table here. This sequence counts not only the number of segments but also the number of triangles added in the next stage (one triangle per segment).

GENERATING THE KOCH SNOWFLAKE

Stage	0	1	2	3	4
# segments at this stage	3	3(4)	$3(4)^2$	$3(4)^3$	$3(4)^4$
# triangles added in making this stage		3	3(4)	$3(4)^2$	$3(4)^3$
Area of one triangle added in making this stage		1/9	$(1/9)^2$	$1(9)^3$	$1(9)^4$
Total area added in making this stage		3(1/9)	$3(4)(1/9)^2$	$3(4)^2(1/9)^3$	$3(4)^3(1/9)^4$

Thus, we get the third line of the table by shifting the previous line over by one. The sizes of the little add-on triangles are in the sequence shown in the fourth line of the table. The total added area for any stage is the product of the number of triangles added (numbers from the third line) times the size of a basic add-on triangle (taken from the fourth line). The sequence of total add-on area, obtained by multiplying terms from the sequences in lines 3 and 4, is given in the last line of the table.

To find the total area of the Koch snowflake, we now add the infinitely many add-on areas to the original area of 1, giving us the infinite series

$$1 + 3\left(\frac{1}{9}\right) + 3(4)\left(\frac{1}{9}\right)^2 + 3(4)^2\left(\frac{1}{9}\right)^3 + 3(4)^3\left(\frac{1}{9}\right)^4 + \cdots$$

$$= 1 + \left(\frac{3}{9}\right) + \left(\frac{3}{9}\right)\left(\frac{4}{9}\right) + \left(\frac{3}{9}\right)\left(\frac{4}{9}\right)^2 + \left(\frac{3}{9}\right)\left(\frac{4}{9}\right)^3 + \cdots$$

$$= 1 + \left(\frac{1}{3}\right) + \left(\frac{1}{3}\right)\left(\frac{4}{9}\right) + \left(\frac{1}{3}\right)\left(\frac{4}{9}\right)^2 + \left(\frac{1}{3}\right)\left(\frac{4}{9}\right)^3 + \cdots.$$

If we ignore the first term, we have a geometric series which we can sum by the usual formula. Thus the entire sum is

$$1 + \frac{\dfrac{1}{3}}{\left[1 - \dfrac{4}{9}\right]} = 1 + \frac{3}{5} = \frac{8}{5}$$

Other beautiful and mysterious fractals, such as the one on the left, have been produced by repetitive processes like the one that generates the Koch Snowflake.

DEFINITION

Given a sequence $\{a_n\}$, if the sequence of partial sums

$$s_n = \sum_{k=1}^{n} a_k$$

converges to a limit value v, then the infinite series

$$\sum_{k=1}^{\infty} a_k$$

is said to *converge* to (sum to) v.

We shall investigate the convergence of Eqs. (1) and (2) numerically in a moment, but first let's think about it without particular numbers.

One condition necessary for a series to converge is that the sequence of the terms of the series must converge to zero. If this does not happen, there is no hope of their sum converging. Even this condition is not sufficient to guarantee convergence. As we add more and more terms, even if the new terms to be added are converging to 0, there are also ever more of them. Convergence or divergence depends on which of these conflicting processes dominates the other, the ever-smaller sizes of the terms or the ever-greater number of them.

Let's look at two similar series in order to get a taste of the subtlety of this conflict. First, consider the series

$$\sum_{k=1}^{\infty} \frac{1}{k}$$

and its partial sum

$$s_n = \sum_{k=1}^{n} \frac{1}{k}.$$

We saw in Example 1 that $s_3 = 11/6$. Some other partial sums are $s_{10} \approx 2.92896$, $s_{100} \approx 5.18737$, $s_{1000} \approx 7.48546$, and $s_{10,000} \approx 9.78760$. Now, consider a similar series

$$\sum_{k=1}^{n} \frac{1}{k^2}$$

and its partial sum

$$t_n = \sum_{k=1}^{n} \frac{1}{k^2}.$$

Some of the values of t_n are $t_3 = 49/36$, $t_{10} \approx 1.54976$, $t_{100} \approx 1.63498$, $t_{1000} \approx 1.64393$, and $t_{10,000} \approx 1.64483$. The question is whether these two partial sums (s_n and t_n) converge to a specific value or keep growing and therefore diverge. The graphs of s_n and t_n for $n = 1, 2, \ldots, 100$ are shown in Figure 5. Unfortunately, graphs of sequences for finite values of n do not reveal any definite information about the convergence or divergence of sequences.

In this case, s_n diverges and t_n converges. These amazing results are usually proved in calculus courses.

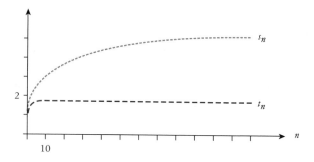

FIGURE 5 Graphs of s_n and t_n ($n = 1, 2, \ldots, 100$).

Geometric Series

Consider the first-order, linear difference equation $a_{n+1} = a_n + r^n$, where $a_0 = 0$ and r is any real number. We see through iteration that $a_1 = 0 + r^0 = 1$, $a_2 = 1 + r^1$, $a_3 = 1 + r + r^2$, and $a_4 = 1 + r + r^2 + r^3$, etc. The terms of this sequence can also be represented in summation notation. We see, for instance, that

$$a_4 = \sum_{k=0}^{3} r^k$$

and, in general,

$$a_n = \sum_{k=0}^{n-1} r^k.$$

It turns out that this difference equation and, therefore, this summation have an analytic solution of the form

$$a_n = \frac{(r^n - 1)}{(r - 1)}, \quad \text{if } r \neq 1. \tag{3}$$

Verifying that this function satisfies the given difference equation is left as an exercise.

The infinite series produced by this summation,

$$\sum_{k=0}^{\infty} r^k,$$

is called the *geometric series*. Analysis of the sequence of partial sums for this series, represented by the solution $a_n = ((r^n - 1)/(r - 1))$, reveals that when $|r| < 1$, $\lim_{n\to\infty} r^n = 0$, and the series converges to $-1/(r - 1)$. This is often written as if $|r| < 1$, then

$$\sum_{k=0}^{\infty} r^k = \frac{1}{1 - r}. \tag{4}$$

The value of r is usually called the *common ratio*.

EXAMPLE 10   Geometric Series for $r = $ 1/2
When $r = 1/2$, the difference equation $a_{n+1} = a_n + (1/2)^n$, $a_0 = 0$, is equivalent to

$$a_n = \sum_{k=0}^{n-1} \left(\frac{1}{2}\right)^k.$$

Since the solution is known from the above discussion, we can find

$$a_{10} = \sum_{k=0}^{9} \left(\frac{1}{2}\right)^k$$

to be equal to $((1/2)10 - 1)/(1/2 - 1) = 1.998$. Similarly, we can evaluate the geometric series:

$$\sum_{k=0}^{\infty} \left(\frac{1}{2}\right)^k = \frac{1}{1 - \left(\frac{1}{2}\right)} = 2. \; \blacktriangle$$

EXAMPLE 11   Series with Hidden Geometric Series   The series $\sum_{k=3}^{\infty} 6(3/4)^{k-4}$ may not immediately look like it fits the form of the geometric series. The sum of the first few terms can be evaluated as $\{8 + 6 + (9/2) + (27/8) + \cdots\}$. However, we can rewrite this series by separating out the first term, factoring out a 6, and establishing a new index to obtain $8 + 6\sum_{k=0}^{\infty}(3/4)^k$. Then we can evaluate this series using Eq. (4) with $r = (3/4)$ to obtain $8 + 6[1/(1 - (3/4))] = 8 + 6(4) = 32. \; \blacktriangle$

Exercises for Section 1.9

---

1. Evaluate the following partial sums:
    (a) $\sum_{k=1}^{6} k^2$
    (b) $\sum_{k=1}^{8} (k + 1)$
    (c) $\sum_{k=1}^{5} (-1)^k/k$
    (d) $\sum_{k=1}^{5} 2^k$
    (e) $\sum_{k=1}^{5} (1/2)^k$
    (f) $\sum_{k=1}^{5} (-1/2)^k$

2. Determine the sequences a_n from the given sequences for Δa_n and the given value for a_1:

    (a) $\Delta a_n = \{0, -2, -2, 0, 2, 2, 0\}$, $a_1 = 2$
    (b) $\Delta a_n = \{4, 4, 4, 4, 4, 4, 4\}$, $a_1 = -8$
    (c) $\Delta a_n = \{0.1, 0.2, 0.3, 0.4, 0.5, 0.6\}$, $a_1 = 0$

3. Determine the sequences a_n from the given sequences for Δa_n and the given term of a_n:

    (a) $\Delta a_n = \{0, -3, -3, 0, 3, 3, 0\}$, $a_2 = 0$

(b) $\Delta a_n = \{1, -1, -3, -5, -3, -1, 1\}$, $a_3 = 3$

4. Use the following hourly average speed data for a seven-hour car trip to determine the hourly odometer readings given that the final odometer reading after completion of the trip was $48,373$: $s_n = \{57, 62, 55, 64, 49, 55, 47\}$.

5. Given $\Delta^2 a_n = \{2, 2, 2, 2, 2, 2, 2, 2, 2\}$, $\Delta a_1 = 2$, and $a_1 = 2$, find Δa_n and a_n.

6. Verify that $a_n = ((r^n - 1)/(r - 1))$ is a solution of the difference equation $a_{n+1} = a_n + r^n$.

7. Explain the concepts of the sum of a difference (Theorem 1) and the difference of a sum.

8. Using Eq. (4) and the method shown in Example 9, evaluate the following geometric series:
   (a) $\sum_{k=0}^{\infty} (1/3)^k$
   (b) $\sum_{k=0}^{\infty} (7/8)^k$
   (c) $\sum_{k=2}^{\infty} 2(1/4)^k$
   (d) $\sum_{k=0}^{\infty} 8(1/10)^k$

9. Using Eq. (4) and the method shown in Example 9, evaluate the following:
   (a) $\sum_{k=1}^{\infty} (3/10^k)$
   (b) $\sum_{k=1}^{\infty} (\pi/3)^{k-1}$

10. Write each of the following repeating decimal numbers as a quotient integer, and determine the series representation:
   (a) $0.222\ldots$
   (b) $0.393939\ldots$
   (c) $0.555\ldots$
   (d) $1.61616\ldots$

SECTION 1.10   *Transition to Calculus*

We have spent this chapter studying functions of a discrete variable. Sequences are one form of such functions, with their domain being the discrete set of positive (or nonnegative) integers. We saw that sequences were graphed by plotting the points formed from the set of ordered pairs (n, a_n). At times, we filled in the intervals (values for all the real numbers) between points with constant values or connected these points with straight lines to form connected graphs.

In this section, we study and analyze functions of a *continuous variable*, specifically functions with domains consisting of intervals of real numbers. No longer are we confined to the domain of positive integers. We often denote such functions by a single letter such as f, reserving the notation $f(x)$ for the value that the function assigns to the domain value x.

For example, f might denote the function that adds 2 to a number. Then $f(3) = 5$ and, in general, $f(x) = x + 2$. However, mathematical usage is sometimes as informal as ordinary English, and we often use the notation $f(x)$ to refer to the function, that is the rule (in this case, a formula) that assigns output values to various input values of x. We can use other letters or even words to denote the function and the variable besides f and x. It is usually advisable to use letters or names that connote the use of the function and variable in the problem being solved. Sometimes, sequences and general functions of *discrete variables* are written using this notation. For example, throughout this chapter, we could have used $a(n)$ instead of a_n to denote sequences.

The graph of $f(x)$ still consists of points formed by the ordered pairs of the form $(x, f(x))$. However, because the x-values come from an interval of real numbers, the graph becomes a "filled-in" curve instead of just discrete points.

EXAMPLE 1 Graphing Functions We graph the two functions $f(x) = 3x + 1$ and $g(x) = x^2 - 1$ over the interval $-4 \leq x \leq 4$. The graphs of these two functions are shown together on the same coordinate axes for the given interval in Figure 1. Notice that the curves on the graph for both functions are filled in. ▲

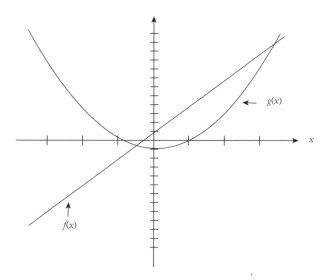

FIGURE 1 Graphs of the functions $f(x)$ and $g(x)$.

Our goals in this section are to analyze functions and determine the behavior of graphs of functions of real variables just as we did for sequences. For this purpose, we will use the familiar operations of the difference operator Δ and the summation operator Σ to analyze functions, just as we used these operations to analyze sequences. Recall from Sections 1.2 and 1.3 that the difference operator determined *slope* by treating sequences as if the points of the graph of the sequence were connected with straight lines.

Let's see how well a sequence can approximate a function by comparing piecewise-linear (or linearly connected) graphs of sequence to graphs of functions.

EXAMPLE 2 Comparing Graphs of Functions and Sequences Compare the graphs of the quadratic function $f(x) = x^2 - 6x + 7$ with the piecewise-linear graph of its corresponding sequence. Note that for a quadratic, there are discrepancies between $f(x) = x^2 - 6x + 7$ in the domain $1 \leq x \leq 6$ and the piecewise-linear graph of the first six terms of $\{a_n\} = \{n^2 - 6n + 7\}$, as shown on the plot of these two functions in Figure 2. ▲

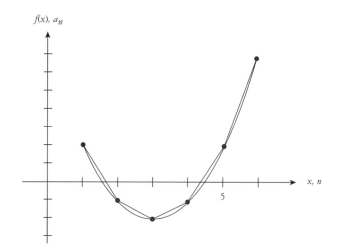

FIGURE 2 Graphs of $f(x)$ and $\{a_n\}$.

Of course, a graph of a linear function of the form $f(x) = mx + b$ is identical to the piecewise-linear graph of the sequence $a_n = mn + b$ over the same domain.

Difference Tables

While there are several obvious differences in these two graphs, in many ways the piecewise-linear graph of a_n, where the discrete points on the graph are connected by straight lines in a piecewise fashion, seems to be a good approximation to $f(x)$. Since we already know how to analyze and determine properties of the sequence a_n using differences and summation, we use this same analysis to approximate the properties of $f(x)$. In particular, Table 1, a difference table for a_n, can be thought of as an analytic tool to approximate the properties of $f(x)$. The annotations on Table 1 provide the slope and concavity analysis for a_n. Table 1 also reflects notation appropriate for $f(x)$ by following the same formulas as those for a_n, in that $\Delta f(x) = f(x + 1) - f(x)$ and $\Delta^2 f(x) = f(x + 2) - 2f(x - 1) + f(x)$.

From the data in Table 1, we determine that sequence a_n has a relative minimum at $n = 3$ and no inflection points over the interval. The properties of slope, concavity, relative extrema, and inflection points for a_n are approximations for these same properties of $f(x)$. Therefore, from this analysis we conjecture that $x^2 - 6x + 7$ has a minimum near $x = 3$, decreases for $0 < x < 3$, increases for $3 < x < 6$, and is concave up over the interval $0 < x < 6$. The closer the piecewise-linear graph of a_n approximates $f(x)$, the better the properties of a_n match those of $f(x)$.

TABLE 1 A Difference Table for a_n.

n or x	a_n or $f(x)$	Δa_n or $\Delta f(x)$	$\Delta^2 a_n$ or $\Delta^2 f(x)$
1	2	-3 decreasing	2 concave up
2	-1	-1 decreasing	2 concave up
3	-2	1 increasing	2 concave up
4	-1	3 increasing	2 concave up
5	2	5 increasing	
6	7		

E X A M P L E 3 Analyzing a Fourth-Degree Polynomial Compare the graph of the fourth-degree polynomial $g(x) = 0.6x^4 - 11.8x^3 +$ $82.95x^2 - 249.9x + 275$ with its corresponding sequence. The plots of these two functions, one continuous $g(x)$ and one discrete $\{b_n\}$, on the same axes showing some of the major differences are given in Figure 3. In Example 2, the relative minimum of the function and sequence occur at the same location. This is not always the case, as this example and the exercise problems will show. The discrete function $\{b_n\}$ completely misses the local minimum of g_x near $x = 3.5$ and the local maximum of g_x near $x = 4.3$. ▲

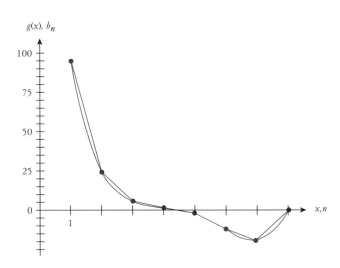

FIGURE 3 Graphs of $g(x)$ and $\{b_n\}$.

Determining Area

Another property of sequences investigated earlier in this chapter is the area of the region between the piecewise-constant graph of the sequence and the horizontal axis. For this property, the sequence was graphed by extending its values as constant in the intervals between the integers, and the operation Σ was used to sum the areas of rectangular regions. Just as we did for the other properties, this area property for sequences can be used as an approximation for area under functions.

E X A M P L E 4 **Area Under the Curve** We approximate the area between the graph of the function $f(x) = -x^2 + 2x + 15$ and the x-axis over the interval $0 \leq x \leq 5$ using the piecewise-constant graph of its associated sequence $\{a_n\} = \{-n^2 + 2n + 15\}$. The plots of $f(x)$ and $\{a_n\}$ and the region of interest are shown on the graph in Figure 4.

The area covered by the rectangles approximates the area under the curve of $f(x)$. The rectangles have a width of 1 unit, with their heights determined by the first five terms of $\{a_n\}$, (16, 15, 12, 7, 0). Therefore, the approximate area between $f(x)$ and the x-axis formed by the rectangles in the Figure 4 is

$$\sum_{n=1}^{5} a_n = 16 + 15 + 12 + 7 = 50$$

square units. We could improve this estimate for the area by using more, but skinnier, rectangles

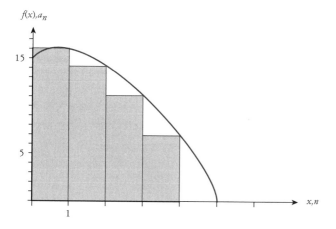

FIGURE 4 Graphs of $f(x)$ and $\{a_n\}$.

in the approximation. It is the subject of calculus that enables us to do exactly that. Using calculus, we can calculate the area exactly through the use of what we might loosely describe as an infinite number of rectangles and summing their areas through a limiting process when the limit exists. ▲

Finding Zeros of Functions

One important property of a function that we have not discussed for a sequence is the location of its *zeros*—that is, the values of x for which $f(x) = 0$. These values are also called the *roots* of the equation $f(x) = 0$. We use these two terms, "zeros" and "roots," interchangeably. This is a standard question to raise whenever we analyze the behavior of a function.

E X A M P L E 5 **Resupply Model** Through experience and analysis, the manager of a storage facility has determined that the function $s(t) = -3t^2 + 12t + 10$ models the approximate amount of a product left in the inventory of a storage facility after t days from the last resupply. We want to find when the supply of this product will be exhausted and a new resupply needed.

To determine this, the facility manager must find the smallest positive root of $s(t)$ or solve for t when $s(t) = 0$. Root-finding techniques produce roots of -0.70801 and 4.70801 for $s(t)$. Therefore, the manager knows that a resupply of this product must be scheduled within 4.7 days of the previous resupply to prevent the inventory of the product from becoming exhausted. How did those root-finding techniques work? Read on! ▲

Over the course of history, mathematicians have devoted considerable effort to developing techniques for finding roots of functions. For some functions, the root-finding techniques are simple and easy to use. Other functions seem to defy simple techniques and require great effort even to approximate their roots. If the function is linear, $f(x) = mx + b$, we can solve directly for the root or x-intercept, $x = -b/m$. If the function is a *quadratic* polynomial, $f(x) = ax^2 + bx + c$, then its two roots are simply given by the *quadratic formula*

$$x = \frac{(-b \pm \sqrt{b^2 - 4ac})}{2a}.$$

Other formulas have been devised for finding the three roots of a *cubic* equation and the four roots of a *quartic* equation, but it has been proven that no such general formula can exist for finding the roots of higher-degree polynomial equations. Moreover, we often face the problem of finding the roots of equations involving functions that are not polynomials.

Suppose we start with a function $f(x)$ on an interval $[a, b]$ such that $f(a)$ and $f(b)$ have opposite signs. That is, the graph starts and ends on opposite sides of the x-axis as we follow it from $x = a$ to $x = b$. Further, we will require that the function $f(x)$ be *continuous* on the interval; that is, its graph can be drawn with no breaks or jumps. Otherwise, it is a *discontinuous* function. For a discontinuous function, the graph of the function can simply "jump" across the x-axis, and no root exists. Figure 5 shows a continuous function with a root in the interval shown and a discontinuous function that jumps over the x-axis and, therefore, has no root in the interval.

Thus, if $f(x)$ is a continuous function on the interval $[a, b]$ such that $f(a)$ and $f(b)$ have opposite signs, then there must be at least one point r within

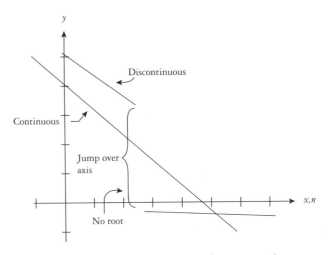

FIGURE 5 Graphs of a continuous function and a discontinuous function.

this interval, $x = r$, where the graph crosses the x-axis. Therefore, $f(r) = 0$, which means $f(x)$ has a root at r. This fact is guaranteed by a property of continuous functions.

D E F I N I T I O N

Suppose a continuous function $f(x)$ is defined on a closed and bounded interval $[a, c]$, where $u = f(a)$ and $w = f(c)$. If v is any value between u and w, then there will be at least one point b between a and c where $f(b) = v$. This is called the *intermediate value property*.

This property, which is little more than a rephrasing of the definition of continuity, ensures that a continuous function assumes all intermediate values on any bounded, closed interval. In particular, if $f(x)$ is positive at one point and negative at another, then it must assume the intermediate value $y = 0$ somewhere in that interval. That is, it must have at least one real root.

Suppose $f(x)$ is a continuous function on the interval $[a, b]$, where $f(a)$ and $f(b)$ have opposite signs. The *bisection method* for approximating a root in this interval is based on the idea that if we find the midpoint m of the interval $[a, b]$, then a root must occur either in the left half-interval $[a, m]$ or in the right half-interval $[m, b]$. If we can find out which, we then bisect the new half-interval at m_1 and so generate an interval one-quarter the size of the original interval. This procedure can be continued indefinitely to produce a sequence of ever-smaller subintervals, each containing a root. Eventually, the endpoints of the subinterval will be close enough to produce the desired accuracy for the location of the root. The geometry of two steps of the bisection method is sketched in Figure 6.

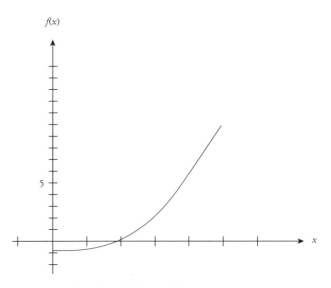

FIGURE 7 Graph of $x^{5/3} - x^{2/3} - 1$.

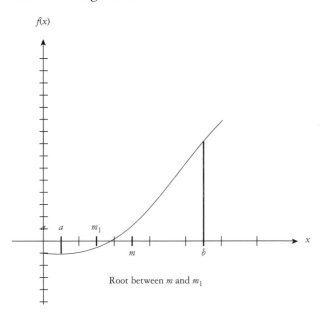

FIGURE 6 Graphic description of two steps of the bisection method.

EXAMPLE 6 **Root Finding** We find a root of $f(x) = x^{5/3} - x^{2/3} - 1 = 0$. The graph of $f(x)$ in the interval $0 \le x \le 5$ is shown in Figure 7.

To apply the bisection method, we start with an initial interval where the function changes sign. If we examine the graph of this function, we see that such an interval is $[a, b] = [1, 2]$, since $f(a) = f(1) = -1$ and $f(b) = f(2) = 0.59$. Consequently, the first midpoint is $m_1 = 1.5$, where $f(1.5) = -0.3448$. Thus, we conclude that a root must be between $x = 1.5$ (where $f(1.5)$ is negative) and $x = 2$ (where $f(2)$ is positive). We then bisect the subinterval $[1.5, 2]$ at $m_2 = 1.75$ and find that $f(1.75) = 0.0891$. Consequently, we conclude that a root is between $x = 1.5$ (where $f(1.5)$ is negative) and $x = 1.75$ (where $f(1.75)$ is positive). We summarize the results of continuing this process in Table 2.

After 10 iterations of the bisection method, we obtain an estimate that a root is between 1.701172 and 1.702145. After 10 more iterations, we find that the root is between 1.701607 and 1.701609. We don't know if the last digit is 7, 8, or 9, but whichever it is, we get a rounded-off estimate of 1.70161.

TABLE 2 The Results of the Bisection Method.

	Left endpoint	Right endpoint
$n = 0$	1	2
$n = 1$	1.5	2
$n = 2$	1.5	1.75
$n = 3$	1.625	1.75
$n = 4$	1.6875	1.75
$n = 5$	1.6875	1.71875
$n = 6$	1.6875	1.703125
$n = 7$	1.695313	1.703125
$n = 8$	1.699220	1.703125
$n = 9$	1.701172	1.703125
$n = 10$	1.701172	1.702145

A graphical portrayal of the steps of the bisection method for this example is given in Figure 8. It shows the intervals containing the root that are produced by each of the first six steps of the method. ▲

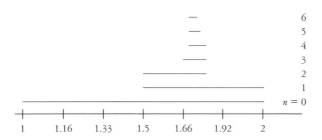

FIGURE 8 Graphical portrayal of the bisection method for Example 6.

The accuracy of the estimation can be increased by simply continuing the process further. Nevertheless, the calculations are tedious to perform by hand, even with the aid of a scientific calculator. However, this type of repetitive computation is ideal for a programmable calculator or a computer.

Analyzing a Function

EXAMPLE 7 Analysis of $x^2/2^x$ Analyze the behavior of the function $f(x) = x^2/2^x$. We begin by examining the graph of this function, which is easily produced using a graphing calculator or a computer graphics program. The graph of $f(x)$ is shown in Figure 9 for the interval $[-1, 9]$.

From the graph, we see that the function decreases from $x = -1$ to $x = 0$, where it passes through the origin and hence has a root at $x = 0$. It then increases from $x = 0$ to approximately $x = 3$ and then decreases thereafter, approaching the x-axis. Moreover, the graph is concave up from $x = -1$ to about $x = 1$, is then concave down until about $x = 5$, and is concave up thereafter. Thus, the function has inflection points near $x = 1$ and near $x = 5$.

We desire to identify more accurately where the function achieves its maximum and where it changes concavity. We know from our previous investigations of sequences and their approximations to functions that $f(x)$ achieves a relative maximum near where the sign of $\Delta f(x)$ changes from positive to negative; it has a point of inflection near where the sign of $\Delta^2 f(x)$ changes from positive to negative or negative to positive.

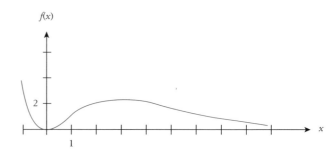

FIGURE 9 Graph of $f(x)$.

First we investigate the relative maximum of $f(x)$ near $x = 3$. We examine a series of values of the function near $x = 3$. We will not use integer values of x, but values near $x = 3$ selected at set intervals. The interval distance is denoted by Δx. In this case we define $\Delta f(x)$ as the change of the function over the interval $(x, x + \Delta x)$. Therefore, $\Delta f(x) = f(x + \Delta x) - f(x)$. This is a generalization of our previous definitions where we always used $\Delta x = 1$. The diagram in Figure 10 shows the relationships for Δx and $\Delta f(x)$. In this case we use $\Delta x = 0.1$ and organize the calculations into Table 3.

From Table 3, we observe that the function increases by 0.0038 from $x = 2.7$ to 2.8, then increases by somewhat less from $x = 2.8$ to 2.9, and then decreases from $x = 2.9$ to 3.0. We

therefore conclude that the relative maximum occurs somewhere near $x = 2.9$, where $f(x) = 1.1267$. We could improve on this estimate by taking values of x that are still closer together by reducing Δx.

We also observe from the graph that there is a point of inflection near $x = 1$. To pin it down more closely, we need to determine where $\Delta^2 f(x)$ changes sign. Once again, the definition of $\Delta^2 f(x)$ must reflect the use of Δx, which is no longer needed to be set as 1. Therefore, we now define $\Delta^2 f(x)$ by $\Delta^2 f(x) = \Delta f(x + \Delta x) - \Delta f(x)$. We use a set of points near $x = 1$ and construct a difference table using $\Delta x = 0.1$, as in Table 4.

From Table 4 we see that the point of inflection occurs near $x = 0.8$, where the signs of the second differences change from positive to negative.

We can refine this estimate of the location of the point of inflection by expanding the table to consider values of x close to 0.8 by reducing the interval distance Δx to 0.01. However, in order for this analysis to be accurate, the Δx must remain fixed or constant for all calculations in each

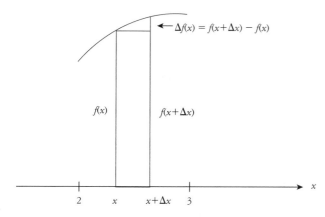

$\Delta f(x) = f(x + \Delta x) - f(x)$

$f(x)$ $f(x + \Delta x)$

2   x $x + \Delta x$ 3

FIGURE 10 Graphical relationship of Δx and $\Delta f(x)$.

TABLE 3 A Difference Table.

x	$f(x)$	$\Delta f(x)$
2.7	1.1219	0.0038
2.8	1.1257	0.0010
2.9	1.1267	−0.0017
3.0	1.125	−0.0042
3.1	1.1208	

TABLE 4 A Difference Table.

x	$f(x)$	$\Delta f(x)$	$\Delta^2 f(x)$	
0.5	0.1768	0.0607	0.0034	Concave up
0.6	0.2375	0.0641	0.0019	Concave up
0.7	0.3016	0.0660	0.0005	Concave up
0.8	0.3676	0.0665	−0.0006	Concave down
0.9	0.4341	0.0659	−0.0014	Concave down
1.0	0.5000	0.0645	−0.0022	Concave down
1.1	0.5645	0.0623	−0.0027	Concave down
1.2	0.6268	0.0596	−0.0033	Concave down
1.3	0.6864	0.0563	−0.0035	Concave down
1.4	0.7427	0.0528	−0.0038	Concave down
1.5	0.7955			

Cars Compute Areas

The speedometer in a car measures how fast the wheels are turning and converts this into mph for the forward motion of the car. The measurement is made continuously through an electrical circuit so the needle on the dashboard changes smoothly. If we were to plot the needle position (speed) against time, we might get the graph shown at the right for a 30-minute trip.

But how is the total trip mileage up to now computed to be displayed on the trip meter or odometer? Mathematically, this is done by finding the area under the curve of the figure between the left end, where time equals 0, to the current time, 20 minutes in the illustration. This computation of area is done by a mechanical device that turns gears from the tires through the transmission to the speedometer. In order for the speedometer and odometer to keep accurate measurements, these gears must be precisely manufactured in the proper gear ratios. It is amazing that the engineering applications in cars, such as the electrical circuits and gear ratios, relate to the mathematical concepts of functions, rates of change, and area under curves.

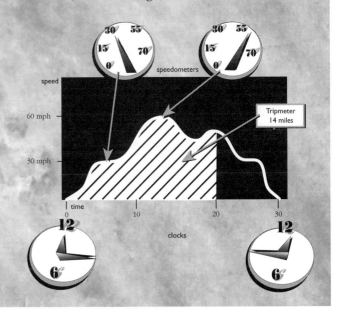

distinct difference table. Then we can use $x = 0.75, 0.76, 0.77, \ldots, 0.85, 0.86$ and find that the second difference is closest to 0 near $x = 0.84$. You can repeat this process still further to estimate the location of the point of inflection as accurately as needed. It turns out that the first point of inflection occurs at $x \approx 0.8451$. ▲

As the previous discussion suggests, in order to obtain exact locations for relative extrema and points of inflection, it is necessary to somehow shrink the distance Δx to zero. Although this seems like a paradox, or perhaps mumbo-jumbo, it turns out that this can be done with advanced tools learned in calculus. There is a second reason, aside from enhanced accuracy, for learning these advanced tools—one can never be sure of finding all relative extrema or points of inflection if we know $f(x)$ only for a finite sample of x-values (multiples of $\Delta x = 0.1$ in the example above).

For example, how do we know that the proper graph of $y = x^2/2^x$ does not have a very high spike between $x = 1.0$ and $x = 1.1$, as shown in Figure 11? It does not have such a spike, but we cannot know this if we do not calculate $f(x)$ for values between $x = 1.0$ and $x = 1.1$. If it had such a spike, there would be a relative maximum between $x = 1.0$ and $x = 1.1$ of which we would have no knowledge. No matter how many points we take, there will always be intervals between points in which spikes can conceivably occur.

Using the power of calculus, we can find exact slopes, exact locations of extrema and inflection points, instantaneous rates of change, and exact areas between curves.

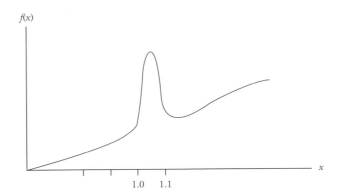

FIGURE 11 Zoomed-in version of part of Figure 9, with a fictitious spike.

All these concepts are based on applying the limit concept to the discrete operations of the difference Δ and the summation Σ.

Exercises for Section 1.10

1. Construct a difference table and use it to estimate the locations of relative extrema to one decimal place for the following functions over the given intervals:
   (a) $x^2 - 6x + 13$, $1 \leq x \leq 6$
   (b) $x^3 - 2x^2 - 4x + 5$, $0 \leq x \leq 5$
   (c) $2^x - x^3$, $0 \leq x \leq 9$
   (d) $x^{5/2} - 4x^{3/2} - 3x^{1/2}$, $0 \leq x \leq 4$

2. Compare the graph of $x^2 - 7x + 10$ with the piecewise-linear graph of its corresponding sequence. In particular, see if the relative minima are located at the same point.

3. Construct a difference table and use it to estimate the locations of inflection points to one decimal place for the following functions over the given intervals:
   (a) $x^2 - 6x + 13$, $1 \leq x \leq 6$
   (b) $x^3 - 2x^2 - 4x + 5$, $0 \leq x \leq 6$
   (c) $2^x - x^3$, $0 \leq x \leq 9$
   (d) $x^2 + 3 - 20/x$, $1 \leq x \leq 7$

4. Approximate the area between the graph of the given function and the x-axis over the given interval.
   (a) $3x + 2$, $0 \leq x \leq 5$
   (b) $x^2 - 8x + 17$, $0 \leq x \leq 6$
   (c) $-x^2 + 6x - 4$, $0 \leq x \leq 2$
   (d) $2x - 8$, $0 \leq x \leq 4$

5. Locate the following roots to two-decimal-place accuracy:
   (a) The root of $x^3 - 3x + 1 = 0$ between 0 and 1
   (b) The root of $x^3 - 3x + 1 = 0$ between 1 and 2
   (c) The root of $x^3 - 3x + 1 = 0$ between -3 and 0
   (d) The largest root of $2x^3 - 4x^2 - 3x + 1 = 0$

(e) The point of intersection of $f(x) = x \sin x$ and $g(x) = \cos x$ between 0 and $\pi/2$

(f) The first positive point of intersection of $f(x) = 2x$ and $g(x) = \tan x$

6. Locate the inflection point of $x^2/2^x$ near $x = 5$ to two decimal places.

7. Given that your car's tires have a 15″ radius:
   (a) How many complete rotations of the tire are needed to travel one mile?
   (b) Since the main gear in your car's odometer makes one rotation every mile, what is the gear ratio needed from the car's axle to the odometer gear?

8. On a car trip you drove 55 mph during the first hour, 62 mph during the second and third hours, and 65 mph for the last hour and a half.
   (a) Make a graph of your car's speed.
   (b) How far did you travel on your trip?
   (c) Find the area under the graph of your car's speed between one and four hours.

9. Explain the differences between a continuous function and a sequence.

Chapter 1 Exercises

1. Determine an expression for the general term for each of the following sequences:
   (a) $\{2, 9, 16, 23, 30, \ldots\}$
   (b) $\{3, 12, 27, 48, 75, \ldots\}$
   (c) $\{1, 3, 11, 25, 45, \ldots\}$
   (d) $\{-2, 1, 4, 7, 10, \ldots\}$

2. Construct a difference table out to the third difference for the first 10 terms of the following sequences:
   (a) $a_n = 10n^2 + 3n - 6$
   (b) $b_n = 4n^3 - 2n$
   (c) $c_n = n^2 - 3/n - 7$

3. Build a difference table for the given terms of the following sequence, and use it to determine the degree of the polynomial that defines the general term: $\{1, -3, -7, -5, 9, 41, 97, \ldots\}$.

4. Find the following limits:
   (a) $\lim_{n \to \infty} n - 2/n + 2$
   (b) $\lim_{n \to \infty} n^2 - 1/n^2 + 1$
   (c) $\lim_{n \to \infty} n + 2/n^2 - 7$

   (d) $\lim_{n \to \infty} n^2 + 8/n$

5. Find the equilibrium values for the following:
   (a) $c_{n+1} = 8c_n - 14$
   (b) $a_{n+1} = a_n^2 + 9a_n + 15$
   (c) $b_{n+1} = b_n^3 - 8b_n$
   (d) $p_{n+1} = 1.3p_n - p_n^2$

6. Draw a cobweb for the first four iterations of the following:
   (a) $a_{n+1} = 1.2a_n - 1, \quad a_0 = 1$
   (b) $b_{n+1} = b_n^2 - 2b_n + 1, \quad b_0 = 2$

7. Iterate the following systems to find the first four values of the components of the system:
   (a) $a_{n+1} = 2a_n - 3b_n, \quad a_0 = 1$
       $b_{n+1} = 3a_n - b_n, \quad b_0 = 2$
   (b) $c_{n+1} = 2c_n - 0.5d_n, \quad c_0 = 2$
       $d_{n+1} = 4.7c_n, \quad d_0 = 1$

8. Approximate the area between the graphs of $f(x) = 3x + 4$ and $g(x) = x^2 - 6x + 4$ between $x = 0$ and $x = 9$.

9. A drug is effective in treating a disease if the concentration remains above 300 mg/l. The safe limit is 1000 mg/l. If the drug decays at a rate of 30% of the amount present at the start of the hour, 100 mg/l is added each hour, and the initial concentration is 500 mg/l, how long will the drug dosage remain within the effective and safe concentration level?

10. Find the solutions to the following difference equations for any n:

(a) $a_{n+1} = 0.5a_n$, $a_0 = 10$

(b) $b_{n+1} = 2b_n$, $b_0 = 4$

(c) $c_{n+1} = 4c_n + 20$, $c_0 = 5$

(d) $d_{n+1} = -0.8d_n + 10$, $d_0 = 20$

(e) $x_{n+1} = 0.6x_n - 8$, $x_0 = 100$

11. Find the balance after five years if $1000 is deposited at 4.5% interest compounded monthly. What is the balance after 20 years?

12. Evaluate the following geometric series:

(a) $\sum_{k=0}^{\infty}(2/3)^k$

(b) $\sum_{k=2}^{\infty}2(1/5)^{k-1}$

2

POSITION

Introduction

The fundamental concerns of geometry are position and shape. This chapter deals mostly with position, although shape appears now and then. Typical questions of position include the following:

1. Two lines are given. Do they intersect? If so, where?

2. In a plane a circle is moving along a straight line. Two points in the plane are specified. Will the circle bump into the line segment between these points? If so, where and when?

3. Two points are on the surface of a ball. How should the points move along the surface of the ball in order to meet in the shortest time?

4. Given a polygon and a line, which point on the polygon is farthest from (or nearest to) the line?

5. In space two points and a plane are given. Are the points on the same side of the plane or on opposite sides?

Questions such as those in the first three examples arise frequently in problems of navigation, object location, collision avoidance, and design. For the first question, think of the design of flight paths near a major airport; for the second, imagine maneuvering a cement truck through a construction site; and for the third, consider a long-distance transport aircraft and its in-flight refueling tanker plane. We will illustrate these questions further in the first five sections of this chapter, with examples from the rapidly developing fields of robotics and remote sensing. Questions of the last two types arise very often in connection with resource allocation and other kinds of management decisions, as we will show in the final section of this chapter.

The two mathematical tools of greatest value in the geometry of position are *vectors* and *coordinate systems.* You have probably done some work with vectors, and you certainly have dealt with coordinates, but we will try to introduce these ideas as if they were entirely new to you.

The mathematical background needed for this chapter is basic algebra, geometry, and trigonometry. Specific results are reviewed when first used. Also, we use the following notation. If A and B are distinct points, the line through them is denoted by AB, the line segment with endpoints A and B is denoted by $[A, B]$, and the length of this segment, that is, the distance between A and B, is denoted by $|AB|$.

SECTION 2.1 *Vectors*

Some quantities can be described by a number alone: Your age, height, and body temperature are examples. Other quantities must be described by both a number and a direction:

a. move two feet to the left;

b. run northwest at six miles per hour;

c. push a 10-pound weight straight up.

Quantity (a) is an example of a *displacement,* (b) is a *velocity,* and (c) is a *force.*

A quantity that is described by a number alone is called a scalar quantity. The mathematical object that describes a scalar quantity—that is, a number—is called a *scalar.* A quantity that is described by a number together with a direction is called a vector quantity, and the mathematical object that describes a vector quantity is called a *vector.* You have worked with scalars all your life; vectors may be less familiar. In this section you'll begin to learn how to do a kind of arithmetic and algebra with vectors, much as you perform these mathematical operations with scalars to solve problems.

Scalar algebra was not created for fun. It developed hundreds of years ago in response to the need to solve practical problems in construction and accounting. So it is with vector algebra and with most of computational geometry. Scalar algebra may have sufficed for the construction of sailing ships, but it won't do for the construction of spaceships. Older mathematics was adequate for double-entry bookkeeping, but it isn't enough for determining how to deploy 100 different limited resources, such as raw materials, labor, and the use of subcontractors, in a way to maximize your company's profits, subject to dozens of government regulations and the pressure of competition.

Vector Notation and Representation

A vector is often depicted by an *arrow,* a term we prefer to the more standard "directed line segment." The arrow points in the direction of the vector, and the length of the arrow represents the scalar information of the vector, which is called the *norm* of the vector. Arrows depicting the vector quantities (a), (b), and (c) above are shown in Figure 1.

Remember: Scalars have sign and size, like -5.4 and 6 (that is, $+6$), while vectors like those shown

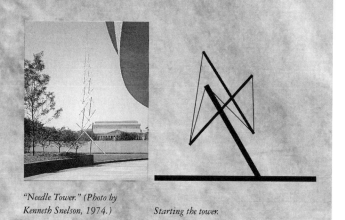

Vectors in Art

The sculpture, "Needle Tower," by the American artist Kenneth Snelson, is made of pieces of rigid aluminum tubing and flexible wire cable. Notice that the pieces of tubing do not touch each other. Why doesn't the sculpture fall down? How would you construct this sculpture? These are essentially problems in the *geometry of position*. Perhaps a schematic sketch of part of the sculpture will help (see drawing), but it might be better to work with pieces of wood dowel and string.

"Needle Tower." (Photo by Kenneth Snelson, 1974.)

Starting the tower.

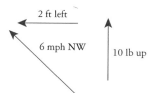

FIGURE 1 Depiction of vectors.

above, have both norm (which is never negative) and direction.

Scalars are often depicted on a *coordinate axis,* which is established on a line *l* by the selection of two points: an *origin O* and a *unit point U.* The half-line from *O* through *U* is the *positive side* of *l,* and the half-line from *O* away from *U* is the *negative side,* as shown in Figure 2. The coordinate of a point *P* on the positive side of the axis is the distance |*OP*| between *O* and *P*, where the length |*OU*| is used as the

unit of measurement. The coordinate of a point *Q* on the negative side is the negative of the distance, $-|OQ|$. Thus, the coordinate of *O* itself is $|OO| = 0$, and the coordinate of the unit point *U* is 1. On the axis in Figure 3, *P* depicts the scalar $\pi = 3.14\ldots$ and *Q* depicts the scalar $\sqrt[3]{-12} = -2.27\ldots.$

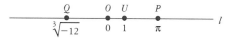

FIGURE 3 Coordinates of points on an axis.

Let *A* and *B* be points on a coordinate axis with respective coordinates *a* and *b*. The distance between *A* and *B*, that is, the length |*AB*| of line segment [*A, B*], is $a - b$ if $b < a$, is 0 if $b = a$, and is $b - a$ if $a < b$. A formula for this distance is conveniently provided by absolute-value notation. Recall that the *absolute value* of a number *x* is

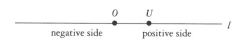

FIGURE 2 A coordinate axis.

$$|x| = \begin{cases} x & \text{if } x > 0, \\ 0 & \text{if } x = 0, \\ -x & \text{if } x < 0. \end{cases} \quad (1)$$

If A has coordinate a and B has coordinate b, then

the distance between A and B = $|a - b|$. $\quad$ (2)

Scalars are usually denoted by lowercase Roman or Greek letters, such as a, b, x, α, θ, etc. We denote vectors by bold lowercase Roman letters, such as **a**, **b**, **v**, **u**, **w**. (In handwritten work, draw a little arrow over the letter: $\vec{u}$, $\vec{v}$, etc.) We denote the arrow from point A to point B by $\overline{AB}$. (You may prefer to write $\overrightarrow{AB}$.) Point A is the *tail* (or the *initial point*) of the arrow and B is the arrow's *head* (or *terminal point*). We will often refer to "*vector $\overline{AB}$*" and write **v** = $\overline{AB}$ in place of the more precise statement, "the vector **v** is represented by the arrow $\overline{AB}$."

There is an important difference between the geometric representations of scalars by points and of vectors by arrows. On an axis, distinct points represent different scalars. But two distinct arrows may represent the same vector. Since a vector is completely specified by its direction and norm, it follows that two arrows that point the same way and have the same length must depict the same vector. Consider parallelogram $ABCD$ of Figure 4. Arrows $\overline{AB}$ and $\overline{DC}$ point the same way and have the same length; hence these arrows depict the same vector, and we may write $\overline{AB} = \overline{DC}$. Similarly, $\overline{CB} = \overline{DA}$. The arrows $\overline{AB}$ and $\overline{CD}$ have equal length but point in opposite directions; hence, $\overline{AB} \neq \overline{CD}$.

FIGURE 4 A parallelogram.

The norm of a vector **v** is denoted by $|\mathbf{v}|$. Confusion of this notation with absolute value is unlikely. (Nevertheless, many authors write norm **v** as $\|\mathbf{v}\|$, because this notation is essential in some parts of advanced mathematics.) If $\mathbf{v} = \overline{AB}$, then $|\mathbf{v}|$ is the distance $|AB|$ between A and B.

EXAMPLE 1 Finding a Norm Suppose **v** = $\overline{AB}$, where the Cartesian coordinates of A and B are (9, 2) and (4, 14), respectively. Then $|\mathbf{v}| = |AB| = \sqrt{(9 - 4)^2 + (2 - 14)^2} = \sqrt{169} = 13$. (Cartesian coordinates and the distance formula are reviewed in Section 2.2). ▲

A *unit vector* is a vector of norm 1. If we are given in a plane a circle K with center C and radius 1, then each unit vector in that plane is represented by one and only one arrow, with tail C and head a point on K, as illustrated in Figure 5.

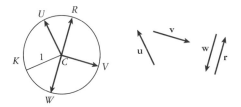

FIGURE 5 $\mathbf{u} = \overline{CU}$, $\mathbf{v} = \overline{CV}$, $\mathbf{w} = \overline{CW}$, $\mathbf{r} = \overline{CR}$ (**u**, **v**, **w**, and **r** are unit vectors).

Arbie the Robot and Vector Addition

As a way to introduce vector algebra, we turn now to *robotics,* a term coined in 1942 by the science and science fiction writer Isaac Asimov (1920–1993). Figure 6 is a two-dimensional rendering of *Arbie,* a sophisticated flying robot dedicated to the location of lost objects. The robot is centered around a camera with lens at point R. The back of the camera is a

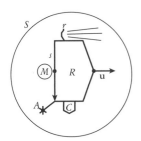

FIGURE 6 Arbie the robot.

FIGURE 8 Command **a** moves Arbie from R to R', where $\mathbf{a} = \overline{RR'}$.

sensitized plate, s, with a light meter, M, in the center. Data are also gathered by a radar antenna, r. Information and commands are sent and received through the antenna, A, and processed by the on-board computer, C. The entire machine is surrounded by a sensing sphere, S, with center R. (The sphere S is not a material object; it represents a warning system established by an onboard short-range radar, not shown in Figure 6).

When instructed to move distance d, where d is a positive scalar, Arbie moves d units in the direction from M to R, which is indicated in Figure 6 by the unit vector $\mathbf{u}$. The robot can move only in the direction of $\mathbf{u}$. To change direction, Arbie rotates around its center, R.

For the purposes of this section we will ignore all systems on the robot and depict Arbie simply by its center, R, and unit direction vector, $\mathbf{u}$, as in Figure 7.

$R \,\overset{\longrightarrow}{\underset{\mathbf{u}}{\bullet}}$ **FIGURE 7** Simplified Arbie.

A *step* for Arbie is a vector $\mathbf{a}$. When step $\mathbf{a}$ is executed, Arbie first orients itself by rotating around R so that $\mathbf{u}$ points in the same direction as $\mathbf{a}$; it then moves distance $|\mathbf{a}|$. The result is a displacement of R to point R', as in Figure 8. After the step the robot direction vector $\mathbf{u}$ points in the direction from R to R'. Just how step command $\mathbf{a}$ is sent and how Arbie

orients itself before moving will be considered in later sections of this chapter.

Suppose Arbie executes two steps, $\mathbf{a}$ followed by $\mathbf{b}$. The robot moves from R to R' and then from R' to R'', as in Figure 9. The total displacement from R to R'' represents the *sum* of the displacements from R to R' and from R' to R'', that is,

$$\overline{RR''} = \overline{RR'} + \overline{R'R''}. \tag{3}$$

This extension of the addition operation from addition of scalars to addition of vectors needs further justification. If Arbie starts at a point S different from R, then the step sequence $\mathbf{a}$, $\mathbf{b}$ will send it to a point S'' different from R''. But arrows $\overline{RR'}$ and $\overline{SS'}$ have the same length and direction since they both depict vector $\mathbf{a}$, and $\overline{R'R''}$ and $\overline{S'S''}$ have the same length and direction since they both depict vector $\mathbf{b}$. Then triangles $SS'S''$ and $RR'R''$ are congruent, by side-angle-side, and consequently arrows $\overline{RR''}$ and $\overline{SS''}$ also have equal length and direction (Figure 10). Therefore, these arrows depict the same

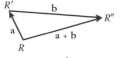

FIGURE 9 Vector addition.

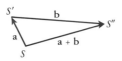
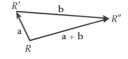

FIGURE 10 Vector addition is well defined.

vector, **a** + **b**. We may thus define vector addition as follows.

D E F I N I T I O N

Given vectors **a** and **b**, their *sum* **a** + **b** is found as follows. Depict **a** by an arrow $\overline{RR'}$ and **b** by an arrow $\overline{R'R''}$ whose tail is the head of the first arrow. Then **a** + **b** is the vector depicted by the arrow $\overline{RR''}$.

If vectors **a** and **b** do not point in the same or in opposite directions, there is another way to picture their sum, which is shown in Figure 11. Depict **a** and **b** by arrows with the same tail. Complete the parallelogram that has these arrows as sides. Then **a** + **b** is the vector with the same tail given by the diagonal of the parallelogram. This depiction of vector addition is called the *parallelogram law.*

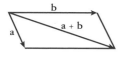

FIGURE 11 The parallelogram law of vector addition.

When we add up a list of scalars, we know that no matter in what order we add them, the sum is always the same. For instance, given 4, 7, 3, and 9, we might take 4, add 7, then add 3, and then add 9, to get 23; or we might take 3, add 4, then add 9, and then add 7, to get 23 again. Does this convenient fact also hold for addition of a list of vectors? The reason why scalar addition behaves so well is that it satisfies the commutative law,

$$a + b = b + a,$$

and the associative law,

$$(a + b) + c = a + (b + c),$$

for all scalars a, b, and c. Now it's easy to see from Figure 12 that vector addition also obeys these laws:

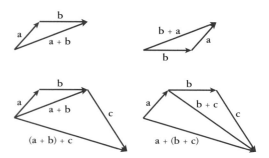

FIGURE 12 The commutative and associative laws of vector addition.

$$\mathbf{a} + \mathbf{b} = \mathbf{b} + \mathbf{a}, \tag{4}$$

$$(\mathbf{a} + \mathbf{b}) + \mathbf{c} = \mathbf{a} + (\mathbf{b} + \mathbf{c}), \tag{5}$$

for all vectors **a**, **b**, and **c**. Hence, the order in which you add a list of vectors does not matter; you always get the same answer. For example, suppose Arbie is at point P and receives commands for steps **a**, **b**, **c**, and **d**, shown in Figure 13. Execution of the steps in alphabetical order or in the order **d**, **a**, **b**, **c** will bring the robot to the same point Q, because

$$\mathbf{a} + \mathbf{b} + \mathbf{c} + \mathbf{d} = \mathbf{d} + \mathbf{a} + \mathbf{b} + \mathbf{c} = \overline{PQ}.$$

(Note, however, that after executing these steps in alphabetical order, Arbie will be facing in the direction of **d**, while execution in the order **d**, **a**, **b**, **c** will leave it facing in the direction of **c**.)

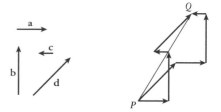

FIGURE 13 Vector addition in different orders.

It could happen that after several steps Arbie returns to its starting point. This would occur, for instance, if the robot started at point O, stayed on the ground, and followed the instructions "turn right 120° and move 2 feet" three times in succession (Figure 14): Arbie would go from O to A to B to O. Then the vector sum $\overline{OA} + \overline{AB} + \overline{BO}$ would describe a displacement of no distance in no direction. To allow for this possibility, we introduce the *zero vector*, denoted **0**, which has no direction and norm zero. The zero vector may be depicted by a single point P, thought of as an "arrow" $\overline{PP}$ of length zero. All such "arrows" are equal, so for any point Q, $\overline{QQ} = \overline{PP} = \mathbf{0}$.

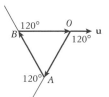

FIGURE 14 Arbie makes a round trip.

By Eq. (3), $\overline{PP} + \overline{PQ} = \overline{PQ}$ and $\overline{PQ} + \overline{QQ} = \overline{PQ}$. Thus, for any vector **v**,

$$\mathbf{0} + \mathbf{v} = \mathbf{0} \quad \text{and} \quad \mathbf{v} + \mathbf{0} = \mathbf{v}. \tag{6}$$

This law of vector addition is analogous to the law of scalar addition that $0 + x = x$ and $x + 0 = x$ for all scalars x. Note also that Eq. (6) is consistent with the definition of vector addition.

The solution of the scalar equation $a + x = 0$ is often written $x = -a$; similarly, we denote the solution to the vector equation $\mathbf{a} + \mathbf{x} = \mathbf{0}$ by $\mathbf{x} = -\mathbf{a}$. By Eq. (3), $\overline{AB} + \overline{BA} = \overline{AA} = \mathbf{0}$. Hence,

$$\text{if } \mathbf{a} = \overline{AB}, \text{ then } -\mathbf{a} = \overline{BA}. \tag{7}$$

The vector $-\mathbf{a}$ has the same norm as **a** but direction opposite to **a**.

Vector addition has been introduced through robotics, in which the vectors appear as displacements. It is a remarkable fact that vector quantities of *all kinds* add in this same way. For example, if you are flying an airplane A on a course and speed given by the vector **c** while the wind is blowing according to the vector **w**, then your actual course and speed will be given by the vector $\mathbf{c} + \mathbf{w}$, as in Figure 15(a). As another example, if you pull on an object O with a force **u** while I pull on it with a force **v**, then the object will actually experience the force $\mathbf{u} + \mathbf{v}$, as in Figure 15(b).

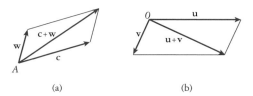

(a) (b)

FIGURE 15 More vector sums.

Notice that in Figure 15 we've used the parallelogram diagonal depiction of vector addition. It would be worth your time to test the reality of the parallelogram addition property of force vectors by using three pieces of rope tied together at one end, spring scales tied to the other ends, and a few fellow students to pull ropes, read the scales, and measure the angles. (Note: The norms of the force vectors will be the readings on the scales when the ropes are at rest; they have nothing to do with the lengths of the ropes.)

Multiplication of Vectors by Scalars

Suppose Arbie takes a step described by the vector **a**. A step in the same direction as **a** but twice as long should surely be denoted 2**a**. A step in the same direction as **a** but half as long should be denoted $(1/2)\mathbf{a}$, or $\mathbf{a}/2$. A step of the same length as **a** in the direction *opposite* to **a** has already been called $-\mathbf{a}$, so a step in the direction opposite to **a** but three times as long should be denoted $-3\mathbf{a}$. We formalize these notations in the following definition.

DEFINITION

The *product* of a scalar k and a vector **v** is the vector, denoted $k\mathbf{v}$, that points in the same direction as **v** if $k > 0$, in the opposite direction to **v** if $k < 0$, and has norm given by

$$|k\mathbf{v}| = |k||\mathbf{v}|. \tag{8}$$

The product $0\mathbf{v}$ is defined to be the zero vector **0**.

E X A M P L E 2 Multiples of a Vector In Figure 16, suppose C is the midpoint of segment $[A, B]$ and D is the midpoint of segment $[A, C]$. Let $\mathbf{v} = \overline{AC}$. Then $\overline{AB} = 2\mathbf{v}$, $\overline{AD} = (1/2)\mathbf{v}$, $\overline{CD} = (-1/2)\mathbf{v}$, and $\overline{BD} = (-3/2)\mathbf{v}$. ▲

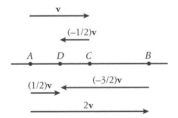

FIGURE 16 Multiples of **v**.

Multiplication by scalars can be used to find the unit vector in the same direction as a given nonzero vector. If $\mathbf{v} \neq 0$, then $|\mathbf{v}| > 0$; so $k = 1/|\mathbf{v}| > 0$. Let $\mathbf{u} = k\mathbf{v}$. Then **u** and **v** point in the same direction and $|\mathbf{u}| = |k||\mathbf{v}| = (1/|\mathbf{v}|)|\mathbf{v}| = 1$. Thus, the

unit vector in the same direction as the nonzero vector **v** is

$$\mathbf{u} = \frac{1}{|\mathbf{v}|}\,\mathbf{v}, \tag{9}$$

which is also written as $\mathbf{v}/|\mathbf{v}|$.

Two nonzero vectors $\mathbf{v} = \overline{AB}$ and $\mathbf{w} = \overline{CD}$ are said to be *parallel* if the lines AB and CD are parallel. (We regard a line as parallel to itself.) If $\mathbf{w} = k\mathbf{v}$ for some scalar k, it follows from the definition that **v** and **w** are parallel. Conversely, suppose the nonzero vectors **v** and **w** are parallel. Then the vectors must point in the same direction or in opposite directions. If they point in the same direction, then $\mathbf{w} = k\mathbf{v}$ where $k = |\mathbf{w}|/|\mathbf{v}|$; if they point in the opposite direction, then $\mathbf{w} = k\mathbf{v}$ where $k = -|\mathbf{w}|/|\mathbf{v}|$. To summarize:

Two nonzero vectors are parallel if and only if one of them is a scalar multiple of the other.

E X A M P L E 3 Parallel Vectors Suppose points A, B, and C have Cartesian coordinates $(7, 4)$, $(2, 7)$, and $(5, 8)$, respectively. Let us find the coordinates of the point D such that $\overline{CD} = -2\overline{AB}$. Let the coordinates of D be (x, y). Introduce the points P with coordinates $(2, 4)$ and Q with coordinates $(x, 8)$. By the statement above this example, lines AB and CD are parallel. Then right triangles ABP and CDQ are similar; so their corresponding sides are proportional. (Plot the points and draw the triangles. Cartesian coordinates are reviewed in Section 2.2.) Since $\overline{CD} = -2\overline{AB}$, we have $|CQ|/|AP| = |DQ|/|BP| = |CD|/|AB| = 2$. Now B lies above and to the left of A and $\overline{CD}$ has direction opposite that of $\overline{AB}$; so D must lie below and to the right of C. Then $|CQ| = x - 5 = 2|AP| = 2 \cdot 5 = 10$, $x = 15$, and $|DQ| = 8 - y = 2|BP| = 2 \cdot 3 = 6$, $y = 2$; so D has coordinates $(15, 2)$. ▲

Vectors in Sports

We're not sure if Ivan Unger and Gladys Roy were really playing tennis on the wing of this flying airplane in 1925. But if they were, just think of the velocity and force vectors with which they would have had to contend. What would have happened if Gladys had served the ball by aiming directly at Roy? Where should she have aimed if she wanted Roy to return the serve? Would it have made a difference if the plane were speeding up or slowing down rather than flying at a steady rate? Draw the vectors involved; first from the standpoint of an observer on the ground and then from the standpoint of the unknown photographer,

who we assume was in an airplane moving with the same velocity vector as the "tennis court."

Tennis, anyone? (Photo from unknown photographer, circa 1925. U.P.I., Fotofolio.)

Here are three important laws of vector algebra, which you are asked to illustrate in the exercises at the end of this section. For all scalars a and b and vectors $\mathbf{u}$ and $\mathbf{v}$,

$$(a + b)\mathbf{v} = a\mathbf{v} + b\mathbf{v}, \tag{10}$$

$$a(\mathbf{u} + \mathbf{v}) = a\mathbf{u} + a\mathbf{v}, \tag{11}$$

$$a(b\mathbf{v}) = (ab)\mathbf{v}. \tag{12}$$

One may deduce, either from these laws or from the definition itself, the following easy-to-believe results for all scalars a and vectors $\mathbf{v}$:

$$a\mathbf{0} = \mathbf{0}, \tag{13}$$

$$1\mathbf{v} = \mathbf{v}, \tag{14}$$

$$(-a)\mathbf{v} = -(a\mathbf{v}). \tag{15}$$

From Eqs. (14) and (15), $(-1)\mathbf{v} = -\mathbf{v}$. In view of

Eqs. (12) and (15), we may write $a(b\mathbf{v})$ simply as $ab\mathbf{v}$ and $(-a)\mathbf{v}$ as $-a\mathbf{v}$.

Armed with these results, we may use vectors to discover and prove many theorems of geometry. Here is an example; others appear in the exercises.

EXAMPLE 4 A Well-Known Fact about Parallelograms   In Figure 17, the diagonals of parallelogram $ABCD$ intersect at E. What can be said about this point?

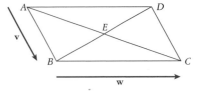

FIGURE 17 Another famous theorem.

Solution. We are interested in the location of point E along segments BD and AC. Thus, we should look at the relation between vectors $\overline{BE}$ and $\overline{BD}$ and between $\overline{AE}$ and $\overline{AC}$. There are scalars a and b such that $\overline{BE} = a\overline{BD}$ and $\overline{AE} = b\overline{AC}$. We would like to evaluate these scalars. Let $\mathbf{v} = \overline{AB}$ and $\mathbf{w} = \overline{BC}$. Then $\overline{BE} = a\overline{BD} = a(\overline{BA} + \overline{AD}) = a(-\mathbf{v} + \mathbf{w})$ and $\overline{AE} = b\overline{AC} = b(\mathbf{v} + \mathbf{w})$. Now $\overline{AB} = \overline{AE} + \overline{EB} = \overline{AE} - \overline{BE}$; so $\mathbf{v} = b(\mathbf{v} + \mathbf{w}) - a(-\mathbf{v} + \mathbf{w})$. Now do some vector algebra:

$$\mathbf{v} = b\mathbf{v} + b\mathbf{w} + a\mathbf{v} - a\mathbf{w}$$
$$= (a + b)\mathbf{v} + (-a + b)\mathbf{w};$$

so $(a + b - 1)\mathbf{v} = (a - b)\mathbf{w}$. Now $(a + b - 1)\mathbf{v}$ is a vector parallel to $\mathbf{v}$, while $(a - b)\mathbf{w}$ is a vector parallel to $\mathbf{w}$. However, $\mathbf{v}$ and $\mathbf{w}$ are obviously *not* parallel. The only way this could happen is for each $(a + b - 1)\mathbf{v}$ and $(a - b)\mathbf{w}$ to be the zero vector! Then we must have

$$\begin{cases} a + b - 1 = 0, \\ a - b = 0. \end{cases}$$

From the second equation, $a = b$. Then, from the first equation, $2a - 1 = 0$, and so $a = 1/2$. Hence, $a = b = 1/2$, and $\overline{BE} = (1/2)\overline{BD}$ *and* $\overline{AE} = (1/2)\overline{AC}$. Thus, E is the midpoint of both diagonals. ▲

You probably already knew the result of Example 4, and you may know a shorter way to prove it. But the idea behind our investigation is tremendously useful in geometry and a host of other applications. Let us state the idea in a more general form. Suppose that the vectors $\mathbf{v}$ and $\mathbf{w}$ are not parallel and that $a\mathbf{v} + b\mathbf{w} = c\mathbf{v} + d\mathbf{w}$ for some scalars a, b, c, and d. Then we have $(a - c)\mathbf{v} = (d - b)\mathbf{w}$. If this vector were not the zero vector, it would have to be parallel to both $\mathbf{v}$ and $\mathbf{w}$, which is impossible. Therefore, $(a - c)\mathbf{v} = \mathbf{0}$ and $(d - b)\mathbf{w} = \mathbf{0}$, from which we conclude that $a = c$ and $b = d$. To summarize:

If $\mathbf{v}$ *and* $\mathbf{w}$ *are not parallel and* $a\mathbf{v} + b\mathbf{w} = c\mathbf{v} + d\mathbf{w}$, *then* $a = c$ *and* $b = d$.

Exercises for Section 2.1

1. Identify each of the following as a vector quantity or a scalar quantity:
   (a) $2000
   (b) what your team is doing in a tug of war
   (c) seven centimeters per hour
   (d) 1800 miles per hour toward Baghdad.

2. Let $\mathbf{u}$ and $\mathbf{v}$ be vectors.
   (a) Explain why $|\mathbf{u} + \mathbf{v}| \leq |\mathbf{u}| + |\mathbf{v}|$.
   (b) Under what conditions for $\mathbf{u}$ and $\mathbf{v}$ does $|\mathbf{u} + \mathbf{v}| = |\mathbf{u}| + |\mathbf{v}|$?

3. True or false?
   (a) For all vectors $\mathbf{u}$ and $\mathbf{v}$, $|\mathbf{u} - \mathbf{v}| \leq |\mathbf{u}| - |\mathbf{v}|$.
   (b) For all vectors $\mathbf{u}$ and $\mathbf{v}$, $|\mathbf{u} - \mathbf{v}| \leq |\mathbf{u}| + |\mathbf{v}|$.

4. Draw a regular hexagon $ABCDEF$, that is, a hexagon with all sides equal and all angles between successive sides equal.

(a) Express $\overline{AB}$ in terms of $\overline{DE}$.

(b) Express $\overline{AB}$ in terms of $\overline{CF}$.

(c) Express $\overline{AC}$ in terms of $\overline{AD}$ and $\overline{AF}$.

(d) Simplify $\overline{AB} + \overline{BC} + \overline{CD} + \overline{DE} + \overline{EF}$.

(e) Let O be the center of the circle through the vertices of the hexagon. Simplify $\overline{OA} + \overline{OB} + \overline{OC} + \overline{OD} + \overline{OE}$.

5. (a) Let $ABCDE$ be a pentagon that is regular in the sense of Exercise 4. Then A, B, C, D, and E lie on a circle, say with center O. Consider the vectors represented by the arrows $\overline{OA}, \overline{OB}, \overline{OC}, \overline{OD}$, and $\overline{OE}$. Show that if these vectors are placed head to tail, as in the definition of vector addition, they form a regular pentagon, and hence $\overline{OA} + \overline{OB} + \overline{OC} + \overline{OD} + \overline{OE} = \mathbf{0}$.

(b) Let P be any point in space such that $\overline{PA} + \overline{PB} + \overline{PC} + \overline{PD} + \overline{PE} = \mathbf{0}$. Prove that P is O. (Hint: Show that $5\overline{PO} = \mathbf{0}$.)

6. (a) Suppose vectors $\mathbf{a}$ and $\mathbf{b}$ have the same direction. Describe the norm and direction of $\mathbf{a} + \mathbf{b}$.

(b) Consider a "flexible" parallelogram $AA'YX$, with $\mathbf{a} = \overline{AA'}$, $|\overline{AX}| = |\mathbf{b}|$, and X free to move. Draw pictures to show that as $\angle XAA'$ decreases toward $0°$, the vector sum $\overline{AA'} + \overline{AX}$, found by the parallelogram law, gets closer and closer to $\mathbf{a} + \mathbf{b}$.

7. (a) Suppose vectors $\mathbf{a}$ and $\mathbf{b}$ have the opposite direction and $|\mathbf{a}| > |\mathbf{b}|$. Describe the norm and direction of $\mathbf{a} + \mathbf{b}$.

(b) Consider a "flexible" parallelogram $AA'YX$, with $\mathbf{a} = \overline{AA'}$, $|\overline{AX}| = |\mathbf{b}|$, and X free to move. Draw pictures to show that as $\angle XAA'$ increases toward $180°$, the vector sum $\overline{AA'} + \overline{AX}$, found by the parallelogram law, gets closer and closer to $\mathbf{a} + \mathbf{b}$.

(c) Draw pictures for the case where $|\mathbf{a}| < |\mathbf{b}|$ and for that where $|\mathbf{a}| = |\mathbf{b}|$.

8. Draw illustrations of the law of vector algebra given in Eq. (10) for each of the following cases:

(a) $a > b > 0$

(b) $a > 0 > b$

(c) $0 > a > b$

9. Draw illustrations of the law of vector algebra given in Eq. (11) for each of the following cases:

(a) $\mathbf{u}$ and $\mathbf{v}$ not parallel and $a > 0$.

(b) $\mathbf{u}$ and $\mathbf{v}$ not parallel and $a < 0$.

(c) $\mathbf{u}$ and $\mathbf{v}$ have the same direction and $a > 0$.

(d) $\mathbf{u}$ and $\mathbf{v}$ have the same direction and $a < 0$.

(e) $\mathbf{u}$ and $\mathbf{v}$ have opposite direction, $\mathbf{u} + \mathbf{v}$ is not $\mathbf{0}$, and $a > 0$.

(f) $\mathbf{u}$ and $\mathbf{v}$ have opposite direction, $\mathbf{u} + \mathbf{v}$ is not $\mathbf{0}$, and $a < 0$.

(g) $\mathbf{u} + \mathbf{v} = \mathbf{0}$.

10. Draw illustrations of the law given in Eq. (12) for the following cases:

(a) $a > b > 0$

(b) $a > 0 > b$

(c) $0 > a > b$

(d) $ab = 0$

11. (a) Let ABC be a triangle, let A' and B' be the midpoints of the sides opposite A and B, and let D be the intersection of AA' and BB'. What can be said about D?

(b) Let C' be the midpoint of the side opposite C. Use part (a) to explain why D is on CC'.

(c) Let O be any point. Express $\overline{OD}$ in terms of $\overline{OA}, \overline{OB}$, and $\overline{OC}$.

12. Let A, B, C, and D be four points in space, not necessarily all in the same plane, with no three points on the same line. Let the midpoints of segments $[A, B]$, $[B, C], [C, D]$, and $[D, A]$ be E, F, G, and H, respectively. Use vectors to prove that not only are the points E, F, G, and H always in a plane, they always form a parallelogram! (Hint: Let O be any point. Let $\overline{AO} = \mathbf{a}$, etc.)

13. Given triangle ABC, let D be the point on $[A, B]$ two-thirds of the way from A to B, and let E be the point on $[C, B]$ two-thirds of the way from C to B. What can be said about $[D, E]$?

14. Arbie must travel from A to B without entering the pool of quicksand shown on the following page. Send the step commands $\mathbf{a, b, c, d, e}$ in a proper order to accomplish this mission.

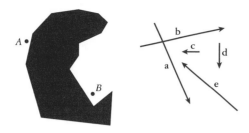

Travel from *A* to *B* on dry land.

15. ESSAY: Use the library, beginning with the *Encyclopaedia Britannica* and *The Oxford English Dictionary,* to discover who invented vectors and vector algebra and to learn what other meanings have been given to the word *vector.* Your paper might concentrate on the life of one of the founders of the subject, or it could compare and contrast the different meanings of the term in modern science and elsewhere.

SECTION 2.2 *Coordinate Systems*

We said that Arbie can fly, but for now let us assume that it stays on the ground—that is, on a fixed plane. Then a step command is a vector **a** in the plane. To respond to command **a**, the robot rotates about its center *R* until its unit direction vector **u** points in the direction of **a**, then it moves distance |**a**|. We adopt the standard convention that a *positive rotation* is one in the counterclockwise sense, as viewed from above the plane, and a *negative* rotation is clockwise. For instance, a 90° rotation is a one-quarter turn counterclockwise, and a −135° rotation is three-eighths of a full clockwise turn, as shown in Figure 1.

FIGURE 1 Positive and negative rotations.

Step commands are prepared for the robot in the form "turn, move" and are beamed to the robot antenna as ordered pairs (r, θ), where r is the (positive or zero) distance to be moved and θ is the (positive, negative, or zero) rotation. Although the robot executes the instructions in the order θ, r, we shall adhere to the traditional notation (r, θ). Arbie's *home base* in the plane consists of a point *O* and a unit vector **i**. The robot is at home base when $R = O$ and

u = **i**. As the robot moves around the plane, the vector **i** remains fixed while the vector **u**, attached to Arbie, varies (in direction, not in length).

If there are no barriers in the plane, then Arbie can go from home base to any point *X* with one command (r, θ). Four examples appear in Figure 2.

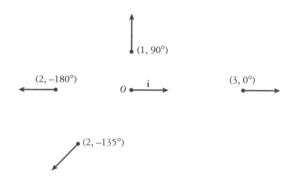

FIGURE 2 Four step commands.

Let's come back for a while from robotics to mathematics. Home base allows us to introduce a *polar coordinate system* in the plane. The point *O* is the *pole,* and the half-line from *O* in the direction of the unit vector **i** is the *initial ray* of the system. A step command (r, θ) to reach a point *X* is a pair of *polar coordinates* for *X*.

Since a positive or negative rotation of 360° around *O* returns every point to itself, the value of θ in the polar coordinates of a point is not uniquely determined by the point. For example the point in

Figure 2 with coordinates $(2, -135°)$ also has coordinates $(2, 225°)$, and the point $(1, 90°)$ also has coordinates $(1, -270°)$. To make polar coordinates unique, the values allowed for θ may be restricted. The two most common restrictions are $0° \leq \theta < 360°$ and $-180° < \theta \leq 180°$. The restriction $-90° \leq \theta < 270°$ appears near the end of this section. Even with such a restriction, the coordinates of the pole are not unique, since at O any value of θ may be used. Despite these minor difficulties, the polar coordinate system has found wide application in science and engineering. For example, when airplanes and tropical storms are detected by radar, the results appear as blips on a polar coordinate system on the screen of a cathode ray tube.

Another important coordinate system derives from the mathematical work of the French philosopher René Descartes (1596–1650) and in his honor is called the *Cartesian coordinate system.* You are already familiar with this system, but we want to review it in order to introduce some slightly different notation, which is suitable for generalization to higher dimensions. In two dimensions, that is, in a plane, take two perpendicular coordinate axes with the same point O as origin. In *standard orientation* we designate these axes as *first* and *second* in such a way that a 90° positive rotation around O carries the positive side of the first axis to the positive side of the second axis. See Figure 3, in which the positive sides

of the axes are indicated by the placement of the labels 1 and 2. This arrangement is called a Cartesian coordinate system in the plane.

Suppose X is a point in a planar Cartesian coordinate system. If the lines through X parallel to the axes cross the first and second axes at points with coordinates x_1 and x_2, then the ordered pair (x_1, x_2) constitutes the Cartesian coordinates of X. As in Figure 4, we often write $X(x_1, x_2)$ as shorthand for "point X, with Cartesian coordinates (x_1, x_2)."

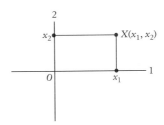

FIGURE 4 Cartesian coordinates of a point.

Clearly, the origin O has coordinates $(0, 0)$, the points on the first axis have coordinates of the form $(x_1, 0)$, and the points on the second axis have coordinates of the form $(0, x_2)$. The axes divide the plane into four *quadrants,* described and numbered as in Figure 5. The first quadrant, in which both coordinates are positive, is also known as the *positive quadrant.*

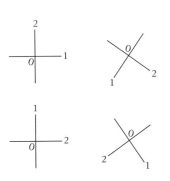

FIGURE 3 Orientations of Cartesian coordinate systems: (a) standard (counterclockwise) orientations; (b) non-standard (clockwise) orientations.

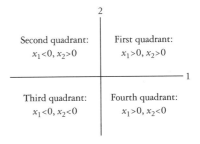

FIGURE 5 The Cartesian quadrants.

Suppose $X(x_1, x_2)$ and $Y(y_1, y_2)$ are points in a plane with a Cartesian coordinate system. We want to obtain a formula for the distance between X and Y, that is, for the length $|XY|$ of line segment $[X, Y]$. Suppose first that $x_1 \neq y_1$ and $x_2 \neq y_2$. Let Z be the point with coordinates (x_1, y_2). Then XYZ is a right triangle (Figure 6); so by the theorem of Pythagoras,

$$|XY|^2 = |ZY|^2 + |XZ|^2. \qquad (1)$$

From Eq. (2), in Section 2.1, $|ZY| = |x_1 - y_1|$ and $|XZ| = |x_2 - y_2|$. Recall that for any number x, $(-x)^2 = x^2$. It then follows from Eq. (1) of Section 2.1 that for all numbers x, $|x|^2 = x^2$. Then Eq. (1) may be written $|XY|^2 = (x_1 - y_1)^2 + (x_2 - y_2)^2$, from which we obtain the *distance formula,*

$$|XY| = \sqrt{(x_1 - y_1)^2 + (x_2 - y_2)^2}. \qquad (2)$$

If $x_1 = y_1$, then (Figure 7) $|XY| = |x_2 - y_2|$; so Eq. (2) also holds in this case. Similarly, the formula holds when $x_2 = y_2$.

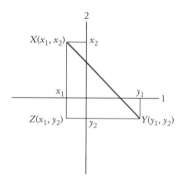

FIGURE 6 Derivation of the distance formula.

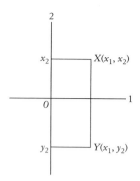

FIGURE 7 A special case.

EXAMPLE 1   Using the Distance Formula

Let's find

a.   all points with distance 6 from each of the points $(-3, 3)$ and $(3, -3)$,

b.   all points with the same distance from each of the points $(1, 2)$, $(3, 5)$, and $(7, 4)$,

c.   all points with the same distance from each of the points $(1, 2)$, $(3, 4)$, and $(5, 6)$.

Solution.

a.   Such points (x_1, x_2) satisfy, by Eq. (2), $(x_1 + 3)^2 + (x_2 - 3)^2 = 36$ and $(x_1 - 3)^2 + (x_2 + 3)^2 = 36$. Subtract the second equation from the first and use the identity $a^2 - b^2 = (a + b)(a - b)$ to get $6(2x_1) - 6(2x_2) = 0$, $x_1 = x_2$. Then the first equation with $x_1 = x_2$ gives $x_1^2 + 6x_1 + 9 + x_1^2 - 6x_1 + 9 = 36$, $2x_1^2 = 18$, $x_1 = \pm 3$. So there are two such points, $(3, 3)$ and $(-3, -3)$. (Plot the points and try to find the answer in an easier way.)

b.   Such points (x_1, x_2) satisfy, by Eq. (2), $(x_1 - 1)^2 + (x_2 - 2)^2 = (x_1 - 3)^2 + (x_2 - 5)^2$ and $(x_1 - 1)^2 + (x_2 - 2)^2 = (x_1 - 7)^2 + (x_2 - 4)^2$, from which we get the equations

(i) $-4x_1 - 6x_2 = -29$
(ii) $12x_1 + 4x_2 = 50$ }.

(From the first equation in part (b), we get

$$x_1^2 - 2x_1 + 1 + x_2^2 - 4x_2 + 4 = x_1^2 - 6x_1$$
$$+ 9 + x_2^2 - 10x_2 + 25,$$
$$-2x_1 - 4x_2 + 5 = -6x_1 - 10x_2 + 34.)$$

Multiply both sides of Eq. (i) by 3 and add to Eq. (ii) to get $-14x_2 = -37$, $x_2 = 37/14$. Then, from Eq. (i), $-4x_1 - 6(37/14) = -29$, $-4x_1 = -166/7$, $x_1 = 83/14$. So the only such point has the coordinates $(83/14, 37/14)$.

c. Proceeding as in part (b) leads to

$$\left.\begin{array}{r} -4x_1 - 4x_2 = -20 \\ 8x_1 + 8x_2 = 56 \end{array}\right\}.$$

Double the first equation and add it to the second to get $0 = 16$, which is false. Thus, there are no such points. (Plot the given points to see why.) ▲

Suppose a plane has both a polar and a Cartesian coordinate system, where the initial ray of the polar system is the positive first axis of the Cartesian system. Let $\mathbf{j}$ be the vector obtained by rotating $\mathbf{i}$ through 90° about the origin, so that $\mathbf{j}$ points in the positive direction of the second axis. Then for any point $X(x_1, x_2)$, we have

$$\overline{OX} = x_1\mathbf{i} + x_2\mathbf{j}. \tag{3}$$

This is illustrated for a point X in the fourth quadrant in Figure 8. Since $x_1 > 0$ and $x_2 < 0$ in this quadrant, the vector $x_1\mathbf{i}$ points in the same direction as $\mathbf{i}$ and the vector $x_2\mathbf{j}$ points in the direction opposite to $\mathbf{j}$, which gives the parallelogram law addition as shown. The vector $\mathbf{x} = x_1\mathbf{i} + x_2\mathbf{j}$ represented by the arrow $\overline{OX}$ is called the *position vector* of point X relative to $\mathbf{i}$ and $\mathbf{j}$.

Suppose a point X in the positive quadrant has Cartesian coordinates (x_1, x_2) and polar coordinates

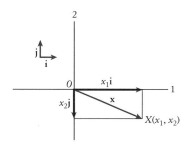

FIGURE 8 Position vector **x** of X.

(r, θ). Let P be the point with Cartesian coordinates $(x_1, 0)$. Then OXP is a right triangle, $\angle XOP = \theta$, $|OX| = r$, $|OP| = x_1$, and $|XP| = x_2$ (see Figure 9). Recall the definitions of *sine* and *cosine*: $\sin \theta = opposite/hypotenuse = |XP|/|OX|$ and $\cos \theta = adjacent/hypotenuse = |OP|/|OX|$. Then

$$x_1 = r \cos \theta, \qquad x_2 = r \sin \theta. \tag{4}$$

(Note: Some calculators and computer software will reject $\sin \theta$, in which case you'll have to enter $\sin(\theta)$.) If $r = 1$, then the point X is on the *unit circle,* that is, the circle with center the origin O that contains the the unit point $U(1, 0)$. (Note that $(1, 0)$ are both the Cartesian and the polar coordinates of U.) As θ goes from 0° to 90°, X moves along the unit circle from U to the point with Cartesian coordinates $(0, 1)$ and polar coordinates $(1, 90°)$. We can use these observations to extend the definitions of sine and cosine to arbitrary angles, as shown in Figure 10. This extension may be expressed by the following definition.

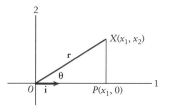

FIGURE 9 Polar and Cartesian coordinates.

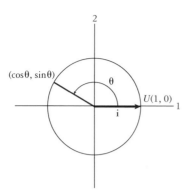

FIGURE 10 Extension of sine and cosine to any angle.

DEFINITION

If θ is any angle (positive, negative, or zero), the *sine* of θ is the second Cartesian coordinate of the point with polar coordinates $(1, \theta)$ and the *cosine* of θ is the first Cartesian coordinate of this point.

EXAMPLE 2 Some Sines and Cosines

Let us find the sine and cosine of the angles in Figure 1. We have already noted that the point with polar coordinates $(1, 90°)$ has Cartesian coordinates $(0, 1)$; so $\sin 90° = 1$ and $\cos 90° = 0$. The point (x_1, x_2) with polar coordinates $(1, -135°)$ is in the third quadrant and is equidistant from both axes. (Plot this point on a copy of Figure 10.) Hence, both coordinates are the same negative number; call it a. By Eq. (2), $\sqrt{(x_1 - 0)^2 + (x_2 - 0)^2} = 1$, from which $2a^2 = 1$. Thus, $a = -\sqrt{1/2} = -0.71$, that is, $\sin(-135°) = \cos(-135°) = -0.71$. (Note: In a statement such as "$-\sqrt{1/2} = -0.71$," it is to be understood that the equality is correct only to as many decimal places as are shown. Thus, for example, $-\sqrt{1/2} = -0.71$, $-\sqrt{1/2} = -0.707$, and $-\sqrt{1/2} = -0.70711$.) ▲

From Figure 5 we see that sine is positive in the first and second quadrants and cosine is positive in the first and fourth quadrants. From Figure 10 we see that sine and cosine are always between -1 and 1, and we may easily find the values of θ for which $\sin \theta$ or $\cos \theta$ is -1 or 1. Many other useful facts about sine and cosine can be deduced from Figure 10 and will appear in the exercises.

Equation (4) for converting polar coordinates (r, θ) to Cartesian coordinates (x_1, x_2) applies to all points in the plane, not just to points in the first quadrant. To see why, suppose the point X has polar coordinates (r, θ), where $r > 0$, and Cartesian coordinates (x_1, x_2). Then the vector $\mathbf{u} = \overline{OX}/r$ is a unit vector in the same direction as $\overline{OX}$. There is a unique point $P(\cos \theta, \sin \theta)$ on the unit circle such that $\mathbf{u} = \overline{OP} = \cos \theta \mathbf{i} + \sin \theta \mathbf{j}$. Then $\overline{OX} = r\mathbf{u} = r \cos \theta \mathbf{i} + r \sin \theta \mathbf{j}$, and from Eq. (3), we have $x_1 = r \cos \theta$ and $x_2 = r \sin \theta$. If $r = 0$, then X is the origin, and these equations still hold. Thus, Eq. (4) holds for all points in the plane.

Next we consider how to convert the other way, from Cartesian coordinates (x_1, x_2) to polar coordinates (r, θ). By the distance formula,

$$r = \sqrt{x_1^2 + x_2^2}. \tag{5}$$

Finding a value of θ (remember that θ is not unique) requires another function from trigonometry. Look again at Figure 9 and recall the definition of *tangent:* $\tan \theta = \textit{opposite/adjacent} = |XP|/|OP|$. Now $|XP|/|OP| = (|XP|/|OX|)/(|OP|/|OX|) = \sin \theta/\cos \theta$; so for $0° < \theta < 90°$, we have

$$\tan \theta = \sin \theta/\cos \theta. \tag{6}$$

This identity is used to extend the definition of the tangent function from the first quadrant to all quadrants. Since division by zero is not defined, we must leave $\tan \theta$ undefined if $\cos \theta = 0$, that is, if $\theta =$

$\pm 90°$. If $x_1 \neq 0$, then $\cos \theta \neq 0$, and from Eq. (4), $x_2/x_1 = r \sin \theta / r \cos \theta = \tan \theta$, that is,

$$x_2/x_1 = \tan \theta. \tag{7}$$

If the point (x_1, x_2) is not on the 2-axis, then $x_1 \neq 0$ and Eq. (7) holds. But this equation is still not sufficient to determine a polar coordinate of the point. Suppose the point (x_1, x_2) has polar coordinates (r, θ). The point $(-x_1, -x_2)$ has the same distance from the origin as (x_1, x_2) but is on the opposite side; thus, polar coordinates for the point $(-x_1, -x_2)$ are $(r, \theta + 180°)$. The points (x_1, x_2) and $(-x_1, -x_2)$ may be called *polar opposites*. Figure 11 shows a case with (x_1, x_2) in the fourth quadrant. Equation (7) cannot distinguish between polar opposite points. To handle this distinction, we define the *principal inverse tangent* of a scalar s, denoted $\tan^{-1} s$, to be the angle θ such that $-90° < \theta < 90°$ and $\tan \theta = s$.

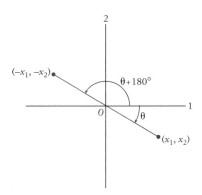

FIGURE 11 Polar opposite points.

EXAMPLE 3 An Inverse Tangent To find $\tan^{-1}(-1) = \theta$, note that $-1 = 1/-1$. Then θ is the angle from $\mathbf{i}$ to the arrow from $(0, 0)$ to $(1, -1)$, which is obviously $-45°$. So $\tan^{-1}(-1) = -45°$. ▲

Most calculators and math software have a built-in principal inverse tangent function. Be sure the machine is in degree mode if you want to get answers in degrees. The radian mode will be discussed later in this chapter.

If $x_1 = 0$, then (x_1, x_2) is on the second axis and $\cos \theta = 0$. In this case it is apparent from Figure 5 that we may take $\theta = 90°$ if $x_2 > 0$ and $\theta = -90°$ if $x_2 < 0$. We may now summarize conversion from Cartesian coordinates (x_1, x_2) to polar coordinates (r, θ). Given Cartesian coordinates (x_1, x_2), use Eq. (5) to calculate r, and use the following equation to find θ:

$$\theta = \begin{cases} \tan^{-1}(x_2/x_1) & \text{if } x_1 > 0, \\ 180° + \tan^{-1}(x_2/x_1) & \text{if } x_1 < 0, \\ 90° & \text{if } x_1 = 0 \text{ and } x_2 > 0, \\ -90° & \text{if } x_1 = 0 \text{ and } x_2 < 0. \end{cases} \tag{8}$$

EXAMPLE 4 Some Conversions from Cartesian to Polar Coordinates The table below shows the values obtained when converting Cartesian coordinates $(4, 3)$, $(-5, -12)$, and $(0, -8)$ to polar coordinates.

Cartesian	r	θ	Polar
$(4, 3)$	$\sqrt{4^2 + 3^2}$	$\tan^{-1}(4/3)$	$(5, 53°)$
$(-5, -12)$	$\sqrt{(-5)^2 + (-12)^2}$	$180° +$ $\tan^{-1}(-12/-5)$	$(13, 247°)$
$(0, -8)$	$\sqrt{0^2 + (-8)^2}$	$-90°$	$(8, -90°)$ ▲

Cartesian and polar coordinates provide useful ways to express vectors in the plane. Let $\mathbf{v}$ be a non-zero vector, and let V be the point such that $\mathbf{v} = \overline{OV}$. Say V has Cartesian coordinates (v_1, v_2). Then from Eq. (3),

$$\mathbf{v} = v_1 \mathbf{i} + v_2 \mathbf{j}. \tag{9}$$

If V has polar coordinates (r, θ), then $r = |\mathbf{v}|$ and

Coordinates and Air Traffic Control

The radar screens pictured here have a polar coordinate system. The origin is the location of the rotating radar antenna. The initial ray may be taken as the 0° direction at the top of the screen. Note that the direction of increase for the θ coordinate is clockwise instead of counterclockwise. The r coordinate gives the distance from the antenna to the object that reflects

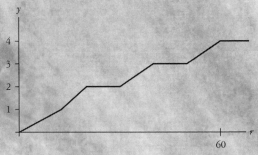

An ideal approach path.

the radar signal. This may be computed from a precise measurement of the time from a radar pulse's transmission until the reflected pulse is received by the antenna. A longer-range version of this type of radar may be used to determine whether an airplane is going in the correct direction to bring it to the airport.

Once the plane is heading in the right direction, it is still necessary to know whether it is at the correct altitude. The correct altitude, y, of the airplane above the level of the runway will depend on the distance coordinate r, as well as on other conditions such as weather and air traffic. Thus, there are ideal altitude functions $y = f(r)$ for various conditions. Above is a graph of one such function, with r in miles and y in thousands of feet. How should a radar system be designed so as to determine whether an airplane is at the correct altitude?

High-resolution, short-range radar display of an airport. (Photo from Henry W. Cole (1992), Understanding Radar, *2nd ed., Oxford: Blackwell, p. 101.)*

θ is the angle between $\mathbf{v}$ and $\mathbf{i}$. Then by Eq. (4), $v_1 = |\mathbf{v}| \cos \theta$ and $v_2 = |\mathbf{v}| \sin \theta$. Hence,

$$\mathbf{v} = |\mathbf{v}|(\cos \theta \mathbf{i} + \sin \theta \mathbf{j}). \qquad (10)$$

Equations (9) and (10) are each called the **ij**-*resolution* of $\mathbf{v}$. Equation (9) is the Cartesian form of the resolution, and Eq. (10) is the polar form.

If $\mathbf{v} = v_1\mathbf{i} + v_2\mathbf{j}$, then $|\mathbf{v}| = |\overline{OV}|$, where $V(v_1, v_2)$. Then by the distance formula,

$$|\mathbf{v}| = \sqrt{v_1^2 + v_2^2}. \qquad (11)$$

We conclude this section with a return to robotics, illustrating vector resolution in both Cartesian and polar coordinate systems.

E X A M P L E 5 Arbie Follows Orders Arbie is at the point A with Cartesian coordinates $(-7.0, 2.0)$ relative to home base and with its direction vector $\mathbf{u}$ rotated $70°$ counterclockwise from $\mathbf{i}$. It receives the step command "turn clockwise $40°$ and move 6." What Cartesian and polar coordinates of the point B does the robot reach after it executes this command?
Solution. The position vector of R before the command is $\mathbf{a} = -7.0\mathbf{i} + 2.0\mathbf{j}$. The rotation from $\mathbf{i}$ to $\mathbf{u}$ before the command is $+70°$, and the command is to rotate $-40°$, so the rotation from $\mathbf{i}$ to $\mathbf{u}$ after execution is $-40° + 70° = 30°$. Then the ij-resolution of the command vector is $\mathbf{c} = 6(\cos 30°\mathbf{i} + \sin 30°\mathbf{j}) = 5.2\mathbf{i} + 3.0\mathbf{j}$. Then the position vector of R after execution is $\mathbf{b} = \mathbf{a} + \mathbf{c} = (-7.0 + 2.0)\mathbf{i} + (5.2 + 3.0)\mathbf{j} = -1.8\mathbf{i} +$

5.0$\mathbf{j}$. The robot is now at the point with Cartesian coordinates $(-1.8, 5.0)$ and polar coordinates (r, θ), where $r = \sqrt{(-1.8)^2 + 5.0^2} = 5.3$ and $\theta = 180° + \tan^{-1}(5.0/-1.8) = 180° + (-70°) = 110°$, as in Figure 12. ▲

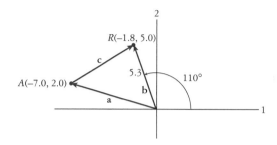

FIGURE 12 Arbie follows orders.

E X A M P L E 6 Preparing a Step Command
Arbie is at $A(8, 10)$ with $\mathbf{u}$ turned $140°$ clockwise from $\mathbf{i}$. Prepare the step command to send it to $B(12, 8)$.
Solution.

$$r = |AB| = \sqrt{(8 - 12)^2 + (10 - 7)^2} = 5,$$

$$\overline{AB} = -\overline{OA} + \overline{OB} = -(8\mathbf{i} + 10\mathbf{j}) + (12\mathbf{i} + 7\mathbf{j})$$
$$= 4\mathbf{i} - 3\mathbf{j}.$$

By Eqs. (9) and (10), $-3 = 5 \sin \theta$, $\sin \theta = -3/5$. From a rough sketch, the clockwise angle from $\mathbf{i}$ to $\overline{AB}$ is acute; hence, $\theta = \sin^{-1}(-3/5) = -37°$. (The function $\sin^{-1}$ is discussed in the next section.) Thus, $\mathbf{u}$ must rotate counterclockwise from $-140°$ to $-37°$, which is a turn of $+103°$. The command, sent as a polar pair in degree mode, is $(5, 103)$. ▲

1. Draw each of the following Arbie journeys, and describe each one by a single "turn and move" step.
    (a) (1) Turn left 90° and move 2 ft; (2) same as (1).
    (b) (1) Turn left 90° and move 2 ft; (2) turn right 90° and move 2 ft.
    (c) (1) Turn left 90° and move 2 ft; (2) turn right 90° and move 2 ft; (3) turn left 90° and move 2 ft.
    (d) (a), then (b), then (c).
    (e) (d), then turn right 90° and move 4 ft; then turn right 90° and move 4 ft.

2. Find the Cartesian coordinates of the points with the following polar coordinates, and plot the points: $A(3, 45°)$, $B(4, 120°)$, $C(3, 210°)$, $D(2, 330°)$, $E(4, 360°)$.

3. Let T be a right triangle with legs a and b and hypotenuse c. Show that four copies of T can be placed around a square of side c to make a square of sides $a + b$, as in the figure below. Also show that the four copies of T can be assembled into a rectangle of sides a and $2b$. Use the formula for the area of a rectangle (or square) to deduce from this the Pythagorean theorem, $a^2 + b^2 = c^2$.

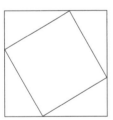

The Pythagorean theorem

4. In right triangle ABC with hypotenuse c and legs a and b opposite A and B, draw CD perpendicular to AB and let $p = |BD|$, as in the following figure.

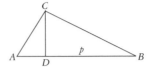

Pythagoras again

    (a) Explain why triangles ABC, ACD, and CBD are similar.
    (b) Obtain from part (a) the proportions $a/p = c/a$ and $b/(c - p) = c/b$.
    (c) Obtain from part (b) the Pythagorean theorem, $a^2 + b^2 = c^2$.

5. In the plane with Cartesian coordinates, suppose $A(2, -4)$ and $B(-2, -1)$. Use the distance formula to find all points C such that ABC is an equilateral triangle.

6. In the plane with Cartesian coordinates, suppose $C(c_1, c_2)$ and $a > 0$. Show that the circle with center C and radius a consists of all points (x_1, x_2) such that $x_1^2 + x_2^2 + b_1 x_1 + b_2 x_2 + d = 0$, where b_1, b_2, and d are constants. Express b_1, b_2, and d in terms of c_1, c_2, and a.

7. Using Exercise 6, describe the set of all points with Cartesian coordinates (x_1, x_2) that satisfy the following equations:
    (a) $x_1^2 + x_2^2 + 4x_1 - 6x_2 + 12 = 0$
    (b) $x_1^2 + x_2^2 - 10x_1 + 9 = 0$
    (c) $x_1^2 + x_2^2 + 2x_1 + 2x_2 + 2 = 0$
    (d) $x_1^2 + x_2^2 - 6x_1 - 8x_2 + 28 = 0$

8. Use the unit circle definitions of sine and cosine to verify the following identities for all angles θ:
    (a) $\cos(\theta \pm 360°) = \cos\theta$,
        $\sin(\theta \pm 360°) = \sin\theta$
    (b) $\cos(\theta \pm 180°) = -\cos\theta$,
        $\sin(\theta \pm 180°) = -\sin\theta$
    (c) $\cos(-\theta) = \cos\theta$, $\sin(-\theta) = -\sin\theta$
    (d) $\cos(180° - \theta) = -\cos\theta$,
        $\sin(180° - \theta) = \sin\theta$

(e) $\cos(90° - \theta) = \sin \theta$, $\sin(90° - \theta) = \cos \theta$

(f) $\sin^2 \theta + \cos^2 \theta = 1$ (Note: The expression $(\sin \theta)^2$ is traditionally shortened to $\sin^2 \theta$, with similar expressions for the other trigonometric functions and for other exponents, *except* for the exponent -1.)

9. On the unit circle in Figure 10, consider the points $U(1, 0)$, $A(\cos \alpha, \sin \alpha)$, $B(\cos \beta, -\sin \beta)$, and $C(\cos(\alpha + \beta), \sin(\alpha + \beta))$. Then $\angle AOB = \angle COU$, so $|AB| = |CU|$. Square both sides of this equation, and use the distance formula and identity (f) of Exercise 8 to obtain the *addition formula for cosine:*

$$\cos(\alpha + \beta) = \cos \alpha \cos \beta - \sin \alpha \sin \beta.$$

10. In the addition formula for cosine (Exercise 9), replace α by $90° - \alpha$ and use identity (e) of Exercise 8 to obtain the *addition formula for sine:*

$$\sin(\alpha + \beta) = \sin \alpha \cos \beta + \cos \alpha \sin \beta.$$

In these addition formulas, replace β by $-\beta$ and then by α to obtain first the *subtraction formulas* and then the *double-angle formulas* for sine and cosine.

11. From Eq. (6) and various identities and formulas derived above obtain the following formulas and identities for tangent:

$$\tan(\theta + 180°) = \tan \theta, \qquad \tan(-\theta) = -\tan \theta,$$

$$\tan(\alpha + \beta) = \frac{\tan \alpha + \tan \beta}{1 - \tan \alpha \tan \beta},$$

$$\tan(\alpha - \beta) = \frac{\tan \alpha - \tan \beta}{1 + \tan \alpha \tan \beta}.$$

12. Find polar coordinates for the points with the following Cartesian coordinates, and plot the points: $A(7, 7)$, $B(0, 5)$, $C(-4, 3)$, $D(-6, -2)$, $E(0, -3)$, $F(5, -1)$, $G(4, 0)$.

13. In each of the following, Arbie's direction vector before the step command is $\mathbf{u} = (-4\mathbf{i} + 3\mathbf{j})/5$. Given the starting point R, find the position vector $\overrightarrow{OR'}$ of the robot after it follows the step command(s) (r, θ):

(a) $R(2, 3)$; $(4, 40°)$

(b) $R(3, 2)$; $(4, 40°)$

(c) $R(-5, 6)$; $(3, 250°)$

(d) $R(3, -4)$; $(5, -70°)$, then $(2, 30°)$

14. Negative values of r are also used in polar coordinates. They are easily understood as "turn and move" instructions to Arbie. If $r < 0$, then, after turning, the robot moves the distance $|r|$, but the motion is *backward,* that is, in reverse. Plot the points with the following polar coordinates, and find positive polar coordinates for them: $(-4, 45°)$, $(4, -45°)$, $(-3, 150°)$, $(5, -60°)$, $(2, -120°)$, $(-7, -45°)$, $(-9, -330°)$.

15. For positive r and θ, express the polar coordinates $(-r, \theta)$, $(r, -\theta)$, and $(-r, -\theta)$ in positive polar coordinates.

16. Explain why the point with polar coordinates $(4, 70°)$ also has polar coordinates $(4, 1870°)$. Suppose a point P has polar coordinates (I), where $r > 0$ and $0° \leq \theta < 360°$.

(a) Find all positive polar coordinates of P.

(b) Find *all* polar coordinates of P.

17. Explain why Eq. (4) holds no matter which polar coordinate pair is used.

18. **ESSAY:** Discover who Pythagoras was, when he lived, where he worked, who he taught, and so forth. Find four, five, or even more additional proofs of the Pythagorean theorem. (There are a great many.) Prove the *converse* of the theorem: For a triangle ABC with sides of length a, b, c opposite A, B, C, if $a^2 + b^2 = c^2$, then the angle at C is a right angle.

SECTION 2.3 *More Vector Algebra in Plane and Space*

This section is intended to expand your ability to work with vectors in a variety of ways. We first recall several results from geometry that are used to extend vector algebra and its applications.

Vectors and Trigonometry

Figure 1 shows an arbitrary triangle ABC with angles α, β, γ at the corners opposite sides a, b, c, respectively. Recall first that

$$\alpha + \beta + \gamma = 180°. \tag{1}$$

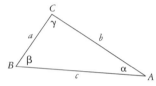

FIGURE 1 An arbitrary triangle.

There are two theorems of trigonometry that are of use in vector algebra: the law of sines and the law of cosines. The law of sines consists of the following proportions, which are easy to remember.

$$\frac{\sin \alpha}{a} = \frac{\sin \beta}{b} = \frac{\sin \gamma}{c}. \tag{2}$$

The law of cosines consists of three equations:

$$\begin{cases} c^2 = a^2 + b^2 - 2ab \cos \gamma, \\ b^2 = a^2 + c^2 - 2ac \cos \beta, \\ a^2 = b^2 + c^2 - 2bc \cos \alpha, \end{cases} \tag{3}$$

which are easier to remember in word form:

The square of any side equals the sum of the squares of the other sides minus twice their product with the cosine of the included angle.

Notice that if an included angle is 90°, then since $\cos 90° = 0$, that case of the law of cosines reduces to the Pythagorean theorem. Derivations of these laws are outlined in the exercises. Of the two laws, the law of cosines is the more important to vector algebra. We will indicate how the law of sines can be avoided if desired.

The first application to vector algebra is to vector addition. Suppose the norms of two nonzero vectors **a** and **b** and the angle θ between **a** and **b** are known. What is the norm and direction of their sum, **c** = **a** + **b**? If θ is either 0° or 180°, then the vectors are parallel and the answer is easy to figure out. Suppose $0° < \theta < 180°$. Then **a**, **b**, and **c** may be drawn as shown in Figure 2 to form a triangle with angle $180° - \theta$ opposite to side **c**. Then by the law of cosines, we have the equation $|\mathbf{c}|^2 = |\mathbf{a}|^2 + |\mathbf{b}|^2 - 2|\mathbf{a}||\mathbf{b}| \cos(180° - \theta)$. Then, since $\cos(180° - \theta) = -\cos \theta$, we have

$$|\mathbf{c}| = \sqrt{|\mathbf{a}|^2 + |\mathbf{b}|^2 + 2|\mathbf{a}||\mathbf{b}| \cos \theta}. \tag{4}$$

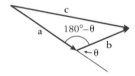

FIGURE 2 Finding **c** = **a** + **b**.

For the direction of **c**, we must find the angle between **c** and one of the other vectors. This may be done with the law of sines. Since, for any angle ϕ, $\sin(180° - \phi) = \sin \phi$, we must be careful to determine whether the angle sought is acute ($<90°$) or obtuse ($>90°$). One safe method is to find the angle ϕ opposite the shorter of **a** or **b**, since that angle is sure to be acute. If s is a scalar between -1 and 1, the *principal inverse sine* of s, denoted $\sin^{-1} s$, is the angle ϕ such that $-90° \le \phi \le 90°$ and $\sin \phi = s$. Calculators and math software usually include the inverse sine function, which is sometimes denoted *arcsin* instead of $\sin^{-1}$.

EXAMPLE 1 Adding Two Commands Arbie receives the command "turn 80.7° and move 5.0 m," followed by the command "turn 160.0° and move 25.0 m." (The letter "m" is the standard abbreviation for *meters*.) Replace these two commands by a single command that would move the robot to the same position.

Solution. Let **a** be the first step vector, let **b** be the second, and let **c** = **a** + **b**. By Eq. (4), $|\mathbf{c}| = \sqrt{5.0^2 + 25.0^2 + 2\cdot 5.0\cdot 25.0 \cos 160°} = 20.4$ m. The angle opposite **c** is $\gamma = 180° - 160° = 20°$. We may use the law of sines to find the angle α opposite the smaller side **a**: $(\sin \alpha)/5.0 = (\sin 20°)/20.4$, $\alpha = \sin^{-1} 0.0838 = 4.8°$. Alternatively, we may use the law of cosines: $a^2 = b^2 + c^2 - 2bc \cos \alpha$, $5.0^2 = 25.0^2 + 20.4^2 - 2\cdot 25.0\cdot 20.4 \cos \alpha$, from which $\cos \alpha = 0.996$, $\alpha = 5.0°$. If we had used the more accurate value $|\mathbf{c}| = 20.27343$, we would get $\alpha = 4.8°$. This more accurate value of $|\mathbf{c}|$ does not change the result—to the nearest tenth of a degree—when the law of sines is used. We will say more about this sort of relative error sensitivity in a later section. In either case, the angle opposite **b** is $\beta = 180° - 20.0° - 4.8° = 155.2°$, by Eq. (1). Then the counterclockwise turn for the single command is $80.7° + 155.2° = 235.9°$. It would be more efficient to turn clockwise and issue the command "turn $-124.1°$ and move 20.4 m." With Arbie preset in the degree–meter mode, this command is sent as the ordered pair $(20.4, -124.1)$. Note that after executing this single command, Arbie would not be facing in the same direction as it would be after executing the two original commands. ▲

Projection and Components

Let **u** be a unit vector and **v** be any vector. The *orthogonal projection* of **v** on **u**, which we shall call simply the *projection* of **v** on **u**, is a vector obtained as follows. Represent **v** by the arrow $\overline{AB}$. Let l be the line through point A that is parallel to **u**. Draw the line through point B that is perpendicular to l; say this line crosses l at point C, as in Figure 3. The projection of **v** on **u** is the vector depicted by an arrow $\overline{AC}$. This vector may be seen as the shadow of vector **v** cast upon line l by the sun high overhead, which may explain the use of the word "projection." Note that this definition implies that the projection of the zero vector on any unit vector is just the zero vector again.

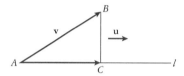

FIGURE 3 Projection of **v** on **u**.

We want to study the relation between the vector **v** and its projection on the vector **u**. Since $\overline{AC}$ is parallel to **u**, we know from the property of parallel nonzero vectors given in Section 2.1 that there is a scalar k such that $\overline{AC} = k\mathbf{u}$. This scalar k is called the *component* of **v** *with respect to* **u**, or briefly the **u**-*component* of **v**. If **u** is a unit vector and **v** is a nonzero vector, there is a useful formula for the **u**-component of **v** in terms of the norm of **v** and the angle θ between **u** and **v**. Since $|\mathbf{u}| = 1$, we have $|\overline{AC}| = |k|$. If $\theta < 90°$, as in Figure 4(a), then $k\mathbf{u}$ and **u** point in the same direction; so $k > 0$, and $|AC| = k$. Then $k = |\mathbf{v}| \cos \theta$. We claim that this result holds for all θ, not just for $\theta < 90°$. If $\theta = 90°$, as in Figure 4(b), then $\overline{AC} = \mathbf{0}$, and $k = 0$. Since $\cos 90° = 0$,

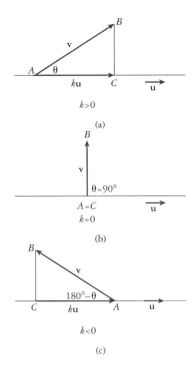

$k>0$

(a)

$\theta=90°$

$k=0$

(b)

$k<0$

(c)

FIGURE 4 Projections and scalar components.

the result holds in this case as well. If $\theta > 90°$, as in Figure 4(c), then $k\mathbf{u}$ and $\mathbf{u}$ point in opposite directions; so $k < 0$, and $|\overline{AC}| = -k$. From right triangle ABC, $|\overline{AC}| = |\mathbf{v}| \cos(180° - \theta)$. But $\cos(180° - \theta) = -\cos\theta$, so we have $-k = -|\mathbf{v}| \cos\theta$, and again the result is true. If $\mathbf{v} = 0$, the angle θ is not well defined. But $|0| \cos\theta = 0$ no matter what θ is; so the result holds for the zero vector as well. To summarize:

If $\mathbf{u}$ is a unit vector and $\mathbf{v}$ is any vector, then *the* $\mathbf{u}$-*component of* $\mathbf{v}$ *is* $|\mathbf{v}| \cos\theta$, where θ is the angle between $\mathbf{u}$ and $\mathbf{v}$. It follows that *the projection of* $\mathbf{v}$ *on* $\mathbf{u}$ *is* $|\mathbf{v}| \cos\theta\,\mathbf{u}$.

The projection of $\mathbf{v}$ on an arbitrary nonzero vector $\mathbf{w}$ is defined as the projection of $\mathbf{v}$ on the unit

vector in the direction of $\mathbf{w}$, which we know from Eq. (9) of Section 2.1 to be the vector $\mathbf{w}/|\mathbf{w}|$. If θ is the angle between $\mathbf{v}$ and $\mathbf{w}$, it follows from the previous paragraph that *the projection of* $\mathbf{v}$ *on* $\mathbf{w}$ *is*

$$|\mathbf{v}| \cos\theta\ \mathbf{w}/|\mathbf{w}|. \qquad (5)$$

EXAMPLE 2 Arbie's Shadow Suppose Arbie is flying west in a straight line at 250 mph, ascending at an angle of $40°$. Let $\mathbf{h}$ be a horizontal velocity vector of norm 10 mph pointing west in the vertical plane containing the robot's line of flight (Figure 5). Let $\mathbf{v}$ be the robot's velocity vector. Then the *ground velocity* vector $\mathbf{g}$, that is, the velocity vector of Arbie's shadow, is the projection of $\mathbf{v}$ on $\mathbf{h}$, which by Eq. (5) is $|\mathbf{v}| \cos 40°\mathbf{h}/|\mathbf{h}|$. Thus, $\mathbf{g} = (250 \cos 40°/10)\mathbf{h} = 19.2\mathbf{h}$. The *ground speed* of the robot is the norm of its ground velocity, $|19.2\mathbf{h}| = (19.2)(10) = 192$ mph. ▲

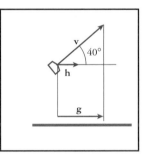

FIGURE 5 Arbie and its shadow.

EXAMPLE 3 Nelly's Wagon Nelly pulls on a wagon with a force of 35 lb at an angle of $30°$ (Figure 6). The wagon is attached to a spring scale attached to a wall. What is the scale's reading?
Solution. Let $\mathbf{f}$ be the force vector applied by Nelly. Place a unit vector $\mathbf{u}$ in the direction in which the spring is stretched. Then the reading

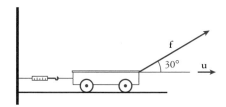

FIGURE 6 Pulling a wagon.

on the scale is the **u**-component of **f**, which by Eq. (5) is $|\mathbf{f}| \cos 30° = (35)(0.87) = 30$ lb. ▲

The Dot Product

We turn next to another operation of vector algebra, which has many uses.

DEFINITION

The *dot product* of two nonzero vectors **v** and **w** is

$$\mathbf{v} \cdot \mathbf{w} = |\mathbf{v}||\mathbf{w}| \cos \theta, \qquad (6)$$

where θ is the angle between **v** and **w**. The dot product of the zero vector with any vector is defined to be the scalar zero: $\mathbf{0} \cdot \mathbf{v} = \mathbf{v} \cdot \mathbf{0} = 0$.

The dot should *not* be omitted from this notation: Write $\mathbf{v} \cdot \mathbf{w}$, not $\mathbf{vw}$. Note that it does not matter whether angle θ is measured in the positive (counterclockwise) or negative (clockwise) direction, because $\cos(360° - \theta) = \cos(-\theta) = \cos \theta$. Thus, we may always assume that $0° \le \theta \le 180°$.

If **u** is a unit vector, the **u**-component of any vector **v** is $|\mathbf{v}| \cos \theta = |\mathbf{u}||\mathbf{v}|\cos \theta$ since $|\mathbf{u}| = 1$. Thus, *the **u**-component of **v** is* $\mathbf{u} \cdot \mathbf{v}$ and *the projection of **v** on **u** is* $(\mathbf{u} \cdot \mathbf{v})\mathbf{u}$. Since the angle between a nonzero vector and itself is $0°$ and $\cos 0° = 1$, it follows that $\mathbf{v} \cdot \mathbf{v} = |\mathbf{v}|^2$ and consequently that $|\mathbf{v}| = \sqrt{\mathbf{v} \cdot \mathbf{v}}$. This

result also holds for the zero vector, since $\mathbf{0} \cdot \mathbf{0} = 0$. From Result (5), the projection of **v** on **w** is $|\mathbf{v}| \cos \theta \; \mathbf{w}/|\mathbf{w}|$, which can be rewritten as $|\mathbf{v}||\mathbf{w}| \cos \theta/|\mathbf{w}|^2$. Thus for any vector **v** and any nonzero vector **w**, *the projection of **v** on **w** is*

$$\frac{\mathbf{v} \cdot \mathbf{w}}{\mathbf{w} \cdot \mathbf{w}} \mathbf{w}. \qquad (7)$$

If **v** and **w** are not zero, then from the definition of dot product, $\cos \theta = \mathbf{v} \cdot \mathbf{w}/|\mathbf{v}||\mathbf{w}|$; hence, *the angle between **v** and **w** is*

$$\cos^{-1} \frac{\mathbf{v} \cdot \mathbf{w}}{|\mathbf{v}||\mathbf{w}|}. \qquad (8)$$

Nonzero vectors are parallel if and only if the angle between them is either $0°$ or $180°$. Now for $0° \le \theta \le 180°$, $\theta = 0°$ *or* $180°$ if and only if $\cos \theta = \pm 1$; so from Result (8) we see that *vectors **v** and **w** are parallel if and only if* $|\mathbf{v} \cdot \mathbf{w}| = |\mathbf{v}||\mathbf{w}|$, that is, if and only if $(\mathbf{v} \cdot \mathbf{w})^2 = (\mathbf{v} \cdot \mathbf{v})(\mathbf{w} \cdot \mathbf{w})$. For $0° \le \theta \le 180°$, we have $\cos \theta = 0$ if and only if $\theta = 90°$; hence, *vectors **v** and **w** are perpendicular if and only if*

$$\mathbf{v} \cdot \mathbf{w} = 0. \qquad (9)$$

These tests for parallel and perpendicular vectors hold for the zero vector as well if we adopt the convention that **0** is both parallel and perpendicular to every vector. We say that two nonzero vectors have the *same general direction* if the angle between them is less than $90°$, and they have the *opposite general direction* if the angle between them is greater than $90°$. Now $\cos \theta > 0$ *for* $0° \le \theta < 90°$ and $\cos \theta < 0$ for $90° < \theta \le 180°$; hence, *vectors **v** and **w** have the same general direction if* $\mathbf{v} \cdot \mathbf{w} > 0$ *and have opposite general direction if* $\mathbf{v} \cdot \mathbf{w} < 0$.

We postpone computational examples using the dot product until later in this section, when we will derive a practical formula for calculating the dot product of two vectors from their components.

Vectors in Space

In some of our work in Section 2.2, we assumed that Arbie did not fly but was confined to one plane. This enabled us to introduce various coordinate systems as an aid to our work with vectors. But throughout Section 2.1 and the present section, *the definitions of vectors and vector operations have made no mention of whether we are in a plane or in space.* Vectors, vector quantities, arrows representing vectors, norm, vector equality, vector addition, multiplication of a vector by a scalar, position vector, projection of one vector on another, components, vector resolution, dot product, and the laws of vector algebra are all the same, not only in two and three dimensions, but in any number of dimensions.

Since the surfaces of paper and chalkboard are two-dimensional, it is harder to draw convincing pictures of some combinations of vectors in space than it was in the plane. For example, consider three mutually perpendicular unit vectors **i**, **j**, and **k**. If we drew **i** and **j** as we did in Section 2.2, you would not see **k**, since it would come straight out of the paper, as in Figure 7(a). Instead, we show **i**, **j**, and **k** from an angle, as in Figure 7(b). You can see this arrangement in space if you stand near a corner of a room with one wall in front of you and another wall to your left. Let O be the corner near your feet. Then the vectors **i** and **j** run along the intersection of the floor with the walls to your left and in front, respec-

tively, and the vector **k** runs up the intersection of these walls.

We use **i**, **j**, and **k** to introduce a Cartesian coordinate system in space with three coordinate axes having the same point O as origin and the positive direction of the first, second, and third axes indicated by the vectors **i**, **j**, and **k**, respectively. As with Cartesian coordinates in the plane, there are two possible orientations. In the *standard* orientation, as in Figure 8(a), if you view the plane containing the first two axes from a point on the positive half of the third axis, the first and second axes will appear in standard orientation, in the sense of Figure 3, in Section 2.2. We will use this orientation.

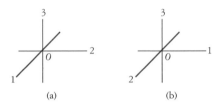

FIGURE 8 Cartesian coordinate systems in space: standard (a) versus nonstandard (b) orientation.

Let X be a point in space with a Cartesian coordinate system, as in Figure 9. Let the **i**-, **j**-, and **k**-components of $\overline{OX}$ be x_1, x_2, and x_3, respectively. Then the Cartesian coordinates of X are defined to be the ordered triple (x_1, x_2, x_3). It follows that the

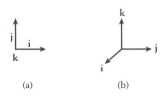

FIGURE 7 Mutually perpendicular unit vectors. (a) Where is **k**? (b) Shown from an angle, **i**, **j**, and **k** are all visible.

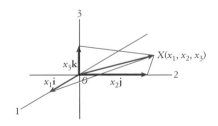

FIGURE 9 Coordinates of a point.

projections of the vector $\mathbf{x} = \overline{OX}$ on the vectors $\mathbf{i}$, $\mathbf{j}$, and $\mathbf{k}$ are $x_1\mathbf{i}$, $x_2\mathbf{j}$, and $x_3\mathbf{k}$, respectively, and that

$$\mathbf{x} = x_1\mathbf{i} + x_2\mathbf{j} + x_3\mathbf{k}. \qquad (10)$$

To clarify Eq. (10), in Figure 10 we redraw the representations of the projected vector. Displacements from 0 to X' to X'' to X result in the displacement from 0 to X, $\overline{OX'} + \overline{X'X''} + \overline{X''X} = \overline{OX}$.

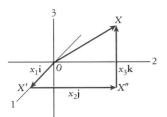

FIGURE 10 Illustrating Eq. (10).

The length of arrow $\overline{OX}$ is, as you are asked to show in Exercise 9,

$$|\overline{OX}| = \sqrt{x_1^2 + x_2^2 + x_3^2}. \qquad (11)$$

Given a vector $\mathbf{x}$ in space with a Cartesian coordinate system, there is a unique point $X(x_1, x_2, x_3)$ such that $\mathbf{x} = \overline{OX}$. Then by Eq. (11), $|\mathbf{x}| = \sqrt{x_1^2 + x_2^2 + x_3^2}$. Suppose $X(x_1, x_2, x_3)$ and $Y(y_1, y_2, y_3)$ are points. Then the distance between X and Y is

$$\begin{aligned}
|\overline{XY}| &= |-\overline{OX} + \overline{OY}| \\
&= |-(x_1\mathbf{i} + x_2\mathbf{j} + x_3\mathbf{k}) + (y_1\mathbf{i} + y_2\mathbf{j} + y_3\mathbf{k})| \\
&= |(x_1 - y_1)\mathbf{i} + (x_2 - y_2)\mathbf{j} + (x_3 - y_3)\mathbf{k}|,
\end{aligned}$$

and so by Eq. (11), *the distance between X and Y is*

$$\sqrt{(x_1 - y_1)^2 + (x_2 - y_2)^2 + (x_3 - y_3)^2}. \qquad (12)$$

We are now able to use the law of cosines to obtain a very useful formula for the dot product of two vectors in terms of their components. Let $\mathbf{a} =$ $a_1\mathbf{i} + a_2\mathbf{j} + a_3\mathbf{k}$ and $\mathbf{b} = b_1\mathbf{i} + b_2\mathbf{j} + b_3\mathbf{k}$ be vectors in space, and let θ be the angle between them. Let $\mathbf{c} = \mathbf{a} - \mathbf{b}$, and apply the law of cosines to the triangle with sides $\mathbf{a}$, $\mathbf{b}$, and $\mathbf{c}$ (Figure 11) to get

$$\begin{aligned}
|\mathbf{c}|^2 &= |\mathbf{a}|^2 + |\mathbf{b}|^2 - 2|\mathbf{a}||\mathbf{b}| \cos \theta \\
&= |\mathbf{a}|^2 + |\mathbf{b}|^2 - 2\mathbf{a} \cdot \mathbf{b}.
\end{aligned}$$

Then

$$\begin{aligned}
2\mathbf{a} \cdot \mathbf{b} &= |\mathbf{a}|^2 + |\mathbf{b}|^2 - |\mathbf{c}|^2 \\
&= (a_1^2 + a_2^2 + a_3^2) + (b_1^2 + b_2^2 + b_3^2) \\
&\quad - [(a_1 - b_1)^2 + (a_2 - b_2)^2 + (a_3 - b_3)^2] \\
&= 2(a_1b_1 + a_2b_2 + a_3b_3),
\end{aligned}$$

from which we get

$$\mathbf{a} \cdot \mathbf{b} = a_1b_1 + a_2b_2 + a_3b_3. \qquad (13)$$

If $\mathbf{a}$ and $\mathbf{b}$ are in the plane through the first two axes, then this formula reduces to

$$\mathbf{a} \cdot \mathbf{b} = a_1b_1 + a_2b_2.$$

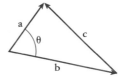

FIGURE 11 The vector triangle for $\mathbf{a} \cdot \mathbf{b}$.

EXAMPLE 4 Measuring an Angle The angle between the vectors $\mathbf{a} = 3\mathbf{i} + 4\mathbf{j} - \mathbf{k}$ and $\mathbf{b} = -2\mathbf{i} + 8\mathbf{j} - 3\mathbf{k}$ is

$$\begin{aligned}
&\cos^{-1} \frac{\mathbf{a} \cdot \mathbf{b}}{|\mathbf{a}||\mathbf{b}|} \\
&= \cos^{-1} \frac{3 \cdot (-2) + 4 \cdot 8 + (-1)(-3)}{\sqrt{[(3^2 + 4^2 + (-1)^2)][(-2)^2 + 8^2 + (-3)^2]}} \\
&= \cos^{-1} \frac{29}{\sqrt{26 \cdot 77}} = \cos^{-1}(0.65) = 50°.
\end{aligned}$$

(Draw the vectors.) ▲

Robots and Geometry

Robots are used for a variety of activities, some very specific, some more general. The photo below shows a small robotic arm. The design is based largely on that of the human arm. Because the joints are rotational, the geometry and linear algebra of rotations is of importance in their design and control.

Early mobile robots used wheels, which restricted their environment to smooth floors with few obstacles. Designs for motion in rough terrain use legs. The

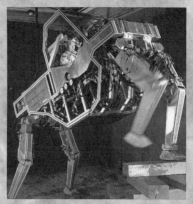

A large climbing robot. (Photo from Michael Brady, ed. (1989), Robotics Science, *Cambridge, MA: MIT Press, p. 584.)*

A robot arm. (Photo from Robert E. Parkin (1991), Applied Robotic Analysis, *Englewood Cliffs, NJ: Prentice Hall, p. 387.)*

GE Walking Truck, shown in the above photo, is a four-legged model. Other types use the six-legged design favored by insects, in which three legs can be raised and moved while the other three maintain the balance. The importance of vectors and of the geometry of position in the design and control of mobile robots is clear.

E X A M P L E 5 **Making a Vector Projection**
The projection of $\mathbf{v} = 12\mathbf{i} - 9\mathbf{k}$ on $\mathbf{w} = -2\mathbf{i} + 5\mathbf{j}$ is

$$\frac{\mathbf{v} \cdot \mathbf{w}}{\mathbf{w} \cdot \mathbf{w}} \mathbf{w} = \frac{(12)(-2) + (0)(5) + (-9)(0)}{(-2)^2 + 5^2} \mathbf{w}$$
$$= \frac{-24}{29} \mathbf{w} = \frac{48}{29} \mathbf{i} - \frac{120}{29} \mathbf{j}.$$

(Again, make a drawing.) ▲

E X A M P L E 6 **Parallel or Perpendicular?**
Given the points $A(-6, 5)$, $B(2, 11)$, $C(-5, -3)$, and $D(2, 1)$, are the lines AB and CD parallel or perpendicular? Are the lines AD and BC parallel or perpendicular?
Solution. We have $\overline{AB} = 8\mathbf{i} + 6\mathbf{j}$, $\overline{CD} = 7\mathbf{i} + 4\mathbf{j}$, $\overline{AD} = 8\mathbf{i} - 4\mathbf{j}$, and $\overline{BC} = -7\mathbf{i} - 14\mathbf{j}$. Now $\overline{AB} \cdot \overline{CD} = 56 + 24 = 80$; so AB and CD aren't perpendicular. Also, $\overline{AB} \cdot \overline{AB} = 100$,

$\overline{CD} \cdot \overline{CD} = 65$, and $80^2 \neq 100 \cdot 65$; so AB and CD aren't parallel either. Now $\overline{AD} \cdot \overline{BC} = -56 + 56 = 0$; so AD is perpendicular, and hence is not parallel, to BC. ▲

With Eq. (13) verifying the following rules of calculation for all vectors $\mathbf{a}$, $\mathbf{b}$, and $\mathbf{c}$ and scalars k is straightforward:

$$\mathbf{a} \cdot \mathbf{b} = \mathbf{b} \cdot \mathbf{a},$$
$$\mathbf{a} \cdot (\mathbf{b} + \mathbf{c}) = \mathbf{a} \cdot \mathbf{b} + \mathbf{a} \cdot \mathbf{b}, \qquad (14)$$
$$(k\mathbf{a}) \cdot \mathbf{b} = \mathbf{a} \cdot (k\mathbf{b}) = k(\mathbf{a} \cdot \mathbf{b}).$$

More applications of this sort of vector algebra will appear in later sections.

Exercises for Section 2.3

1. (a) Consider a parallelogram with base b and height h. By cutting it in two and reassembling the pieces into a rectangle, show that the area of the parallelogram is bh.
   (b) Consider a triangle with base b and height h. By assembling two copies of the triangle into a parallelogram, show that the area of the triangle is $bh/2$.
   (c) Consider the triangle in Figure 1. Show that its area is $(1/2)bc \sin \alpha$.
   (d) Find two more expressions like that of part (c) for the area of the triangle.
   (e) From the fact that the expressions found in parts (c) and (d) give the same area, deduce the law of sines.

2. For the law of cosines, it suffices to prove for the triangle in Figure 1 that $c^2 = a^2 + b^2 - 2ab \cos \gamma$ for all possible angles γ. If $\gamma = 90°$, the result is the Pythagorean theorem.
   (a) Suppose $\gamma > 90°$. Draw the altitude $[A, D]$, let h be its length, and let $d = |CD|$. Use the Pythagorean theorem to find two expressions for

h^2 in terms of a, b, c, and d, and deduce that $c^2 = a^2 + b^2 + 2ad$. Express d in terms of b and γ to complete this case.
   (b) Suppose $\gamma < 90°$. If D is on $[B, C]$, draw the altitude $[A, D]$; otherwise, draw the altitude $[B, D]$. Let h be the length of the altitude drawn, and let $d = |CD|$. Complete this case similarly to that of part (a).

3. In each part, replace the sequence of step commands by a single step command.
   (a) "Turn $90°$ and move 2 ft," then "turn $-90°$ and move 2 ft."
   (b) "Turn $90°$ and move 2 ft," then "turn $90°$ and move 2 ft."
   (c) Execute the step command "turn $30°$ and move 3 m" three times.
   (d) Execute the step command "turn $-45°$ and move 8 m" seven times.
   (e) "Turn $290.33°$ and move 2.45 cm," then "turn $-342.66°$ and move 2.54 cm."

4. A gas pipe has a plug in it at ground level that I must pull out. The plug is attached to a four-foot length of rope on which I can pull with a force of 70 lb. I know

that a force of 50 lb *exerted straight up* will pull the plug, and the noxious gas will then shoot out. How close to the plug do I have to get?

5. A hawk is diving down at an angle of $35°$ below the horizon. Its shadow cast by the sun directly overhead is moving along the ground at 40 ft/sec. What is the hawk's speed?

6. Let $\mathbf{a} = a_1\mathbf{i} + a_2\mathbf{j} + a_3\mathbf{k}$, $\mathbf{b} = b_1\mathbf{i} + b_2\mathbf{j} + b_3\mathbf{k}$, and $\mathbf{c} = c_1\mathbf{i} + c_2\mathbf{j} + c_3\mathbf{k}$. Use Eq. (13) to verify the laws in Eq. (14).

7. Measure $\angle ABC$, where $A(1, 2, 3)$, $B(2, 1, 3)$, and $C(3, 1, 2)$.

8. For the points of Exercise 7, calculate $|\overline{AB} + \overline{AC}|$.

9. In Figure 10, draw segment $[0, X'']$. Use the Pythagorean theorem twice to deduce Result (11).

10. In space with Cartesian coordinates, consider the points $A(6, -2, 5)$, $B(4, 5, -2)$, $C(2, 1, 6)$, $D(4, 6, -1)$, $E(-2, 8, -6)$, and $F(-3, 8, -5)$. Which four points are the vertices of a rectangle?

11. Calculate the $\mathbf{w}$-component of $\mathbf{v}$, where $\mathbf{v} = -3\mathbf{i} + \mathbf{j} + 4\mathbf{k}$ and $\mathbf{w} = 2\mathbf{i} - 3\mathbf{j} + \mathbf{k}$.

12. Find the projection of $\overline{PQ}$ on $\overline{RS}$, where $P(6, 5, 4)$, $Q(3, 2, 1)$, $R(1, 2, 3)$, and $S(4, 5, 6)$. Draw the vectors to see why the result is true.

13. Calculate the projection of $\mathbf{i} - \mathbf{k}$ on $\mathbf{i} + \mathbf{j} + \mathbf{k}$. Draw the vectors to see why the result is true.

14. Show that the sphere with center (c_1, c_2, c_3) and radius a consists of all points (x_1, x_2, x_3) that satisfy an equation of the form $x_1^2 + x_2^2 + x_3^2 + b_1x_1 + b_2x_2 + b_3x_3 + d = 0$. Express b_1, b_2, b_3, and d in terms of c_1, c_2, c_3, and a.

15. In triangle ABC let D be the foot of the altitude from A to BC, let E be the foot of the altitude from B to AC, and let H be the intersection of AD and BE.

(a) Draw the triangle and explain why there are scalars x, y such that $\overline{CH} = \overline{CA} + x(\overline{CA} - \overline{CD})$, $\overline{CH} = \overline{CB} + y(\overline{CB} - \overline{CE})$.

(b) Let $\mathbf{a} = \overline{CA}$ and $\mathbf{b} = \overline{CB}$. Note that $\overline{CD}$ is the projection of $\mathbf{a}$ on $\mathbf{b}$ and $\overline{CE}$ is the projection of $\mathbf{b}$ on $\mathbf{a}$. From part (a), $\overline{CA} + x(\overline{CA} - \overline{CD}) = \overline{CB} + y(\overline{CB} - \overline{CE})$. Write this equation in terms of $\mathbf{a}$, $\mathbf{b}$, x, and y.

(c) Using the independence of $\mathbf{a}$ and $\mathbf{b}$, solve the equation of part (b) for x. From this and the first equation of part (a), obtain an expression for $\overline{CH}$ in terms of $\mathbf{a}$ and $\mathbf{b}$.

(d) Use part (c) to show that CH is perpendicular to AB. Explain why this proves the famous theorem that the three altitudes of any triangle always meet in a point.

16. Draw the analog in three dimensions to the parallelogram law of vector addition (Figure 11 in Section 2.1). Begin with arrows $\overline{OA}$, $\overline{OB}$, and $\overline{OC}$, where points 0, A, B, and C are not all in the same plane.

17. Three vectors are said to be *independent* if they are not all parallel to the same plane.

(a) Explain why if three vectors are independent, then no two of them are parallel.

(b) Explain the following analog to a result given near the end of Section 2.1. If vectors $\mathbf{u}$, $\mathbf{v}$, $\mathbf{w}$ are independent and $p\mathbf{u} + q\mathbf{v} + r\mathbf{w} = s\mathbf{u} + t\mathbf{v} + x\mathbf{w}$ for some scalars p, q, r, s, t, x, then it must be that $p = s$, $q = t$, and $r = x$.

18. **ESSAY:** We mentioned in Section 2.1 that the word *robotics* was coined by Isaac Asimov. But he did not coin the word *robot*. Find out who did. Then read and comment on the stage play that you will discover in the course of your research.

SECTION 2.4 *Lines, Planes, and Circles*

As Arbie flies around, it often moves in a straight line, as when proceeding to a selected site; it sometimes flies in a circle, as when surveying the site; and it occasionally approaches a plane, as when landing at a site. In this section we'll use vector algebra to describe these geometrical objects. Of course, the world is filled with curves and surfaces that are more complicated than lines, circles, and planes. But the objects of this section are a sensible place to start. Moreover, as you will learn in more advanced courses, lines, circles, and planes are essential tools used to study *all* "smooth" curves and surfaces. Thus, while the subject of this section may be introductory, it is also fundamental.

Parametric Description of Lines

Let A be a point, let $\mathbf{e}$ be a nonzero vector, and let l be the line through A parallel to $\mathbf{e}$. Then point X is on l if and only if the vectors $\overline{AX}$ and $\mathbf{e}$ are parallel (Figure 1), that is, if and only if $\overline{AX} = t\mathbf{e}$ for some scalar t. (If $A = X$, then $t = 0$.) Introduce a Cartesian coordinate system with origin O and let $\mathbf{a} = \overline{OA}$. Then $\overline{AX} = \overline{OX} - \overline{OA} = \overline{OX} - \mathbf{a}$, so X is on l if and only if

$$\overline{OX} = \mathbf{a} + t\mathbf{e}. \qquad (1)$$

Equation (1) is a *point-direction* vector equation for line l.

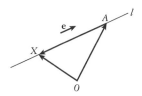

FIGURE 1 Point-direction description of a line.

Let B be a point on l different from A, and let $\mathbf{b} = \overline{OB}$. Then $\mathbf{b} - \mathbf{a}$ is a nonzero vector parallel to l, so we may set $\mathbf{e} = \mathbf{b} - \mathbf{a}$ in Eq. (1) to get $\overline{OX} = \mathbf{a} + t(\mathbf{b} - \mathbf{a})$, that is,

$$\overline{OX} = (1 - t)\mathbf{a} + t\mathbf{b}. \qquad (2)$$

This is a *point-point* vector equation for line AB. The variable t in each of these equations is called a *parameter,* and both equations are *parametric vector equations* for l.

EXAMPLE 1 Two Lines The line through the point A with Cartesian coordinates $(2, -3, 4)$ parallel to the vector $\mathbf{e} = 3\mathbf{i} + 6\mathbf{j} - 5\mathbf{k}$ has the point-direction equation $\overline{OX} = (2\mathbf{i} - 3\mathbf{j} + 4\mathbf{k}) + t(3\mathbf{i} + 6\mathbf{j} - 5\mathbf{k})$. For $X(x_1, x_2, x_3)$, this is equivalent to the system of parametric scalar equations

$$\left. \begin{aligned} x_1 &= 2 + 3t \\ x_2 &= -3 + 6t \\ x_3 &= 4 - 5t \end{aligned} \right\}.$$

The line through A and $B(7, -8, 2)$ has the point-point equation $\overline{OX} = (1 - t)(2\mathbf{i} - 3\mathbf{j} + 4\mathbf{k}) + t(7\mathbf{i} - 8\mathbf{j} + 2\mathbf{k})$, that is,

$$\left. \begin{aligned} x_1 &= 2(1 - t) + 7t = 2 + 5t \\ x_2 &= -3(1 - t) - 8t = -3 - 5t \\ x_3 &= 4(1 - t) + 2t = 4 - 2t \end{aligned} \right\}. \quad \blacktriangle$$

As the parameter t in Eq. (2) increases from 0 to 1, the point X moves along the line segment $[A, B]$ from A to B. Think of t as time; then in one unit of time the point X moves a distance of $|\mathbf{b} - \mathbf{a}|$ in the direction of vector $\mathbf{b} - \mathbf{a}$. Thus, the speed of point X is $|\mathbf{b} - \mathbf{a}|$, and the velocity vector of point X is $\mathbf{v} = \mathbf{b} - \mathbf{a}$. If $t < 0$, then X is on the side of point

A opposite point B; if $t > 1$, then X is on the side of point B opposite point A, as in Figure 2.

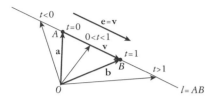

FIGURE 2 Point-point vector representation of line AB.

EXAMPLE 2 The Flying Robot Arbie is flying in a straight line at a constant speed of 250 km/hr. At 10:07 A.M. it is at the point $A(-50, 200, -150)$, in Cartesian coordinates calibrated in kilometers, headed toward the point $B(100, 150, 0)$. When does the robot reach point B? Where is it at 10:39 A.M.?

Solution. $|AB| = \sqrt{150^2 + (-50)^2 + 150^2} = 218$ km, a distance that Arbie covers in $218/250 = 0.87$ hr $= 52$ min. Hence, the robot reaches B at 10:59 A.M. Let $\mathbf{a} = \overline{OA}$ and $\mathbf{b} = \overline{OB}$. At 10:39 we have $t = 32$ min $= 0.53$ hr, and from Eq. (2), $\overline{OX} = 0.47(-50\mathbf{i} + 200\mathbf{j} - 150\mathbf{k}) + 0.53(100\mathbf{i} + 150\mathbf{j}) = 29.5\mathbf{i} + 173.5\mathbf{j} - 70.5\mathbf{k}$; so at 10:39 the robot is at the point $X(29.5, 173.5, -70.5)$. ▲

Let P be a point with position vector $\mathbf{p} = \overline{OP}$. We seek a formula for the distance d from P to the line l with point-direction equation (1). Suppose $\mathbf{p} - \mathbf{a}$ is not perpendicular to l. Let $\overline{AQ}$ be the projection of $\overline{AP}$ on $\mathbf{e}$. Then APQ is a right triangle with $|PQ| = d$ and $|AQ| =$ the length of the projection of $\mathbf{p} - \mathbf{a}$ on $\mathbf{e}$ (Figure 3), which by Eq. (7) of Section 2.3, is

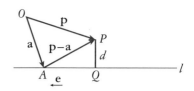

FIGURE 3 Finding distance (P, l).

$$\left| \frac{(\mathbf{p} - \mathbf{a}) \cdot \mathbf{e}}{\mathbf{e} \cdot \mathbf{e}} \mathbf{e} \right| = \frac{|(\mathbf{p} - \mathbf{a}) \cdot \mathbf{e}||\mathbf{e}|}{\mathbf{e} \cdot \mathbf{e}} = \frac{|(\mathbf{p} - \mathbf{a}) \cdot \mathbf{e}|}{|\mathbf{e}|},$$

since $\mathbf{e} \cdot \mathbf{e} = |\mathbf{e}|^2$. Then by the Pythagorean theorem, $|\mathbf{p} - \mathbf{a}|^2 = d^2 + [(\mathbf{p} - \mathbf{a}) \cdot \mathbf{e}/|\mathbf{e}|]^2$, so *the distance d from P to l is given by*

$$d^2 = (\mathbf{p} - \mathbf{a}) \cdot (\mathbf{p} - \mathbf{a}) - \frac{[(\mathbf{p} - \mathbf{a}) \cdot \mathbf{e}]^2}{\mathbf{e} \cdot \mathbf{e}}. \qquad (3)$$

If $\mathbf{p} - \mathbf{a}$ is perpendicular to l, then $d = |\mathbf{p} - \mathbf{a}|$ and $(\mathbf{p} - \mathbf{a}) \cdot \mathbf{e} = 0$, and so Eq. (3) still holds.

EXAMPLE 3 Arbie Approaches a Line Arbie is on the ground at $R(19, 36)$, facing $T(23, 33)$, where coordinates are in kilometers. We want the robot to go to either the line l_1 through $A(2, 9)$ and $B(4, 8)$ or the line l_2 through $C(20, 1)$ and $D(26, 3)$, whichever is closer. What step command should be sent?

Solution. (Draw a sketch first.) By Eq. (3) with $\mathbf{p} - \mathbf{a} = (19\mathbf{i} + 36\mathbf{j}) - (2\mathbf{i} + 9\mathbf{j}) = 17\mathbf{i} + 27\mathbf{j}$ and $\mathbf{e} = (4\mathbf{i} + 8\mathbf{j}) - (2\mathbf{i} + 9\mathbf{j}) = 2\mathbf{i} - \mathbf{j}$, the distance d_1 from R to l_1 satisfies $d_1^2 = (17^2 + 27^2) - [17 \cdot 2 + 27 \cdot (-1)]^2/(2^2 + (-1)^2) = 1008.2$. By Eq. (3) with $\mathbf{p} - \mathbf{a} = (19\mathbf{i} + 36\mathbf{j}) - (20\mathbf{i} + \mathbf{j}) = -\mathbf{i} + 35\mathbf{j}$ and $\mathbf{e} = (26\mathbf{i} + 3\mathbf{j}) - (20\mathbf{i} + \mathbf{j}) = 6\mathbf{i} + 2\mathbf{j}$, the distance d_2 from R to l_2 satisfies $d_2^2 = 1328.4$; so we should send Arbie to l_1. The length of the step is $d_1 = \sqrt{1008.2} = 31.75$ km. A direction vector $\mathbf{f}$ for the step must be perpendicular to the vector $\mathbf{e} = 2\mathbf{i} - \mathbf{j}$.

By inspection, $(\mathbf{i} + 2\mathbf{j}) \cdot (2\mathbf{i} - \mathbf{j}) = (1)(2) + (2)(-1) = 0$, so $\mathbf{i} + 2\mathbf{j}$ is perpendicular to $\mathbf{e}$. But Arbie is above l_1, so $\mathbf{f}$ must have general direction opposite to $\mathbf{j}$. Thus we take $\mathbf{f} = -\mathbf{i} - 2\mathbf{j}$. Arbie's current heading is $\overline{RT} = (23\mathbf{i} + 33\mathbf{j}) - (19\mathbf{i} + 36\mathbf{j}) = 4\mathbf{i} - 3\mathbf{j} = \mathbf{h}$. The angle between $\mathbf{h}$ and $\mathbf{f}$ is $\cos^{-1}(\mathbf{h} \cdot \mathbf{f}/(|\mathbf{h}||\mathbf{f}|)) = \cos^{-1}(-2/5\sqrt{5}) = \cos^{-1}(-0.18) = 100.3°$. Since the general direction of $\mathbf{f}$ is opposite to $\mathbf{i}$, the $100.3°$ rotation from $\mathbf{h}$ to $\mathbf{f}$ is clockwise. The step command is "rotate clockwise $100.3°$, then move 31.75 km," which is beamed to the robot in the polar form $(31.75, -100.3°)$. ▲

Planes in Space

We next present two ways to describe a plane $\mathbf{P}$ in space: first, a parametric vector equation, and second, a scalar equation. In each description assume $A(a_1, a_2, a_3)$ is a point on $\mathbf{P}$, and let $\mathbf{a} = \overline{OA} = a_1\mathbf{i} + a_2\mathbf{j} + a_3\mathbf{k}$.

For the first description, we establish a sort of generalized Cartesian coordinate system in $\mathbf{P}$. Let $\mathbf{e}_1$ and $\mathbf{e}_2$ be any two vectors that are parallel to $\mathbf{P}$ but not parallel to each other. Let l_1 and l_2 be the lines through A parallel to $\mathbf{e}_1$ and $\mathbf{e}_2$. For an arbitrary point X in $\mathbf{P}$, suppose that the line through X parallel to l_2 intersects l_1 at the point X_1 and that the line through X parallel to l_1 intersects l_2 at the point X_2, as in Figure 4. Then by the parallelogram law, $\overline{AX} = \overline{AX_1} + \overline{AX_2}$. Since l_1 and l_2 are parallel to $\mathbf{e}_1$ and $\mathbf{e}_2$, there are unique scalars s and t such that $\overline{AX_1} = s\mathbf{e}_1$ and $\overline{AX_2} = t\mathbf{e}_2$. Thus, each point X is associated with a unique ordered pair (s, t), called the *coordinates of X relative to $\mathbf{e}_1$ and $\mathbf{e}_2$ with origin A.* Ordinary Cartesian coordinates result from the choices $A = 0$, $\mathbf{e}_1 = \mathbf{i}$, and $\mathbf{e}_2 = \mathbf{j}$, as in Figure 8 of Section 2.2. We may now describe $\mathbf{P}$ as consisting of all points X such that

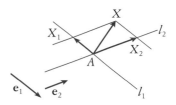

FIGURE 4 Generalized Cartesian coordinates.

$$\overline{OX} = \mathbf{a} + s\mathbf{e}_1 + t\mathbf{e}_2. \tag{4}$$

This is a *parametric vector equation* for $\mathbf{P}$, with two parameters, s and t.

EXAMPLE 4 Stuck in a Plane Arbie has been confined to the plane $\mathbf{P}$ containing the points $A(-1, -2, -4)$, $B(1, -1, -1)$, and $C(2, 0, -3)$. Can the robot reach the points $D(-14, -10, -13)$ and $E(4, 1, 1)$?

Solution. Let $\mathbf{e}_1 = \overline{AB} = 2\mathbf{i} + \mathbf{j} + 3\mathbf{k}$ and $\mathbf{e}_2 = \overline{AC} = 3\mathbf{i} + 2\mathbf{j} + \mathbf{k}$. Note that these vectors are not parallel. Point D is reachable if and only if there are scalars s and t such that $\overline{OD} = \mathbf{a} + s\mathbf{e}_1 + t\mathbf{e}_2$, that is, if and only if the vector equation $-14\mathbf{i} - 10\mathbf{j} - 13\mathbf{k} = -\mathbf{i} - 2\mathbf{j} - 4\mathbf{k} + s(2\mathbf{i} + \mathbf{j} + 3\mathbf{k}) + t(3\mathbf{i} + 2\mathbf{j} + \mathbf{k})$ can be solved for s and t. Equating the $\mathbf{i}$-, $\mathbf{j}$-, and $\mathbf{k}$-components yields the system

$$\left. \begin{array}{l} -14 = -1 + 2s + 3t \\ -10 = -2 + s + 2t \\ -13 = -4 + 3s + t \end{array} \right\},$$

which we simplify to

$$\left. \begin{array}{ll} \text{(i)} & 2s + 3t = -13 \\ \text{(ii)} & s + 2t = -8 \\ \text{(iii)} & 3s + t = -9 \end{array} \right\}.$$

Subtract two times Eq. (ii) from Eq. (i) to get $-t = 3$, $t = -3$. Then from Eq. (ii), $s - 6 = -8$, $s = -2$. These values satisfy (i) too, but in order to be a solution of the system, they must also satisfy (iii): $3(-2) + (-3) = -9$; thus, they do satisfy. So the robot can reach point D within the plane $\mathbf{P}$.

In order to reach point E, we must be able to solve the equation $\overline{OE} = \mathbf{a} + s\mathbf{e}_1 + t\mathbf{e}_2$, which leads to the simplified system

$$\begin{array}{rl} \text{(i)} & 2s + 3t = 5 \\ \text{(ii)} & s + 2t = 3 \\ \text{(iii)} & 3s + t = 5 \end{array}\Bigg\}.$$

Again, subtract twice Eq. (ii) from Eq. (i) to get $-t = -1$, $t = 1$, and then use (ii), to get $s + 2 = 3$, $s = 1$, which also satisfies Eq. (i). But $3 \cdot 1 + 1 \neq 5$, so Eq. (iii) is not satisfied, and hence this system does not have a solution. Thus, Arbie cannot get to point E until it can leave plane $\mathbf{P}$. ▲

For the second description of a plane in space, suppose $\mathbf{n} = n_1\mathbf{i} + n_2\mathbf{j} + n_3\mathbf{k}$ is a nonzero vector that is perpendicular to plane $\mathbf{P}$. Such a vector is called a *normal* to $\mathbf{P}$. The normal is perpendicular to every line in $\mathbf{P}$ (Figure 5). Thus, a point $X(x_1, x_2, x_3)$ is on $\mathbf{P}$ if and only if $\overline{AX}$ is perpendicular to $\mathbf{n}$, that is, if and only if

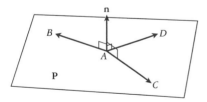

FIGURE 5 If A is on $\mathbf{P}$ and n is a normal to $\mathbf{P}$, then B, C, and D are on $\mathbf{P}$.

$$(\overline{OX} - \overline{OA}) \cdot \mathbf{n} = 0. \tag{5}$$

By Eq. (13) in Section 2.3, this gives the scalar equation $n_1(x_1 - a_1) + n_2(x_2 - a_2) + n_3(x_3 - a_3) = 0$, which can also be written as $n_1x_1 + n_2x_2 + n_3x_3 = b$, where the constant $b = n_1a_1 + n_2a_2 + n_3a_3$. To summarize:

> *Every plane perpendicular to a nonzero vector* $\mathbf{n} = n_1\mathbf{i} + n_2\mathbf{j} + n_3\mathbf{k}$ *has an equation of the form* $n_1x_1 + n_2x_2 + n_3x_3 = b$, *where b is a constant. If the plane contains the point $A(a_1, a_2, a_3)$, then* $b = n_1a_1 + n_2a_2 + n_3a_3$.

Conversely, consider any scalar equation of the form $n_1x_1 + n_2x_2 + n_3x_3 = b$, where n_1, n_2, and n_3 are not all zero. Let $\mathbf{n} = n_1\mathbf{i} + n_2\mathbf{j} + n_3\mathbf{k}$ and $\mathbf{a} = (b/\mathbf{n} \cdot \mathbf{n})\mathbf{n}$. Then $n_1a_1 + n_2a_2 + n_3a_3 = \mathbf{a} = (b/\mathbf{n} \cdot \mathbf{n})\mathbf{n} \cdot \mathbf{n} = b$, and so by the above statement, $n_1x_1 + n_2x_2 + n_3x_3 = b$ is an equation for the plane through $A(a_1, a_2, a_3)$ perpendicular to $\mathbf{n} = n_1\mathbf{i} + n_2\mathbf{j} + n_3\mathbf{k}$.

EXAMPLE 5 Intersecting Planes and Perpendiculars Let l be the line of intersection of the planes with scalar equations $2x_1 - 3x_2 + x_3 = 7$ and $-4x_1 + x_2 + 2x_3 = -9$. Find a parametric vector equation and also a scalar equation for the plane $\mathbf{P}$ through $A(2, -5, 3)$ that is perpendicular to l.
Solution. The vectors $\mathbf{e}_1 = 2\mathbf{i} - 3\mathbf{j} + \mathbf{k}$ and $\mathbf{e}_2 = -4\mathbf{i} + \mathbf{j} + 2\mathbf{k}$, which are not parallel, are each perpendicular to one of the given planes; hence both vectors are perpendicular to line l. Then, since $\mathbf{P}$ is perpendicular to l, both $\mathbf{e}_1$ and $\mathbf{e}_2$ are parallel to $\mathbf{P}$. (Draw a sketch.) Then a parametric vector equation for $\mathbf{P}$ is $\overline{OX} = \mathbf{a} + s\mathbf{e}_1 + t\mathbf{e}_2$, where $\mathbf{a} = 2\mathbf{i} - 5\mathbf{j} + 3\mathbf{k}$. Equating compo-

nents, we may write this parametric vector equation as a system of three parametric scalar equations:

$$
\begin{aligned}
\text{(i)} \quad & x_1 = 2 + 2s - 4t \\
\text{(ii)} \quad & x_2 = -5 - 3s + t \\
\text{(iii)} \quad & x_3 = 3 + s + 2t
\end{aligned}
$$

But this is not a single scalar equation without parameters. (Such an equation can, however, be obtained from these equations. See Exercise 15.) To get such an equation, we need a nonzero vector $\mathbf{n} = n_1\mathbf{i} + n_2\mathbf{j} + n_3\mathbf{k}$ perpendicular to $\mathbf{P}$. Since $\mathbf{e}_1$ and $\mathbf{e}_2$ are parallel to $\mathbf{P}$, we must have $\mathbf{n} \cdot \mathbf{e}_1 = 0$ and $\mathbf{n} \cdot \mathbf{e}_2 = 0$, that is,

$$
\begin{aligned}
2n_1 - 3n_2 + n_3 &= 0 \\
-4n_1 + n_2 + 2n_3 &= 0
\end{aligned}
$$

Add twice the first equation to the second to get $-5n_2 + 4n_3 = 0$, $n_2 = (4/5)n_3$. Then from the first equation, $2n_1 = 3n_2 - n_3 = (12/5)n_3 - n_3 = (7/5)n_3$, $n_1 = (7/10)n_3$. Any nonzero value for n_3 will give a nonzero perpendicular vector $\mathbf{n}$. A convenient value is $n_3 = 10$, which yields $\mathbf{n} = 7\mathbf{i} + 8\mathbf{j} + 10\mathbf{k}$. Since $7 \cdot 2 + 8(-5) + 10 \cdot 3 = 4$, the corresponding scalar equation for $\mathbf{P}$ is $7x_1 + 8x_2 + 10x_3 = 4$. ▲

Slope of a Line

For the rest of this section we will confine ourselves to a plane with a Cartesian coordinate system and unit vectors $\mathbf{i}$ and $\mathbf{j}$ on the positive first and second axes. Equations (1) and (2) are still of use here, but there are other helpful ideas for lines in the plane, which we introduce through the following example.

E X A M P L E 6 Do the Lines Cross? Consider the points $A(-1, 5)$, $B(5, 9)$, $C(6, 1)$, and $D(9, 3)$. Where does line AB cross line CD?

First Solution. Line AB has the parametric vector equation $\overline{OX} = -\mathbf{i} + 5\mathbf{j} + t(6\mathbf{i} + 4\mathbf{j})$, and line CD has the equation $\overline{OX} = 6\mathbf{i} + \mathbf{j} + s(3\mathbf{i} + 2\mathbf{j})$, so for a point of intersection we must solve the system $-1 + 6t = 6 + 3s$, $5 + 4t = 1 + 2s$, that is,

$$
\begin{aligned}
\text{(i)} \quad & 6t - 3s = 7 \\
\text{(ii)} \quad & 4t - 2s = -4
\end{aligned}
$$

Multiply (i) by 2, (ii) by -3, and add the resulting equations, to get the result $0 = 26$, which is false. Think of the logic of what has been shown: If this system has a solution, then $0 = 26$. It follows that the system does not have a solution, that is, that lines AB and CD are parallel and hence never cross.

Second Solution. Draw the lines in the coordinate plane (Figure 6). They look like they might be parallel. If we can show they make the same angle with the first axis, then they *are* parallel. Introduce the points $E(5, 5)$ and $F(9, 1)$ to form right triangles ABE and CDF. We have $|BE|/|AE| = 4/6 = 2/3$ and $|DF|/|CF| = 2/3$. Since the legs of these right triangles are in proportion, the triangles are similar. Then $\angle BAE = \angle DCF$. Since lines AB and CE make the same angle with the first axis, they are parallel. ▲

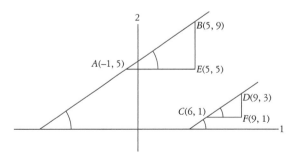

FIGURE 6 Parallel lines.

The second solution to this problem contains two useful, closely related ideas about lines in a planar Cartesian coordinate system.

DEFINITION

Let l be a nonvertical line, that is, a line that is not parallel to the second axis. For any two points $A(a_1, a_2)$ and $B(b_1, b_2)$ on l, the ratio

$$m = \frac{a_2 - b_2}{a_1 - b_1} \qquad (7)$$

is called the *slope* of l.

By similar triangles, the same ratio is obtained no matter which two points on l are used. Also, since $(b_2 - a_2)/(b_1 - a_1) = (a_2 - b_2)/(a_1 - b_1)$, it does not matter which of the two points is named (a_1, a_2) and which is named (b_1, b_2). If $m > 0$, then l slopes up to the right; if $m = 0$, then l is horizontal, that is, parallel to the first axis; if $m < 0$, then l slopes down to the right. The farther m is from zero (positive or negative), the steeper is the slope (up or down). The *angle of inclination* of l is the angle α between l and the first axis, with the restriction $-90° < \alpha \leq 90°$ that makes α uniquely determined by l. The angle of inclination of a horizontal line is defined to be $0°$. (See Figure 7.)

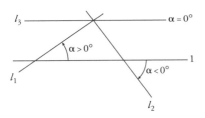

FIGURE 7 Angles of inclination.

Suppose l is a nonvertical line with the positive angle of inclination α. Let $A(a_1, a_2)$ and $B(b_1, b_2)$ be distinct points on l, and let $C(b_1, a_2)$, as in Figure 8. Then in right triangle ABC, $\tan \alpha = \tan \angle BAC = opposite/adjacent = |BC|/|AC| = (b_2 - a_2)/(b_1 - a_1) =$ the slope of l, by Definition (7). If the angle of inclination is negative, then the slope of l and $\tan \alpha$ are both negative and, as you should discover from drawing the triangle, we again have $\tan \alpha =$ slope l. Finally, a horizontal line has slope 0, and $\tan 0° = 0$. Thus, *if a nonvertical line has slope m and angle of inclination α, then*

$$m = \tan \alpha \quad \text{and} \quad \alpha = \tan^{-1} m. \qquad (8)$$

If l is vertical, that is, parallel to the second axis, then $\alpha = 90°$ and $\tan \alpha$ is not defined. On the other hand, if $A(a_1, a_2)$ and $B(b_1, b_2)$ are any two points on a vertical line, then $a_1 = b_1$, so the right side of Definition (7) is not defined. Thus, each of the equations in Eq. (8) holds whenever either side exists.

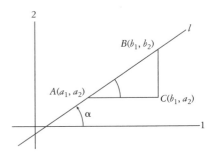

FIGURE 8 Angle of inclination α of line l.

Let l_1 and l_2 be nonvertical lines with slopes m_1 and m_2 and angles of inclination α_1 and α_2. We can use Eq. (8) to obtain useful ways to determine if these lines are parallel, perpendicular, or neither. First, these lines are parallel if and only if their angles of inclination are equal. Then by Eq. (8),

$$l_1 \parallel l_2 \text{ if and only if } m_1 = m_2. \qquad (9)$$

The slope of the line through the points $O(0, 0)$ and $M(1, m)$ is $(m - 0)/(1 - 0) = m$. Thus, the vector $\overline{OM} = \mathbf{i} + m\mathbf{j}$ is a nonzero vector parallel to every line of slope m. If l_1 and l_2 are nonvertical lines with slopes m_1 and m_2, then these lines are perpendicular if and only if the vectors $\mathbf{m}_1 = \mathbf{i} + m_1\mathbf{j}$ and $\mathbf{m}_2 = \mathbf{i} + m_2\mathbf{j}$ are perpendicular. From what we learned about perpendicular vectors in Section 2.3, these vectors are perpendicular if and only if $\mathbf{m}_1 \cdot \mathbf{m}_2 = 0$, that is, if and only if $1 + m_1 m_2 = 0$. Thus:

Two nonvertical lines with slopes m_1 and m_2 are perpendicular if and only if

$$m_1 m_2 = -1. \tag{10}$$

E X A M P L E 7 Arbie Approaches Slowly
Arbie starts at $A(30, 72)$, with coordinates in miles, and creeps at 2 mph straight toward the line joining $B(-40, 0)$ and $C(0, -15)$. After 24 hours, how far is the robot from home base? *Solution.* Slope $BC = (0 - (-15))/(-40 - 0) = -3/8$. Arbie's path is perpendicular to BC, so by Eq. (11) its slope is $m = 8/3$. The unit vectors parallel to this path are $\pm(\mathbf{i} + (8/3)\mathbf{j})/|\mathbf{i} + (8/3)\mathbf{j}| = \pm(3\mathbf{i} + 8\mathbf{j})/\sqrt{73}$. From a rough sketch we see that Arbie is moving in a general direction opposite to $\mathbf{j}$; so we choose the minus sign. The robot's speed is 2, so its velocity is $\mathbf{v} = (-2/\sqrt{73}) \cdot (3\mathbf{i} + 8\mathbf{j})$. Its displacement in 24 hours is $24\mathbf{v} = (-48/\sqrt{73})(3\mathbf{i} + 8\mathbf{j})$, and its position vector is then $\overline{OR} = 30\mathbf{i} + 72\mathbf{j} + (-48/\sqrt{73}) \cdot (3\mathbf{i} + 8\mathbf{j}) = 13.15\mathbf{i} + 27.06\mathbf{j}$. Arbie's distance from $O(0, 0)$ is then $|\overline{OR}| = \sqrt{13.15^2 + 27.06^2} = 30.08$ miles. ▲

Circles and Radian Measure of Angles

So far we have measured angles and rotations in degrees. This is the most commonly used system of angle measure, but other systems do exist. The *grad* system divides the circle into 400 equal parts; so a right angle measures 90°, and a 90° angle measures 100 grad. Measurement systems such as these are artificial, in the sense that they are based on an arbitrarily chosen number of subdivisions. There is another way to measure angles that does not use an arbitrary choice, namely, measure by radius, which is known as *radian* measure. Many calculations and results in mathematics, especially in calculus, are best done using radian measure of angles.

Given an angle, choose any radius r and draw the circular arc s of radius r within the angle, as in Figure 9. The radian measure of the angle is the number of *radii* (the plural of *radius*) in s, that is,

$$\text{radian measure of an angle} = \text{included arc/radius} \tag{11}$$
$$= s/r.$$

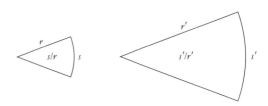

FIGURE 9 Radian measure of angle.

Two observations about Eq. (11) should be made at once. First, the ratio s/r is a pure number, without any units such as degrees or grad. Nevertheless, people often speak, for example, of "an angle of 1.36 radians," rather than the correct phrase, "an angle of radian measure 1.36." Second, it does not matter which radius is used. If for a given angle you use radius r and circular arc s, while I use r' and s', as in Figure 7, then s'/r' will be equal to s/r. Another way to express this observation is to note that it does not matter in what units r and s are measured (feet, meters, etc.); the ratio s/r will be the same number. This fact may seem obvious from the similarity of the shapes in Figure 7, but an actual proof that it is true uses some rather sophisticated ideas.

We turn next to a redefinition of sine, cosine, and the other functions of trigonometry as functions of numbers rather than of angles. It is in this form that these functions assume their crucial importance to applied mathematics. Figure 10 is essentially a copy of Figure 10 in Section 2.2, in which sine and cosine were defined for general angles. If t is the counterclockwise distance along the unit circle from unit point U to point W, then the radian measure of $\angle UOW$ is $t/1 = t$, so the Cartesian coordinates of W are (cos t, sin t). This defines sine and cosine as functions of an arbitrary positive number t rather than as functions of an angle or a rotation, and it makes clear how the two definitions are related. For $t < 0$, travel clockwise along the circle a distance of $-t$ from U to reach W. This extends the definitions of sin t and cos t to all real numbers t. You can picture these definitions with the following model. Represent all numbers t on a coordinate axis. Imagine that this axis is a flexible tape. Attach the origin, $t = 0$, of the tape to the unit point U. Wrap the positive half of the tape counterclockwise around and around the circle and the negative half clockwise around and around the circle. Then any number t on the tape is wrapped to the point W(cos t, sin t) (which is why we denoted this point by W).

The other functions of trigonometry are now defined as functions of real numbers, rather than angles or rotations, by standard identities. We have already introduced the defining identity for *tangent*: tan $t = $ sin t/cos t. The remaining functions, which we shall not use in this chapter, are *cotangent*: cot $t = $ cos t/sin t, *secant*: sec $t = 1$/cos t, and *cosecant*: csc $t = 1$/sin t. Because the general definitions of these functions rely upon the unit circle rather than upon a right triangle, they are often called the *circular functions* instead of the trigonometric functions.

For a positive right angle placed as described above, the distance t is one-fourth of the circumference of the unit circle. Recall that the circumference of a circle of radius r is $2\pi r$. Then $t = (1/4)2\pi(1) = \pi/2$. Thus, the radian measure of a 90° angle is $\pi/2$. Then the radian measure of a 1° angle is $(\pi/2)/90 = \pi/180 = 0.01745$, and an angle of radian measure 1 has degree measure $(90/(\pi/2))° = (180/\pi)° = 57.30°$. In general, an $x°$ angle has radian measure $0.01745x$, and an angle of radian measure y has $(57.30y)°$. Be sure that your calculator or math software is set in the mode you want for your calculations, either degree or radian.

In Figure 11 Arbie, represented as usual by the point R, is moving on a circle with radius a and center $C(c_1, c_2)$. Let θ be the radian measure of the counterclockwise angle from **i** to $\overline{CR}$. Then the unit vector in the direction from C to R is cos θ**i** + sin θ**j**, $\overline{CR} = a($cos θ**i** + sin θ**j**), and $\overline{OR} = \overline{OC} + \overline{CR}$, that is,

$$\overline{OR} = (c_1 + a \cos \theta)\mathbf{i} + (c_2 + a \sin \theta)\mathbf{j}. \quad (12)$$

Let point F have position vector $\overline{OF} = \overline{OC} +$

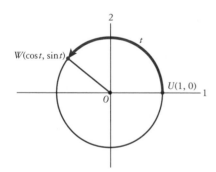

FIGURE 10 Definition of sine and cosine of any number.

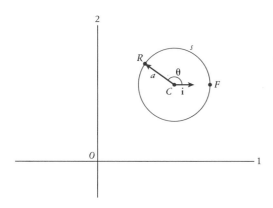

FIGURE 11 Circular motion.

$a\mathbf{i}$, and let s be the distance along the circle counterclockwise from F to R. By Eq. (11), $\theta = s/a$; so

$$s = a\theta. \tag{13}$$

The *angular velocity* of R is the rate ω at which θ is changing with time. (Note that the letter ω, lowercase Greek omega, is not the letter w.) If R is circling in the counterclockwise direction, then θ is increasing and so $\omega > 0$; if R is moving in the clockwise direction, then θ is decreasing and $\omega < 0$. The *circular velocity* of R is the rate v at which s is changing. For counterclockwise motion, s is increasing and $v > 0$; for clockwise motion, s is decreasing and $v < 0$. From Eq. (13), the rate at which s is changing is a times the rate at which θ is changing, that is,

$$v = a\omega. \tag{14}$$

EXAMPLE 8 Big Ben? Let M be the tip of the 15-foot-long minute hand of a giant clock. Since the hand makes 1 clockwise revolution in 60 minutes, the angular velocity of M is $\omega = -2\pi/60 = -\pi/30 = -0.105$ rad/min. Then by Eq. (14), the circular speed of M is $v = (15)(-\pi/30) = -\pi/2$ ft/min. In 20 minutes, M travels a distance of $s = |(20)(-\pi/2)| = 10\pi = 31.4$ ft. ▲

EXAMPLE 9 Arbie Circles Arbie is crawling counterclockwise at 2 meters per hour around the circle with center C, at coordinates (5, 6) in meters, and radius 4 meters. Will the robot hit a wall on segment $[A, B]$, where $A(3, 4)$ and $B(6, 10)$? If so, where and when?
Solution. First, note that $|AC| = \sqrt{8} < 4$, while $|BC| = \sqrt{17} > 4$; so A is inside the circle and B is outside. Thus, Arbie will eventually hit wall $[A, B]$. To answer the question "when?" requires us to choose a point of reference on the circle from which to measure time. Since the standard angle θ is measured from $\mathbf{i}$, the most convenient point to choose is the point F on the circle such that $\overline{CF}$ has the same direction as $\mathbf{i}$ (see Figure 11). Point F has coordinates (9, 6), since $(5\mathbf{i} + 6\mathbf{j}) + 4\mathbf{i} = 9\mathbf{i} + 6\mathbf{j}$. Set time $t = 0$ when Arbie is at point F. This makes θ a function of t, with $\theta = 0$ when $t = 0$. Then $\theta = \omega t$. By Eq. (14), the circular speed is $2 = 4\omega$; so $\omega = 1/2$ and $\theta = t/2$. By Eq. (12), a parametric vector equation for the circle is $\overline{OR} = 5\mathbf{i} + 6\mathbf{j} + 4 (\cos \theta \mathbf{i} + \sin \theta \mathbf{j})$. By Eq. (2), a parametric vector equation for the wall is $\overline{OX} = (1 - u)(3\mathbf{i} + 4\mathbf{j}) + u(6\mathbf{i} + 10\mathbf{j})$, $0 \le u \le 1$. When Arbie hits the wall, we have $R = X$, from which $5 + 4 \cos \theta = 3(1 - u) + 6u$ and $6 + 4 \sin \theta = 4(1 - u) + 10u$. Rewrite these equations as $4 \cos \theta = 3u - 2$ and $4 \sin \theta = 6u - 2$. We may use the identity $\sin^2 \theta + \cos^2 \theta = 1$—which, as you may recall, follows from the distance formula and the definition of sine and cosine—to eliminate θ from these equations by squaring both sides of each equation and adding to get $16 = 45u^2 - 36u + 8$, that is, $45u^2 - 36u - 8 = 0$. The quadratic formula then gives $u = 0.98$ or $u = -0.18$. Since we must have $0 \le u \le 1$, impact occurs for $u = 0.98$, at the point $Q(5.9, 9.9)$. Now

The Fathers of Analytic Geometry

The history of the development of mathematical ideas, like all intellectual history, is complicated and often obscured by controversy, oversimplification, and fantasy. The creation of analytic geometry is a case in point. The contributions of René Descartes appear in a 100-page appendix, entitled *La Géometrie,* to his most famous work, the *Discours de la Méthode,* published in 1637. This work is a powerful, original contribution to mathematics, but it does not fully describe what we would now call a Cartesian coordinate system. It is told that Descartes' ideas on analytic geometry came to him as he lay in bed watching a fly walk up a wall and along the ceiling. He saw that the distances of the fly from the walls would determine its position on the ceiling.

At about the same time—actually rather earlier—many ideas in analytic geometry were developed by Pierre de Fermat (1601?–1665), a French lawyer who is widely regarded as the greatest mathematician of his time. Fermat obtained the general equations for straight lines and circles and discussed ellipses, parabolas, and hyperbolas. His results were communicated mainly by letter and were not published until after his death.

The introduction of coordinate axes in something like their modern form may be due to the extraordinary German philosopher and mathematician, Gottfried Wilhelm Leibniz (1646–1716), who along with Isaac Newton (1642–1727) was the creator of much of calculus.

René Descartes (1596–1650). (Photo from Howard Eves (1976), Introduction to the History of Mathematics, *4th ed. Holt, Rinehart, Winston, p. 280.)*

$\cos \theta = (3u - 2)/4 = 0.24$ and Q is in the upper half of the circle; hence, $0 < \theta < \pi$ and $\theta = \cos^{-1} 0.24 = 1.33$. (In case it's needed, the degree measure of $\angle FCR$ is $(1.33 \cdot 57.30)° = 76.2°$.) Since $\theta = t/2$, it follows that Arbie hits the wall at $t = 2 \cdot 1.33 = 2.66$ hours, that is, 2 hours 40 minutes after it left point D. (Sketch the circle and wall.) ▲

1. Arbie starts at $(-4, -7, -9)$, where the Cartesian coordinates are in kilometers, and moves at 12 km/hr in the direction of the vector $7\mathbf{i} + 10\mathbf{k}$. Where is Arbie after 6-1/4 hours?

2. Arbie has been moving with constant velocity for exactly four hours and is now at $(5, -1, -1)$. If it was at $(3, -3, 4)$ one hour ago, where did it start its trip?

3. In each part, find the distance from point P to line l.
   (a) $P(7, 3, 12)$, $l = \overline{OX} = \mathbf{a} + t\mathbf{e}$, where $\mathbf{a} = -\mathbf{i} + \mathbf{j} - \mathbf{k}$ and $\mathbf{e} = \mathbf{i} + \mathbf{j} + 3\mathbf{k}$
   (b) $P(0, 5, -4)$, $l = AB$, where $A(9, -5, 7)$ and $B(-2, 4, 4)$

4. In the plane, Arbie is circling $C(5, -3)$ clockwise at a distance of 9. How close does it get to the line l through $(-10, 0)$ and $(0, 12)$?

5. Suppose in Exercise 4 that when the robot reaches the point on the circle nearest l it stops. Calculate the step command that will send it directly to the line.

6. Consider the points in Cartesian coordinates $A(6, -5, 3)$, $B(20, -6, -13)$, $C(15, 0, -8)$, and $D(17, 1, -11)$. Do the lines AB and CD intersect? If so, where?

7. Consider the points with the Cartesian coordinates $A(9, -5, 7)$, $B(6, -5, 3)$, $C(17, 1, -11)$, $D(15, 0, -8)$, and $E(-2, 5, 4)$. Does the plane containing A, B, and C intersect line DE? If so, where?

8. The *angle between two planes* in space may be defined as follows. Declare the angle between parallel planes to be 0. If planes $\mathbf{P}_1$ and $\mathbf{P}_2$ are not parallel, they will intersect in a line l. Let $\mathbf{Q}$ be a plane perpendicular to l. Plane $\mathbf{Q}$ intersects planes $\mathbf{P}_1$ and $\mathbf{P}_2$ in two lines, l_1 and l_2. Define the angle between $\mathbf{P}_1$ and $\mathbf{P}_2$ to be the smaller angle between l_1 and l_2. Draw a picture of two planes making a 60° angle. Now suppose vectors $\mathbf{n}_1$ and $\mathbf{n}_2$ are perpendicular to $\mathbf{P}_1$ and $\mathbf{P}_2$. Use your drawing to explain why the angle between $\mathbf{P}_1$ and $\mathbf{P}_2$ equals the angle between $\mathbf{n}_1$ and $\mathbf{n}_2$. Use this to find the angle between the planes with scalar equations $3x_1 - 5x_2 + 2x_3 = 6$ and $3x_1 + 5x_2 + 2x_3 = 8$.

9. Using Exercise 8, find a scalar equation for the plane through $(1, 1, 1)$ and the origin that makes a 60° angle with the plane through $(-1, -1, 2)$, $(2, 2, 1)$, and $(3, -2, -1)$.

10. Arbie 1 starts at $(1, 5)$, with coordinates in meters, and circles the point $(4, 5)$ clockwise. At the same moment, Arbie 2 starts at $(10, 9)$ and circles the point $(7, 9)$ counterclockwise. If Arbie 1 is moving at 2 m/min, how fast should Arbie 2 move in order to meet Arbie 1 as soon as possible? There is an infinite number of correct answers to this problem (explain why); give the smallest one.

11. Is the angle θ of the polar coordinates of a point P the same as the angle of inclination of line OP?

12. Since every point in the plane has an infinite number of different polar coordinate pairs (Section 2, Exercises 13–15), we must be careful about the definition of the graph of an equation in polar coordinates. Let $F(r, \theta)$ be an expression involving r or θ (or both). We say that a point is on the graph of the polar coordinate equation $F(r, \theta) = 0$ if it has *at least one* polar coordinate pair that satisfies the equation. (In this context, θ is virtually always expressed in radian measure.) Consider, for example, the polar equation $4r\theta + 15\pi = 0$. The point with polar coordinates $(3, \pi/4)$ is on the graph of this equation even though $4(3)(\pi/4) + 15\pi$ is not zero, because another polar coordinate pair for this point is $(-3, 5\pi/4)$, which *does* satisfy the equation. Graph the following polar coordinate equations:
   (a) $r = 5$
   (b) $r = -5$
   (c) $\theta = \pi/3$ (careful!)
   (d) $r = \theta$

13. This exercise concerns the polar equation $r = \cos\theta$.
   (a) Begin to graph the equation by carefully plot-

ting the points for which $\theta = n\pi/10$, $n = 0, 1, 2, 3, 4, 5$.

(b) Obtain four more points on the graph by means of the identity $\cos(-\theta) = \cos\theta$. Sketch a smooth curve through the 10 plotted points.

(c) Consider a few points with $\pi/2 < \theta < 3\pi/2$, and come to the realization that you already had the whole graph in part (b).

(d) Multiply both sides of the polar equation by r, and use results from Section 2.2 to see what the graph really is.

14. Graph the following daisies, which some authors call "roses":

(a) $r = \sin(4\theta)$
(b) $r = \cos(3\theta)$
(c) $r = \sin(5\theta)$
(d) $r = \cos(6\theta)$

15. In Example 5, solve Eqs. (i) and (ii) for s and t in terms of x_1, x_2. Then substitute these expressions into Eq. (iii), and rewrite the result to obtain the scalar equation for **P** that was found in the example.

16. Here's a way to see why the area A of a disk of radius a is given by the formula $A = \pi a^2$. (This is not a formal proof.) Imagine the disk as a piece of cloth formed from concentric rings of thread. With scissors, cut from the bottom B of the disk straight to the center C. Unwrap the outer ring of thread, and lay it out straight away from the cut. Unwrap the next ring and lay it on top of the first, with the left ends next to each other. Continue like

this all the way to the central "ring," which is really just a dot. The resulting stack of threads would look like the triangle in this figure.

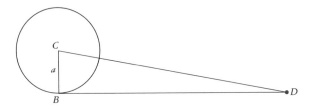

Finding the area of a disk.

(a) Explain why $[C, D]$ is a straight line segment.
(b) Find $|BD|$.
(c) Since triangle BCD consists of the same threads that made the disk, A must equal the area of the triangle. Find A.

17. Derive a parametric vector equation for the curve that is the intersection of the sphere with center $(2, 3, 4)$ and radius 5 with the sphere with center $(4, 2, 3)$ and radius 6.

18. ESSAY: You know that, for $a > 0$, $\mathbf{r} = (a \cos t)\,\mathbf{i} + (a \sin t)\,\mathbf{j}$, $0 \le t \le 2\pi$, describes a circle. What curve is described by $\mathbf{r} = (a \cos t)\,\mathbf{i} + (b \sin t)\,\mathbf{j}$, $0 \le t \le 2\pi$, where a and b are different positive numbers? Draw some graphs for various values of a and b. (Appropriate computer software would be helpful here, but it is not required.) Look up Kepler's laws of planetary motion, and describe how they are relevant to your graphs.

SECTION 2.5 *Remote Sensing and Location*

Why is Arbie moving around? It is seeking lost objects, such as crashed airplanes or lost hikers. In some cases, the object is attached to a *beacon,* represented here by a point B, that sends out ultraviolet light in all directions. When Arbie sees the light, it can lo-

cate the object. In this section we examine how the robot "sees" and then "locates" an object. In the first part, with Arbie confined to a plane, we present rather detailed descriptions of the geometry of point location and barrier avoidance. In the second part we introduce the most commonly used coordinate systems in space and give some indication of the variety of geometric ideas used there to solve problems of imaging, guidance, and location.

Location in the Plane

As we mentioned in Section 2.1, the point R to which we have often referred when discussing the robot is actually the lens of a camera. To begin we assume that the robot is confined to the plane. Then the back of the camera may be modeled by a sensitized coordinate axis, called the s-axis, of length $2a$ with origin M at its midpoint, as shown in Figure 1. Let b be the focal length of the camera, that is, the distance from M to R. If the beacon B falls within the field of view of the camera, as shown in the figure, then it produces an image with coordinate s on the sensitized axis, where $-a \leq s \leq a$. In addition, a light meter located at M measures the intensity i of the image of B. Once the image has been recorded, the robot has two ways to locate the beacon: *variation of intensity* and *triangulation*.

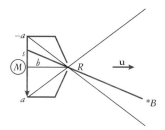

FIGURE 1 Arbie's camera and light meter.

The variation of intensity method of location depends on the *inverse square law,* according to which the intensity i of the image of B is inversely proportional to the square of the distance d from the image to B, that is,

$$i = k/d^2, \qquad (1)$$

where k is a constant. This method proceeds as follows. First, Arbie rotates around R through an angle

of $\tan^{-1}(s/b)$, where as usual a positive rotation is counterclockwise and a negative rotation is clockwise. This will move the image of B onto the light meter so that the direction to B will be given by the vector $\overline{MR}$. (Draw a sketch to see how this works.) Since the image of B is now on the light meter, its intensity i_1 will be measured. Next, the robot moves a fixed distance c toward B and measures the intensity i_2 of the image of B. If the distance from M to B after this move is d, then the distance before the move is $d + c$. Then from Eq. (1),

$$\frac{i_2}{i_1} = \frac{\dfrac{k}{d^2}}{\dfrac{k}{(d + c)^2}} = \frac{(d + c)^2}{d^2},$$

$$\sqrt{\frac{i_2}{i_1}} = \frac{d + c}{d} = 1 + \frac{c}{d},$$

$$d = \frac{c}{\sqrt{\dfrac{i_2}{i_1}} - 1}.$$

Then the distance from R to B is $d - b$, that is,

$$|RB| = \frac{c}{\sqrt{\dfrac{i_2}{i_1}} - 1} - b. \qquad (2)$$

E X A M P L E   1   A Distant Beacon   Suppose c is 20 meters and b is 2 centimeters, that is, 0.02 m. Say Arbie measures $i_1 = 0.000301$ and $i_2 = 0.000304$. Then $i_2/i_1 = 304/301$, and Eq. (2) gives $|RB| = 4023$ meters. (As is usually the case, b is negligible.) ▲

The trouble with this result is that it is highly sensitive to slight errors in the measurements of image intensity. For example, if the true value of i_1 were

0.000001 more than the result reported above, then $i_2/i_1 = 304/302$, and Eq. (2) would give $|RB| = 6050$ meters. Thus, an error of $1/3$ of 1% in one reading led to an error of 34% in the measurement of position. With any formula or method of measurement in applied mathematics, it is of great importance to carry out a *sensitivity analysis* to see whether small errors in the input measurements are likely to create much larger errors in the output. A variety of theorems from algebra, calculus, and probability are used in such analyses, which are beyond the scope of this chapter. The present example certainly suggests that, unless the light meter is extremely accurate, it would be good to have an alternative method of location—which brings us to triangulation.

Triangulation is based on the law of sines, Eq. (2) of Section 2.3. In this method of triangulation, Arbie travels a preset distance c from its present location R, in a direction *not* parallel to line RB, to location S and then reacquires an image of B. From the record of the robot's rotations since leaving home base, the angles between the home base vector $\mathbf{i}$ and $\overline{BR}$ and between $\mathbf{i}$ and $\overline{BS}$ are known, from which the angles $\alpha_1 = \angle BRS$ and $\alpha_2 = \angle BSR$ are found. Let $|SB| = d$ and $\beta = \angle RBS$. Now $\alpha_1 + \alpha_2 + \beta = 180°$; so $\sin \beta = \sin[180° - (\alpha_1 + \alpha_2)] = \sin(\alpha_1 + \alpha_2)$. Then by the law of sines, we have $d/\sin \alpha_1 = c/\sin(\alpha_1 + \alpha_2)$, from which

$$d = c \sin \alpha_1/\sin(\alpha_1 + \alpha_2). \qquad (3)$$

EXAMPLE 2 A Triangulation If $c = 20$ m, $\alpha_1 = 120.3°$, and $\alpha_2 = 58.4°$, then Eq. (3) gives $d = 761$ m (see Figure 2). ▲

Just as with the first method, this one is subject to serious consequences of small errors. For instance, if in Example 2 the true value of α_1 is $120.7°$, then

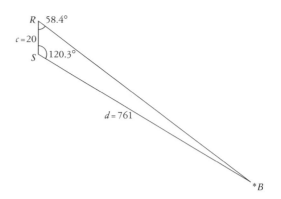

FIGURE 2 Location of a beacon.

Eq. (3) gives $d = 1095$ m. Let's analyze this a bit more. Since the measurements of Arbie's rotations are directly linked to its mechanical robotic drive, we shall assume that they are highly accurate and hence that the main source of error is in the determination of the coordinates s of the images of B before and after the robot's last step. Small errors in these measurements are likely to produce small errors in the value of angle β and consequently small errors in the value of $\sin \beta = \sin(\alpha_1 + \alpha_2)$. But if B is a long way away, then β will be close to zero, and consequently $\sin(\alpha_1 + \alpha_2)$ will also be close to zero. Unfortunately, $\sin(\alpha_1 + \alpha_2)$ appears in the *denominator* of Eq. (3). Thus, even a small change in its value can make a great change in the value of d. As you will see in the exercises, the situation is much the same with the first method of location.

The choice of location method might appear to depend solely on which is more accurate—the image intensity measurement or the image coordinate measurement. But there could be other considerations. With the variation of intensity method, we are sure to get a result after one step; but one step in the triangulation method might place Arbie behind a wall where it could not see B. In this event, the robot would have to keep making steps until it reacquired the beacon. Thus, even if triangulation is the more accurate method, it might be wise to use variation of intensity to get a rough answer before triangulating. This discussion is an illustration of the kind of interplay between analysis and real-world requirements that characterizes much of the work of an applied mathematician.

If Arbie is likely to bump into walls, as discussed in previous sections, isn't that likely to hurt the camera or light meter? The designers have taken care of that. The robot is surrounded by a *sensing shield,* represented in Figure 3 by a circle with center R and radius k. As we mentioned in Section 2.1, the shield is not a physical object but rather a representation of the outer range of a special short-range onboard radar system.

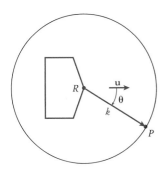

FIGURE 3 The sensing shield.

If Arbie is in the "move" part of a "turn and move" instruction when the shield encounters a wall at point P, the *avoidance program* is triggered. The robot stops, the angle θ from the direction vector $\mathbf{u}$ to the vector $\overline{RP}$ is found, and the step command "turn $\theta + 180°$ and move k" is issued. This results in a displacement by $\overline{PR} = -\overline{RP}$, that is, one sensory radius back from P. Arbie then resumes its search. Of course, it is possible to have a configuration of walls that will defeat this simple avoidance program (see the exercises). As with the location method discussion, this is just an illustration of the continuing process of applied mathematics.

How should Arbie proceed to look for B? Suppose we design the camera so that the sensitized axis is a little more than twice as long as the focal length, that is, with a slightly greater than b. Then the field of view covers an angle of $2 \tan^{-1}(a/b)$, which is a little more than $\pi/2$ radians. Then with four counterclockwise rotations of $\pi/2$, the robot can look in every direction and return to its previous orientation. If during this look-around Arbie sees B, it then switches to the location program, using one or both of the methods described above. Otherwise, having returned to its previous position, it moves forward by a fixed amount d and looks around again, unless the avoidance program is triggered during the move.

This search program is likely to need modification in order to stop Arbie from wandering off or bouncing around without ever finding B. We might, for example, include an instruction that if a certain number N of consecutive unsuccessful look-arounds have been done, the robot should rotate $\pi/2$ before continuing. You will be asked to explore this in the exercises.

Should we turn Arbie off once it has located B? Probably not, considering the possibility of errors in measurement. It might be prudent to turn the robot toward B, travel a fixed distance, and run the location program again. This could be continued either until there is no further variation in the reported location of B or until the sensing shield radar acquires it.

Coordinates, Orientation, and Imaging in Space

For the rest of this section we return to the true environment of a flying robot: space. For this purpose we will use the classical notation for the variables x_1, x_2, and x_3, namely, x, y, and z. Then the axes in a Cartesian coordinate system are called the x-axis, y-axis, and z-axis, and the planes containing pairs of these axes are the *xy*-plane, the *xz*-plane, and the *yz*-plane, as in Figure 4.

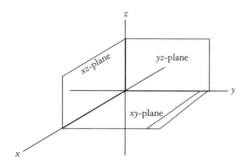

FIGURE 4 Classical Cartesian coordinate system in space.

For some problems it is more convenient to use a coordinate system other than the Cartesian. In the plane we introduced polar coordinates. Suppose that in space we use polar coordinates in the *xy*-plane but retain the *z*-coordinate axis, as in Figure 5. In other words, we replace the first two Cartesian coordinates *x*, *y* of a point *P* by the polar coordinates *r*, θ of the point at the foot of the line through *P* perpendicular to the *xy*-plane. This mixed system of coordinates is called the *cylindrical coordinate system.*

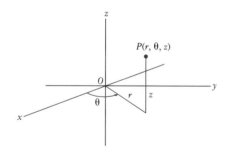

FIGURE 5 Cylindrical coordinates.

In Section 2.2 we learned how to convert between Cartesian and polar coordinates in the plane with Eqs. (4), (5), and (8). We can use this to convert between Cartesian and cylindrical coordinates in

space. In particular, *If a point has cylindrical coordinates* (*r*, θ, *z*), *then its Cartesian coordinates are*

$$(r \cos \theta, r \sin \theta, z). \tag{4}$$

EXAMPLE 2 A Spiral Flight Suppose Arbie takes off 30 yds from the base of a 1200-ft TV tower and circles around it at the rate of 1 revolution every 2 minutes, while ascending at the rate of 10 yds/min. Where is the robot after 11-1/2 minutes, and how far is it from the top of the tower?

Solution. Set up Cartesian coordinates in yards with the tower as the *z*-axis and the takeoff point at (30, 0, 0). Then the top of the tower is at (0, 0, 400). Let *t* be the elapsed time in minutes from takeoff, and measure rotation in radians. In cylindrical coordinates, $r = 30$ and $\theta = \pi t$, since Arbie goes through a full revolution of 2π radians in 2 minutes. Since *z* is increasing at the rate of 10 yds/min, we have $z = 10t$. Then the cylindrical coordinates of *R* at time *t* are (30, πt, 10*t*), and from Eq. (6) its Cartesian coordinates are (30 cos πt, 30 sin πt, 10*t*). After 11.5 min the cylindrical coordinates of *R* are (30, 11.5π, 115) and its Cartesian coordinates are (0, −30, 115) since we have cos 11.5π = cos[11.5π − 6(2π)] = cos(−0.5π) = 0 and sin 11.5π = sin(−0.5π) = −1. The distance of this point from the top of the tower is
$$\sqrt{0^2 + (-30)^2 + (400 - 115)^2} =$$
286.6 yds. ▲

For global and interplanetary travel there is another coordinate system that is often most effective. In a Cartesian coordinate system, let *P* be a point different from the origin. Let φ be the angle from **k** to $\overline{OP}$, as in Figure 6. Then $0 \leq \phi \leq \pi$. If *P* is on the positive *z*-axis, then $\phi = 0$; if *P* is in the *xy*-

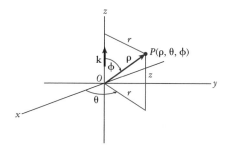

FIGURE 6 Spherical coordinates.

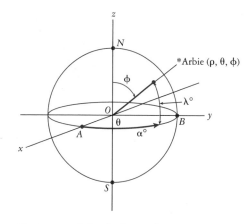

FIGURE 7 Spherical coordinates, latitude, and longitude.

plane, then $\phi = \pi/2$; if P is on the negative z-axis, then $\phi = \pi$. Let $\rho = |OP|$, and let the cylindrical coordinates of P be (r, θ, z). The *spherical coordinates* of P are (ρ, θ, ϕ). (Note: ρ is the lowercase Greek letter rho, not the letter p.) From Figure 6 we see that $z = \rho \cos \phi$ and that $r = \rho \cos(\pi/2 - \phi) = r\sin \phi$. Then from Eq. (4), $x = \rho \sin \phi \cos \theta$ and $y = \rho \sin \phi \sin \theta$. To summarize, *If a point has Cartesian coordinates (x, y, z), cylindrical coordinates (r, θ, z), and spherical coordinates (ρ, θ, ϕ), then*

$$x = r \cos \theta = \rho \sin \phi \cos \theta, \tag{5}$$

$$y = r \sin \theta = \rho \sin \phi \sin \theta, \tag{6}$$

$$z = \rho \cos \phi, \tag{7}$$

$$\rho = \sqrt{r^2 + z^2} = \sqrt{x^2 + y^2 + z^2}, \tag{8}$$

$$\phi = \cos^{-1}(z/r). \tag{9}$$

EXAMPLE 3 Arbie Visits Another World

Our robot is hovering above a spherical asteroid of radius 900 m (Figure 7). Let O be the center of the asteroid, let N be its north pole, let A be a point on the equator, and let B be the point on the equator reached by traveling east from point A one-fourth of the way around the equator. Lines ON, OA, and OB are mutually perpendicular, so we may set up Cartesian coordinates with origin O and with A, B, N on the positive x-, y-, and z-axes.

Establish latitude and longitude lines on the asteroid with the Greenwich Meridian (the zero longitude line) running from N through A to the south pole, S. If Arbie is at the point with spherical coordinates (ρ, θ, ϕ), where $0 \leq \theta \leq 2\pi$, then its altitude is $\rho - 900$. The point on the asteroid's surface that is directly below the robot has longitude θ east and latitude $\pi/2 - \phi$, where a positive value is north of the equator and a negative value is (in absolute value) south. Conversely, suppose the degree coordinates of the point are $\alpha°$ east and $\lambda°$ north. Then its second spherical coordinate is $\theta = \alpha\pi/180$ radians, and its third spherical coordinate is $\phi = \pi/2 - \lambda(\pi/180)$ radians. See Figure 7.

Suppose Arbie is 100 m above the point at $70°$ east, $40°$ north when it begins to ascend at 12 m/min while decreasing longitude at $3°$/min and decreasing latitude at $2°$/min. Then its spherical coordinates after t minutes are (ρ, θ, ϕ), where $\rho = 100 + 12t$, $\theta = 70\pi/180 - 3\pi t/180$, and $\phi = \pi/2 - 40\pi/180 + (2\pi t/180)$. (Notice that as latitude decreases in the northern hemisphere, the spherical coordinate ϕ increases.)

Computer Imaging: An Interview with Paul Nagin

At the time of this writing, Paul Nagin is Chairman of Computer Science at Hofstra University on Long Island. Before coming to Hofstra, he was a professor in the department of ophthalmology at the Tufts University Medical School, where he worked on the application of computer imaging technology

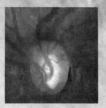

Digitized images (a) with region processing (b) and edge processing (c). (Source: Paul Nagin and John Impagliazzo (1995), Computer Science, A Breadth-First Approach with C, *New York: Wiley, pp. 654, 656, 657.)*

to the diagnosis of glaucoma, cataract, and other diseases of the eye. With John Impagliazzo, he is the co-author of *Computer Science, A Breadth-First Approach with C* (Wiley, 1995), which represents an attempt to do for first-year computer science students what this book is attempting for mathematics students.

RB: How does geometry come into the diagnosis of glaucoma?

PN: Evidence of excessive pressure within the eyeball is shown by changes in the appearance of the end of the optic nerve, which is at the back of the eyeball in the retina. The physician needs to track changes in the shape of this area over time. These can be quite subtle; a 5% change in area may be important. An accurate, consistent record of quantitative observations, in combination with other clinical data, is necessary for a good diagnosis of the progress of the condition. In practice a special camera is used to make a series of im-

ages of the retina. The images obtained will vary, not just because of clinically significant changes in the nerve end but due to involuntary movements of the eye and possible blurring of parts of the image by cataract formation, among other sources of difficulty.

RB: So the raw data obtained by the camera needs to be . . . ?

PN: Processed. And for this purpose digitization and computer technology are well suited.

RB: How does digitization work?

PN: The image region is divided into small regions called *pixels*. In the standard video image there are 512 rows of 512 pixels each, a $2^9 \times 2^9$ array. Some high-resolution systems use $2^{11} \times 2^{11}$ or even finer arrays. The camera scans the array and records for each pixel the *pixel vector* (x, y, i), where (x, y) are the Cartesian coordinates of the pixel and, for a monochromatic picture, i is the median intensity (also called the brightness or the gray level) of the image in that pixel. For a color picture, the pixel vector is (x, y, r, g, b), where r, g, b are the median intensities of red, green, and blue.

RB: So coordinates and vectors—at least as data storage arrays—make their appearance! Once the computer has all this information, what does it do with it?

PN: Using pattern recognition, the computer may perform coordinate transformations in order to bring each image to a standard position so that certain fixed reference spots are in alignment. Comparison of intensities for these and other spots may cause the recommended rejection of some images.

RB: What do you do with the retained data?

PN: Two useful procedures are *region processing* and *edge processing*. Region processing seeks out connected regions over which there is relatively small variation of intensity (or, in some cases, of intensity of one particular color) and redraws them with a constant intensity. This is useful for determining the areas of various regions. Edge processing isolates *boundary pixels* around which there are relatively large variations of intensity. This helps determine the precise shape and degree of irregularity of a region. Here's an example (shown on previous page) of the same data from an optic nerve scan processed in each way.

RB: Very interesting. Are there other geometrical ideas that are of use in this work?

PN: Yes. For example, by taking two simultaneous photographs of the back of the eye from different angles, a stereographic image is obtained from which it is possible to deduce the degree of "cupping" of the optic nerve, that is, its change from being flat. Calculation of cup volume is an important clinical measure of excessive pressure in the eyeball due to diseases such as glaucoma. Before the introduction of this geometrical technique, physicians tried to deduce the degree of cupping from the configuration of local blood vessels. But this is often quite misleading.

RB: So electronic technology combines with computer science and geometry to enhance our ability to deal with serious disease.

PN: Yes, and there's still a lot more to be done.

Let us calculate the distance from the robot to the south pole after 10 minutes. If $t = 10$, then $\rho = 220$, $\theta = 2\pi/9$, and $\phi = 7\pi/18$. Then from Eqs. (5)–(7), Arbie is at the point with Cartesian coordinates (158, 133, 75). The Cartesian coordinates of the south pole are (0, 0, -900); so the distance from the robot to the south pole is
$$\sqrt{158^2 + 133^2 + 975^2} = \sqrt{993{,}278} = 966 \text{ m.} \ \blacktriangle$$

If Arbie is to fly in space, the simple "turn and move" instructions for plane travel will not suffice. The direction vector **u** must be extended to a *frame* consisting of mutually perpendicular vectors **u**, **v**, **w**

in standard orientation attached to the robot as in Figure 8. What we have been calling rotations of the robot are now seen to be rotations around the vertical axis, **w**. A rotation around this vector is called a *yaw,* from an old sailing term. A rotation around **v** is a *pitch,* and a rotation around **u** is a *roll.* Thus a typical step instruction to the robot might be "roll $+10°$, pitch $-35°$, yaw $+20°$, and move 30 meters." Cal-

FIGURE 8 Arbie in 3-D.

culations with such steps are best done with the help of more machinery of linear algebra than we have introduced in this chapter. An excellent beginning is to follow this chapter with the study of the next one in this book.

The sensitized *s*-axis introduced earlier in this section for the two-dimensional robot must now be replaced by a sensitized *plate,* in the form of a square plane section. Use the vectors **v**, **w**, in that order, to introduce Cartesian coordinates in this plane, with origin at the center of focus *M* and *s*- and *t*-axes, with $-a \leq s \leq a$ and $-a \leq t \leq a$. If the beacon *B* is within the field of view of the camera with lens *R*, then it produces an image with coordinates (s, t), as in Figure 9. Again, it requires more math than we have covered here to deal with some of the problems this suggests. For example, suppose that, as in the previous section, there is a light meter at the origin *M*. If an image of *B* appears at coordinates (s, t), what roll, pitch, yaw instructions will bring the image to *M* and bring **u** parallel to $\overline{RB}$?

Now that we have a two-dimensional sensitized plate, we should consider the images not merely of single points but of entire solid bodies within the camera's field. The formation of such an image is known as *central projection* from the object through the *center* (the lens *R* of the camera) to the *image plane* (the plane containing the sensitized plate.) Of course, the image is likely to be a considerable distortion of the actual shape of the body. Suppose, for example, that the body is not a three-dimensional solid but merely a square *ABCD*. The image is a quadrangle $A'B'C'D'$, but it is not likely to be a square. What

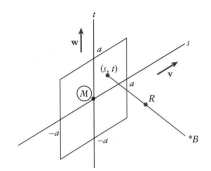

FIGURE 9 Arbie's camera plate.

are conditions on the position of *ABCD* that will ensure that $A'B'C'D'$ is a square? If $A'B'C'D'$ is not a square, is it at least a parallelogram? a trapezoid?

Questions of this sort are answered in the mathematical field called *projective geometry.* Some of the computational tools of projective geometry, notably the use of so-called homogeneous coordinates, are of prime importance in the development of computer graphics and imaging systems. A related question is the following. Given a number of projected images of a solid body taken from different positions in space, to what extent can we deduce the actual shape of the body? When x-rays or other high-frequency radiation rather than visible or ultraviolet light rays are used, this question has led to some extremely effective applications of sophisticated mathematics to the diagnostic technique known as *tomography,* including the type of medical imaging known as a CAT scan. This is not the place to begin a systematic study of these fascinating areas of geometry, but we hope you will want to pursue them.

1. When Arbie is at $(60, -25)$, Cartesian coordinates in meters, and the counterclockwise angle from **i** to **u** is $40°$, it picks up an image of the beacon B at coordinate $s = 1.47$ cm. The focal length of the camera is $b = 2.5$ cm, and the standard distance of movement is $c = 3$ m. The robot executes the variation of intensity method of location and obtains $i_1 = 0.0268$ and $i_2 = 0.0346$. Calculate the coordinates of the beacon.

2. When Arbie is at $(-56, 40)$, Cartesian coordinates in meters, and the counterclockwise angle from **i** to **u** is $310°$, it picks up an image of the beacon B at coordinate $s = -0.78$ cm. The focal length is $b = 2.5$ cm. Arbie begins the triangulation method of location but encounters obstacles that result in execution of the following three step commands: rotate $+60°$ and move 20; rotate $+30°$ and move 20; rotate $-40°$ and move 30. It then picks up an image at $s = 2.14$ cm. Calculate the coordinates of the beacon.

3. As part of its internal guidance system program, Arbie obtains a "command ratio" h from three measured voltages v_1, v_2, and v_3 by means of the formula $h = \sqrt{3v_1 + v_2^2}/v_3$. Typical ranges for these measurements are $1.00 \leq v_1 \leq 9.00$, $5.00 \leq v_2 \leq 7.00$, and $0.50 \leq v_3 \leq 4.50$. Carry out a sensitivity analysis.

4. Suppose Arbie starts its search program from the position and direction shown in the following figure with these parameter settings: field of view $\pi/2$; sensing shield radius $k = $ length of wall shown in the figure, fixed move distance $c = k$; after three unsuccessful look-arounds, turn $\pi/2$ and continue. Let R_0 be the initial position of the camera lens, and let R_n be its position after n steps. Plot R_0, R_1, R_2, etc., up to the point at which Arbie first sees the beacon B.

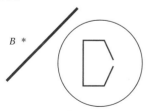

5. Repeat Exercise 4, but with the turn of $\pi/2$ after three unsuccessful look-arounds replaced by $-\pi/2$.

6. Suppose the wall of Exercise 4 is replaced by the wall shown in this figure.

    (a)   Explain why with the settings of Exercise 4 Arbie will never find B.

    (b)   Change the setting "after *three* unsuccessful look-arounds" in Exercise 4 to a sufficiently large number so that Arbie will eventually find B, and plot the points R_0, R_1, R_2, and L showing the acquisition.

7. Return to Exercise 4 with the original settings, and place an additional wall as shown in the figure below. Will Arbie find B? If not, modify the settings so that it does.

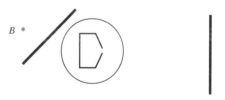

8. Consider the triple $(3, 2, 1)$. It may be regarded as the coordinates of a point P in a Cartesian, a cylindrical, or a spherical system in space, with radian measure of angles.

    (a)   If $(3, 2, 1)$ are the Cartesian coordinates of P, what are its cylindrical and spherical coordinates?

    (b)   If $(3, 2, 1)$ are the cylindrical coordinates of P, what are its Cartesian and spherical coordinates?

(c) If (3, 2, 1) are the spherical coordinates of P, what are its Cartesian and cylindrical coordinates?

9. Assume the Earth is a perfect sphere of radius 4000 miles. Let P_1 and P_2 be points on the surface with latitude and longitude (λ_1, ϕ_1) and (λ_2, ϕ_2), in degrees, respectively. Find the shortest distance between P_1 and P_2. You may assume both points are in the eastern hemisphere north of the equator. (Hint: Consider $\overline{OP}_1$ and $\overline{OP}_2$, where O is the center of the earth.

10. When Arbie is at home base in three dimensions, the lens R is at the origin O and the frame vectors $\mathbf{u}, \mathbf{v}, \mathbf{w}$ (Figure 8) coincide with $\mathbf{i}, \mathbf{j}, \mathbf{k}$, respectively. Starting at home base, the robot is instructed to roll $60°$, pitch $45°$, yaw $30°$, and move 20 m.

    (a) Find the coordinates of the points on the unit sphere, with center O and radius 1, to which $\mathbf{u}, \mathbf{v}$, and $\mathbf{w}$ point after the roll has been executed.

    (b) Find the coordinates of the points on the unit sphere to which $\mathbf{u}, \mathbf{v}$, and $\mathbf{w}$ point after the roll and pitch have been executed.

    (c) Find the coordinates of the points on the unit sphere to which $\mathbf{u}, \mathbf{v}$, and $\mathbf{w}$ point after the roll, pitch, and yaw have been executed.

    (d) Find the coordinates of R after all instructions have been executed.

11. Arbie, in 3-space, has on its screen the complete image of a circle with center C. The circle is in a plane perpendicular to the frame vector $\mathbf{u}$ and the points M, R, and C are aligned. Suppose the robot begins to pitch forward. Draw the images on the screen at several stages from the circle image to the image consisting of a single point.

12. ESSAY: Use the library, especially some of its computer data bases, to learn about the satellite-based *global positioning system.* Report on some of the mathematics involved. Find out which car makers are planning to offer the system as an option, and when.

SECTION 2.6 *Optimization and N-Space*

In the introduction to this chapter we mentioned the position problems, "which point on a given polygon is farthest from a given line?" and, "are two given points in space on the same side of a given plane?" On the first page of Section 2.1 we referred to the need for modern mathematics to meet the challenge of finding a best possible way to deploy a large number of different resources in order to attain a given objective under many restrictions. In this section we will establish connections between these problems in geometry and challenges from the worlds of business and government, and we will describe, in outline, one of the most effective mathematical methods used to deal with them. Much of this material can fairly be described as modern mathematics, since it was developed during and after the second world war (1939–1945).

Half-Planes

In a plane with Cartesian coordinates, let l be the line with slope m through the point (a_1, a_2). By the definition of slope, a point (x_1, x_2) different from (a_1, a_2) is on l if and only if $(x_2 - a_2)/(x_1 - a_1) = m$. Multiply both sides of this equation by $x_1 - a_1$ to get

$$x_2 - a_2 = m(x_1 - a_1), \qquad (1)$$

which is satisfied by all points on l, including (a_1, a_2), and by no other points. Equation (1) is the famous *point-slope* equation for l. It may be rewritten as

$$x_2 = mx_1 + b, \qquad (2)$$

where $b = a_2 - ma_1$. The constant b is called the *vertical intercept* of l, because l crosses the second axis at the point $(0, b)$. (In classical x, y notation, Eq. (2) is $y = mx + b$ and b is called the y-intercept.)

Line l divides the plane into two *half-planes*, as shown in Figure 1. The upper half-plane consists of all points (x_1, x_2) for which $x_2 \geq mx_1 + b$, and the lower half-plane consists of all points (x_1, x_2) for which $x_2 \leq mx_1 + b$. The line $x_2 = mx_1 + b$ itself is included in both half-planes.

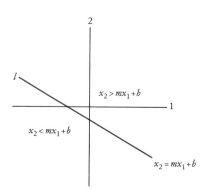

FIGURE 1 Half-planes.

The vertical line through (a_1, a_2) does not have a slope. It consists of all points (a_1, x_2), where x_2 ranges over all numbers. Thus, an equation for the vertical line through (a_1, a_2) is simply

$$x_1 = a_1. \qquad (3)$$

This line splits the plane into the right half-plane $x_1 \geq a_1$ and the left half-plane $x_1 \leq a_1$.

Every line has an equation of the form

$$c_1 x_1 + c_2 x_2 = d, \qquad (4)$$

where c_1 and c_2 are not both zero. If $c_2 \neq 0$, we may rewrite Eq. (4) in the form $x_2 = (-c_1/c_2)x_1 + (d/c_2)$, which we recognize from Eq. (2) as an equation for the line with slope $-c_1/c_2$ and vertical intercept d/c_2. If $c_2 = 0$, rewrite Eq. (4) as $x_1 = d/c_1$, which is an equation for the vertical line with *horizontal intercept* d/c_1, that is, which crosses the first axis at $(d/c_1, 0)$.

The half-planes bounded by the line with Eq. (4) are, of course, given by $c_1 x_1 + c_2 x_2 \geq d$ and $c_1 x_1 + c_2 x_2 \leq d$. (Note: If $c_2 > 0$, then $c_1 x_1 + c_2 x_2 \geq d$ is the upper half-plane; but if $c_2 < 0$, then $c_1 x_1 + c_2 x_2 \geq d$ is the *lower* half-plane.)

EXAMPLE 1 Arbie's Last Appearance
Our faithful robot is stuck in a Cartesian coordinate plane at $(-100, -7)$ and needs to go directly to one of two refueling stations, ALPHA at $(370, -32)$ and BRAVO at $(250, -27)$. Enemy agents have set up a powerful and destructive x-ray laser beam that starts very far away but is known to pass through the points $(-800, 30)$ and $(500, -40)$, where other stations have been vaporized. What should Arbie do? (And what did it actually do?)

Solution. The line l of the laser beam has slope $m = (30 + 40)/(-800 - 500) = -7/130$ and point-slope equation $x_2 - 30 = (-7/130)(x_1 + 800)$, which we rewrite as $7x_1 + 130x_2 = -1700$.

Note that Arbie and the refueling stations are not in immediate danger, since their coordinates do not satisfy this equation of destruction. The robots and stations are safely in one of the two half-planes **H**, given by $7x_1 + 130x_2 < -1700$, or **K**, given by $7x_1 + 130x_2 > -1700$. For Arbie's coordinates we have $(7)(-100) + (130)(-7) = -1610 > -1700$; so the robot is in half-plane **K**. Since $(7)(370) + (130)(-32) =$

$-1570 > -1700$, ALPHA is also in **K**, but since $(7)(250) + (130)(-27) = -1760 < -1700$, BRAVO is in **H**. Thus, Arbie should go to the farther station, ALPHA (Figure 2). (Alas! The robot headed toward BRAVO because it was closer. The sensing shield did not pick up the laser beam, and Arbie was destroyed. But don't despair; a new and improved model, Arbie 2, is in preparation.) ▲

FIGURE 2 Arbie's last mission.

Convexity

The property behind the conclusion that Arbie should have gone to station ALPHA is one of great importance for geometry and many of its applications.

D E F I N I T I O N

A region **C** is *convex* if, given any two points in **C**, every point on the line segment joining those points is also in **C**.

Half-planes are convex; that's why Arbie would have been safe if it had remained in half-plane **K**. Examples of other convex regions include disks and (solid) rectangles. But, for example, stars and rings are not convex (see Figure 3).

The intersection of any number of half-planes, even an infinite number of them, is always a convex region. To see why, suppose **S** is the intersection of

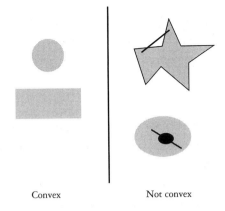

Convex   Not convex

FIGURE 3 Convex and nonconvex sets.

a family of half-planes. Let A and B be points in **S**, and let P be a point on the segment $[A, B]$. Since A and B are in **S**, these points are in every half-plane in the family. But all half-planes are convex, so P is in every half-plane of the family. Thus, P is in the intersection **S**. Then by the definition above, **S** is convex.

A region that equals the intersection of a *finite* number of half-planes is called a *convex polygonal region*. Every such region is the graph of all points that satisfy a finite set of *linear inequalities*—inequalities of the form $c_1x_1 + c_2x_2 \geq d$ or $c_1x_1 + c_2x_2 \leq d$.

E X A M P L E   2   A Convex Polygonal Region
Figure 4 contains the graph of the convex polygonal region *ABCDEF,* given by Inequalities (i)–(vi) in the figure caption. The points A through F are called the *vertices* (or *corner points*) of the region. Their coordinates may be found from the equations for their boundary lines. For example, vertex C is on lines (iii) and (iv); so its coordinates (x_1, x_2) must satisfy the equations $x_1 + 3x_2 = 18$, $x_1 + 2x_2 = 13$. Subtract the second equation from the first to get $x_2 = 5$. Then the second equation gives $x_1 + 10 = 13$, $x_1 = 3$; so $C(3, 5)$.

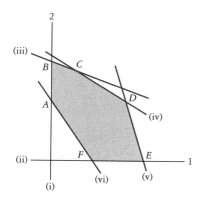

FIGURE 4 Graph of the system of inequalities. (i) $x_1 \geq 0$; (ii) $x_2 \geq 0$; (iii) $x_1 + 3x_2 \leq 18$; (iv) $x_1 + 2x_2 \leq 13$; (v) $4x_1 + x_2 \leq 24$; (vi) $4x_1 + 3x_2 \geq 12$.

Similarly (check these), $A(0, 4)$, $B(0, 6)$, $D(5, 4)$, $E(6, 0)$, and $F(3, 0)$. ▲

Linear Programming

Convex polygonal regions arise in a branch of applied mathematics called *linear programming*. Here is an example.

EXAMPLE 3 Vitamin Dosage I need to take some vitamins. There are two types of vitamin capsules, Energee and Pepp. Both types contain vitamins A, C, D, and E, as well as the newly discovered Z, which interests me. Each Energee capsule has 1 g of vitamin A, 1 g C, 4 g D, 4 g E, and 5 g Z. Each capsule of Pepp has 3 g A, 2 g C, 1 g D, 3 g E, and 2 g Z. Each day I must take at most 18 g A, at most 13 g C, at most 24 g D, and at least 12 g E.

a. How many pills of each type should I take in order to satisfy my vitamin requirements and get the greatest amount of vitamin Z? (There are no restrictions on the amount of Z that I may take.)

b. Energee capsules cost \$6 each and Pepp capsules cost \$4 each. How can I satisfy my vitamin requirements at lowest cost; and if that's what I do, how much vitamin Z will I get?

Solution. Let x_1 be the number of Energee capsules I take per day, and let x_2 be the number of Pepp capsules I take per day. Then the number of grams of vitamin A that I get per day is $x_1 + 3x_2$, which must be at most 18. Thus, Inequality (iii) must hold. The requirements for C, D, and E lead to Inequalities (iv), (v), and (vi). Since I cannot take a negative number of capsules of either type, Inequalities (i) and (ii) also hold. Inequalities (i) through (vi) are called the *constraints* for this problem. Every point (x_1, x_2) in the convex polygonal region of Figure 4 is a feasible solution to part (a) or (b) of the problem. The region in Figure 4 is called the *feasible region* (or the *region of feasibility*) of the problem.

Part (a). Let z be the number of grams of vitamin Z in a dose of x_1 Energee capsules and x_2 Pepp capsules. Then $z = 5x_1 + 2x_2$. Consider, for example, the feasible point (4, 1), which is circled in Figure 5. If I took 4 Energee capsules and 1

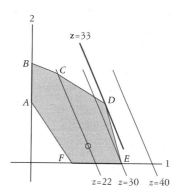

FIGURE 5 The vitamin region with z-lines.

Pepp capsule, I would get $z = 5(4) + 2(1) = 22$ g of Z. The line $z = 22$, that is, the line with equation $5x_1 + 2x_2 = 22$, intersects the feasible region in many points, as shown in Figure 5. Thus, there are many feasible ways to get 22 g of Z. Can we get more? Yes: The line $z = 30$ contains, among others, the feasible point (6, 0). The lines $z = 22$ and $z = 30$ are parallel, of course, since a point of intersection would have to yield $z = 22$ and $z = 30$ for the same dosage. Since the line $z = 30$ is to the right of the line $z = 22$, we see that as larger values of z are tried, the z-line moves to the right. So to get the greatest amount of Z, we move the z-line to the right as far as we can without leaving the feasible region. It appears from Figure 5 that the last z-line that intersects the feasible region is the z-line through the vertex $D(5,4)$, for which $z = 5(5) + 2(4) = 33$. (Note that, for instance, the line $z = 40$ contains no points of the feasible region; so there is no way that I can safely get as much as 40 g of Z.) Thus, the greatest amount of Z that I can take safely is 33 g, and the only way I can do this is to take 5 Energee capsules and 4 Pepp capsules. *Part (b).* Let w be the dollar cost of a dose of x_1 Energee capsules and x_2 Pepp capsules. Then $w = 6x_1 + 4x_2$. The lines $w = 30$ and $w = 20$ in Figure 6 show that we are looking for the lowest w-line that still intersects the feasible region. It appears from the figure that this line is the w-line through vertex A, for which $w = 6(0) + 4(4) = 16$. Thus, the least expensive way to satisfy my vitamin requirements is to take no Energee capsules and 4 Pepp capsules, for a cost of $16 per day. This will provide me with $z = 5(0) + 2(4) = 8$ g of vitamin Z. (Note from Figure 6 that I cannot, for instance, satisfy my requirements for as little as $10.) ▲

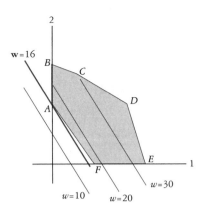

FIGURE 6 The vitamin region with cost lines.

The geometric fact behind the solutions of the vitamin problem is as follows. Let **C** be a convex polygonal region and let l_0 be a line. Move a line l across **C**, keeping l parallel to l_0, until a line l_1 is reached such that any further movement of l will cause the line to leave **C** (Figure 7). Then l_1 must contain a vertex V of **C**. If an equation for l_0 is $c_1 x_1 + c_2 x_2 = d$, then each line l parallel to l_0 has an equation of the form $c_1 x_1 + c_2 x_2 = k$ for some constant k. If the vertex V on l_1 has coordinates (v_1, v_2), then l_1 has the equation $c_1 x_1 + c_2 x_2 = e$, where $e = c_1 v_1 + c_2 v_2$.

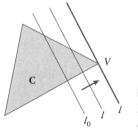

FIGURE 7 The optimal line goes through a corner.

In part (a) of the vitamin problem, we wanted the maximum value of $5x_1 + 2x_2$ in the feasible region; in part (b) we wanted the minimum value of

$6x_1 + 4x_2$ in this region. An important part of modern applied mathematics, engineering, and business decisions consists of finding, or trying to find, the maximum or the minimum value of an expression on a region. This field of study, known generally as *optimization,* is a large and growing subject that uses many results of mathematics. The specific problem of optimizing an expression of the form $c_1x_1 + c_2x_2$ in a convex, planar, polygonal region is called *two-dimensional linear programming.*

The result illustrated in Figure 7 implies that the optimum values of the expression $c_1x_1 + c_2x_2$ occur at the vertices of the region. This yields a simple but effective method to solve any two-dimensional linear programming problem. *To optimize $z = c_1x_1 + c_2x_2$, subject to a set of constraints, each of the form $a_1x_1 + a_2x_2 \leq b$ or $a_1x_1 + a_2x_2 \geq d$:*

1. Sketch the feasible region defined by the constraints in order to find which pairs of boundary lines intersect at the vertices.

2. For each pair of lines found in step 1, solve the system of two linear equations to obtain the coordinates of that vertex.

3. Substitute the coordinates of each vertex into $c_1x_1 + c_2x_2$ to get a value of z. The maximum (minimum) occurs at the vertex with the largest (smallest) value.

For part (a) of our vitamin problem, we have

vertex: $A(0, 4)$ $B(0, 6)$ $C(3, 5)$ $D(5, 4)$ $E(6, 0)$ $F(3, 0)$
z-value: 8   12   25   33   30   15

So the maximum does occur at D, as we thought we saw in Figure 5. Suppose you have reversed your belief in the value of vitamin Z and want to avoid it as much as possible while satisfying all the constraints. Then it is obvious from this table that you

should go to vertex A: Take no Energee capsules and 4 Pepp capsules to receive the minimum feasible dose of 8 g Z.

Notes on the Simplex Algorithm

The vitamin example serves well as an introduction to linear programming, but the method given here for finding the optimum values of z is of limited practical use because linear programming problems in practice often involve not two variables but ten, a hundred, or even several thousand. Fortunately, extraordinarily efficient methods have been developed for solving such practical problems. We shall describe briefly the most important of these methods, known as the *simplex algorithm,* for part (a) of the vitamin dosage problem. Consider the boundary of the feasible region in Figure 4, which consists of the vertices A, B, C, D, E, F and the edges AB, BC, CD, DE, EF, FA joining them. (In this discussion we write the line segment $[X, Y]$ as XY.) This boundary is known as the *simplex* (more precisely, the *1-simplex*) of the region. The idea is to locate a vertex and then to move along edges around the simplex so as to increase the value of the objective variable z. Eventually you will reach a vertex such that if you move away from it along the simplex in any direction, the value of z will not increase. Then that vertex gives the optimum value. For example, suppose you begin at vertex A. Traveling along edge AF will decrease the value of z, but traveling along AB will increase it; so you go to B. In the same way, you then go on to C, and then to D. Going from D along either DC or DE will cause z to decrease; so the answer is at D.

Why is the simplex algorithm sure to work? *Because the feasible region is convex.* If the region were not convex, the algorithm might terminate at the wrong vertex. This is illustrated in Figure 8. Suppose that

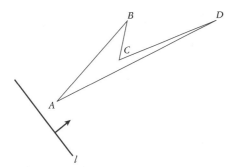

FIGURE 8 Optimization on a nonconvex simplex.

the z-line is parallel to l and that increasing z moves l in the direction shown. If the simplex algorithm were at vertex B, it would terminate, since moving along the edges from B to either A or C would decrease z. But the maximum value of z is really at vertex D. Vertex B is called a *local maximum* because it gives a larger value of z than any *nearby* point in the feasible region. But D is the genuine answer, called the *global maximum*. If the feasible region is convex, then this cannot occur.

It should be emphasized that the simplex algorithm does not require any graphs to be drawn. Our instructions above to "find a vertex of the feasible region" and to "move along an edge to the next vertex" are actually carried out by pure algebra, using some of the ideas from linear algebra that will appear in the next chapter. Thus, this algorithm is very well suited for implementation on computers. There are, in fact, several good linear programming software packages available for students and for various groups of professional users.

Why is the simplex algorithm so efficient? This is not obvious from our two-dimensional example, but even it contains a hint. Our path from A to B to C to the answer D never had to consider vertices E or F. For problems that involve a great number of variables, the simplex algorithm usually has to consider only a small fraction of all the vertices of the region of feasibility. In fact, it will usually not be necessary to calculate where most of the vertices are.

In order to demonstrate more convincingly the efficiency of the simplex algorithm, we consider a hypothetical problem involving three variables x_1, x_2, x_3, in which we are to maximize $z = c_1 x_1 + c_2 x_2 + c_3 x_3$ subject to six constraints of the form $a_1 x_1 + a_2 x_2 + a_3 x_3 \leq or \geq b$. As we learned in Section 2.5, each equation $a_1 x_1 + a_2 x_2 + a_3 x_3 = b$ represents a plane in space; so each constraint is a *half-space* bounded by one of six planes.

The feasible region for our problem is the intersection of six half-spaces. It would be tedious to draw this region. A simple example of such a region would be a solid cube; the actual feasible region would likely be much less regular than that. But even without drawing the region we know that it must be convex, because a half-space is obviously convex and the intersection of any number of convex regions is convex.

In general, each vertex of our feasible region is the intersection of three boundary planes. (Think for example of the corners of a cube.) As you will learn in a later chapter, there are in general 20 such intersections made by sets of 3 of our 6 boundary planes. To find each such point we would have to solve a system of three equations. In many cases, the point found by all this work would not be a vertex of the feasible region, because it would not belong to the other three half-spaces. Of course, the vertices could eventually all be found and substituted into $z = a_1 x_1 + a_2 x_2 + a_3 x_3$ to find the answer. Typically, with a problem of this sort, the simplex algorithm would visit three to five vertices without ever considering a nonvertex or bothering to find the other vertices, and it would reach the answer with much less effort.

A Six-Dimensional Robot

The Cincinnati Milarcon T^3 robot illustrates two ideas from this section. First, the positioning of the robot wrist is given in terms of *pitch, roll,* and *yaw,* as with the orientation of Arbie in three dimensions. Second, the "command space" of this robot is six-dimensional. Each position command is a 6-tuple of turn commands, $(\theta_a, \theta_s, \theta_e, \theta_p, \theta_r, \theta_y)$, consisting of the changes in angle for the arm sweep, shoulder swivel, elbow extension, pitch, roll, and yaw, respectively. The commands could instead be interpreted as vector components in 6-space and written as 6×1 column matrices.

For practical use, one (or more) of many different devices is attached to the robot's wrist, such as a temperature probe, an arc welding point, or a gripper. Instructions to the device are sent after the robot has been positioned by a command 6-tuple. Thus, the full instruction space for the T^3 would have dimension greater than six.

Beginning at a predetermined initial setting, the robot is given a series of actions, each consisting of a position command followed by a device instruction. The last position command returns the robot to the initial setting. In the setup phase, the position commands and device instructions are sent by pressing buttons or moving a mouse or tracker ball on the ACRAMATIC unit (see the illustration). Early results may be modified after viewing program execution on slow-motion videotape. The goal is to achieve maximum efficiency of motion consistent with the immediate environment. As with other applications involving mathematics and engineering, this is not a mere series of calculations, but a continuing dialog—part science, part art.

Once the desired actions have been made efficient enough, the entire set of instructions may be placed in the ACRAMATIC memory unit. The Milarcon T^3 is then ready to begin, for example, a complicated set of assembly instructions for the rear door of a hatchback, with tireless accuracy, considerable speed, and minimal supervision. It is through devices such as this that many modern large plant and factory operations, ranging from sawmills to computer manufacturing, are run with remarkably few human workers.

While the robotic revolution has resulted in job displacement, it has not resulted in overall job loss. It has, however, added to the need for more education, especially in topics such as those that you are studying in this book!

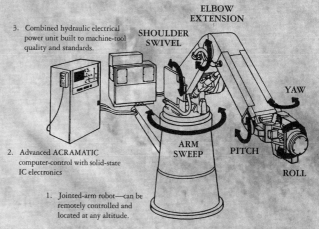

The T^3's robotic arm. (Source: V. Daniel Hunt (1985), Smart Robots, New York: Chapman and Hall, p. 237.)

A Glimpse into Higher Dimensions

In previous sections we introduced coordinate systems on the line, in the plane, and in space to help us use vectors to understand and solve problems. By these means, problems involving one, two, or three unknown quantities could be modeled on a line, in a plane, or in space. Now it is apparent that there are a great many problems, such as problems of optimization, that involve far more than three variables. Is it possible to imagine spaces of dimension greater than that of ordinary space and to introduce coordinate systems and vectors in these spaces in order to solve at least some of these multivariable problems? Of course it is!

The difference between our approach to high-dimensional spaces and our view of low-dimensional spaces (lines, planes, and space) is implicit in the language we have just used. We have an intuitive understanding of low-dimensional spaces, and we "introduce" coordinates in them to help with calculation. With high-dimensional spaces, we first create the coordinate system and then use them to *define the space.*

The coordinate systems that we will use for this purpose are the obvious analogs of the Cartesian coordinate systems in low-dimensional space. With these systems we identify points on a line with real numbers x, points in the plane with ordered pairs (x_1, x_2), and points in space with ordered triples (x_1, x_2, x_3). The analogs for more dimensions are ordered *4-tuples* (x_1, x_2, x_3, x_4), *5-tuples* $(x_1, x_2, x_3, x_4, x_5)$, and so forth.

DEFINITION

Let n be a positive integer. *Euclidean n-space,* denoted $\Re^n$, is the set of all n-tuples $(x_1, x_2, \ldots, x_n)$ of real numbers. The n-tuples are called the *points* of $\Re^n$.

Note the difference in viewpoint. In two dimensions, for example, we "know" what a point X is, we introduce coordinates, and we write $X(x_1, x_2)$, that is, "X has coordinates the ordered pair (x_1, x_2)." In five dimensions, for example, we say the point X *is* the 5-tuple $(x_1, x_2, x_3, x_4, x_5)$ and we write $X = (x_1, x_2, x_3, x_4, x_5)$.

The letter $\Re$ is used in the notation $\Re^n$ because the coordinates $x_1, x_2, \cdots x_n$ come from the set of all *real numbers,* that is, numbers that can be expressed as decimals. In view of this definition, what we have up to this point in this chapter called "space" should now be called *3-space,* a plane is a *2-space,* and a line is an *1-space.* (In fact, to the truly pedantic, a single point constitutes a *0-space,* and nothing at all is a *−1-space.*) As in low dimensions, the point $(0, 0, \ldots, 0)$ is usually denoted by O.

In the plane we introduced the vectors $\mathbf{i}$ and $\mathbf{j}$ such that the vector represented by the arrow $\overline{OX}$, where the coordinates of X are (x_1, x_2), has components x_1 and x_2. Once the vectors $\mathbf{i}, \mathbf{j}$ have been specified, in that order, the vector represented by $\overline{OX}$ may be identified with the ordered pair of scalar components x_1, x_2. In order to distinguish this pair from the coordinate pair (x_1, x_2), it is customary to write it as a vertical array called a *column matrix,*

$$\begin{bmatrix} x_1 \\ x_2 \end{bmatrix}.$$

Similarly, in 3-space with $\mathbf{i}, \mathbf{j}, \mathbf{k}$ specified in that order, the vector represented by $\overline{OX}$, where $X(x_1, x_2, x_3)$, is identified with the column matrix

$$\begin{bmatrix} x_1 \\ x_2 \\ x_3 \end{bmatrix}.$$

This prompts the following definition.

DEFINITION

The *vector* in $\Re^n$ with *components* $x_1, x_2, \ldots, x_n$ is the column matrix

$$\mathbf{x} = \begin{bmatrix} x_1 \\ x_2 \\ \vdots \\ x_n \end{bmatrix}.$$

The definitions of *n*-space and vector in *n*-space are keys to the study of geometry in any finite number of dimensions. Among many other applications, they allow the simplex algorithm to be made into an algebraic algorithm suitable for use with a computer.

Some beautiful and surprising discoveries are being made in high-dimensional geometry. For example, recent work by John Conway (Princeton University) and Neil Sloane (Bell Labs) on very efficient ways to pack eight-dimensional spheres into eight-dimensional space—ways that do not work in lower dimensions—have provided precise information on the best way to transmit large quantities of data on telephone lines!

As you continue to study high-dimensional geometry, it will often be helpful to keep in mind the source of the ideas, which is *low*-dimensional geometry. If a concept or computation seems dauntingly abstract or complicated, always try to see what it would look like in one, two, or three dimensions. We hope you'll enjoy it.

Exercises for Section 2.6

---

1. Find equations of the form $a_1 x_1 + a_2 x_2 = b$ for each of the following lines:
   (a) with slope 2/3 and vertical intercept 4
   (b) with slope -3 and passing through $(-1, -4)$
   (c) with slope 7 and horizontal intercept 8
   (d) with undefined slope and passing through $(-2, -9)$
   (e) horizontal, through $(-2, -9)$
   (f) through $(7, -2)$ and $(-3, 8)$
   (g) through $(19, 4)$ and $(19, -5)$
   (h) through $(-5, -2)$ and $(71, -2)$
   (i) the first axis
   (j) the second axis

2. Let a and b be nonzero constants. Show that an equation for the line with horizontal intercept a and vertical intercept b is

$$\frac{x_1}{a} + \frac{x_2}{b} = 1.$$

3. Draw (separate) graphs of each of the following inequalities:
   (a) $2x_1 + 3x_2 \geq 12$
   (b) $2x_1 + 3x_2 \leq 12$
   (c) $2x_1 - 3x_2 \geq 12$
   (d) $2x_1 - 3x_2 \leq 12$
   (e) $-2x_1 + 3x_2 \geq 12$
   (f) $-2x_1 + 3x_2 \leq 12$
   (g) $x_2 \geq 5$
   (h) $x_2 \leq -2$
   (i) $x_1 \geq -1$
   (j) $x_1 \leq 2$

4. Let K be a circle with radius 3. Which of the follow-

ing sets of points in the plane of K are convex? (Draw each set.)

(a) K itself
(b) K and its center
(c) all points with distance at most 3 from K
(d) all points with distance less than 3 from K
(e) all points with distance more than 3 from K

5. Graph each of the following regions:
(a) $x_1 \geq 0, x_2 \geq 0, x_1 + x_2 \leq 3, x_1 + 2x_2 \leq 4$
(b) $5x_1 - x_2 \geq 8, x_1 - 5x_2 \leq 8, x_1 + x_2 \leq 2$
(c) $x_1 + x_2 \geq 4, 2x_1 + x_2 \leq 4, x_1 + 2x_2 \leq 4$

6. (a) Find the maximum and minimum values of $z = 6x_1 + x_2$, subject to constraints $4x_1 + 3x_2 \geq 22, 2x_1 - 2x_2 \leq 8, x_1 + 5x_2 \leq 31$.
(b) Find the maximum and minimum values of $z = x_1 + 7x_2$, subject to the same constraints an in part (a).

7. The Titan Conglomerate needs 9 million gallons of fuel oil and 7 million gallons of gasoline. The Supergas refinery will deliver 40,000 gal oil and 80,000 gal gasoline per day for $110,000 per day. The Exxtrabig refinery will deliver 60,000 gal oil and 30,000 gal gasoline per day for $70,000 per day. How does Titan get the best deal, and what does it cost?

8. A mineral company owns two mines that it can operate for any number of shifts. Each shift of the ALPHA mine results in the production of 1 ton high-grade ore, 3 tons medium-grade, and 5 tons low-grade; each shift of the BETA mine produces 2 tons of each type of ore. The company can sell all the ore it produces, but the mines have different operating costs. Each shift for which the ALPHA mine is run will eventually net the company a $200 profit; each BETA shift will net $208. Storage restrictions limit production to at most 88 tons high-grade, 160 tons medium-grade, and 200 tons low-grade. How does the company maximize its net, and what will that net be?

9. Consider the region in 3-space defined by the inequalities $x_1 \geq 0, x_2 \geq 0, x_3 \geq 0, 6x_1 + 2x_2 + 3x_3 \geq 6, 4x_1 + 2x_2 + 3x_3 \leq 12$. If the inequalities were equalities, this would describe a set of five planes in space. For

each set of three of these planes, find the coordinates of the point in which the three planes intersect. How many points are thus found? For each of these points, test whether or not the inequalities are all true, in which case the point is a corner point. How many corner points are there? Draw the region.

10. Maximize each of $y = 6x_1 + 4x_2 + 3x_3$, $z = 5x_1 + 2x_2 + 3x_3$, and $w = 6x_1 + 3x_2 + 5x_3$, subject to the constraints of Exercise 9. For each problem, trace on the feasible region the path that the simplex algorithm would follow to the optimal corner point if it began at the corner point $(1, 0, 0)$.

11. Let $\mathbf{a}$ and $\mathbf{b}$ be vectors in 2-space, and let k be a scalar. Let the column matrix representations of these vectors with respect to the ordered basis $\mathbf{i}, \mathbf{j}$ be

$$\mathbf{a} = \begin{bmatrix} a_1 \\ a_2 \end{bmatrix}, \qquad \mathbf{b} = \begin{bmatrix} b_1 \\ b_2 \end{bmatrix}.$$

Write the column matrix representations for $k\mathbf{a}$, for $-\mathbf{a}$, and for $\mathbf{a} + \mathbf{b}$. Also write the expressions for the norm $|\mathbf{a}|$ and for the dot product $\mathbf{a} \cdot \mathbf{b}$ in terms of the entries in these matrices. Repeat this for vectors in 3-space. Based on this, write the obvious definitions of $k\mathbf{a}$, $-\mathbf{a}$, and $\mathbf{a} \cdot \mathbf{b}$ for $\mathbf{a}$ and $\mathbf{b}$ in n-space. Verify that all the laws of vector algebra hold for these definitions. (See Section 2.1, Eqs. (4)–(6), (8), and (11)–(16), and Section 2.3, Eq. (14).)

12. Consider the set E of all points in 3-space with coordinates (x, y, z) such that x, y, and z are integers satisfying $x^2 + (y - 2)^2 + (z + 1)^2 \leq 3$. Find an equation of the form $ax + by + cz = d$ for a plane that contains exactly one point of E and has half of the remaining points of E in each half-space.

13. ESSAY: Discuss the need for sensitivity analysis in linear programming. Create a "story problem" that leads to a two-dimensional linear programming problem in which $z = 100x_1 + 200x_2$ is to be maximized, subject to the constraints $x_1 \geq 0, x_2 \geq 0$ and to one constraint of the form $ax_1 + bx_2 \leq c$ such that the answer to the problem occurs at one vertex, but if the objective function is changed to $z = 100.01x_1 + 200x_2$, the answer occurs

at a different vertex. Also, discuss linear programming problems in which the values of x_1, x_2, K are required to be integers, giving examples of practical problems in which this requirement would be necessary. This topic, which is called *integer programming,* has an extensive literature about it.

Chapter 2 Exercises

1. Let *ABC* be an arbitrary triangle. Let *D*, *E*, *F* be points outside the triangle such that triangles *ABD*, *ACE*, and *BCF* are each equilateral. What can be said about triangle *DEF*?

2. In a plane with Cartesian and polar coordinates suppose there are mirrors along the line segments from $(-1, 0)$ to $(-1, 1)$ and from $(-1, 1)$ to $(0, 1)$. If a laser beam is sent from the origin toward the point with polar coordinates $(1, \theta)$, find the Cartesian coordinates of the point on the horizontal axis, if any, that the beam strikes.

3. In 3-space with Cartesian coordinates let **P** be the plane through the points A $(1, 1, -1)$, B $(1, -1, 1)$, and C $(-1, 1, 1)$. Consider the points on the two lines in **P** through A that make a $60°$ angle with line *AB*. Among these points, find the one that is closest to the point D $(1, 1, 1)$.

4. Suppose a spherical orange of radius 2 cm is resting on the surface of a globe of the earth of radius 20 cm, with the navel of the orange on the surface of the globe at $10°$ North, $20°$ East. Roll the orange along the globe, following the shortest path toward the point at $60°$ North, $100°$ East, until the navel first touches the globe again. Find the latitude and longitude of the navel now.

CHAPTER

3

LINEAR ALGEBRA

SECTION 3.1 *Linear Systems as Models*

Linear algebra may be the most widely used mathematical subject in the modern world, beyond simple arithmetic. This chapter will give you some glimpses why the preceding statement is true. The problems to which people apply mathematical models today tend to involve large, complex situations, such as

a.   measuring the effect of price changes in oil on our national economy with its thousands of interdependent industries,

b.   deciding how much of each consumer product will be made next month at each of a company's 20 production facilities around the country, or

c.   developing a battery of tests that can predict how well applicants will perform at a given job.

Linear algebra provides the language and methods for organizing and analyzing such problems involving many variables. It extends the single-variable algebra learned in high school to an algebra that handles large arrays of numbers just as easily.

First of all, what does one mean by the term "linear algebra"? Or even more simply, what does the term "linear" mean? One good answer is that "linear" means "like a line." If a variable y is a linear function $f(x)$ of a variable x, then the graph of $y = f(x)$ will be a straight line, such as $y = 2x - 1$, as in Figure 1.

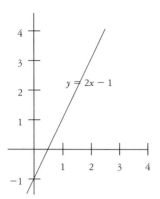

FIGURE 1 y is a linear function of x.

A *linear combination* of variables, or *linear expression,* such $3x + 4y - 6z$, has linear (line-like) terms for each variable. A linear expression cannot have quadratic terms such as x^2 or xy; these terms are *nonlinear.*

As in all mathematical subjects, there are two basic types of mathematical building blocks in linear algebra. There are the *objects*—arrays of numbers—and there are the *operations* that are performed on these objects—such as adding or multiplying two arrays together. In this section we introduce the basic objects. Subsequent sections in this chapter present the basic operations. We shall also introduce in this section some mathematical models involving arrays of numbers. These models will be used repeatedly throughout this chapter to illustrate new concepts.

Vectors and Matrices

The simplest array of numbers in linear algebra is a *vector.* Our definition of a vector in this chapter will be different from the definition given near the beginning of Chapter 2.

> **DEFINITION**
>
> An ordered list of n numbers is called a *vector,* or an *n-vector.*

It is common to represent a vector in coordinate space as an arrow pointing from the origin to a point whose coordinates are the entries of the vector. For example, Figure 2 displays the vector $[2, 1]$ as an arrow from the origin to the point with coordinates $(2, 1)$. In Chapter 2 a vector was defined to be a number and a direction. In the language of Chapter 2, the vector $[2, 1]$ has a magnitude of $\sqrt{5}$ and an angle (with the x-axis) of $26.6°$.

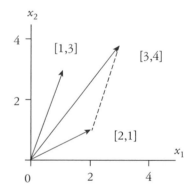

FIGURE 2 Vector addition with arrows.

EXAMPLE 1 Vectors and Motion In Chapter 2 we used vectors to represent the location and motion of the robot Arbie. For example, if Arbie moved 2 units right and 1 unit up, and then moved 1 unit right and 3 units up, we can represent each motion with a vector—$\begin{bmatrix} 2 \\ 1 \end{bmatrix}$ and then $\begin{bmatrix} 1 \\ 3 \end{bmatrix}$. The total motion is $\begin{bmatrix} 3 \\ 4 \end{bmatrix}$, which equals the sum $\begin{bmatrix} 2 \\ 1 \end{bmatrix} + \begin{bmatrix} 1 \\ 3 \end{bmatrix}$. (Vector addition and other vector operations will be discussed in the next section.) In Figure 2, vector $\begin{bmatrix} 1 \\ 3 \end{bmatrix}$ is displayed in standard form as an arrow from the origin to the point (1, 3). Then in the vector sum $\begin{bmatrix} 2 \\ 1 \end{bmatrix} + \begin{bmatrix} 1 \\ 3 \end{bmatrix}$, the arrow for vector $\begin{bmatrix} 1 \\ 3 \end{bmatrix}$ is displaced and now goes from point (2, 1) to point (3, 4). Such vector displacements were discussed in Chapter 2. ▲

We shall use lowercase, boldface letters, such as $\mathbf{v}$, to denote vectors, as was done in Chapter 2. We write v_1 for the first entry in vector $\mathbf{v}$, v_2 for the second entry in $\mathbf{v}$, and v_i for the ith entry in $\mathbf{v}$. Two examples of vectors are

$$\mathbf{v} = [1, 2, 3, 4] \qquad \text{and} \qquad \mathbf{c} = \begin{bmatrix} 7 \\ 8 \\ 9 \end{bmatrix}.$$

Here $\mathbf{v}$ is a 4-vector and $\mathbf{c}$ is a 3-vector. For example, $v_2 = 2$ and $c_2 = 8$.

A vector can be treated as a row of numbers or as a column of numbers. For vectors alone, the choice of row or column format is unimportant, but when vectors and matrices are multiplied together, it is very important to distinguish clearly whether a vector is to be treated as a row vector or as a column vector. *The standard convention in this chapter* (as in most mathematics texts) *is to assume that a vector is a column vector unless explicitly stated otherwise.*

DEFINITION

A *matrix* is a rectangular array of numbers.

We refer to a matrix as an $m \times n$ *matrix* when the matrix has m rows and n columns, and we use capital, boldface letters, such as $\mathbf{A}$, to denote matrices. (A common handwritten way to indicate a matrix is with a wavy line under the letter, such as $\underset{\sim}{A}$.) We use the notation a_{ij} to denote the entry in matrix $\mathbf{A}$ occurring in row i and column j. Examples of matrices are

$$\mathbf{A} = \begin{bmatrix} 4 & 3 & 8 \\ 1 & 2 & 3 \\ 4 & 5 & 6 \end{bmatrix} \qquad \text{and} \qquad \mathbf{M} = \begin{bmatrix} 9 & 5 & 1 \\ 2 & 7 & 6 \end{bmatrix}, \quad (1)$$

where $a_{23} = 3$, $a_{31} = 4$, and $m_{13} = 1$.

A column n-vector is also an $n \times 1$ matrix, and a row n-vector is a $1 \times n$ matrix. Conversely, an $m \times n$ matrix $\mathbf{A}$ can be thought of as a set of n column vectors (each of length m) or as a set of m row vectors (each of length n). We will use the following notation:

$\mathbf{a}_j$ denotes the jth column vector in $\mathbf{A}$; and

$\mathbf{a}'_i$ denotes the ith row vector in $\mathbf{A}$.

(It is common in linear algebra to use the letter i to refer to the number of some row in a matrix and the letter j to refer to the number of some column.) For example, in the matrix $\mathbf{A}$ in Eq. (1),

$$\mathbf{a}_2 = \begin{bmatrix} 3 \\ 2 \\ 5 \end{bmatrix} \qquad \text{and} \qquad \mathbf{a}'_1 = [4\ 3\ 8].$$

Models Involving Systems of Linear Equations

The models introduced in the following three examples will be used repeatedly throughout this chapter to illustrate concepts and techniques of linear algebra.

EXAMPLE 2   An Oil Refinery Model

We consider a company that runs three oil refineries. Each refinery produces three petroleum-based products: heating oil, diesel oil, and gasoline. Suppose that from 1 barrel of petroleum, the first refinery produces 16 gallons of heating oil, 8 gallons of diesel oil, and 4 gallons of gasoline. The second and third refineries produce different amounts of these three products, as described in the following matrix $\mathbf{A}$.

$$
\mathbf{A} = \begin{matrix} \text{Heating oil} \\ \text{Diesel oil} \\ \text{Gasoline} \end{matrix}
\begin{matrix} \text{Refinery 1} & \text{Refinery 2} & \text{Refinery 3} \end{matrix} \\
\begin{bmatrix} 16 & 8 & 8 \\ 8 & 20 & 8 \\ 4 & 10 & 20 \end{bmatrix} \quad (2)
$$

Each column of $\mathbf{A}$ is a vector of outputs by a refinery. For example, from 1 barrel of oil, refinery 3 produces 8 gallons of heating oil, 8 gallons of diesel oil, and 20 gallons of gasoline. We can represent refinery 3's production with an output

vector $\mathbf{a}_3 \, 3 = \begin{bmatrix} 8 \\ 8 \\ 20 \end{bmatrix}$. Each row of $\mathbf{A}$ is a vector of

the amounts of some product produced by different refineries. The row vector for gasoline is $\mathbf{a}_3' = [4, 10, 20]$.

Let x_i denote the number of barrels of petroleum used by the ith refinery. Suppose there is a demand for 9600 gallons of heating oil, 12,800 gallons of diesel oil, and 16,000 gallons of gasoline. Note that these demands are unrealistically small for a real refinery, where the demands would be in the millions of gallons.

The x_i's need to satisfy the following system of linear equations:

$$
\begin{aligned}
16x_1 + 8x_2 + 8x_3 &= 9600, \\
8x_1 + 20x_2 + 8x_3 &= 12800, \quad (3) \\
4x_1 + 10x_2 + 20x_3 &= 16000,
\end{aligned}
$$

or, as a single-vector equation in the column vectors of the matrix (we define vector addition formally in the next section),

$$
x_1 \begin{bmatrix} 16 \\ 8 \\ 4 \end{bmatrix} + x_2 \begin{bmatrix} 8 \\ 20 \\ 10 \end{bmatrix} + x_3 \begin{bmatrix} 8 \\ 8 \\ 20 \end{bmatrix} = \begin{bmatrix} 9600 \\ 12800 \\ 16000 \end{bmatrix}. \quad (4)
$$

If $\mathbf{b}$ is the column vector of the right-side demands in Eq. (4) and $\mathbf{x}$ is a (column) vector of the x_i's, then matrix algebra should give us a way to write the system of equations concisely in terms of $\mathbf{A}$, $\mathbf{b}$, and $\mathbf{x}$. Indeed, we shall learn how to do this in the next section. In Section 3.4 we shall learn how to solve any linear system of three equations in three variables. ▲

Observe how the concept of a vector allows us to recast the system of three equations in Eq. (3) as a single equation involving vectors in Eq. (4). Collecting information into appropriate groupings to reduce the number or complexity of equations is often just as important in mathematics as developing ways of solving the equations.

EXAMPLE 3   Markov Chain for Weather

Suppose we categorize weather in our city into three states: sunny, cloudy, or rainy. If it is cloudy today, then the probability is 1/2 that it will be sunny tomorrow, 1/4 that it will be cloudy tomorrow, and 1/4 that it will be rainy tomorrow. Other probabilities for tomorrow's weather apply if it is cloudy today or if it is rainy today. These probabilities would be determined by looking at several years of past day-to-day weather trends in our city. It is convenient to display these probabilities in a matrix **A**. (Note that probability will be discussed at greater length in Chapter 8; we will just make informal use of some probability ideas in this example.)

$$
\mathbf{A} = \begin{array}{c} \\ \text{Tomorrow} \\ \text{Sunny} \\ \text{Cloudy} \\ \text{Rainy} \end{array}
\begin{array}{ccc} \text{Today} \\ \text{Sunny} \quad \text{Cloudy} \quad \text{Rainy} \\
\begin{bmatrix} \frac{3}{4} & \frac{1}{2} & \frac{1}{4} \\ \frac{1}{8} & \frac{1}{4} & \frac{1}{2} \\ \frac{1}{8} & \frac{1}{4} & \frac{1}{4} \end{bmatrix} \end{array} \quad (5)
$$

This is a simplified version of a probability-based approach to predicting the weather. A more realistic (but mathematically more complex) probabilistic approach would predict tomorrow's weather by considering the weather today at several locations to the west of our city (since weather in North America tends to move from west to east). For example, the weather today in Buffalo, Pittsburgh, and Washington, D.C., should give a good basis for predicting tomorrow's weather in New York City.

The probabilities in matrix (5) are called *transition probabilities,* and the matrix is called a *transition matrix.* Each column corresponds to the type of weather today. Each row corresponds to the type of weather tomorrow. For example, entry a_{23}, which is found in the cloudy row and rainy column and is 1/2, gives the probability that it will be cloudy tomorrow given that it is rainy today.

A model like matrix (5) for predicting the probabilities of being in different states in the next period in terms of the current states is called

a *Markov chain.* In a Markov chain the probabilities in each column of the transition matrix must add up to 1. A convenient way to display the information in a Markov chain is with a *transition diagram.* The diagram for the weather Markov chain is drawn in Figure 3. There is a node for each state and an arrow for each transition probability.

The data in the transition matrix **A** are used to compute the probabilities of being sunny or cloudy or rainy tomorrow given the probabilities of being sunny or cloudy or rainy today. Let p_1, p_2, and p_3 denote today's probability of sun, clouds, and rain, respectively, and let p_1', p_2', and p_3' denote tomorrow's probability of sun, clouds, and rain, respectively. For this Markov chain, the following probability formulas are used to compute tomorrow's probabilities:

$$
\begin{aligned}
p_1' &= \tfrac{3}{4} p_1 + \tfrac{1}{2} p_2 + \tfrac{1}{4} p_3, \\
p_2' &= \tfrac{1}{8} p_1 + \tfrac{1}{4} p_2 + \tfrac{1}{2} p_3, \\
p_3' &= \tfrac{1}{8} p_1 + \tfrac{1}{4} p_2 + \tfrac{1}{4} p_3.
\end{aligned} \quad (6)
$$

For example, in this weather Markov chain, if early in the morning we hear a weather report

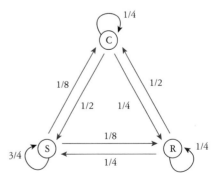

FIGURE 3 Markov transition diagram.

that there is a 50–50 chance that it will be cloudy or rainy today, then $p_1 = 0$, $p_2 = 1/2$, and $p_3 = 1/2$. From these probabilities for today's weather, we can use the Markov chain to predict tomorrow's weather probabilities, p_1', p_2', and p_3':

$$p_1' = \tfrac{3}{4} p_1 + \tfrac{1}{2} p_2 + \tfrac{1}{4} p_3 = 0 + \tfrac{1}{2} \cdot \tfrac{1}{2} + \tfrac{1}{4} \cdot \tfrac{1}{2} = \tfrac{3}{8},$$
$$p_2' = \tfrac{1}{8} p_1 + \tfrac{1}{4} p_2 + \tfrac{1}{2} p_3 = 0 + \tfrac{1}{4} \cdot \tfrac{1}{2} + \tfrac{1}{2} \cdot \tfrac{1}{2} = \tfrac{3}{8},$$
$$p_3' = \tfrac{1}{8} p_1 + \tfrac{1}{4} p_2 + \tfrac{1}{4} p_3 = 0 + \tfrac{1}{4} \cdot \tfrac{1}{2} + \tfrac{1}{4} \cdot \tfrac{1}{2} = \tfrac{1}{4}.$$
$$\tag{7}$$

We explain the formula for p_1' in Eqs. (6) and (7) intuitively as follows. To compute the probability of a sequence of two events, such as the probability p_2 of now being in state 2 along with the probability 1/2 of switching from state 2 to state 1, we multiply these two probabilities together, to get $(1/2)p_2$. We can be in state 1 (sunny) tomorrow either because we are in state 1 today and we stay in state 1—this is the probability $(3/4)p_1$—or because we are in state 2 (cloudy) today and then switch to state 1—this is the probability $(1/2)p_2$—or because we are in state 3 (rainy) today and then switch to state 1—this is the probability $(1/4)p_3$.

We can employ the probability formulas in Eq. (6) to predict the weather probabilities two days ahead, p_1'', p_2'', and p_3'', based on tomorrow's weather probabilities, and $p_1' = 3/8$, $p_2' = 3/8$, and $p_3' = 1/4$, obtained from Eq. (7):

$$p_1'' = \tfrac{3}{4} p_1' + \tfrac{1}{2} p_2' + \tfrac{1}{4} p_3' = \tfrac{3}{4} \cdot \tfrac{3}{8} + \tfrac{1}{2} \cdot \tfrac{3}{8} + \tfrac{1}{4} \cdot \tfrac{1}{4} = \tfrac{34}{64},$$
$$p_2'' = \tfrac{1}{8} p_1' + \tfrac{1}{4} p_2' + \tfrac{1}{2} p_3' = \tfrac{1}{8} \cdot \tfrac{3}{8} + \tfrac{1}{4} \cdot \tfrac{3}{8} + \tfrac{1}{2} \cdot \tfrac{1}{4} = \tfrac{17}{64},$$
$$p_3'' = \tfrac{1}{8} p_1' + \tfrac{1}{4} p_2' + \tfrac{1}{4} p_3' = \tfrac{1}{8} \cdot \tfrac{3}{8} + \tfrac{1}{4} \cdot \tfrac{3}{8} + \tfrac{1}{4} \cdot \tfrac{1}{4} = \tfrac{13}{64}.$$
$$\tag{8}$$

From probabilities for two days hence, we could predict three days ahead, and so on. The computations give the following table assuming today's weather probabilities are 0 sunny, 1/2 cloudy, 1/2 rainy.

	Sunny	Cloudy	Rainy
Today:	0	$\tfrac{1}{2}$	$\tfrac{1}{2}$
1 day ahead:	$\tfrac{3}{8}$	$\tfrac{3}{8}$	$\tfrac{2}{8}$
2 days ahead:	$\tfrac{34}{64}$	$\tfrac{17}{64}$	$\tfrac{13}{64}$
3 days ahead:	$\tfrac{149}{256}$	$\tfrac{60}{256}$	$\tfrac{47}{256}$
5 days ahead:	$\sim\tfrac{14}{23}$	$\sim\tfrac{5}{23}$	$\sim\tfrac{4}{23}$
10 days ahead:	$\sim\tfrac{14}{23}$	$\sim\tfrac{5}{23}$	$\sim\tfrac{4}{23}$
100 days ahead:	$\tfrac{14}{23}$	$\tfrac{5}{23}$	$\tfrac{4}{23}$

$$\tag{9}$$

Observe that after several days, the probabilities stabilize at 14/23, 5/23, and 4/23 for sunny, cloudy, and rainy weather, respectively. These probabilities are predictions this model made for the more distant future when all we know is today's weather and the day-to-day probabilities of weather contained in the Markov chain transition matrix. These probabilities say, in effect, that on an average day in the distant future, the chances are 14/23 for sunny weather, 5/23 for cloudy weather, and 4/23 for rainy weather. ▲

Later in this chapter we will formulate a system of three linear equations in three variables and solve it to determine these "average-day" probabilities directly. For any Markov chain, it will turn out to be mathematically easier to determine the stable probabilities dozens of periods away than to determine the probabilities four days away. Mathematics is often better at discerning long-term trends that are inherent in a model than in finding intermediate-term results that vary from day to day.

Markov chains are widely used in science because linear algebra provides effective mathematical tools to answer virtually any question one can pose about a Markov chain.

E X A M P L E   4   Industrial Growth Model   We consider a possible industrial growth model for pollution and industrial development in a third-world country. Let P be the current level of pollution and D the current level of industrial development. (Each will be measured in units that combine a variety of appropriate indices, such as carbon monoxide levels in air, contaminants in rivers, etc., for pollution.) Let P' and D' be the levels of pollution and industrial development, respectively, in five years. Suppose that based on experiences in similar developing countries, an international development agency believes that the following simple linear model should be a useful predictor of pollution and industrial development over successive five-year periods.

$$P' = P + 2D$$
$$D' = 2P + D \tag{10}$$

We use Eq. (10) to project the pollution and development levels for the next 50 years, as shown with the following table, where initially $P = 4$ and $D = 2$.

	P	D
Now	4	2
5 years	8	10
10 years	28	26
15 years	80	82
20 years	244	242
25 years	728	730
30 years	2188	2186
⋮	⋮	⋮
50 years	177,148	177,146

(11)

Notice that the difference between the values of P and D each year is always 2 in Table (11) (in odd multiples of 5 years, D is larger; in even multiples of 5 years, P is larger). Observe also that after the first two 5-year periods, both quantities grow by a factor of about 3 each period. We give tables for two other initial values for P and D; one for $P = 1, D = 1$ and the other for the nonsense pair $P = 1, D = -1$.

	P	D	P	D
Now	1	1	1	-1
5 years	3	3	-1	1
10 years	9	9	1	-1
15 years	27	27	-1	1
20 years	81	81	1	-1
25 years	243	243	-1	1

(12)

For the first $[P, D]$-vector in Table (12), the values of pollution and industrial development exactly triple each year, while for the second (P, D)-vector, the "nonsense" one, the vector is multiplied by -1 each period. In Section 3.7 we shall develop tools from linear algebra to explain the behavior in Table (11) in terms of the values in Table (12). The nonsense pair $P = 1, D = -1$ will explain why in Table (11) P is alternately 2 larger than D and then 2 smaller than D over successive periods. ▲

In the language of difference equations from Chapter 1, the industrial model in Example 4 would have been expressed as a pair of difference equations:

$$P_n = P_{n-1} + 2D_{n-1},$$
$$D_n = 2P_{n-1} + D_{n-1}. \tag{13}$$

While the numbers that come from Eqs. (10) and (13) will be the same, the two approaches will provide different insights. One of the important assets of mathematical analysis is its ability to gain insight

Mathematics at Work: Didon Pachner, Mathematician at the National Security Agency

Walter Meyer: Is working in a mathematical job anything like studying mathematics?

Didon Pachner: In some ways it is. A lot of what I learned in school is coming in handy in my job! I still have to read math books when

Didon Pachner

I am not familiar with a topic. However, there are some differences. When I read now, it is because I really need to learn the material—it is no longer just for a grade. Learning new things is much more enjoyable because now I really want to learn and can read on my own schedule.

Walter Meyer: What kind of mathematics do you use most?

Didon Pachner: Recently I have been using a lot of linear algebra, and I like it much more than I did when I was in school. The types of mathematics that I use vary a lot. I am in an intern program where I rotate to different departments for six-month tours. Each department has its own needs and specialties.

Other types of math I have used include abstract algebra, probability, and statistics.

Walter Meyer: Sounds a little intense.

Didon Pachner: It can be. There are certainly pressures to work hard. I try to maintain a balance by having outside interests: weight training, playing the piano, and working in a soup kitchen.

Walter Meyer: What does your employer do?

Didon Pachner: I work for the National Security Agency, whose mission is to contribute to secure communications. This means, for example, encoding messages so they can't be read by other countries.

Walter Meyer: How did you get into this job?

Didon Pachner: I was a mathematics major at the University of Maryland and got involved with a co-op program at NSA. It grew into a full-time job.

Walter Meyer: Any advice for today's students?

Didon Pachner: I think that trying to really understand what you are learning is important. It will take you farther than just memorizing formulas.

---

into a problem by looking at the problem several different ways and then combining the results of the different approaches to obtain an even deeper understanding of the problem.

In all the models presented thus far, the number of linear equations equals the number of variables. We finish our introduction of linear models with two models where the number of linear equations is greater than the number of variables. There are also models where the number of linear equations is less than the number of variables, although we will not consider any such models in this chapter.

EXAMPLE 5 Overdetermined System

Suppose the second refinery in the oil refinery model in Example 2 is not operating and we must try to meet the demand with just two refineries. The original refinery system in Eq. (3) now becomes

	Refinery 1		Refinery 3	Demand	
Heating oil	$16x_1$	$+$	$8x_2$	$=$	9600
Diesel oil	$8x_1$	$+$	$8x_2$	$=$	12800 (14)
Gasoline	$4x_1$	$+$	$20x_2$	$=$	16000

It is unlikely we can meet the three demands exactly with just two refineries. If we choose, we can probably meet two of the demands with two refineries, but not all three. Such a problem is called *overdetermined* because there are more constraints than variables.

In this problem, our goal is usually to find the best approximate solution. Some trial and error leads to a possible approximate solution of $x_1 = 300$ and $x_2 = 800$. This yields outputs of

	Refinery 1	Refinery 3	Output	Demand	Error
Heating oil	16(300) +	8(800) =	11200	9600	+1600
Diesel oil	8(300) +	8(800) =	8800	12800	−4000
Gasoline	4(300) +	20(800) =	17200	16000	+1200 ▲

EXAMPLE 6 Fitting a Line to Given Points

A situation that arises frequently in statistics is the problem of "fitting" a line $y = ax + b$ as closely as possible to a set of data points (x_i, y_i). For example, take the four points $(0, 1)$, $(2, 1)$, $(4, 4)$, and $(6, 5)$, where x_i might be the ith student's score on a college entrance exam and y_i might be the ith student's subsequent GPA in college (see Figure 4). Here the system of equations has four equations and just two variables:

$$\begin{aligned} 1 &= 0a + b, \\ 1 &= 2a + b, \\ 4 &= 4a + b, \\ 5 &= 6a + b. \ \blacktriangle \end{aligned} \qquad (15)$$

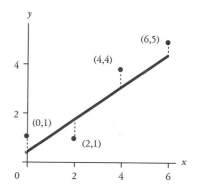

FIGURE 4 Fitting a line to points.

Note that Example 6 is just a more extreme example of an overdetermined system, a concept introduced in Example 5. In Section 3.8 we develop a method to solve a simplified version of the problem in Example 6. Matrix theory has a general method for efficiently solving all overdetermined problems, like those in Examples 5 and 6, although it is beyond the scope of this text.

1. Given the matrix $\mathbf{A} = \begin{bmatrix} 7 & 4 & 3 & 1 \\ 2 & -1 & 5 & 8 \\ 1 & 6 & 7 & 3 \end{bmatrix}$, write out the following row and column vectors and entries.

   (a) $\mathbf{a}_1$ (b) $\mathbf{a}_1'$ (c) $\mathbf{a}_2$ (d) a_{32} (e) a_{21}

2. Plot the following vectors as arrows in the $x_1 x_2$-plane, as in Figure 1.

   (a) $\mathbf{a} = \begin{bmatrix} 1 \\ 2 \end{bmatrix}$ (d) $\mathbf{d} = \begin{bmatrix} 1 \\ -2 \end{bmatrix}$

   (b) $\mathbf{b} = \begin{bmatrix} 0 \\ 3 \end{bmatrix}$ (e) $\mathbf{e} = \begin{bmatrix} 2 \\ 5 \end{bmatrix}$

   (c) $\mathbf{c} = \begin{bmatrix} -2 \\ 3 \end{bmatrix}$

3. The vectors named in this exercise refer to the vectors in Exercise 2.

   (a) Plot the effect of taking vector $\mathbf{a}$ and adding to it vector $\mathbf{b}$.
   (b) Plot the effect of taking vector $\mathbf{b}$ and adding to it vector $\mathbf{a}$. Is this the same final vector as obtained in part (a)?
   (c) Plot the effect of taking vector $\mathbf{c}$ and adding to it vector $\mathbf{d}$.
   (d) Plot the effect of taking vector $\mathbf{b}$ and adding to it vector $\mathbf{e}$.

4. (a) In the matrix $\mathbf{A}$ for the oil refinery model in Example 2, state in words what the entries a_{13} and entries a_{21} represent. What do the numbers in column $\mathbf{a}_3$ of $\mathbf{A}$ represent?
   (b) Suppose refinery 2 is modernized and its output for each barrel of oil is doubled. What is the new matrix $\mathbf{A}$ of production amounts?

5. In Example 2, suppose refinery 1 processes 20 barrels of petroleum, refinery 2 processes 30 barrels, and refinery 3 processes 50 barrels. With this production schedule, for which product does production deviate the most from the set of demands 500, 700, 1000?

6. Consider the following refinery model. There are three refineries, 1, 2, and 3, and from each barrel of crude petroleum, the different refineries produce the following amounts (measured in gallons) of heating oil, diesel oil, and gasoline.

	Refinery 1	Refinery 2	Refinery 3
Heating oil	24	15	9
Diesel oil	6	15	12
Gasoline	9	12	18

   Suppose that we have the following demand: 300 gallons of heating oil, 450 gallons of diesel oil, and 600 gallons of gasoline. Write a system of equations whose solution would determine production levels to yield the desired amounts of heating oil, diesel oil, and gasoline. As in Example 2, let x_i be the number of barrels processed by the ith refinery.

7. A textile company runs three clothing factories. Each factory produces three types of men's clothing: shirts, pants, and coats. Suppose that from one roll of cloth the first factory produces 20 shirts, 10 pants, and 5 coats. The second and third factories produce different amounts of these three products as described in the following matrix $\mathbf{A}$. The demands are 500 shirts, 850 pants, and 1000 coats.

		Factory 1	Factory 2	Factory 3
	Shirts	20	4	4
$\mathbf{A} =$	Pants	10	14	5
	Coats	5	5	12

   Make a mathematical model of this textile problem with a system of three equations, similar to the refinery model.

8. A furniture manufacturer makes tables, chairs, and sofas. One month the company has available 300 units of wood, 350 units of labor, and 225 units of upholstery. The manufacturer wants a production schedule that uses

all of these resources in the month. The different products require the following amounts of the resources:

	Table	Chair	Sofa
Wood	4	1	3
Labor	3	2	5
Upholstery	0	2	4

Make a mathematical model of this production problem, similar to the refinery model.

9. For the weather Markov chain in Example 3, determine the probability distribution for tomorrow's weather, using Eq. (6), when
  (a) $p_1 = 1/3, p_2 = 1/3, p_3 = 1/3$
  (b) $p_1 = 1/4, p_2 = 1/4, p_3 = 1/2$
  (c) $p_1 = 14/23, p_2 = 5/23, p_3 = 4/23$

10. Consider the following Markov chain model involving the states of mind of Professor Mindthumper. The states are alert (A), hazy (H), and stupor (S). If in state A or H today, then tomorrow the professor has a 1/3 chance of being in each of the three states. If in state S today, then tomorrow with probability 1 the professor will still be in state S.

Write the transition matrix $\mathbf{A}$ for this Markov chain.

11. If the local professional basketball team, the Sneakers, wins today's game, they have a 2/3 chance of winning their next game. If they lose this game, they have a 1/2 chance of winning their next game.
  (a) Make a Markov chain for this problem: Give the matrix of transition probabilities, and draw the transition diagram.
  (b) If there is a 50–50 chance of the Sneakers winning today's game, what are the chances they win their next game?
  (c) If they won today, what are the chances of winning the game after the next?

12. If the stock market went up today, historical data show that tomorrow it has a 60% chance of going up, a 20% chance of staying the same, and a 20% chance of going down. If the market were unchanged today, then tomorrow it has a 20% chance of being unchanged, a 40% chance of going up, and a 40% chance of going down. If the market goes down today, then tomorrow it has a 20% of going up, a 20% chance of being unchanged, and a 60% chance of going down.
  (a) Make a Markov chain for this problem: Give the matrix of transition probabilities and the transition diagram.
  (b) If today there is a 30% chance the market goes up, a 10% chance it is unchanged, and a 60% chance it goes down, what is the probability distribution for the market tomorrow?

13. The following model for learning a concept over a set of lessons identifies four states of learning: I = ignorance, E = exploratory thinking, S = superficial understanding, and M = mastery. If now in state I, after one lesson you have 1/2 probability of still being in I and 1/2 probability of being in E. If now in state E, you have 1/4 probability of being in I, 1/2 in E, and 1/4 in S. If now in state S, you have 1/4 probability of being in E, 1/2 in S, and 1/4 in M. If in M, you always stay in M (with probability 1).
  (a) Make a Markov chain model of this learning model.
  (b) If you start in state I, what is your probability distribution after two lessons? After three lessons?

14. Suppose the equations in the pollution/development model in Example 4 were

$$P' = 3P + D,$$
$$D' = 2P + 2D.$$

For the following initial $[P, D]$-vectors, produce a table like that in (11) for P and D over six successive five-year periods. In each case, estimate how much the numbers grow each period.
  (a) $P = 4, D = 2$ (c) $P = 1, D = -2$
  (b) $P = 1, D = 1$

15. (a) By trial and error, try to find production levels for the first and second refineries in Example

2, assuming that refinery 3 is not operating, which come as close as possible to the demand.

(b) Repeat part (a) now with refinery 1 not operating.

16. **(Computer Project)** Use a computer program to follow the Markov chains in the following exercises for 5, 10, 25, and 100 periods by iterating the next-period formula (6) as done in Example 3.

(a) Exercise 10 with Markov chain initially in the state of alert

(b) Exercise 11 with Markov chain initially in winning state

(c) Exercise 12 with Markov chain initially .5 in "going up" and .5 in "going down."

(d) Exercise 13 with Markov chain initially .5 in ignorance and .5 in exploratory thinking.

SECTION 3.2 *Basic Operations on Vectors and Matrices*

Scalar Multiplication and Addition of Matrices

In the next two sections, we develop the increasingly complex computations for multiplying vectors and matrices together. We will start simply and develop complicated operations as collections of simpler operations. The ultimate goal is to define the multiplication of two matrices. While critically important in many linear models, matrix products are numerically messy and so we also need a mathematical language for expressing matrix operations that avoids this messiness. This language is called *matrix algebra.*

The simplest operation in matrix algebra is *matrix addition,* which is performed by adding the corresponding entries of matrices together. The same method is used to add vectors. For addition to make sense, the matrices or vectors must have the same size.

EXAMPLE 1 Matrices of Test Scores

Suppose we are recording the test scores of four students in three subjects. We will call the students A, B, C, and D and the subjects 1, 2, and 3. The students have two one-hour exams and a final exam in each course, each graded out of 10 points. For each of the three tests we form a matrix of test scores with rows for students and columns for subjects. Call the matrices S_1, S_2, and S_3.

$$
S_1 = \begin{array}{c} \\ A \\ B \\ C \\ D \end{array}
\begin{array}{ccc} 1 & 2 & 3 \end{array}
\left[\begin{array}{ccc} 6 & 8 & 9 \\ 8 & 5 & 8 \\ 8 & 7 & 8 \\ 4 & 6 & 6 \end{array}\right]
\qquad
S_2 = \left[\begin{array}{ccc} 5 & 9 & 8 \\ 6 & 7 & 9 \\ 7 & 8 & 8 \\ 5 & 6 & 7 \end{array}\right]
$$

$$
S_3 = \left[\begin{array}{ccc} 6 & 7 & 9 \\ 8 & 6 & 9 \\ 8 & 7 & 8 \\ 6 & 5 & 6 \end{array}\right]
$$

We want a matrix T of the total course scores of each student in each course (without any weighting to make the final exam count more). T is given by the matrix algebra expression

$$
T = S_1 + S_2 + S_3.
$$

Summing the corresponding entries in S_1, S_2, and S_3, we obtain the matrix T:

$$T = S_1 + S_2 + S_3 = \begin{bmatrix} 6 & 8 & 9 \\ 8 & 5 & 8 \\ 8 & 7 & 8 \\ 4 & 6 & 6 \end{bmatrix} + \begin{bmatrix} 5 & 9 & 8 \\ 6 & 7 & 9 \\ 7 & 8 & 8 \\ 5 & 6 & 7 \end{bmatrix}$$

$$+ \begin{bmatrix} 6 & 7 & 9 \\ 8 & 6 & 9 \\ 8 & 7 & 8 \\ 6 & 5 & 6 \end{bmatrix} = \begin{bmatrix} 17 & 24 & 26 \\ 22 & 18 & 26 \\ 23 & 22 & 24 \\ 15 & 17 & 19 \end{bmatrix}. \ \blacktriangle$$

A single number in matrix algebra is called a *scalar*. The operation of multiplying a vector or matrix by a scalar c is called *scalar multiplication*. Scalar multiplication is performed by multiplying each entry in the vector or matrix by the scalar. For example,

$$\text{if } b = \begin{bmatrix} 2 \\ 5 \\ -1 \end{bmatrix},$$

$$\text{then } 2b = \begin{bmatrix} 4 \\ 10 \\ -2 \end{bmatrix}$$

and

$$\text{if } D = \begin{bmatrix} 2 & 4 & 5 & 1 \\ 3 & 9 & 2 & 5 \\ 1 & 6 & 6 & 2 \end{bmatrix},$$

$$\text{then } 3D = \begin{bmatrix} 6 & 12 & 15 & 3 \\ 9 & 27 & 6 & 15 \\ 3 & 18 & 18 & 6 \end{bmatrix}.$$

We used scalar multiplication of vectors in Example 2 of the previous section when we rewrote the oil refinery system of equations,

$$16x_1 + 8x_2 + 8x_3 = 9600,$$
$$8x_1 + 20x_2 + 8x_3 = 12800, \quad (1)$$
$$4x_1 + 10x_2 + 20x_3 = 16000,$$

as a vector equation by factoring out the different x_i's:

$$x_1 \begin{bmatrix} 16 \\ 8 \\ 4 \end{bmatrix} + x_2 \begin{bmatrix} 8 \\ 20 \\ 10 \end{bmatrix} + x_3 \begin{bmatrix} 8 \\ 8 \\ 20 \end{bmatrix} = \begin{bmatrix} 9600 \\ 12800 \\ 16000 \end{bmatrix}. \quad (2)$$

The vector version of this system of equations emphasizes that x_1 is the production level for a vector of oil products from refinery 1, and x_2 and x_3 represent the same for refineries 2 and 3, respectively.

EXAMPLE 1 (continued) Matrices of Test Scores Suppose that the final exam should be weighted twice as much as each hour test. Furthermore, we want the overall course score to be out of 10 points (like each test). That is, the overall course score is a weighted average of the three tests, with the first two tests counting for 25% each and the final exam counting for 50%. Then the matrix W of weighted course scores is given by the matrix expression

$$W = \tfrac{1}{4} S_1 + \tfrac{1}{4} S_2 + \tfrac{1}{2} S_3. \quad (3)$$

We compute W by using the formula in Eq. (3) for each entry. For example, the entry w_{12}, student A's weighted overall course score in course 2, is

$$w_{12} = \tfrac{1}{4} \cdot 8 + \tfrac{1}{4} \cdot 9 + \tfrac{1}{2} \cdot 7 = 7\tfrac{3}{4}.$$

Altogether, we have

$$\mathbf{W} = \frac{1}{4}\begin{bmatrix} 6 & 8 & 9 \\ 8 & 5 & 8 \\ 8 & 7 & 8 \\ 4 & 6 & 6 \end{bmatrix} + \frac{1}{4}\begin{bmatrix} 5 & 9 & 8 \\ 6 & 7 & 9 \\ 7 & 8 & 8 \\ 5 & 6 & 7 \end{bmatrix}$$

$$+ \frac{1}{2}\begin{bmatrix} 6 & 7 & 9 \\ 8 & 6 & 9 \\ 8 & 7 & 8 \\ 6 & 5 & 6 \end{bmatrix} = \begin{bmatrix} 6 & 8 & 9 \\ 8 & 6 & 9 \\ 8 & 7 & 8 \\ 5 & 6 & 6 \end{bmatrix}.$$

(4)

Note that only whole numbers appear in the answer. Fractions were rounded off, with those .5 or greater being rounded up (e.g., 3.6 becomes 4). ▲

Matrix addition is defined only for matrices of the same size. It makes no sense to add matrices that do not contain similar information in an array of similar size. So far, we are just using matrices to organize information, like a spreadsheet.

Scalar Product of Vectors

We now present a form of vector multiplication that is more complex than any sort of multiplication done with single numbers. It is called a *scalar product* because the result of this operation is a scalar (a single number). The following example illustrates the natural way that scalar products arise in calculations with arrays of numbers. Suppose we have a vector $\mathbf{p}$ of prices for a set of three vegetables, celery, broccoli, and squash, $\mathbf{p} = [.80, 1.00, .50]$. Suppose we are also given a vector $\mathbf{d} = [5, 3, 4]$ of the weekly demand in a household for these three vegetables. Then the scalar product of $\mathbf{p}$ and $\mathbf{d}$, written $\mathbf{p} \cdot \mathbf{d}$, equals the cost of the household's weekly demand for these three vegetables. In this case,

$$\mathbf{p} \cdot \mathbf{d} = [.80, 1.00, 50] \cdot [5, 3, 4] = (.80) \cdot 5 \\ + (1.00) \cdot 3 + (.50) \cdot 4 = 4 + 3 + 2 = 9.$$

DEFINITION

Let $\mathbf{a} = [a_1, a_2, \ldots, a_n]$ and $\mathbf{b} = [b_1, b_2, \ldots, b_n]$ be vectors of the same size n. Each vector can be either a row or a column vector. Then the *scalar product* $\mathbf{a} \cdot \mathbf{b}$ of $\mathbf{a}$ and $\mathbf{b}$ is a scalar equal to the sum of the products $a_i b_i$, that is, $\mathbf{a} \cdot \mathbf{b} = a_1 b_1 + a_2 b_2 + \cdots + a_n b_n$.

The scalar product $\mathbf{a} \cdot \mathbf{b}$ makes sense only when $\mathbf{a}$ and $\mathbf{b}$ have the same length.

E X A M P L E   2   Scalar Products   Suppose oranges are 30 cents each, apples 50 cents each, and bananas 40 cents each at a store. Anne wants to get 3 oranges, 2 apples, and 4 bananas, while Bill wants to get 2 oranges, 3 apples, and 2 bananas. We form a price vector $\mathbf{p} = [.30, .50, .40]$ for the respective fruits and form demand vectors for Anne, $\mathbf{a} = [3, 2, 4]$, and for Bill, $\mathbf{b} = [2, 3, 2]$. Then the scalar products $\mathbf{p} \cdot \mathbf{a}$ and $\mathbf{p} \cdot \mathbf{b}$ give the costs of Anne's and Bill's purchases, respectively:

$$\mathbf{p} \cdot \mathbf{a} = [.30, .50, .40] \cdot [3, 2, 4] \\ = (.30) \cdot 3 + (.50) \cdot 2 + (.40) \cdot 4 \\ = .90 + 1.00 + 1.60 = 3.50,$$

$$\mathbf{p} \cdot \mathbf{b} = [.30, .50, .40] \cdot [2, 3, 2] \\ = (.30) \cdot 2 + (.50) \cdot 3 + (.40) \cdot 2 \\ = .60 + 1.50 + .80 = 2.90. \; ▲$$

E X A M P L E   3   A Geometric View of Scalar Products   When two vectors point in approximately the same general direction (in their geometric depiction with arrows), their scalar product is positive. When two vectors point in approximately opposite directions, their scalar product is negative. Most interestingly, when two vectors form a right angle, their scalar product will always be zero. This property of scalar products will be discussed in Section 3.8. Figure 1 illustrates these assertions. ▲

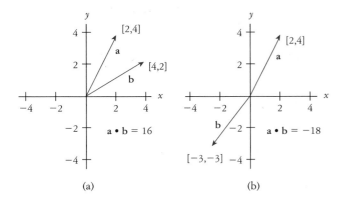

(a)   (b)

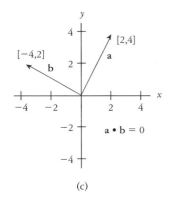

(c)

FIGURE 1 The scalar product of two vectors pointing in the same general direction is positive (a). The scalar product of two vectors pointing in opposite directions is negative (b). The scalar product of two vectors that form a right angle is zero (c).

Observe that the scalar product $\mathbf{p} \cdot \mathbf{a}$ in Example 2 is a linear combination of the entries in each vector. For example, if the numbers of oranges, apples, and

bananas were a vector of unknowns $\mathbf{a} = [x, y, z]$, then $\mathbf{p} \cdot \mathbf{a} = .30x + .50y + .40z$. Conversely, *any linear combination of variables can be expressed as a scalar product.*

E X A M P L E 4 The Oil Refinery Equations as Scalar Products Consider the first linear equation of the oil refinery model (see Eq. (3)):

$$16x_1 + 8x_2 + 8x_3 = 9600. \tag{5}$$

The left side of this equation is a linear combination of the variables. If

$$\mathbf{a} = [16, 8, 8] \quad \text{and} \quad \mathbf{x} = \begin{bmatrix} x_1 \\ x_2 \\ x_3 \end{bmatrix}$$

(the reason for writing $\mathbf{a}$ as a row vector and $\mathbf{x}$ as a column vector will be explained shortly), then the left side of Eq. (5) can be written as a scalar product:

$$\mathbf{a} \cdot \mathbf{x} = [16, 8, 8] \cdot \begin{bmatrix} x_1 \\ x_2 \\ x_3 \end{bmatrix} = 16x_1 + 8x_2 + 8x_3.$$

Any system of linear equations can be written in terms of a system of scalar products. For example, the left sides of the equations in (3) for the oil refinery model are

$$\text{Heating oil:} \quad 16x_1 + 8x_2 + 8x_3 = [16, 8, 8] \cdot \begin{bmatrix} x_1 \\ x_2 \\ x_3 \end{bmatrix}$$

$$= \mathbf{a}_1' \cdot \mathbf{x},$$

Diesel oil: $8x_1 + 20x_2 + 8x_3 = [8, 20, 8] \cdot \begin{bmatrix} x_1 \\ x_2 \\ x_3 \end{bmatrix}$

$$= \mathbf{a}_2' \cdot \mathbf{x},$$

Gasoline: $4x_1 + 10x_2 + 20x_3 = [4, 10, 20] \cdot \begin{bmatrix} x_1 \\ x_2 \\ x_3 \end{bmatrix}$

$$= \mathbf{a}_3' \cdot \mathbf{x}, \tag{6}$$

where $\mathbf{a}_i'$ is the vector formed by coefficients in the ith equation in (6). These coefficient vectors together form a matrix of coefficients associated with (6):

$$\mathbf{A} = \begin{matrix} & \text{Refinery 1} & \text{Refinery 2} & \text{Refinery 3} \\ \text{Heating oil} & & & \\ \text{Diesel oil} & & & \\ \text{Gasoline} & & & \end{matrix} \begin{bmatrix} 16 & 8 & 8 \\ 8 & 20 & 8 \\ 4 & 10 & 20 \end{bmatrix}. \quad \blacktriangle$$

Our original numerical setting of the scalar product involved a short-hand way of expressing the arithmetic computations that arose in, say, determining the cost of shopping for vegetables. Now we have extended the use of scalar products to a symbolic setting where they have the ability to express linear equations concisely.

The following special vectors play a useful role in scalar products.

D E F I N I T I O N

The *coordinate vector* $\mathbf{i}_k$ is a vector with a 1 in the kth position and 0s elsewhere.

For example, the coordinate 2-vector $\mathbf{i}_1$ is $\begin{bmatrix} 1 \\ 0 \end{bmatrix}$. Coordinate vectors are as important as they are simple. They are what their name implies. They point in the direction of one of the coordinate axes. They were discussed in Section 2.4, where they were given special names; in two-dimensional space, $\mathbf{i}$ was used for $\begin{bmatrix} 1 \\ 0 \end{bmatrix}$, and $\mathbf{j}$ for $\begin{bmatrix} 0 \\ 1 \end{bmatrix}$. (Coordinate vectors can be written as either row or column vectors.) Any 2-vector $\mathbf{v}$ is a linear combination of the coordinate 2-vectors $\mathbf{i}_1$ and $\mathbf{i}_2$. For example, $\begin{bmatrix} 5 \\ 3 \end{bmatrix} = 5 \begin{bmatrix} 1 \\ 0 \end{bmatrix} + 3 \begin{bmatrix} 0 \\ 1 \end{bmatrix}$. Thus, we have

$$\mathbf{v} = \begin{bmatrix} v_1 \\ v_2 \end{bmatrix} = v_1 \mathbf{i}_1 + v_2 \mathbf{i}_2 = v_1 \begin{bmatrix} 1 \\ 0 \end{bmatrix} + v_2 \begin{bmatrix} 0 \\ 1 \end{bmatrix}. \tag{7}$$

Observe also that $\mathbf{v} \cdot \mathbf{i}_2 = \begin{bmatrix} v_1 \\ v_2 \end{bmatrix} \cdot \begin{bmatrix} 0 \\ 1 \end{bmatrix} = v_1 \cdot 0 + v_2 \cdot 1 = v_2$. For example $\begin{bmatrix} 5 \\ -6 \end{bmatrix} \cdot \begin{bmatrix} 0 \\ 1 \end{bmatrix} = 5 \cdot 0 + (-6) \cdot 1 = -6$. In general, we see that *the scalar product $\mathbf{v} \cdot \mathbf{i}_k$ of any n-vector $\mathbf{v}$ with the kth coordinate n-vector $\mathbf{i}_k$ is equal to v_k, the kth entry of $\mathbf{v}$.*

Matrix–Vector Products

While there are many important uses of single scalar products that we shall encounter in linear algebra, the most important use of scalar products is as a building block for defining the multiplication of a matrix times a vector and, in the next section, a matrix times another matrix.

DEFINITION

The *matrix–vector product* of an $m \times n$ matrix $\mathbf{A}$ and a column n-vector $\mathbf{c}$ is a column vector of scalar products $\mathbf{a}_i' \cdot \mathbf{c}$, of the *rows* $\mathbf{a}_i'$ of $\mathbf{A}$ with $\mathbf{c}$. If $\mathbf{A} = \begin{bmatrix} a_{11} & a_{12} & a_{13} \\ a_{21} & a_{22} & a_{23} \end{bmatrix}$ is a 3×2 matrix and $\mathbf{c} = \begin{bmatrix} c_1 \\ c_2 \\ c_3 \end{bmatrix}$ is a column 3-vector, then the matrix–vector product $\mathbf{Ac}$ is

$$\mathbf{Ac} = \begin{bmatrix} \mathbf{a}_1' \cdot \mathbf{c} \\ \mathbf{a}_2' \cdot \mathbf{c} \end{bmatrix} = \begin{bmatrix} a_{11}c_1 + a_{12}c_2 + a_{13}c_3 \\ a_{21}c_1 + a_{22}c_2 + a_{23}c_3 \end{bmatrix}.$$

In scalar products involving matrices, the first vector in the scalar product must be a row vector and the second vector must be a column vector.

For example, if

$$\mathbf{A} = \begin{bmatrix} -2 & 1 & 2 \\ 3 & 4 & 5 \end{bmatrix} \quad \text{and} \quad \mathbf{c} = \begin{bmatrix} 1 \\ 2 \\ 3 \end{bmatrix},$$

then $\quad \mathbf{Ac} = \begin{bmatrix} -2 & 1 & 2 \\ 3 & 4 & 5 \end{bmatrix} \begin{bmatrix} 1 \\ 2 \\ 3 \end{bmatrix}$

$$= \begin{bmatrix} -2 \cdot 1 + 1 \cdot 2 + 2 \cdot 3 \\ 3 \cdot 1 + 4 \cdot 2 + 5 \cdot 3 \end{bmatrix} = \begin{bmatrix} 6 \\ 29 \end{bmatrix}.$$

Remember that the number of columns in $\mathbf{A}$ must equal the length of $\mathbf{c}$.

EXAMPLE 4 (continued) The Oil Refinery Model

Returning to the left side of the oil refinery equations given in (6), we see that if we make a vector of the left sides, we have

$$\begin{bmatrix} 16x_1 + 8x_2 + 8x_3 \\ 8x_1 + 20x_2 + 8x_3 \\ 4x_1 + 10x_2 + 20x_3 \end{bmatrix} = \begin{bmatrix} [16, 8, 8] \cdot \mathbf{x} \\ [8, 20, 8] \cdot \mathbf{x} \\ [4, 10, 20] \cdot \mathbf{x} \end{bmatrix}$$

$$= \begin{bmatrix} \mathbf{a}_1' \cdot \mathbf{x} \\ \mathbf{a}_2' \cdot \mathbf{x} \\ \mathbf{a}_3' \cdot \mathbf{x} \end{bmatrix} = \mathbf{Ax}. \quad (8)$$

If we let $\mathbf{b} = \begin{bmatrix} 9600 \\ 12800 \\ 16000 \end{bmatrix}$ be the column vector of demands for the different products, then using Eq. (8), the system of refinery equations becomes

$\mathbf{Ax} = \mathbf{b}$, in expanded form:

$$\begin{bmatrix} 16x_1 + 8x_2 + 8x_3 \\ 8x_1 + 20x_2 + 8x_3 \\ 4x_1 + 10x_2 + 20x_3 \end{bmatrix} = \begin{bmatrix} 9600 \\ 12800 \\ 16000 \end{bmatrix}. \ \blacktriangle \quad (9)$$

Thus, a system of simultaneous linear equations can be written very compactly in matrix notation as $\mathbf{Ax} = \mathbf{b}$. This is an example of the notational power of matrix algebra. In a few pages we have come a long way from our original cost-of-vegetables introduction to scalar products.

It is always useful to test new mathematical concepts with simple examples. The simplest type of vector is a coordinate vector (defined earlier). Let us see what happens when we multiply a matrix times a coordinate vector.

EXAMPLE 5 Coordinate Vectors in Matrix–Vector Products

Earlier we introduced the coordinate vectors $\mathbf{i}_k$, which have a 1 in the kth po-

Linear Programming

What are the odds that someone who nearly flunked high school algebra would develop one of the most valuable advances in applied mathematics in the 20th century? Well, it happened to George Dantzig. After getting angry with himself for his poor high school performance, Dantzig became a mathematician and found himself working for the Air Force in 1946. He was trying to find the most efficient way to move soldiers and material around to meet military objectives. Such problems had been solved by non-mathematical means and Dantzig's insight was that the problems could be set up mathematically in terms of linear inequalities such as $x + y + 3z <= 17$. The problems he dealt with are now called linear programming problems and are similar to the ones we studied in Chapter 2.

George Dantzig

After setting up the problem, Dantzig was stumped about solving it. The geometric approach described in Chapter 2 wasn't powerful enough. Dantzig saw that an important step toward being able to manipulate such a system of inequalities is to convert it to a system of equations. For example, $x + y + 3z <= 17$ becomes $x + y + 3z + d = 17$ where d is the non-negative difference between the right and left sides of the inequality. Now, Dantzig was able to use the tools of linear algebra, such as Gaussian elimination, to manipulate such linear equations.

By combining geometry and linear algebra, Dantzig was able to find the simplex algorithm for solving linear programming problems. This method is still in widespread use today. Economists use it in their economic theories and managers of business have put it to work to save untold amounts of time and money. Not bad for a kid who struggled with algebra.

---

sition and 0s elsewhere. Suppose we multiply the matrix $\mathbf{A} = \begin{bmatrix} -2 & 1 & 2 \\ 3 & 4 & 5 \end{bmatrix}$ times the first coordinate 3-vector $\mathbf{i}_1$:

$$\mathbf{Ai}_1 = \begin{bmatrix} -2 & 1 & 2 \\ 3 & 4 & 5 \end{bmatrix} \begin{bmatrix} 1 \\ 0 \\ 0 \end{bmatrix}$$
$$= \begin{bmatrix} -2 \cdot 1 + 1 \cdot 0 + 2 \cdot 0 \\ 3 \cdot 1 + 4 \cdot 0 + 5 \cdot 0 \end{bmatrix} = \begin{bmatrix} -2 \\ 3 \end{bmatrix} \quad (10)$$

The result is the first column of $\mathbf{A}$. Recall from the definition of matrix–vector products that $\mathbf{Ai}_1$ is a column vector of scalar products $\mathbf{a}'_i$ of the rows $\mathbf{a}'_i$ of $\mathbf{A}$ with $\mathbf{i}_1$. As observed earlier, the scalar product of any vector with $\mathbf{i}_1$ is equal to the first entry of that vector. Looking at Eq. (10), we see that $\mathbf{Ai}_1$ is a column vector consisting of the first entry in each row of $\mathbf{A}$. The set of first entries in each row yields the first *column* of $\mathbf{A}$. ▲

There is another way to view matrix–vector products. In the definition before, $\mathbf{Ac}$ is defined in terms of the *rows* of $\mathbf{A}$, each of which forms a scalar product with $\mathbf{c}$. However, recall that the refinery equations in Eq. (8) were expressed at the beginning of this section in Eq. (2) as a linear combination of the *columns* of the refinery matrix:

$$x_1 \begin{bmatrix} 16 \\ 8 \\ 4 \end{bmatrix} + x_2 \begin{bmatrix} 8 \\ 20 \\ 10 \end{bmatrix} + x_3 \begin{bmatrix} 8 \\ 8 \\ 20 \end{bmatrix} = \begin{bmatrix} 9600 \\ 12800 \\ 16000 \end{bmatrix}. \quad (11)$$

The same representation is available for any matrix–vector product. For the product $\mathbf{Ai}_1$ in Eq. (10), we have

$$\mathbf{Ai}_1 = 1 \begin{bmatrix} -2 \\ 3 \end{bmatrix} + 0 \begin{bmatrix} 1 \\ 4 \end{bmatrix} + 0 \begin{bmatrix} 2 \\ 5 \end{bmatrix} = \begin{bmatrix} -2 \\ 3 \end{bmatrix}. \quad (12)$$

By grouping the entries in a matrix into columns, we recast the matrix–vector product in a way that gives us better insight into "how it works." This is much easier to understand than Eq. (10). In general, we have,

---

The product $\mathbf{Ac}$ can be viewed as a linear combination of the columns of $\mathbf{A}$:

$$\mathbf{Ac} = c_1 \mathbf{a}_1 + c_2 \mathbf{a}_2 + \cdots + c_n \mathbf{a}_n, \quad (13)$$

where $\mathbf{a}_1, \mathbf{a}_2, \ldots, \mathbf{a}_n$ are the columns of matrix $\mathbf{A}$ and $c_1, c_2, \ldots, c_n$ are the entries of column vector $\mathbf{c}$.

A constant theme in mathematics is to develop ever more complex operations and then to find ways of simplifying their complexity.

EXAMPLE 6 Markov Chain for Weather

In the Markov chain model introduced in Example 3 of the previous section, the equations for determining the probabilities p_1', p_2', and p_3' of sunny, cloudy, or rainy weather tomorrow given the probabilities p_1, p_2, and p_3 of sunny, cloudy, or rainy weather today were

$$\begin{aligned} p_1' &= \tfrac{3}{4} p_1 + \tfrac{1}{2} p_2 + \tfrac{1}{4} p_3, \\ p_2' &= \tfrac{1}{8} p_1 + \tfrac{1}{4} p_2 + \tfrac{1}{2} p_3, \\ p_3' &= \tfrac{1}{8} p_1 + \tfrac{1}{4} p_2 + \tfrac{1}{4} p_3. \end{aligned} \quad (14)$$

If $\mathbf{p} = \begin{bmatrix} p_1 \\ p_2 \\ p_3 \end{bmatrix}$ is the vector of today's probabilities, $\mathbf{p}' = \begin{bmatrix} p_1' \\ p_2' \\ p_3' \end{bmatrix}$ is the vector of tomorrow's probabilities, and $\mathbf{A}$ is the matrix of transition probabilities:

	Tomorrow	Today Sunny	Cloudy	Rainy
$\mathbf{A} =$	Sunny	$\tfrac{3}{4}$	$\tfrac{1}{2}$	$\tfrac{1}{4}$
	Cloudy	$\tfrac{1}{8}$	$\tfrac{1}{4}$	$\tfrac{1}{2}$
	Rainy	$\tfrac{1}{8}$	$\tfrac{1}{4}$	$\tfrac{1}{4}$

then the system of equations in (14) can be written in matrix algebra as $\mathbf{p}' = \mathbf{Ap}$ since

$$\begin{aligned} \mathbf{Ap} &= \begin{bmatrix} \tfrac{3}{4} & \tfrac{1}{2} & \tfrac{1}{4} \\ \tfrac{1}{8} & \tfrac{1}{4} & \tfrac{1}{2} \\ \tfrac{1}{8} & \tfrac{1}{4} & \tfrac{1}{4} \end{bmatrix} \begin{bmatrix} p_1 \\ p_2 \\ p_3 \end{bmatrix} \\ &= \begin{bmatrix} \tfrac{3}{4} p_1 + \tfrac{1}{2} p_2 + \tfrac{1}{4} p_3 \\ \tfrac{1}{8} p_1 + \tfrac{1}{4} p_2 + \tfrac{1}{2} p_3 \\ \tfrac{1}{8} p_1 + \tfrac{1}{4} p_2 + \tfrac{1}{4} p_3 \end{bmatrix}. \ \blacktriangle \quad (15) \end{aligned}$$

Multiplication of a vector followed by a matrix $c\mathbf{A}$ can be defined similarly to the product of a matrix followed by a vector. In $c\mathbf{A}$, one forms the scalar products of the *row* vector $\mathbf{c}$ with each column of $\mathbf{A}$. The result is a row vector of scalar products: $c\mathbf{A} = [\mathbf{c}\cdot\mathbf{a}_1, \ \mathbf{c}\cdot\mathbf{a}_2, \ \mathbf{c}\cdot\mathbf{a}_3]$. The size of $\mathbf{c}$ must equal the number of rows in $\mathbf{A}$. Details are left to the exercises at the end of this section.

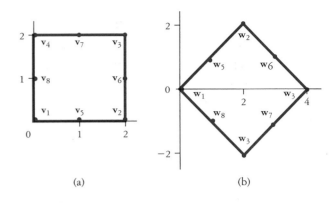

(a)   (b)

EXAMPLE 7 A Geometric View of Matrix–Vector Products

In geometric terms, when we multiply a matrix $\mathbf{A}$ times a vector $\mathbf{v}$, then that vector $\mathbf{v}$ is transformed by the multiplication into another vector, $\mathbf{w}$. We can view multiplication by $\mathbf{A}$ as a function: $\mathbf{w} = f(\mathbf{v})$, where $f(\mathbf{v}) = \mathbf{A}\mathbf{v}$. This type of transformation is used in computer graphics, for example, to produce creative lettering and moving images in television commercials. It provides a way to visualize the effect on a vector of multiplying it by a matrix.

Rather than using arrows, this time we will simply represent 2-vectors as points in the plane. Consider the vectors

$$\mathbf{v}_1 = \begin{bmatrix} 0 \\ 0 \end{bmatrix}, \quad \mathbf{v}_2 = \begin{bmatrix} 2 \\ 0 \end{bmatrix}, \quad \mathbf{v}_3 = \begin{bmatrix} 2 \\ 2 \end{bmatrix}, \quad \mathbf{v}_4 = \begin{bmatrix} 0 \\ 2 \end{bmatrix},$$

$$\mathbf{v}_5 = \begin{bmatrix} 1 \\ 0 \end{bmatrix}, \quad \mathbf{v}_6 = \begin{bmatrix} 2 \\ 1 \end{bmatrix}, \quad \mathbf{v}_7 = \begin{bmatrix} 1 \\ 2 \end{bmatrix}, \quad \mathbf{v}_8 = \begin{bmatrix} 0 \\ 1 \end{bmatrix},$$

where $\mathbf{v}_1$, $\mathbf{v}_2$, $\mathbf{v}_3$, and $\mathbf{v}_4$ are the corners of a square of side 2 and $\mathbf{v}_5$, $\mathbf{v}_6$, $\mathbf{v}_7$, and $\mathbf{v}_8$ are the midpoints of the sides of this square (see Figure 2a). If we multiply these $\mathbf{v}_i$'s by the matrix $\mathbf{A} = \begin{bmatrix} 1 & 1 \\ 1 & -1 \end{bmatrix}$, we obtain the vectors $\mathbf{w}_i = \mathbf{A}\mathbf{v}_i$, which form a square on its side, as shown in Figure 2b:

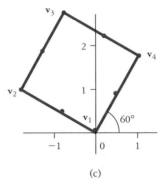

(c)

FIGURE 2 (a) The original square; (b) the rotated square; (c) the original square $\times \ \mathbf{R}_{60°}$.

$$\mathbf{w}_1 = \begin{bmatrix} 1 & 1 \\ 1 & -1 \end{bmatrix}\begin{bmatrix} 0 \\ 0 \end{bmatrix} = \begin{bmatrix} 0 \\ 0 \end{bmatrix},$$

$$\mathbf{w}_2 = \begin{bmatrix} 1 & 1 \\ 1 & -1 \end{bmatrix}\begin{bmatrix} 2 \\ 0 \end{bmatrix} = \begin{bmatrix} 2 \\ 2 \end{bmatrix},$$

$$\mathbf{w}_3 = \begin{bmatrix} 1 & 1 \\ 1 & -1 \end{bmatrix}\begin{bmatrix} 2 \\ 2 \end{bmatrix} = \begin{bmatrix} 4 \\ 0 \end{bmatrix},$$

$$\mathbf{w}_4 = \begin{bmatrix} 1 & 1 \\ 1 & -1 \end{bmatrix}\begin{bmatrix} 0 \\ 2 \end{bmatrix} = \begin{bmatrix} 2 \\ -2 \end{bmatrix},$$

$$\mathbf{w}_5 = \begin{bmatrix} 1 & 1 \\ 1 & -1 \end{bmatrix}\begin{bmatrix} 1 \\ 0 \end{bmatrix} = \begin{bmatrix} 1 \\ 1 \end{bmatrix},$$

$$\mathbf{w}_6 = \begin{bmatrix} 1 & 1 \\ 1 & -1 \end{bmatrix}\begin{bmatrix} 2 \\ 1 \end{bmatrix} = \begin{bmatrix} 3 \\ 1 \end{bmatrix},$$

$$\mathbf{w}_7 = \begin{bmatrix} 1 & 1 \\ 1 & -1 \end{bmatrix}\begin{bmatrix} 1 \\ 2 \end{bmatrix} = \begin{bmatrix} 3 \\ -1 \end{bmatrix},$$

$$\mathbf{w}_8 = \begin{bmatrix} 1 & 1 \\ 1 & -1 \end{bmatrix}\begin{bmatrix} 0 \\ 1 \end{bmatrix} = \begin{bmatrix} 1 \\ -1 \end{bmatrix}.$$

(16)

Note how the midpoint vectors $\mathbf{v}_5$, $\mathbf{v}_6$, $\mathbf{v}_7$, and $\mathbf{v}_8$ in the original square are transformed into the midpoint vectors $\mathbf{w}_5$, $\mathbf{w}_6$, $\mathbf{w}_7$, and $\mathbf{w}_8$ in the rotated square. We have assumed, without proof, that the line segment between $\mathbf{v}_1$ and $\mathbf{v}_2$ is transformed into the line segment between $\mathbf{w}_1$ and $\mathbf{w}_2$; the same assumption holds for the other line segments. Determining how the points along the line segment between $\mathbf{v}_1$ and $\mathbf{v}_2$ are transformed would be very tedious if they are not transformed into the line segment between $\mathbf{w}_1$ and $\mathbf{w}_2$. We shall prove shortly that line segments are mapped into line segments in such transformations. ▲

EXAMPLE 8 Rotation Matrices Consider the matrix $\mathbf{R}_\theta = \begin{bmatrix} \cos\theta & -\sin\theta \\ \sin\theta & \cos\theta \end{bmatrix}$. We claim that multiplying the matrix $\mathbf{R}_\theta$ times a vector $\mathbf{v}$ has the same effect as rotating the vector $\mathbf{v}$ $\theta°$ counterclockwise around the origin. For example,

$\mathbf{R}_\theta \mathbf{i}_1 = \begin{bmatrix} \cos\theta & -\sin\theta \\ \sin\theta & \cos\theta \end{bmatrix} \begin{bmatrix} 1 \\ 0 \end{bmatrix} = \begin{bmatrix} \cos\theta \\ \sin\theta \end{bmatrix}$. Figure 2c shows the effect of multiplying the square in Figure 2a by $\mathbf{R}_{60°} = \begin{bmatrix} \cos 60° & -\sin 60° \\ \sin 60° & \cos 60° \end{bmatrix} = \begin{bmatrix} 1/2 & -2/\sqrt{3} \\ 2/\sqrt{3} & 1/2 \end{bmatrix}$. ▲

We now show that any transformation that takes a 2-vector $\mathbf{v}$ to the 2-vector $\mathbf{w} = \mathbf{Av}$, for some 2×2 matrix $\mathbf{A}$, always maps lines into lines. The proof given here is meant to illustrate the connections between geometry and linear algebra and the power of matrix algebra. The details of the proof are not that important.

This theorem has great practical value, as noted at the end of Example 7, for it simplifies computer graphics computations by only requiring us to compute where endpoints are mapped.

Theorem 1. For any 2×2 matrix $\mathbf{A}$, the mapping of 2-space $\mathbf{v} \rightarrow \mathbf{w} = \mathbf{Av}$ takes lines into lines, except for special cases where it takes a line into a point.

Proof (Optional). The line segment L between vectors $\mathbf{v}_1$ and $\mathbf{v}_2$ can be represented as the set of vectors

$$L = \{\mathbf{u} : \mathbf{u} = \mathbf{v}_1 + c(\mathbf{v}_2 - \mathbf{v}_1), 0 \leq c \leq 1\}.$$

When $c = 0$, we obtain $\mathbf{v}_1$, since $\mathbf{v}_1 + 0(\mathbf{v}_2 - \mathbf{v}_1) = \mathbf{v}_1$. As c increases, an increasing fraction of the vector $\mathbf{v}_2 - \mathbf{v}_1$ is added to $\mathbf{v}_1$, yielding a vector that moves from $\mathbf{v}_1$ toward $\mathbf{v}_2$. Let $\mathbf{w}_1 = \mathbf{Av}_1$ and $\mathbf{w}_2 = \mathbf{Av}_2$. We need to show that this mapping takes L into the line segment

$$L' = \{\mathbf{y} : \mathbf{y} = \mathbf{w}_1 + c(\mathbf{w}_2 - \mathbf{w}_1), 0 \leq c \leq 1\}.$$

We shall show that the vector $\mathbf{u} = \mathbf{v}_1 + c(\mathbf{v}_2 - \mathbf{v}_1)$ is mapped to the vector $\mathbf{y} = \mathbf{w}_1 + c(\mathbf{w}_2 - \mathbf{w}_1)$, that is, $\mathbf{y} = \mathbf{Au}$:

$$\begin{aligned}
\mathbf{Au} &= \mathbf{A}(\mathbf{v}_1 + c(\mathbf{v}_2 - \mathbf{v}_1)) & \text{(17a)} \\
&= \mathbf{Av}_1 + \mathbf{A}(c(\mathbf{v}_2 - \mathbf{v}_1)) & \text{(17b)} \\
&= \mathbf{Av}_1 + c\mathbf{A}(\mathbf{v}_2 - \mathbf{v}_1) & \text{(17c)} \\
&= \mathbf{w}_1 + c(\mathbf{Av}_2 - \mathbf{Av}_1) & \\
&= \mathbf{w}_1 + c(\mathbf{w}_2 - \mathbf{w}_1) = \mathbf{y}. & \text{(17d)}
\end{aligned}$$

In the special case where $\mathbf{w}_1 = \mathbf{w}_2$ (this happens occasionally), L' collapses to the single point $\mathbf{w}_1$.

▲ ▲ ▲

Surveying and Navigation

Linear algebra provides a means to calculate the longitude and latitude of a yet unknown mountain summit S from measurements at known locations A and B. If A, B, and S lie close together, then the plane through them approximates the surface of the earth, as a map does, and the longitude and latitude play the roles of Cartesian x and y coordinates.

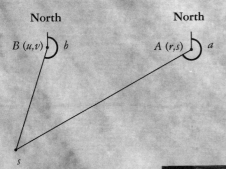

North

$B\ (u,v)$ b

North

$A\ (r,s)$ a

s

Measuring azimuths from Goat Peak fire-lookout (photograph ©1995 by Yves Nievergelt).

Thus, a compass or a theodolite may measure the directions a and b of that summit S from the two reference locations $A = (r, s)$ and $B = (u, v)$. The directions a and b refer to the angles, called "azimuths," from due North to the lines of sight from A through S, and from B through S. Then linear algebra gives the coordinates (x, y) of S. Indeed, (x, y) lies on the line through A in the direction a, which has the

equation $x \cdot \cos(a) - y \cdot \sin(a) = r \cdot \cos(a) - s \cdot \sin(a)$, and also on the line through B in the direction b, which has the equation $x \cdot \cos(b) - y \cdot \sin(b) = u \cdot \cos(b) - v \cdot \sin(b)$. Thus, computing the coordinates (x, y) from the measurements of a and b reduces to solving the linear system

$$\begin{bmatrix} \cos(a) & -\sin(a) \\ \cos(b) & -\sin(b) \end{bmatrix} \begin{bmatrix} x \\ y \end{bmatrix} = \begin{bmatrix} r \cdot \cos(a) - s \cdot \sin(a) \\ u \cdot \cos(b) - v \cdot \sin(b) \end{bmatrix}.$$

For example, from Goat Peak at

$$A = (r, s) = (-120°24'19'', 48°37'51''),$$

one sees a summit S in the azimuth $a = 242°$. Similarly, from Driveway Butte at

$$B = (u, b) = (-120°31'59'', 48°38'03''),$$

one sees the same summit S in the azimuth $b = 198°$. Substituting such values in the preceding linear system and solving for (x, y) yields the coordinates of the summit,

$$S = (x, y) = (-120°33'40'', 48°32'53'').$$

Hikers may employ the same computations to identify the summit, by looking up the location $(x, y) = (-120°33'40'', 48°32'53'')$ on a map, which reveals the name of the summit: Silver Star Mountain.

In Eq. lines (17b) and (17d) we used the distributive law of matrix–vector products—$A(b_1 + b_2) = Ab_1 + Ab_2$ [in Eq. (17b) $b_1 = v_1$ and $b_2 = c(v_2 - v_1)$]—and in Eq. (17c) we used the law of scalar factoring, $A(cb) = cAb$. We assume these laws without proof (laws of matrix algebra will be discussed at the end of the next section). Exercise 15 asks the reader to confirm these laws for particular matrices and vectors.

These two laws together give us the following important property: the *linearity* of matrix–vector products.

Theorem 2. For any scalar numbers r and q, any $m \times n$ matrix A, and any column n-vectors b and c:

$$A(rb + qc) = rAb + qAc.$$

Because of this property, a mapping like $v \to w = Av$ is called a *linear transformation*.

Exercises for Section 3.2

1. Suppose in Example 1 that the final exam counted four times as much as a one-hour exam so that the weights on the three tests should be 1/6, 1/6, 2/3, respectively. Recompute the course score matrix W with these weights.

2. Let $A = \begin{bmatrix} 3 & 1 & 2 \\ 5 & 3 & 6 \end{bmatrix}$ and $B = \begin{bmatrix} 4 & 2 & 3 \\ 1 & 5 & 0 \end{bmatrix}$. Determine

   (a) $2A$ (c) $A + 2B$
   (b) $5B$ (d) $2A - 3B$

3. Let $A = \begin{bmatrix} 4 & 1 & 5 & 4 \\ 0 & 3 & 4 & 9 \\ 2 & 3 & 1 & 6 \end{bmatrix}$ and $B = \begin{bmatrix} 2 & 3 & 1 & 4 \\ -2 & -3 & 4 & 1 \\ 3 & 5 & 0 & 4 \end{bmatrix}$. Determine

   (a) $3A$ (d) $A + B$
   (b) $2B$ (e) $2A + 3B$
   (c) $-3B$ (f) $3A - 2B$

4. The following matrix A gives the price (in cents) of three different kinds of candies bought at three different stores:

	Candy A	Candy B	Candy C
Store 1	10	20	20
$A =$ Store 2	25	30	20
Store 3	30	40	35

(a) Suppose the price of candy doubles. What will the matrix of candy prices be?

(b) Suppose the price of candy increases by 50% and there is a tax of 5 cents on each piece of candy. What will the matrix of candy prices be now?

5. Let all matrices in this exercise be 4×4. Let $\mathbf{I}$ denote the 4×4 matrix with 1s in the main diagonal (that is, entries whose row number equals their column number) and 0s elsewhere. Let $\mathbf{J}$ denote the matrix with a 1 in each entry. Let

$$\mathbf{A} = \begin{bmatrix} 1 & 0 & 1 & 0 \\ 0 & 1 & 0 & 1 \\ 1 & 0 & 1 & 0 \\ 0 & 1 & 0 & 1 \end{bmatrix}.$$

Express the following matrices in terms of $\mathbf{I}$, $\mathbf{J}$, and $\mathbf{A}$:

(a) $\begin{bmatrix} 6 & 2 & 2 & 2 \\ 2 & 6 & 2 & 2 \\ 2 & 2 & 6 & 2 \\ 2 & 2 & 2 & 6 \end{bmatrix}$ (c) $\begin{bmatrix} 5 & 3 & 1 & 3 \\ 3 & 5 & 3 & 1 \\ 1 & 3 & 5 & 3 \\ 3 & 1 & 3 & 5 \end{bmatrix}$

(b) $\begin{bmatrix} 0 & 1 & 0 & 1 \\ 1 & 0 & 1 & 0 \\ 0 & 1 & 0 & 1 \\ 1 & 0 & 1 & 0 \end{bmatrix}$

6. Let $\mathbf{a} = \begin{bmatrix} 1 \\ 2 \end{bmatrix}$, $\mathbf{b} = \begin{bmatrix} -2 \\ 3 \end{bmatrix}$, $\mathbf{c} = \begin{bmatrix} -4 \\ 2 \end{bmatrix}$, and $\mathbf{d} = \begin{bmatrix} 3 \\ -2 \end{bmatrix}$. Plot the following pairs of vectors, and compute their scalar products:

(a) $\mathbf{a}$, $\mathbf{c}$ (d) $\mathbf{a}$, $\mathbf{d}$
(b) $\mathbf{b}$, $\mathbf{c}$ (e) $\mathbf{c}$, $\mathbf{d}$
(c) $\mathbf{b}$, $\mathbf{d}$

7. Let $\mathbf{a} = \begin{bmatrix} 3 \\ 1 \\ 2 \end{bmatrix}$, $\mathbf{b} = \begin{bmatrix} 0 \\ 2 \\ -2 \end{bmatrix}$, and $\mathbf{c} = \begin{bmatrix} 1 \\ 4 \\ 8 \end{bmatrix}$. Compute

(a) $\mathbf{a} \cdot \mathbf{b}$ (c) $\mathbf{a} \cdot (\mathbf{b} + \mathbf{c})$
(b) $\mathbf{b} \cdot \mathbf{c}$ (d) $\mathbf{a} \cdot \mathbf{a}$

8. Let $\mathbf{a}$, $\mathbf{b}$, and $\mathbf{c}$ be as in Exercise 7. Let

$$\mathbf{A} = \begin{bmatrix} 4 & 1 & 5 & 4 \\ 0 & 3 & 4 & 9 \\ 2 & 3 & 1 & 6 \end{bmatrix}, \quad \mathbf{B} = \begin{bmatrix} 2 & 1 & -1 \\ 3 & -1 & 4 \\ 2 & 1 & 3 \end{bmatrix},$$

and $\mathbf{C} = \begin{bmatrix} -2 & 3 & -2 \\ 0 & 3 & 1 \\ 4 & 5 & 3 \\ -1 & 2 & 3 \end{bmatrix}.$

Which of the following matrix calculations are well-defined (the sizes match)? If the computation makes sense, perform it.

(a) $\mathbf{Aa}$ (d) $\mathbf{Ab}$
(b) $\mathbf{Bb}$ (e) $\mathbf{Ba}$
(c) $\mathbf{Cc}$

9. Calculate the following expressions, unless the sizes do not match. The vectors $\mathbf{a}$, $\mathbf{b}$, $\mathbf{c}$ are as defined in Exercise 7, and matrices $\mathbf{A}$, $\mathbf{B}$, $\mathbf{C}$ are as in Exercise 8.

(a) $\mathbf{a} \cdot (\mathbf{Bc})$ (d) $\mathbf{Ba} + \mathbf{Ab}$
(b) $\mathbf{B}(\mathbf{a} + \mathbf{c})$ (e) $(\mathbf{Bc}) \cdot (\mathbf{Bc})$
(c) $(\mathbf{A} + \mathbf{B})\mathbf{b}$

10. Suppose you want to have a party catered and will need 10 hero sandwiches, 6 quarts of fruit punch, 3 quarts of potato salad, and 2 plates of hors d'oeuvres. The following data give the unit costs of these supplies from three different caterers:

	Caterer A	Caterer B	Caterer C
Hero sandwich	$4.00	$6.00	$5.00
Fruit punch	$2.00	$1.00	$0.85
Potato salad	$0.65	$0.85	$1.00
Hors d'oeuvres	$6.00	$5.00	$7.00

(a) Express the problem of determining the costs of catering the party by each caterer as a matrix–vector product (be careful whether you place the vector first or second in the product).

(b) Determine the costs of catering with each caterer.

11. Write out the following systems of equations. Here **x** denotes a column vector of variables $x_1, x_2, \ldots$, where the number of variables equals the number of columns in **A**.

(a) $\mathbf{Ax} = \mathbf{b}$, where $\mathbf{A} = \begin{bmatrix} 1 & 3 \\ 2 & 3 \end{bmatrix}$, $\mathbf{b} = \begin{bmatrix} 4 \\ -2 \end{bmatrix}$

(b) $\mathbf{Ax} = \mathbf{b}$, where $\mathbf{A} = \begin{bmatrix} 4 & -1 \\ 1 & 3 \end{bmatrix}$, $\mathbf{b} = \begin{bmatrix} 2 \\ 1 \end{bmatrix}$

(c) $\mathbf{Ax} = \mathbf{b}$, where $\mathbf{A} = \begin{bmatrix} 2 & 1 & 0 \\ 5 & -1 & 4 \\ 5 & 2 & 3 \end{bmatrix}$,

$\mathbf{b} = \begin{bmatrix} 2 \\ 3 \\ 4 \end{bmatrix}$

(d) $\mathbf{Ax} = \mathbf{b}$, where $\mathbf{A} = \begin{bmatrix} -1 & 0 & 0 \\ 0 & 0 & 2 \\ 0 & 1 & 0 \end{bmatrix}$,

$\mathbf{b} = \begin{bmatrix} 0 \\ 0 \\ 0 \end{bmatrix}$

12. Write the following systems of equations in matrix notation. Define any matrices or vectors you use.

(a) $2x_1 + 4x_2 = 7$
$1x_1 - 3x_2 = 4$

(b) $3x_1 + x_2 + 3x_3 = 6$
$2x_1 + 4x_3 = 2$
$x_1 - 2x_2 = 5$

(c) $x_1 = 3x_1 - 2x_2$
$x_2 = 4x_1 + 3x_2$
$x_3 = 5x_1 + 1x_2$

13. Consider the system of equations

$2x_1 + 3x_2 - 2x_3 = 5y_1 + 2y_2 - 3y_3 + 200,$

$x_1 + 4x_2 + 3x_3 = 6y_1 - 4y_2 + 4y_3 - 120,$

$5x_1 + 2x_2 - x_3 = 2y_1 - 2y_3 + 350.$

(a) Write this system of equations in matrix form. Define the vectors and matrices you introduce.

(b) Rewrite in matrix form with all the variables on the left side (and just scalars on the right).

14. The vector–matrix product of row vector **b** with matrix **A** is a row vector whose ith entry is the scalar product of **b** with the ith column of **A**. Compute the following vector–matrix products, if defined, where $\mathbf{a} = [1, 2, 3]$, $\mathbf{b} = [2, 4, 0]$, and $\mathbf{c} = [3, 9, 1]$, and **A**, **B**, and **C** are the matrices defined in Exercise 8.

(a) $\mathbf{aA}$ (d) $(\mathbf{a} + \mathbf{c})\mathbf{B}$
(b) $\mathbf{bB}$ (e) $\mathbf{aB} + \mathbf{bA}$
(c) $\mathbf{cC}$

15. This example asks the reader to confirm the distributive law—$\mathbf{A}(\mathbf{b} + \mathbf{c}) = \mathbf{Ab} + \mathbf{Ac}$—for matrix–vector products. Let

$$\mathbf{a} = \begin{bmatrix} 2 \\ 0 \\ 5 \end{bmatrix}, \qquad \mathbf{b} = \begin{bmatrix} 1 \\ 3 \\ -2 \end{bmatrix}, \qquad \text{and } \mathbf{c} = \begin{bmatrix} 2 \\ 3 \\ 4 \end{bmatrix},$$

and let **A** be the matrix in Exercise 11c, **B** the matrix in Exercise 8, and **C** the matrix in Exercise 16.

(a) Confirm that $\mathbf{A}(\mathbf{a} + \mathbf{c}) = \mathbf{Aa} + \mathbf{Ab}$.
(b) Confirm that $\mathbf{B}(\mathbf{a} + \mathbf{b}) = \mathbf{Ba} + \mathbf{Bb}$.
(c) Confirm that $\mathbf{C}(\mathbf{b} + \mathbf{c}) = \mathbf{Cb} + \mathbf{Cc}$.

16. For

$$\mathbf{a} = [1, 2, 3], \qquad \mathbf{b} = \begin{bmatrix} 2 \\ 3 \\ 4 \end{bmatrix},$$

$$\text{and } \mathbf{C} = \begin{bmatrix} 2 & 1 & 0 \\ 5 & -1 & 4 \\ 5 & 2 & 3 \end{bmatrix},$$

compute $(\mathbf{aC}) \cdot \mathbf{b}$ and $\mathbf{a} \cdot (\mathbf{Cb})$ to confirm that these two matrix expressions are equal.

17. Plot the eight transformed vectors $\mathbf{w}_i = f(\mathbf{v}_i) = \mathbf{Av}_i$ of the eight vectors $\mathbf{v}_1, \mathbf{v}_2, \ldots, \mathbf{v}_8$ in Example 7 for the following matrices. In each case, draw lines segments from $\mathbf{w}_1$ to $\mathbf{w}_2$, from $\mathbf{w}_2$ to $\mathbf{w}_3$, from $\mathbf{w}_3$ to $\mathbf{w}_4$, and from

w_4 to w_1. Also, check that the vectors w_5, w_6, w_7, and w_8 lie at the middle of these respective line segments.

(a) $A = \begin{bmatrix} 1 & 0 \\ 1 & 1 \end{bmatrix}$ (c) $A = \begin{bmatrix} -2 & 0 \\ 1 & -1 \end{bmatrix}$

(b) $A = \begin{bmatrix} .7 & -.7 \\ .7 & .7 \end{bmatrix}$

18. (a) Write the rotation matrix for a rotation of $90°$ and for $135°$.
    (b) Plot the eight transformed vectors $w_i = f(v_i) = Av_i$ of the eight vectors v_1, v_2, . . . , v_8 in Example 7 for the matrices in part (a).

19. Show that for any vector a, $a \cdot a$ is the sum of the squares of the entries of a.

20. Prove the assertion in this section that the scalar product $c \cdot i_k$ of any n-vector v with the kth coordinate n-vector i_k is equal to v_k, the kth entry of v.

21. Let 1 denote a row n-vector of all 1s. Let A be the $n \times n$ transition matrix for some Markov chain.
    (a) Show that for each column a_j of A, $1 \cdot a_j = 1$.
    (b) Using part (a), show that $1A = 1$.

22. (a) One can express polynomial multiplication in terms of a matrix–vector product as follows: To multiply the quadratic $3x^2 + 2x + 4$ by $3x^2 + 1x - 4$, we multiply

$$\begin{bmatrix} 3 & 0 & 0 \\ 2 & 3 & 0 \\ 4 & 2 & 3 \\ 0 & 4 & 2 \\ 0 & 0 & 4 \end{bmatrix} \begin{bmatrix} 3 \\ 1 \\ -4 \end{bmatrix}.$$

The resulting vector will give the coefficients in the product. Confirm this.

(b) For the polynomial multiplication $(4x^3 - 3x^2 + 1x + 2)(3x^2 - 2x + 5)$, write the associated matrix–vector product.

23. (Computer Project) Write a computer program to add two matrices A and B where both are $m \times n$. Assume that m and n are given and that the entries of the matrices are stored in arrays $A(I, J)$ and $B(I, J)$.

24. (Computer Project) Write a computer program to read in scalars r and s and then compute the linear combination $rA + sB$ of the $m \times n$ matrices A and B. Assume that m and n are given and that the entries of the matrices are stored in arrays $A(I, J)$ and $B(I, J)$.

25. (Computer Project) Write a computer program (or use one supplied by the instructor) to do the following:
    (a) enter a set of points p by their coordinates in the plane, and specify which pairs are to be connected by lines (to form a letter or other simple figure you choose);
    (b) enter a 2×2 matrix A that will be used to transform the figure, by applying the mapping $p' = Ap$ to the points and then connecting with lines the designated pairs of transformed points.
        (i) Let A be a matrix performing a $45°$ rotation.
        (ii) Describe in words the effect of the matrix $A = \begin{bmatrix} 1 & 1 \\ 1 & 0 \end{bmatrix}$.
        (iii) Create a matrix that expands the size of a figure by a factor of 2 (twice as large) and rotates the figure $90°$.

SECTION 3.3 *Matrix Multiplication*

In the previous section we introduced the concept of the scalar product $\mathbf{a} \cdot \mathbf{b}$ of two vectors $\mathbf{a}$ and $\mathbf{b}$. By treating a matrix as a sequence of column vectors, we used this scalar product to define the product $\mathbf{Ab}$ of a matrix $\mathbf{A}$ times a vector $\mathbf{b}$. In this section we extend this process the final step, to define the product $\mathbf{AB}$ of two matrices.

We compute the matrix product $\mathbf{AB}$ of two matrices $\mathbf{A}$ and $\mathbf{B}$ by forming the scalar product of each row of $\mathbf{A}$ with each column of $\mathbf{B}$. For example, if

$$\mathbf{A} = \begin{bmatrix} 1 & 2 \\ -1 & 0 \end{bmatrix} \qquad \mathbf{B} = \begin{bmatrix} 4 & 5 & 1 \\ 6 & 7 & 2 \end{bmatrix},$$

then the entry $(1, 2)$ in $\mathbf{AB}$ is the scalar product of the first row of $\mathbf{A}$ with the second column of $\mathbf{B}$:

$$\mathbf{a}_1' \cdot \mathbf{b}_2 = [1, 2] \cdot \begin{bmatrix} 5 \\ 7 \end{bmatrix} = 1 \cdot 5 + 2 \cdot 7 = 19.$$

In general, the scalar product of row i of $\mathbf{A}$ with column j of $\mathbf{B}$ forms entry (i, j) in $\mathbf{AB}$:

$$\mathbf{AB} = \begin{bmatrix} 1 & 2 \\ -1 & 0 \end{bmatrix} \begin{bmatrix} 4 & 5 & 1 \\ 6 & 7 & 2 \end{bmatrix}$$

$$= \begin{bmatrix} 1 \cdot 4 + 2 \cdot 6 & 1 \cdot 5 + 2 \cdot 7 & 1 \cdot 1 + 2 \cdot 2 \\ -1 \cdot 4 + 0 \cdot 6 & -1 \cdot 5 + 0 \cdot 7 & -1 \cdot 1 + 0 \cdot 2 \end{bmatrix}$$

$$= \begin{bmatrix} 16 & 19 & 5 \\ -4 & -5 & -1 \end{bmatrix}. \tag{1}$$

Matrix Multiplication

Let $\mathbf{A}$ be an $m \times r$ matrix and $\mathbf{B}$ an $r \times n$ matrix. The number of columns in $\mathbf{A}$ must equal the number of rows in $\mathbf{B}$. The matrix product $\mathbf{AB}$ is an $m \times n$ matrix obtained by forming the scalar product of each row $\mathbf{a}_i'$ in $\mathbf{A}$ with each column $\mathbf{b}_j$ in $\mathbf{B}$. The (i, j)th entry in $\mathbf{AB}$ is $\mathbf{a}_i' \cdot \mathbf{b}_j$:

$$\mathbf{AB} = \begin{bmatrix} \mathbf{a}_1' \cdot \mathbf{b}_1 & \mathbf{a}_1' \cdot \mathbf{b}_2 & \cdots & \mathbf{a}_1' \cdot \mathbf{b}_n \\ \mathbf{a}_2' \cdot \mathbf{b}_1 & \mathbf{a}_2' \cdot \mathbf{b}_2 & \cdots & \mathbf{a}_2' \cdot \mathbf{b}_n \\ \vdots & \vdots & \vdots & \vdots \\ \mathbf{a}_m' \cdot \mathbf{b}_1 & \mathbf{a}_m' \cdot \mathbf{b}_2 & \cdots & \mathbf{a}_m' \cdot \mathbf{b}_n \end{bmatrix}. \tag{2}$$

Matrix multiplication is a very complicated operation when viewed in terms of the arithmetic operations on the individual entries. Not surprisingly, not all properties of scalar (single-number) multiplication also hold for matrix multiplication. What is surprising is that most properties of scalar multiplication do hold (they are summarized near the end of this section). One property of scalar multiplication that does not hold for matrix multiplication is commutativity.

E X A M P L E 1 Matrix Multiplication Is Not Commutative. Let $\mathbf{A} = \begin{bmatrix} 1 & 1 \\ 3 & 4 \end{bmatrix}$ and $\mathbf{B} = \begin{bmatrix} 1 & -1 \\ 0 & 2 \end{bmatrix}$. Then

$$\mathbf{AB} = \begin{bmatrix} 1 \cdot 1 + 1 \cdot 0 & 1 \cdot (-1) + 1 \cdot 2 \\ 3 \cdot 1 + 4 \cdot 0 & 3 \cdot (-1) + 4 \cdot 2 \end{bmatrix} = \begin{bmatrix} 1 & 1 \\ 3 & 5 \end{bmatrix}$$

and

$$\mathbf{BA} = \begin{bmatrix} 1 \cdot 1 + (-1) \cdot 3 & 1 \cdot 1 + (-1) \cdot 4 \\ 0 \cdot 1 + 2 \cdot 3 & 0 \cdot 1 + 2 \cdot 4 \end{bmatrix}$$

$$= \begin{bmatrix} -2 & -3 \\ 6 & 8 \end{bmatrix}.$$

Thus $\mathbf{AB} \neq \mathbf{BA}$. ▲

E X A M P L E 2 Special Matrices That Do Commute—The Identity Matrix An *identity matrix* $\mathbf{I}$ is a square matrix that has 1s on the main diagonal and 0s elsewhere:

$$I = \begin{bmatrix} 1 & 0 & 0 & \cdots & 0 \\ 0 & 1 & 0 & \cdots & 0 \\ 0 & 0 & 1 & \cdots & 0 \\ \vdots & \vdots & \vdots & \vdots & \vdots \\ 0 & 0 & 0 & \cdots & 1 \end{bmatrix}. \tag{3}$$

When it is important to emphasize the size of the identity matrix, we write I_n, for the $n \times n$ identity matrix. The kth column or kth row of I is the kth coordinate vector i_k (with a 1 in the kth entry and 0s elsewhere). Coordinate vectors were introduced in the previous section.

I is the identity element in matrix multiplication (it plays a role similar to that of the number 1 in scalar multiplication). That is, for any square matrix A,

$$IA = AI = A. \tag{4}$$

For the matrix A in Example 1, we have

$$IA = \begin{bmatrix} 1 & 0 \\ 0 & 1 \end{bmatrix} \begin{bmatrix} 1 & 1 \\ 3 & 4 \end{bmatrix}$$

$$= \begin{bmatrix} 1 \cdot 1 + 0 \cdot 3 & 1 \cdot 1 + 0 \cdot 4 \\ 0 \cdot 1 + 1 \cdot 3 & 0 \cdot 1 + 1 \cdot 4 \end{bmatrix} = \begin{bmatrix} 1 & 1 \\ 3 & 4 \end{bmatrix}.$$

The reader should check that $AI = A$ for this matrix A. ▲

E X A M P L E   2 (c o n t i n u e d)   Special Matrices That Do Commute—Inverses If $ac = 1$, for scalar numbers a and c, then c is called the *reciprocal* of a, that is, $c = a^{-1}$ (and $a = c^{-1}$). For example, 1/3 is the reciprocal of 3. The term *inverse* is used in matrix multiplication. If A^{-1} is the inverse of A, then $A^{-1}A = I$; also $AA^{-1} = I$, where I denotes the identity matrix introduced

in Eq. (3). The matrix $A = \begin{bmatrix} 1 & 1 \\ 3 & 4 \end{bmatrix}$ has the inverse $A^{-1} = \begin{bmatrix} 4 & -1 \\ -3 & 1 \end{bmatrix}$. We confirm this as follows:

$$AC = \begin{bmatrix} 1 & 1 \\ 3 & 4 \end{bmatrix} \begin{bmatrix} 4 & -1 \\ -3 & 1 \end{bmatrix}$$

$$= \begin{bmatrix} 1 \cdot 4 + 1 \cdot (-3) & 1 \cdot (-1) + 1 \cdot 1 \\ 3 \cdot 4 + 4 \cdot (-3) & 3 \cdot (-1) + 4 \cdot 1 \end{bmatrix}$$

$$= \begin{bmatrix} 1 & 0 \\ 0 & 1 \end{bmatrix} \tag{5a}$$

and

$$CA = \begin{bmatrix} 4 & -1 \\ -3 & 1 \end{bmatrix} \begin{bmatrix} 1 & 1 \\ 3 & 4 \end{bmatrix}$$

$$= \begin{bmatrix} 4 \cdot 1 + (-1) \cdot 3 & 4 \cdot 1 + (-1) \cdot 4 \\ (-3) \cdot 1 + 1 \cdot 3 & (-3) \cdot 1 + 1 \cdot 4 \end{bmatrix}$$

$$= \begin{bmatrix} 1 & 0 \\ 0 & 1 \end{bmatrix}. \tag{5b}$$

Observe that a matrix and its inverse always commute, since their product (in either order) is always the identity matrix I. On the other hand, not all matrices have inverses. We learn how to compute the inverse A^{-1} of a matrix A—if the inverse exists—in Section 3.5. Note that the inverse of a matrix is just another matrix (no fractions are required). ▲

Matrix multiplication involves a lot of tedious arithmetic. After some pencil-and-paper practice, it is appropriate to let a computer multiply matrices for you. Indeed, tedious numerical chores in matrix multiplication were a prime motivation for the development of digital computers. With three simple loops, the following short computer program performs the matrix multiplication $\mathbf{C} = \mathbf{AB}$.

```
input:    positive integers m, r, n, m × r
          matrix A, and r × n matrix B
output:   m × n matrix C, which is the
          matrix product AB
```

```
For i from 1 to m
    For j from 1 to n
        C[i, j] ← 0
        For k from 1 to r                    (6)
            C[i, j] ← C[i, j]
                + A[i, k]*B[k, j]
        end for
end for
```

Use of Matrix Multiplication in Organizing Data

In this chapter we will see many different uses for matrix multiplication. We start with an example in which matrix multiplication is used to compile a table of scalar product computations.

EXAMPLE 3 A Collection of Computer Computation Times A MEGACRUNCH computer requires 3 minutes to do a type-1 job (say, a statistics problem). 4 minutes to do a type-2 job, and 2 minutes to do a type-3 job. The computer has 6 type-1 jobs, 8 type-2 jobs, and 10 type-3 jobs. How long will the computer take to perform all these jobs?

If $\mathbf{t} = [3, 4, 2]$ is the vector of the times to do the various jobs and $\mathbf{n} = \begin{bmatrix} 6 \\ 8 \\ 10 \end{bmatrix}$ is the vector

of the numbers of each type of job (the reason $\mathbf{t}$ is written as a row vector and $\mathbf{n}$ as a column vector will be clear shortly), then the total time required will be the value of the scalar product $\mathbf{t} \cdot \mathbf{n}$:

$$\text{Total time} = \mathbf{t} \cdot \mathbf{n} = [3, 4, 2] \cdot \begin{bmatrix} 6 \\ 8 \\ 10 \end{bmatrix}$$
$$= 3 \cdot 6 + 4 \cdot 8 + 2 \cdot 10$$
$$= 18 + 32 + 20 = 70. \quad \blacktriangle$$

Suppose instead of one supercomputer, we have four supercomputers: MEGACRUNCH BITBUSTER, WHOPPER, and ULTIMA. Then instead of a vector of computation times for MEGACRUNCH to perform different jobs, we need a matrix $\mathbf{A}$ of computation times for the different supercomputers to perform the different types of jobs:

$$\mathbf{A} = \begin{array}{c} \\ \text{MEGACRUNCH} \\ \text{BITBUSTER} \\ \text{WHOPPER} \\ \text{ULTIMA} \end{array} \overset{\begin{array}{ccc} \text{Type of} \\ \text{job} \\ 1 \ \ 2 \ \ 3 \end{array}}{\begin{bmatrix} 3 & 4 & 2 \\ 5 & 7 & 3 \\ 1 & 2 & 1 \\ 3 & 3 & 3 \end{bmatrix}} \text{ matrix of times.}$$

To calculate how long it would take each supercomputer to do 6 type-1, 8 type-2, and 10 type-3 jobs, we multiply $\mathbf{A}$ times the column vector $\mathbf{n} = \begin{bmatrix} 6 \\ 8 \\ 10 \end{bmatrix}$:

$$\mathbf{An} = \begin{bmatrix} 3 & 4 & 2 \\ 5 & 7 & 3 \\ 1 & 2 & 1 \\ 3 & 3 & 3 \end{bmatrix} \begin{bmatrix} 6 \\ 8 \\ 10 \end{bmatrix}$$
$$= \begin{bmatrix} 3 \cdot 6 + 4 \cdot 8 + 2 \cdot 10 \\ 5 \cdot 6 + 7 \cdot 8 + 3 \cdot 10 \\ 1 \cdot 6 + 2 \cdot 8 + 1 \cdot 10 \\ 3 \cdot 6 + 3 \cdot 8 + 3 \cdot 10 \end{bmatrix} = \begin{bmatrix} 70 \\ 116 \\ 32 \\ 72 \end{bmatrix}. \quad (7)$$

Next let us do this calculation not just for one

set of jobs, but for three sets of jobs. Set A will be the previous set $\mathbf{n} = \begin{bmatrix} 6 \\ 8 \\ 10 \end{bmatrix}$. Sets B and C will be $\begin{bmatrix} 2 \\ 5 \\ 5 \end{bmatrix}$ and $\begin{bmatrix} 4 \\ 4 \\ 4 \end{bmatrix}$, respectively. Let us calculate the times required to do each set on each computer by expanding the vector $\mathbf{n}$ into a matrix $\mathbf{N}$ of three column vectors:

$$
\mathbf{N} = \begin{array}{c} \text{Sets} \\ \text{Type } 1 \\ \text{Type } 2 \\ \text{Type } 3 \end{array}
\begin{array}{ccc} A & B & C \end{array}
\begin{bmatrix} 6 & 2 & 4 \\ 8 & 5 & 4 \\ 10 & 5 & 4 \end{bmatrix} \quad \text{matrix of jobs.}
$$

The calculation of $\mathbf{An}$ in Eq. (7) required us to multiply each row of $\mathbf{A}$ times the column vector $\mathbf{n}$. Now we need to multiply each row of $\mathbf{A}$ (one for each computer) times each column of $\mathbf{N}$ (one for each set of jobs):

$$
\mathbf{AN} = \begin{bmatrix} 3 & 4 & 2 \\ 5 & 7 & 3 \\ 1 & 2 & 1 \\ 3 & 3 & 3 \end{bmatrix} \begin{bmatrix} 6 & 2 & 4 \\ 8 & 5 & 4 \\ 10 & 5 & 4 \end{bmatrix}
$$

$$
= \begin{bmatrix} 3\cdot6 + 4\cdot8 + 2\cdot10 & 3\cdot2 + 4\cdot5 + 2\cdot5 \\ 5\cdot6 + 7\cdot8 + 3\cdot10 & 5\cdot2 + 7\cdot5 + 3\cdot5 \\ 1\cdot6 + 2\cdot8 + 1\cdot10 & 1\cdot2 + 2\cdot5 + 1\cdot5 \\ 3\cdot6 + 3\cdot8 + 3\cdot10 & 3\cdot2 + 3\cdot5 + 3\cdot5 \end{bmatrix}
$$

(8)

$$
\begin{bmatrix} 3\cdot4 + 4\cdot4 + 2\cdot4 \\ 5\cdot4 + 7\cdot4 + 3\cdot4 \\ 1\cdot4 + 2\cdot4 + 1\cdot4 \\ 3\cdot4 + 3\cdot4 + 3\cdot4 \end{bmatrix}
$$

$$
= \begin{array}{c} \text{Sets} \\ \text{MEGACRUNCH} \\ \text{BITBUSTER} \\ \text{WHOPPER} \\ \text{ULTIMA} \end{array}
\begin{array}{ccc} A & B & C \end{array}
\begin{bmatrix} 70 & 36 & 36 \\ 116 & 60 & 60 \\ 32 & 172 & 16 \\ 72 & 36 & 36 \end{bmatrix}
\begin{array}{l} \text{matrix of total} \\ \text{computation} \\ \text{times.} \end{array}
$$

We now consider a more practical version of the previous example.

EXAMPLE 4 Building New Data Matrices

Suppose we are given the following matrices: matrix $\mathbf{A}$ gives the amounts of raw material required to build different products; matrix $\mathbf{B}$ tells how many of the products are needed to build two types of houses; and matrix $\mathbf{C}$ gives the demand for houses in two countries.

A	Raw material		
	Wood	Labor	Steel
Item A	5	20	10
Item B	4	25	8
Item C	10	10	5

B	Items needed in house		
	A	B	C
House I	4	8	3
House II	5	5	2

C	Demand for houses	
	House I	House II
Spain	50,000	200,000
Italy	80,000	500,000

a.   Which matrix product tells the amounts of each raw material needed to build each type of house?

b.   Which matrix product tells how much of each item is needed to build all houses (types I and II combined) in each different country?

c.   Which matrix product tells how much of each raw material is needed to build all houses (types I and II combined) in each different country?

For part (a) we need a matrix of amounts of

each raw material for each house. Each of these amounts will be the scalar product of the amount of a raw material used by each item (the columns of **A**) and the number of each item needed to make a particular house (the rows of **B**). By the rules of matrix multiplication, to get a matrix of scalar products of the columns of **A** times the rows of **B**, we form the matrix product **BA**. (The rows of **BA** stand for the types of house, and the columns for the different raw materials.)

$$\mathbf{BA} = \begin{bmatrix} 4 & 8 & 3 \\ 5 & 5 & 2 \end{bmatrix} \begin{bmatrix} 5 & 20 & 10 \\ 4 & 25 & 8 \\ 10 & 10 & 5 \end{bmatrix}$$

$$= \begin{bmatrix} 4\cdot5 + 8\cdot4 + 3\cdot10 & 4\cdot20 + 8\cdot25 + 3\cdot10 \\ 5\cdot5 + 5\cdot4 + 2\cdot10 & 5\cdot20 + 5\cdot25 + 2\cdot10 \end{bmatrix}$$

$$\begin{array}{c} 4\cdot10 + 8\cdot8 + 3\cdot5 \\ 5\cdot10 + 5\cdot8 + 2\cdot5 \end{array} = \begin{bmatrix} 82 & 310 & 119 \\ 65 & 470 & 100 \end{bmatrix}$$

For part (b) we need a matrix of amounts of each item in all houses in a country. The number of each item needed for different houses is given in the columns of **B**, and the number of each type of house needed in each country is given by the rows of **C**. So we must form scalar products of the rows of **C** with the columns of **B**. We obtain these amounts from the matrix product **CB**.

$$\mathbf{CB} = \begin{bmatrix} 50{,}000 & 200{,}000 \\ 80{,}000 & 500{,}000 \end{bmatrix} \begin{bmatrix} 4 & 8 & 3 \\ 5 & 5 & 2 \end{bmatrix}$$

$$= \begin{bmatrix} 50{,}000\cdot4 + 200{,}000\cdot5 & 50{,}000\cdot8 + 200{,}000\cdot5 \\ 80{,}000\cdot4 + 500{,}000\cdot5 & 80{,}000\cdot8 + 500{,}000\cdot5 \end{bmatrix}$$

$$\begin{array}{c} 50{,}000\cdot3 + 200{,}000\cdot2 \\ 80{,}000\cdot3 + 500{,}000\cdot2 \end{array}$$

$$= \begin{bmatrix} 1{,}200{,}000 & 1{,}400{,}000 & 550{,}000 \\ 2{,}820{,}000 & 3{,}140{,}000 & 1{,}240{,}000 \end{bmatrix}$$

For part (c) we need a matrix of amounts of each raw material needed for all houses in each country. The product **BA** in (a) gives the amounts of each raw material needed for each house. The scalar product of the columns of **BA** with the rows of **C** will give the desired result. Thus the required matrix product is **C(BA)**. The calculation of the matrix is left as an exercise. ▲

Interpretations of Powers of Matrices

In the Markov chain model for weather used in previous sections, the equations for determining the probabilities p_1', p_2', and p_3' of sunny, cloudy, or rainy weather tomorrow given the probabilities p_1, p_2, and p_3 of sunny, cloudy, or rainy weather today were as follows:

$$\begin{aligned} p_1' &= \tfrac{3}{4}p_1 + \tfrac{1}{2}p_2 + \tfrac{1}{4}p_3, \\ p_2' &= \tfrac{1}{8}p_1 + \tfrac{1}{4}p_2 + \tfrac{1}{2}p_3, \\ p_3' &= \tfrac{1}{8}p_1 + \tfrac{1}{4}p_2 + \tfrac{1}{4}p_3. \end{aligned} \tag{9}$$

If $\mathbf{p} = \begin{bmatrix} p_1 \\ p_2 \\ p_3 \end{bmatrix}$ is the vector of current probabilities,

$\mathbf{p}' = \begin{bmatrix} p_1' \\ p_2' \\ p_3' \end{bmatrix}$ is the vector of tomorrow's probabilities,

and **A** is the matrix of transition probabilities:

	Tomorrow	Today Sunny	Cloudy	Rainy
A =	Sunny	$\tfrac{3}{4}$	$\tfrac{1}{2}$	$\tfrac{1}{4}$
	Cloudy	$\tfrac{1}{8}$	$\tfrac{1}{4}$	$\tfrac{1}{2}$
	Rainy	$\tfrac{1}{8}$	$\tfrac{1}{4}$	$\tfrac{1}{4}$

then we showed in Example 6 of Section 3.2 that Eq. (9) can be written as $\mathbf{p}' = \mathbf{Ap}$. As a review, we check this claim as follows:

$\mathbf{p}' = \mathbf{Ap}$ expands out as $\begin{bmatrix} p_1' \\ p_2' \\ p_3' \end{bmatrix} = \begin{bmatrix} \frac{3}{4} & \frac{1}{2} & \frac{1}{4} \\ \frac{1}{8} & \frac{1}{4} & \frac{1}{2} \\ \frac{1}{8} & \frac{1}{4} & \frac{1}{4} \end{bmatrix} \begin{bmatrix} p_1 \\ p_2 \\ p_3 \end{bmatrix}$

$$= \begin{bmatrix} \frac{3}{4} p_1 + \frac{1}{2} p_2 + \frac{1}{4} p_3 \\ \frac{1}{8} p_1 + \frac{1}{4} p_2 + \frac{1}{2} p_3 \\ \frac{1}{8} p_1 + \frac{1}{4} p_2 + \frac{1}{4} p_3 \end{bmatrix}. \tag{10}$$

A Markov chain can be used to predict probabilities for many periods. A table of such calculations for the weather Markov chain was presented in Example 3 of Section 3.1.

Just as tomorrow's probability vector $\mathbf{p}'$ can be computed by the matrix expression $\mathbf{p}' = \mathbf{Ap}$, so the probability vector $\mathbf{p}''$ for two days hence is given by the formula

$$\begin{aligned} \mathbf{p}'' &= \mathbf{Ap}' \\ &= \mathbf{A}(\mathbf{Ap}) \\ &= (\mathbf{A}^2)\mathbf{p}. \end{aligned} \tag{11}$$

The last equality in Eq. (11)—the key step—uses the associative law for matrix multiplication: $\mathbf{A}(\mathbf{Ap}) = (\mathbf{AA})\mathbf{p}$. Equation (11) says that instead of using Eq. (10) twice to compute $\mathbf{p}''$—by first using it to determine $\mathbf{p}'$ and then from $\mathbf{p}'$ using Eq. (10) to determine $\mathbf{p}''$—we can instead compute $\mathbf{p}''$ directly from $\mathbf{p}$ by multiplying $\mathbf{A}^2$ times $\mathbf{p}$.

Similarly, the distribution $\mathbf{p}^{(3)}$ three days hence should be given by

$$\mathbf{p}^{(3)} = \mathbf{Ap}'' = \mathbf{A}(\mathbf{A}^2\mathbf{p}) = (\mathbf{A}^3)\mathbf{p}. \tag{12}$$

Again, in Eq. (12) the critical step is the associative law of matrix multiplication.

EXAMPLE 5 Powers of Markov Transition Matrices To illustrate the preceding discussion, let us compute $\mathbf{A}^2$ and $\mathbf{A}^3$ for the weather Markov transition matrix $\mathbf{A}$:

$$\mathbf{A}^2 = \begin{bmatrix} \frac{3}{4} & \frac{1}{2} & \frac{1}{4} \\ \frac{1}{8} & \frac{1}{4} & \frac{1}{2} \\ \frac{1}{8} & \frac{1}{4} & \frac{1}{4} \end{bmatrix} \begin{bmatrix} \frac{3}{4} & \frac{1}{2} & \frac{1}{4} \\ \frac{1}{8} & \frac{1}{4} & \frac{1}{2} \\ \frac{1}{8} & \frac{1}{4} & \frac{1}{4} \end{bmatrix}$$

$$= \begin{bmatrix} \frac{3}{4}\cdot\frac{3}{4}+\frac{1}{2}\cdot\frac{1}{8}+\frac{1}{4}\cdot\frac{1}{8} & \frac{3}{4}\cdot\frac{1}{2}+\frac{1}{2}\cdot\frac{1}{4}+\frac{1}{4}\cdot\frac{1}{4} \\ \frac{1}{8}\cdot\frac{3}{4}+\frac{1}{4}\cdot\frac{1}{8}+\frac{1}{2}\cdot\frac{1}{8} & \frac{1}{8}\cdot\frac{1}{2}+\frac{1}{4}\cdot\frac{1}{4}+\frac{1}{2}\cdot\frac{1}{4} \\ \frac{1}{8}\cdot\frac{3}{4}+\frac{1}{4}\cdot\frac{1}{8}+\frac{1}{4}\cdot\frac{1}{8} & \frac{1}{8}\cdot\frac{1}{2}+\frac{1}{4}\cdot\frac{1}{4}+\frac{1}{4}\cdot\frac{1}{4} \end{bmatrix}$$

$$\begin{matrix} \frac{3}{4}\cdot\frac{1}{4}+\frac{1}{2}\cdot\frac{1}{2}+\frac{1}{4}\cdot\frac{1}{4} \\ \frac{1}{8}\cdot\frac{1}{4}+\frac{1}{4}\cdot\frac{1}{2}+\frac{1}{2}\cdot\frac{1}{4} \\ \frac{1}{8}\cdot\frac{1}{4}+\frac{1}{4}\cdot\frac{1}{2}+\frac{1}{4}\cdot\frac{1}{4} \end{matrix}$$

$$= \begin{bmatrix} \frac{21}{32} & \frac{18}{32} & \frac{16}{32} \\ \frac{6}{32} & \frac{8}{32} & \frac{9}{32} \\ \frac{5}{32} & \frac{6}{32} & \frac{7}{32} \end{bmatrix}. \tag{13}$$

Next we compute $\mathbf{A}^3$ (the arithmetic details are left as an exercise):

$$\mathbf{A}^3 = \mathbf{AA}^2 = \begin{bmatrix} \frac{3}{4} & \frac{1}{2} & \frac{1}{4} \\ \frac{1}{8} & \frac{1}{4} & \frac{1}{2} \\ \frac{1}{8} & \frac{1}{4} & \frac{1}{4} \end{bmatrix} \begin{bmatrix} \frac{21}{32} & \frac{18}{32} & \frac{16}{32} \\ \frac{6}{32} & \frac{8}{32} & \frac{9}{32} \\ \frac{5}{32} & \frac{6}{32} & \frac{7}{32} \end{bmatrix}$$

$$= \begin{bmatrix} \frac{160}{256} & \frac{152}{256} & \frac{146}{256} \\ \frac{53}{256} & \frac{58}{256} & \frac{62}{256} \\ \frac{43}{256} & \frac{46}{256} & \frac{48}{256} \end{bmatrix}. \tag{14}$$

The entries in $\mathbf{A}^2$ are transition probabilities for two days, and the entries in $\mathbf{A}^3$ are transition probabilities for three days. For example, the value 6/32 in entry (2, 1) of $\mathbf{A}^2$ means that if we are now in state 1 (sunny), then the chance is 6/32 that in two days we will be in state 2 (cloudy). And the value of 53/256 in entry (2, 1) of $\mathbf{A}^3$ tells us that if it is now sunny, the probability is 53/256 that in three days it will be cloudy. The values we obtained in computing $\mathbf{A}^2$ and $\mathbf{A}^3$ look reasonable. In particular, the numbers in each column of $\mathbf{A}^2$ and $\mathbf{A}^3$ sum to 1.

Note that the three columns of $\mathbf{A}^3$ all have about the same values. That is, the first entries in each column in $\mathbf{A}^3$ are all close to 150/256, the second entries are all close to 58/256, and the third entries are all close to 46/256. Recall that the first column in $\mathbf{A}^3$ gives the probabilities, if sunny today, of different types of weather in three days. The second column in $\mathbf{A}^3$ gives the probabilities, if cloudy today, of different types of weather in three days, similarly for the third column if rainy today. The interpretation of the similarity of the columns of $\mathbf{A}^3$ is that, no matter what today's weather is, the chances are about the same of getting sunny weather in three days, or getting cloudy weather, or getting rainy weather. Intuitively, as we predict further into the future, what today's weather is will become unimportant. ▲

Example 5 illustrates how, with very concise notation, matrix algebra allows us to express quite complex calculations. It also shows how raising a matrix to a power corresponds to repeating a Markov chain calculation for several periods. The following example gives a geometric interpretation of how powers of a matrix correspond to repeating an operation.

EXAMPLE 6 Powers of a Rotation Matrix

In Example 8 of the previous section, we introduced the rotation matrix $\mathbf{R}_\theta =$
$\begin{bmatrix} \cos\theta & -\sin\theta \\ \sin\theta & \cos\theta \end{bmatrix}$. We noted that multiplying $\mathbf{R}_\theta$ by a 2-vector $\mathbf{v}$ had the effect of rotating the vector $\theta°$ counterclockwise about the origin. Then $\mathbf{R}_\theta^2$ will rotate $\mathbf{v}$ twice by $\theta°$ counterclockwise about the origin. $(\mathbf{R}_\theta^2)\mathbf{v} = \mathbf{R}_\theta(\mathbf{R}_\theta\mathbf{v})$. That is, $\mathbf{R}_\theta^2$ equals $\mathbf{R}_{2\theta}$. We confirm this claim with a little trigonometric calculation:

$$
\begin{aligned}
\mathbf{R}_\theta\mathbf{R}_\theta &= \begin{bmatrix} \cos\theta & -\sin\theta \\ \sin\theta & \cos\theta \end{bmatrix}\begin{bmatrix} \cos\theta & -\sin\theta \\ \sin\theta & \cos\theta \end{bmatrix} \\
&= \begin{bmatrix} (\cos\theta)^2 - (\sin\theta)^2 & -2\sin\theta\cos\theta \\ 2\sin\theta\cos\theta & (\cos\theta)^2 - (\sin\theta)^2 \end{bmatrix} \\
&= \begin{bmatrix} \cos 2\theta & -\sin 2\theta \\ \sin 2\theta & \cos 2\theta \end{bmatrix}.
\end{aligned}
$$

Here we have used the trigonometric double-angle formulas: $\cos 2\theta = (\cos\theta)^2 - (\sin\theta)^2$ and $\sin 2\theta = 2\sin\theta\cos\theta$. Observe that the preceding matrix multiplication could be used to *derive* the double-angle formulas! As an exercise, the reader should try using this approach to derive the formulas for $\sin(\theta + \Psi)$ and $\cos(\theta + \Psi)$.

As a concrete illustration of powers of a rotation matrix, consider

$$
\mathbf{R}_{60°} = \begin{bmatrix} \cos 60° & -\sin 60° \\ \sin 60° & \cos 60° \end{bmatrix} = \begin{bmatrix} \dfrac{1}{2} & \dfrac{-2}{\sqrt{3}} \\ \dfrac{2}{\sqrt{3}} & \dfrac{1}{2} \end{bmatrix}.
$$

Then $(\mathbf{R}_{60°})^3$ will equal $\mathbf{R}_{180°} = \begin{bmatrix} -1 & 0 \\ 0 & -1 \end{bmatrix}$
(check this). ▲

Graphs and Matrices

We now introduce a new application of matrix multiplication, one that involves graphs and symmetric matrices. A *graph* $G = (N, E)$ consists of a set N of nodes and a collection E of edges that are pairs of nodes. There is a natural way to "draw" a graph. We make a point for each node and draw lines linking each pair of nodes forming an edge. For example, the graph G with node set $N = \{a, b, c, d, e\}$ and edge set $E = \{(a, b), (a, c), (a, d), (b, c), (b, d), (d, e),$ $(e, e)\}$ is drawn in Figure 1. An edge may link a node with itself, as at node e in Figure 1. Such an edge is called a *loop*.

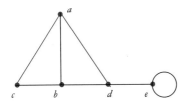

FIGURE 1

This type of "graph" is different from the more familiar "graph of a function," but these graphs model a wide variety of scientific and industrial activities. Organizational charts, electrical circuits, telephone networks, and road maps are examples of graphs. Industry spends billions of dollars every year solving applied graph problems. A flowchart for a computer program is a form of graph.

Many questions about graphs concern paths. A *path* is a sequence of nodes with edges linking consecutive nodes. The length of a path is the number of edges on it. For example, in Figure 1, (a, b, d, e) is a path of length 3 between a and e. A single edge is a path of length 1. We may want to find the shortest path between two nodes, or to determine if a path exists between a given pair of nodes. The latter question arises over and over again in dozens of practical settings—for example, in routing telephone calls or in studying the effect on networks of random disruption, say, due to lightning. We shall now show that matrix multiplication can be used to answer questions about the existence of various types of paths in a graph.

Before we can work with graphs systematically, we need a way to represent them. The *adjacency ma-trix* of a graph is a common way of representing a graph with numbers.

The adjacency matrix $\mathbf{A}(G)$ of the graph G in Figure 1 is

$$
\mathbf{A}(G) = \begin{array}{c} a \\ b \\ c \\ d \\ e \end{array} \begin{bmatrix} 0 & 1 & 1 & 1 & 0 \\ 1 & 0 & 1 & 1 & 0 \\ 1 & 1 & 0 & 0 & 0 \\ 1 & 1 & 0 & 0 & 1 \\ 0 & 0 & 0 & 1 & 1 \end{bmatrix}. \tag{15}
$$

Matrix $\mathbf{A}(G)$ is *symmetric*, that is, entry (i, j) equals entry (j, i) since node i being adjacent to node j is equivalent to node j being adjacent to node i.

We claim that the ith and jth nodes, for $i \neq j$, can be joined by a path of length 2 if and only if entry (i, j) in $\mathbf{A}^2(G)$, the square of the adjacency matrix $\mathbf{A}(G)$, is positive. First let us consider $\mathbf{A}^2(G)$ for the graph in Figure 1. To do this, we must find the scalar product of each row of $\mathbf{A}(G)$ with each column of $\mathbf{A}(G)$. Since $\mathbf{A}(G)$ is symmetric, this is equivalent to finding the scalar product of each row with every other row. Consider the scalar product of a's row and b's row in $\mathbf{A}(G)$:

$$
\begin{array}{c} a \\ b \\ \, \end{array} \begin{bmatrix} 0 & 1 & 1 & 1 & 0 \\ 1 & 0 & 1 & 1 & 0 \\ \cdot & \cdot & \cdot & \cdot & \cdot \end{bmatrix} \tag{16}
$$

Moving a Robot Arm

When a robot arm with many rotational joints changes its joint angles, the tool at the tip (a gripper

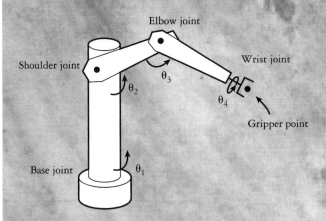

Elbow joint

Shoulder joint

Wrist joint

θ_3

θ_2

θ_4

Gripper point

Base joint

θ_1

in the figure) also moves. If we know the new values, θ_j, of the joint angles after the motion, we ought to be able to predict exactly where the tip will be and the direction in which it will point. But how? One common solution to this problem involves matrix multiplication. For each joint, there is a 4×4 matrix $\mathbf{M}_i$ whose entries are computed by plugging θ_i into easily obtained formulas. Then we multiply to compute

$$\mathbf{E} = \mathbf{M}_1\mathbf{M}_2 \cdots \mathbf{M}_n,$$

where n is the number of joints. For the robot pictured left, there would be four matrices to multiply.

The first three entries of the last column of $\mathbf{E}$ are the third coordinates of the gripper point. The other entries show the gripper's orientation—how it is pointing.

In practice, one often wants to do this mathematics in reverse. The robot determines where it wants its gripper to be and how it is to be oriented in space. Then it needs to find the joint angles that will do this. For this "inverse kinematics problem," we also need the matrix multiplication equation.

Animals and human beings have the same problem to solve for their own arms. If you want to ring a doorbell, you need to bring your finger to the bell and have it pointing the right way. Somehow the brain knows how to move the arm joints to achieve this. No one knows exactly how the brain does this, even though we can solve the problem for robots with the help of the above matrix multiplication equation.

$$a \cdot b = [0, 1, 1, 1, 0] \cdot [1, 0, 1, 1, 0] = 0 \cdot 1 + 1 \cdot 0 + 1 \cdot 1 + 1 \cdot 1 + 0 \cdot 0 = 2.$$

The product of two entries in this scalar product will be 1 if and only if the two entries are both 1. Thus, the value of the scalar product is simply the

number of positions where the two vectors both have a 1.

We now interpret the computation of the scalar product in (16) in terms of adjacencies in the graph. In (16), when rows a and b have a 1 in q's column, this means that a and b are both adjacent to q. For example, from (16) we see that a and b are both adjacent to nodes c and d. Then (a, c, b) and (a, d, b) are paths of length 2 between a and b. In general, when two nodes n_i and n_j are adjacent to a common node n_k, then (n_i, n_k, n_j) will be a path of length 2 between n_i and n_j. Summarizing, we have the following result.

Theorem 1. If $\mathbf{A}(G)$ is the adjacency matrix for graph G, then the (i, j) entry in $\mathbf{A}^2(G)$ equals the number of paths of length 2 between the ith and jth nodes.

E X A M P L E 7 Paths in Graphs Find which pairs of different nodes in the graph in Figure 1 are joined by a path of length 2. By the preceding discussion, we can answer this question by computing $\mathbf{A}^2(G)$:

$$
\mathbf{A}^2(G) = \begin{array}{c} a \\ b \\ c \\ d \\ e \end{array} \begin{array}{c} \begin{array}{ccccc} a & b & c & d & e \end{array} \\ \begin{bmatrix} 3 & 2 & 1 & 1 & 1 \\ 2 & 3 & 1 & 1 & 1 \\ 1 & 1 & 2 & 2 & 0 \\ 1 & 1 & 2 & 3 & 1 \\ 1 & 1 & 0 & 1 & 2 \end{bmatrix} \end{array}. \quad (17)
$$

The positive off-diagonal entries tell us which pairs of different nodes are joined by a path of length 2. The answer from Eq. (17) is all pairs of different nodes except c, e. Note that the matrix in Eq. (17) not only tells us if there is a path of length 2, but additionally how many length-2 paths join each pair of nodes. ▲

Basic Laws of Matrix Algebra

Having introduced matrix addition and multiplication, we now summarize the laws of matrix algebra that apply to these operations. With the exception of commutativity, the laws are basically the same that applied in single-variable (scalar) algebra. It requires some effort to verify them (which is beyond the scope of this chapter). *We assume here that the matrices in each law have the proper size for the matrix operations of addition and multiplication to make sense.*

Basic Laws of Matrix Algebra

Associative Law: Matrix addition and multiplication are associative:

$$
(\mathbf{A} + \mathbf{B}) + \mathbf{C} = \mathbf{A} + (\mathbf{B} + \mathbf{C}) \text{ and } (\mathbf{AB})\mathbf{C} = \mathbf{A}(\mathbf{BC}).
$$

Commutative Law: Matrix addition is commutative: $\mathbf{A} + \mathbf{B} = \mathbf{B} + \mathbf{A}$. Matrix multiplication is not commutative: $\mathbf{AB} \neq \mathbf{BA}$ (except in special cases).

Distributive Law: $\mathbf{A}(\mathbf{B} + \mathbf{C}) = \mathbf{AB} + \mathbf{AC}$ *and* $(\mathbf{B} + \mathbf{C})\mathbf{A} = \mathbf{BA} + \mathbf{CA}$.

Law of Scalar Factoring: $(r\mathbf{A})\mathbf{B} = \mathbf{A}(r\mathbf{B}) = r(\mathbf{AB})$.

Since a vector is just a $1 \times n$ matrix or an $n \times 1$ matrix, these laws also apply to vectors in expressions containing matrix–vector products. For example, *our definition of a matrix–vector product is a special case of matrix multiplication.* In matrix–vector products, the vector was a column vector, which is the same as an $n \times 1$ matrix of 1 column. If $\mathbf{A}$ is an $m \times n$ matrix and $\mathbf{b}$ is an $n \times 1$ matrix (a column n-vector), then as matrix multiplication, $\mathbf{Ab}$ is an $m \times 1$ matrix (a column m-vector).

We close this section by noting that there are several ways to interpret matrix multiplication. First, we can view it as the scalar product of each row of $\mathbf{A}$ with each column of $\mathbf{B}$. Second, we can view it as a sequence of matrix–vector products, that is,

$$AB = A[b_1, b_2, \ldots, b_n] = [Ab_1, Ab_2, \ldots, Ab_n].$$

For $A = \begin{bmatrix} 1 & 2 \\ -1 & 0 \end{bmatrix}$, $B = \begin{bmatrix} 4 & 5 & 1 \\ 6 & 7 & 2 \end{bmatrix}$, with product

$AB = \begin{bmatrix} 16 & 19 & 5 \\ -4 & -5 & -1 \end{bmatrix}$ (the matrices used in the example at the very start of this section), we check that the first column of AB is the matrix–vector product Ab_1:

$$Ab_1 = \begin{bmatrix} 1 & 2 \\ -1 & 0 \end{bmatrix}\begin{bmatrix} 4 \\ 6 \end{bmatrix} = \begin{bmatrix} 1 \cdot 4 + 2 \cdot 6 \\ -1 \cdot 4 + 0 \cdot 6 \end{bmatrix} = \begin{bmatrix} 16 \\ -4 \end{bmatrix}.$$

Finally, we can also view AB as an extension of the vector–matrix product $a_i'B$ to the vector–matrix products of each row of A with B:

$$AB = \begin{bmatrix} a_1' \\ a_2' \\ \vdots \\ a_m' \end{bmatrix} B = \begin{bmatrix} a_1'B \\ a_2'B \\ \vdots \\ a_m'B \end{bmatrix}. \tag{18}$$

For A and B above, we check that the first row of AB is the vector–matrix product $a_1'B$:

$$a_1'B = [1\ 2]\begin{bmatrix} 4 & 5 & 1 \\ 6 & 7 & 2 \end{bmatrix}$$
$$= [1 \cdot 4 + 2 \cdot 6,\ 1 \cdot 5 + 2 \cdot 7,\ 1 \cdot 1 + 2 \cdot 2]$$
$$= [16, 19, 5].$$

Equivalent Definitions of Matrix Multiplication

a. Entry (i, j) of AB is the scalar product $a_i' \cdot b_j$.

b. Column j of AB is the matrix–vector product Ab_j.

c. Row i of AB is the vector–matrix product $a_i' \cdot B$.

Recall from the column approach to the matrix–vector product described in the previous section, that the column vector Ab_j is a linear combination of the columns of A. By the column version of matrix multiplication (item b above), the next theorem follows.

Theorem 2. Each column of the product matrix AB is a linear combination of the columns of A. Similarly, each row of AB is a linear combination of the rows of B. For example, in the above product

$$AB: \begin{bmatrix} 1 & 2 \\ -1 & 0 \end{bmatrix}\begin{bmatrix} 4 & 5 & 1 \\ 6 & 7 & 2 \end{bmatrix} = \begin{bmatrix} 16 & 19 & 5 \\ -4 & -5 & -1 \end{bmatrix},$$

column 2 of AB is the following linear combination of A's columns:

$$5\begin{bmatrix} 1 \\ -1 \end{bmatrix} + 7\begin{bmatrix} 2 \\ 0 \end{bmatrix} = \begin{bmatrix} 19 \\ 5 \end{bmatrix}.$$

Exercises for Section 3.3

1. Indicate which pairs of the following matrices can be multiplied together, and give the size of the resulting product:

   (a) a 3×5 matrix A

   (b) a 3×3 matrix B

   (c) a 2×2 matrix C

   (d) a 4×2 matrix D

   (e) a 4×2 matrix E

2. Let

$$A = \begin{bmatrix} 1 & 2 \\ 3 & 4 \end{bmatrix}, \qquad B = \begin{bmatrix} 3 & 1 \\ 2 & 5 \end{bmatrix},$$

$$C = \begin{bmatrix} 1 & -1 & 0 \\ 2 & 10 & -2 \end{bmatrix}.$$

Compute the following matrix products (if possible):
- (a) **AC**
- (b) **AB**
- (c) **CB**
- (d) **BA**
- (e) **(BA)C**

3. Let

$$A = \begin{bmatrix} 1 & 0 & -1 \\ 2 & -2 & 0 \\ 0 & 1 & -1 \end{bmatrix}, \qquad B = \begin{bmatrix} 1 & 2 & 3 & 4 \\ 2 & 4 & 6 & 8 \\ 3 & 5 & 7 & 9 \end{bmatrix},$$

$$C = \begin{bmatrix} 5 & 4 & 1 \\ 1 & 0 & 2 \\ 3 & 2 & 1 \\ 0 & 1 & 3 \end{bmatrix}.$$

Compute these matrix products (if possible):
- (a) **BA**
- (b) **AB**
- (c) **BC**
- (d) **CB**
- (e) **CA**

4. Compute just one row or column, as requested, in the following matrix products and (**A**, **B**, and **C** are as defined in Exercise 3):
- (a) row 1 in A^2
- (b) column 2 in **BC**
- (c) column 3 in **CA**

5. Show that $AB = BA$ for the matrices $A = \begin{bmatrix} 1 & 2 \\ 3 & 7 \end{bmatrix}$ and $B = \begin{bmatrix} 7 & -2 \\ -3 & 1 \end{bmatrix}$.

6. For **A**, **B**, and **C** in Exercise 3, compute entry (2, 3) in (**AB**)C. Explain first why it is not necessary to multiply out **AB** fully to determine entry (2, 3) in (**AB**)C.

7. Suppose we are given the following matrices involving the costs of fruits at different stores, the amounts of

fruit different types of people want, and the numbers of people of different types in different towns.

	Store A	Store B
Apple	.10	.15
Orange	.15	.20
Pear	.10	.10

	Apple	Orange	Pear
Person A	5	10	3
Person B	4	5	5

	Person A	Person B
Town 1	1000	500
Town 2	2000	1000

Call the first matrix **A**, the second matrix **B**, and the third matrix **C**.
- (a) Compute a matrix that tells how much each person's fruit purchases cost at each store.
- (b) Compute a matrix telling how many of each fruit is purchased in each town.

8. Suppose we are given the following matrices: Matrix **A** gives the amount of time each of three jobs requires of I/O (input/output), of execution time, and of system overhead; matrix **B** gives the charges (per unit of time) of different computer activities under two different charging plans; matrix **C** (actually a vector) tells how many jobs of each type there are; and matrix **D** tells the fraction of the time that each time-charging plan (the columns in matrix **B**) is used each day.

A	I/O	Time Execution	System
Job A	5	20	10
Job B	4	25	8
Job C	10	10	5

B	Time Charges Plan I	Plan II
I/O	2	3
Execution	6	5
System	3	4

	Number of jobs of each type			Fraction of time
C		D		
Job A	$\begin{bmatrix} 4 \\ 5 \\ 3 \end{bmatrix}$	Plan I		$\begin{bmatrix} .3 \\ .7 \end{bmatrix}$
Job B		Plan II		
Job C				

Find matrix products for the following arrays using **A**, **B**, **C**, and **D**. Compute the numbers in these arrays.

(a) Compute the matrix product **AB**.

(b) Find the total cost of each type of job for each charge plan.

(c) Compute the total amount of I/O, execution, and system overhead time for all the jobs (all jobs are summarized in matrix **C**).

(d) Find the total cost of all jobs when run under plan I and under plan II.

(e) Calculate the average cost of one unit of I/O, of execution, and of system overhead time. (Hint: Use matrix **D**).

9. Consider a growth model for the numbers of computers (C) and dogs (D) from year to year:

$$D' = 3C + D,$$

$$C' = 2C + 2D.$$

Let $\mathbf{x} = \begin{bmatrix} C \\ D \end{bmatrix}$ be the initial vector, and let $\mathbf{x}^{(k)}$ denote the vector of computers and dogs after k years. Let **A** be the matrix of coefficients in this system. Write an expression for $\mathbf{x}^{(k)}$ in terms of **A** and **x**.

10. Consider a variation on the Markov chain for weather. Now there is just sunny and cloudy weather. The new transition matrix **A** is

	Sunny	Cloudy
A = Sunny	$\frac{2}{3}$	$\frac{1}{3}$
Cloudy	$\frac{1}{3}$	$\frac{2}{3}$

(a) Compute $\mathbf{A}^2$. What probability does entry $(1, 2)$ in $\mathbf{A}^2$ represent?

(b) Compute $\mathbf{A}^3$. What is the probability if sunny today that it is sunny in three days?

(c) Compute $\mathbf{A}^4$. What vector do the columns of $\mathbf{A}^k$, for $k = 2, 3, 4$, seem to be approaching?

11. Compute $\mathbf{A}^3$ in Example 5 by multiplying $\mathbf{A}^2$ times **A**.

12. Consider the following transition matrices A for weather Markov chains.

	Sunny	Cloudy	Rainy
Sunny	$\frac{1}{2}$	$\frac{1}{4}$	$\frac{1}{4}$
(i) Cloudy	$\frac{1}{4}$	$\frac{1}{2}$	$\frac{1}{4}$
Rainy	$\frac{1}{4}$	$\frac{1}{4}$	$\frac{1}{2}$

	Sunny	Cloudy	Rainy
Sunny	$\frac{1}{2}$	$\frac{1}{2}$	$\frac{1}{4}$
(ii) Cloudy	$\frac{1}{2}$	$\frac{1}{4}$	$\frac{1}{4}$
Rainy	0	$\frac{1}{4}$	$\frac{1}{2}$

(a) Compute $\mathbf{A}^2$. What probability does entry $(3, 2)$ in $\mathbf{A}^2$ represent?

(b) Compute $\mathbf{A}^3$. What is the probability if sunny today that it is rainy in three days?

13. Use the method suggested in Example 6 to derive the trigonometric formulas for $\sin(\theta + \psi)$ and $\cos(\theta + \psi)$.

14. Confirm that $(\mathbf{R}_{60°})^3 = \mathbf{R}_{180°} = \begin{bmatrix} -1 & 0 \\ 0 & -1 \end{bmatrix}$ in Example 6.

15. Draw the graphs with the following adjacency matrices:

(a) $\begin{bmatrix} 0 & 0 & 0 & 1 \\ 0 & 0 & 1 & 0 \\ 0 & 1 & 0 & 0 \\ 1 & 0 & 0 & 0 \end{bmatrix}$

(b) $\begin{bmatrix} 0 & 1 & 1 & 1 \\ 1 & 0 & 0 & 1 \\ 1 & 0 & 0 & 1 \\ 1 & 1 & 1 & 0 \end{bmatrix}$

(c) $\begin{bmatrix} 0 & 0 & 1 & 1 & 0 & 0 \\ 0 & 1 & 1 & 0 & 1 & 1 \\ 1 & 1 & 0 & 0 & 0 & 1 \\ 1 & 0 & 0 & 1 & 1 & 1 \\ 0 & 1 & 0 & 1 & 0 & 0 \\ 0 & 1 & 1 & 1 & 0 & 1 \end{bmatrix}$

16. Write the adjacency matrices for the following graphs:

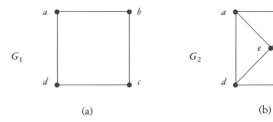

(a)

(b)

(c)

(d)

17. Compute the square of the adjacency matrix for the following graphs from the previous exercise:

(a) G_1 (c) G_3

(b) G_2 (d) G_4

SECTION 3.4 *Gaussian Elimination*

While solving scalar (single-number) linear equations, such as $2x + 3 = 9$, is easy, solving matrix equations of the form $\mathbf{Ax} = \mathbf{b}$ is fairly complicated and requires some careful thought. In this section we develop the procedure of *Gaussian elimination* for solving any system of m linear equations in n variables—to find the unique solution, if one exists, or to find all solutions, or to show that no solution exists. This method of elimination was systematized by Karl Frederic Gauss around 1820 to solve systems of linear equations that arose in astronomical and land-surveying computations (30 years before matrix multiplication was defined). High school students learn Gauss's elimination method to solve two linear equations in two unknowns. Here we present the method in a systematic form applicable to larger systems of equations.

Initially we shall assume that the number of equations equals the number of variables—n linear equations in n variables. Gaussian elimination involves two stages. The first stage transforms the given system of equations into *upper triangular form,* with only 0s below the main diagonal of the matrix, such as

$$2x_1 + x_2 - x_3 = 1,$$
$$x_2 + 4x_3 = 5,$$
$$x_3 = 2,$$

or $\mathbf{Ax} = \mathbf{b}$, where

$$\mathbf{A} = \begin{bmatrix} 2 & 1 & -1 \\ 0 & 1 & 4 \\ 0 & 0 & 1 \end{bmatrix}, \quad \mathbf{x} = \begin{bmatrix} x_1 \\ x_2 \\ x_3 \end{bmatrix}, \quad \mathbf{b} = \begin{bmatrix} 1 \\ 5 \\ 2 \end{bmatrix}. \quad (1)$$

Computed Tomography

Computed Tomography scanners *compute* pictures of a patient's brain from X-rays measured only outside the patient's head.

The accompanying diagram illustrates the role of linear algebra in Computed Tomography. In the triangle, the three discs represent three small internal organs with yet unknown masses $X1$, $X2$, and $X3$, while the straight lines represent X-rays. The small organs lie at yet unknown locations, which makes it impossible to aim each X-ray through only one organ. Here, the X-ray along $L12$ traverses the total mass $X1 + X2$ of the organs with masses $X1$ and $X2$, which absorb some of the intensity of the X-ray. By measuring the intensity absorbed, we can tell the total mass the ray passed through. So $X1 + X2$ is a known quantity $B12$—i.e., $X1 + X2 = B12$. The same applies to the other lines and we get the following system of 3 equations in 3 unknowns.

$$\begin{aligned} X1 + X2 \quad\;\; &= B12, \\ X2 + X3 &= B23, \\ X1 \quad\;\; + X3 &= B31. \end{aligned}$$

Gaussian elimination reveals the masses ($X1$, $X2$, $X3$).

For medical diagnostics, Computed Tomography calculates the density of tissues not at only three loca-

tions, but at thousands of points in each organ, and each X-ray passes through many such points. Hence we get linear systems of thousands of equations with thousands of unknowns, with the unknown XN de-

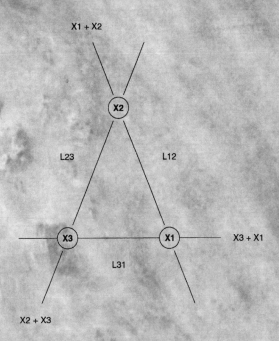

noting the density (or, more accurately, the "coefficient of linear absorption") at the N-th selected point. The solution of such problems involves not only linear algebra, but also mathematics at the graduate level.

The second stage uses *back substitution* to obtain values for the variables. That is, knowing from the third equation of (1) that $x_3 = 2$, we can solve for x_2 in the second equation:

$$x_2 + 4(2) = 5 \quad \text{or} \quad x_2 = -3.$$

And now knowing x_2 and x_3, we can solve for x_1 in the first equation:

$$2x_1 + (-3) - 2 = 1 \quad \text{or} \quad x_1 = 3.$$

The best way to show how Gaussian elimination works is with some examples. Then we summarize the steps in the procedure. While our examples involve two or three equations, the method works for any number of equations. Since the numbers, not the variables, are what matter in the equations, we often use the coefficient matrix $\mathbf{A}$, with an added column for the right-side vector $\mathbf{b}$. This matrix $[\mathbf{A}|\mathbf{b}]$ of coefficients plus the right-side vector is called the *augmented coefficient matrix*. We initially show both the system of equations $\mathbf{Ax} = \mathbf{b}$ and the augmented matrix $[\mathbf{A}|\mathbf{b}]$.

E X A M P L E 1 Gaussian Elimination We start with a simple system of two equations in two unknowns.

$$
\begin{array}{ll}
\text{(a)} & x + y = 4 \\
\text{(b)} & 2x - y = -1
\end{array}
\qquad
\left[\begin{array}{cc|c} 1 & 1 & 4 \\ 2 & -1 & -1 \end{array}\right]
\quad (2)
$$

To eliminate the $2x$-term from (b), we subtract 2 times (a) from (b), and obtain the following new second equation.

$$
\begin{array}{lr}
\text{(b)} & 2x - y = -1 \\
-2\text{(a)} & -(2x + 2y = 8) \\
\hline
\text{(b$'$)} = \text{(b)} - 2\text{(a)} & 0 - 3y = -9
\end{array}
$$

Our new system of equations is

$$
\begin{array}{ll}
\text{(a)} & x + y = 4, \\
\text{(b$'$)} & -3y = -9,
\end{array}
\qquad
\left[\begin{array}{cc|c} 1 & 1 & 4 \\ 0 & -3 & -9 \end{array}\right].
$$

Observe that any solution to (a) and (b) is also a solution to (a) and (b$'$). We can reverse the step creating (b$'$). That is, (b$'$) = (b) − 2(a) implies (b) = (b$'$) + 2(a). Thus, (b) is formed from (b$'$) and a multiple of (a), and so any solution to (a) and (b$'$) is a solution to (a) and (b). But (b$'$) is trivial to solve and gives

$$y = 3.$$

Substituting $y = 3$ in (a), we have

$$x + 3 = 4 \quad \Rightarrow \quad x = 4 - 3 = 1.$$

The reader should verify that $x = 1, y = 3$ satisfy (a) and (b). *When solving a system of linear equations, one can always verify that the answer is correct by substituting the values for x and y in the original equations and checking for equality.* ▲

E X A M P L E 2 Gaussian Elimination for Refinery Problem Recall the refinery problem introduced in Section 3.1 with three refineries whose production levels had to be chosen to meet the demands for heating oil, diesel oil, and gasoline.

Heating oil (a) $16x_1 + 8x_2 + 8x_3 = 9600$
Diesel oil (b) $8x_1 + 20x_2 + 8x_3 = 12800$
Gasoline (c) $4x_1 + 10x_2 + 20x_3 = 16000$

$$\begin{bmatrix} 16 & 8 & 8 & 9600 \\ 8 & 20 & 8 & 12800 \\ 8 & 10 & 20 & 16000 \end{bmatrix} \quad (3)$$

We use multiples of equation (a) to eliminate x_1 from (b) and (c). First, subtract 1/2 times (a) from (b) to eliminate the $8x_1$-term from (b) and obtain a new second equation (b′).

$$\begin{array}{ll} \text{(b)} & 8x_1 + 20x_2 + 8x_3 = 12800 \\ -\frac{1}{2}\text{(a)} & -(8x_1 + 4x_2 + 4x_3 = 4800) \\ \hline \text{(b′)} = \text{(b)} - \frac{1}{2}\text{(a)} & 0 + 16x_2 + 4x_3 = 8000 \end{array}$$

In a similar fashion we subtract 1/4 times (a) from (c) to eliminate the $4x_1$-term from (c) and obtain a new equation (c′).

$$\begin{array}{ll} \text{(c)} & 4x_1 + 10x_2 + 20x_3 = 16000 \\ -\frac{1}{4}\text{(a)} & -(4x_1 + 2x_2 + 2x_3 = 2400) \\ \hline \text{(c′)} = \text{(c)} - \frac{1}{4}\text{(a)} & 8x_2 + 18x_3 = 13600 \end{array}$$

Our new system of equations is as shown below.

$$\begin{array}{ll} \text{(a)} & 16x_1 + 8x_2 + 8x_3 = 9600 \\ \text{(b′)} = \text{(b)} - \frac{1}{2}\text{(a)} & 16x_2 + 4x_3 = 8000 \\ \text{(c′)} = \text{(c)} - \frac{1}{4}\text{(a)} & 8x_2 + 18x_3 = 13600 \end{array}$$

$$\begin{bmatrix} 16 & 8 & 8 & 9600 \\ 0 & 16 & 4 & 8000 \\ 0 & 8 & 18 & 13600 \end{bmatrix} \quad (4)$$

Next we use equation (b′) to eliminate the $8x_2$-term from (c′) and obtain a new third equation (c″).

$$\begin{array}{ll} \text{(c′)} & 8x_2 + 18x_3 = 13600 \\ -\frac{1}{2}\text{(b′)} & -(8x_2 + 2x_3 = 4000) \\ \hline \text{(c″)} = \text{(c′)} - \frac{1}{2}\text{(b′)} & 16x_3 = 9600 \end{array}$$

Our final system of equations is as follows.

$$\begin{array}{ll} \text{(a)} & 16x_1 + 8x_2 + 8x_3 = 9600 \\ \text{(b′)} & 16x_2 + 4x_3 = 8000 \\ \text{(c″)} = \text{(c′)} - \frac{1}{2}\text{(b′)} & 16x_3 = 9600 \end{array}$$

$$\begin{bmatrix} 16 & 8 & 8 & 9600 \\ 0 & 16 & 4 & 8000 \\ 0 & 0 & 16 & 9600 \end{bmatrix} \quad (5)$$

Any solution to the original system in (3) is a solution to the new system in (5). Furthermore, by reversing the steps in going from (3) to (5) (so that (3) is formed from linear combinations of the equations in (5)), we see that any solution to (5) is a solution to (3).

System (5) is in upper triangular form, and we can solve using back substitution. From (c″), we have

$$x_3 = \frac{9600}{16} = 600.$$

Substituting this value for x_3 in (b′), we have

$$16x_2 + 4(600) = 8000 \Rightarrow 16x_2 = 8000 - 2400$$
$$\Rightarrow x_2 = \frac{5600}{16} = 350.$$

Next, substituting the values for x_3 and x_2 in (a), we have

$16x_1 + 2(350) + 2(600) = 9600$ or

$$x_1 = \frac{(9600 - 700 - 1200)}{16} = 125.$$

So the vector of the production levels of the three refineries is [125, 350, 600]. ▲

EXAMPLE 3   System of Equations Without Unique Solution   Suppose we change the third equation in (3) in the previous example and use the following values.

(a) $16x_1 + 8x_2 + 8x_3 = 9600$
(b) $8x_1 + 20x_2 + 8x_3 = 12800$
(c) $12x_1 + 14x_2 + 8x_3 = 11200$

$$\begin{bmatrix} 16 & 8 & 8 & | & 9000 \\ 8 & 20 & 8 & | & 12800 \\ 12 & 14 & 8 & | & 11200 \end{bmatrix} \quad (6)$$

After eliminating x_1 from (b) and (c), we have the following system.

(a) $16x_1 + 8x_2 + 8x_3 = 9600$
(b') = (b) − $\frac{1}{2}$ (a) $16x_2 + 4x_3 = 8000$
(c') = (c) − $\frac{3}{4}$ (a) $8x_2 + 2x_3 = 4000$

$$\begin{bmatrix} 16 & 4 & 4 & | & 9600 \\ 0 & 16 & 4 & | & 8000 \\ 0 & 8 & 2 & | & 4000 \end{bmatrix} \quad (7)$$

Next we subtract 1/2 (b') from (c') to eliminate the $8x_2$-term, but this eliminates all of (c').

(a) $16x_1 + 8x_2 + 8x_3 = 9600$
(b') $16x_2 + 4x_3 = 8000$
(c'') = (c') − $\frac{1}{2}$ (b') $0 = 0$

$$\begin{bmatrix} 16 & 8 & 8 & | & 9600 \\ 0 & 16 & 4 & | & 8000 \\ 0 & 0 & 0 & | & 0 \end{bmatrix} \quad (8)$$

That is, equation (c') is just 1/2 (b'). We now have only two equations in three unknowns. This system has an infinite number of solutions, since we can make up any value for x_3 and then use back substitution to determine x_2 and x_1. ▲

EXAMPLE 4   System of Equations With No Solution   Let us reconsider (6) with the third equation having 16000 on the right-hand side.

(a) $16x_1 + 8x_2 + 8x_3 = 9600$
(b) $8x_1 + 20x_2 + 8x_3 = 12800$
(c) $12x_1 + 14x_2 + 8x_3 = 16000$

$$\begin{bmatrix} 16 & 8 & 8 & | & 9600 \\ 8 & 20 & 8 & | & 12800 \\ 12 & 14 & 8 & | & 16000 \end{bmatrix} \quad (9)$$

Eliminating x_1 as before, we get the following system.

(a) $16x_1 + 8x_2 + 8x_3 = 9600$
(b') $16x_2 + 4x_3 = 8000$
(c') $8x_2 + 2x_3 = 8800$

$$\begin{bmatrix} 16 & 8 & 8 & | & 9600 \\ 0 & 16 & 4 & | & 8000 \\ 0 & 8 & 2 & | & 8800 \end{bmatrix} \quad (10)$$

When we use (b') to eliminate the $8x_2$-term in (c'), we obtain the following equations.

$$
\begin{array}{ll}
\text{(a)} & 16x_1 + 8x_2 + 8x_3 = 9600 \\
\text{(b}'\text{)} & \phantom{16x_1 + {}} 16x_2 + 4x_3 = 8000 \\
\text{(c}''\text{)} = \text{(c}'\text{)} - \tfrac{1}{2}\text{(b}'\text{)} & \phantom{16x_1 + 8x_2 + {}} 0 = 4800
\end{array}
$$

$$
\left[
\begin{array}{ccc|c}
16 & 8 & 8 & 9600 \\
0 & 16 & 4 & 8000 \\
0 & 0 & 0 & 4800
\end{array}
\right] \quad (11)
$$

In (11), (c$''$) is an inconsistent equation. Therefore, this system has no solution. ▲

We now summarize the steps of Gaussian elimination.

Method of Gaussian Elimination for Solving n Linear Equations in n Variables

A. Elimination Stage

1. Subtract multiples of the first equation from other equations to eliminate x_1 from other equations.

2. Subtract multiples of the second equation from the following equations to eliminate x_2 from those equations.

3. Continue this elimination process with x_3 in the third equation, x_4 in the fourth equation, and so on, until, for all i, the first term in the ith equation involves x_i (all x_j's, $j < i$, have been eliminated).

B. Back Substitution

Use back substitution on the upper triangular system of equations resulting from the elimination stage to solve for each variable.

There are two complications that may occur:

Complications with Gaussian Elimination

1. If the variable x_i does not appear in the ith equation (that is, $a_{ii} = 0$), then we cannot use the ith equation to eliminate x_i from the remaining equations. In this case, we interchange the ith row with a row that does have a nonzero coefficient of x_i and then continue with the algorithm. We illustrate this case in Example 6 ahead.

2. If all the variables are eliminated from one (or more) equations, as happened in Examples 3 and 4, then either some variables can be assigned any value—as illustrated in Example 3—or no solution exists—as illustrated in Example 4. Theorem 2 ahead discusses the problem of a zeroed-out equation more fully.

A Little Theory About Gaussian Elimination

Underlying the method of Gaussian elimination is the assumption that the steps performed in the elimination process produce a reduced system of equations with the same solutions as the original system of equations. Perhaps the process of elimination could introduce or eliminate a solution. We need a little mathematical theory to resolve this potential problem. The first stage of Gaussian elimination transforms a coefficient matrix into upper triangular form using the following three row operations.

Elementary Row Operations

M. Multiply or divide a row of a matrix (or an equation) by a nonzero number.

S. Subtract a multiple of one row (or equation) from another row (equation).

I. Interchange two rows (equations).

Theorem 1. If $Ax = b$ is a system of equations to which elementary row operations M, S, I are applied to obtain a new system of equations $A'x = b'$, then x^* is a solution to $Ax = b$ if and only if it is also a solution to $A'x = b'$.

Explanation of Theorem. Applying operation M to an equation will not change the set of x-vectors that satisfy the equation. For example, an $[x, y, z]$-vector satisfies $2x + y - z = 3$ if and only if it satisfies this equation multiplied by 2, $4x + 2y - 2z = 6$ (details are left as an exercise). If the vectors satisfying an individual equation are unchanged, then the solution(s) of the system of equations is (are) unchanged.

Next consider row operation S. Suppose a vector $x^* = [x^*, y^*, z^*]$ satisfies two equations, such as $x + y - z = 3$ and $3x + 4y + 5z = 10$. Then x^* will also satisfy any equation obtained by subtracting a multiple of the first equation from the second equation, such as

$$
\begin{array}{r}
3x + 4y + 5z = 10 \\
-2\{x + y - z = 3\} \\
\hline
x + 2y + 7z = 4
\end{array}
$$

Again, details are left as an exercise. Row operation I, interchanging the order of two equations, clearly will not change the solution(s).

Thus, performing one of the three elementary row operations does not change the solution(s) x^* to a system of linear equations. It follows that x^* will continue to be a solution after any number of elementary row operations. So if x^* is a solution to $Ax = b$, then it is also a solution of the reduced system of equations $A'x = b'$ obtained by Gaussian elimination.

Observe that reversing (undoing) an elementary row operation is also an elementary row operation.

If we multiply an equation by the nonzero number r, then we reverse this action by dividing by r (or by multiplying by $1/r$). If we had subtracted r times the first row from the second row, then we reverse by subtracting $-r$ times the first row from the second row. Interchanging the two rows twice puts them in original order. Then if $A'x = b'$ is obtained from $Ax = b$ by elementary row operations, $Ax = b$ can be obtained from $A'x = b'$ by elementary row operations. So if $x°$ is a solution of the reduced system $Ax' = b$, obtained by Gaussian elimination through elementary row operations, then $x°$, is also a solution to the original system $Ax = b$.

▲ ▲ ▲

The examples we have worked show that there are three possible outcomes when a system of n linear equations in n variables is solved: either a unique solution, as in Example 2; or an infinite number of solutions, as in Example 3; or no solution, as in Example 4. While we cannot easily tell in advance which of these three outcomes will occur, each outcome is associated with a different final form of upper triangular system that results from Gaussian elimination. The following theorem describes this association. In this theorem we assume no column of matrix A is all 0s. The proof is technical and is therefore omitted here.

Theorem 2. One of the following three alternatives must result when Gaussian elimination (or Gauss–Jordan elimination is applied to $Ax = b$, a system of n linear equations in n variables.

a. *Unique solution:* The coefficient matrix A is reduced by Gaussian elimination to upper triangular form, and then a unique solution is obtained by back substitution. Or, equivalently, Gauss–Jordan elimination pivots along the main diagonal (with row interchanges, if necessary, to remove 0s along the main diagonal) to reduce the coefficient matrix to an identity matrix.

b. *Infinite number of solutions:* In Gaussian elimination, the last rows in the augmented coefficient matrix $[\mathbf{A}|\mathbf{b}]$ become all 0s, as in Example 3. If the last k rows are all 0s, then the last k variables (and possibly some other variables) can be given any values, and then the remaining variables can be determined by back substitution.

c. *No solution:* In Gaussian elimination, some row in the coefficient matrix becomes all 0s while the associated right-side entry is nonzero, as in Example 4.

Gauss–Jordan Elimination

We next present a variation on Gaussian elimination, known as *Gauss–Jordan elimination,* that is a little slower than Gaussian elimination but eliminates the need for back substitution. Gauss–Jordan elimination uses the equation i to eliminate x_i from all other equations *before,* as well as after, equation i (Gaussian elimination only eliminates x_i in equations after equation i). It also divides equation i by the coefficient of x_i so that the coefficient of x_i in the new equation i is 1. Gauss–Jordan elimination will be used in the next section to find inverses of matrices.

We use the term *pivot on entry* a_{ij} (the coefficient of x_j in equation i) to denote the process of using equation i to eliminate x_j from all other equations (and making 1 be the new coefficient of x_j in equation i).

E X A M P L E 5 Gauss–Jordan Elimination
Let us rework Example 2 using Gauss–Jordan elimination. Now we show only the augmented

coefficient matrix $[\mathbf{A}|\mathbf{B}]$, without the individual equations.

$$
\begin{array}{c}
\text{(a)}\\
\text{(b)}\\
\text{(c)}
\end{array}
\left[
\begin{array}{ccc|c}
16 & 8 & 8 & 9600\\
8 & 20 & 8 & 12800\\
4 & 10 & 20 & 16000
\end{array}
\right]
\tag{12}
$$

First we pivot on entry (1, 1).

$$
\begin{array}{l}
(a') = \frac{1}{16}(a)\\
(b') = (b) - \frac{1}{2}(a)\\
(c') = (c) - \frac{1}{4}(a)
\end{array}
\left[
\begin{array}{ccc|c}
1 & \frac{1}{2} & \frac{1}{2} & 600\\
0 & 16 & 4 & 8000\\
0 & 8 & 18 & 13600
\end{array}
\right]
\tag{13}
$$

Next we pivot on entry (2, 2).

$$
\begin{array}{l}
(a'') = (a') - \frac{1}{2}(b'')\\
(b'') = \frac{1}{16}(b')\\
(c'') = (c') - 8(b'')
\end{array}
\left[
\begin{array}{ccc|c}
1 & 0 & \frac{3}{8} & 350\\
0 & 1 & \frac{1}{4} & 500\\
0 & 0 & 16 & 9600
\end{array}
\right]
\tag{14}
$$

Finally, we pivot on entry (3, 3).

$$
\begin{array}{l}
(a''') = (a') - \frac{3}{8}(c''')\\
(b''') = (b'') - \frac{1}{4}(c''')\\
(c''') = \frac{1}{16}(c'')
\end{array}
\left[
\begin{array}{ccc|c}
1 & 0 & 0 & 125\\
0 & 1 & 0 & 350\\
0 & 0 & 1 & 600
\end{array}
\right]
\tag{15}
$$

Now the upper triangular system of equations corresponding to (15) yields a solution directly, without back substitution:

$$
\begin{aligned}
x_1 &= 125,\\
x_2 &= 350,\\
x_3 &= 600. \ \blacktriangle
\end{aligned}
\tag{16}
$$

Note that in (15) the coefficient matrix has been reduced by Gauss–Jordan elimination to the identity matrix $\mathbf{I}$ (identity matrices were introduced in Ex-

On the Road

A trucking company can enhance its service and its earnings if it can quickly make changes in routing its trucks, to accommodate new pickups and deliveries and other changes in plans. Day & Night Transportation Services, based in Noblesville, Indiana, tried putting cellular phones in its trucks; but the bills were high, service was unavailable in some areas, and there was no way to verify a truck's location.

The company found a solution in equipping the trucks to access the Global Positioning System (GPS),

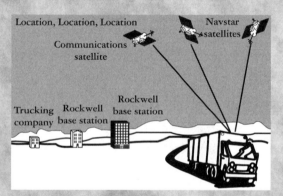

A truck communicates with three GPS satellites and automatically communicates its location to its home base by means of a separate communications satellite.

a collection of 24 high-altitude satellites. The truck receives signals from three of the satellites, and software in the receiver uses techniques of linear algebra to determine the truck's location. The location is determined to within a few feet and automatically relayed to the dispatching office.

The geometry of the intersections of spheres tells why three satellites are needed. When the truck con-

tacts a satellite, the receiver determines, from the time for a signal to go and return, what the distance is from the truck to the satellite—say 14,000 miles. From the point of view of the satellite, the truck is thus known to be located somewhere on the surface of a sphere centered at the satellite with a radius of 14,000 miles. If the truck is also 17,000 miles from a second satellite, it is on the surface of a sphere centered at that satellite with a radius of 17,000 miles. The two spheres intersect in a circle. The truck is also on the surface of a sphere centered at the third satellite, with a radius of say 16,000 miles. This third sphere cuts the circle at precisely two points: one on the surface of the earth and the other thousands of miles above it. It's not hard to tell which of the two is the location of the truck!

The details of the linear algebra are as follows. Let the truck be located at (x, y, z), let the satellites be located at (a_1, b_1, c_1), (a_2, b_2, c_2), and (a_3, b_3, c_3), and let the distances from the truck to the satellites be r_1, r_2, and r_3. From the distance formula in three dimensions, we have

$$(x - a_1)^2 + (y - b_1)^2 + (z - c_1)^2 = r_1^2$$
$$(x - a_2)^2 + (y - b_2)^2 + (z - c_2)^2 = r_2^2$$
$$(x - a_3)^2 + (y - b_3)^2 + (z - c_3)^2 = r_3^2$$

Oh-oh! These equations aren't linear in x, y, and z. However—subtract the second equation from the first, getting

$$(2a_2 - 2a_1)x + (2b_2 - 2b_1)y + (2c_2 - 2c_1)z = A,$$

where $A = r_1^2 - r_2^2 + a_2^2 - a_1^2 + b_2^2 - b_1^2 + c_2^2 - c_1^2$ does not involve x, y, or z. This equation is linear in x, y, and z. Subtracting the third equation from the first, we get another linear equation:

$$(2a_3 - 2a_1)x + (2b_3 - 2b_1)y + (2c_3 - 2c_1)z = B.$$

We may rewrite the two equations as

$$(2a_2 - 2a_1)x + (2b_2 - 2b_1)y = A - (2c_2 - 2c_1)z$$

$$(2a_3 - 2a_1)x + (2b_3 - 2b_1)y = B - (2c_3 - 2c_1)z$$

and solve for x and y in terms of z by means of Gaussian elimination. Substituting these expressions into any one of the original distance equations gives a quadratic equation for z. We find the roots and substitute those into the expressions for x and y. Each root gives a point: One is the location of the truck, and the other is far off the earth.

The actual calculations carried out by the GPS software are modified for better numerical precision, using a technique called a Kalman filter. The values of x, y, and z that are found provide the least-squares fit to the distance data.

—You can find further details about GPS and its use in trucking in "On-Road, On-Time, and On-Line," by Christine White, Byte (April 1995) 60–66.

ample 2 in Section 3.3). A pivot in Gauss–Jordan elimination can be summarized with the following matrices, where a is the pivot entry, b is another entry in the pivot row, c is another entry in the pivot column, and d is an entry in the same column as b and the same row as c. The matrix on the left shows the situation before pivoting, and the matrix on the right shows the situation after pivoting.

$$\begin{bmatrix} a & \cdots & b \\ \vdots & & \\ c & & d \end{bmatrix} \rightarrow \begin{bmatrix} 1 & \cdots & \dfrac{b}{a} \\ \vdots & & \\ 0 & & d - \dfrac{bc}{a} \end{bmatrix}$$

As noted when we discussed complications with Gaussian elimination earlier, if the variable x_i does not appear in the ith equation, then we cannot use the ith equation to eliminate x_i from the remaining equations. In this case, we interchange the ith row with a row that does have a nonzero coefficient of x_i (this is elementary row operation I) and then continue with the algorithm. We illustrate the idea with Gauss–Jordan elimination, but it also applies to Gaussian elimination.

EXAMPLE 6 Solution with Interchange of Rows Let us repeat the previous example but with the numbers in equations (b) and (c) changed:

$$\begin{array}{c} \text{(a)} \\ \text{(b)} \\ \text{(c)} \end{array} \quad \begin{bmatrix} 16 & 8 & 8 & 9600 \\ 8 & 4 & 8 & 12800 \\ 4 & 10 & 10 & 8800 \end{bmatrix} \quad (17)$$

As before, we want to make entry $(1, 1)$ equal to 1 and the rest of the first column 0s.

$$(a') = \tfrac{1}{16}(a)$$
$$(b') = (b) - \tfrac{1}{2}(a)$$
$$(c') = (c) - \tfrac{1}{4}(a)$$
$$\begin{bmatrix} 1 & .5 & .5 & | & 600 \\ 0 & 0 & 4 & | & 8000 \\ 0 & 8 & 8 & | & 6400 \end{bmatrix} \quad (18)$$

Since entry $(2, 2)$ is 0, we cannot pivot on it. Now we need elementary row operation I, interchange of two rows. We interchange the second and third rows, (b') and (c').

$$(a') \qquad \begin{bmatrix} 1 & .5 & .5 & | & 600 \\ 0 & 8 & 8 & | & 6400 \\ 0 & 0 & 4 & | & 8000 \end{bmatrix} \quad (19)$$

We can now pivot on entry $(2, 2)$ in this new matrix (which was entry $(3, 2)$ in (18)).

$$(a'') = (a') - \tfrac{1}{2}(c'')$$
$$(c'') = \tfrac{1}{8}(c')$$
$$(b'') = (b')$$
$$\begin{bmatrix} 1 & 0 & 0 & | & 200 \\ 0 & 1 & 1 & | & 800 \\ 0 & 0 & 4 & | & 8000 \end{bmatrix} \quad (20)$$

Finally we pivot on entry $(3, 3)$ in (20) (which was originally entry $(2, 3)$).

$$(a''') = (a'')$$
$$(c''') = (c'') - (b''')$$
$$(b''') = \tfrac{1}{4}(b'')$$
$$\begin{bmatrix} 1 & 0 & 0 & | & 200 \\ 0 & 1 & 0 & | & -1200 \\ 0 & 0 & 1 & | & 2000 \end{bmatrix} \quad (21)$$

We read the solution from the matrix: $x_1 = 200$; $x_2 = -1200$; $x_3 = 2000$. ▲

Next we consider an example where the number of equations is less than the number of variables.

EXAMPLE 7 Two-Product Refinery

Problem Suppose in our refinery model that the production of gasoline is not important (there is

an excess supply in storage tanks). We are concerned with the demands for heating oil and diesel oil only. The system of equations and augmented coefficient matrix from Example 2 become

Heating oil (a) $\quad 16x_1 + 8x_2 + 8x_3 = 9600,$
Diesel oil (b) $\quad 8x_1 + 20x_2 + 8x_3 = 12800,$

$$\begin{bmatrix} 16 & 8 & 8 & | & 9600 \\ 8 & 20 & 8 & | & 12800 \end{bmatrix}. \quad (22)$$

Using Gauss–Jordan elimination, we eliminate x_1 from row 2 (and divide the first row by 16).

$$(a') = \tfrac{1}{16}(a)$$
$$(b') = (b) - \tfrac{1}{2}(a)$$

$$1x_1 + \tfrac{1}{2}x_2 + \tfrac{1}{2}x_3 = 600$$
$$16x_2 + 4x_3 = 8000$$

$$\begin{bmatrix} 1 & \tfrac{1}{2} & \tfrac{1}{2} & | & 600 \\ 0 & 16 & 4 & | & 8000 \end{bmatrix} \quad (23)$$

Next we pivot on entry $(2, 2)$.

$$(a'') = (a') - \tfrac{1}{2}(b'')$$
$$(b'') = \tfrac{1}{16}(b')$$

$$1x_1 \qquad + \tfrac{3}{8}x_3 = 350$$
$$1x_2 + \tfrac{1}{4}x_3 = 500$$

$$\begin{bmatrix} 1 & 0 & \tfrac{3}{8} & | & 350 \\ 0 & 1 & \tfrac{1}{4} & | & 500 \end{bmatrix} \quad (24)$$

Bringing x_3 over to the right side of the equations in (24), we have the general solution for (22) in terms of x_3:

$$x_1 = 350 - \tfrac{3}{8}x_3, \qquad \begin{bmatrix} x_1 \\ x_2 \end{bmatrix} = \begin{bmatrix} 350 \\ 500 \end{bmatrix} - x_3 \begin{bmatrix} \tfrac{3}{8} \\ \tfrac{1}{4} \end{bmatrix}. \; ▲ \quad (25)$$
$$x_2 = 500 - \tfrac{1}{4}x_3,$$

We see that when there are more variables than equations, one or more of the variables will not be specified by Gaussian elimination and we are free to assign it any value we want. This is similar to case b of Theorem 2, illustrated in Example 3, when an equation is zeroed out during elimination (the zeroing out means that effectively there are then fewer equations than variables).

Gaussian Elimination and Stable Probabilities in Markov Chains

We conclude this section by using the equation-solving tools we have developed to investigate the structure of Markov chains.

E X A M P L E 8 Steady State of the Weather Markov Chain In the weather Markov chain introduced in Section 3.1 with transition matrix,

$$
\begin{array}{c}
\quad\quad\quad\quad\quad \text{Today}\\
\begin{array}{cccc}
\text{Tomorrow} & \text{Sunny} & \text{Cloudy} & \text{Rainy}\\
\end{array}\\
\mathbf{A} = \begin{array}{c}\text{Sunny}\\\text{Cloudy}\\\text{Rainy}\end{array}
\begin{bmatrix}
\frac{3}{4} & \frac{1}{2} & \frac{1}{4}\\
\frac{1}{8} & \frac{1}{4} & \frac{1}{2}\\
\frac{1}{8} & \frac{1}{4} & \frac{1}{4}
\end{bmatrix},
\end{array}
$$

it was noted that over many periods, the probability distribution stabilized at 14/23 sunny, 5/23 cloudy, 4/23 rainy (see (9) in Section 3.1). We confirm that if today's distribution $\mathbf{p}$ is $\begin{bmatrix}\frac{14}{23}\\\frac{5}{23}\\\frac{4}{23}\end{bmatrix}$,

then tomorrow's distribution $\mathbf{p}'\ (= \mathbf{Ap})$ is also $\begin{bmatrix}\frac{14}{23}\\\frac{5}{23}\\\frac{4}{23}\end{bmatrix}$:

$$
\mathbf{p}' = \mathbf{Ap} = \begin{bmatrix}
\frac{3}{4} & \frac{1}{2} & \frac{1}{4}\\
\frac{1}{8} & \frac{1}{4} & \frac{1}{2}\\
\frac{1}{8} & \frac{1}{4} & \frac{1}{4}
\end{bmatrix}\begin{bmatrix}\frac{14}{23}\\\frac{5}{23}\\\frac{4}{23}\end{bmatrix}
$$

$$
= \begin{bmatrix}
\frac{3}{4}\cdot\frac{14}{23} + \frac{1}{2}\cdot\frac{5}{23} + \frac{1}{4}\cdot\frac{4}{23}\\
\frac{1}{8}\cdot\frac{14}{23} + \frac{1}{4}\cdot\frac{5}{23} + \frac{1}{2}\cdot\frac{4}{23}\\
\frac{1}{8}\cdot\frac{14}{23} + \frac{1}{4}\cdot\frac{5}{23} + \frac{1}{4}\cdot\frac{4}{23}
\end{bmatrix} = \begin{bmatrix}\frac{14}{23}\\\frac{5}{23}\\\frac{4}{23}\end{bmatrix}. \quad (26)
$$

We call this special vector $\mathbf{p}^* = \begin{bmatrix}\frac{14}{23}\\\frac{5}{23}\\\frac{4}{23}\end{bmatrix}$ the

stable distribution of the Markov chain. We now show how to solve directly for the stable distribution $\mathbf{p}^*$. As checked in (26), $\mathbf{p}^*$ satisfies the matrix equation, $\mathbf{p}^* = \mathbf{Ap}^*$:

$$
\begin{aligned}
p_1^* &= \tfrac{3}{4}p_1^* + \tfrac{1}{2}p_2^* + \tfrac{1}{4}p_3^*,\\
p_2^* &= \tfrac{1}{8}p_1^* + \tfrac{1}{4}p_2^* + \tfrac{1}{2}p_3^*,\\
p_3^* &= \tfrac{1}{8}p_1^* + \tfrac{1}{4}p_2^* + \tfrac{1}{4}p_3^*.
\end{aligned} \quad (27)
$$

Collecting the p^*'s on the left, we get

$$
\begin{aligned}
\tfrac{1}{4}p_1^* - \tfrac{1}{2}p_2^* - \tfrac{1}{4}p_3^* &= 0,\\
-\tfrac{1}{8}p_1^* + \tfrac{3}{4}p_2^* - \tfrac{1}{2}p_3^* &= 0,\\
-\tfrac{1}{8}p_1^* - \tfrac{1}{4}p_2^* + \tfrac{3}{4}p_3^* &= 0.
\end{aligned} \quad (28)
$$

Solving by Gaussian elimination, we obtain (using the augmented coefficient matrix)

$$
\begin{bmatrix}
\frac{1}{4} & -\frac{1}{2} & -\frac{1}{4} & 0\\
-\frac{1}{8} & \frac{3}{4} & -\frac{1}{2} & 0\\
-\frac{1}{8} & -\frac{1}{4} & \frac{3}{4} & 0
\end{bmatrix} \Rightarrow
\begin{bmatrix}
\frac{1}{4} & -\frac{1}{2} & -\frac{1}{4} & 0\\
0 & \frac{1}{2} & -\frac{5}{8} & 0\\
0 & -\frac{1}{2} & \frac{5}{8} & 0
\end{bmatrix}
$$

$$
\Rightarrow
\begin{bmatrix}
\frac{1}{4} & -\frac{1}{2} & -\frac{1}{4} & 0\\
0 & \frac{1}{2} & -\frac{5}{8} & 0\\
0 & 0 & 0 & 0
\end{bmatrix}. \quad (29)
$$

Rewriting the reduced system of equations with 1s on the main diagonal, we have

$$\begin{bmatrix} 1 & -2 & -1 & | & 0 \\ 0 & 1 & -\frac{5}{4} & | & 0 \\ 0 & 0 & 0 & | & 0 \end{bmatrix} \quad \text{or} \qquad (30a)$$

$$\begin{aligned} p_1^* - 2p_2^* - p_3^* &= 0, \\ p_2^* - \tfrac{5}{4} p_3^* &= 0. \end{aligned} \qquad (30b)$$

Solving (30b), we get

$$p_2^* = \tfrac{5}{4} p_3^* \quad \text{and}$$
$$p_1^* = 2p_2^* + p_3^* = 2(\tfrac{5}{4} p_3^*) + p_3^* = \tfrac{7}{2} p_3^*. \quad (31)$$

Where is our unique vector of stable probabilities? In (31) we obtained an infinite number of solutions to (30), one for each value of p_3^*. This difficulty has occurred because we omitted one important fact, namely that these probabilities must sum to 1:

$$p_1^* + p_2^* + p_3^* = 1. \qquad (32)$$

Using (31), we express p_1^* and p_2^* in terms of p_3^* to obtain

$$1 = p_1^* + p_2^* + p_3^* = \tfrac{7}{2} p_3^* + \tfrac{5}{4} p_3^* + p_3^* = \tfrac{23}{4} p_3^*.$$

Hence, $p_3^* = 4/23$, and then from (31), $p_2^* = 5/23$ and $p_1^* = 14/23$. These are the probabilities of the stable distribution of the weather Markov chain given above. ▲

We could have substituted Eq. (32) for the last equation in (28), since that last equation is zeroed out during Gaussian elimination. Or we could have added (32) to the three equations in (28) to have a system of four equations in three unknowns. Either of these two new systems could be solved by elimination to determine the stable probabilities.

Systems of equations $\mathbf{Ax} = \mathbf{0}$ with zero right sides, such as (28), arise frequently in linear algebra. They have a special name, as our next definition describes.

DEFINITION

A system of linear equations $\mathbf{Ax} = \mathbf{0}$ with zero right sides is called *homogeneous*.

When solving a homogeneous system, we usually are interested in a nonzero solution, as occurred with the system in (30). This will also be the case in the homogeneous system of equations that arises in Section 3.7 to find eigenvectors of a matrix. Note that $\mathbf{x} = \mathbf{0}$ is always a solution to any homogeneous system $\mathbf{Ax} = \mathbf{0}$. Thus, we need multiple solutions to get a nonzero solution. If $\mathbf{A}$ is an $n \times n$ matrix, then by Theorem 2 there will be multiple solutions if and only if at least one row of $\mathbf{A}$ is zeroed out during elimination (note that for homogeneous systems, case c in Theorem 2 cannot happen).

1. In each of the following sets of three equations, show that the third equation equals the second equation minus some multiple of the first equation: (c) = (b) − r(a) for some r.

(a) (i) $x + 2y = 4$
   (ii) $3x + y = 9$
   (iii) $x - 3y = 1$

(b) (i) $2x + y - 2z = -5$
   (ii) $3x - y + z = 8$
   (iii) $6x + .5y - 2z = .5$

2. Solve the following systems of equations using Gaussian elimination.

(a) $2x_1 - 3x_2 + 2x_3 = 0$
    $x_1 - x_2 + x_3 = 7$
    $-x_1 + 5x_2 + 4x_3 = 4$

(b) $-x_1 - x_2 + x_3 = 2$
    $2x_1 + 2x_2 - 4x_3 = -4$
    $x_1 - 2x_2 + 3x_3 = 5$

(c) $x_1 + x_2 + 4x_3 = 4$
    $2x_1 + x_2 + 3x_3 = 5$
    $5x_1 + 2x_2 + 5x_3 = 11$

(d) $2x_1 - 3x_2 - x_3 = 2$
    $3x_1 - 5x_2 - 2x_3 = -1$
    $9x_1 + 6x_2 + 4x_3 = 1$

3. Solve the systems of equations in Exercise 2 using Gauss–Jordan elimination.

4. Use Gaussian elimination to solve the following variations on the refinery problem in Example 2. Sometimes the variation will have no solution, sometimes it will have multiple solutions (express such an infinite family of solutions in terms of x_3), and sometimes the solution will involve negative numbers (a real-world impossibility).

(a) $6x_1 + 2x_2 + 2x_3 = 500$
   $3x_1 + 6x_2 + 3x_3 = 300$
   $3x_1 + 2x_2 + 6x_3 = 1000$

(b) $8x_1 + 4x_2 + 3x_3 = 500$
   $4x_1 + 8x_2 + 5x_3 = 500$
   $12x_2 + 6x_3 = 500$

(c) $6x_1 + 5x_2 + 6x_3 = 500$
   $10x_1 + 10x_2 = 850$
   $2x_1 + 12x_3 = 1000$

5. Solve the following systems of equations using Gaussian elimination.

(a) $x_1 + 3x_2 + 2x_3 - x_4 = 7$
   $x_1 + x_2 + x_3 + x_4 = 3$
   $2x_1 - 2x_2 + x_3 - x_4 = -5$
   $x_1 - 3x_2 - x_3 + 2x_4 = -4$

(b) $3x_1 + 2x_2 + x_3 = 3$
   $x_1 + x_2 - x_4 = 2$
   $2x_1 + x_2 - x_3 + x_4 = -3$
   $x_1 + x_2 + x_3 + x_4 = 0$

6. Determine whether each of the following systems of equations has a unique solution, has multiple solutions, or is inconsistent.

(a) $x_1 - x_2 + x_3 = 5$
   $x_1 + 3x_2 + 6x_3 = 9$
   $-x_1 + 5x_2 + 4x_3 = 10$

(b) $x_1 + x_2 + 2x_3 = 0$
   $2x_1 + x_2 - 3x_3 = 0$
   $-x_1 + 2x_2 + x_3 = 0$

(c) $x_1 + x_2 + 2x_3 = 3$
   $-x_1 - 2x_2 + x_3 = 8$
   $x_1 - x_2 + 8x_3 = 25$

(d) $x_1 + 2x_2 + 3x_3 = 5$
   $3x_1 - x_2 - 2x_3 = -3$
   $-5x_1 + 4x_2 + 10x_3 = 14$

7. Solve each system of equations in Exercise 2 with Gauss–Jordan elimination in which off-diagonal pivots are used. To be exact, pivot on entry (2, 1), then on (3, 2), and finally on (1, 3).

8. The staff dietician at the California Institute of Trigonometry has to make up a meal with 1200 calories, 30 grams of protein, and 300 milligrams of vitamin C. There are three food types to choose from: rubbery jello, dried fish sticks, and mystery meat. They have the following nutritional content per ounce.

	Jello	Fish sticks	Mystery meat
Calories	20	100	200
Protein	1	3	2
Vitamin C	30	20	10

Set up and solve a system of equations to determine how much of each food should be used.

9. A furniture manufacturer makes tables, chairs, and sofas. In one month, the company has available 550 units of wood, 475 units of labor, and 220 units of fabric. The manufacturer wants a production schedule for the month that uses all of these resources. The different products require the following amounts of the resources.

	Table	Chair	Sofa
Wood	4	2	5
Labor	3	2	5
Fabric	0	2	4

Set up and solve a system of equations to determine how much of each product should be manufactured.

10. An investment analyst is trying to find out how much business a secretive TV manufacturer has. The company makes three brands of TV: Brand A, Brand B, and Brand C. The analyst learns that the manufacturer has ordered from suppliers 450,000 type-1 circuit boards, 300,000 type-2 circuit boards, and 350,000 type-3 circuit boards. Brand A uses 2 type-1 boards, 1 type-2 board, and 2 type-3 boards. Brand B uses 3 type-1 boards, 2 type-2 boards, and 1 type-3 board. Brand C uses 1 board of each type. How many TVs of each brand are being manufactured?

11. Find the stable distribution (as done in Example 8) for Markov chains with the following transition matrices.

(a) $\begin{bmatrix} 2/3 & 1/3 & 1/3 \\ 0 & 1/3 & 0 \\ 1/3 & 1/3 & 2/3 \end{bmatrix}$ (c) $\begin{bmatrix} 0 & 1/2 & 1/2 \\ 1/2 & 0 & 1/2 \\ 1/2 & 1/2 & 0 \end{bmatrix}$

(b) $\begin{bmatrix} 1/2 & 1/4 & 1/4 \\ 1/4 & 1/2 & 1/4 \\ 1/4 & 1/4 & 1/2 \end{bmatrix}$

12. Show in Theorem 1 that row operations M and S do not change the set of solutions to a system of linear equations.

13. Describe the row operations that are needed to convert the reduced system of equations in (5) back into the original refinery system in (3).

14. For what values of k does the following refinery-type system of equations have a unique solution with all x_1 nonnegative?

$$6x_1 + 5x_2 + 3x_3 = 500$$
$$4x_1 + x_2 + 7x_3 = 600$$
$$5x_1 + kx_2 + 5x_3 = 1000$$

15. (Computer Project) Solve the following refinery problems as a function of a and b using computer algebra software.

(a)
$$16x_1 + ax_2 + 8x_3 = 600$$
$$8x_1 + 20x_2 + 8x_3 = 800$$
$$bx_1 + 10x_2 + 20x_3 = 1000$$

(b)
$$16x_1 + ax_2 + bx_3 = 600$$
$$8x_1 + 20x_2 + bx_3 = 800$$
$$4x_1 + 10x_2 + 20x_3 = 1000$$

16. (Computer Project) Solve the following Markov chains for the stable probability vector using computer algebra software (add the additional constraint $p_1^* + p_2^* = 1$ or $p_1^* + p_2^* + p_3^* = 1$ to the two or three equations of the type appearing in (28)).

(a) $\begin{bmatrix} p & 1-p \\ 1-p & p \end{bmatrix}$

(b) $\begin{bmatrix} p & 1/2 & 0 \\ 1-p & 0 & 1-p \\ 0 & 1/2 & p \end{bmatrix}$

(c) $\begin{bmatrix} p & 1/3 & 1-p-q \\ q & 1/3 & q \\ 1-p-q & 1/3 & p \end{bmatrix}$

17. **(Computer Project)** Write a computer program to perform Gaussian elimination on a system of n equations in n unknowns (watch out for 0s on the main diagonal).

18. **(Computer Project)** Write a computer program to perform Gauss-Jordan elimination on a system of n equations in n unknowns (watch out for 0 pivots).

SECTION 3.5 *The Inverse of a Matrix*

In this section we study a general method for solving a system $Ax = b$ of n linear equations in n variables for *any* b, instead of for one particular b, as in the previous section. Our tool is the (multiplicative) inverse A^{-1} of a matrix A. As noted in Example 2 of Section 3.3, the inverse A^{-1} of an $n \times n$ matrix A is an $n \times n$ matrix with the property

$$AA^{-1} = I \quad \text{and} \quad A^{-1}A = I,$$

where I is the $n \times n$ identity matrix (with 1s on the main diagonal and 0s elsewhere).

E X A M P L E 1 **Matrix with an Inverse** Matrix $A = \begin{bmatrix} 3 & 1 \\ 4 & 2 \end{bmatrix}$ has the inverse $A^{-1} = \begin{bmatrix} 1 & -1/2 \\ -2 & 3/2 \end{bmatrix}$. (For now, do not worry how this inverse was found.) Multiplying A times A^{-1}, we have

$$AA^{-1} = \begin{bmatrix} 3 & 1 \\ 4 & 2 \end{bmatrix} \begin{bmatrix} 1 & -\frac{1}{2} \\ -2 & \frac{3}{2} \end{bmatrix}$$

$$= \begin{bmatrix} 3 \cdot 1 + 1 \cdot (-2) & 3 \cdot (-\frac{1}{2}) + 1 \cdot \frac{3}{2} \\ 4 \cdot 1 + 2 \cdot (-2) & 4 \cdot (-\frac{1}{2}) + 2 \cdot \frac{3}{2} \end{bmatrix}$$

$$= \begin{bmatrix} 1 & 0 \\ 0 & 1 \end{bmatrix}.$$

The reader can verify that if A^{-1} precedes A, again $A^{-1}A = I$. ▲

Note that the inverse of a matrix is just another matrix with nothing about it to indicate that it is an inverse, unlike the situation with numbers, where inverses are often written as reciprocals; for example, the multiplicative inverse of 11 is 1/11.

Properties of Inverse Matrices

A matrix A is said to be *invertible* if it has an inverse. Some matrices are invertible and some are not. Examples of matrices without inverse will be given shortly. Inverses allow us to "solve" a system of equations symbolically the way one solves the scalar equation $ax = b$ by dividing both sides by a to obtain $x = a^{-1}b$.

Theorem 1. If the $n \times n$ matrix A has an inverse A^{-1}, then the system of equations $Ax = b$ has the solution $x = A^{-1}b$.

Before proving Theorem 1, we illustrate it. Let us consider the system

$$3x + y = 4$$
$$4x + 2y = 6$$ or

$$\mathbf{Ax} = \mathbf{b}, \text{ where } \mathbf{A} = \begin{bmatrix} 3 & 1 \\ 4 & 2 \end{bmatrix} \text{ and } \mathbf{b} = \begin{bmatrix} 4 \\ 6 \end{bmatrix}.$$

From Example 1, we have seen that $\mathbf{A}^{-1} = \begin{bmatrix} 1 & -1/2 \\ -2 & 3/2 \end{bmatrix}$ is the inverse of matrix $\mathbf{A}$. Then by Theorem 1,

$$\mathbf{x} = \mathbf{A}^{-1}\mathbf{b} = \begin{bmatrix} 1 & -\frac{1}{2} \\ -2 & \frac{3}{2} \end{bmatrix}\begin{bmatrix} 4 \\ 6 \end{bmatrix} = \begin{bmatrix} 1 \\ 1 \end{bmatrix}.$$

The reader should check that $\mathbf{x} = \begin{bmatrix} 1 \\ 1 \end{bmatrix}$ is indeed the solution to this system of equations.

Proof of Theorem 1. As in the scalar case, we divide both sides of $\mathbf{Ax} = \mathbf{b}$ by $\mathbf{A}$, that is, multiply both sides of the matrix equation $\mathbf{Ax} = \mathbf{b}$ by $\mathbf{A}^{-1}$:

$$\mathbf{A}^{-1}(\mathbf{Ax}) = \mathbf{A}^{-1}\mathbf{b}. \tag{1}$$

Using the associative law for matrix multiplication and the fact that $\mathbf{A}^{-1}\mathbf{A} = \mathbf{I}$, we can rewrite the left side of Eq. (1) as follows:

$$\mathbf{A}^{-1}(\mathbf{Ax}) = (\mathbf{A}^{-1}\mathbf{A})\mathbf{x} = \mathbf{Ix} = \mathbf{x}. \tag{2}$$

(Recall that the identity matrix $\mathbf{I}$ times any matrix or vector just equals that matrix or vector; see Example 2 of Section 3.3.) Combining Eqs. (1) and (2), we have the desired result: $\mathbf{x} = \mathbf{A}^{-1}\mathbf{b}$.

▲ ▲ ▲

Reviewing the proof in Theorem 1, we first multiply both sides of matrix equation $\mathbf{Ax} = \mathbf{b}$ by $\mathbf{A}^{-1}$ and then rewrite the matrix expression $\mathbf{A}^{-1}(\mathbf{Ax})$ us-

ing three simple matrix facts: (i) the associative law for matrix multiplication; (ii) $\mathbf{A}^{-1}\mathbf{A} = \mathbf{I}$; and (iii) $\mathbf{Ix} = \mathbf{x}$. With these steps we were able to derive a general matrix formula for solving systems of equations. We do not know how to compute inverses, yet we already proved a very important result about inverses. This is the sort of powerful yet simple-to-obtain result that characterizes linear algebra theory.

Let us look at two matrices without inverses.

E X A M P L E   2   Matrix Without an Inverse

We claim that the matrix $\mathbf{B} = \begin{bmatrix} 1 & 4 \\ 2 & 8 \end{bmatrix}$ has no inverse. The key to our claim is the observation that the second row of $\mathbf{B}$ is twice the first row:

$$\mathbf{b}_2' = 2\mathbf{b}_1', \tag{3}$$

where $\mathbf{b}_1'$ and $\mathbf{b}_2'$ are the first and second rows of $\mathbf{B}$, respectively.

Suppose

$$\mathbf{C} = \begin{bmatrix} c_{11} & c_{12} \\ c_{21} & c_{22} \end{bmatrix}$$

were the inverse of $\mathbf{B}$, so that

$$\mathbf{BC} = \mathbf{I} = \begin{bmatrix} 1 & 0 \\ 0 & 1 \end{bmatrix}.$$

If $\mathbf{c}_1$ and $\mathbf{c}_2$ are the two columns of $\mathbf{C}$, then the matrix product $\mathbf{BC}$ is the following collection of scalar products:

$$\mathbf{BC} = \begin{bmatrix} \mathbf{b}_1' \cdot \mathbf{c}_1 & \mathbf{b}_1' \cdot \mathbf{c}_2 \\ \mathbf{b}_2' \cdot \mathbf{c}_1 & \mathbf{b}_2' \cdot \mathbf{c}_2 \end{bmatrix} = \begin{bmatrix} 1 & 0 \\ 0 & 1 \end{bmatrix}. \tag{4}$$

From Eq. (3), $\mathbf{b}'_2 = 2\mathbf{b}'_1$, implying that $\mathbf{b}'_2 \cdot \mathbf{c}_1 = 2(\mathbf{b}'_1 \cdot \mathbf{c}_1)$. We check this: $\mathbf{b}'_2 \cdot \mathbf{c}_1 = 2 \cdot c_{11} + 8 \cdot c_{21} = 2(1 \cdot c_{11} + 4 \cdot c_{21}) = 2(\mathbf{b}'_1 \cdot \mathbf{c}_1)$. Similarly, $\mathbf{b}'_2 \cdot \mathbf{c}_2 = 2(\mathbf{b}'_1 \cdot \mathbf{c}_2)$. Then we see that row 2 of $\mathbf{BC}$, $[\mathbf{b}'_2 \cdot \mathbf{c}_1, \mathbf{b}'_2 \cdot \mathbf{c}_2]$, must be twice row 1, $[\mathbf{b}'_1 \cdot \mathbf{c}_1, \mathbf{b}'_1 \cdot \mathbf{c}_2]$. However, the second row of $\mathbf{I}$ is obviously not twice its first row. This contradiction shows that no inverse can exist.

Generalizing this reasoning, one can show that if one row of $\mathbf{A}$ is a multiple of another row, then no inverse can exist. ▲

EXAMPLE 3 **Matrix with a Partial Inverse**

By the definition of an inverse, a matrix must be square to have an inverse. But we claim that the matrix

$$C = \begin{bmatrix} -2 & 1 \\ 0 & 1 \\ 2 & 1 \end{bmatrix}$$

has a partial (left-sided) inverse

$$C^+ = \begin{bmatrix} -\frac{1}{4} & 0 & \frac{1}{4} \\ \frac{1}{3} & \frac{1}{3} & \frac{1}{3} \end{bmatrix};$$

$$C^+C = \begin{bmatrix} -\frac{1}{4} & 0 & \frac{1}{4} \\ \frac{1}{3} & \frac{1}{3} & \frac{1}{3} \end{bmatrix} \begin{bmatrix} -2 & 1 \\ 0 & 1 \\ 2 & 1 \end{bmatrix} = \begin{bmatrix} 1 & 0 \\ 0 & 1 \end{bmatrix} = I.$$

On the other hand,

$$CC^+ = \begin{bmatrix} -2 & 1 \\ 0 & 1 \\ 2 & 1 \end{bmatrix} \begin{bmatrix} -\frac{1}{4} & 0 & \frac{1}{4} \\ \frac{1}{3} & \frac{1}{3} & \frac{1}{3} \end{bmatrix} = \begin{bmatrix} \frac{5}{6} & \frac{1}{3} & -\frac{1}{6} \\ \frac{1}{3} & \frac{1}{3} & \frac{1}{3} \\ -\frac{1}{6} & \frac{1}{3} & \frac{5}{6} \end{bmatrix}.$$

The matrix C^+ is called the pseudo-inverse of C and has an important role in many applica-

tions. We note, however, that *for no 3×2 matrix C can there exist a 2×3 matrix D such that* $CD = I$. (Proving this claim requires linear algebra beyond the scope of this text.) ▲

Unlike the situation in Example 3, if a matrix is square, then a left inverse is also a right inverse, and vice versa (the proof of this fact is beyond the scope of this chapter). A related concern is, if an inverse exists for a matrix A, is it unique? Might there be two different matrices that are both inverses of A? The following theorem provides the answer.

Theorem 2. If an $n \times n$ matrix A has an inverse A^{-1}, then this inverse must be unique.

Proof. Suppose that Q is another matrix with the property that $AQ = QA = I$. Then by the associative law for matrix algebra, we can compute the triple product QAA^{-1} in two ways:

$$QAA^{-1} = Q(AA^{-1}) = QI = Q \quad \text{and}$$
$$QAA^{-1} = (QA)A^{-1} = IA^{-1} = A^{-1}.$$

Thus, we conclude that $Q = A^{-1}$.

▲ ▲ ▲

We presented Theorem 2 and its proof to give the reader a second example (along with Theorem 1) of matrix algebra at work and the very critical role again played by the associative law of matrix multiplication.

Computing the Inverse of a Matrix

Now we turn to the problem of determining the inverse of a matrix, if it exists. The following example shows how to compute the inverse of a 2 × 2 matrix. The key is to find a system of linear equations that the entries in the inverse must satisfy.

E X A M P L E 4 Computing the Inverse of a 2 × 2 Matrix Consider the 2 × 2 matrix $\mathbf{A}$ presented in Example 1. We seek to find its inverse $\mathbf{X}$:

$$\mathbf{A} = \begin{bmatrix} 3 & 1 \\ 4 & 2 \end{bmatrix}, \qquad \mathbf{X} = \begin{bmatrix} x_{11} & x_{12} \\ x_{21} & x_{22} \end{bmatrix}.$$

We require that $\mathbf{AX} = \mathbf{I}$:

$$\mathbf{AX} = \begin{bmatrix} 3 & 1 \\ 4 & 2 \end{bmatrix}\begin{bmatrix} x_{11} & x_{12} \\ x_{21} & x_{22} \end{bmatrix} = \begin{bmatrix} 1 & 0 \\ 0 & 1 \end{bmatrix} = \mathbf{I}. \quad (5)$$

We determine $\mathbf{X}$ $(= \mathbf{A}^{-1})$ one column at a time from (5). We find the first column $\mathbf{x}_1 = \begin{bmatrix} x_{11} \\ x_{21} \end{bmatrix}$ of matrix $\mathbf{X}$ by setting $\mathbf{Ax}_1$—the first column in product $\mathbf{AX}$ of (5)—equal to the $\mathbf{i}_1$, the first column of $\mathbf{I}$:

$$\mathbf{Ax}_1 = \mathbf{i}_1: \qquad \begin{bmatrix} 3 & 1 \\ 4 & 2 \end{bmatrix}\begin{bmatrix} x_{11} \\ x_{21} \end{bmatrix} = \begin{bmatrix} 1 \\ 0 \end{bmatrix},$$

or

$$\begin{aligned} 3x_{11} + x_{21} &= 1, \\ 4x_{11} + 2x_{21} &= 0. \end{aligned} \qquad (6a)$$

Similarly, the second column of (5) yields the system

$$\begin{aligned} 3x_{12} + x_{22} &= 0, \\ 4x_{12} + 2x_{22} &= 1. \end{aligned} \qquad (6b)$$

Using Gauss–Jordan elimination on the augmented coefficient matrix for (6a), we obtain

$$\left[\begin{array}{cc|c} 3 & 1 & 1 \\ 4 & 2 & 0 \end{array}\right] \Rightarrow \left[\begin{array}{cc|c} 1 & \frac{1}{3} & \frac{1}{3} \\ 0 & \frac{2}{3} & -\frac{4}{3} \end{array}\right] \Rightarrow \left[\begin{array}{cc|c} 1 & 0 & 1 \\ 0 & 1 & -2 \end{array}\right],$$
$$(7a)$$

and so $x_{11} = 1$, $x_{21} = -2$. For (6b) we obtain

$$\left[\begin{array}{cc|c} 3 & 1 & 0 \\ 4 & 2 & 1 \end{array}\right] \Rightarrow \left[\begin{array}{cc|c} 1 & \frac{1}{3} & 0 \\ 0 & \frac{2}{3} & 1 \end{array}\right] \Rightarrow \left[\begin{array}{cc|c} 1 & 0 & -\frac{1}{2} \\ 0 & 1 & \frac{3}{2} \end{array}\right], \quad (7b)$$

and so $x_{12} = -1/2$, $x_{22} = 3/2$.

Substituting these values for x_{ij} back into $\mathbf{X}$ $(= \mathbf{A}^{-1})$, we have

$$\mathbf{X} = \begin{bmatrix} 1 & -\frac{1}{2} \\ -2 & \frac{3}{2} \end{bmatrix}. \ \blacktriangle$$

The method in Example 4 can be extended to determine the inverse, when it exists, for a matrix of any size. The right-side vectors in (6a) and (6b) will become the columns of the appropriate-sized identity matrix $\mathbf{I}$, column vectors with a 1 in one position and 0s in all the other positions. The column vectors $\mathbf{i}_1, \mathbf{i}_2, \mathbf{i}_3, \ldots$ of $\mathbf{I}$ are coordinate vectors.

Theorem 3. Let $\mathbf{A}$ be an $n \times n$ matrix and $\mathbf{i}_1, \mathbf{i}_2, \mathbf{i}_3, \ldots, \mathbf{i}_n$ be the coordinate n-vectors (the columns of the $n \times n$ identity matrix $\mathbf{I}$). Let the n-vectors $\mathbf{x}_1, \mathbf{x}_2, \mathbf{x}_3, \ldots, \mathbf{x}_n$ be the solutions to

$$\mathbf{Ax}_j = \mathbf{i}_j,$$

for $j = 1, 2, \ldots, n$. Then the $n \times n$ matrix $\mathbf{X}$ with column vectors $\mathbf{x}_j$ is the inverse of $\mathbf{A}$:

$$\mathbf{A}^{-1} = \mathbf{X} = [\mathbf{x}_1, \mathbf{x}_2, \ldots, \mathbf{x}_n]. \qquad (8)$$

Note: If any of the systems $\mathbf{A}x_j = \mathbf{i}_j$ does not have a solution, then $\mathbf{A}$ does not have an inverse.

Recall that when we solve a system of equations by elimination, the right sides play a passive role. That is, using a different right side $\mathbf{b}$ does not change any of the calculations involving the coefficients. It only affects the final values that appear on the right side. Thus when we performed Gauss–Jordan elimination on the coefficient matrix in (7a) and (7b) of Example 3, we could have simultaneously applied the elimination steps to an augmented coefficient matrix $[\mathbf{A}\ \mathbf{I}]$ that contained both right-side vectors. The computations would be

$$\begin{bmatrix} 3 & 1 & 1 & 0 \\ 4 & 2 & 0 & 1 \end{bmatrix} \Rightarrow \begin{bmatrix} 1 & \frac{1}{3} & \frac{1}{3} & 0 \\ 0 & \frac{2}{3} & -\frac{4}{3} & 1 \end{bmatrix} \Rightarrow \begin{bmatrix} 1 & 0 & 1 & -\frac{1}{2} \\ 0 & 1 & -2 & \frac{3}{2} \end{bmatrix}. \tag{9}$$

Applying Gauss–Jordan elimination to $[\mathbf{A}\ \mathbf{I}]$ to obtain $[\mathbf{I}\ \mathbf{A}^{-1}]$ is an algorithm for determining $\mathbf{A}^{-1}$.

```
input:   n × (2n) augmented matrix [A I]
output:  n × n matrix A⁻¹ or finding no
         inverse of A
```

```
For i from 1 to n
    If A[i, i] = 0 then
        if another row j is nonzero in ith
            position, then interchange rows i
            and j otherwise STOP (no inverse)
    Pivot on entry A[i, i]
end for
```

E X A M P L E 5 The Inverse of a 3 × 3 Matrix
Let

$$A = \begin{bmatrix} 1 & 0 & 2 \\ 2 & 4 & 2 \\ 1 & 2 & 6 \end{bmatrix}.$$

We compute the inverse by pivoting along the main diagonal of the augmented matrix $[\mathbf{A}\ \mathbf{I}]$.

$$[\mathbf{A}\ \mathbf{I}] = \begin{matrix} (a) \\ (b) \\ (c) \end{matrix} \begin{bmatrix} 1 & 0 & 2 & 1 & 0 & 0 \\ 2 & 4 & 2 & 0 & 1 & 0 \\ 1 & 2 & 6 & 0 & 0 & 1 \end{bmatrix}$$

$$\begin{matrix} (a') = (a) \\ (b') = (b') - 2(a) \\ (c') = (c) - (a) \end{matrix} \begin{bmatrix} 1 & 0 & 2 & 1 & 0 & 0 \\ 0 & 4 & -2 & -2 & 1 & 0 \\ 0 & 2 & 4 & -1 & 0 & 1 \end{bmatrix}$$

$$\begin{matrix} (a'') = (a') \\ (b'') = \frac{1}{4}(b) \\ (c'') = (c') - 2(b'') \end{matrix} \begin{bmatrix} 1 & 0 & 2 & 1 & 0 & 0 \\ 0 & 1 & -.5 & -.5 & .25 & 0 \\ 0 & 0 & 5 & 0 & -.5 & 1 \end{bmatrix}$$

$$\begin{matrix} (a''') = (a'') - 2(c''') \\ (b''') = (b'') + \frac{1}{2}(c''') \\ (c''') = \frac{1}{5}(c') \end{matrix} \begin{bmatrix} 1 & 0 & 0 & 1 & .2 & -.4 \\ 0 & 1 & 0 & -.5 & .2 & .1 \\ 0 & 0 & 1 & 0 & -.1 & .2 \end{bmatrix} \tag{10}$$

Thus

$$A^{-1} = \begin{bmatrix} 1 & \frac{1}{5} & -\frac{2}{5} \\ -\frac{1}{2} & \frac{1}{5} & \frac{1}{10} \\ 0 & -\frac{1}{10} & \frac{1}{5} \end{bmatrix}. \ \blacktriangle$$

E X A M P L E 6 Use of an Inverse in Multiple Right-Hand Sides In Example 2 of Section 3.4, we solved the refinery system of equations by Gauss–Jordan elimination. Let us use the same sequence of pivots with the augmented matrix $[\mathbf{A}\ \mathbf{I}]$ to compute the inverse.

$$\begin{matrix} (a) \\ (b) \\ (c) \end{matrix} \begin{bmatrix} 16 & 8 & 8 & 1 & 0 & 0 \\ 8 & 20 & 8 & 0 & 1 & 0 \\ 4 & 10 & 20 & 0 & 0 & 1 \end{bmatrix}$$

$$\begin{array}{l} (a') = \tfrac{1}{16}(a) \\ (b') = (b) - \tfrac{1}{2}(a) \\ (c') = (c) - \tfrac{1}{4}(a) \end{array} \qquad \left[\begin{array}{ccc|ccc} 1 & \tfrac{1}{2} & \tfrac{1}{2} & \tfrac{1}{16} & 0 & 0 \\ 0 & 16 & 4 & -\tfrac{1}{2} & 1 & 0 \\ 0 & 8 & 18 & -\tfrac{1}{4} & 0 & 1 \end{array}\right]$$

$$\begin{array}{l} (a'') = (a') - \tfrac{1}{2}(b'') \\ (b'') = \tfrac{1}{16}(b') \\ (c'') = (c') - 2(b'') \end{array} \qquad \left[\begin{array}{ccc|ccc} 1 & 0 & \tfrac{3}{8} & \tfrac{5}{64} & -\tfrac{1}{32} & 0 \\ 0 & 1 & \tfrac{1}{4} & -\tfrac{1}{32} & \tfrac{1}{16} & 0 \\ 0 & 0 & 16 & 0 & -\tfrac{1}{2} & 1 \end{array}\right]$$

$$\begin{array}{l} (a''') = (a'') - \tfrac{3}{8}(c''') \\ (b''') = (b'') - \tfrac{1}{4}(c''') \\ (c''') = \tfrac{1}{16}(c'') \end{array} \qquad \left[\begin{array}{ccc|ccc} 1 & 0 & 0 & \tfrac{5}{64} & -\tfrac{5}{256} & -\tfrac{3}{128} \\ 0 & 1 & 0 & -\tfrac{1}{32} & \tfrac{9}{128} & -\tfrac{1}{64} \\ 0 & 0 & 1 & 0 & -\tfrac{1}{32} & \tfrac{1}{16} \end{array}\right]$$

(11)

The inverse is thus

$$\mathbf{A}^{-1} = \left[\begin{array}{ccc} \tfrac{5}{64} & -\tfrac{5}{256} & -\tfrac{3}{128} \\ -\tfrac{1}{32} & \tfrac{9}{128} & -\tfrac{1}{64} \\ 0 & -\tfrac{1}{32} & \tfrac{1}{16} \end{array}\right]. \tag{12}$$

If we were given a new right-hand side vector for the refinery system, say

$$\mathbf{b}^* = \left[\begin{array}{c} 3200 \\ 3200 \\ 1600 \end{array}\right],$$

then the new solution can be obtained by computing $\mathbf{x}^* = \mathbf{A}^{-1}\mathbf{b}^*$.

$$\begin{aligned} \mathbf{x}^* = \mathbf{A}^{-1}\mathbf{b}^* &= \left[\begin{array}{ccc} \tfrac{5}{64} & -\tfrac{5}{256} & -\tfrac{3}{128} \\ -\tfrac{1}{32} & \tfrac{9}{128} & -\tfrac{1}{64} \\ 0 & -\tfrac{1}{32} & \tfrac{1}{16} \end{array}\right]\left[\begin{array}{c} 3200 \\ 3200 \\ 1600 \end{array}\right] \\ &= \left[\begin{array}{c} \tfrac{5}{64}\cdot 3200 - \tfrac{5}{256}\cdot 3200 - \tfrac{3}{128}\cdot 1600 \\ -\tfrac{1}{32}\cdot 3200 + \tfrac{9}{128}\cdot 3200 - \tfrac{1}{64}\cdot 1600 \\ 0\cdot 3200 - \tfrac{1}{32}\cdot 3200 + \tfrac{1}{16}\cdot 1600 \end{array}\right] \\ &= \left[\begin{array}{c} 150 \\ 100 \\ 0 \end{array}\right] \blacktriangle \end{aligned} \tag{13}$$

In many economic applications, finding production levels that meet the given demand—the solution $\mathbf{x}$ in (13) represents the productions levels needed to meet the petroleum demands of vector $\mathbf{b}$—is only half the story. One is also interested in determining how small changes in demand affect the production levels. This topic is called *sensitivity analysis* in economics textbooks.

E X A M P L E 7 Sensitivity Analysis with Inverses Consider what happens to a solution
$$\mathbf{x}^* = \left[\begin{array}{c} 150 \\ 100 \\ 0 \end{array}\right]$$
in the previous refinery example when we change the right side $\mathbf{b}$ a little, say, we increase the first component (heating oil) by one gallon. So $\mathbf{b}$ changes to $\mathbf{b} + \Delta\mathbf{b}$, where $\Delta\mathbf{b} = \left[\begin{array}{c} 1 \\ 0 \\ 0 \end{array}\right]$. Now instead of the solution $\mathbf{x}^*$ ($= \mathbf{A}^{-1}\mathbf{b}$), we have a new solution:

$$\mathbf{x}^0 = \mathbf{A}^{-1}(\mathbf{b} + \Delta\mathbf{b}) = \mathbf{A}^{-1}\mathbf{b} + \mathbf{A}^{-1}\Delta\mathbf{b} = \mathbf{x} + \Delta\mathbf{x}. \tag{14}$$

Then we claim that the solution will change by

$$\Delta\mathbf{x} = \mathbf{A}^{-1}\Delta\mathbf{b} = \mathbf{A}^{-1}\left[\begin{array}{c} 1 \\ 0 \\ 0 \end{array}\right] = (\mathbf{A}^{-1})_1;$$

in this problem, $(\mathbf{A}^{-1})_1 = \left[\begin{array}{c} \tfrac{5}{64} \\ -\tfrac{1}{32} \\ 0 \end{array}\right]$, (15)

where $(\mathbf{A}^{-1})_1$ denotes the first column of $\mathbf{A}^{-1}$.

(Recall that by Example 5 of Section 3.2, for any matrix C, $Ci_1 = c_1$, the first column of C).

Thus, the change Δx in the solution to produce one more unit of the first product is the first column of A, $(A^{-1})_1$. Similarly, the second and third columns of A^{-1} tell how the solution will change if we need one more unit of the second or third product. In sum, the columns of A^{-1} show us how the solution x changes when the right-side vector b changes. As a specific example, let us consider how our solution x^* in (13) changes when we change from

$$b^* = \begin{bmatrix} 3200 \\ 3200 \\ 1600 \end{bmatrix} \quad \text{to} \quad b^0 = \begin{bmatrix} 4800 \\ 3200 \\ 1600 \end{bmatrix}.$$

We take the solution $x^* = A^{-1}b^* = \begin{bmatrix} 150 \\ 100 \\ 0 \end{bmatrix}$

computed in (13) for b^* and change it by $\Delta x = A^{-1}\Delta b$, where $\Delta b = b^0 - b^* = \begin{bmatrix} 1600 \\ 0 \\ 0 \end{bmatrix}$. By

(15), Δx will be $1600(A^{-1})_1$, 1600 times the first column of A^{-1}. So,

$$
x^0 = A^{-1}\begin{bmatrix} 4800 \\ 3200 \\ 1600 \end{bmatrix} = A^{-1}\begin{bmatrix} 3200 \\ 3200 \\ 1600 \end{bmatrix} + A^{-1}\begin{bmatrix} 1600 \\ 0 \\ 0 \end{bmatrix}
$$

$$
= \begin{bmatrix} 150 \\ 100 \\ 0 \end{bmatrix} + 1600\begin{bmatrix} \frac{5}{64} \\ -\frac{1}{32} \\ 0 \end{bmatrix}
$$

$$
= \begin{bmatrix} 150 \\ 100 \\ 0 \end{bmatrix} + \begin{bmatrix} 125 \\ -50 \\ 0 \end{bmatrix} = \begin{bmatrix} 275 \\ 50 \\ 0 \end{bmatrix}. \ \blacktriangle \qquad (16)
$$

We now give a theorem linking inverses to solutions of systems of linear equations. This theorem illustrates that theory in linear algebra is usually easy to prove and gives useful results.

Theorem 4. The following three statements are equivalent for any $n \times n$ matrix A:

a. For some particular vector b, the system of equations $Ax = b$ has a unique solution.

b. For all b, the system of equations $Ax = b$ always has a unique solution.

c. A has an inverse.

Proof.

(a) $\Rightarrow$ (b) By Theorem 2 of the previous section, Gaussian elimination on A either produces a unique solution, multiple solutions, or no solution. Looking at the argument in proving Theorem 2, we see that the outcome depends only on the coefficients of A, not on b. Thus, if Gaussian elimination produces a unique solution for a particular b, it will produce a unique solution for all b.

(b) $\Rightarrow$ (c) By Theorem 3, the inverse is found by solving $Ax = i_j$, for $j = 1, 2, \ldots, n$. By (b), these systems all have a unique solution. Their solutions determine the entries of the inverse.

(c) $\Rightarrow$ (a) If A^{-1} exists, then by Theorems 1 and 2, there is a unique solution to $Ax = b$, namely, $x = A^{-1}b$.

$\blacktriangle \ \blacktriangle \ \blacktriangle$

Note that if a matrix A does *not* have an inverse, then Gauss–Jordan elimination will fail to reduce

Linear Algebra on the Gridiron

Have you ever enjoyed a marching band at a football game? With many band members executing individualized maneuvers, designing shows is very time-consuming. Formations are usually planned by placing labelled dots on a map of the field, forming "stuntsheets."

Soon after student Martin Haye joined the University of California Marching Band in 1989, he envi-

The Cal Band in its pregame "Wedge" formation.

sioned a computer program to print stuntsheets and also yield an animation of the show. He recruited me, a mathematics major, to help him. Haye and I realized that data entry would be a bottleneck, as each stuntsheet involved placing 150 points using the mouse, with 30 stuntsheets per show and at least six shows per football season.

The Cal Band has a traditional marching style, with formations of blocks, rectangles, parallelograms, straight lines, and diagonals. We recognized that many of the band's movements can be described as

affine transformations, which consist of linear transformations composed with translations. For example, a line of band members may rotate 90° clockwise about its midpoint (linear transformation) and then move north 30 steps (translation). We used affine transformations to place points automatically.

We consider the field as lying in the plane $P = \{(x, y, z) | z = 1\}$, identifying a point (x, y) on the field with the point $(x, y, 1)$ in P. A linear transformation $L(x, y, 1) = (A_{11}x + A_{12}y, A_{21}x + A_{22}y, 1)$ can be represented by the matrix

$$\mathbf{L} = \begin{bmatrix} A_{11} & A_{12} & 0 \\ A_{21} & A_{22} & 0 \\ 0 & 0 & 1 \end{bmatrix},$$

and a translation $T(x, y, 1) = (x + A_{13}, y + A_{23}, 1)$ can be represented by the matrix

$$\mathbf{T} = \begin{bmatrix} 1 & 0 & A_{13} \\ 0 & 1 & A_{23} \\ 0 & 0 & 1 \end{bmatrix}.$$

Then the matrix of the affine transformation $A = T \circ L$ is

$$\mathbf{A} = \mathbf{T}\,\mathbf{L} = \begin{bmatrix} A_{11} & A_{12} & A_{13} \\ A_{21} & A_{22} & A_{23} \\ 0 & 0 & 1 \end{bmatrix}.$$

Hence, we can implement a band movement by matrix multiplication.

The show designer defines an affine transformation by manually setting the first three points. The equations for the points,

$$A(x_1, y_1, 1) = (a_1, b_1, 1)$$
$$A(x_2, y_2, 1) = (a_2, b_2, 1)$$
$$A(x_3, y_3, 1) = (a_3, b_3, 1),$$

can be expressed together as

$$A \begin{bmatrix} x_1 & x_2 & x_3 \\ y_1 & y_2 & y_3 \\ 1 & 1 & 1 \end{bmatrix} = \begin{bmatrix} a_1 & a_2 & a_3 \\ b_1 & b_2 & b_3 \\ 1 & 1 & 1 \end{bmatrix}, \quad \text{or} \quad \mathbf{A}\,\mathbf{B} = \mathbf{C}.$$

If the points are not collinear, the matrix $[B]$ has full rank and is invertible, so we find $[A] = [C][B]^{-1}$ and use $[A]$ to place all the other points.

Our computer program greatly reduces the time and data-entry errors of producing a show. The new stuntsheets are easier to read, and animation verifies that there will be no collisions.

The interesting and innovative feature of our story is that *as students,* we were able to visualize marching-band movements in terms of linear algebra and apply what we knew to a practical problem. The next time that you are enjoying the excitement of a marching band, remember the hidden role that linear algebra plays!

—Daniel C. Isaksen

[A I] to [I $\mathbf{A}^{-1}$]. This is the fastest way in general to determine if $\mathbf{A}^{-1}$ exists.

We conclude this section with examples showing the natural role inverses play in geometry and cryptanalysis.

EXAMPLE 8 Testing a Matrix for Invertibility Consider matrix $\begin{bmatrix} 1 & 4 \\ 2 & 8 \end{bmatrix}$ from Example 2, which we showed is not invertible. To test for invertibility, apply Gauss–Jordan elimination to the augmented matrix $\begin{bmatrix} 1 & 4 & 1 & 0 \\ 2 & 8 & 0 & 1 \end{bmatrix}$. We need to subtract twice the first row from the second row, yielding $\begin{bmatrix} 1 & 4 & 1 & 0 \\ 0 & 0 & -2 & 1 \end{bmatrix}$. There is no nonzero entry in the second row (on the left side) on which to pivot, and so this matrix has no inverse. ▲

EXAMPLE 9 The Inverse of Rotation Matrices In Example 8 of Section 3.2, we introduced the rotation matrix $\mathbf{R}_\theta = \begin{bmatrix} \cos\theta & -\sin\theta \\ \sin\theta & \cos\theta \end{bmatrix}$. We noted that multiplying a 2-vector $\mathbf{v}$ by $\mathbf{R}_\theta$ had the effect of rotating the vector $\theta°$ counterclockwise about the origin. We now look at the inverse $\mathbf{R}_\theta^{-1}$ of a rotation matrix $\mathbf{R}_\theta$. Observe that the identity matrix $\mathbf{I}$ is the rotation matrix for a rotation of $0°$: $\mathbf{R}_{0°} = \mathbf{I}$. Observe also that a rotation of $\theta°$ followed by a rotation of $-\theta°$ will produce a net rotation of $0°$. That is, $\mathbf{R}_{-\theta}\mathbf{R}_\theta = \mathbf{R}_{0°} (= \mathbf{I})$, and so $\mathbf{R}_{-\theta}$ is the inverse of

$\mathbf{R}_\theta$. We confirm this by multiplying $\mathbf{R}_{-\theta}\mathbf{R}_\theta$ out (recall: $\cos(-\theta) = \cos\theta$; $\sin(-\theta) = -\sin\theta$; and $(\cos\theta)^2 + (\sin\theta)^2 = 1$).

$$
\begin{aligned}
\mathbf{R}_{-\theta}\mathbf{R}_\theta &= \begin{bmatrix} \cos\theta & -\sin\theta \\ \sin\theta & \cos\theta \end{bmatrix}\begin{bmatrix} \cos-\theta & -\sin-\theta \\ \sin-\theta & \cos-\theta \end{bmatrix} \\
&= \begin{bmatrix} (\cos\theta)^2 + (\sin\theta)^2 & 0 \\ 0 & (\cos\theta)^2 + (\sin\theta)^2 \end{bmatrix} \\
&= \begin{bmatrix} 1 & 0 \\ 0 & 1 \end{bmatrix} \ \blacktriangle
\end{aligned}
$$

EXAMPLE 10 Cryptograms Using Matrix Encryption A common approach to constructing a cryptogram for sending a secret message is to treat each letter in the message as a number between 1 and 26: $A \leftrightarrow 1$, $B \leftrightarrow 2$, $C \leftrightarrow 3$, ..., $Z \leftrightarrow 26$. So MAYDAY would be the numeric sequence 13, 1, 25, 4, 1, 25. Then we apply some algebraic formula to encrypt each letter (number) as some other letter (number). To ensure that the result of some calculation is a number between 1 and 26, one usually assumes that all arithmetic is done mod 26. As an example, suppose letter L_x is encrypted as cryptogram letter C_x using the formula $C_x = 7L_x + 6$. Then the letter M (13) is encrypted as follows:

$$
\begin{aligned}
C_M = 7L_M + 6 &= 7\cdot 13 + 6 \\
&= 97 \equiv 19 \quad (\text{mod } 26) \leftrightarrow S,
\end{aligned}
$$

since: $97 = 3\cdot 26 + 19$.

Such linear encrypting schemes are easy to break. A scheme that is simple to use but fairly hard to break is to group letters into pairs L_1, L_2 and encrypt them into a pair of cryptogram letters C_1, C_2 with two linear equations, such as

$$
\begin{aligned}
C_1 &\equiv 9L_1 + 17L_2 \quad (\text{mod } 26), \\
C_2 &\equiv 7L_1 + 2L_2 \quad (\text{mod } 26).
\end{aligned} \tag{17}
$$

If $L_1 = $ E (5) and $L_2 = $ C (3), then (17) would encrypt them as

$$
C_1 = 9\cdot 5 + 17\cdot 3 = 96 \equiv 18 \quad (\text{mod } 26) \leftrightarrow R,
$$

$$
C_2 = 7\cdot 5 + 2\cdot 3 = 41 \equiv 15 \quad (\text{mod } 26) \leftrightarrow O.
$$

In matrix form, with $\mathbf{c} = \begin{bmatrix} C_1 \\ C_2 \end{bmatrix}$, $\mathbf{l} = \begin{bmatrix} L_1 \\ L_2 \end{bmatrix}$, and $\mathbf{E} = \begin{bmatrix} 9 & 17 \\ 7 & 2 \end{bmatrix}$, (17) becomes

$$
\mathbf{c} \equiv \mathbf{El} \quad (\text{mod } 26). \tag{18}
$$

This encrypting scheme was used by the U.S. government many years ago, although it can easily be broken today with the aid of computers. The appeal of this scheme is that how an individual letter is encrypted depends on the letter just before or after it. If there is a one-to-one encrypting scheme of individual letters, then the frequent letters in the cryptogram will correspond to frequent letters in English, such as E, I, and T, making the cryptogram easy for anyone to break. In this scheme, which encrypts letters in pairs, frequencies of individual cryptogram letters have no useful meaning.

The person who receives the encrypted pair $\mathbf{c}$ will decrypt $\mathbf{c}$ back into the original message pair $\mathbf{I}$ by using the inverse of $\mathbf{E}$. Observe that

$$
\mathbf{E}^{-1}\mathbf{c} \equiv \mathbf{E}^{-1}(\mathbf{El}) = (\mathbf{E}^{-1}\mathbf{E})\mathbf{l} = \mathbf{Il} = \mathbf{l},
$$

and so

$$1 \equiv \mathbf{E}^{-1}\mathbf{c}. \tag{19}$$

The following discussion is optional, and the reader can skip to the inverse of $\mathbf{E}$ *given in (20).* We compute the inverse by performing Gauss–Jordan elimination on $[\mathbf{E}|\mathbf{I}]$, except that now all computations must be performed mod 26. Subtraction mod 26 is performed by adding; for example, $-1 \equiv 25$. Division mod 26 is more complicated. The inverse mod 26 of a number k, $1 \le k \le 26$, must be an integer between 1 and 26. Further, not all integers have an inverse. The situation is similar to the way that the inverse of a matrix, when it exists, is another matrix. For example, the inverse of 9 mod 26 is 3, since $9 \cdot 3 = 27 \equiv 1$. The integer 2 has no inverse mod 26, since the product of a number and its inverse must be 1 (mod 26), an odd number, while 2 times any number is even.

The following augmented matrices show how Gaussian elimination proceeds mod 26 to find the inverse of $\mathbf{E} = \begin{bmatrix} 9 & 17 \\ 7 & 2 \end{bmatrix}$. The reader should check the modular computations.

$$\begin{bmatrix} 9 & 17 & | & 1 & 0 \\ 7 & 2 & | & 0 & 1 \end{bmatrix} \Rightarrow \begin{bmatrix} 1 & 25 & | & 3 & 0 \\ 0 & 9 & | & 5 & 1 \end{bmatrix} \pmod{26}$$

$$\Rightarrow \begin{bmatrix} 1 & 0 & | & 18 & 3 \\ 0 & 1 & | & 15 & 3 \end{bmatrix} \pmod{26}$$

Thus

$$\mathbf{E}^{-1} = \begin{bmatrix} 18 & 3 \\ 15 & 3 \end{bmatrix} \pmod{26}. \tag{20}$$

The decrypting equations are

$$\begin{aligned} L_1 &\equiv 18C_1 + 3C_2 \pmod{26}, \\ L_2 &\equiv 15C_1 + 3C_2 \pmod{26}. \end{aligned} \tag{21}$$

For example, the pair R, O (18, 15) is decrypted using (20) as

$$\begin{aligned} L_1 &= 18 \cdot 18 + 3 \cdot 15 = 324 + 45 \\ &= 369 \equiv 5 \pmod{26} \leftrightarrow \text{E}, \\ L_2 &= 15 \cdot 18 + 3 \cdot 15 = 270 + 45 \\ &= 315 \equiv 3 \pmod{26} \leftrightarrow \text{C}. \end{aligned}$$

So R, O translate back to the original pair E, C, as required. A more sophisticated (unbreakable) encryption scheme is discussed in Section 9.2. ▲

1. Verify for the matrix A in Example 1 that $A^{-1}A = I$.

2. Write the system of equations that entries in the inverse of the following matrices must satisfy. Then find inverses (as in Example 4) or show that none can exist (following the reasoning in Example 2).

(a) $\begin{bmatrix} 0 & 1 \\ 1 & 0 \end{bmatrix}$

(b) $\begin{bmatrix} -1 & 3 \\ 2 & -6 \end{bmatrix}$

(c) $\begin{bmatrix} 1 & 1 & 0 \\ 0 & 1 & 1 \\ 1 & 2 & 1 \end{bmatrix}$

(d) $\begin{bmatrix} 1 & 2 & 1 \\ 2 & 4 & 2 \\ 2 & 5 & 1 \end{bmatrix}$

3. (a) Write out the system of equations that the first column of the inverse of A must satisfy, where

$$A = \begin{bmatrix} 1 & 0 & 2 \\ 0 & 1 & 3 \\ 1 & 0 & 4 \end{bmatrix}.$$

(b) Determine the first column of A^{-1}.

4. Use Gauss–Jordan elimination to find the inverse of the following matrices.

(a) $\begin{bmatrix} 2 & -3 & 2 \\ 1 & -1 & 1 \\ -1 & 5 & 4 \end{bmatrix}$

(b) $\begin{bmatrix} -1 & -3 & 2 \\ 2 & 1 & 3 \\ 5 & 4 & 6 \end{bmatrix}$

(c) $\begin{bmatrix} 1 & 1 & 4 \\ 2 & 1 & 3 \\ 5 & 2 & 5 \end{bmatrix}$

(d) $\begin{bmatrix} 2 & -3 & -1 \\ 3 & -5 & -2 \\ 9 & 6 & 4 \end{bmatrix}$

5. For each matrix A in Exercise 4, solve $Ax = b$, where $b = \begin{bmatrix} 20 \\ 20 \\ 20 \end{bmatrix}$.

6. For each matrix A in Exercise 4, how much will the solution of $Ax = b$ change if b is changed from

(a) the vector $\begin{bmatrix} b_1 \\ b_2 \\ b_3 \end{bmatrix}$ to the vector $\begin{bmatrix} b_1 \\ b_2 + 1 \\ b_3 \end{bmatrix}$?

(b) the vector $\begin{bmatrix} b_1 \\ b_2 \\ b_3 \end{bmatrix}$ to the vector $\begin{bmatrix} b_1 \\ b_2 + 1 \\ b_3 - 1 \end{bmatrix}$?

7. Use Gauss–Jordan elimination to find the inverse of the following matrices.

(a) $\begin{bmatrix} 1 & 1 & -1 & -1 \\ 2 & 0 & 0 & 1 \\ 3 & 0 & 0 & -2 \\ 4 & -2 & 1 & 3 \end{bmatrix}$

(b) $\begin{bmatrix} 3 & 2 & 1 & 0 \\ 1 & 1 & 0 & -1 \\ 2 & 1 & -1 & 1 \\ 1 & 1 & 1 & 1 \end{bmatrix}$

(c) $\begin{bmatrix} 1 & 3 & 2 & -1 \\ 1 & 1 & 1 & 1 \\ 2 & -2 & 1 & -1 \\ 1 & -3 & -1 & 2 \end{bmatrix}$

8. This exercise gives a "picture" of how the inverse of A nearly does not exist when two columns of A are almost the same. For the following matrices A, solve the system $A\begin{bmatrix} u_1 \\ u_2 \end{bmatrix} = \begin{bmatrix} 1 \\ 0 \end{bmatrix}$. Then plot u_1a_1 and u_2a_2 in a two-dimensional coordinate system and show geometrically how the sum of vectors u_1a_1 and u_2a_2 is $\begin{bmatrix} 1 \\ 0 \end{bmatrix}$.

(a) $\begin{bmatrix} 2 & 3 \\ 1 & 2 \end{bmatrix}$

(b) $\begin{bmatrix} 8 & 9 \\ 7 & 7 \end{bmatrix}$

9. (Continuation of Exercise 8 in Section 3.4) The staff dietician at the California Institute of Trigonometry has to make up a meal with 1200 calories, 30 grams of pro-

tein, and 300 milligrams of vitamin C. There are three food types to choose from: rubbery jello, dried fish sticks, and mystery meat. They have the following nutritional content per ounce.

	Jello	Fish sticks	Mystery meat
Calories	20	100	200
Protein	1	3	2
Vitamin C	30	20	10

(a) Find the inverse of this data matrix, and use it to compute the amount of jello, fish sticks, and mystery meat required.

(b) If the protein requirement is increased by 4, how will this change the number of units of jello in the meal?

(c) If the vitamin C requirement is decreased by k milligrams, how much will this change the number of fish sticks in a meal?

10. (Continuation of Exercise 9 in Section 3.4) A furniture manufacturer makes tables, chairs, and sofas. In one month, the company has available 550 units of wood, 475 units of labor, and 270 units of upholstery. The manufacturer wants a production schedule for the month that uses all of these resources. The different products require the following amounts of the resources.

	Table	Chair	Sofa
Wood	4	2	5
Labor	3	2	5
Upholstery	2	0	4

(a) Find the inverse of this data matrix, and use it to determine how much of each product should be manufactured.

(b) If the amount of wood is increased by 30 units, how will this change the number of sofas produced?

(c) If the amount of labor is decreased by k, how will this change your answer in part (a)?

11. (Continuation of Exercise 10 in Section 3.4) An investment analyst is trying to find out how much business

a secretive TV manufacturer has. The company makes three brands of TV: Brand A, Brand B, and Brand C. The analyst learns that the manufacturer has ordered from suppliers 450,000 type-1 circuit boards, 300,000 type-2 circuit boards, and 350,000 type-3 circuit boards. Brand A uses 2 type-1 boards, 1 type-2 board, and 2 type-3 boards. Brand B uses 3 type-1 boards, 2 type-2 boards, and 1 type-3 board. Brand C uses 1 board of each type.

(a) Set up this problem as a system $\mathbf{Ax} = \mathbf{b}$. Find the inverse of $\mathbf{A}$, and use it to determine how many TVs of each brand are being manufactured.

(b) If the number of type-2 boards used is increased by 100,000, how will this change your answer in part (a)?

(c) If the number of type-1 boards is decreased by 10,000k, how much will this change your answer in part (a)?

12. Why must a matrix be square if it has an inverse?

13. Determine the inverse of the following matrix.

$$\begin{bmatrix} 1 & 0 & 0 & 0 \\ 0 & 1 & 0 & 0 \\ a & 0 & 1 & 0 \\ 0 & 0 & 0 & 1 \end{bmatrix}$$

(Hint: The inverse has a simple form; try trial-and-error guesswork.)

14. Suppose we want to run a Markov chain backward—earlier in time—so that the relation in $\mathbf{p}' = \mathbf{Ap}$ becomes reversed and $\mathbf{p}'$ is used to determine $\mathbf{p}$. For example, suppose you were out of town Sunday and returned early Monday morning to read a newspaper giving the chances of sun, clouds, and rain on Monday; now you want to look backward and determine the chances of sun, clouds, or rain Sunday. Using the matrix equation $\mathbf{p}' = \mathbf{Ap}$ to solve for $\mathbf{p}$, we get $\mathbf{p} = A^{-1}\mathbf{p}'$, with the new transition matrix being A^{-1}. (Warning: A little thought should convince the reader that there is no probability

distribution for today that can guarantee it will be sunny tomorrow, and so if $\mathbf{p'} = \begin{bmatrix} 1 \\ 0 \\ 0 \end{bmatrix}$, $\mathbf{p}\ (= \mathbf{A}^{-1}\mathbf{p'})$ will be something artificial (or perhaps $\mathbf{A}^{-1}$ will not exist). Reverse the following Markov chains by finding the inverse (if it exists) of the transition matrix, and then find the "distribution" today if tomorrow's distribution is $\begin{bmatrix} .5 \\ 0 \\ .5 \end{bmatrix}$. Is this distribution really a probability distribution?

(a) the weather Markov chain

(b) $\begin{bmatrix} .5 & 0 & 0 \\ .5 & 1 & .5 \\ 0 & 0 & .5 \end{bmatrix}$

(c) $\begin{bmatrix} .5 & .25 & 0 \\ .5 & .5 & .5 \\ 0 & .25 & .5 \end{bmatrix}$

(d) $\begin{bmatrix} .4 & .3 & .3 \\ .3 & .4 & .3 \\ .3 & .3 & .4 \end{bmatrix}$

15. Use Gauss–Jordan elimination to find the inverse of the following rotation matrices.

(a) $\mathbf{R}_{60°} = \begin{bmatrix} 1/2 & -\sqrt{3}/2 \\ \sqrt{3}/2 & 1/2 \end{bmatrix}$

(b) $\mathbf{R}_\theta = \begin{bmatrix} \cos\theta & -\sin\theta \\ \sin\theta & \cos\theta \end{bmatrix}$

16. Which integers have no inverse mod 26?

17. (a) Find the system of equations for decrypting the following encryption schemes.

(i) $C_1 = 3L_1 + 5L_2$
$C_2 = 5L_1 + 8L_2$

(ii) $C_1 = 11L_1 + 6L_2$
$C_2 = 8L_1 + 5L_2$

(iii) $C_1 = 2L_1 + 3L_2$
$C_2 = 7L_1 + 5L_2$

(Hint: the inverse of 7 is -11, the inverse of -11 is 7.)

(b) Decrypt the cryptogram pair EF in each of these schemes.

18. (Computer Project) Use a computer algebra system to determine the inverse of the following matrices.

(a) $\begin{bmatrix} 1 & 2 & 0 \\ 1 & a & 3 \\ 0 & 1 & 2 \end{bmatrix}$

(b) $\begin{bmatrix} a & 1 & 1 \\ 1 & a & 1 \\ 1 & 1 & a \end{bmatrix}$

(c) $\begin{bmatrix} a & b & c \\ 1 & 0 & 1 \\ c & b & a \end{bmatrix}$

19. (Computer Project) For a scalar a, with $|a| < 1$, $1/(1 - a) = 1 + a + a^2 + a^3 + \dots$. The same identity is true for matrices $\mathbf{A}$ whose "size" is less than 1:

$$(\mathbf{I} - \mathbf{A})^{-1} = \mathbf{I} + \mathbf{A} + \mathbf{A}^2 + \mathbf{A}^2 + \cdots \quad (*)$$

One simple measure of the size of $\mathbf{A}$ is the sum of the entries in each column. If all column sums are less than 1 (for negative entries, use their absolute value), then (*) is valid. Verify this by computing the inverse of $(\mathbf{I} - \mathbf{A})$ by (*), and confirm the result by computing the inverse by the method in this section. Approximate (*) by using a computer to sum $\mathbf{I}$ and the first 8 powers of $\mathbf{A}$ in (*) for the following matrices.

(a) $\begin{bmatrix} .4 & .3 \\ .1 & .3 \end{bmatrix}$

(b) $\begin{bmatrix} .3 & .2 & 0 \\ .2 & .1 & .2 \\ 0 & .2 & .1 \end{bmatrix}$

SECTION 3.6 *Determinant of a Matrix*

In the 1700's when mathematicians first looked systematically at solving systems of linear equations, they wanted to get a formula for their solution in the spirit of the solution to a quadratic equation, $ax^2 + bx + c = 0$, given by the quadratic formula,

$$x = \frac{-b \pm \sqrt{b^2 - 4ac}}{2a}.$$

What grew out of this objective was the theory of determinants and Cramer's rule (given ahead), which does provide the desired "formula" for solving a system of linear equations. The easier approach (with less computation) for solving a system of linear equations using Gaussian elimination was not developed by Gauss until a century later.

Determinant Formula for Solving 2 × 2 Systems

We want a general formula for solving a system of two linear equations in two variables. This formula will express the variables in terms of the coefficients and right-side entries. To obtain such a formula, we need to solve the following system where all coefficients are left as parameters to be specified.

$$\mathbf{Ax} = \mathbf{b}: \quad \begin{matrix} a_{11}x_1 + a_{12}x_2 = b_1 \\ a_{21}x_1 + a_{22}x_2 = b_2 \end{matrix} \tag{1}$$

Multiplying the first equation by a_{22} and the second by a_{12} and then subtracting, we obtain

$$\begin{aligned} a_{11}a_{22}x_1 + a_{12}a_{22}x_2 &= a_{22}b_1 \\ -(a_{12}a_{21}x_1 + a_{12}a_{22}x_2 &= a_{12}b_2) \\ \hline (a_{11}a_{22} - a_{12}a_{21})x_1 &= a_{22}b_1 - a_{12}b_2 \end{aligned}$$

Solving for x_1, we have

$$x_1 = \frac{a_{22}b_1 - a_{12}b_2}{a_{11}a_{22} - a_{12}a_{21}}. \tag{2}$$

Substituting (2) in the first equation of (1) and simplifying, we obtain

$$x_2 = \frac{a_{11}b_2 - a_{21}b_1}{a_{11}a_{22} - a_{12}a_{21}}. \tag{3}$$

Formulas (2) and (3) are messy, but they give answers for solving any 2 × 2 system. Note that (2) and (3) will fail to give answers if their denominator is zero. Zero denominators in (2) and (3) will become a very important issue in the next section, where we do not want a unique solution to a certain system of linear equations.

On the other hand, if the denominator $a_{11}a_{22} - a_{12}a_{21}$ is not zero, then the system of equations has a solution that is uniquely given by (2) and (3).

EXAMPLE 1 Solving a 2 × 2 System by Formula Consider the following system of two linear equations in two variables.

$$\begin{matrix} 2x_1 - 3x_2 = 4 \\ x_1 + 2x_2 = 9 \end{matrix} \tag{4}$$

Let us use formulas (2) and (3) to solve for x_1 and x_2.

$$x_1 = \frac{a_{22}b_1 - a_{12}b_2}{a_{11}a_{22} - a_{12}a_{21}} = \frac{2 \cdot 4 - (-3) \cdot 9}{2 \cdot 2 - (-3) \cdot 1} = \frac{35}{7} = 5$$

$$x_2 = \frac{a_{11}b_2 - a_{21}b_1}{a_{11}a_{22} - a_{12}a_{21}} = \frac{2 \cdot 9 - 4 \cdot 1}{2 \cdot 2 - (-3) \cdot 1} = \frac{14}{7} = 2 \; \blacktriangle \tag{5}$$

It is possible to extend these formulas to obtain expressions for the solutions to three linear equations in three variables, and more generally to n linear equations in n variables. However, these expressions become huge, and evaluating them normally takes far longer than solving by Gaussian elimination.

The common denominator, $a_{11}a_{22} - a_{12}a_{21}$, in (2) and (3) is called the determinant of system (1).

DEFINITION

The *determinant, det*($\mathbf{A}$), of a 2×2 matrix $\mathbf{A} = \begin{bmatrix} a_{11} & a_{12} \\ a_{21} & a_{22} \end{bmatrix}$ is defined to be

$$\det(\mathbf{A}) = a_{11}a_{22} - a_{12}a_{21}.$$

The determinant of $\mathbf{A} = \begin{bmatrix} a_{11} & a_{12} \\ a_{21} & a_{22} \end{bmatrix}$ is often written $\begin{vmatrix} a_{11} & a_{12} \\ a_{21} & a_{22} \end{vmatrix}$. In the 2×2 case, det($\mathbf{A}$) is simply the product of the two main diagonal entries minus the product of the two off-diagonal entries. A complex recursive formula for determinants of larger matrices will be given shortly. As in the case of 2×2 matrices, the determinant of any square matrix $\mathbf{A}$ will turn out to be the denominator in the algebraic expressions for the solution of the system $\mathbf{Ax} = \mathbf{b}$.

We can write the numerators in formulas (2) and (3) for x_1 and x_2 as determinants of the matrices obtained by replacing the first and second columns, respectively, of $\mathbf{A}$ by the right-side vector $\mathbf{b}$. That is, let

$$\mathbf{A}_1 = \begin{bmatrix} b_1 & a_{12} \\ b_2 & a_{22} \end{bmatrix} \quad \text{and} \quad \mathbf{A}_2 = \begin{bmatrix} a_{11} & b_1 \\ a_{21} & b_2 \end{bmatrix}. \quad (6)$$

Then

$$\det(\mathbf{A}_1) = \begin{vmatrix} b_1 & a_{12} \\ b_2 & a_{22} \end{vmatrix} = b_1 a_{22} - a_{12} b_2,$$

$$\hspace{10cm} (7)$$

$$\det(\mathbf{A}_2) = \begin{vmatrix} a_{11} & b_1 \\ a_{21} & b_2 \end{vmatrix} = a_{11} b_2 - b_1 a_{21}.$$

The expressions in (7) are exactly the numerators in (2) and (3). So using $\det(\mathbf{A}_1)$ and $\det(\mathbf{A}_2)$, our formulas for x_1 and x_2 are

$$x_1 = \frac{\det(\mathbf{A}_1)}{\det(\mathbf{A})} \quad \text{and} \quad x_2 = \frac{\det(\mathbf{A}_2)}{\det(\mathbf{A})}.$$

The numerators in the solutions of the systems of n linear equations in n variables turn out to have the same form. That is, if we define $\mathbf{A}_i$ to be the matrix obtained from $\mathbf{A}$ by replacing the ith column $\mathbf{a}_i$ of $\mathbf{A}$ by the right-side vector $\mathbf{b}$:

$$\mathbf{A}_i = [\mathbf{a}_1, \mathbf{a}_2, \dots, \mathbf{a}_{i-1}, \mathbf{b}, \mathbf{a}_{i+1}, \dots, \mathbf{a}_n], \quad (8)$$

then the solution to $\mathbf{Ax} = \mathbf{b}$ is given by the following determinant-based formula, called *Cramer's rule.*

Theorem 1. Let $\mathbf{Ax} = \mathbf{b}$ be a system of n linear equations in n variables. If $\det(\mathbf{A}) \neq 0$, then the solution is given by $x_i = \det(\mathbf{A}_i)/\det(\mathbf{A})$, $i = 1, 2, \dots, n$.

Gabriel Cramer was a Swiss mathematician who lived from 1704 to 1752. We shall not prove Cramer's rule. (The interested reader can refer to H. Anton (1995), *Elementary Linear Algebra,* 7th ed., New York: John Wiley & Sons.)

Applying Cramer's rule to the system of equations in Example 1,

$$2x_1 - 3x_2 = 4,$$
$$x_1 + 2x_2 = 9,$$

we obtain

$$x_1 = \frac{\begin{vmatrix} 4 & -3 \\ 9 & 2 \end{vmatrix}}{\begin{vmatrix} 2 & -3 \\ 1 & 2 \end{vmatrix}} = \frac{4 \cdot 2 - (-3) \cdot 9}{2 \cdot 2 - (-3) \cdot 1} = \frac{35}{7} = 5,$$

$$x_2 = \frac{\begin{vmatrix} 2 & 4 \\ 1 & 9 \end{vmatrix}}{\begin{vmatrix} 2 & -3 \\ 1 & 2 \end{vmatrix}} = \frac{2 \cdot 9 - 4 \cdot 1}{2 \cdot 2 - (-3) \cdot 1} = \frac{14}{7} = 2.$$

The same solution we obtained earlier in (5).

Cramer's rule also yields an easy-to-remember formula for the entries in the inverse $\mathbf{A}^{-1} = \begin{bmatrix} a_{11}^* & a_{12}^* \\ a_{21}^* & a_{22}^* \end{bmatrix}$ of a 2 × 2 matrix $\mathbf{A}$. Recall that since $\mathbf{A}\mathbf{A}^{-1} = \mathbf{I}$, then the first column $\mathbf{a}_1^*$ of the inverse satisfies $\mathbf{A}\mathbf{a}_1^* = \begin{bmatrix} 1 \\ 0 \end{bmatrix}$, and the second column $\mathbf{a}_2^*$ of the inverse satisfies $\mathbf{A}\mathbf{a}_2^* = \begin{bmatrix} 0 \\ 1 \end{bmatrix}$. (See Theorem 3 in Section 3.5 for a discussion of these equations satisfied by the column vectors of the inverse.) For any 2 × 2 matrix $\mathbf{A}$, the system of equations $\mathbf{A}\mathbf{a}_1^* = \begin{bmatrix} 1 \\ 0 \end{bmatrix}$ has the solution by Cramer's rule:

$$a_{11}^* = \frac{\begin{vmatrix} 1 & a_{12} \\ 0 & a_{22} \end{vmatrix}}{\det(\mathbf{A})} = \frac{a_{22}}{\det(\mathbf{A})},$$

$$a_{21}^* = \frac{\begin{vmatrix} a_{11} & 1 \\ a_{21} & 0 \end{vmatrix}}{\det(\mathbf{A})} = \frac{-a_{21}}{\det(\mathbf{A})}. \tag{9}$$

The simple form of the numerator comes from having a right-side vector of $\begin{bmatrix} 1 \\ 0 \end{bmatrix}$. The same simplification occurs in solving $\mathbf{A}\mathbf{a}_2^* = \begin{bmatrix} 0 \\ 1 \end{bmatrix}$ by Cramer's rule, yielding

$$a_{12}^* = \frac{-a_{12}}{\det(\mathbf{A})}, \qquad a_{22}^* = \frac{a_{11}}{\det(\mathbf{A})}.$$

These single-number numerators lead to the following general formula for the inverse of a 2 × 2 matrix.

Formula for Inverse of a 2 × 2 Matrix

If $\mathbf{A} = \begin{bmatrix} a_{11} & a_{12} \\ a_{21} & a_{22} \end{bmatrix}$, then

$$\mathbf{A}^{-1} = \frac{1}{\det(\mathbf{A})} \begin{bmatrix} a_{22} & -a_{12} \\ -a_{21} & a_{11} \end{bmatrix}. \tag{10}$$

In words, the inverse of a 2 × 2 matrix $\mathbf{A}$ is obtained as follows: Interchange the two diagonal entries, change the sign of the two off-diagonal entries, and then divide all entries of $\mathbf{A}$ by the determinant. No such simple construction of the inverse is possible for larger matrices.

E X A M P L E   2   Inverse of a 2 × 2 Matrix
Let us use (10) to obtain the inverse of the 2 × 2 matrix $\mathbf{A} = \begin{bmatrix} 3 & 1 \\ 4 & 2 \end{bmatrix}$. (This matrix's inverse was computed by elimination in Example 4 of Section 3.5.) We first compute $\det(\mathbf{A}) = 3 \cdot 2 - 1 \cdot 4 = 2$. Then by (10),

$$\mathbf{A}^{-1} = \tfrac{1}{2} \begin{bmatrix} 2 & -1 \\ -4 & 3 \end{bmatrix} = \begin{bmatrix} 1 & -\tfrac{1}{2} \\ -2 & \tfrac{3}{2} \end{bmatrix}. \ \blacktriangle \tag{11}$$

Determinants for 3 × 3 and Larger Systems

We now present a general recursive formula for the determinant of an $n \times n$ matrix. We need to introduce two terms. The (i, j)-*minor* $\mathbf{A}_{ij}$ of a matrix $\mathbf{A}$ is a submatrix of $\mathbf{A}$ obtained by deleting the ith row and jth column of $\mathbf{A}$. The (i, j)-*cofactor* C_{ij} of $\mathbf{A}$ is the determinant of the minor $\mathbf{A}_{ij}$ multiplied by $(-1)^{i+j}$:

$$C_{ij} = (-1)^{i+j}\det(\mathbf{A}_{ij}). \tag{12}$$

General Definition of a Determinant

The determinant of an $n \times n$ matrix $\mathbf{A}$ is defined to be

$$\det(\mathbf{A}) = a_{11}C_{11} + a_{12}C_{12} + a_{13}C_{13} + \cdots + a_{1n}C_{1n}. \tag{13}$$

Actually, in formula (13) the terms a_{11}, a_{12}, $a_{13}, \ldots, a_{1n}$ (and associated cofactors) can be replaced by any row (or column) so that our definition generalizes to

$$\det(\mathbf{A}) = a_{i1}C_{i1} + a_{i2}C_{i2} + a_{i3}C_{i3} + \cdots + a_{in}C_{in}, \tag{14}$$

for some i, $1 \leq i \leq n$.

In the case of the determinant of a 3×3 matrix $\mathbf{A}$, the cofactors involve 2×2 determinants, which we already know how to compute. Thus, we have

$$
\det(\mathbf{A}) = \begin{vmatrix} a_{11} & a_{12} & a_{13} \\ a_{21} & a_{22} & a_{23} \\ a_{31} & a_{32} & a_{33} \end{vmatrix} = a_{11}\begin{vmatrix} a_{22} & a_{23} \\ a_{32} & a_{33} \end{vmatrix}
$$

$$
- a_{12}\begin{vmatrix} a_{21} & a_{23} \\ a_{31} & a_{33} \end{vmatrix} + a_{13}\begin{vmatrix} a_{21} & a_{22} \\ a_{31} & a_{32} \end{vmatrix}
$$

$$
= a_{11}(a_{22}a_{33} - a_{23}a_{32}) - a_{12}(a_{21}a_{33} - a_{23}a_{31}) + a_{13}(a_{21}a_{32} - a_{22}a_{31}). \tag{15}
$$

EXAMPLE 3   Determinant of a 3 × 3 Matrix

Using (15), we can compute the following 3×3 determinant:

$$
\begin{vmatrix} 1 & 2 & 3 \\ 4 & 5 & 6 \\ 7 & 8 & 0 \end{vmatrix} = 1\begin{vmatrix} 5 & 6 \\ 8 & 0 \end{vmatrix} - 2\begin{vmatrix} 4 & 6 \\ 7 & 0 \end{vmatrix} + 3\begin{vmatrix} 4 & 5 \\ 7 & 8 \end{vmatrix}
$$

$$
= 1(5 \cdot 0 - 6 \cdot 8) - 2(4 \cdot 0 - 6 \cdot 7) + 3(4 \cdot 8 - 5 \cdot 7) = 1(-48)
$$

$$
- 2(-42) + 3(-3)
$$

$$
= -48 + 84 - 9 = 27. \ \blacktriangle
$$

Knowing how to compute the determinant of a 3×3 matrix allows us to compute the determinant of a 4×4 matrix, since the right side in formula (13) involves 3×3 cofactors. In general, with enough stamina (or better, a computer program) we can compute the determinant of a matrix of any size. We note that there exists a simpler way using Gaussian elimination to compute large determinants.

For the determinant of a 3×3 matrix $\mathbf{A}$, there is a picture that can be used in which one multiplies the numbers lying on the 6 "diagonals" in the augmented 3×5 array shown below (no such approach is available for larger matrices). The products marked by solid lines have plus signs, and the products marked by dashed lines have minus signs.

$$
det(\mathbf{A}) = \begin{vmatrix} a_{11} & a_{12} & a_{13} & a_{11} & a_{12} \\ a_{21} & a_{22} & a_{23} & a_{21} & a_{22} \\ a_{31} & a_{32} & a_{33} & a_{31} & a_{32} \end{vmatrix}
$$

$$
= a_{11}a_{22}a_{33} + a_{12}a_{23}a_{31} + a_{13}a_{21}a_{32}
$$

$$
- a_{13}a_{22}a_{31} - a_{11}a_{23}a_{32} - a_{12}a_{21}a_{33} \tag{16}
$$

EXAMPLE 4   Solving the Refinery Equations by Cramer's Rule   In Section 3.1, we presented a system of equations for controlling

the production of three oil refineries. Let us solve this system with Cramer's rule.

Heating oil	$16x_1 + 8x_2 + 8x_3 = 9600$
Diesel oil	$8x_1 + 20x_2 + 8x_3 = 12800$
Gasoline	$4x_1 + 10x_2 + 20x_3 = 16000$

The determinant of the coefficient matrix for this system of equations is, by (15),

$$\det(\mathbf{A}) = \begin{vmatrix} 16 & 8 & 8 \\ 8 & 20 & 8 \\ 4 & 10 & 20 \end{vmatrix} = 16 \begin{vmatrix} 20 & 8 \\ 10 & 20 \end{vmatrix}$$
$$- 8 \begin{vmatrix} 8 & 8 \\ 4 & 20 \end{vmatrix} + 8 \begin{vmatrix} 8 & 20 \\ 4 & 10 \end{vmatrix}$$
$$= 16 \cdot (20 \cdot 20 - 8 \cdot 10) - 8(8 \cdot 20 - 8 \cdot 4)$$
$$+ 8(8 \cdot 10 - 20 \cdot 4) = 16(320)$$
$$- 8(128) + 8(0) = 4096. \tag{17}$$

To determine x_1 using Cramer's rule, we need $\det(\mathbf{A}_1)$. Recall that $\mathbf{A}_1$ is obtained from $\mathbf{A}$ by replacing the first column of $\mathbf{A}$ by the numbers on the right side of the equations. This time we use (16) to compute $\det(\mathbf{A}_1)$:

$$\det(\mathbf{A}_1) = \begin{vmatrix} 9600 & 8 & 8 \\ 12800 & 20 & 8 \\ 16000 & 10 & 20 \end{vmatrix} = 9600 \cdot 20 \cdot 20$$
$$+ 8 \cdot 8 \cdot 16000 + 8 \cdot 12800 \cdot 10$$
$$- 8 \cdot 20 \cdot 16000 - 8 \cdot 12800 \cdot 20$$
$$- 9600 \cdot 8 \cdot 10$$
$$= 3,840,000 + 1,024,000 + 1,024,000$$
$$- 2,560,000 - 2,048,000 - 768,000$$
$$= 512,000. \tag{18}$$

We thus have

$$x_1 = \frac{\det(\mathbf{A}_1)}{\det(\mathbf{A})} = \frac{512000}{4096} = 125.$$

In a similar fashion we use (16) to compute $\det(\mathbf{A}_2)$:

$$\det(\mathbf{A}_2) = \begin{vmatrix} 16 & 9600 & 8 \\ 8 & 12800 & 8 \\ 4 & 16000 & 20 \end{vmatrix} = 16 \cdot 12800 \cdot 20$$
$$+ 9600 \cdot 8 \cdot 4 + 8 \cdot 8 \cdot 16000$$
$$- 8 \cdot 12800 \cdot 4 - 8 \cdot 16000 \cdot 16$$
$$- 20 \cdot 8 \cdot 9600$$
$$= 4,096,000 + 307,200 + 1,024,000$$
$$- 409,600 - 2,048,000 - 1,536,000$$
$$= 14,336,000. \tag{19}$$

And thus

$$x_2 = \frac{\det(\mathbf{A}_2)}{\det(\mathbf{A})} = \frac{1433600}{4096} = 350.$$

It is left as an exercise for the reader to use compute $\det(\mathbf{A}_3)$ and to determine x_3. ▲

Some Theory of Determinants

We conclude this section with a little theory about determinants. The first result is stated without proof.

Theorem 2. The determinant of a matrix product is the product of the determinants:

$$\det(\mathbf{AB}) = \det(\mathbf{A})\det(\mathbf{B}).$$

EXAMPLE 5 Determinants of Products

Consider the matrix $\mathbf{A} = \begin{bmatrix} 3 & 1 \\ 4 & 2 \end{bmatrix}$ and the matrix $\mathbf{B} = \begin{bmatrix} 3 & 1 \\ 1 & 1 \end{bmatrix}$. $\det(\mathbf{A}) = 6 - 4 = 2$ and $\det(\mathbf{B}) = 3 - 1 = 2$. Then $\mathbf{AB} = \begin{bmatrix} 10 & 4 \\ 14 & 6 \end{bmatrix}$ and $\det(\mathbf{AB}) = 60 - 56 = 4$, confirming that $\det(\mathbf{AB}) = \det(\mathbf{A})\det(\mathbf{B})$. ▲

Theorem 3. Let $\mathbf{A}$ be a square matrix, and suppose that the inverse $\mathbf{A}^{-1}$ exists. Then

a. $\det(\mathbf{A}) \neq 0$,

b. $\det(\mathbf{A}^{-1}) = 1/\det(\mathbf{A})$.

Proof. Since $\mathbf{AA}^{-1} = \mathbf{I}$ and, as noted earlier, $\det(\mathbf{I}) = 1$, then by Theorem 2 we have

$$\det(\mathbf{A})\det(\mathbf{A}^{-1}) = \det(\mathbf{I}) = 1. \qquad (20)$$

If the product of two real numbers is 1, both numbers must be nonzero. Thus, part (a) follows. Dividing both sides of Eq. (20) by $\det(\mathbf{A})$ gives part (b).

▲ ▲ ▲

Theorem 4. Our next result is known as the fundamental theorem for solving $\mathbf{Ax} = \mathbf{b}$. The following four statements are equivalent for any $n \times n$ matrix $\mathbf{A}$.

a. For some particular vector $\mathbf{b}$, the system of equations $\mathbf{Ax} = \mathbf{b}$ has a unique solution.

b. For all $\mathbf{b}$, the system of equations $\mathbf{Ax} = \mathbf{b}$ always has a unique solution.

c. $\mathbf{A}$ has an inverse.

d. $\det(\mathbf{A}) \neq 0$.

Proof. In Theorem 4 of Section 3.5, we proved the equivalence of parts (a), (b), and (c). By Theorem 3, (c) implies (d). Theorem 1 (Cramer's rule) says that (d) implies (a).

▲ ▲ ▲

Theorem 4 is an example of how mathematics likes to look at a problem—in this case, solving a system of linear equations—from many different points of view. We take Theorem 4 of the previous section, which relates solutions using Gaussian elimination to the existence of an inverse. Now we add another condition, that the determinant is nonzero. No matter what the applied setting from which a system of linear equations arises, this theorem is waiting to help scientists determine if the system has a unique solution. Sometimes the system will need to be solved by elimination to determine if there is a solution and if it is unique. But in some cases the existence of a unique solution for another $\mathbf{b}$-vector, say, may be known, and Theorem 4 then guarantees a unique solution and the existence of an inverse.

Notice that this theorem is only concerned with unique solutions. If there is no inverse or if the determinant is zero, it is possible that there could be multiple solutions to $\mathbf{Ax} = \mathbf{b}$. Conversely, if we want multiple solutions, then the determinant must be zero. This consequence of Theorem 4 plays an important role in the next section when we require multiple solutions to a system of linear equations.

EXAMPLE 6 Determinants of Inverses and Powers

To illustrate the previous theorems, consider the matrix $\mathbf{A} = \begin{bmatrix} 3 & 1 \\ 4 & 2 \end{bmatrix}$ whose inverse was seen to be

$$A^{-1} = \begin{bmatrix} 1 & -\frac{1}{2} \\ -2 & \frac{3}{2} \end{bmatrix}$$

in Example 2. Observe that $\det(A) = 3 \cdot 2 - 1 \cdot 4 = 2$.

(i) *Inverses.* By Theorem 3, $\det(A^{-1}) = 1/\det(A) = 1/2$. We confirm this by computing the determinant of A^{-1} directly: $\det(A^{-1}) = 1 \cdot (3/2) - (-1/2) \cdot (-2) = 1/2$.

(ii) *Powers.* Consider next $A^2 = AA$. By Theorem 2, $\det(A^2) = \det(A)\det(A) = (\det(A))^2$. Since $\det(A) = 2$, then $\det(A^2) = 2^2 = 4$. We confirm this by computing A^2:

$$A^2 = \begin{bmatrix} 3 & 1 \\ 4 & 2 \end{bmatrix}\begin{bmatrix} 3 & 1 \\ 4 & 2 \end{bmatrix} = \begin{bmatrix} 13 & 5 \\ 20 & 8 \end{bmatrix} \quad \text{and} \quad \det(A^2)$$
$$= 13 \cdot 8 - 5 \cdot 20 = 104 - 100 = 4. \qquad (21)$$

By iterating the product rule for determinants, we see that $\det(A^k) = (\det(A))^k = 2^k$. ▲

Exercises for Section 3.6

1. Compute the determinant of the following matrices.

(a) $\begin{vmatrix} 2 & 4 \\ -3 & -6 \end{vmatrix}$ (c) $\begin{vmatrix} 1 & 3 \\ 5 & -2 \end{vmatrix}$

(b) $\begin{vmatrix} 3 & 2 \\ 1 & 0 \end{vmatrix}$

2. Find the (unique) solution to the following systems of equations, if possible, using Cramer's rule.

(a) $3x + y = 7$
    $2x - 2y = 7$

(b) $x + y = 34$
    $2x - y = 30$

3. Consider the two-refinery production of diesel oil and gasoline. The second refinery, which has yet to be built, will produce 15 gallons of gasoline and k gallons of diesel oil from each barrel of crude oil. We have

Diesel oil $10x_1 + kx_2 = D,$

Gasoline $5x_1 + 15x_2 = G,$

where k is to be determined, D is the demand for diesel oil, G is the demand for gasoline, and x_i is number of barrels of crude oil processed by refinery $i, i = 1, 2$. Solve this system of equations to determine x_1 and x_2 in terms

of k, D, and G using Cramer's rule. What values of k yield a nonunique solution? In practical terms, what does this nonuniqueness mean?

4. Does either of the following systems of equation have a nonzero solution? If the solution is not unique, give the set of all possible solutions.

(a) $2x - 6y = 0$
    $-x + 3y = 0$

(b) $3x + 4y = 0$
    $6x + 2y = 0$

5. Compute the following determinants.

(a) $\begin{vmatrix} 2 & 0 & -1 \\ 0 & 1 & 3 \\ -4 & 0 & 2 \end{vmatrix}$ (c) $\begin{vmatrix} 2 & 1 & 0 \\ 0 & 0 & 2 \\ 2 & 2 & 2 \end{vmatrix}$

(b) $\begin{vmatrix} 1 & 0 & 1 \\ 0 & 2 & 2 \\ 1 & 2 & 3 \end{vmatrix}$ (d) $\begin{vmatrix} 1/6 & 1/7 & 1/8 \\ 1/7 & 1/8 & 1/9 \\ 1/8 & 1/9 & 1/10 \end{vmatrix}$

6. Use Cramer's rule to solve for x_3 in Example 4.

7. If you double the first column in the system

$$ax + by = e,$$
$$cx + dy = f,$$

show using Cramer's rule that the value of x is half as large and the value of y is unchanged.

8. If **A**, **B**, and **C** are 4×4 matrices such that $\mathbf{C} = \mathbf{AB}$, $\det(\mathbf{A}) = 3$, and $\det(\mathbf{C}) = 6$, determine
   (a) $\det(\mathbf{B})$ (c) $\det(\mathbf{ABC})$
   (b) $\det(4\mathbf{B})$ (d) $\det(\mathbf{B}^{-1}\mathbf{A})$

9. (**Computer Project**) Use a computer algebra system to find the determinant of $\mathbf{A}^{-1}$, where

$$\mathbf{A} = \begin{bmatrix} a_{11} & a_{12} \\ a_{21} & a_{22} \end{bmatrix}.$$

Show that $\det(\mathbf{A}^{-1}) = 1/\det(\mathbf{A})$.

10. (**Computer Project**) Use a computer algebra system to determine the inverse of the matrix

$$\begin{bmatrix} a & b & c \\ d & e & f \\ g & h & i \end{bmatrix},$$

and confirm that the common denominator in all terms of the inverse is the determinant.

11. (**Computer Project**) Let

$$\mathbf{A} = \begin{bmatrix} a_{11} & a_{12} \\ a_{21} & a_{22} \end{bmatrix} \quad \text{and} \quad \mathbf{B} = \begin{bmatrix} b_{11} & b_{12} \\ b_{21} & b_{22} \end{bmatrix}$$

be arbitrary 2×2 matrices. Use a computer algebra system to multiply **AB** symbolically and to determine $\det(\mathbf{AB})$. Then multiply out $\det(\mathbf{A})\det(\mathbf{B})$. Show that $\det(\mathbf{AB}) = \det(\mathbf{A})\det(\mathbf{B})$.

SECTION 3.7 *Eigenvectors and Eigenvalues*

When **A** is a square matrix, multiplying a vector by **A** occasionally has exactly the same effect as multiplying it by a simple scalar λ. When $\mathbf{Au} = \lambda\mathbf{u}$ (where $\mathbf{u} \neq 0$), the vector **u** is called an *eigenvector* of **A** (*eigen* is the German word for proper) and the scalar λ is called an *eigenvalue* of **A**. While such an occurrence might seem unlikely, it is actually common in many matrix models. For example, in the weather Markov chain introduced in Section 3.1, we saw as we experimented with computing successive probability distributions over many periods that eventually the probability vector converged to a stable distribution **p*** that stayed the same from one period to the next, that is, $\mathbf{Ap^*} = \mathbf{p^*}$.

EXAMPLE 1 Stable Probability Vector for Weather Markov Chain In Section 3.1 we introduced the weather Markov chain with transition matrix,

$$
\mathbf{A} = \begin{array}{c} \\ \text{Sunny} \\ \text{Cloudy} \\ \text{Rainy} \end{array}
\begin{array}{c} \text{Today} \\ \begin{array}{ccc} \text{Sunny} & \text{Cloudy} & \text{Rainy} \end{array} \\ \begin{bmatrix} \frac{3}{4} & \frac{1}{2} & \frac{1}{4} \\ \frac{1}{8} & \frac{1}{4} & \frac{1}{2} \\ \frac{1}{8} & \frac{1}{4} & \frac{1}{4} \end{bmatrix} \end{array}.
$$

We showed in Section 3.4 that

$$\mathbf{p^*} = \begin{bmatrix} \frac{14}{23} \\ \frac{5}{23} \\ \frac{4}{23} \end{bmatrix}$$

is a stable probability distribution for this transition matrix $\mathbf{A}$. That is,

$$\mathbf{p^*} = \mathbf{Ap^*} = \begin{bmatrix} \frac{3}{4} & \frac{1}{2} & \frac{1}{4} \\ \frac{1}{8} & \frac{1}{4} & \frac{1}{2} \\ \frac{1}{8} & \frac{1}{4} & \frac{1}{4} \end{bmatrix} \begin{bmatrix} \frac{14}{23} \\ \frac{5}{23} \\ \frac{4}{23} \end{bmatrix}$$

$$= \begin{bmatrix} \frac{3}{4}\cdot\frac{14}{23} + \frac{1}{2}\cdot\frac{5}{23} + \frac{1}{4}\cdot\frac{4}{23} \\ \frac{1}{8}\cdot\frac{14}{23} + \frac{1}{4}\cdot\frac{5}{23} + \frac{1}{2}\cdot\frac{4}{23} \\ \frac{1}{8}\cdot\frac{14}{23} + \frac{1}{4}\cdot\frac{5}{23} + \frac{1}{4}\cdot\frac{4}{23} \end{bmatrix} = \begin{bmatrix} \frac{14}{23} \\ \frac{5}{23} \\ \frac{4}{23} \end{bmatrix}.$$

Thus $\mathbf{Ap^*} = \mathbf{p^*}$, and $\mathbf{p^*}$ is an eigenvector of $\mathbf{A}$ with eigenvalue 1. ▲

This section introduces eigenvalues and eigenvectors and tries to give the reader some sense of their great usefulness. The most important use of eigenvectors in science and engineering arises in differential equations, which are beyond the scope of this book. However, eigenvalues and eigenvectors also play a major role in analyzing growth models.

E X A M P L E 2 Eigenvalues in a Growth Model We consider an industrial growth model for pollution and industrial development in a developing country, introduced in Example 4 of Section 3.1. Let P be the current level of pollution and D the current level of industrial development. (Each will be measured in units that combine a variety of appropriate indices, such as carbon monoxide levels in air, contaminants in rivers, etc., for pollution.) Let P' and D' be the levels of pollution and industrial development, respectively, in five years. Suppose that based on experiences in similar developing countries, an in-

ternational development agency believes that the following simple linear model should be a useful predictor of pollution and industrial development over successive five-year periods:

$$\begin{aligned} P' &= P + 2D, \\ D' &= 2P + D. \end{aligned} \tag{1}$$

If initially we had $P = 1$, $D = 1$, then we compute $P' = 1(1) + 2(1) = 3$, $D' = 2(1) + 1(1) = 3$. Letting $P = 3$, $D = 3$, we obtain $P' = 1(3) + 2(3) = 9$, $D' = 2(3) + 1(3) = 9$. Extending these calculations, we see that for $P = a$ and $D = a$, we will obtain $P' = 3a$ and $D' = 3a$. That is, if

$$\begin{bmatrix} P \\ D \end{bmatrix} = \begin{bmatrix} a \\ a \end{bmatrix}, \quad \text{then} \quad \begin{bmatrix} P' \\ D' \end{bmatrix} = \begin{bmatrix} 3a \\ 3a \end{bmatrix} = 3\begin{bmatrix} a \\ a \end{bmatrix}.$$

So 3 is an eigenvalue of $\mathbf{A} = \begin{bmatrix} 1 & 2 \\ 2 & 1 \end{bmatrix}$, the coefficient matrix in (1), and any multiple of $\begin{bmatrix} 1 \\ 1 \end{bmatrix}$, that is, a vector of the form $\begin{bmatrix} a \\ a \end{bmatrix}$, is an associated eigenvector of $\mathbf{A}$. Looking at powers of $\mathbf{A}$, we have

$$\mathbf{A}^2\begin{bmatrix} a \\ a \end{bmatrix} = \mathbf{A}\left(\mathbf{A}\begin{bmatrix} a \\ a \end{bmatrix}\right) = \mathbf{A}\left(3\begin{bmatrix} a \\ a \end{bmatrix}\right) = 3\mathbf{A}\begin{bmatrix} a \\ a \end{bmatrix} = 3^2\begin{bmatrix} a \\ a \end{bmatrix},$$

and in general, one obtains

$$\mathbf{A}^k\begin{bmatrix} a \\ a \end{bmatrix} = 3^k\begin{bmatrix} a \\ a \end{bmatrix}. \tag{2}$$

Note that if we initially had the (nonsense) vector

Eigenvalues and the Stability of Buckeyballs

For a long time, chemists thought that carbon atoms could attach themselves to other carbon atoms in fixed patterns in only two ways: diamond or graphite. But in 1985 it was discovered that 60 carbon atoms could take up positions at the corners of a truncated icosahedron, as in the Buckeyball shown to the right. The lines connecting the corners represent chemical bonds between the atoms. This is the same pattern we often see in soccer balls.

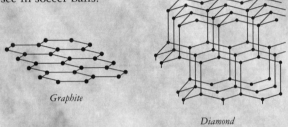

Graphite

Diamond

This discovery was considered revolutionary, and a new field of chemistry was born. Scientists began to speculate about related ways of hooking up carbon atoms and what applications such molecules might have in medicine and other areas. Theoretical chemists started by drawing other patterns of carbon molecules on paper and wondering which ones could actually exist. Before trying to build one of these theoretical possibilities, chemists like to have an idea of what its properties might turn out to be. One of the most im-

portant is the degree of stability—are the bonds strong enough to hold the atoms despite their natural vibrations and gyrations? It turns out that this can be predicted before building the molecule. Start by drawing the graph of the molecule with vertices representing atoms and edges representing chemical bonds (for the Buckeyball this would be the figure shown below). Then work out the adjacency matrix of

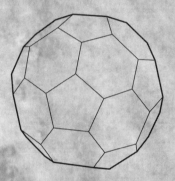

the graph (a matrix with 60 rows and 60 columns in the case of the Buckeyball). Next work out the eigenvalues of this matrix. These eigenvalues are plugged into a simple arithmetic formula, and the size of the resulting number indicates the stability of the potential molecule. This is based on the elegant and powerful theory of Erich Hückel, a chemist with an interest in mathematics.

$$\begin{bmatrix} P \\ D \end{bmatrix} = \begin{bmatrix} 1 \\ -1 \end{bmatrix}, \quad \text{then} \quad \begin{bmatrix} P' \\ D' \end{bmatrix} = \begin{bmatrix} -1 \\ 1 \end{bmatrix}.$$

So -1 is also an eigenvalue of $\mathbf{A}$ with eigenvector $\begin{bmatrix} 1 \\ -1 \end{bmatrix}$ (or any multiple of $\begin{bmatrix} 1 \\ -1 \end{bmatrix}$). ▲

Even though the eigenvector $\begin{bmatrix} 1 \\ -1 \end{bmatrix}$ makes no practical sense in this model, this eigenvector will play just as important a role as the "practical" eigenvector $\begin{bmatrix} 1 \\ 1 \end{bmatrix}$.

This property of matrix multiplication acting like scalar multiplication for certain vectors happens for all matrices. It is the key to understanding the behavior of many linear models. An $n \times n$ matrix usually has n eigenvalues, each with an infinite collection of eigenvectors.

Observe that if $\mathbf{e}$ is an eigenvector of $\mathbf{A}$ with $\mathbf{Ae} = \lambda\mathbf{e}$, then any multiple $r\mathbf{e}$ of $\mathbf{e}$ is also an eigenvector, as illustrated in Example 2.

The following example shows how eigenvectors provide a simplifying way to carry out matrix–vector computations.

**EXAMPLE 3 Using Eigenvectors to
Evaluate Powers of a Matrix** The pollution/development growth model from Example 2 has the form

$$\mathbf{c}' = \mathbf{Ac}, \quad \text{where } \mathbf{A} = \begin{bmatrix} 1 & 2 \\ 2 & 1 \end{bmatrix},$$

$$\mathbf{c} = \begin{bmatrix} P \\ D \end{bmatrix}, \quad \mathbf{c}' = \begin{bmatrix} P' \\ D' \end{bmatrix}.$$

In Example 2 we saw that the two eigenvalues

and associated eigenvectors of this matrix $\mathbf{A}$ are $\lambda = 3$ with $\mathbf{e} = \begin{bmatrix} 1 \\ 1 \end{bmatrix}$ and $\lambda' = 1$ with $\mathbf{e}' = \begin{bmatrix} 1 \\ -1 \end{bmatrix}$.

Suppose we want to determine the effects of this growth model over 20 periods with the starting vector $\mathbf{c} = \begin{bmatrix} 1 \\ 7 \end{bmatrix}$. In order to take advantage of the fact that it is much easier to compute $\mathbf{A}^{20}\mathbf{e}$ and $\mathbf{A}^{20}\mathbf{e}'$ (see (2)) than $\mathbf{A}^{20}\mathbf{c}$, let us express $\mathbf{c}$ as a linear combination of $\mathbf{e}$ and $\mathbf{e}'$.

$$\mathbf{c} = a\mathbf{e} + b\mathbf{e}': \begin{bmatrix} 1 \\ 7 \end{bmatrix} = a\begin{bmatrix} 1 \\ 1 \end{bmatrix} + b\begin{bmatrix} 1 \\ -1 \end{bmatrix}$$

$$\text{or} \quad \begin{aligned} 1 &= 1a + 1b \\ 7 &= 1a - 1b \end{aligned} \quad (3)$$

The system in (3) can be solved by elimination to yield $a = 4$, $b = -3$. So $\mathbf{c} = 4\mathbf{e} - 3\mathbf{e}'$.

Using the linearity of matrix–vector products, we can write

$$\begin{aligned} \mathbf{Ac} = \mathbf{A}(4\mathbf{e} - 3\mathbf{e}') &= 4(\mathbf{Ae}) - 3(\mathbf{Ae}') \\ &= 4(3\mathbf{e}) - 3(1\mathbf{e}') \\ &\quad \text{(since } \mathbf{e}, \mathbf{e}' \text{ are eigenvectors)} \\ &= 12\mathbf{e} - 3\mathbf{e}'. \end{aligned} \quad (4)$$

For 20 periods, we have

$$\begin{aligned} \mathbf{A}^{20}\mathbf{c} = \mathbf{A}^{20}(4\mathbf{e} - 3\mathbf{e}') &= 4(\mathbf{A}^{20}\mathbf{e}) - 3(\mathbf{A}^{20}\mathbf{e}') \\ &= 4(3^{20}\mathbf{e}) - 3[(-1)^{20}\mathbf{e}')] = 4 \cdot 3^{20}\begin{bmatrix} 1 \\ 1 \end{bmatrix} \\ &- 3\begin{bmatrix} 1 \\ -1 \end{bmatrix} = \begin{bmatrix} 4 \cdot 3^{20} \\ 4 \cdot 3^{20} \end{bmatrix} - \begin{bmatrix} 3 \\ -3 \end{bmatrix}. \end{aligned} \quad (5)$$

Observe that we needed both the "nonsense" eigenvector $\begin{bmatrix} 1 \\ -1 \end{bmatrix}$ and the "real" eigenvector $\begin{bmatrix} 1 \\ 1 \end{bmatrix}$

to express **c** as a linear combination of eigenvectors.

Note how the eigenvector with the larger eigenvalue swamps the other eigenvector in the final answer in (5). The relative effect of the other eigenvector is so small that it can be neglected. So after n periods we have

$$\mathbf{A}^n\mathbf{c} \approx \mathbf{A}^n(4\mathbf{e}) = 4 \cdot 3^n\mathbf{e} = 4 \cdot 3^n\begin{bmatrix} 1 \\ 1 \end{bmatrix}. \qquad (6)$$

This is much easier than multiplying $\mathbf{A}^n\mathbf{c}$ out directly for various values of n. ▲

As shown in Example 3, one important use of eigenvectors is to simplify computations of the form $\mathbf{A}^n\mathbf{c}$. We express **c** as a linear combination of eigenvectors, $\mathbf{c} = a\mathbf{e} + b\mathbf{e}'$. Then the messy matrix calculation $\mathbf{A}^n\mathbf{c}$ can be rewritten as the much easier $a\mathbf{A}^n\mathbf{e} + b\mathbf{A}^n\mathbf{e}'$. We give two more examples of this process.

E X A M P L E 4   Expressing a Vector in Terms of Eigenvectors   Consider the following model for two interacting species C and D from year to year:

$$\mathbf{c}' = \mathbf{Ac}, \quad \text{where } \mathbf{A} = \begin{bmatrix} 1 & 2 \\ 3 & 2 \end{bmatrix},$$

$$\mathbf{c} = \begin{bmatrix} C \\ D \end{bmatrix}, \qquad \mathbf{c}' = \begin{bmatrix} C' \\ D' \end{bmatrix}.$$

We are given the two eigenvalues and associated eigenvectors of this matrix $\mathbf{A}$: $\lambda = 4$ with $\mathbf{e} = \begin{bmatrix} 2 \\ 3 \end{bmatrix}$ and $\lambda' = -1$ with $\mathbf{e}' = \begin{bmatrix} 1 \\ -1 \end{bmatrix}$.

Suppose we want to determine the effects of

this growth model after 20 periods with the starting vector $\mathbf{c} = \begin{bmatrix} 6 \\ 4 \end{bmatrix}$. We first express **c** as a linear combination of **e** and **e'**:

$$\mathbf{c} = a\mathbf{e} + b\mathbf{e}': \begin{bmatrix} 6 \\ 4 \end{bmatrix} = a\begin{bmatrix} 2 \\ 3 \end{bmatrix} + b\begin{bmatrix} 1 \\ -1 \end{bmatrix} \quad \text{or}$$

$$\begin{bmatrix} 2 & 1 \\ 3 & -1 \end{bmatrix}\begin{bmatrix} a \\ b \end{bmatrix} = \begin{bmatrix} 6 \\ 4 \end{bmatrix}. \qquad (7)$$

We can solve (7) by Gaussian elimination or by Gauss–Jordan elimination (or by computing the inverse of $\begin{bmatrix} 2 & 1 \\ 3 & -1 \end{bmatrix}$). Gauss–Jordan elimination on the augmented coefficient matrix yields

$$\begin{bmatrix} 2 & 1 & | & 6 \\ 3 & -1 & | & 4 \end{bmatrix} \Rightarrow \begin{bmatrix} 1 & \frac{1}{2} & | & 3 \\ 0 & -\frac{5}{2} & | & -5 \end{bmatrix} \Rightarrow \begin{bmatrix} 1 & 0 & | & 2 \\ 0 & 1 & | & 2 \end{bmatrix}. \quad (8)$$

Thus $\mathbf{c} = 2\mathbf{e} + 2\mathbf{e}'$, that is, $\begin{bmatrix} 6 \\ 4 \end{bmatrix} = 2\begin{bmatrix} 2 \\ 3 \end{bmatrix} + 2\begin{bmatrix} 1 \\ -1 \end{bmatrix}$.

To find the population vector after 20 periods, we need to compute $\mathbf{A}^{20}\mathbf{c}$.

$$\begin{aligned}
\mathbf{A}^{20}\mathbf{c} &= \mathbf{A}^{20}(2\mathbf{e} + 2\mathbf{e}') = 2(\mathbf{A}^{20}\mathbf{e}) + 2(\mathbf{A}^{20}\mathbf{e}') \\
&= 2(4^{20}\mathbf{e}) + 2((-1)^{20}\mathbf{e}') \\
&\qquad \text{(note that } (-1)^{20} = 1) \\
&= 2 \cdot 4^{20}\begin{bmatrix} 2 \\ 3 \end{bmatrix} + 2\begin{bmatrix} 1 \\ -1 \end{bmatrix} \approx 2 \cdot 4^{20}\begin{bmatrix} 2 \\ 3 \end{bmatrix}. \quad (9)
\end{aligned}$$

Notice again how the term associated with the larger eigenvalue dominates the result in (9). ▲

E X A M P L E 5   Eigenvalues and Eigenvectors of a 3 × 3 Matrix   Consider the following growth model for species C, D, and E:

$$\mathbf{c}' = \mathbf{Ac}, \quad \text{where } \mathbf{A} = \begin{bmatrix} 5 & 4 & 2 \\ 4 & 5 & 2 \\ 2 & 2 & 2 \end{bmatrix},$$

$$\mathbf{c} = \begin{bmatrix} C \\ D \\ E \end{bmatrix}, \quad \mathbf{c}' = \begin{bmatrix} C' \\ D' \\ E' \end{bmatrix}.$$

We are given the three eigenvalues and associated eigenvectors of this matrix $\mathbf{A}$:

$$\lambda_1 = 10 \quad \text{with } \mathbf{e}_1 = \begin{bmatrix} 2 \\ 2 \\ 1 \end{bmatrix}, \quad \lambda_2 = 1 \text{ with}$$

$$\mathbf{e}_2 = \begin{bmatrix} 1 \\ 0 \\ -2 \end{bmatrix}, \quad \text{and } \lambda_3 = 1 \text{ with } \mathbf{e}_3 = \begin{bmatrix} 0 \\ 1 \\ -2 \end{bmatrix}.$$

Note that $\lambda_2 = \lambda_3$.

Suppose now we want to determine the effects of this growth model after 10 periods with the starting vector $\mathbf{c} = \begin{bmatrix} 6 \\ 9 \\ 6 \end{bmatrix}$. We first express $\mathbf{c}$ as a linear combination of $\mathbf{e}_1$, $\mathbf{e}_2$, and $\mathbf{e}_3$.

$\mathbf{c} = a\mathbf{e}_1 + b\mathbf{e}_2 + c\mathbf{e}_3$:

$$\begin{bmatrix} 6 \\ 9 \\ 6 \end{bmatrix} = a\begin{bmatrix} 2 \\ 2 \\ 1 \end{bmatrix} + b\begin{bmatrix} 1 \\ 0 \\ -2 \end{bmatrix} + c\begin{bmatrix} 0 \\ 1 \\ -2 \end{bmatrix}$$

$$\text{or} \quad \begin{bmatrix} 2 & 1 & 0 \\ 2 & 0 & 1 \\ 1 & -2 & -2 \end{bmatrix}\begin{bmatrix} a \\ b \\ c \end{bmatrix} = \begin{bmatrix} 6 \\ 9 \\ 6 \end{bmatrix} \quad (10)$$

Gaussian elimination on the augmented coefficient matrix yields

$$\begin{bmatrix} 2 & 1 & 0 & 6 \\ 2 & 0 & 1 & 9 \\ 1 & -2 & -2 & 6 \end{bmatrix} \Rightarrow \begin{bmatrix} 2 & 1 & 0 & 6 \\ 0 & -1 & 1 & 3 \\ 0 & -\frac{5}{2} & -2 & 3 \end{bmatrix}$$

$$\Rightarrow \begin{bmatrix} 2 & 1 & 0 & 6 \\ 0 & -1 & 1 & 3 \\ 0 & 0 & -\frac{9}{2} & -\frac{9}{2} \end{bmatrix}. \quad (11)$$

So $c = 1$, and back substitution yields $b = (3 - 1)/-1 = -2$ and $a = (6 - 1(-2))/2 = 4$. Then

$$\mathbf{c} = 4\mathbf{e}_1 - 2\mathbf{e}_2 + \mathbf{e}_3, \quad \text{or}$$

$$\begin{bmatrix} 6 \\ 9 \\ 6 \end{bmatrix} = 4\begin{bmatrix} 2 \\ 2 \\ 1 \end{bmatrix} - 2\begin{bmatrix} 1 \\ 0 \\ -2 \end{bmatrix} + \begin{bmatrix} 0 \\ 1 \\ -2 \end{bmatrix}.$$

To find the population vector after 10 periods, we need to compute $\mathbf{A}^{10}\mathbf{c}$.

$$\mathbf{A}^{10}\mathbf{c} = \mathbf{A}^{10}(4\mathbf{e}_1 - 2\mathbf{e}_2 + \mathbf{e}_3)$$
$$= 4(\mathbf{A}^{10}\mathbf{e}_1) - 2(\mathbf{A}^{10}\mathbf{e}_2) + \mathbf{A}^{10}\mathbf{e}_3 = 4(10^{10}\mathbf{e}_1)$$

$$- 2(1^{10}\mathbf{e}_2) + 1(1^{10}\mathbf{e}_3) = 4 \cdot 10^{10}\begin{bmatrix} 2 \\ 2 \\ 1 \end{bmatrix}$$

$$- 2\begin{bmatrix} 1 \\ 0 \\ -2 \end{bmatrix} + \begin{bmatrix} 0 \\ 1 \\ -2 \end{bmatrix} \approx 4 \cdot 10^{10}\begin{bmatrix} 2 \\ 2 \\ 1 \end{bmatrix} \blacktriangle \quad (12)$$

Let us generalize the pattern after many periods in the previous examples. Suppose we are given eigenvectors $\mathbf{e}$, $\mathbf{e}'$ of a 2×2 matrix $\mathbf{A}$ with eigenvalues λ and λ', respectively, where $\lambda > \lambda'$. We express the vector $\mathbf{c}$ as a linear combination of $\mathbf{e}$ and $\mathbf{e}'$ by determining coefficients a and b so that $\mathbf{c} = a\mathbf{e} + b\mathbf{e}'$. Then by the linearity of the matrix–vector products, $\mathbf{Ac}$ and $\mathbf{A}^2\mathbf{c}$ can be calculated as

$$\mathbf{Ac} = \mathbf{A}(a\mathbf{e} + b\mathbf{e}') = a\mathbf{Ae} + b\mathbf{Ae}' = a\lambda\mathbf{e} + b\lambda'\mathbf{e}',$$
$$\mathbf{A}^2\mathbf{c} = \mathbf{A}^2(a\mathbf{e} + b\mathbf{e}') = a\mathbf{A}^2\mathbf{e} + b\mathbf{A}^2\mathbf{e}' \qquad (13)$$
$$= a\lambda^2\mathbf{e} + b\lambda'^2\mathbf{e}'.$$

More generally,

$$\mathbf{A}^n\mathbf{c} = \mathbf{A}^n(a\mathbf{e} + b\mathbf{e}') = a\mathbf{A}^n\mathbf{e} + b\mathbf{A}^n\mathbf{e}'$$
$$= a\lambda^n\mathbf{e} + b\lambda'^n\mathbf{e}'. \qquad (14)$$

As noted in Examples 3 and 4, for large n, λ^n will be much larger than λ'^n, since $\lambda > \lambda'$, and so we have

$$\mathbf{A}^n\mathbf{c} \approx a\lambda^n\mathbf{e}. \qquad (15)$$

To give a geometric picture of the convergence of vectors toward multiples of $\lambda^n\mathbf{e}$ in Example 3, Figure 1 shows how $\mathbf{A}$, $\mathbf{A}^2$, and $\mathbf{A}^3$ map the positive quadrant of the plane toward vectors close to $\mathbf{e}$. (While eigenvectors might have seemed a bit mysterious at first, readers hopefully now realize that eigenvectors are our "friends.")

Eigenvectors clearly provide a very simple way to follow growth models over many periods. In many other types of problems in linear algebra, such as systems of linear differential equations, eigenvalues and eigenvectors provide the simplest way to obtain solutions.

We call an eigenvalue λ^* of the square matrix $\mathbf{A}$ the *dominant eigenvalue* of $\mathbf{A}$ if λ^* is larger in absolute value than any other eigenvalue of $\mathbf{A}$. An eigenvector $\mathbf{e}^*$ associated with the dominant eigenvalue of $\mathbf{A}$ is called the *dominant eigenvector* of $\mathbf{A}$ when all eigenvectors associated with λ^* are scalar multiples of one another. For example, the identity matrix $\mathbf{I}$ has one eigenvalue, 1, but all vectors are eigenvectors ($\mathbf{Ic} = \mathbf{c}$ for all $\mathbf{c}$) and so $\mathbf{I}$ has no dominant eigenvector. However, most square matrices have a dominant eigenvalue and dominant eigenvector, and

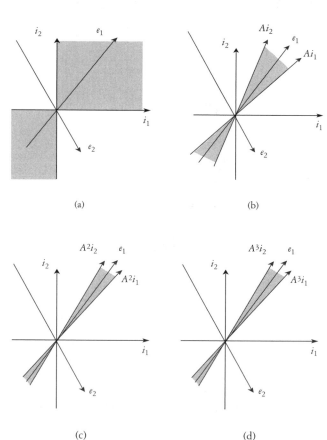

FIGURE 1 Vectors converging toward dominant eigenvector.

the nice behavior we saw in the previous examples is what normally happens.

Principle for Long-Term Behavior of $\mathbf{A}^n\mathbf{c}$

Let the square matrix $\mathbf{A}$ have a dominant eigenvector $\mathbf{e}^*$. Then for any vector $\mathbf{c}$, the expression $\mathbf{A}^n\mathbf{c}$ normally approaches a multiple of $\mathbf{e}^*$ as n becomes large.

Corollary to Principle. Let the square matrix $\mathbf{A}$ have a dominant eigenvector $\mathbf{e}^*$. Then the columns of $\mathbf{A}^n$ normally approach multiples of $\mathbf{e}^*$ as n becomes large.

Proof of Corollary. Recall that $\mathbf{i}_k$ is the vector with 1 in the kth position and 0s elsewhere. By the above principle, $\mathbf{A}^n\mathbf{i}_k$ normally approaches a multiple of $\mathbf{e}^*$ as n becomes large. But $\mathbf{A}^n\mathbf{i}_k$ equals the kth column of $\mathbf{A}^n$ (refer to Example 5 in Section 3.2).

▲ ▲ ▲

As an example of this corollary, for $\mathbf{A} = \begin{bmatrix} 1 & 2 \\ 2 & 1 \end{bmatrix}$ with $\lambda = 3$ and $\mathbf{e} = \begin{bmatrix} 1 \\ 1 \end{bmatrix}$, we have

$$\mathbf{A}^2 = \begin{bmatrix} 5 & 4 \\ 4 & 5 \end{bmatrix}, \qquad \mathbf{A}^4 = \begin{bmatrix} 41 & 40 \\ 40 & 41 \end{bmatrix},$$

$$\mathbf{A}^8 = \begin{bmatrix} 3281 & 3280 \\ 3280 & 3281 \end{bmatrix}.$$

Exercise 10 gives an example of a matrix $\mathbf{A}$ in which one column in the powers of $\mathbf{A}$ does not approach a multiple of the dominant eigenvector.

The principle states that $\mathbf{A}^n\mathbf{c}$ *normally* approaches a multiple of $\mathbf{e}^*$. The reason for the word *normally* is that if $\mathbf{c}$ is a multiple of another eigenvector $\mathbf{e}^\circ$, not the dominant eigenvector, then $\mathbf{A}^n\mathbf{c}$ will always be a multiple of $\mathbf{e}^\circ$. (A similar difficulty arises if $\mathbf{c}$ is a linear combination of nondominant eigenvectors.)

Our matrix algebra analysis has revealed that not only do population vectors in growth models of the form $\mathbf{p}' = \mathbf{A}\mathbf{p}$ converge to a multiple of the dominant eigenvector (in the spirit of the way Markov chain probabilities converge to a stable probability distribution), but more interestingly the columns of every square matrix typically converge to multiples of the dominant eigenvector when the matrix is raised to increasing powers. Squaring a matrix is a fairly complicated operation, and computing the fifth power of a 3×3 matrix is very tedious. Now we learn that the higher the power, the more predictable the result in the sense that the columns will become multiples of the dominant eigenvector. Note the ease with which matrix algebra allowed us to derive these important results in Eqs. (13)–(15) and in the proof of the corollary.

Mathematics has revealed a simple pattern in this mess of computations. It is often the case in mathematics that the long-term behavior can be nicely predicted by theory while the short-term behavior requires tedious calculations.

The reader should look back at the powers of the weather Markov chain matrix of Example 2, which were computed in Eqs. (13) and (14) of Section 3.3, and check that the columns of these powers were converging toward multiples of the dominant eigenvector $\begin{bmatrix} \frac{14}{23} \\ \frac{5}{23} \\ \frac{4}{23} \end{bmatrix}$ of this matrix (actually converging to this dominant eigenvector itself).

The determination of the eigenvalues and eigenvectors of a square matrix is one of the most carefully studied computational problems in all of mathematics. For 2×2 matrices the theory of determinants, introduced in the previous section, yields a simple computational procedure for computing eigenvalues, which we present just ahead. The problem is more complicated for larger matrices. While determinants still work in theory for larger matrices, in practice other methods (beyond the scope of this chapter) are used.

Determining the Eigenvalues of a 2 × 2 Matrix

The defining equation for an eigenvalue λ of a square matrix $\mathbf{A}$ and its associated eigenvector $\mathbf{e}$ is

$$\mathbf{A}\mathbf{e} = \lambda\mathbf{e} \quad \text{or} \quad \mathbf{A}\mathbf{e} - \lambda\mathbf{e} = 0. \qquad (16)$$

Writing $\lambda\mathbf{e}$ as $\lambda\mathbf{I}\mathbf{e}$, we have

$(\mathbf{A} - \lambda\mathbf{I})\mathbf{e} = \mathbf{0}$ for the model in Example 1:

$$(1 - \lambda)e_1 + 2e_2 = 0,$$
$$2e_1 + (1 - \lambda)e_2 = 0. \quad (17)$$

Once an eigenvalue λ is found, we can obtain an associated eigenvector $\mathbf{e}$ by solving the system of equations in (17). More importantly, we can use (17) to determine the eigenvalues themselves. The key to finding the eigenvalues is to remember that each eigenvalue has an infinite set of associated eigenvectors, for if $\mathbf{Ae} = \lambda\mathbf{e}$, then for any scalar r, $\mathbf{A}(r\mathbf{e}) = r\mathbf{Ae} = \lambda r\mathbf{e}$. Thus, when λ is an eigenvalue, the system in (17) has an infinite number of solutions. Using previous information about when systems of equations have multiple solutions, we obtain the following theorem.

Theorem 1. The values λ that make $\det(\mathbf{A} - \lambda\mathbf{I}) = 0$ are the set of eigenvalues of $\mathbf{A}$. The associated eigenvector(s) for λ are the nonzero solutions to $(\mathbf{A} - \lambda\mathbf{I})\mathbf{x} = \mathbf{0}$.

Proof. Let us think back to what we learned in earlier sections about systems of equations with an infinite number of solutions. In particular, recall Theorem 4 in Section 3.6, which said that (17) has only one solution, namely $\mathbf{e} = \mathbf{0}$ (the zero vector), if and only if the determinant of the system in (17) is nonzero: $\det(\mathbf{A} - \lambda\mathbf{I}) \neq 0$. If (17) has another solution besides $\mathbf{e} = \mathbf{0}$, then it follows that $\det(\mathbf{A} - \lambda\mathbf{I}) = 0$. Further, if (17) has multiple solutions, it must have an infinite number of solutions (by Theorem 2 of Section 3.4). So eigenvalues must be values of λ that make $\det(\mathbf{A} - \lambda\mathbf{I}) = 0$.

▲ ▲ ▲

For any $n \times n$ matrix $\mathbf{A}$, $\det(\mathbf{A} - \lambda\mathbf{I})$ will be a polynomial of degree n in λ, called the *characteristic* *polynomial* of $\mathbf{A}$. A polynomial of degree n can have up to n zeros. These zeros of the characteristic polynomial of $\mathbf{A}$ are the eigenvalues of $\mathbf{A}$.

EXAMPLE 6 Determining Eigenvalues and Eigenvectors Consider the system of pollution/development equations again:

$$C' = C + 2D, \quad \text{or} \quad \mathbf{c}' = \mathbf{Ac},$$
$$D' = 2C + D,$$

where $\mathbf{A} = \begin{bmatrix} 1 & 2 \\ 2 & 1 \end{bmatrix}$.

Earlier the eigenvalues and eigenvectors for $\mathbf{A}$ were given to us. Now we calculate them by the theorem. We must find the zeros of the characteristic polynomial $\det(\mathbf{A} - \lambda\mathbf{I})$.

$$\det(\mathbf{A} - \lambda\mathbf{I}) = \begin{vmatrix} 1 - \lambda & 2 \\ 2 & 1 - \lambda \end{vmatrix}$$
$$= (1 - \lambda)(1 - \lambda) - 2 \cdot 2$$
$$= (1 - 2\lambda + \lambda^2) - 4 = \lambda^2 - 2\lambda$$
$$- 3 = (\lambda - 3)(\lambda + 1). \quad (18)$$

The zeros of $\det(\mathbf{A} - \lambda\mathbf{I}) = (\lambda - 3)(\lambda + 1)$ are thus 3 and -1.

To find an eigenvector $\mathbf{e}$ for the eigenvalue 3, we must solve the system $\mathbf{Ae} = 3\mathbf{e}$ or, as in (17), $(\mathbf{A} - 3\mathbf{I})\mathbf{e} = \mathbf{0}$, where

$$\mathbf{A} - 3\mathbf{I} = \begin{bmatrix} 1 & 2 \\ 2 & 1 \end{bmatrix} - \begin{bmatrix} 3 & 0 \\ 0 & 3 \end{bmatrix} = \begin{bmatrix} -2 & 2 \\ 2 & -2 \end{bmatrix}.$$

Writing out $(\mathbf{A} - 3\mathbf{I})\mathbf{e} = \mathbf{0}$, we have

$$-2e_1 + 2e_2 = 0 \quad \Rightarrow \quad e_1 = e_2,$$
$$2e_1 - 2e_2 = 0. \quad (19)$$

Face Space

A "technology of the future" that is here today is *computer facial recognition.* The Commonwealth of Massachusetts started using it in 1995 to help eliminate fraudulent applications for driver's licenses. Soon you may identify yourself to an automatic teller machine not just by a magnetic bankcard but with your face as well.

Several competing systems for facial recognition are under development and testing. One, developed at MIT under the name Photobook, uses linear algebra.

Photobook works from a database of faces to identify the 40 best facial features for distinguishing one face from another; surprisingly, these do not correspond to what we think of as recognizable facial characteristics. The distinguishing features, called *eigenfaces,* serve as basis vectors for *face space.* Any face can be represented as a linear combination of eigenfaces, just as a vector is represented in terms of basis vectors. Two faces are compared by determining the difference between them (subtracting their corresponding coordinates relative to the basis of eigenfaces) and calculating the length (magnitude) of this difference vector. It takes only 80 bytes to store a face, in terms of the coordinates of its projections onto the eigenfaces.

Just as we seek orthonormal vectors for a basis, Photobook selects orthonormal eigenfaces, for more precise identification. It uses the method of *principal components,* which involves finding eigenvalues and eigenvectors of the matrix of covariances (squares of correlations) between eigenfaces. Photobook finds the 40 largest eigenvalues (and accompanying eigenvectors) of a $16{,}384 \times 16{,}384$ matrix, based on a coarse scan of the face consisting of 128×128 pixels of grayscale. The resulting eigenvectors are the eigenfaces.

Curiously, Photobook is not fooled by small changes in appearance, such as a change in hairstyle or the addition or subtraction of a hat, glasses, a beard, or a mustache! Nor do the lighting or the angle of the head affect recognition. The software achieves a 99.9% recognition rate in verification and a 95% rate in recognizing whether a face is in the database.

—You can find further details in "Face value," by Edmund X. de Jesus, *Byte* (February 1995), 85–90.

The second equation in (19) is the same as the first equation (and can be ignored). Then **e** is an eigenvector if $e_1 = e_2$ or, equivalently, if **e** is a multiple of $\begin{bmatrix} 1 \\ 1 \end{bmatrix}$.

It is left as an exercise for the reader to verify that $\mathbf{e}' = \begin{bmatrix} 1 \\ -1 \end{bmatrix}$ is an eigenvector for $\lambda = -1$ by showing that this $\mathbf{e}'$ is a solution to $(\mathbf{A} - \mathbf{I})\mathbf{e}' = 0$. ▲

Note that in Example 8 of Section 3.4 when we solved for the steady-state population distribution of the weather Markov chain, we were really just solving for the dominant eigenvector (associated with eigenvalue 1).

Finding the zeros of polynomials generally requires some numerical iterative scheme, but for second-degree polynomials the quadratic formula can be used. To find a dominant eigenvector and associated dominant eigenvalue directly, we can also use the principle for long-term behavior of $\mathbf{A}^n\mathbf{c}$: For any vector $\mathbf{c}$, the expression $\mathbf{A}^n\mathbf{c}$ normally approaches a multiple of the dominant eigenvector $\mathbf{e}^*$ as n becomes large; and from $\mathbf{e}^*$, the associated eigenvalue λ^* is obtained (since $\mathbf{A}\mathbf{e}^* = \lambda^*\mathbf{e}^*$).

Eigenvalue/Eigenvector Analysis of Growth Models

Next we completely analyze two linear growth models using our new knowledge of eigenvectors and eigenvalues.

EXAMPLE 7 Eigenvalues and Eigenvectors for Rabbit/Fox Population Model We consider a simple ecological model for the survival of rabbits and foxes. Suppose the current number of rabbits (R) naturally grows by 10% a year in the absence of foxes. So next year's number of rabbits (R') follows the growth law: $R' = 1.1R$. Also, suppose the current number of foxes (F) decreases by 15% a year in the absence of rabbits, so that $F' = .85F$. However, when foxes and rabbits are in the same habitat, the foxes eat the rabbits, decreasing the number of rabbits and allowing foxes to increase. The model we propose is as follows:

$$R' = 1.1R - .15F, \qquad \text{or}$$
$$F' = .1R + .85F,$$
$$\mathbf{r}' = \mathbf{Ar}: \quad \begin{bmatrix} R' \\ F' \end{bmatrix} = \begin{bmatrix} 1.1 & -.15 \\ .1 & .85 \end{bmatrix}\begin{bmatrix} R \\ F \end{bmatrix}. \quad (20)$$

First, to get some numerical picture of the model, we use (20) to compute the following table of rabbit and fox populations over many periods when we start with the sample population vector $\begin{bmatrix} R \\ F \end{bmatrix} = \begin{bmatrix} 10 \\ 8 \end{bmatrix}$.

$$
\begin{array}{lll}
\text{0 years:} & \text{10 rabbits,} & \text{8 foxes} \\
\text{1 year:} & \text{9.8 rabbits,} & \text{7.8 foxes} \\
\text{2 years:} & \text{9.6 rabbits,} & \text{7.6 foxes} \\
\text{3 years:} & \text{9.4 rabbits,} & \text{7.4 foxes} \\
\quad \vdots & \quad \vdots & \quad \vdots \\
\text{10 years:} & \text{8.4 rabbits,} & \text{6.4 foxes} \quad (21) \\
\quad \vdots & \quad \vdots & \quad \vdots \\
\text{20 years:} & \text{7.4 rabbits,} & \text{5.4 foxes} \\
\quad \vdots & \quad \vdots & \quad \vdots \\
\text{50 years:} & \text{6.3 rabbits,} & \text{4.3 foxes} \\
\quad \vdots & \quad \vdots & \quad \vdots \\
\text{100 years:} & \text{6.02 rabbits,} & \text{4.02 foxes}
\end{array}
$$

Let us try to obtain a mathematical analysis of the behavior shown in (21). As in Example 3, we can simplify the computation of $\mathbf{A}^n\mathbf{r}$ by finding the eigenvalues λ, λ' and eigenvectors $\mathbf{e}, \mathbf{e}'$ for this matrix and using them to write our starting population vector $\mathbf{r} = \begin{bmatrix} 10 \\ 8 \end{bmatrix}$ in terms of $\mathbf{e}$ and $\mathbf{e}'$, $\mathbf{r} = a\mathbf{e} + b\mathbf{e}'$.

We first compute $\det(\mathbf{A} - \lambda\mathbf{I})$, the characteristic polynomial of $\mathbf{A}$:

$$
\begin{aligned}
\det(\mathbf{A} - \lambda\mathbf{I}) &= \begin{bmatrix} 1.1 - \lambda & -.15 \\ .1 & .85 - \lambda \end{bmatrix} \\
&= (1.1 - \lambda)(.85 - \lambda) - .1(-.15) \\
&= \lambda^2 - 1.95\lambda + .95 \\
&= (\lambda - 1)(\lambda - .95). \quad (22)
\end{aligned}
$$

The eigenvalues are the zeros of $\lambda^2 - 1.95\lambda + .95$: $\lambda = 1$ and $\lambda' = .95$.

To find an eigenvector $\mathbf{e}$ associated with $\lambda = 1$, we solve $(\mathbf{A} - \mathbf{I})\mathbf{e} = \mathbf{0}$, where

$$\mathbf{A} - \mathbf{I} = \begin{bmatrix} 1.1 & -.15 \\ .1 & .85 \end{bmatrix} - \begin{bmatrix} 1 & 0 \\ 0 & 1 \end{bmatrix} = \begin{bmatrix} .1 & -.15 \\ .1 & -.15 \end{bmatrix},$$

$$(\mathbf{A} - \mathbf{I})\mathbf{e} = \mathbf{0}: \quad .1e_1 - .15e_2 = 0$$
$$\Rightarrow \quad e_1 = \tfrac{3}{2}e_2 \quad \Rightarrow \quad \mathbf{e} = \begin{bmatrix} 3 \\ 2 \end{bmatrix},$$
$$.1e_1 - .15e_2 = 0.$$

The nonzero multiples of $\mathbf{e} = \begin{bmatrix} 3 \\ 2 \end{bmatrix}$ are eigenvectors associated with $\lambda = 1$. Note that an eigenvector associated with an eigenvalue of 1 will be a "stable" population vector that stays the same period after period, that is, $\mathbf{e} = \mathbf{A}\mathbf{e}$.

Next we determine the eigenvector of $\lambda' = .95$. We solve $(\mathbf{A} - .95\mathbf{I})\mathbf{e}'$, where

$$\mathbf{A} - .95\mathbf{I} = \begin{bmatrix} 1.1 & -.15 \\ .1 & .85 \end{bmatrix} - \begin{bmatrix} .95 & 0 \\ 0 & .95 \end{bmatrix}$$
$$= \begin{bmatrix} .15 & -.15 \\ .1 & -.1 \end{bmatrix},$$

$$(\mathbf{A} - .95\mathbf{I})\mathbf{e}' = \mathbf{0}: \quad .15e_1' - .15e_2' = 0$$
$$\Rightarrow \quad e_1' = e_2' \quad \Rightarrow \quad \mathbf{e}' = \begin{bmatrix} 1 \\ 1 \end{bmatrix},$$
$$.1e_1' - .1e_2' = 0.$$

The nonzero multiples of $\mathbf{e}' = \begin{bmatrix} 1 \\ 1 \end{bmatrix}$ are eigenvectors associated with $\lambda' = .95$.

We now express the starting vector $\mathbf{r} = \begin{bmatrix} 10 \\ 8 \end{bmatrix}$ used in (21) in terms of eigenvectors $\mathbf{e} = \begin{bmatrix} 3 \\ 2 \end{bmatrix}$ and $\mathbf{e}' = \begin{bmatrix} 1 \\ 1 \end{bmatrix}$:

$\mathbf{r} = a\mathbf{e} + b\mathbf{e}'$:

$$\begin{bmatrix} 10 \\ 8 \end{bmatrix} = a\begin{bmatrix} 3 \\ 2 \end{bmatrix} + b\begin{bmatrix} 1 \\ 1 \end{bmatrix} \quad \text{or} \quad \begin{matrix} 3a + 1b = 10, \\ 2a + 1b = 8. \end{matrix} \quad (23)$$

By elimination, we find that $a = 2$, $b = 4$, and so $\mathbf{r} = 2\mathbf{e} + 4\mathbf{e}'$.

Then using (23) to compute the population sizes in (21) gives

$$\mathbf{r}^{(n)} = \mathbf{A}^n\mathbf{r} = 2\mathbf{A}^n\mathbf{e} + 4\mathbf{A}^n\mathbf{e} = 2(1^n\mathbf{e}) + 4(.95^n\mathbf{e}')$$
$$= 2\begin{bmatrix} 3 \\ 2 \end{bmatrix} + 4 \cdot .95^n\begin{bmatrix} 1 \\ 1 \end{bmatrix} = \begin{bmatrix} 6 \\ 4 \end{bmatrix} + .95^n\begin{bmatrix} 4 \\ 4 \end{bmatrix}. \quad (24)$$

So $\mathbf{r}^n$ is composed of a stable population term $\begin{bmatrix} 6 \\ 4 \end{bmatrix}$ and a second term of $.95^n\begin{bmatrix} 4 \\ 4 \end{bmatrix}$ that slowly decays away (the vectors in the second term lie on a line with slope $4/4 = 1$). With (24), the behavior in (21) is completely explained!

Generalizing the calculation in (24), if the starting vector $\mathbf{r}$ has the eigenvector representation $\mathbf{r} = a\mathbf{e} + b\mathbf{e}'$, then

$$\mathbf{r}^{(n)} = \mathbf{A}^n\mathbf{r} = a\mathbf{A}^n\mathbf{e} + b\mathbf{A}^n\mathbf{e}' = a\mathbf{e} + b(.95^n\mathbf{e}')$$
$$= \begin{bmatrix} 3a \\ 2a \end{bmatrix} + .95^n\begin{bmatrix} b \\ b \end{bmatrix}. \quad (25)$$

And the long-term stable population is $\begin{bmatrix} 3a \\ 2a \end{bmatrix}$. The critical number is a. To find a for any starting vector $\mathbf{r} = \begin{bmatrix} R \\ F \end{bmatrix}$, we substitute the general vector $\begin{bmatrix} R \\ F \end{bmatrix}$ for $\begin{bmatrix} 10 \\ 8 \end{bmatrix}$ in (23) and solve by Cramer's rule. We obtain

$$a = \frac{\det(\mathbf{A}_1)}{\det(\mathbf{A})} = \frac{\begin{vmatrix} R & 1 \\ F & 1 \end{vmatrix}}{\begin{vmatrix} 3 & 1 \\ 2 & 1 \end{vmatrix}} = \frac{R - F}{1} = R - F.$$

So the long-term stable population is

$$\begin{bmatrix} 3(R - F) \\ 2(R - F) \end{bmatrix}. \ \blacktriangle$$

E X A M P L E 8   Rental Cars Revisited   In Example 1 of Section 1.8, the following difference-equation model was introduced for a_n, the number of rental cars available in city A on day n, and b_n, the number of rental cars available in city B on day n:

$$\begin{aligned} a_{n+1} &= 0.9a_n + 0.12b_n, \\ b_{n+1} &= 0.1a_n + 0.88b_n. \end{aligned} \qquad (26)$$

This difference equation model can be converted to a matrix model:

$$\mathbf{c}' = \mathbf{Ac}, \quad \text{where } \mathbf{A} = \begin{bmatrix} 0.9 & 0.12 \\ 0.1 & 0.88 \end{bmatrix},$$

$$\mathbf{c} = \begin{bmatrix} a_n \\ b_n \end{bmatrix}, \quad \text{and} \quad \mathbf{c}' = \begin{bmatrix} a_{n+1} \\ b_{n+1} \end{bmatrix}. \qquad (27)$$

In Section 1.8 the initial number of rental cars at cities A and B was $a_0 = 120$, $b_0 = 150$. The analytic solution to system (26) was given, without explanation in Eq. (2) of Section 1.8 to be

$$\begin{aligned} a_n &= -27.3(0.78)^n + 147.3, \\ b_n &= 27.3(0.78)^n + 122.7. \end{aligned} \qquad (28)$$

Our eigenvalue/eigenvector analysis can be applied to the model (27) to verify the formulas in (28). The preceding methods show that eigenvalues and eigenvectors of (27) are $\lambda = 1$, $\mathbf{e} = \begin{bmatrix} 6 \\ 5 \end{bmatrix}$, and $\lambda' = .78$, $\mathbf{e}' = \begin{bmatrix} 1 \\ -1 \end{bmatrix}$ (details are left as an exercise). Writing the initial vector as a linear combination of the eigenvectors, we obtain the system of equations:

$$\mathbf{c} = g\mathbf{e} + h\mathbf{e}': \begin{bmatrix} 120 \\ 150 \end{bmatrix} = g\begin{bmatrix} 6 \\ 5 \end{bmatrix} + h\begin{bmatrix} 1 \\ -1 \end{bmatrix} \quad \text{or}$$

$$\begin{aligned} 6g + 1b &= 120, \qquad (29) \\ 5g - 1b &= 150. \end{aligned}$$

Solving (29), we find that $g = -270/11$ and $b = 300/11$. Then

$$\begin{aligned} \mathbf{c}^{(n)} = \mathbf{A}^n\mathbf{c} &= g\lambda^n\mathbf{e} + h\lambda'^n\mathbf{e}' \\ &= -\tfrac{270}{11}\begin{bmatrix} 6 \\ 5 \end{bmatrix} + \tfrac{300}{11}\begin{bmatrix} 1 \\ -1 \end{bmatrix} = \begin{bmatrix} 147.3 \\ 122.7 \end{bmatrix} \\ &\quad + (.78)^n\begin{bmatrix} -27.3 \\ 27.3 \end{bmatrix}, \qquad (30) \end{aligned}$$

confirming the formulas from Section 1.8 given in (28). $\blacktriangle$

1. (a) The matrix $\begin{bmatrix} 1 & 6 \\ -2 & -6 \end{bmatrix}$ has eigenvectors $\mathbf{e} = \begin{bmatrix} -2 \\ 1 \end{bmatrix}$ and $\mathbf{e}' = \begin{bmatrix} -3 \\ 2 \end{bmatrix}$. What are the corresponding eigenvalues for these eigenvectors?

    (b) The matrix $\begin{bmatrix} 1 & -2 \\ -2 & 1 \end{bmatrix}$ has eigenvectors $\mathbf{e} = \begin{bmatrix} 1 \\ 1 \end{bmatrix}$ and $\mathbf{e}' = \begin{bmatrix} 1 \\ -1 \end{bmatrix}$. What are the corresponding eigenvalues for these eigenvectors?

    (c) The matrix

    $$\begin{bmatrix} -4 & 4 & 4 \\ -1 & 1 & 2 \\ -3 & 2 & 4 \end{bmatrix}$$

    has eigenvectors

    $$\mathbf{e}_1 = \begin{bmatrix} 2 \\ 0 \\ 1 \end{bmatrix}, \qquad \mathbf{e}_2 = \begin{bmatrix} 6 \\ 4 \\ 5 \end{bmatrix}, \qquad \mathbf{e}_3 = \begin{bmatrix} 4 \\ 3 \\ 2 \end{bmatrix}.$$

    What are the corresponding eigenvalues for these eigenvectors?

2. Verify for each Markov transition matrix $\mathbf{A}$ that the given vector is a stable probability vector.

    (a) $\mathbf{A} = \begin{bmatrix} 1/4 & 1/2 \\ 3/4 & 1/2 \end{bmatrix}$, $\mathbf{p} = \begin{bmatrix} 2/5 \\ 3/5 \end{bmatrix}$

    (b) $\mathbf{A} = \begin{bmatrix} 1/3 & 1/2 & 0 \\ 2/3 & 0 & 2/3 \\ 0 & 1/2 & 1/3 \end{bmatrix}$, $\mathbf{p} = \begin{bmatrix} 3/10 \\ 4/10 \\ 3/10 \end{bmatrix}$

3. The matrix $\begin{bmatrix} 2 & 5 \\ 6 & 1 \end{bmatrix}$ has eigenvalue $\lambda = 7$ with eigenvector $\mathbf{e} = \begin{bmatrix} 1 \\ 1 \end{bmatrix}$ and eigenvalue $\lambda' = -4$ with eigenvector $\mathbf{e}' = \begin{bmatrix} -5 \\ 6 \end{bmatrix}$.

    (a) We want to compute $\mathbf{A}^3\mathbf{v}$, where $\mathbf{v} = \begin{bmatrix} -2 \\ 9 \end{bmatrix}$. Writing $\mathbf{v}$ as $\mathbf{v} = 3\mathbf{e}_1 + \mathbf{e}_2$, compute $\mathbf{A}^3\mathbf{v}$ indirectly as in Example 3.

    (b) Give an approximate formula for $\mathbf{A}^n\mathbf{v}$.

    (c) Determine a and b so that the vector $\mathbf{v} = \begin{bmatrix} 2 \\ 13 \end{bmatrix}$ can be written as $\mathbf{v} = a\mathbf{e}_1 + b\mathbf{e}_2$. Then use this representation of $\mathbf{v}$ to compute $\mathbf{A}^3\mathbf{v}$.

4. The matrix $\begin{bmatrix} 1 & 1 \\ 0 & 2 \end{bmatrix}$ has eigenvectors $\mathbf{e} = \begin{bmatrix} 1 \\ 1 \end{bmatrix}$ and $\mathbf{e}' = \begin{bmatrix} 1 \\ 0 \end{bmatrix}$.

    (a) We want to compute $\mathbf{A}^4\mathbf{v}$, where $\mathbf{v} = \begin{bmatrix} 3 \\ 1 \end{bmatrix}$. Writing $\mathbf{v}$ as $\mathbf{v} = \mathbf{e} + 2\mathbf{e}'$, compute $\mathbf{A}^4\mathbf{v}$ indirectly as in Example 3.

    (b) Give an approximate formula for $\mathbf{A}^n\mathbf{v}$.

    (c) Determine a and b so that the vector $\mathbf{v} = \begin{bmatrix} 6 \\ 9 \end{bmatrix}$ can be written as $\mathbf{v} = a\mathbf{e}_1 + b\mathbf{e}_2$, and use this representation of $\mathbf{v}$ to compute $\mathbf{A}^5\mathbf{v}$.

5. (a) Determine the eigenvalues of each of the following matrices (part (vi) requires computer assistance).

    (i) $\begin{bmatrix} 1 & 2 \\ 5 & 4 \end{bmatrix}$

    (ii) $\begin{bmatrix} 4 & -1 \\ 1 & 2 \end{bmatrix}$

    (iii) $\begin{bmatrix} 2 & 5 \\ 1 & -2 \end{bmatrix}$

$$(iv) \quad \begin{bmatrix} 3 & 2 & 4 \\ 2 & 0 & 2 \\ 4 & 2 & 3 \end{bmatrix}$$

$$(v) \quad \begin{bmatrix} 1 & 1 & 2 \\ 1 & 3 & 1 \\ 2 & 1 & 1 \end{bmatrix}$$

$$(vi) \quad \begin{bmatrix} 3 & 2 & 0 & 1 \\ 1 & 4 & 0 & 1 \\ 1 & -2 & 2 & -3 \\ -1 & 2 & 0 & 5 \end{bmatrix}$$

(b) Determine an eigenvector associated with the dominant eigenvalue for the matrices shown in part (a).

6. The matrix $\mathbf{B} = \begin{bmatrix} 2/3 & -1/3 \\ -1/3 & 2/3 \end{bmatrix}$ is the inverse of

$\mathbf{A} = \begin{bmatrix} 2 & 1 \\ 1 & 2 \end{bmatrix}$.

(a) Verify that $\mathbf{e} = \begin{bmatrix} 1 \\ 1 \end{bmatrix}$ and $\mathbf{e}' = \begin{bmatrix} 1 \\ -1 \end{bmatrix}$ are eigenvectors of both $\mathbf{A}$ and $\mathbf{B}$.

(b) Determine the eigenvalues of $\mathbf{A}$ and $\mathbf{B}$.

7. (a) For the following rabbit/fox models, determine both eigenvalues.
   (i) $R' = 1.1R - .3F$
       $F' = .2R + .4F$
   (ii) $R' = 1.3R - .2F$
       $F' = .15R + .9F$
   (iii) $R' = 1.1R + .1F$
       $F' = .2R + 1.1F$

(b) Determine an eigenvector $\mathbf{e}$ associated with the largest eigenvalue in each system in part (a).

(c) Determine the other eigenvector $\mathbf{e}'$ (associated with the smaller eigenvalue) for each system in part (a).

(d) If the initial population is $\mathbf{x} = \begin{bmatrix} 10 \\ 10 \end{bmatrix}$, then express $\mathbf{x}^{(k)}$ as a linear combination of $\mathbf{e}$ and $\mathbf{e}'$, as in Eq. (25), for each system in part (a). Use this expression to describe in words the behavior of this model over time.

8. Verify the eigenvalues and eigenvectors for the car rental model in (27).

9. The following growth model for computer science teachers (T), computer operations (O), and computer programmers (P) predicts population changes from decade to decade.

$$T' = T - O$$
$$O' = -T + 20 - P$$
$$P' = \qquad -O + P$$

(a) Determine the eigenvalues and associated eigenvectors for this system. (This part requires computer assistance).

(b) Suppose initially we have

$$\mathbf{p} = \begin{bmatrix} T \\ O \\ P \end{bmatrix} = \begin{bmatrix} 50 \\ 0 \\ 10 \end{bmatrix}.$$

Write $\mathbf{p}$ as a linear combination of the eigenvectors.

(c) Use the information in part (b) to determine an approximate value for the population sizes 12 decades from now.

10. (a) Determine the dominant eigenvalue and an associated eigenvector for the following system of equations by computing successive (x', y') pairs for many periods. Use $\begin{bmatrix} 1 \\ 0 \end{bmatrix}$ as your starting vector.

$$x' = .707x - .707y$$
$$y' = .707x + .707y$$

(b) Plot the successive iterates on graph paper. Try other starting vectors. State in words, the effect in (x, y)-coordinates of this linear model.

(c) Solve the characteristic equation $\det(\mathbf{A} - \lambda\mathbf{I}) = 0$ to determine the eigenval-

ues for this matrix of coefficients. You are finding out that imaginary eigenvalues correspond
to rotations. Note that $.707 = \sin 45° = \cos 45°$.

11. Consider the matrix

$$A = \begin{bmatrix} 2 & 0 & 0 \\ 0 & 3 & 1 \\ 0 & 2 & 2 \end{bmatrix}.$$

Its dominant eigenvector is

$$\begin{bmatrix} 0 \\ 1 \\ 1 \end{bmatrix}.$$

Show that the first column in powers of A does not converge to a multiple of this dominant eigenvector. Why
does this result not contradict the corollary to the principle of long-term behavior of $A^n c$?

12. **(Computer Project)** Use computer software to find
the 5th, 10th, and 20th powers of the following matrices,
and show that the columns in these powers are converging to multiples of the dominant eigenvector (found in
Exercise 5b).

(a) $\begin{bmatrix} 1 & 1 & 2 \\ 1 & 3 & 1 \\ 2 & 1 & 1 \end{bmatrix}$

(b) $\begin{bmatrix} 3 & 2 & 0 & 1 \\ 1 & 4 & 0 & 1 \\ 1 & -2 & 2 & -3 \\ -1 & 2 & 0 & 5 \end{bmatrix}$

13. **(Computer Project)** Use a computer algebra system to determine the eigenvalues and associated eigenvectors for the general 3×3 matrix

$$\begin{bmatrix} a & b & 1 \\ d & e & 1 \\ 1 & 1 & 1 \end{bmatrix}.$$

14. **(Computer Project)** A growth model for a species
can be split into three age groups: baby (b), youth (y),
adult (a). We use a matrix model of the form $p' = Ap$,
where

$$p = \begin{bmatrix} b \\ y \\ a \end{bmatrix}, \quad p' = \begin{bmatrix} b' \\ y' \\ a' \end{bmatrix}, \quad A = \begin{bmatrix} 0 & 0 & c \\ s_1 & 0 & 0 \\ 0 & s_2 & s_3 \end{bmatrix},$$

and c = number of children born each period by an adult,
s_1 = probability of a baby surviving one period to become
a youth, s_2 = probability of a youth surviving one period
to become an adult, and s_3 = probability of an adult
surviving the current period. Such a growth model is
called a *Leslie model*. Explore some Leslie growth models
for different values of c, s_1, s_2, and s_3 of your choice. Use
software packages to determine the eigenvectors and eigenvalues. Initially set $s_3 = 0$.

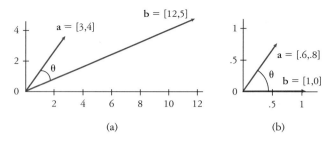

FIGURE 1 (a) Example 1(ii); (b) Example 1(iii).

SECTION 3.8 *Angles, Orthogonality, and Projections*

One of the great achievements of mathematics that set the stage for Newton's calculus and the scientific world as we know it today was the combination of geometry, developed systematically by the Greeks beginning in 300 B.C., with algebra, which was evolving more slowly starting from the time of Diophantus in Alexandria in 200 A.D. The resulting subject—analytic geometry—was developed in the seventeenth century by the French mathematicians Descartes, Fermat, and Pascal. This section brings analytic geometry ideas to bear on some vector and matrix problems.

Angles Between Vectors

We start by deriving a formula for the angle between two vectors. We find several unexpected applications of it. In the subsequent discussion we will be particularly interested in vectors that are perpendicular to each other. We then consider the projection of one vector onto another and apply this concept to the statistical problem of fitting a line through a set of points. By the length $|\mathbf{a}|$ of a vector $\mathbf{a}$, we mean the distance from the origin to the point with coordinates given by $\mathbf{a}$. That is,

$$|\mathbf{a}| = \sqrt{a_1^2 + a_2^2 + \cdots + a_n^2}.$$

Our initial goal is a formula for the cosine of the angle between two vectors $\mathbf{a}$ and $\mathbf{b}$. This is the angle formed at the origin by the line segments to the two vector points (the angle is measured in the plane formed by the two line segments; see Figure 1). The formula is simple, and its proof is fairly short and

not hard to follow. However, the proof uses an indirect algebraic argument that gives one no feel for why the formula is true.

Theorem 1.   The cosine of the angle θ between two nonzero vectors $\mathbf{a}$ and $\mathbf{b}$ is

$$\cos \theta = \frac{\mathbf{a} \cdot \mathbf{b}}{|\mathbf{a}||\mathbf{b}|}. \tag{1}$$

If $\mathbf{a}$ and $\mathbf{b}$ are unit-length vectors, Eq. (1) becomes $\cos \theta = \mathbf{a} \cdot \mathbf{b}$. The angle between two vectors is the angle whose cosine is $(\mathbf{a} \cdot \mathbf{b})/|\mathbf{a}||\mathbf{b}|$.

Proof.   Our proof involves expressing the square of the difference between $\mathbf{a}$ and $\mathbf{b}$, $|\mathbf{a} - \mathbf{b}|^2$, in two very different ways. The result will follow immediately when we set the two expressions for $|\mathbf{a} - \mathbf{b}|^2$ equal to one another. The first way we express $|\mathbf{a} - \mathbf{b}|^2$ uses the law of cosines from trigonometry:

$$|\mathbf{a} - \mathbf{b}|^2 = |\mathbf{a}|^2 + |\mathbf{b}|^2 - 2|\mathbf{a}||\mathbf{b}| \cos \theta. \tag{2}$$

For the second way, we note that the square of the length $|\mathbf{c}|^2$ of any vector $\mathbf{c}$ is simply $\mathbf{c} \cdot \mathbf{c}$, since both $|\mathbf{c}|^2$ and $\mathbf{c} \cdot \mathbf{c}$ equal $c_1^2 + c_2^2 + \cdots + c_n^2$. Letting $\mathbf{c} =$

$\mathbf{a} - \mathbf{b}$, we use the distributive law for scalar products to obtain

$$
\begin{aligned}
|\mathbf{a} - \mathbf{b}|^2 &= (\mathbf{a} - \mathbf{b}) \cdot (\mathbf{a} - \mathbf{b}) \\
&= \mathbf{a} \cdot \mathbf{a} + \mathbf{b} \cdot \mathbf{b} - 2\mathbf{a} \cdot \mathbf{b} \\
&= |\mathbf{a}|^2 + |\mathbf{b}|^2 - 2\mathbf{a} \cdot \mathbf{b}. \quad\quad (3)
\end{aligned}
$$

The right sides of (3) and (2) must be equal. Moreover, the right side of (3) is the same as the right side of (2) except for the last term. So the last terms must be equal:

$$
-2|\mathbf{a}||\mathbf{b}| \cos \theta = -2\mathbf{a} \cdot \mathbf{b}.
$$

Solving for $\cos \theta$ yields Theorem 1. Wasn't that a sneaky way to verify Eq. (1)! A slightly longer proof using only the Pythagorean theorem and the concept of a projection (introduced shortly) is given in the appendix at the end of the section.

▲ ▲ ▲

E X A M P L E 1   Examples of Angles Between Vectors

(i)   If $\mathbf{a} = [1, 0, 0]$ and $\mathbf{b} = [0, 1, 0]$, then $\cos \theta = \mathbf{a} \cdot \mathbf{b} = 0$. Since $\cos 90° = 0$, we conclude that $\mathbf{a}$ and $\mathbf{b}$ form a $90°$ angle.

(ii)   If $\mathbf{a} = [3, 4]$ and $\mathbf{b} = [12, 5]$, then $|\mathbf{a}| = 5 (= \sqrt{9 + 16})$ and $|\mathbf{b}| = 13 (= \sqrt{144 + 25})$. So $\cos \theta = (\mathbf{a} \cdot \mathbf{b})/|\mathbf{a}||\mathbf{b}| = (3 \cdot 12 + 4 \cdot 5)/5 \cdot 13 = 56/65 \approx .86$. The angle with a cosine of .86 is approximately $36°$ (see Figure 1a).

(iii)   If $\mathbf{a} = [.6, .8]$, with $|\mathbf{a}| = 1$ and $\mathbf{b} = [1, 0]$, then $\cos \theta = \mathbf{a} \cdot \mathbf{b} = .6 \cdot 1 + .8 \cdot 0 = .6$—just the first coordinate (see Figure 1b).

(iv)   If $\mathbf{a} = [\cos \theta, \sin \theta]$ and $\mathbf{b} = [1, 0]$, with $|\mathbf{a}| = \sqrt{(\cos \theta)^2 + (\sin \theta)^2} = 1$, then $\cos \theta = \mathbf{a} \cdot \mathbf{b} = \cos \theta \cdot 1 + \sin \theta \cdot 0 = \cos \theta.$ ▲

Linear Correlation

Before pursuing the mathematical uses of Theorem 1, we note that it has an interesting application in statistics. The cosine of the angle $\theta(\mathbf{x}, \mathbf{y})$ between two vectors $\mathbf{x}$ and $\mathbf{y}$ tells us if the vectors are close together (when $\cos \theta(\mathbf{x}, \mathbf{y})$ is near 1), or opposites of one another (when $\cos \theta(\mathbf{x}, \mathbf{y})$ is near -1), or unrelated, i.e., close to perpendicular (when $\cos \theta(\mathbf{x}, \mathbf{y})$ is near 0).

Suppose $\mathbf{x}$ and $\mathbf{y}$ are vectors of data from an experiment. Perhaps $\mathbf{x}$ is the vector of scores of 10 students on a math test and $\mathbf{y}$ is the vector of scores of the same 10 students on a language test. Let 0 be the average score on each test so that a positive score is an above-average result and a negative score is below average. Although when plotted these are vectors in 10-dimensional space and cannot be visualized, the vectors and the angle between them behave just like the vectors in two dimensions plotted in Figure 1.

Thus $\cos \theta(\mathbf{x}, \mathbf{y})$ tells us how closely related these two data vectors are and helps us predict future relations between math and language scores. If $\cos \theta(\mathbf{x}, \mathbf{y})$ is .6 (similar to vectors $\mathbf{a}$ and $\mathbf{b}$ in Figure 1b), then performance on these two tests is closely related and we can view a student's score on one test as a reasonably good predictor of how he/she will do on the other test. If $\cos \theta(\mathbf{x}, \mathbf{y})$ is $-.7$, then the score vectors point in almost opposite directions, and a high score on one test is very likely to produce a below-average score on the other test. If $\cos \theta(\mathbf{x}, \mathbf{y}) = 0$ (the vectors $\mathbf{x}$ and $\mathbf{y}$ are at right angles), then performance on one test tells us nothing about the likely performance on the other test (in statistics, one says that the two data variables are uncorrelated).

DEFINITION

Let $\mathbf{x} = [x_1, x_2, \ldots, x_n]$ and $\mathbf{y} = [y_1, y_2, \ldots y_n]$ be two sets of observations with the property that the average x-value and the average y-value are each 0. Then the *correlation coefficient, Cor* $(\mathbf{x}, \mathbf{y})$, of $\mathbf{x}$ and $\mathbf{y}$ is defined to be $\cos \theta(\mathbf{x}, \mathbf{y})$.

$$\text{Cor}(\mathbf{x}, \mathbf{y}) = \frac{\mathbf{x} \cdot \mathbf{y}}{|\mathbf{x}||\mathbf{y}|} = \frac{\Sigma x_i y_i}{\sqrt{\Sigma x_i^2} \sqrt{\Sigma y_i^2}} \qquad (4)$$

Recall that the average x-value is $\bar{x} = (1/n)\Sigma x_i$. If $\bar{x} \neq 0$, then one can subtract $\bar{x}$ from each x_i to get a revised vector that does have an average value of 0. One can do the same for y-values. We need an average value of 0 so that the opposite of a high score (a positive value) will be a low score (a negative value). This way the terms $x_i y_i$ in Eq. (4) for pairs of oppositely correlated entries x_i, y_i will be negative (when $x_i y_i$ is the product of a positive and a negative number), leading to a negative correlation.

EXAMPLE 2 Correlation Coefficient Suppose that we ask the faculty in the Mythology Department at KnowItAll U. to rate the quality of their students and that we poll the students to get a rating of the quality of each of the eight faculty. The results of our experiment are presented in the following table (where we have processed the data to make the average value 0 in each category).

Faculty	x_i Quality of students	y_i Student rating
1. Aristotle	+5	+2
2. Galileo	−5	−7
3. Goldbrick	−2	0
4. Hasbeen	+3	−1
5. Leadbottom	−4	−5
6. Mercury	+5	+3
7. Merlin	+5	0
8. Midas	−7	+8

Applying Eq. (4) to this data, we obtain

$$\begin{aligned}
\text{Cor}(\mathbf{x}, \mathbf{y}) &= \frac{\Sigma x_i y_i}{\sqrt{\Sigma x_i^2} \sqrt{\Sigma y_i^2}} \\
&= \frac{5 \cdot 2 + (-5)(-7) + \cdots + (-7)8}{\sqrt{178} \cdot \sqrt{152}} \\
&= \frac{21}{(13.3)(12.3)} \approx .1.
\end{aligned}$$

Looking back at the data in the table, we are a little surprised to see such a low correlation since the numbers in the two columns correspond fairly well for most faculty with the glaring exception of Midas. Statisticians would call Midas's data pair $(-7, 8)$ an *outlier,* an observation that fits poorly with the rest of the data. (A little investigation reveals that Midas is a terrible teacher but is still well liked because he gives the students pieces of gold.)

Let us throw out Midas's numbers and recompute the correlation coefficient. This requires us to adjust the data so that the averages in each column are again 0. The new numbers are as follows.

Faculty	Quality of students	Student rating
1. Aristotle	+4	+3
2. Galileo	−6	−6
3. Goldbrick	−3	+1
4. Hasbeen	+2	0
5. Leadbottom	−5	−4
6. Mercury	+4	+4
7. Merlin	+4	+1

$$\text{Cor}(\mathbf{x}, \mathbf{y}) = \frac{85}{\sqrt{122} \cdot \sqrt{79}} \approx \frac{85}{(11)(8.9)} \approx .9,$$

which shows a high degree of correlation. ▲

The concept of a correlation coefficient gives a good example of the power of applying geometric interpretations to vectors that arise from nongeometric settings.

Orthogonal Vectors

The most interesting angle between two vectors is a right angle (90°), that is, the angle formed when the two vectors are perpendicular to one another. There is a special mathematical name for perpendicular vectors.

DEFINITION

Two nonzero vectors are called *orthogonal* when the angle between them is 90°.

Since the cosine of 90° is 0, from Theorem 1 we have the following result.

Theorem 2. Two nonzero vectors **a** and **b** are orthogonal if and only if $\mathbf{a} \cdot \mathbf{b} = 0$.

Note that in Theorem 2 we can omit the denominator in the formula for the cosine because the fraction is zero if and only if its numerator is zero (recall that the denominator cannot be 0 since both vectors are nonzero and hence have positive lengths). To illustrate the role of orthogonality in linear algebra in unexpected places, let us look at its role in matrix inverses.

EXAMPLE 3 Inverse Matrices and Orthogonality When we multiply two matrices **A** and **B** together, entry (i, j) in the product **AB** is the scalar product of the ith row a_i' of **A** and the jth column b_j of **B**. In the case of a matrix **A** and its inverse $\mathbf{A}^{-1}$, their product $\mathbf{A}^{-1}\mathbf{A}$ equals the identity matrix **I** in which all entries are 0 except on the main diagonal. Thus, the ith row of $\mathbf{A}^{-1}$ has a scalar product of 0 with all columns of **A** except the ith column and is hence orthogonal to all columns of **A** except the ith column. Since $\mathbf{A}\mathbf{A}^{-1}$ also equals **I**, the jth column of $\mathbf{A}^{-1}$ is orthogonal to all rows of **A** except the jth row. As an example, the 2×2 matrix $\mathbf{A} = \begin{bmatrix} 3 & 1 \\ 4 & 2 \end{bmatrix}$ has the inverse

$$\mathbf{A}^{-1} = \begin{bmatrix} 1 & -\frac{1}{2} \\ -2 & \frac{3}{2} \end{bmatrix}.$$

Figure 2 plots the columns of **A** and the rows of $\mathbf{A}^{-1}$. ▲

Because of the central role of orthogonality in matrix inverses, *if a matrix **A** has columns that are mutually orthogonal, then its inverse $\mathbf{A}^{-1}$ is very easy to determine* (with no elimination computation needed). Let **A** have orthogonal columns. Form a matrix **B**

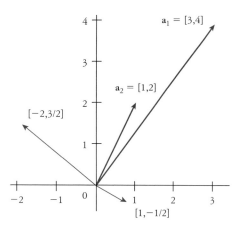

FIGURE 2 Graph for Example 2.

whose ith row $\mathbf{b}'_i$ is a vector equal to the ith column $\mathbf{a}_i$ of $\mathbf{A}$. Then $\mathbf{BA}$ will have 0s everywhere except on the main diagonal, since entry (i, j) of $\mathbf{BA}$ is $\mathbf{b}'_i \cdot \mathbf{a}_j = \mathbf{a}_i \cdot \mathbf{a}_j = 0$ (by the orthogonality of columns $\mathbf{a}_i$, $\mathbf{a}_j$). So $\mathbf{BA}$ is almost $\mathbf{I}$. Next look at diagonal entry (i, i) of $\mathbf{BA}$: $\mathbf{b}'_i \cdot \mathbf{a}_i = \mathbf{a}_i \cdot \mathbf{a}_i (= |\mathbf{a}_i|^2)$. To make this entry equal to 1, we modify $\mathbf{B}$ by dividing the ith row of $\mathbf{B}$ by $|\mathbf{a}_i|^2$. We have thus proved the next theorem, which deals with the inverse of a matrix with orthogonal columns.

Theorem 3. Let $\mathbf{A}$ be a square matrix with orthogonal columns $\mathbf{a}_j$. Then the ith row of $\mathbf{A}^{-1}$ equals $(1/|\mathbf{a}_i|^2)\mathbf{a}_i$.

E X A M P L E   4   Inverse of a Matrix with Orthogonal Columns   The matrix

$$\mathbf{A} = \begin{bmatrix} 2 & 1 & -2 \\ 1 & 2 & 2 \\ 2 & -2 & 1 \end{bmatrix}$$

has orthogonal columns. Then

$$\mathbf{A}^{-1} = \begin{bmatrix} \frac{2}{9} & \frac{1}{9} & \frac{2}{9} \\ \frac{1}{9} & \frac{2}{9} & -\frac{2}{9} \\ -\frac{2}{9} & \frac{2}{9} & \frac{1}{9} \end{bmatrix}$$

by Theorem 3. For example, the second row of $\mathbf{A}^{-1}$ is the vector formed by taking the second column $\mathbf{a}_2$ of $\mathbf{A}$ and dividing its entries by $|\mathbf{a}_2|^2 = \mathbf{a}_2 \cdot \mathbf{a}_2 = 1 \cdot 1 + 2 \cdot 2 + (-2) \cdot (-2) = 9$. ▲

A common theme in mathematics is that when an object—in this case, a matrix—has nice theoretical properties—in this case, orthogonal columns—then theory can supply shortcuts for computations involving this object. As objects (and operations) get more complex, the pressure grows to find ways to simplify the computations. Much of the development of mathematical theory has been motivated by this need to find special properties to simplify complex computations. We saw this situation in the previous section, where the computation of $\mathbf{A}^n\mathbf{c}$ was greatly simplified in the case where $\mathbf{c}$ was an eigenvector of $\mathbf{A}$.

Projections and Orthogonal Decomposition

Now we consider a vector closely related to the angle between two vectors. The *projection* $\mathbf{p}$ of vector $\mathbf{b}$ onto vector $\mathbf{a}$ is a multiple of $\mathbf{a}$, that is, $\mathbf{p} = q\mathbf{a}$, for some scalar q, and is defined as follows. Form a right triangle with one corner at the origin, the hypotenuse being vector $\mathbf{b}$ (that is, the line segment from the origin to point $\mathbf{b}$), the projection $\mathbf{p}$ being the side adjacent to the origin, and the third side being $\mathbf{b} - \mathbf{p}$ (see Figure 3a). The following mental image can

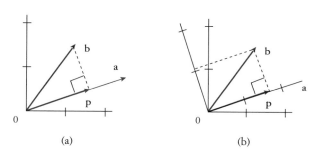

FIGURE 3 Projections of **b** onto **a**.

be associated with Figure 3a. (Refer to Chapter 2 for a geometric discussion of projections.) Rotate Figure 3a slightly clockwise so that the line representing vector **a** is horizontal; think of this line as the ground level. Suppose the sun is directly overhead. Then the shadow that the line representing vector **b** casts on the ground (i.e., on vector **a**) is the projection of **b** onto **a**. The length of the projection of **b** onto **a** can be interpreted as the value of **b**'s coordinate along the **a**-axis, if a coordinate system were used in which **a** was one of the axis directions (see Figure 3b). Note that the entries of any 2-vector $\mathbf{b} = [b_1, b_2]$ are the lengths of projections of **b** onto the coordinate vectors $\mathbf{i}_1 = [1, 0]$ and $\mathbf{i}_2 = [0, 1]$.

Observe from Figure 3 that the projection $q\mathbf{a}$ of **b** onto **a** is *the multiple of* **a** *that is closest to* **b**. That is, along the line formed by all multiples $r\mathbf{a}$ of **a**, the projection $q\mathbf{a}$ is the vector that minimizes the distance $|\mathbf{b} - r\mathbf{a}|$. We summarize the key properties we shall need of the projection **p** of a vector **b** onto a vector **a**.

Property 1. The projection **p** is a multiple of **a**: $\mathbf{p} = q\mathbf{a}$, for some scalar q.

Property 2. The projection **p** is orthogonal to the vector $\mathbf{b} - \mathbf{p}$.

Property 3. The projection **p** is the multiple of **a** closest to **b**.

By property 1, the projection of **b** onto **a** is totally characterized by the scalar q. We now give a simple formula for q.

Theorem 4. The projection $q\mathbf{a}$ of vector **b** onto vector **a** is given by the formula

$$q = \frac{\mathbf{a} \cdot \mathbf{b}}{\mathbf{a} \cdot \mathbf{a}}. \tag{5}$$

When **a** is a unit-length vector, $q = \mathbf{a} \cdot \mathbf{b}$.

Proof. Combining properties 1 and 2, the projection **p**, which equals $q\mathbf{a}$, is orthogonal to $\mathbf{b} - \mathbf{p}$, that is, to $\mathbf{b} - q\mathbf{a}$. If $q\mathbf{a}$ is orthogonal to $\mathbf{b} - q\mathbf{a}$, then **a** itself is also orthogonal to $\mathbf{b} - q\mathbf{a}$. The orthogonality of **a** and $\mathbf{b} - q\mathbf{a}$ means that $(\mathbf{b} - q\mathbf{a}) \cdot \mathbf{a} = 0$ (by Theorem 2). A little vector algebra then yields

$$(\mathbf{b} - q\mathbf{a}) \cdot \mathbf{a} = 0 \quad \Rightarrow \quad \mathbf{b} \cdot \mathbf{a} - q\mathbf{a} \cdot \mathbf{a} = 0 \quad \Rightarrow$$
$$\mathbf{b} \cdot \mathbf{a} = q\mathbf{a} \cdot \mathbf{a} \quad \Rightarrow \quad q = \frac{\mathbf{a} \cdot \mathbf{b}}{\mathbf{a} \cdot \mathbf{a}}. \tag{6}$$

▲ ▲ ▲

Since the projection **p** of vector **b** onto vector **a** is orthogonal to $\mathbf{b} - \mathbf{p}$ (property 2), **b** is decomposed into a part **p** that is a multiple of **a** and a part $\mathbf{b} - \mathbf{p}$ orthogonal to **a**. (A proof of (5) from first principles, without referring to Theorem 2, is given in the appendix at the end of the section.)

E X A M P L E 5 **Projections and Orthogonal Decompositions**

a. Find the projection **p** of **b** = $[-1, 3]$ onto **a** = $\mathbf{i}_2$ = $[0, 1]$, and use it to decompose **b** into orthogonal parts, one part parallel to **a** and one part orthogonal to **a**. This projection of **b** is $q\mathbf{i}_2$, where

$$q = \frac{\mathbf{a} \cdot \mathbf{b}}{\mathbf{a} \cdot \mathbf{a}} = \frac{0 \cdot (-1) + 1 \cdot 3}{0 \cdot 0 + 1 \cdot 1} = \frac{3}{1} = 3.$$

So **p** = $3\mathbf{i}_2$ = $[0, 3]$, and the orthogonal part is **b** − **p** = $[-1, 3] - [0, 3] = [-1, 0]$ (see Figure 4a).

b. Find the projection **p** of **b** = $[-1, 3]$ onto **a** = $[2, 4]$, and use it to decompose **b** into orthogonal parts. This projection is $q\mathbf{a}$, where

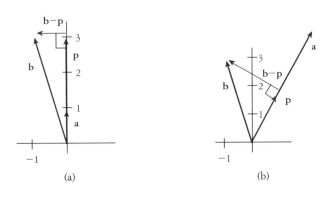

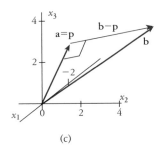

(c)

FIGURE 4 Example 5 graphs.

$$q = \frac{\mathbf{a} \cdot \mathbf{b}}{\mathbf{a} \cdot \mathbf{a}} = \frac{2 \cdot (-1) + 4 \cdot 3}{2 \cdot 2 + 4 \cdot 4} = \frac{10}{20} = \frac{1}{2}.$$

So **p** = $q\mathbf{a}$ = $.5[2, 4]$ = $[1, 2]$, and the orthogonal part is **b** − **p** = $[-1, 3] - [1, 2] = [-2, 1]$ (see Figure 4b).

c. Find the projection **p** of **b** = $[2, -4]$ onto **a** = $[2, 1]$, and use it to decompose **b** into orthogonal parts. Now

$$q = \frac{\mathbf{a} \cdot \mathbf{b}}{\mathbf{a} \cdot \mathbf{a}} = \frac{2 \cdot 2 + 1 \cdot (-4)}{2 \cdot 2 + 1 \cdot 1} = \frac{0}{5}.$$

Then **a** and **b** are orthogonal. So the projection **p** is the zero vector $0\mathbf{a}$ = $[0, 0]$, and the orthogonal part is **b** − **p** = **b** = $[2, -4]$.

d. Find the projection **p** of **b** = $[-1, 5, 3]$ onto **a** = $[-2, 0, 2]$, and use it to decompose **b** into orthogonal parts. Here

$$q = \frac{\mathbf{a} \cdot \mathbf{b}}{\mathbf{a} \cdot \mathbf{a}} = \frac{-2 \cdot -1 + 0 \cdot 5 + 2 \cdot 3}{-2 \cdot -2 + 0 \cdot 0 + 2 \cdot 2} = \frac{8}{8} = 1.$$

So the projection is **p** = $1\mathbf{a}$ = $1[-2, 0, 2]$ = $[-2, 0, 2]$, and the orthogonal part is **b** − **p** = $[-1, 5, 3] - [-2, 0, 2] = [1, 5, 1]$ (see Figure 4c). ▲

Theorem 3 discussed the form of the inverse of a matrix **A** whose columns are orthogonal. The key fact was that the ith row of $\mathbf{A}^{-1}$ equals the vector $(1/|\mathbf{a}_i|^2)\mathbf{a}_i$ (the ith column of **A** divided the square of the column's length). Suppose that for such an **A**, we wanted to solve the system of equations **Ax** = **b**. Then **x** = $\mathbf{A}^{-1}\mathbf{b}$, and the ith component x_i of **x** is the scalar product of the ith row of $\mathbf{A}^{-1}$ times $\mathbf{b}$. That is,

$$x_i = \left(\frac{1}{|\mathbf{a}_i|^2}\right)\mathbf{a}_i \cdot \mathbf{b} = \frac{\mathbf{a}_i \cdot \mathbf{b}}{\mathbf{a}_i \cdot \mathbf{a}_i}. \tag{7}$$

Thus, x_i is just the projection coefficient of $\mathbf{b}$ onto $\mathbf{a}_i$—small world!

EXAMPLE 6 Solving Ax = b for a Matrix with Orthogonal Columns Consider $\mathbf{Ax} = \mathbf{b}$, where $\mathbf{A} = \begin{bmatrix} 3 & 4 \\ -4 & 3 \end{bmatrix}$ has orthogonal columns and $\mathbf{b} = \begin{bmatrix} 3 \\ 6 \end{bmatrix}$. As a linear combination of columns problem, this case is

$$x_1\begin{bmatrix} 3 \\ -4 \end{bmatrix} + x_2\begin{bmatrix} 4 \\ 3 \end{bmatrix} = \begin{bmatrix} 3 \\ 6 \end{bmatrix}. \tag{8}$$

Figure 5 shows the situation graphically. Since $\mathbf{A}$'s columns $\mathbf{a}_1 = \begin{bmatrix} 3 \\ -4 \end{bmatrix}$ and $\mathbf{a}_2 = \begin{bmatrix} 4 \\ 3 \end{bmatrix}$ are or-

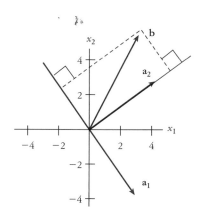

FIGURE 5 Example 6 graph.

thogonal, in the decomposition of $\mathbf{b}$ into its projection $\mathbf{p}$ onto $\mathbf{a}_1$ and an orthogonal part $\mathbf{b} - \mathbf{p}$ (as performed in Example 5), the orthogonal part $\mathbf{b} - \mathbf{p}$ will be a multiple of $\mathbf{a}_2$. (We note that all vectors orthogonal to $\mathbf{a}_1$ in the plane will be multiples of one another.) So $\mathbf{b} - \mathbf{p}$ will be a projection of $\mathbf{b}$ on $\mathbf{a}_2$. Thus, we have argued that Eq. (8) can be solved by letting x_1 be the projection coefficient of $\mathbf{b}$ onto $\mathbf{a}_1$ and x_2 be the projection coefficient of $\mathbf{b}$ onto $\mathbf{a}_2$. Using Eq. (7), we have

$$x_1 = \frac{\mathbf{a}_1 \cdot \mathbf{b}}{\mathbf{a}_1 \cdot \mathbf{a}_1} = \frac{3 \cdot 3 + (-4) \cdot 6}{3 \cdot 3 + (-4) \cdot (-4)} = -\frac{15}{25} = -\frac{3}{5},$$
$$x_2 = \frac{\mathbf{a}_2 \cdot \mathbf{b}}{\mathbf{a}_2 \cdot \mathbf{a}_2} = \frac{4 \cdot 3 + 3 \cdot 6}{4 \cdot 4 + 3 \cdot 3} = \frac{30}{25} = \frac{6}{5}. \ \blacktriangle \tag{9}$$

The following theorem states our shortcut for solving $\mathbf{Ax} = \mathbf{b}$ when $\mathbf{A}$ has orthogonal columns.

Theorem 5. The solution of the system of equations $\mathbf{Ax} = \mathbf{b}$, where $\mathbf{A}$ is a square matrix with orthogonal columns, is the vector of projection coefficients when $\mathbf{b}$ is projected onto each of the columns of $\mathbf{A}$. Thus,

$$x_i = \frac{\mathbf{a}_i \cdot \mathbf{b}}{\mathbf{a}_i \cdot \mathbf{a}_i}.$$

EXAMPLE 7 Solving Ax = b for a 3 × 3 Matrix with Orthogonal Columns Consider the matrix

$$\mathbf{A} = \begin{bmatrix} 2 & 1 & -2 \\ 1 & 2 & 2 \\ 2 & -2 & 1 \end{bmatrix}$$

with orthogonal columns. Let

$$\mathbf{b} = \begin{bmatrix} 1 \\ 4 \\ 6 \end{bmatrix}.$$

Then the solution of $\mathbf{Ax} = \mathbf{b}$ is, by Eq. (7),

$$
\begin{aligned}
x_1 &= \frac{\mathbf{a_1} \cdot \mathbf{b}}{\mathbf{a_1} \cdot \mathbf{a_1}} = \frac{2 \cdot 1 + 1 \cdot 4 + 2 \cdot 6}{2 \cdot 2 + 1 \cdot 1 + 2 \cdot 2} = \frac{18}{9} = 2, \\
x_2 &= \frac{\mathbf{a_2} \cdot \mathbf{b}}{\mathbf{a_2} \cdot \mathbf{a_2}} = \frac{1 \cdot 1 + 2 \cdot 4 + (-2) \cdot 6}{1 \cdot 1 + 2 \cdot 2 + (-2) \cdot (-2)} \\
&= \frac{-3}{9} = -\frac{1}{3}, \\
x_3 &= \frac{\mathbf{a_3} \cdot \mathbf{b}}{\mathbf{a_3} \cdot \mathbf{a_3}} = \frac{(-2) \cdot 1 + 2 \cdot 4 + 1 \cdot 6}{(-2) \cdot (-2) + 2 \cdot 2 + 1 \cdot 1} = \frac{12}{9} \\
&= \frac{4}{3}. \ \blacktriangle
\end{aligned}
\tag{10}
$$

Linear Regression

We next apply the concept of projections to a very important problem in statistics, one seemingly unrelated to projections and angles. The problem is statistical regression. In regression, one is given a set of data points (x_i, y_i) and one seeks to fit a line $\hat{y} = cx + d$ as closely as possible to a set of data points. The variable $\hat{y}$ is an estimator for the true y-value based the assumption that $\hat{y}$ is a linear function of x. The value $\hat{y}_i$, where $\hat{y}_i = cx_i + d$, would be the regression estimate for y_i. The name *regression,* which means movement back to a less developed state, comes from the idea that our model recaptures a simple relationship between the x_i and the y_i which randomness has obscured (the variables regress to a linear relationship). While the regression line seeks to fit past data, its important use is to predict the future. If we have several hundred data points in which x is the number of years a student has studied French and y is the rating on an hour-long interview in French of the person's fluency in French, then a regression line for this data can give an employer a general estimate of how fluent in French future job applicants are likely to be without a lengthy test.

For the present, we shall consider the simple case where the line has the form $\hat{y} = cx$.

EXAMPLE 8   A Simple Regression Problem

Suppose we want to fit the three points $(0, 1)$, $(2, 1)$, and $(4, 4)$ to a line of the form

$$\hat{y} = cx. \tag{11}$$

Look at Figure 6. The x-value might represent the number of semesters of college mathematics a student has taken, and the y-value the student's score on a graduate admissions test. There are thousands of other settings that might give rise to these values. The estimate in Eq. (11) would help us predict the y-values for other x-values, for example, predict how future students might do on the test based on the amount of math they have taken.

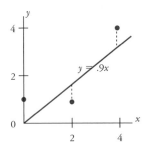

FIGURE 6 A simple regression problem, $y = cx$.

The three points in this problem are readily seen not to lie on a common line, much less a line through the origin (any line of the form in Eq. (11) passes through the origin). So we have to find a choice of c that gives the best possible fit, that is, a line $\hat{y} = cx$ passing as close to these three points as possible.

What do "best possible fit" and "as close as possible" mean? A first thought would be to minimize the sum of the errors, $\Sigma(\hat{y}_i - y_i)$. However, this approach does not work since the sum might be nearly zero because some large negative errors canceled some large positive errors. Instead, we should look at absolute errors, $|\hat{y}_i - y_i|$, the absolute difference between the value $\hat{y}_i (= cx_i)$ predicted by Eq. (11) and the true value y_i. On the other hand, absolute values are not easy to use in mathematical equations. The most common approach used in such problems is to minimize $\Sigma(\hat{y}_i - y_i)^2$, the sum of the squares of the errors. Taking the squares of differences yields positive numbers without using absolute values. There is also a geometric reason, which we shall give shortly, for using squares.

For the points $(0, 1)$, $(2, 1)$, $(4, 4)$, the expression $\Sigma(\hat{y}_i - y_i)^2$ for the sum of squares of the errors (SSE) is

$$
\begin{aligned}
SSE &= (0c - 1)^2 + (2c - 1)^2 + (4c - 4)^2 \\
&= 1 + (4c^2 - 4c + 1) + (16c^2 - 32c + 16) \\
&= 20c^2 - 36c + 18.
\end{aligned}
\tag{12}
$$

One can use calculus to find the value of c that minimizes the expression in (12). One can also plot the graph to estimate the minimum point. However, there is a geometric interpretation of the sum of squares of errors that will al-low us to convert the minimization of (12) into a projection problem. Let $\mathbf{x}$ be the vector of our x-values and $\mathbf{y}$ the vector of our corresponding y-values. Here

$$
\mathbf{x} = \begin{bmatrix} 0 \\ 2 \\ 4 \end{bmatrix} \quad \text{and} \quad \mathbf{y} = \begin{bmatrix} 1 \\ 1 \\ 4 \end{bmatrix}.
$$

Further, let $\hat{y}$ be the vector of estimates for $\mathbf{y}$. Equation (11) can now be rewritten as

$$
\hat{y} = c\mathbf{x}.
\tag{13}
$$

That is, the estimates

$$
\hat{y} = \begin{bmatrix} \hat{y}_1 \\ \hat{y}_2 \\ \hat{y}_3 \end{bmatrix}
$$

from Eq. (11) will be c times the x-values $\begin{bmatrix} 0 \\ 2 \\ 4 \end{bmatrix}$.

Think of $\mathbf{x}$, $\mathbf{y}$, and $\hat{y}$ as vectors in three-dimensional space, where $\hat{y}$ is a multiple of $\mathbf{x}$. Then the obvious strategy for approximating $\mathbf{y}$ by $\hat{y}$ is to pick the value of c that makes $c\mathbf{x} (= \hat{y})$ as geometrically close as possible to $\mathbf{y}$ (see Figure 7). That is, we want to minimize the distance $|c\mathbf{x} - \mathbf{y}|$ in three-dimensional space between $c\mathbf{x}$ and $\mathbf{y}$.

This distance between

$$
c\mathbf{x} = \begin{bmatrix} 0c \\ 2c \\ 4c \end{bmatrix} \quad \text{and} \quad \mathbf{y} = \begin{bmatrix} 1 \\ 1 \\ 4 \end{bmatrix}
$$

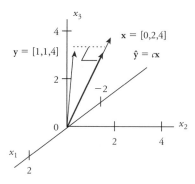

FIGURE 7 The graph of $\hat{y} = cx$.

is simply

$$|c\mathbf{x} - \mathbf{y}| = \sqrt{(0c - 1)^2 + (2c - 1)^2 + (4c - 4)^2}.$$

(14)

Comparing (12) and (14), we see that $|c\mathbf{x} - \mathbf{y}|$ is the square root of the SSE. Then minimizing SSE is equivalent to minimizing the distance $|c\mathbf{x} - \mathbf{y}|$ since minimizing the value of a positive expression also minimizes the value of the square root of the expression.

But the multiple of $\mathbf{x}$ that is closest to $\mathbf{y}$ is the projection of $\mathbf{y}$ onto $\mathbf{x}$ (property 3 of projections). By Theorem 4,

$$c = \frac{\mathbf{y} \cdot \mathbf{x}}{\mathbf{x} \cdot \mathbf{x}} = \frac{1 \cdot 0 + 1 \cdot 2 + 4 \cdot 4}{0 \cdot 0 + 2 \cdot 2 + 4 \cdot 4} = \frac{18}{20} = .9.$$

Then the regression line is $\hat{y} = .9x$, and our estimates $\hat{y}_i$ are (see Figure 6):

$$\hat{y}_1 = .9x_1 = .9 \cdot 0 = 0,$$
$$\hat{y}_2 = .9x_2 = .9 \cdot 2 = 1.8,$$
$$\hat{y}_3 = .9x_3 = .9 \cdot 4 = 3.6. \;\blacktriangle$$

Appendix to Section 3.8

Alternative Proof of Theorems 1 and 4

We seek to prove Theorem 1, that $\cos \theta = (\mathbf{a} \cdot \mathbf{b})/|\mathbf{a}||\mathbf{b}|$, and Theorem 4, that the length q of the projection of $\mathbf{b}$ onto $\mathbf{a}$ is given by $q = (\mathbf{a} \cdot \mathbf{b})/|\mathbf{a}|^2$ or, equivalently, $q = (\mathbf{a} \cdot \mathbf{b})/(\mathbf{a} \cdot \mathbf{a})$. Referring to Figure 8, θ is the angle formed by $\mathbf{b}$ and $\mathbf{a}$ or, equivalently, the angle formed by $\mathbf{b}$ and $\mathbf{p} = q\mathbf{a}$, the projection of $\mathbf{b}$ onto $\mathbf{a}$. In the right triangle in Figure 8,

$$\cos \theta = \frac{|q\mathbf{a}|}{|\mathbf{b}|} = \frac{q|\mathbf{a}|}{|\mathbf{b}|}. \qquad (*)$$

It remains to determine q. By the Pythagorean theorem,

$$|\mathbf{p}|^2 + |\mathbf{b} - \mathbf{p}|^2 = |\mathbf{b}|^2, \qquad \text{or}$$
$$|q\mathbf{a}|^2 + |\mathbf{b} - q\mathbf{a}|^2 = |\mathbf{b}|^2, \qquad \text{or}$$
$$q\mathbf{a} \cdot q\mathbf{a} + (\mathbf{b} - q\mathbf{a}) \cdot (\mathbf{b} - q\mathbf{a}) = \mathbf{b} \cdot \mathbf{b}.$$

Using the distributive and commutative laws for scalar products,

$$q^2\mathbf{a} \cdot \mathbf{a} + \mathbf{b} \cdot \mathbf{b} - 2q\mathbf{a} \cdot \mathbf{b} + q^2\mathbf{a} \cdot \mathbf{a} = \mathbf{b} \cdot \mathbf{b}.$$

After simplifying, we have

$$2q^2\mathbf{a} \cdot \mathbf{a} = -2q\mathbf{a} \cdot \mathbf{b} \quad \text{or} \quad q = \frac{\mathbf{a} \cdot \mathbf{b}}{|\mathbf{a}|^2}.$$

This proves Theorem 4. Substituting for q in (*) yields Theorem 1.

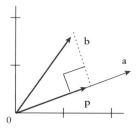

FIGURE 8 The graph of proof that $\cos \theta = (\mathbf{a} \cdot \mathbf{b})/|\mathbf{a}||\mathbf{b}|$.

1. Compute the cosine of and determine the angle made by the following pairs of vectors.
   (a) $[3, 4], [-3, 4]$
   (b) $[1, 2], [3, 1]$
   (c) $[1, 1, 1], [1, -1, 2]$
   (d) $[1, 1, 1], [2, -1, 3]$

2. If a triangle has corners at the following three points, find the cosine of each of the angles of the triangle.
   (a) $[2, 0, 1], [3, -1, 1]$, and $[4, 2, 0]$
   (b) $[3, 0], [-2, 1]$, and $[5, 6]$

3. Determine which of the following pairs of vectors are parallel (the angle between them is $0°$ or $180°$), orthogonal, or neither.
   (a) $[3, 2], [3, 2]$
   (b) $[6, 4], [-2, 3]$
   (c) $[1, 2, -3], [2, -1, 3]$
   (d) $[1, 0, 2], [0, 4, 0]$

4. Consider the following data of student performance, where the x-value is a scaled score (to have average value of 0) of high school grades, the y-value is a scaled score of SAT scores, and the z-value is an unscaled score of college grades.

Student	A	B	C	D	E
x	-4	-2	0	2	4
y	2	-1	-2	-1	2
z	3	6	7	7	6

Compute the correlation coefficient between the vectors of $\mathbf{x}$- and $\mathbf{y}$-values and that between the vectors of $\mathbf{x}$- and $\mathbf{z}$-values.

5. The following data show scores that three students received on a battery of six different tests.

	George	Ronnie	Jimmy
General IQ	12	20	10
Math	8	22	4
Reading	16	14	10
Running	24	16	12
Speaking	12	10	30
Watching	12	14	8

Compute the correlation coefficient between (a) George and Ronnie, (b) Ronnie and Jimmy. (Hint: Remember first to subtract the average value from each number.)

6. (a) Compute the correlation coefficient of the following readings from seven students of their IQs and their scores at the arcade game Zaxxon.

Student	A	B	C	D	E	F	G
IQ	120	130	105	90	125	120	110
Zaxxon	11,000	7000	10,000	12,000	8000	100,000	8000

   (b) Delete student F, and recompute the correlation coefficient. (Hint: Remember first to subtract the average value from each number.)

7. For each of the following 2×2 matrices $\mathbf{A}$, find the inverse $\mathbf{A}^{-1}$ and confirm graphically that the ith column of $\mathbf{A}$ is orthogonal to the jth row of $\mathbf{A}^{-1}$ for $i \neq j$.
   (a) $\begin{bmatrix} 2 & -1 \\ -4 & 3 \end{bmatrix}$ (b) $\begin{bmatrix} 1 & 2 \\ 2 & 3 \end{bmatrix}$

8. For the following matrices $\mathbf{A}$ with orthogonal columns, determine the inverse $\mathbf{A}^{-1}$ using Theorem 3.
   (a) $\begin{bmatrix} -6 & 4 \\ 3 & 8 \end{bmatrix}$ (c) $\begin{bmatrix} 2 & -3 & 6 \\ -6 & 2 & 3 \\ 3 & 6 & 2 \end{bmatrix}$
   (b) $\begin{bmatrix} 1 & -4 \\ 2 & 2 \end{bmatrix}$

9. Determine the length q of the projection of **b** onto **a**.
   (a) **a** = [0, 1], **b** = [4, 0]
   (b) **a** = [3, 1], **b** = [−3, 5]
   (c) **a** = [−3, −4], **b** = [3, 7]
   (d) **a** = [3, 2, 1], **b** = [1, −3, 3]
   (e) **a** = [−2, 3, 1], **b** = [6, 3, 3]

10. Decompose each vector **b** in Exercise 9 into two orthogonal components, one component parallel to **a** and one component orthogonal to **a**.

11. Use Theorem 5 to solve the system of linear equations **Ax** = **b**, where the columns of **A** are orthogonal. Plot the columns of **A** and **b**, and plot the projections of **b** onto the columns of **A**.

(a) $\mathbf{A} = \begin{bmatrix} -3 & 3 \\ 1 & 9 \end{bmatrix}, \mathbf{b} = \begin{bmatrix} -5 \\ 1 \end{bmatrix}$

(b) $\mathbf{A} = \begin{bmatrix} -6 & 4 \\ 3 & 8 \end{bmatrix}, \mathbf{b} = \begin{bmatrix} 7 \\ 2 \end{bmatrix}$

(c) $\mathbf{A} = \begin{bmatrix} 2 & -3 & 6 \\ -6 & 2 & 3 \\ 3 & 6 & 2 \end{bmatrix}, \mathbf{b} = \begin{bmatrix} -2 \\ 5 \\ -1 \end{bmatrix}$

(d) $\mathbf{A} = \begin{bmatrix} -1 & 4 & -1 \\ 2 & 1 & -2 \\ 1 & 2 & 3 \end{bmatrix}, \mathbf{b} = \begin{bmatrix} 4 \\ 1 \\ 2 \end{bmatrix}$

12. The following data give the number of accidents bus drivers had in one year as a function of the number of years on the job.

Years on job	2	4	6	8	10	12
Accidents	10	8	3	8	4	5

Fit this data with a regression model of the form $\hat{y} = qx$, where x is the number of years of experience and y is the number of bus accidents. Plot the observed numbers of accidents and the predicted numbers. How good is the fit?

13. Seven students earned the following scores on a test after studying for different numbers of weeks:

Student	A	B	C	D	E	F	G
Weeks of study	0	1	2	3	4	5	6
Test score	3	4	7	6	10	6	10

Fit this data with a regression model of the form $\hat{y} = qx$, where x is the number of weeks studied and y is the test score. Plot the observed scores and the predicted scores.

14. The following data give the numbers of fish a fisher caught on successive weekends during a fishing season.

Weekend number	1	2	3	4	5	6	7
Fish caught	4	9	13	14	18	22	28

Fit this data with a regression model of the form $\hat{y} = qx$, where x is the number of the weekend and y is the number of fish caught.

Summary of Chapter 3

Chapter 3 introduced basic concepts and methods of linear algebra and used them to solve a variety of problems associated with linear models. Two key examples were used to illustrate many of these concepts and methods: the oil refinery production model and the weather Markov chain model.

The first section introduced vectors and matrices and presented several linear models, including the oil refinery and weather models, that are expressed mathematically in terms of vectors and matrices.

The second section presented the building block of matrix algebra operations, the scalar product, from which matrix–vector products were built. Matrix–vector products were used to represent models introduced in Section 3.1. They were also given interpretations as geometric transformations.

The third section introduced matrix multiplication and gave applications of this operation to data organization, Markov chains, and graphs.

The fourth section presented Gaussian elimination for solving a system of linear equations. It examined the mathematical and practical interpretations of the various possible outcomes of the elimination algorithm. Gaussian elimination was used to determine the steady-state weather distribution in the weather Markov chain model.

The fifth section introduced the concept of a matrix inverse and showed how to compute inverses using Gauss–Jordan elimination, a variation of Gaussian elimination introduced in the previous section. Additional theory was presented, and inverses were applied to sensitivity analysis of the oil refinery model, rotation matrices (introduced in Section 3.2), and cryptography.

The sixth section gave a brief overview of determinants. While originally developed to obtain a closed-form formula for the solution of a system of linear equations, their primary use in this chapter is in finding eigenvalues.

The seventh section introduced eigenvalues and eigenvectors. The initial objective of the section was to illustrate the usefulness of these concepts in computations involving linear growth models. Then we showed how to determine the eigenvalues and eigenvectors of a matrix and applied these methods to analyze the long-term behavior of two growth models.

The last section had a more geometric focus, looking at angles between vectors and the projection of one vector onto another. Of particular interest was the situation where two or more vectors are mutually perpendicular to one another, what are called orthogonal vectors. It was shown that systems of linear equations are very easy to solve if in matrix format they have orthogonal columns. The section closed with an important statistical application of projections called linear regression.

It is hoped that this chapter helped the reader gain an appreciation of the power and usefulness of linear algebra and linear models.

Chapter 3 Exercises

1. Let $\mathbf{a} = \begin{bmatrix} 3 \\ 1 \\ 2 \end{bmatrix}$, $\mathbf{b} = \begin{bmatrix} 0 \\ 2 \\ -2 \end{bmatrix}$, $\mathbf{c} = \begin{bmatrix} 1 \\ 4 \\ 8 \end{bmatrix}$,

$\mathbf{A} = \begin{bmatrix} 4 & 1 & 5 & 4 \\ 0 & 3 & 4 & 9 \\ 2 & 3 & 1 & 6 \end{bmatrix}$, $\mathbf{B} = \begin{bmatrix} 2 & 1 & -1 \\ 3 & -1 & 4 \\ 2 & 1 & 3 \end{bmatrix}$,

$\mathbf{C} = \begin{bmatrix} -2 & 3 & -2 \\ 0 & 3 & 1 \\ 4 & 5 & 3 \\ -1 & 2 & 3 \end{bmatrix}$. Calculate the following expressions unless the sizes do not match.

(a) **aB**　　　　　　　(e) **AB**
(b) **Ca**　　　　　　　(f) **BA**
(c) **(a + c)A**　　　　(g) **aBc**
(d) **(c − d)C**　　　　(h) **ACB**

2. Suppose we are given the following matrices: matrix **A** gives the amounts of raw material required to build different products; matrix **B** gives the costs of these raw materials in two different countries; matrix **C** tells how many of the products are needed to build two types of houses; and matrix **D** gives the demand for houses in the two countries.

Raw material

A	Wood	Labor	Steel
Item A	5	20	10
Item B	4	25	8
Item C	10	10	5

Cost by country

B	Spain	Italy
Wood	$2	$3
Labor	$6	$5
Steel	$3	$4

Items needed in house

C	A	B	C
House I	4	8	3
House II	5	5	2

D	House I	House II
Spain	50,000	200,000
Italy	80,000	500,000

(a) Which matrix product tells how many of each item is needed to build each type of house?
(b) Which matrix product tells how much it costs to make each item in each country?
(c) Which matrix product gives the cost of building each type of house in each country?

3. Consider a variation on the Markov chain for weather. Now there is just Sunny and Cloudy weather. The new transition matrix **A** is

$$
A = \begin{array}{c} \text{Sunny} \\ \text{Cloudy} \end{array} \begin{array}{cc} \text{Sunny} & \text{Cloudy} \\ \left[\begin{array}{cc} \frac{3}{4} & \frac{1}{2} \\ \frac{1}{4} & \frac{1}{2} \end{array} \right. & \left. \vphantom{\frac{1}{2}} \right] \end{array}
$$

(a) Compute A^2. What probability does entry (1, 2) in A^2 represent?
(b) Compute A^3. What is the probability if sunny today that it is sunny in three days?

4. A furniture manufacturer makes tables, chairs and sofas. In one month, the company has available 270 units of wood, 520 units of labor, and 290 units of fabric. The manufacturer wants a production schedule for the month that uses all of these resources. The different products require the following amounts of the resources.

	Table	Chair	Sofa
Wood	5	3	4
Labor	4	2	8
Fabric	0	1	5

Set up and solve a system of equations to determine how much of each product should be manufactured.

5. Re-solve problem 4 by first computing the inverse for the data matrix. If the amount of wood is increased by 10 units, how will this change the numbers of chairs produced?

6. Consider the two-refinery production of heating oil and gasoline. The second refinery has not been built, but when it is built it will produce 10 gallons of heating oil and k gallons of gasoline from each barrel of crude oil. We have

$$
\begin{array}{ll} \text{Heating oil} & 6x_1 + 10x_2 = H \\ \text{Gasoline} & 4x_1 + 4kx_2 = G \end{array}
$$

where k is to be determined, H is the demand for heating oil, G is the demand for gasoline (and x_i is number of barrels of crude oil processed by refinery i, $i = 1, 2$).

Solve this system of equations to determine x_1 and x_2 in terms of k, H, G using Cramer's Rule. What values of k yield a non-unique solution?

7. (a) For the following rabbit/fox model, determine both eigenvalues.

$$R' = 1.2R - .2F$$
$$F' = .2R + .8F$$

(b) Determine an eigenvector $\mathbf{e}$ associated with the largest eigenvalue.

(c) Determine the other eigenvector $\mathbf{e}'$ (associated with the smaller eigenvalue).

(d) If the initial population is $\mathbf{x} = \begin{bmatrix} 10 \\ 10 \end{bmatrix}$, then express $\mathbf{x}^{(k)}$ as a linear combination of $\mathbf{e}$ and $\mathbf{e}'$. Use this expression to describe in words the behavior of this model over time.

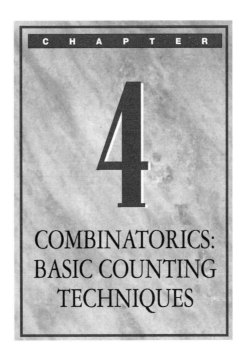

CHAPTER 4

COMBINATORICS: BASIC COUNTING TECHNIQUES

Introduction

In 1992, Bellcore, the research arm of the local telephone operating companies, announced that it would soon lift the restriction of requiring "0" or "1" as the middle digit of an area code. According to Bellcore, more area codes were needed. Can we be more precise? How many area codes were possible under the restriction? How many new area codes are possible now that the restriction has been lifted? These and similar problems require us to *enumerate* possibilities, or to count.

One of the first mathematical activities early humans engaged in was counting. Among the most primitive societies discovered by anthropologists, there are typically words for "one," "two," and "many." Values higher than two are often represented in an additive fashion; for example, five might be conveyed as "two-two-one." In most societies, numeration systems have been developed to express large numbers more conveniently. For example, Figure 1 shows fragments of ancient Egyptian cubits, the equivalent of modern rulers, showing both length markings and the Egyptian numerals.

Our concern in this chapter is not how to express

FIGURE 1 Fragments of Egyptian royal cubits, c. 250 B.C.

a large number, but how to determine that number. We certainly do not want to write out all possible 1992 area codes and then count our list by ones! (We look at area codes in Section 4.1, Examples 12 and 13, and again in Section 4.2.)

Students in elementary school learn two techniques, other than counting by ones, for combining "small" numbers and arriving at a "larger" answer, namely addition and multiplication. These techniques have been expanded by mathematicians into two basic principles of advanced counting, or enumeration: the addition principle and the multiplication principle. We will use these principles, and some related formulas, to answer our area code questions and to solve other enumeration problems. When we study probability in Chapter 8, Chance, we will be building on what we learn in this chapter.

SECTION 4.1 *Addition and Multiplication Principles*

Combinatorics is the branch of mathematics that studies enumeration problems. We start by learning the two basic principles underlying the solution of all enumeration problems. The basic principles are not particularly difficult to apply in simple cases, such as those we will encounter in this section. But in later sections, the problems become more complex, requiring thoughtful analysis and often the use of both principles.

The Addition Principle

We use the addition principle to count the total number of possibilities when we are given several nonoverlapping sets of possibilities. For example, suppose a high school senior has been accepted at three colleges, which we will call X, Y, and Z, but of course will only attend one, either X or Y or Z. Let's suppose that each of the colleges has its own strengths, so the student's possible majors depend on the college in which the student enrolls, as shown in the following table.

College X	College Y	College Z
chemistry	mathematics	astronomy
biology	accounting	physics
	pre-law	engineering
	business	
(2 possibilities)	(4 possibilities)	(3 possibilities)

It does not surprise us that the number of possible majors for this student is 9, since $9 = 2 + 4 + 3$.

Let's look at the features of this example to extract those central to the proper application of the addition principle. Perhaps the most important feature, in terms of the typical language used to present combinatorics problems, is that the word "or" occurs or is implied in the problem in the guise of mutually exclusive sets of possibilities. In our example there are three colleges: The student will attend College X *or* College Y *or* College Z. At each college the student has a set of possible majors. The student cannot major both in chemistry at College X and in astronomy at College Y; the student will attend just one college, and only one major is possible. (For simplicity, we require the student to have a single major, not a double major.)

We note that there are no shared elements among the three sets of possible majors; in set theory these three sets are called *pairwise-disjoint sets,* since no possible major is common to any two of the colleges. (Appendix 1 contains a brief introduction to set theory.) The addition principle is used to count the total number of elements in the collection of pairwise-disjoint sets.

DEFINITION

The *addition principle* says that given a collection C of k pairwise-disjoint sets, $S_1, S_2, \ldots, S_k$, with the number of elements of the set S_i given by n_i. Suppose we want to find the number of elements in the union $S_1 \cup S_2 \cup \ldots \cup S_k$. Since no element of this union is a member of more than one of the component sets, the sum we want is $n_1 + n_2 + \cdots + n_k = \sum_{i=1}^{k} n_i$. (The symbol on the right-hand side, the Greek capital letter sigma, is used to denote a sum.)

E X A M P L E   1   An Easy Use of the Addition Principle   A state government official wants to know how many automobile license plates are possible if the plates are formed using one of two arrangements: two letters followed by three numbers or four numbers followed by one letter. The disjoint sets here are $S_1 = \{$plates with two letters followed by three numbers$\}$ and $S_2 = \{$plates with four numbers followed by one letter$\}$. The resident mathematician, using the multiplication principle, which we will learn soon, tells the official that there are 676,000 possible plates in set S_1 and 260,000 possible plates in set S_2. The government official knows the addition principle and thus determines that there are 676,000 + 260,000 = 936,000 types of plates possible using just those two arrangements of numbers and letters. Although this is presented as an example of the addition principle, solving the government official's problem requires analysis and the use of both principles. ▲

The pairwise-disjoint requirement for the sets is crucial to the addition principle. Suppose that the head of a Chamber of Commerce wishes to enumerate the businesses in a community and asks one person to make a list of all businesses engaged in sales, including, for example, shoe stores, and another person to make a list of all businesses providing services, such as repairing shoes. This may appear to be a fine way to proceed, until we discover that Local City Shoe Shop both sells and repairs shoes. Thus Local City is on both lists, it is an element of both sets, and we no longer have pairwise-disjoint sets. It is clear that by mistakenly following the addition principle to determine the number of businesses, the head of the Chamber of Commerce would count Local City twice, not once, as is proper. When the sets are not pairwise disjoint, we need a generalization of the addition principle. Here is that generalization for the case of two overlapping sets.

One generalization of the *addition principle* is this: Given two sets S_1 and S_2, with the number of elements of the set S_i given by n_i, and the number of elements in the intersection $S_1 \cap S_2$ being k, then the number of elements in the union $S_1 \cup S_2$ is $n_1 + n_2 - k$.

EXAMPLE 2 If the community above has 573 businesses engaged in sales (the first set) and 249 businesses providing services (the second set), and 117 of the businesses engage in both activities, then the total number of distinct businesses in the union of these two sets is $573 + 249 - 117 = 705$. ▲

EXAMPLE 3 **Testing a Program Module—First View** A computer programmer wishes to provide data to test every execution path of a module exactly once. (We use the term *module* to mean any identifiable subsection of computer code; procedures, functions, and subroutines are examples of modules.) An execution path is the set of individual steps or instructions actually followed in carrying out the program, in going from its start to its end. Most programs contain many alternative routes for going from start to end; each route is a possible execution path. The particular execution path followed is determined by the data set being processed by the module. For example, suppose the code contains the equivalent of this pseudo-code:

```
If a < b then
        Follow execution path one
else
        Follow execution path two
end if
```

The relative sizes of the variables a and b determine which execution path will be followed. Thus, in order to test both paths, two sets of data would be required, one with $a < b$ and one with $a \geq b$. If execution paths one and/or two contain further decisions, additional "if" structures, then the module would have more than two execution paths and more than two data sets would be needed.

The programmer needs to know how many execution paths there are in order to know how many sets of test data to provide. A flowchart of a module, shown in Figure 2, will help us find the number of execution paths for that module.

An execution path for the module corresponds to a route along the directed edges of the flowchart, from the oval labeled "Start of Module" to the oval labeled "End of Module." The module in Figure 2 selects exactly one of three submodules to execute; the first submodule has 7 different execution paths, the second submodule has 5, and the third submodule has 8. The submodules are shown as multiway decisions with several internal execution paths. In this example each submodule corresponds to a set of execution paths, and since only one of the three submodules

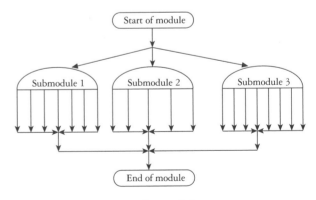

FIGURE 2 Flowchart of a module.

can be executed each time the module is executed, the three sets of execution paths are disjoint. The sum of the numbers of elements (execution paths) in the three submodules is $7 + 5 + 8 = 20$, so the programmer must provide 20 sets of test data in order to test all the execution paths. Again, we see an easy calculation; in the next section we will see how combinatoric considerations can be very important in deciding how to test large programs, consisting of many modules. ▲

The Multiplication Principle and Independent Arrangements

Now we learn the multiplication principle, which is used to count the number of potential "arrangements" of several "objects," each of which comes from a specific set of elements.

E X A M P L E   4   Ice Cream Sundaes   Suppose you are entertaining five friends and want to present each one with a unique ice cream sundae consisting of one scoop of ice cream and one topping. You check your kitchen and find 3 ice creams—vanilla (V), chocolate (C), and strawberry (S)—and 2 toppings—fudge (F) and butterscotch (B). In this situation, the objects you are arranging are the scoop of ice cream and the topping. The tree diagram in Figure 3 displays the possibilities.

The figure suggests that multiplication will tell us the number of possible sundaes, each of which corresponds to an endpoint (known as a leaf) of the tree: Multiply the number of possibilities for the first object (ice cream scoop) by the number of possibilities for the second object (topping). We get $3 \cdot 2 = 6$ possibilities for the number of sundaes. You could provide each of

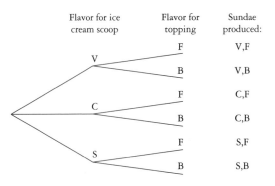

FIGURE 3   Tree diagram for an ice cream sundae.

your 5 friends with a unique sundae, and there is one arrangement left for you. ▲

There is an order in which one physically creates an ice cream sundae, but it does not affect our calculations. We are really planning the arrangements of flavors for the ice cream and the topping, not for physically making the sundaes. We could plan the flavor for the topping before we plan the flavor for the ice cream, but of course we could not make the actual sundae in that way.

An essential aspect of Example 4 is that we are creating an arrangement of "objects," for example scoops, and for each position in the arrangement there is a fixed number of possibilities available for the object. That number of possibilities must not depend on which object is in any other position in the arrangement. This last condition is called *independence*. In our example if no one likes vanilla ice cream with butterscotch sauce, then the branch of the tree showing V as the ice cream would have only leaf, namely F, while the two branches starting with C and S would each have two leaves. The number of leaves would depend on the flavor used on the first branch, so we would not have independence. This more complex, nonindependent situation would require us to carefully use both the addition and multiplication principles; in Section 4.2 we study such examples.

DEFINITION

We call an arrangement of objects an *independent arrangement* if the number of possibilities for any specific position in the arrangement does not depend on which objects are in the other positions of the arrangement.

If we are making an independent arrangement consisting of two objects and we have a set of n_1 possibilities for the first position in the arrangement and a set of n_2 possibilities for the second, then the number of possible arrangements we can make is the product $n_1 \cdot n_2$. If we are making an independent arrangement of k objects and we have a set of n_i possibilities for the ith position in the arrangement, then the number of possible arrangements we can make is the product

$$n_1 \cdot n_2 \cdot \cdots \cdot n_k = \prod_{i=1}^{k} n_i.$$

The symbol on the right-hand side is the Greek capital letter pi, which is used here to denote a product.

By contrast with the addition principle, associated with pairwise-disjoint sets of possibilities and the word "or," the multiplication principle is often associated with the word "and." We use the multiplication principle when we are simultaneously putting several objects into an arrangement; that arrangement consists of the object in the first position, *and* the object in the second position, *and* the one in the third position, *and* so on. The multiplication principle is very flexible, as the following examples show.

EXAMPLE 5 Counting Possible RNA Molecules   RNA molecules are chemicals found in living cells. An RNA molecule is a long string of hundreds or even thousands of positions; the exact length depends on the species being examined. Each position in the string is occupied by one of four different chemicals called bases and represented by the letters A, C, G, and U. Since the bases in an RNA molecule can occur in any order at all, the base in any one position is independent of the bases in any other. How many different RNA molecules could there be if the molecule is exactly 100 bases long? We have 100 positions, with a set of 4 choices for each position. A tree for this situation would have 100 "levels" of branches between the "root" of the tree (shown at the left in the tree in Figure 4) and the "leaves" (shown at the right in the figure). Instead

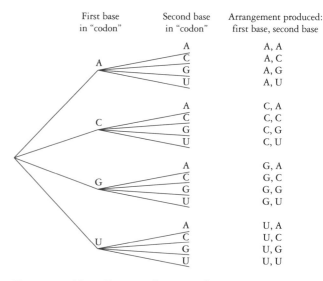

FIGURE 4 Tree diagram for a two-base "codon."

of drawing such a large tree, we proceed directly to the multiplication principle. That principle says that the number of possible RNA molecules of length 100 is $4 \cdot 4 \cdot 4 \cdot \cdots \cdot 4$, 100 factors each of which is 4, giving 4^{100}, or approximately $1.6 \cdot 10^{60}$! This tree has an enormous, almost unimaginable, number of branches. ▲

The number $1.6 \cdot 10^{60}$, which is 16 followed by 59 zeros, tells us the size of the set of possible RNA molecules of length 100. It is a very large number. In other examples and in some of the exercises in this chapter, we encounter other very large numbers. In the Spotlight "Beyond Finite Sets," we consider some sets that are so large that ordinary numbers are not adequate to describe their sizes.

E X A M P L E 6   How Long Must Codons Be?
We now consider codons (pronounced "code-ons"), which are short parts, just a few bases, of an RNA molecule. In the biology of cell activity, each codon acts as the blueprint for making an amino acid. There are 20 different amino acids. If we assume that all codons have the same number of bases—the actual case—then what is the minimum number of bases per codon so that each amino acid is represented by at least one codon?

Clearly, one base is not enough. Since there are only 4 bases, there could only be 4 codons, but we have 20 amino acids. A tree diagram for two bases is shown in Figure 4.

With only 2 bases per codon, there are $4 \cdot 4 = 4^2 = 16$ possibilities; again not enough codons for all 20 amino acids. If we were to draw the tree for a codon with 3 bases, we would have $4 \cdot 4 \cdot 4 = 4^3 = 64$ possibilities. This number is clearly sufficient to code for just 20 different amino acids. In fact, codons do have three bases, exactly the minimum number we calculated. Of

the 64 possibilities, 3 serve as chain termination signals, and the other 61 are codons and do code for amino acids. Most amino acids are represented by more than one codon. ▲

E X A M P L E 7   Ice Cream Cones   Suppose a customer in an ice cream store wants a 2-scoop cone and requires that the bottom scoop be some variety of chocolate, of which the store has 4, and that the top scoop be some variety of fruit, of which the store has 6. The multiplication principle applies, giving $4 \cdot 6 = 24$ possible 2-scoop cones.

Note that if our store has raspberry chocolate ice cream, that flavor could be considered as both a variety of chocolate and a variety of fruit. In applying the multiplication principle, we require independence of choices, but we can have sets of choices that overlap, that is, are not disjoint. By contrast, the addition principle requires that the sets of choices be disjoint. If the customer wants a single scoop of either a variety of chocolate or a variety of fruit, then the existence of flavors like raspberry chocolate would require us to use the generalized addition principle to determine the number of choices. ▲

Did you notice the uses of the words "and" and "or" in the previous example? The "or" helps us to find disjoint sets and usually leads to the addition principle. The "and" helps us to see independent arrangements of objects and usually leads to the multiplication principle.

E X A M P L E 8   Cone With Distinct Flavors
Suppose that you and I go to the ice cream store, and we require only that the bottom scoop and the top scoop be of different flavors. Today the store has 7 flavors available. So the server may put

Beyond Finite Sets

Some of the numbers we consider in this chapter are so large that they are beyond imagination. How large do numbers get? That is an easy question to ask but not such an easy question to answer. Suppose we think about the set of posi-

Georg Cantor

tive integers: $Z^+ = \{1, 2, 3, 4, \ldots\}$. What number corresponds to the size of Z^+? That number can't be 100!, because when we list all the positive integers in order, 100! is on the list, and it is followed by 100! + 1, then 100! + 2, and so on. The same argument tells us that any number, no matter how large, is also in the "middle" of the list.

We need a new concept, a "number" that denotes the size of the set Z^+. Mathematicians have studied the kinds of "numbers" that we need here. They are not finite values, values that get bigger when we add 1 and smaller when we divide by 2. The concept we need here is that of *infinity*. Infinity does not behave like an ordinary, finite number. A few examples will introduce this mind-boggling concept.

First, we clarify a relationship between set size and order relations between finite numbers. Suppose we have two sets, A and B, and they are of the same size, which we call n. We understand that we can make pairings—each pairing joining an element of A with an element of B. Every element of A will be in exactly one of these pairings, and every element of B will also be in

exactly one. Mathematicians call this "establishing a one-to-one correspondence" between the elements of A and those of B. Two sets can be put into one-to-one correspondence exactly when they are the same size.

Now let us return to the infinite set Z^+, and let us form a new set $Z^+ \cup \{0\}$, which we will call Z^{nonneg}. Now it seems obvious that Z^{nonneg} is bigger than Z^+, since we created the former by adding one element to the latter. What happens if we use the idea of one-to-one correspondence? For every element of Z^+, for example, z, we will correspond to it the element $z - 1$ from Z^{nonneg}. That is,

Element of Z^+	1	2	3	...	100	...	1234	...	z
Element of Z^{nonneg}	0	1	2	...	99	...	1233	...	$z-1$

It would appear that the two sets Z^+ and Z^{nonneg} are of the same size, because we have shown that there is a one-to-one correspondence between them.

And it gets worse! Suppose we consider the set $2Z^+ = \{2, 4, 6, 8, \ldots\}$, the set of all positive multiples of 2. Does it not seem reasonable to think that $2Z^+$ must be half the size of Z^+, since the odd positive integers are in Z^+ but not in $2Z^+$? But again, a one-to-one correspondence shows us a different story:

Element of Z^+	1	2	3	...	100	...	1234	...	z
Element of $2Z^+$	2	4	6	...	200	...	2468	...	$2z$

By setting up a one-to-one correspondence between the element z of Z^+ and the element $2z$ of $2Z^+$, we seem to show that those two sets are of the same size.

In dealing with infinite sets, mathematicians today define set size based on one-to-one correspondence, so Z^+, Z^{nonneg}, and $2Z^+$ are all the same size. (There are bigger infinities.) The name of that size is *aleph null,* written $\aleph_0$; it was named by Georg Cantor, who studied infinite sets in the 1870s and noted their paradoxical behaviors. When Cantor first introduced his ideas, the mathematical community rejected his ideas as nonsense, and rejected him for his faith in his mathematical beliefs.

Mathematicians have wrestled with the concept of infinity since the times of the ancient Greeks, in the fifth century B.C. Even today, there is a difference of opinion in the mathematical community about how to treat infinity. A nice overview of this history is in "A brief history of infinity," by A. W. Moore, in the April 1995 issue of *Scientific American.*

any of 7 flavors on the bottom, but may not repeat that flavor on the top, so there are only 6 choices for that scoop. This situation is shown in Figure 5, with the details of the second scoop shown in only one case.

The number of possibilities for the second scoop is 6, even though the exact flavors in that set of 6 is different for you and me if my first scoop is vanilla and your first scoop is chocolate. Because the number is 6 for both of us, we can use the multiplication principle. Thus, for each of us, there are $7 \cdot 6 = 42$ possible cones available. ▲

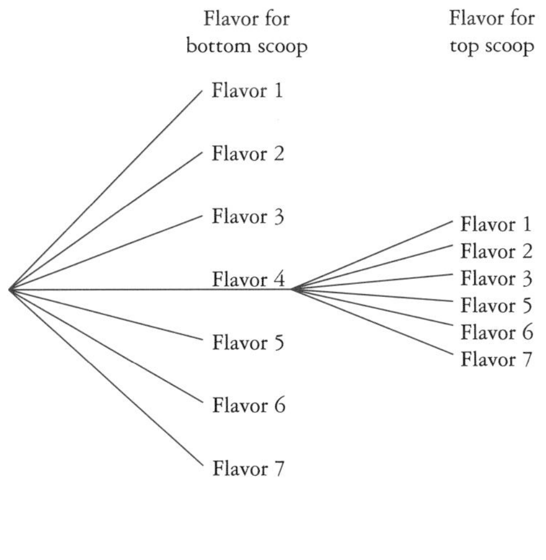

FIGURE 5 A partial tree for a two-scoop ice cream cone.

EXAMPLE 9 Binary Strings In many mathematical abstractions of real-world situations, we have an arrangement of objects that were each chosen from a set of just two elements. Examples of this include the answers to a true/false test and binary strings (arrangements of 0s and 1s). We are interested in knowing how many arrangements are possible if there are k objects in the arrangement. In our examples this would mean k questions on the test or k symbols in the string. Using the multiplication principle, we see that we have $2 \cdot 2 \cdot 2 \cdots \cdot 2 = 2^k$ possible arrangements. Specifically, this means that there are $2^3 = 8$ binary strings of length 3, namely 000, 001, 010, 011, 100, 101, 110, and 111. In computers, one use of binary strings is as addresses of memory locations. If a computer memory has one million locations, then how long must the address strings be so that every location has its own address? Based on your calculations, do you think that exactly one million is a sensible memory size? ▲

EXAMPLE 10 Counting Subsets The collection of all the subsets of a set S is called the *power set* of the set S. For example, the set $S = \{a, b, c\}$ has three elements. It is standard to consider both S and the empty set, $\varnothing$, as subsets of S. This may not seem very natural to some of us at first glance; as we progress in this example, the reason for this standard should become clear. After listing S and $\varnothing$, we list the other subsets of S: $\{a\}, \{b\}, \{c\}, \{a, b\}, \{a, c\}$, and $\{b, c\}$. Our listing puts all of the one-element subsets first, then lists all of the two-element subsets. We see that there are eight subsets of a three-element set, but the relationship between the eight and the three is not at all obvious from this presentation.

We need to look at this problem from a different point of view in order to see the relationship. If we can view the problem as counting arrangements of objects, where each arrangement corresponds to a set in the power set of S, then we can use the multiplication principle. Suppose we think of creating a subset of S by examining each element of S and deciding whether or not the element is to be in the subset. We will use "1" to mean "yes" and "0" to mean "no" (more about this choice in Chapter 7). Thus the binary string "001" means a and b are not in the subset but c is. We summarize this scheme in the following table.

Element a	Element b	Element c	Resulting Subset
0	0	0	$\varnothing$
0	0	1	$\{c\}$
0	1	0	$\{b\}$
0	1	1	$\{b, c\}$
1	0	0	$\{a\}$
1	0	1	$\{a, c\}$
1	1	0	$\{a, b\}$
1	1	1	$\{a, b, c\} = S$

The order in which the table lists the eight subsets is not the order that comes naturally to some of us, but using it gives us insight into the original problem. The original problem now corresponds to an arrangement of three symbols, each of which is a response to a question having two possible answers:

Is *a* in the subset?	Is *b* in the subset?	Is *c* in the subset?
two possibilities: "0" (no) or "1" (yes)	two possibilities: "0" (no) or "1" (yes)	two possibilities: "0" (no) or "1" (yes)

The multiplication principle tells us that there are $2 \cdot 2 \cdot 2 = 2^3 = 8$ possible subsets of a set of 3 elements. We now see a link between the 3 and the 8; the 3 is the power of the base 2, which gives us the answer 8. Now we understand the decision to include the original set itself and the empty set as subsets. Without them, the formula we just developed would not work! ▲

Where did the particular order of the arrangements of 0s and 1s in our table come from? Why is the first line all 0s, the last line all 1s, the second line 001, the third 010, and so on? If you are familiar with binary numeration, you know the answer: We are listing the 8 3-bit binary numerals in their numeric order, from 000, which is binary for 0, to 111, which is binary for 7. By using a well-organized order, we not only get the eight different arrangements, but we get them in a standard pattern without having to be concerned about leaving out any arrangement.

We can generalize the specific result from Example 10.

Theorem 1. The number of distinct subsets of a set S of k elements is 2^k.

Proof. Each subset of S corresponds to an arrangement of k symbols, each symbol being a "0" or a "1." A "0" in position p of the arrangement means that the pth element of the set is not in the subset being described; a "1" means it is. The multiplication principle tells us that there are 2^k possible arrangements, so there are 2^k distinct subsets of a set having k elements.

▲ ▲ ▲

EXAMPLE 11 Testing a Module—A Second Look Let us return to the problem of finding the number of execution paths in a module. We recall that an execution path in a flowchart is a path from start to end, obeying the arrows on the line segments. This time the module is broken into two submodules, one following another, as in the flowchart in Figure 6.

We assume that the choice of execution path made in one submodule is independent of the choice made in the other one. Thus an execution path for the entire module can be thought of as an arrangement with two positions: the first telling us the path followed in submodule 1, and the other telling us the path followed in submodule 2.

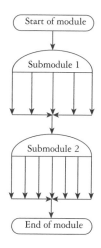

FIGURE 6 Flowchart with two submodules invoked in succession.

Number of possible paths in first submodule	Number of possible paths in second submodule
5	7

The multiplication principle applies here, giving $5 \cdot 7 = 35$ possible execution paths. ▲

We end this section by starting an analysis of the announcement mentioned in the introduction to this chapter. In 1992, Bellcore announced that soon the restriction of requiring a "0" or a "1" as the middle digit of an area code would be lifted, because more area codes were needed. To understand this more fully, we need to understand how telephone numbers are constructed and the restrictions on the digits occupying the different positions in acceptable arrangements for telephone numbers.

We need a bit of terminology for our discussion. A telephone number, such as 708-555-4321, consists of three parts. The 708 is the area code, the 555 is the exchange, and the 4321 we will call the location.

When area codes and direct dialing of long-distance calls were introduced (prior to their use, all long-distance calls were made through an operator), there was no requirement to dial a "1" first if you were dialing a number out of your area code. It was necessary to distinguish area codes from exchanges, because the numbers dialed were each processed immediately, so there was no way of counting the number of digits in the telephone number being dialed and thus knowing if it started with an area code or not. The method chosen was to require that, for the second digit, area codes used either 0 or 1 and exchanges used one of the other eight digits.

There were some additional restrictions on the possible digits for an area code. First, an initial string of one or more 0s and/or 1s was reserved to signal various situations, such as operator intervention (and eventually, long distance), so an area code may not begin with 0 or 1, leaving eight possibilities for that position. Second, codes like 800 and 411 are reserved for special purposes, so the third digit cannot be a repeat of the second; thus there are nine possibilities for the third digit. All these restrictions are incorporated into the following table.

Old area code restrictions

First digit	*Second digit*	*Third digit*
8 possibilities: 2, 3, . . . , 9	2 possibilities: 0 or 1	9 possibilities: 0, 1, . . . , 9, but not the same as used for the second digit

E X A M P L E 1 2 Old Area Codes and Old Exchanges The multiplication principle tells us that there are $8 \times 2 \times 9 = 144$ possibilities for area codes that conform to the original restrictions, as detailed in the above table. As you will calculate in Exercise 7, for any one area code, there are 6,400,000 possible "old-style" phone numbers, those not having 0 or 1 as the second digit of the exchange. Thus, using original area codes and exchanges, there could be $144 \cdot 6,400,000 = 921,600,000$ (from the multiplication principle) different telephone numbers

for the United States, Canada, and the Caribbean, the territory covered by area codes. ▲

Although 921 million is bigger than the combined populations of this territory, many people now have more than one phone number. A person could have a home number, a work number, a fax number, a car-phone number, and so forth. Even with sharing some of those numbers with other people, the limit is being pushed. Thus the phone company needs more area codes and could even make use of more exchanges.

Some time after the introduction of area codes, the telephone company introduced the convention of using a 1 or a 0 to begin an out-of-area phone call. Since the initial "1" or "0" signals that the number being dialed includes an area code, the restriction on the middle digits of area codes and exchanges is no longer needed. The restriction is no longer used for exchanges, so more of them are possible.

E X A M P L E 1 3 Old Area Codes and New-style Exchanges In Exercise 8, you will calculate that there are 8 million possible "new-style" phone numbers in any one area code, those having any digit as the second digit of the exchange. Using these new-style phone numbers, we get $144 \cdot 8,000,000 = 1,152,000,000$ possible phone numbers in any one area code. This is 25% more than the 921 million we had before, but not a big enough increase to satisfy the anticipated needs. ▲

In order for us to consider the effects of lifting the restriction that the middle digit of area codes be a "0" or a "1," we will need to use both the addition and multiplication principles in one problem. Problems requiring both principles are our focus in Section 4.2.

(Calculator availability is assumed for these exercises.)

1. As part of preparing the evening meal, Jen finds four kinds of frozen vegetables, five varieties of canned vegetables, and three different fresh vegetables available. If Jen will serve just one vegetable at dinner, how many possible choices are there?

2. A student finds six interesting history courses in the college catalog and four interesting philosophy courses. If the student decides to take either a history course or a philosophy course as an elective next semester, what is the total number of courses from which the student will choose an elective? If the student decides to take two electives, one history course and one philosophy course, what is the total number of possibilities the student has for the pair of electives?

3. Some friends have just surprised Chris with a visit, and it is time for dinner. A quick check of the pantry and freezer gives this inventory: four varieties of meat, seven varieties of vegetables, three types of bread, and two sorts of potatoes. How many different menus consisting of one each of meat, vegetable, bread, and potato could Chris create?

4. A very nonmusical student has learned that there are 12 different notes on a keyboard and that a typical piece of music contains notes of 7 different durations (half notes, quarter notes, dotted quarter notes, etc.). How many different choices would this student calculate there are available to a composer who is writing the next note of a melody?

The results of next two exercises were used in Example 1.

5. How many automobile license plates are possible if the plates are formed using two letters followed by three numbers? (This is the format used for license plates when the author of this chapter was young. Why is this format no longer popular?)

6. How many automobile license plates are possible if the plates are formed using four numbers followed by one letter? (Have you ever seen a license plate in this format? Why is this not likely to be a standard format?)

The results of the next two exercises were used in Examples 12 and 13.

7. How many old-style phone numbers can an area code support? There are seven digits in a phone number, but for the same reason as with area codes, the first digit may not be a 0 or a 1. Old-style individual phone numbers never have a 0 or a 1 as the second digit, which is the middle digit of the exchange. The other five digits can be any of 0, 1, . . . , 9.

8. How many new-style phone numbers can an area code support? There are seven digits in a phone number, but for the same reason as with area codes, the first digit may not be a 0 or a 1. New-style individual phone numbers may have any digit as the second digit, which is the middle digit of the exchange. The other five digits can be any of 0, 1, . . . , 9.

9. A survey of 5000 people found that 3020 of them had a furry pet, such as a dog or cat, 705 of them had a nonfurry pet, such as a bird or fish, and 345 people had both kind of pets. How many of the 5000 had no pets?

10. A college has 4357 students playing intramural basketball or soccer. If 3240 play basketball and 2875 play soccer, then how many students play both sports?

11. Suppose that in some town, 20,450 people have allergies and 13,700 people have arthritis. What can you say about the size of the town? Can you tell its minimum size? Can you tell its maximum size? Explain.

12. A magazine description of how a typical U.S. adult spends the day says that the adult spends 2.5 hours eat-

ing, 1.5 hours commuting, 8 hours working, 3 hours listening to the radio or watching TV, 7 hours sleeping, 2 hours socializing, 2.5 hours on household chores, 1 hour on hobbies. Explain how this description could be true, even though the sum of the allotted times adds up to more than the 24 hours in a day.

SECTION 4.2 *Using Both Principles in One Problem*

In this section we will see how to use both the addition and multiplication principles in one problem. Our first two examples are continuations of examples from Section 4.1.

Revisiting Two Examples

EXAMPLE 1 More Ice Cream Sundaes
In Example 4 in Section 4.1, we were concerned with ice cream sundaes made from three possible flavors of ice cream and two possible toppings. We noted that if nobody would be happy with a vanilla and butterscotch sundae, then the arrangement of flavors would not be independent. Here's how we can count the number of possible sundaes, allowing for this dislike of vanilla–butterscotch. We think of two disjoint sets of arrangements: arrangements with vanilla ice cream and arrangements with some other flavor of ice cream. In the first type of arrangement, there is only 1 possibility for the ice cream and 1 possibility, namely fudge, for the topping, for a total of $1 \cdot 1 = 1$ possible arrangement. In the second type of arrangement, there are 2 possibilities for the bottom scoop, chocolate and strawberry, and either of the 2 flavors is possible for the topping, giving us $2 \cdot 2 = 4$ possible arrangements. We used the multiplication principle to get the numbers 1 and 4. Now, since the two types of ar-

rangements are disjoint, we use the addition principle to find the total number of acceptable arrangements: $1 + 4 = 5$. We can also look back to Figure 2 of Section 4.1 and see that if we omit the leaf of the tree labeled V,B, there remain 5 leaves on that tree which correspond to acceptable cones. ▲

EXAMPLE 2 More About Area Codes We continue the area code discussion from the end of Section 4.1. We will determine the total number of area codes after the lifting of the restriction requiring the middle digit of the code to be a "0" or "1." We look at a pair of disjoint sets of area codes: the original ones with 0 or 1 as the middle digit, and the new ones with one of 2 through 9 as the middle digit. We have already calculated that there are $8 \cdot 2 \cdot 9 = 144$ original area codes. Once we calculate the number of possibilities for new area codes, the addition principle will give us the total number of possible area codes.

What happens if the restriction on the second digit is lifted? The restriction on the first digit (not a "0" or a "1") remains, so there are 8 possible first digits, as before. The second digit can now be any of the 8 digits formerly banned. For the old codes, there were only 2 possibilities for the second digit, and that digit could not be repeated as the third digit, leaving 9 choices. Now there are 8 additional possibilities for the second digit, and if we use any of them, there are no limitations (10 choices) on the third digit. There are thus $8 \cdot 8 \cdot 10 = 640$ new area codes (the multiplication principle). The following table summarizes the situation.

First digit	Second digit	Third digit
8 choices: 2, 3, . . . , 9	8 choices: 2, 3, . . . , 9	10 choices: 0, 1, . . . , 9

With the 640 new codes plus the 144 old ones, there are 784 codes altogether (the addition principle), more than 5 times as many as were available before. This is a much more dramatic increase in potential telephone numbers than was achieved by allowing the unrestricted use of any digit for the middle digit of an exchange. Do you see why? ▲

It is important to note that a straightforward application of the multiplication principle would not work to determine the number of possible area codes because the choice of second digit determines the *number of possibilities* for the third digit. So we divided our problem into two disjoint cases—original codes and new ones—calculated the number of possibilities for each, and then used the addition principle to give us the total number of area codes.

The Importance of Careful Analysis

An essential part of any counting problem is careful analysis before beginning our computations. Recall how important the analysis was in Example 10 and Theorem 1 of Section 4.1, counting the number of subsets of a set of k elements. If we fail to make a good analysis, we could get the wrong answer. Even if our analysis is correct, we could wind up doing unnecessarily long and complicated calculations if there is a better analysis, a simpler way of looking

at the situation. Sometimes, as in Examples 1 and 2 here, we break up the problem into two or more disjoint cases, each of which will be *enumerated* using the multiplication principle. Then the cases are *combined* using the addition principle. At times, however, our problem at its highest level has the characteristics of a multiplication principle, but in order to determine the number of possibilities for each object, we need to use the addition principle. Here is an example of using both principles in which we use the multiplication principle second, not first.

E X A M P L E 3 Scheduling Courses A student is selecting courses to fill in the 10:00 A.M. and 2:00 P.M. time slots. The 10:00 course will be either one of four music courses or one of three art courses. The 2:00 course will be either one of six psychology courses or one of five sociology courses. In how many ways could the student fill the two time slots? In this example each time slot corresponds to an object in an arrangement, and we assume that the 10:00 course is independent of the 2:00 course. Using the addition principle, we find that there are 7 possible 10:00 courses and 11 possible 2:00 courses. Then applying the multiplication principle, we get $7 \cdot 11 = 77$ ways for the student to choose the two courses. ▲

Testing Modules: Avoiding Unnecessary Work

E X A M P L E 4 Testing Modules—Avoiding a Combinatorial Explosion As a final example of using both principles, we return to counting the number of execution paths in a module. This module is more complex than ones in the previous examples, and the submodules it invokes are also more complex, as shown in Figure 1.

Design of the Space Shuttle Engine

On January 28, 1986, the space shuttle *Challenger* exploded just after liftoff. The "blue-ribbon" presidential commission formed to investigate the disaster found that the immediate cause of the accident was an "O" ring that lost its necessary flexibility in the cold conditions prevailing that morning. Commission member Dr. Richard Feynman, Nobel Prize-winning physicist from the California Institute of Technology, in Appendix F of the Commission's report, pointed out problems in the way that the shuttle engine was designed and tested. As we read Dr. Feynman's words, we see a relationship to our discussion of testing program modules:

> The usual way that such engines are designed . . . may be called the component system, or bottom up design. First it is necessary to thoroughly understand the properties and limitations of the materials to be used . . . and tests are begun in experimental rigs to determine those. With this knowledge, larger component parts . . . are designed and tested individually. . . . Finally one works up to the final design of the entire engine. . . .

This bottom-up method for testing engines is similar to testing a program using the black-box method. The testing of the materials and simple components is analogous to the testing of individual modules in a program. Then after the simple components check out, more complex components are created from them, just as complex programs are built from simpler modules. Feynman continued:

> The space shuttle main engine was handled in a different manner—top down, we might say. The engine was designed and put together all at once with relatively little detailed preliminary study of the materials and components. But now, when troubles are found . . . it is more expensive and difficult to discover the causes and make changes. . . .

The top-down testing of the shuttle engine is like testing a program like a glass box, looking at the entire construction all at once. So when a shuttle engine problem arose, it was a problem in a complex machine with many components. As we learned in this chapter, the number of tests required to ensure correctness is significantly larger in glass-box, or top-down, testing than in black-box, or bottom-up, testing because we are multiplying instead of adding to get the number of tests required. There may not be enough time to do all the tests required in the glass-box scenario. Feynman continued:

> A further disadvantage of the top-down method is that if an understanding of a fault is obtained, a simple fix . . . may be impossible to implement without a redesign of the entire engine.

In fact, there was a major delay in the space shuttle program while the shuttle engine was redesigned following the *Challenger* explosion.

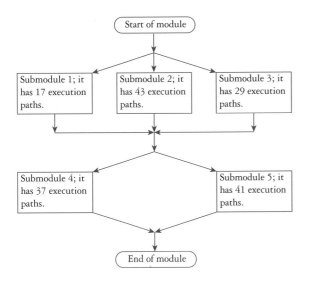

FIGURE 1 A complex module with many execution paths.

We see from the flowchart that any execution path for the entire module is an arrangement of two subpaths: exactly one execution subpath from the union of submodules 1, 2, and 3; and exactly one execution subpath from the union of submodules 4 and 5. As before, we assume that the choice of subpaths is independent. We will use the multiplication principle here, but first we need to use the addition principle to find the number of subpaths for each of the positions in the arrangement. The number of execution paths for this module is the product of the number of execution paths for the union of submodules 1, 2, and 3 and the number of execution paths for the union of submodules 4 and 5. Specifically, this means $(17 + 43 + 29) \cdot (37 + 41) = 6942$. Since the programmer would need to create a set of test data for each of these execution paths, and time would be needed to run the module for each of those data sets, this is a daunting number of different sets of test data required in order to test this module! And of course, this module is by no means as complex as modules get. So what do programmers do?

There are two points of view possible when testing: A submodule can be considered a *glass box,* allowing the programmer to look inside it and see all the execution paths. Or the submodule can be viewed as a *black box,* a section of code that produces specific outputs from specific inputs, but without the programmer being concerned, in this view, about the number of execution paths within the module. Instead of viewing the entire module and its submodules as a glass box, giving 6942 execution paths to test, the programmer can view the submodules as black boxes, so that their execution paths are not visible. We can test the module just to see if the correct submodule is invoked. At that level, the module itself has $3 \cdot 2 = 6$ execution paths. If we have previously tested each of the five submodules separately as glass boxes and are sure that each works, at a cost of $17 + 43 + 29 + 37 + 41 = 167$ data sets, then the number of data sets needed to test the entire system is now $167 + 6 = 173$, a far cry from the 6942 we calculated originally. ▲

This dramatic reduction in the number of data sets was achieved in Example 4 by changing the testing situation from one in which the multiplication principle was needed, because there were several modules being tested simultaneously, to a situation in which the addition principle was applicable since at any one time only one module was under scrutiny. Of course, if the code had been written without submodules, there would be no possibility of viewing submodules as black boxes, and we would be forced to test all 6942 execution paths. This is an important reason why programmers use submodules.

Testing every path in every module, using a combination of glass-box and black-box views, is one way to verify that a computer program is correct.

There are other approaches to verifying program correctness. One involves creating proofs, based on the structure of the program and the individual statements. Advanced computer science courses study program verification methods.

Program verification is a branch of software engineering. As an engineering discipline, it is thus related to other forms of engineering, in particular, those dealing with physical materials. Testing methods used in engineering include rigorous testing of individual components before assembling multicomponent systems. The motivation behind this methodology is similar to the use of black-box testing of programs, and when it is not followed, major problems can occur (see the Spotlight "Design of the Space Shuttle Engine").

The results obtained when testing physical materials are *probabilities,* not exact predictions. For example, a series of tests would determine that a certain type of steel beam would withstand a certain weight 95% of the time. This sort of information is very different from saying that a module gives the correct answer when a particular execution path is followed. In the case of the module, we expect 100% correct answers on that path. What exactly could we expect of the steel beam? In Chapter 8 of this text we study probability theory and discuss the interpretation of a number like 95%.

Exercises for Section 4.2

(Calculator availability is assumed for these exercises.)

1. A customer in an ice cream store requires that one scoop be some variety of chocolate, of which the store has 4, and the other scoop be some variety of fruit, of which the store has 6. The customer will accept either a cone with fruit as the top scoop and chocolate as the bottom scoop, or a cone with fruit as the bottom scoop and chocolate as the top scoop. How many possible cones would the customer find acceptable?

2. Another hungry customer in the same store as in

Exercise 1 wants a three-scoop cone with anything on the bottom, and either fruit in the middle and chocolate on the top or chocolate in the middle and fruit on top. How many possible cones are there for this customer?

3. How many license plates are possible if each plate must have three letters and three numbers and all the letters must be in one group, either at the left or the right end of the plate?

4. Suppose automobile license plates consist either of a single letter followed by three digits or a pair of letters followed by two digits. How many different license plates are possible?

5. How many different tests could a professor construct from a test bank if the first question is to be chosen either from a set of four essay questions or from a set of nine matching problems; the second question is to be chosen either from a set of seven fill-in-the-blank questions or from a set of three compare-and-contrast questions; and each of the next five questions is to be one of 37 short-answer questions, with the condition that none of the 37 be used more than once?

6. A college offers a choice of majors in 25 different subjects. A student can have a single major or combine any two subjects into a double major. If a student is considering either a single or a double major, how many different possibilities does he have?

7. A graduating senior has many options in the geographical area in which she wants to live. She has four full-time job offers and three acceptances at graduate schools for full-time study, including living stipend, leading to a master's degree in two years. The student can postpone starting any full-time job for as long as two years, and once having started the job, she can take one two-year leave of absence for graduate work but must return to the same job. The student can also postpone entering graduate school for up to two years. In addition, the student has five acceptances at graduate school for part-time study and seven possible part-time job offers; any combination of part-time study and part-time work will provide the student with enough money for living expenses and a master's degree in four years.

How many options does the student have for arranging the next four years of her life and having a master's degree at the end of that time? (Hint: Try drawing a tree showing the different choices for the student; for example, during the first year the student could either work full time, be a full-time student, or combine part-time work and part-time study. Make some reasonable assumptions, such as limiting the number of job switches and eliminating the possibility of switching schools or switching between part-time and full-time study.)

8. Suppose the module in Example 4 has the same structure as in Figure 2, but the submodules 1, 2, 3, 4, and 5 have 24, 39, 44, 27, and 48 execution paths, respectively. What is the difference in the number of data sets needed to test the module if we use the black-box or the glass-box methods?

9. A state currently has only one form of license plates, three numbers followed by three letters. The state is considering two schemes for generating more plates. Scheme 1 is to use the current form and its reverse, three letters followed by three numbers. Scheme 2 is to have three types of plates: Type 1 has two letters followed by four numbers; type 2 has two numbers followed by two letters followed by two more numbers; and type 3 has four numbers followed by two letters. Which scheme gives more possible plates, and how many more plates does it allow?

10. The written Hawaiian language has five vowels (a, e, i, o, u) and seven consonants (h, k, l, m, n, p, w). The basic rules for forming an acceptable word in Hawaiian are that consonants may not occur in consecutive positions in the word and that any number of vowels may occur in consecutive positions. For example, the name "Hawaiian" has four consecutive vowels, but "opliau" would be unacceptable because the "p" and "l" are consecutive. How many three-letter Hawaiian words can be formed using these rules? What about the number of four-letter words?

11. **Writing project.** In this section we have seen that the way a computer program is written and tested can have a major effect on how many different sets of data

would be needed to completely test for errors. The task of creating errorfree computer programs is becoming more and more important in our world, as computers are used not only for maintaining our bank balances, but also to assist pilots, operate shutdown systems in nuclear power plants, and control our telephone systems.

Read the two articles by Ivars Peterson listed below, and write a report explaining the issues to an intelligent friend who has neither taken this course nor read the articles.

"Finding fault. The formidable task of eradicating software bugs." *Science News* **139**(2/16/91):104–106.

"Warning: This software may be unsafe." *Science News* **130**(9/13/86):171–173.

SECTION 4.3 *Factorials, Permutations, and Combinations*

Arrangements "Without Replacement"

In this section we look at a special situation in which the multiplication principle applies. That situation is an arrangement of objects all coming from the same initial set of objects, with the additional condition that once an object is chosen, it cannot be chosen again. A professor would not repeat a question twice on a test; in any one round of a card game, a particular card can only be dealt once; in planting shrubs in the garden, once a bush has been planted, it is not available to be planted elsewhere. Mathematicians use the term "without replacement" to describe such arrangements; once an object is "used" in the arrangement, it is not "replaced" in the set of possibilities for use in another position. In Example 8 of Section 4.1 we had an arrangement "without replacement": We wanted an ice cream cone with any of the store's 7 flavors for the bottom scoop of ice cream, but we wanted the second scoop to be different from the first, thus limiting the number of possibilities for the second scoop to 6. By contrast, if we had allowed the top scoop of the two-scoop cone to be a repeat of the bottom scoop, the arrangements would have been "with replacement": We would have "replaced" the flavor into the set of possibilities so that it could be "used" again.

Factorials and Permutations

E X A M P L E   1   Using a Question Bank to Make a Test   Let us consider a professor who wishes to select 4 questions from a test bank of 45 questions, without repeating any question. The professor considers the order of presenting the four questions as very important in constructing the test, so differing orders of those questions are counted as different arrangements. Putting the hardest question first makes for a different sort of test than putting the questions in increasing order of difficulty. How many different tests can the professor construct? The following table presents the possibilities.

First question	*Second question*	*Third question*	*Fourth question*
45 possibilities	44 possibilities	43 possibilities	42 possibilities

It is important to note that regardless of which particular question is used as the first question, there are 44 choices left for the second question. We have the type of independence we require in order to use the multiplication principle, namely that the *number of possibilities* at any one place in the arrangement is not related to the specific choices already made. We would not have the independence we need if the data bank had 10 essentially similar questions so that using one of them would cause the professor to eliminate the other 9 from the pool of possibilities. Applying the multiplication principle gives the professor $45 \cdot 44 \cdot 43 \cdot 42 = 3,575,880$ possible tests. ▲

EXAMPLE 2 Editorial Decisions Sometimes problems of arrangements "without replacement" require that all the objects be eventually chosen. For example, an editor has five short stories to include in a book. Suppose that the editor has no literary reason for preferring that any one of the stories precede any of the others. Thus we have

First story	Second story	Third story	Fourth story	Fifth story
5 choices	4 choices	3 choices	2 choices	1 choice

The editor has $5 \cdot 4 \cdot 3 \cdot 2 \cdot 1 = 120$ possible arrangements for the order in which the short stories appear in the book. ▲

The name for a product of consecutive positive integers starting with 1, such as the product above, is *factorial*.

DEFINITION

n factorial, written as $n!$, is defined for a positive integer n as the product $1 \cdot 2 \cdot 3 \cdot \cdots \cdot n$. We define $0!$ as 1; the reason for this seemingly strange definition will become clear in just a few paragraphs when we introduce permutations.

We can also define $n!$ as a function f, using a difference equation and initial condition, such as you learned in Chapter 1:

The initial condition, for $n = 0$, is $f(0) = 0! = 1$.

The difference equation, for $n > 0$, is $f(n) = n! = n \cdot (n - 1)! = n \cdot f(n - 1)$.

In Exercise 16 at the end of this section, you are asked to verify that the formula given for $n!$ in the definition does satisfy the difference equation.

EXAMPLE 3 A Shopping Trip Suppose that you are doing some holiday gift shopping and need to visit 10 stores. Let us initially make the obviously unrealistic assumption that the order in which you visit these stores is totally unimportant to you; any order is as good as any other. Then you have a situation in which you have 10 choices for the first store to visit, 9 choices for the second store, and so on, down to 2 choices for the ninth store and just 1 choice for the tenth store. Thus, the number of choices you have is the product $10 \cdot 9 \cdot \cdots \cdot 2 \cdot 1$ or, using the commutative law of multiplication, $1 \cdot 2 \cdot \cdots \cdot 9 \cdot 10$, which is $10!$, or 3,628,800. Perhaps your reaction to this number is to wonder how anyone ever decides which order is best; "best" often means the least amount of walking between stores, but it could be based on some other criterion. In Chapter 5 this classic problem, known as the Traveling Salesperson Problem, is discussed. ▲

DEFINITION

A *permutation* is an arrangement of objects, without replacement, from a fixed set of choices.

Permutations include the cases of using all the objects at hand, as with the editor or the holiday shopping, and the cases of using just some of the objects, as with the professor's test bank. If there are n objects and all the objects are used, then the number of permutations is $n!$. When there are only k positions in the arrangement, but still n objects to choose from, then the number of permutations is represented by the symbol $_nP_k$ and is the product of k integers:

$$_nP_k = n \cdot (n - 1) \cdot (n - 2) \cdot \cdots \cdot (n - (k - 1)). \quad (1)$$

This product can be expressed in terms of factorials. The number of permutations of n objects into k positions is given by the formula in Eq. (2).

$$_nP_k = \frac{n!}{(n - k)!}. \quad (2)$$

We see that expression on the right of Eq. (1) is really $n!$ with the smallest $n - k$ factors of that product "missing." Thus, the denominator of expression on the right of Eq. (2) is $(n - k)!$, which cancels out the undesired factors. We also now see why there was the somewhat strange definition of $0!$ as 1, since when $k = n$, we have $0!$ in the denominator of the formula and want the result to be $n!$.

Since $_nP_k$ is defined using factorials, we suspect that there is a way to define it using a difference equation and initial condition, just as we did for factorials. But since $_nP_k$ is a function of two variables, n and k, we need to treat one variable as a fixed quantity in order to use this technique. For example, suppose we decide to fix k in the range $0 \leq k \leq n$ and treat $_nP_k$ as a function, f, of n. Then we would have $n = k$ as the initial value of n, giving us the initial condition

$$f(k) = _kP_k = \frac{k!}{(k - k)!} = \frac{k!}{0!} = k!,$$

and for $n > k$ we would have the difference equation

$$f(n) = _nP_k = \frac{n!}{(n - k)!} = \frac{n \cdot (n - 1)!}{(n - k) \cdot (n - k - 1)!}$$
$$= \frac{n \cdot (n - 1)!}{(n - k) \cdot (n - 1 - k)!} = \frac{n}{n - k} \cdot {_{n-1}P_k}$$
$$= \frac{n}{n - k} \cdot f(n - 1).$$

In Exercises 17–19 of this section, you will be asked to verify that our factorial definition does satisfy this difference equation and initial condition and to develop a difference equation and initial condition for a different function, this time assuming that n is fixed and k is the variable.

EXAMPLE 4 Many Photos Suppose that we want a photograph for a yearbook having four people chosen from these seven people: Pedro, Bobbi, John, Claudia, Priscilla, Bruce, and David. Note that not only does it matter which four people are in the photo, but the caption in the yearbook will depend on the order in which the people are arranged. How many different captions are possible? We are making a permutation of 4 of the 7 people, so we have $_7P_4$, which is

$$\frac{7!}{(7 - 4)!} = \frac{7!}{3!} = \frac{7 \cdot 6 \cdot 5 \cdot 4 \cdot 3 \cdot 2 \cdot 1}{3 \cdot 2 \cdot 1}$$
$$= 7 \cdot 6 \cdot 5 \cdot 4 = 840. \; \blacktriangle$$

Combinations

As we think about the 840 different photographs in Example 4, we realize that in one of these photographs, we will see, from left to right, Pedro, Priscilla, Claudia, and Bobbi. In a different photo, these same four appear arranged in a different order: Bobbi, Priscilla, Pedro, and Claudia. In how many distinct photos does the same quartet appear? The answer is easy: the number of ways that 4 people can be arranged in a line; that is, 4! or 24. If we were to consider another group of four (say Bobbi, Claudia, David, and Bruce), there would also be 24 distinct photos containing them.

We see then that the 840 photographs can be broken up into piles of 24, with a separate pile for each quartet. How many piles (quartets) would there be? Simply divide the total number of photos by the number of photos in each pile:

$$\frac{840}{24} = 35.$$

It's useful to rewrite this result in a way that recalls its history:

Number of different groups of 4 that can be chosen from a set of 7 $= \dfrac{7 \cdot 6 \cdot 5 \cdot 4}{4 \cdot 3 \cdot 2 \cdot 1} = \dfrac{7!}{4!3!}.$

Many mathematical problems turn out to be of the form "find the number of different groups of k individuals that can be made/chosen from a set of n." What is the general formula for this number? We see that it is $_nP_k/k!$, the number of permutations of k individuals selected from a set of n, divided by $k!$, the number of different rearrangements, or permutations, of any one group of k individuals. Mathematicians use the name *combination* in this situation.

D E F I N I T I O N

A *combination* of k elements from a set of n elements is a subset of k of the n elements.

We recall that the order in which we list the elements of a set does not matter in defining the set. Similarly, the order in which we list the elements of the combination, really a subset, does not matter in defining the combination. Thus, the combination {Bruce, Priscilla} is the same combination as {Priscilla, Bruce}.

The formula we developed for the number of combinations of k elements from a set of n, $_nP_k/k!$, is symbolized in one of two ways:

$$\frac{_nP_k}{k!} = {}_nC_k = \binom{n}{k} = \frac{n!}{(n-k)!k!} \qquad (3)$$

The new symbols, the middle two expressions of Eq. (3), are both read "n choose k," meaning the number of different ways we can choose k objects from a set of n objects. The expression at the right end of (3) is the usual definition of the number of combinations, "n choose k."

As we did in the discussion of permutations, we could look for a difference equation and initial condition to represent the number of combinations, either with k held as fixed or with n as fixed. These will be left as exercises (Exercises 20 and 21), as will be the development of some of the many interesting relationships that are true for $\binom{n}{k}$ (Chapter Exercises 30–35). In particular, Chapter Exercise 35 will explain why the formula for combinations is also that for *binomial coefficients*.

E X A M P L E 5 Selecting is Like Discarding
Suppose we want to count the number of ways to choose 5 letters from the set of 8 letters: {A, B, C, D, E, F, G, H}. Our formula says the answer is "8 choose 5." Another way to look at the problem is that the task of selecting 5 letters to keep is equivalent to the job of picking 3 letters to dis-

card: Every choice of 3 to discard defines a collection of 5 to keep. Thus, the number of ways to select 5 objects from a collection of 8 should be equal to the number of ways to select 3 objects from 8. Since this latter number is "8 choose 3," we should have

$$8 \text{ choose } 5 = 8 \text{ choose } 3, \quad \text{or}$$

$$\binom{8}{5} = \binom{8}{3}, \quad \text{which is}$$

$$\frac{8!}{5!3!} = \frac{8!}{3!5!},$$

which we see is true. ▲

Let us consider the generalization of Example 5. In choosing k objects from a set of n, we are simultaneously picking $n - k$ objects not to be in our collection. For example, in choosing 6 executive committee members from an organization of 10 members, we can either designate the specific 6 committee members or designate 4 organization members who will not be on the committee. Including k members on a committee is equivalent to excluding $n - k$ members. Thus, the number of distinct ways to choose k objects from a set of n should be equal to the number of ways of choosing $n - k$ objects from a set of n. This gives us the following theorem.

Theorem 1. For $n \geq 0$ and $0 \leq k \leq n$, "n choose k" and "n choose $(n - k)$" are equal; that is,

$$\binom{n}{n - k} = \binom{n}{k}.$$

Proof. By formula (3), the number of ways of choosing $n - k$ objects from a set of n is

$$\binom{n}{n - k} = \frac{n!}{(n - k)!(n - (n - k))!}$$

$$= \frac{n!}{(n - k)!k!} = \frac{n!}{k!(n - k)!} = \binom{n}{k}.$$

▲ ▲ ▲

EXAMPLE 6 Checking the Extremes It's always useful to check a formula in extreme cases. For example, does our formula give the correct answer for the number of ways of choosing 1 object from a set of 10? Of course, we don't need any formula to answer this question. Since any of the 10 can be selected as the unique object, there are 10 distinct ways. The formula gives

$$\binom{10}{1} = \frac{10!}{1!(10 - 1)!} = \frac{10 \cdot 9!}{1 \cdot 9!} = 10.$$

What about an even more extreme case? How many ways can we select 10 objects from a collection of 10? Obviously, only 1 way since all 10 must be included. The formula gives

$$\binom{10}{10} = \frac{10!}{10!(10 - 10)!} = \frac{10!}{10!0!} = \frac{10!}{10!} = 1.$$

The only way this formula gives the correct answer is if we define, as we did earlier, 0! to be equal to 1. ▲

In Exercises 31 and 32 of the Chapter Exercises, you are asked to prove the general cases from Example 6.

Some Practical Considerations

It is not always clear whether a combination or a permutation is being modeled, because it is not al-

Combinatorics Tells Which Algorithm Is Better

Henry Pollak, a retired mathematician, worked for many years at Bell Laboratories, the research arm of AT&T. He spoke with Rochelle Meyer about how combinatoric considerations helped determine which method to use to find minimal cost spanning trees. You will study trees and the two algorithms he mentions ("algorithm" means "method to solve a problem") in Chapter 5 and learn about analyzing algorithms in Chapter 6.

Rochelle: Why was AT&T interested in minimal cost spanning trees?

Henry: In 1956, AT&T used minimal cost spanning trees to find the price to charge business customers for linking up several

Henry Pollak

locations with a private network, called Private Wire Service. To find the trees, clerks physically modeled various possibilities using a large scale map of the United States embedded in the floor of one of their office buildings in New York City. Each location was marked on the map, and the clerks looked for a network which linked up all the locations without ever creating two ways to link up any one pair of locations. Finding the network with the smallest sum of lengths of individual links was the goal. Reasonably intelligent people can in fact get very good results in this

way. But there are limits to what this hands-on (and on-knees) approach can accomplish.

Rochelle: What was the first algorithm the mathematicians at Bell Labs considered?

Henry: Kruskal's algorithm had just been published. As its first step, a sorted list of all distances between pairs of locations had to be created. We know that when there are n locations, there are $\binom{n}{2} = n(n-1)/2$ pairs. For small problems Kruskal's algorithm is fine. But by 1956, AT&T needed to find a minimal cost spanning tree of 500 locations for one customer, Dun and Bradstreet. No one wanted to sort all $500 \cdot 499/2 = 124{,}750$ edges for such a network by hand, and the computers of the day did not have the capacity to process a data set of that size. Another algorithm was needed.

Rochelle: How long did it take to develop an alternative?

Henry: By 1958, Prim, who headed the Mathematics Department at Bell at the time, had his algorithm. One advantage of his algorithm over Kruskal's is that Prim's never deals with more than n data items at a time, so Prim's needs less memory than Kruskal's. Prim's does need to have access to all $\binom{n}{2}$ lengths. But in 1958, this large data set was kept in "auxiliary storage," not in the more limited space of the computer's memory.

ways clear whether we are merely choosing a subset or whether we are both choosing a subset and making an arrangement of it. The former situation is a combination, the latter is a permutation. Here is an example to help you tell the two apart.

EXAMPLE 7 Distinguishing Between Combinations and Permutations In how many ways could we select 6 different people from an original group of 10 to form an executive committee? This problem certainly involves selection without replacement, so the number of ways is either a permutation or a combination. How can we tell which? We need to know whether all 6 people will be essentially interchangeable within the executive committee or whether each one will serve in a different capacity.

If the 6 people will be essentially interchangeable, then we are concerned only with whether or not someone is chosen to be on the executive committee, that is, whether or not the person is in the subset. So we are counting combinations:

Number of distinct ways to choose 6 objects from a
$$\text{set of 10 objects} = \binom{10}{6} = \frac{10!}{6!4!} = \frac{10 \cdot 9 \cdot 8 \cdot 7}{1 \cdot 2 \cdot 3 \cdot 4}$$
$$= 210.$$

On the other hand, if each person will serve on the executive committee in a different capacity—e.g., President, Vice-President, Recording Secretary, Corresponding Secretary, Treasurer, and Auditor—then we are concerned with not only which six people are on the committee, but where in the committee they sit, so to speak. One person sits in the President's chair, one in the Vice-President's, and so on. Each such choice of six

people is a permutation. The number of permutations is

$$_{10}P_6 = \frac{10!}{(10-6)!} = \frac{10!}{4!} = 10 \cdot 9 \cdot 8 \cdot 7 \cdot 6 \cdot 5$$
$$= 151{,}200.$$

In the second situation we are not simply choosing six people to form a group, but rather selecting six to fill particular, distinguishable positions. ▲

Mathematicians often speak of selection *ignoring order* or *considering order*. In selecting an executive committee of six persons, all serving in the same capacity, we are "ignoring order"; it doesn't matter which person is chosen first, which one second, and so on. Ignoring order means selecting a subset, and the number of ways to do that is a combination. But in filling specific offices, we are "considering order." If the first person selected will be president, the second one vice-president, and so on, then the role you play in the organization very much depends on when you were selected. Considering order, as opposed to ignoring it, gives more possibilities, because for any pair of integers, n and k, there are more permutations than combinations. Typically, thinking about the number of ways of *choosing* a group of objects of a certain size from a larger collection means that we are *ignoring order*.

Example 7 reminds us that for fixed values of n and k, $_nP_k > {_nC_k}$. The number of combinations can, however, become very large. The Spotlight "Combinatorics Tells Which Algorithm is Better," relates the true story of how large values for $_nC_k$ prompted search for a better way to solve a problem.

The numbers represented by $_nP_k$ and $\binom{n}{k}$ can be calculated by hand for small values of n and k. For larger values it is quicker and more accurate to use some technology. Many calculators and mathematics computer utilities have built-in functions to generate permutations and combinations and also to calculate $_nP_k$, the number of permutations and $\binom{n}{k} = {}_nC_k$, the number of combinations. The estimates of large factorials in this chapter were obtained using a mathematics utility.

This chapter is an introduction to combinatorics. Interested readers may want to continue learning about this topic in other books. A classic in the field is by Ivan Niven (1965), *Mathematics of Choice: How to Count Without Counting,* New York: Random House.

Exercises for Section 4.3

(Calculator availability is assumed for these exercises.)

In Exercises 1 and 2, we consider two kinds of families: Fussy families consider the left–right order of a display of objects on a shelf to be important aesthetically, whereas nonfussy families are concerned only with which objects are displayed and which are stored in the closet.

1. A well-traveled family has collected 37 souvenirs. Unfortunately, it has only one shelf for souvenirs; that shelf will hold any 10 of the souvenirs. How many different displays of souvenirs could the family consider for its shelf if it were a fussy family? How many if it were nonfussy?

2. The family in Exercise 1 divides its souvenirs into two sets—fragile and nonbreakable. There are 21 fragile souvenirs. If it decides to put 10 fragile souvenirs on the shelf and put a row of 5 nonbreakable souvenirs on the mantel above the fireplace, how many different schemes does it now have for displaying souvenirs? Again, make calculations for both fussy and nonfussy families.

3. A student has 20 books. In how many different ways can all these books be arranged on a single shelf? Why does the answer to this exercise help justify library cataloging systems?

4. A homeowner needs 5 new shrubs and wants to purchase five different types of shrub. The local nursery has 27 types of shrubs that the homeowner finds attractive. How many different collections of 5 shrubs could the homeowner make?

5. The homeowner from Exercise 4 can plant the 5 shrubs in any of 5 spots. A nongardener cares only that 5 new shrubs are on the property. A gardener, however, carefully considers which shrub is planted in which spot. Clearly, a nongardener sees only one possibility—plant the shrubs. How many planting possibilities does a gardener see?

6. For both kinds of homeowners in Exercise 5, the total number of possibilities for the successive tasks of selecting shrubs and then planting them is the product of the calculation in Exercise 4 times the appropriate calculation from Exercise 5. Suppose the gardener-homeowner had viewed the entire project as one task—go to the nursery and select a (different) plant for each of the 5 bare spots in the yard. Calculate the number of ways this task could be accomplished, and compare your answer with the result of multiplying the answers to the two previous exercises.

7. The homeowner of Exercise 4 brings home five different plants and then remembers that two of the locations for new plants have been troublesome—all previous

plantings there have died. In how many ways could the homeowner divide the set of 5 plants into two sets, those going into troublesome locations and those going into other locations? Does it matter in this question whether we are talking about the gardener or the nongardener? Explain.

8. A man has 6 different hats, which he displays in a row in the rear window of his car. How many different displays of all his hats can he create?

9. The family of the man in Exercise 8 gives him 2 new hats to add to his collection. But he can only fit six hats in the rear window. His son claims that the father can now create 28 different hat displays in the window, but the daughter claims the true number of different displays is 20,160. What assumptions are each of these children making in order to arrive at their answers?

10. In a round-robin sporting tournament, each of 16 teams will play each other once. How many games will be played?

Exercises 11–14 are not solved by just finding a permutation or a combination, although a permutation or a combination may be part of the answer.

11. In order to run down the clock, the five members of the basketball team are passing the ball back and forth, delaying taking a shot until the last possible moment. If 7 passes are made before the shot is taken, in how many different orders can the ball visit the players?

12. In a single elimination tournament, like the U.S. Tennis Open, a person (team) who loses one match does not play in any subsequent rounds. How many games are played in such a tournament if there are 16 players in the first round? Can you solve this problem in more than one way? Often this type of tournament is depicted in a newspaper by a diagram showing a column of the 16 original players, paired up and feeding into a column of the 8 winners of round 1, who are then paired up in round 2, and so on, until there is a single winner of the final round. This diagram is a *tree,* a mathematical object that we study in Chapter 5, Section 4. What would happen if the

original number of players/teams in single elimination is not a power of 2 (note that $16 = 2^4$)?

13. A small New England college divides its curriculum into four areas. Certain courses in each division have been designated as "foundation courses." The divisions and the respective number of foundation courses are as follows:

Humanities	20
Foreign language	16
Social sciences	8
Natural sciences	17

Each student must take a foundation course in at least three of the four divisions. In how many different ways can a student satisfy this requirement? Note that as soon as the student has three courses, one in each of three divisions, the requirement is fulfilled.

14. You decide to invest a large inheritance in a portfolio of eight stocks and four bonds. Your broker recommends 12 different stocks and 7 bonds as good investments. In how many ways can you create the portfolio?

15. The test to become a licensed taxicab driver in London, England, involves learning the locations of 25,000 different places and then being able to describe an efficient route for getting from any one of these places to any other one. Explain how the number of different routes a prospective taxicab driver must know might be considered as counting combinations. Why might the number of routes be considered as counting permutations?

16. Verify by substitution that $f(n) = n!$ is a solution to the difference equation and initial condition $f(n) = n \cdot f(n)$, for $n > 0$, and $f(0) = 1$, for $n = 0$.

17. Verify by substitution that $f(n) = {}_nP_k = n!/(n - k)!$ is a solution to the difference equation and initial condition developed in this section: $f(n) = n/(n - k) \cdot f(n - 1)$, for $n > k$, and $f(k) = k!$, for $n = k$, k fixed, $0 \leq k \leq n$.

18. Develop a difference equation and initial condition

for $f(k) = {}_nP_k = n!/(n - k)!$ when n is fixed and k is the variable, $0 \leq k \leq n$. The initial condition will be $k = 0$, and the recurrence equation will relate $f(k) = {}_nP_k$ to $f(k - 1) = {}_nP_{k-1}$.

19. Verify by substitution that $f(n) = {}_nP_k = n!/(n - k)!$ is a solution to the difference equation and initial condition you developed in Exercise 18.

20. Verify by substitution that $f(n) = {}_nC_k = \binom{n}{k} = n!/(n - k)!k!$ is a solution to the difference equation $f(n) = n/(n - k) \cdot f(n - 1)$, for $n > k$, and initial condition

$f(k) = 1$ for $n = k$, k fixed, $0 \leq k \leq n$. Note that this is for the case of k fixed and n being the variable. Compare this difference equation and initial condition with those in Exercise 17.

21. Verify by substitution that $f(k) = {}_nC_k = \binom{n}{k} = n!/(n - k)!k!$ is a solution to the difference equation $f(k) = (n - k + 1)/k \cdot f(k - 1)$, for n fixed, $n > k$, and initial condition $f(0) = 1$, for $k = 0$. Note that this is for the case of n fixed and k being the variable, $0 \leq k \leq n$.

Chapter Summary

Combinatorics is the branch of mathematics in which we analyze and count. In simple situations, we can use either the Addition Principle or the Multiplication Principle; in most of the interesting problems, we need to use a combination of these Principles. An important subset of problems for which the Multiplication Principle applies, often called selection without replacement, leads us to the special situations called permutations and combinations. Today, with modern technology available to us, the actual calculations are not difficult, but the analysis telling us what calculations to perform remains, as it always has been, an essential part of combinatorics.

Chapter 4 Exercises

(Calculator availability is assumed for these exercises.)

1. An $n \times n$ matrix has n different entries in the n columns of the top row. Can each of the remaining rows be formed of the same n entries, but in a different order, without any two rows having the same order? Suppose the matrix has m rows. How large could m be before we are forced to repeat a row?

2. In Chapter 2 we learned that the final position of a robot is independent of the order in which the robot fol-

lows a set of vector directions. If the set of vector directions had eight vectors, in how many ways could the robot arrive at the destination? Why might some of these orders be preferable to others?

3. A groom-to-be must choose a best man. He has no brothers, so he decides that his best man will be either a cousin (he has 7), a close high school friend (he has 3), or a college buddy (he has 5). How many possible choices does he have for a best man?

4.   The Department of Labor wishes to determine the number of people who either became unemployed in the past month or were looking for work in the last month. If they wish to arrive at their final answer by using the addition principle, how would you suggest they define the sets of people they are counting?

5.   A list of people to be invited to a wedding is formed by combining the list of 54 people provided by the groom's parents, the list of 61 people made by the bride's parents, and the list of 44 friends submitted by the happy couple. How many people will be invited to the wedding (state any assumptions you need to make in order to get your answer)?

6.   An absentminded professor is greeted at the local mall by a female student. The professor recognizes that the student is in one of her four classes. The classes have 17, 19, 21, and 37 female students, and no student is in more than one of the professor's classes. As the professor madly tries to mentally scan the names of her female students and attach a name to the one pleasantly chatting in the mall, how many names must the desperate professor consider?

Exercises 7–9 involve DNA, the double-helix molecule that forms the chromosomes in all living things. Like RNA (see Examples 5 and 6 of Section 4.1), DNA is made from chemicals called bases. DNA looks like two long parallel spirals, with bridges from the bases on one spiral to those on the other. Only four types of bases occur in DNA, represented by A, C, G, and T. The bases in DNA can occur in any order, just like the bases of RNA. The bridges between the spirals link either A to T or C to G; no other linkages occur. Thus, if we know the order of the bases on one of the two spirals, we also know the order of the bases on the other. Any two bases linked by a bridge are called a base-pair.

7.   A typical bacterial gene is a section of DNA that has 1500 base-pairs. Calculate the number of possible bacterial genes. (Note: The answer you get represents more genes than could have existed in all the chromosomes that have existed since the beginning of life on our planet. What does this suggest about the future potential for

variation among living things? If viral DNA has a similar length, what can we say about the possibility of new viral diseases occurring in the future—has the human race seen all possible diseases or is there the potential for others?)

8.   A typical mammal gene is a section of DNA that has 2000 base-pairs. Calculate the number of possible mammalian genes. Comparing the lengths of the bacterial genes from Exercise 7 and mammalian genes, we see that the length of a mammalian gene is 4/3 times the length of a bacterial gene. Are there 4/3 as many potential mammalian genes as bacterial genes?

9.   A restriction enzyme recognizes a special sequence of base-pairs and cuts a DNA molecule where the sequence occurs. Some restriction enzymes look for a sequence of four base-pairs; other restriction enzymes seek out sequences that are six base-pairs in length. Assuming that each base-pair sequence corresponds to exactly one restriction enzyme, how many possible restriction enzymes can there be? Assume that there are two distinct sets of restriction enzymes, defined by the length of the sequence they seek. Why is this assumption needed?

10.   In Example 6 of Section 4.1 we learned about codons. How many "codons" could there be if they could not contain two A's side by side?

11.   How many k-digit numerals are possible in a numeration system that is base 5 and has five possible symbols, namely 0, 1, 2, 3, and 4? Consider a numeral such as 421 as a k-digit numeral in which the first $k - 3$ digits are 0s.

12.   How many bits would be necessary in a binary number if we wish to represent 0 through 63? 0 through 100? 0 through 1000? 0 through 50,000? Here is a hint: The four binary numerals possible with just two bits are 00, 01, 10, and 11, which represent the numbers more familiarly known in base 10 as 0, 1, 2, and 3. Thus, with four possibilities we represent the numbers 0 to 3. If we have five-bit binary numerals, we have $2^5 = 32$ possibilities, which would represent 0 through 31.

13.   The League of Concerned Citizens in a certain community wishes to sponsor a debate on environmental pro-

tection issues between two officeholders, one a Democrat and the other a Republican. There are 14 Democrats holding local offices, 5 Democrats representing the community at the state level, and 1 Democrat representing the community at the national level. The corresponding figures for Republicans are 17, 4, and 3. Assuming that no one person holds more than one office, how many different debates could the League sponsor?

14. A state used to have license plates of exactly one form—three letters followed by three numbers. Lately, new forms of license plates have been appearing: Exactly one of the positions formerly held by a letter is now held by a number. How many extra license plates can the state generate using these new forms?

15. A crossword-puzzle fan wants to list all the anagrams (rearrangements) of the word "merchants." Including the original arrangement, how many anagrams are there? (This problem becomes much more difficult if the original word has repeats of some letters. Could we tell if the fourth and fifth letters of "crossword" have been switched? Should we count such an arrangement once or twice?)

16. Ten people are invited to a party; all will attend, and each will arrive alone. The host plans to give the first to arrive a prize of a CD; the second will receive a poster; the third will receive a pair of concert tickets. In how many ways can the prizes be distributed among the party invitees?

17. Suppose the host of the party in Exercise 16 had very little time to shop for prizes, so she bought three pairs of tickets to an upcoming concert and will give out one pair to each of the first three arriving guests. How many different winning subsets of three guests are possible?

18. A coach has 17 first-grade students on a soccer team, none of whom has ever played the game before. Only 11 can play at any one time. For the purposes of this exercise, we will assume that there are 11 recognizable positions on the field and that the students will play the positions the coach assigns them. How many different

assignments of 11 players could the coach form from the 17 members of the team?

19. Continuing Exercise 18, suppose that first-grade soccer players do not play their positions at all, except for the goalie. So the number of different teams the coach can put into the game really boils down to the number of ways the coach can choose a goalie and the number of ways the coach can choose the set of 10 other players. In how many ways can the coach do this? Does it make a difference to the number of possibilities if we consider the choice for goalie as being made before or after the choice of the other 10 players?

20. The coach in Exercise 18 has no initial reasons to assign certain players to certain positions. However, professional coaches do have prior knowledge of the strengths of their team members. How does that prior knowledge help decrease the number of possibilities? Demonstrate your answer with a hypothetical team of 22 experienced soccer players and 4 positions: goalie, defender, midfielder, and striker. State your assumptions, and calculate the resulting number of possible squads.

21. In 1994 the United States hosted its first World Cup soccer tournament. In the first round of World Cup competition, 24 teams are divided into 6 groups of 4. Each group plays a round-robin tournament (see Section 4.3, Exercise 10). Sixteen teams proceed to the next round, which is the start of a single-elimination tournament (see Section 4.3, Exercise 12). There is one additional game to determine third (and thus fourth) place. How many games must be scheduled in a World Cup?

22. The United States assigns a permanent nine-digit Social Security number to each resident, including both citizens and resident aliens. In theory, each person gets exactly one Social Security number, although mistakes can occur. Social Security numbers are not recycled after a person dies or emigrates. Is the United States likely to run out of Social Security numbers? Justify your answer with calculations and census data.

23. At one time, the exchange portions of local phone numbers were formed in the pattern of two letters followed by one digit. For example, "BUtterfield 8" was how

a person thought of the exchange, but you would dial BU8. Then the phone company phased out these exchanges in favor of exchanges expressed by three digits. Why do you think they did this? (Hint: Look at a telephone.)

24. Sometimes a telephone "number" is all digits, as in 846-1234, and sometimes the "number" contains letters, as in 256-CARS. Explain whether the telephone company gets more local phone "numbers" by using both letters and digits than by just using digits. (Hint: Look at a telephone.)

25. Suppose you use a mathematics computer utility to list all the possible two-scoop ice cream cones from a store with 28 flavors. Would you use permutations or combinations for this list? Why? Are there any possible cones not listed by either permutations or combinations? Why?

26. Combinatorics helps explain why simple-looking games have so much appeal. In Bridge each of the four players is dealt a hand consisting of one-quarter, or 13, of the deck of 52 cards. How many different subsets of cards could a player be dealt? Suppose a player played 100 Bridge games per day every day of the year. How many years could the person play without seeing the same set of cards as her hand?

27. A player in a Bridge game plays his 13 cards one at a time, in 13 "tricks." The order in which the player plays the cards is affected by information obtained during the bidding, which precedes the tricks, and the rules of the game. How many different ways can a Bridge player play out a hand, assuming that he pays no attention to the bidding and game strategy?

28. Many medicines, both prescription drugs and over-the-counter ones, can cause undesirable reactions if used at the same time as other medicines. A typical handbook of prescription drugs contains information on 1500 medicines. If one typically looks at a pair of drugs for drug interactions, then how many pairs of drugs would the handbook need to consider?

29. As people age, they tend to take a greater number of medicines per day. Suppose a person is taking three medicines. Further suppose that it is possible for there to be an adverse reaction to the use of three medicines together but not an adverse reaction to taking any two of them. Using the figure of 1500 prescription medicines from Exercise 28, calculate how many different trios of drugs would need to be considered when looking for adverse reactions.

30. After cancellation of the $(n - k)!$ in the denominator, the formula for combinations, $_nC_k$, becomes

$$\frac{n \cdot (n - 1) \cdot \cdots \cdot (n - (k - 1))}{1 \cdot 2 \cdot 3 \cdot \cdots \cdot k}.$$

There is still a denominator, $k!$, in that fraction, yet a combination is supposed to count the number of subsets of size k which can be formed from a set of size n, and we know that a fractional number of subsets is not acceptable. Can you give an intuitive argument explaining why the formula for $_nC_k$ always results in an integer, never a fraction?

31. Show that the extreme values of k in the formula for $_nC_k$ give the desired value of 1, namely that $\binom{n}{0} = \binom{n}{n} = 1$.

32. Show that $\binom{n}{1} = n$.

33. Show that $\binom{n}{k} = \binom{n - 1}{k} + \binom{n - 1}{k - 1}$, for $0 < k < n$.

34. Show that for $n \geq 1$, $\sum_{k=0}^{n} \binom{n}{k} = 2^n$. (Hint: Use the results of Exercises 31 and 33.)

35. When the binomial $(x - y)$ is raised to the power n, each term of the result is of the form of a coefficient times $x^k y^{n-k}$, $0 \leq k \leq n$. Show that the coefficient of the $x^k y^{n-k}$-term of $(x - y)^n$ is $\binom{n}{k}$. This result explains why $\binom{n}{k}$ is often called a *binomial coefficient*.

36. **Writing project.** Recently telephone companies have been offering personal 800 numbers to families. They suggest these numbers could be used by college students to call home without needing to have a charge card or pay for the call. Are there enough 800 numbers to allow for such a new use? If not, then how are the phone companies going to provide this service? What sort of data, other than just calculations, might you need to answer this question? Gather that additional data, and write up the results of your investigation.

37. **Writing project.** How do cellular phone companies keep calls separated? How many separate frequencies would be needed so that each cellular phone could have its own permanent frequency? Are there enough frequen-

cies, or is some other method, such as some sort of coded handshaking, being used? You could get answers to questions like these from companies that sell cellular phones or operate cellular phone services. Write up the results of your investigation, being sure to include the combinatorics involved.

38. **Writing project.** Reread the Spotlight "Beyond Finite Sets," and then read about Cantor's "diagonalization proof" that there are more real numbers than integers. You can find that proof in many books on discrete mathematics and in some math books written for general audiences. Write up the results of your reading as a letter to a friend interested in mathematics.

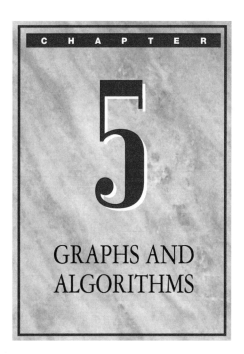

C H A P T E R

5

GRAPHS AND ALGORITHMS

Introduction

In this chapter we investigate what it means to have a method to solve a problem, and we introduce the concept of a graph (different from the familiar concept of a plot of y vs. x) as a useful way to model many situations.

SECTION 5.1 *Problems, Pictures, Procedures*

The Line Painter's Problem

A feature about streets that we take for granted is the line painted down the middle to divide the two directions of traffic. The paint gradually wears off, so trucks need to repaint the lines periodically. For a particular group of streets, an efficient route for the truck would involve passing exactly once along each block, so that there is no wasted travel, such as having to drive back along a street whose center line had already been done. Three questions arise naturally:

➤ Is there an optimal route (one as efficient as possible)?
➤ If so, how can we find it?
➤ If not, what's the best that we can do?

This is an optimization problem; to make it into a mathematical problem, to which we can apply mathematical tools, we need to model the situation with mathematical concepts.

The streets can be represented as line segments, which we call edges, and they meet in intersections, which we call nodes (see Figure 1).

Figure 1 is an example of a *graph*. Although the name is the same as for the graph of a function, this kind of graph is entirely different. See the accompanying Spotlight, "Why Are They Called Graphs?", for the history of "graph" as we use it here.

Graph Concepts

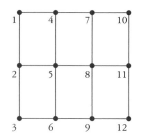

FIGURE 1 A graph to represent the streets for the line painter's route.

In drawings we represent nodes as enlarged points and edges as line segments or arcs joining nodes. However, in graphs that are not as orderly as maps of streets, there may be no way to draw a graph to avoid line intersections that are not nodes of the graph. In Figure 2 the dots labeled $v_1, v_2, \ldots, v_6$ represent nodes, and the line joining v_3 to v_4 represents the edge (v_3, v_4) that is incident to nodes v_3 and v_4. The point P where the edge from v_2 to v_5 crosses the edge from v_3 to v_4 is not a node. We use enlarged points to show which points do represent nodes.

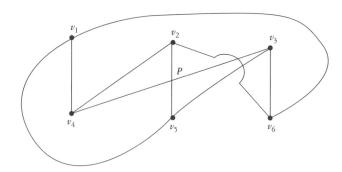

FIGURE 2 In any drawing of this graph in the plane, some edges have to cross at points that are not nodes.

It is convenient to label the nodes $v_1, v_2, \ldots, v_n$, as in Figure 2, because we can then use a matrix to represent which nodes are joined by edges. Such a matrix is called an *adjacency matrix* for the graph.

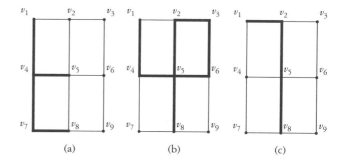

FIGURE 3 Different kinds of paths from v_1 to v_8.
a. $v_1v_4v_5v_4v_7v_8$: An edge and a node are repeated.
b. $v_1v_4v_5v_6v_3v_2v_5v_8$: No edge is repeated, but one node is.
c. $v_1v_2v_5v_8$: No edge is repeated, nor is any node repeated.

The line painter is concerned with covering the route, which in our terms corresponds to traversing edges in the graph, preferably without having to go down any block more than once. At this point we introduce three definitions, which take words common in your experience and give them specific technical meanings to distinguish closely related concepts.

In Section 5.3 we will be concerned with whether a path traverses an edge more than once and whether it visits a node more than once. In Figure 3 we show some paths on the same graph, using thickened edges to denote the edges included in a path. In Figure 3a the pair of nodes v_4 and v_5 occurs twice, first in the order v_4v_5 and then in the order v_5v_4 (v_5 occurs only once, but it can be paired both with its predecessor and with its successor). Hence the edge between v_4 and v_5 is traversed twice. In the path of Figure 3b, each edge that is part of the path is traversed once. Figure 3c shows a path in which no nodes are repeated.

In terms of a path, the line painter's problem translates into the question:

> Is there a path that traverses every edge exactly once?

The line painter does not mind returning to the same point (for example, an intersection of streets), but would like to avoid retracing any blocks. A route that would satisfy the painter is called an *Eulerian path,* after Leonhard Euler (1707–1783), pronounced "oiler." See the accompanying Spotlight for biographical information about this remarkable man.

Eulerian Paths and Cycles

> ### DEFINITION
>
> An *Eulerian path* is a path that traverses every edge exactly once.

In terms of the drawings that we use to represent graphs, an Eulerian path corresponds to tracing all of the edges without lifting your pencil from the paper and without going over any edge more than once. There can be an Eulerian path only if the graph is connected, meaning that for any pair of distinct nodes, there is a path from one to the other (see Figure 4).

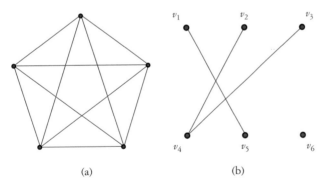

(a)   (b)

FIGURE 4 a. This graph has an Eulerian path. It is a connected graph, but not all connected graphs have Eulerian paths. b. This graph is not connected and hence cannot have an Eulerian path. The graph has three components, none of which is connected to the others by edges.

> ### DEFINITIONS
>
> A graph is *connected* if, for any pair of distinct nodes, there is a path from one to the other. The graph of Figure 4a is connected: Each pair of nodes is connected by an edge, but the necessary path between a pair of nodes may pass through many edges. The graph of Figure 4b is not connected. A graph with just two nodes and no edges is not connected. A graph with just one node (and no edges) is connected, because it cannot fail to be—there are not two distinct nodes without a path between them.
>
> Any graph can be partitioned into one or more subgraphs that are connected and are as large as possible. These subgraphs are called *components.* In Figure 4b there are three components. A connected graph has only one component. A graph with just two nodes and no edges has two components.

The painting truck must start from the city garage and return there after painting lines on a route of streets. The truck must go to a starting point for the route and proceed from there. A good route would be one whose starting and finishing nodes are the same, so that the truck winds up back where it started.

> ### DEFINITION
>
> An *Eulerian cycle* is a cycle that traverses every edge exactly once. In other words, it is an Eulerian path that is a cycle. Figure 5 shows graphs with and without Eulerian cycles.

In 1736 Euler investigated the conditions under which a graph has an Eulerian path or an Eulerian cycle. These conditions involve the degree of a node.

Why Are They Called Graphs?

The term "graph" may seem to you to be overloaded with meanings in mathematics. In Chapter 1 it meant the "graph of a function"; you are also familiar with the term "bar graph" for a particular kind of chart. In this chapter, however, "graph" means a collection of nodes and edges connecting them.

Why yet another meaning? What we now call graph theory was originally a part of mathematics useful in solving puzzles, mazes, and other recreations. The first serious application of those ideas was to chemical molecules. For example, the common "graphic notation" for a molecule of methane, whose chemical formula is CH_4 follows.

The first great mathematician to work in the United States (at Johns Hopkins University), the Englishman James Joseph Sylvester (1814–1897), recog-

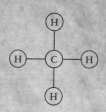

Graphic notation for a molecule of methane.

nized the connection between chemistry and graph theory. It was he who shortened the term "graphical notation" for a chemical molecule to become the mathematical "graph" of nodes and edges. (His main area of research was linear algebra, the subject of Chapter 3, to which he was a major contributor.)

(Adapted with permission of the author and the publisher from Drawing Pictures with One Line: Exploring Graph Theory, by Darrah Chavey. Lexington, MA: COMAP, Inc., 1992.)

DEFINITION

The *degree* of a node is the number of edges incident at that node. For example, the degree of each node in Figure 4a is 4.

Theorem 1. A graph has a Eulerian path if and only if the graph is connected and the number of vertices with odd degree is either 0 or 2. If the number is 0, then any Eulerian path is an Eulerian cycle; if it is 2, then an Eulerian path must begin at one of the vertices of odd degree and end at the other.

You should compare what the theorem predicts for the graphs in Figure 5 and then try to find an Eulerian cycle in Figure 5a and an Eulerian path in Figure 5b. Although Euler developed this theorem in 1736, you probably could have discovered the theorem on your own, after trying a number of examples. It is easy in practice to find whether a Eulerian path exists, as Exercise 13 demonstrates.

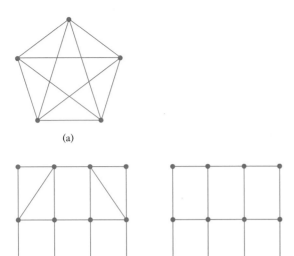

(a)

(b)   (c)

FIGURE 5 a. A graph with an Eulerian cycle. b. A graph with an Eulerian path but no Eulerian cycle. c. A graph with neither an Eulerian path nor an Eulerian cycle.

Euler's theorem tells whether or not a graph has Eulerian paths or cycles. If some do exist, we want to find one. The best proof of Euler's theorem would be one that constructs an Eulerian path. We seek a *procedure* to generate an Eulerian path, a procedure that works on any graph that has an Eulerian path. Mathematical scientists have investigated in detail the nature of such procedures or "recipes," which are called *algorithms*. It turns out to be easy in practice to find whether an Eulerian path exists; Exercises 13 to 15 investigate one such algorithm.

The accompanying Spotlight suggests a generalization of the concept of "graph" that is a better model for some other situations.

We leave graph theory for the time being and now proceed to examine carefully the concept of an algorithm, with examples from other realms of mathematics.

What Is an Algorithm?

DEFINITION

An *algorithm* for a problem is an outline of the steps of a procedure to solve the problem. The procedure must be

deterministic—the nature and order of the steps are described precisely and unambiguously;

effective—it gives a correct solution to the problem; and

finite—it terminates after a finite number of steps.

You are already familiar with several kinds of arithmetical algorithms, the rules that you learned for doing pencil-and-paper addition, subtraction, multiplication, and division of whole numbers, fractions, and decimals. For example, you probably learned an algorithm for adding positive integers that involved first summing the 1's column, posting a carry digit to the 10's column (if necessary), and so on. In algebra you learned algorithms for these operations on polynomials and rational functions.

Another broad class of algorithms consists of mathematical formulas. For example, the values of x that solve the equation $ax^2 + bx + c = 0$ are given by

$$x_1 = \frac{-b + \sqrt{b^2 - 4ac}}{2a}, \qquad x_2 = \frac{-b - \sqrt{b^2 - 4ac}}{2a},$$

provided $a \neq 0$. Try to envision what you would do on a calculator to evaluate one of these formulas for a particular set of values for a, b, and c. The formula is deterministic: It tells what steps to perform and

Leonhard Euler (1707–1783)

Leonhard Euler was born in Basel, Switzerland, in 1707. He learned the basics of mathematics from his father, a minister. His father arranged for him to study further with Johann Bernoulli, one of the great early masters of calculus, at the University of Basel. At age 20, Euler was recruited to a position in St. Petersburg, Russia, at the research academy founded by Tsar Peter the Great. After 14 years in Russia, Prussian ruler Frederick the Great convinced him to go to Berlin to head the Prussian Academy in 1741.

Leonhard Euler
(1707–1783)

Euler's unsophisticated character never really fit in at the Prussian Academy. He had lost the sight of one eye in 1735, and he lost the sight of the other in 1766. The warmth of the Russians' regard for him, and possibly their willingness to aid him in his blindness, convinced him to return to St. Petersburg for the last 17 years of his life.

Blindness might seem an overwhelming loss for a mathematician, but like Beethoven's hearing loss, Euler's blindness did nothing to impair his work. Aided by a tremendous memory, he dictated mathematical papers to a secretary.

Euler was an extremely prolific writer, influencing almost every field of mathematics. He published 530 books and papers during his lifetime, and the *Proceedings of the St. Petersburg Academy* continued publishing his remaining manuscripts for 47 years after his death!

Euler was quite fond of children and had 13 of his own, only 5 of whom survived childhood. It is said that he would sit with children in his lap while performing difficult calculations. His ability to do mathematics amid loud disturbances—and much more quickly than his contemporaries—led a later astronomer to say that "Euler calculated without any apparent effort, just as men breathe and as eagles sustain themselves in the air."

(Adapted with permission of the author and the publisher from Drawing Pictures with One Line: Exploring Graph Theory, by Darrah Chavey. Lexington, MA: COMAP, Inc., 1992.)

---

in what order; it is effective since the results are guaranteed by algebra to be correct solutions (assuming that you press the right numbers and perform the right steps in the right order, and that the limited precision of your calculator does not produce round-off error); and it is finite, since there are only finitely many parts to the formula.

In Chapter 3 you encountered the Gaussian elim-

Generalizing a Graph

How would a letter carrier react to the graph model of Figure 1 for the line painter's problem? The carrier would likely point out that the graph oversimplifies the problem that a carrier faces. For blocks with deliveries on both sides of the street, the carrier usually would not crisscross back and forth in a single trip down the street. The width of the street is not so small that the carrier can totally neglect it. Instead, the carrier goes up one side of the street and (perhaps after making deliveries on other streets) up or down the other.

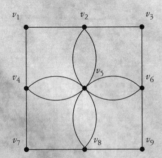

Multigraph for a small delivery route.

We could model the carrier's problem better if a graph could have a separate edge for each side of the street, that is, two edges along each block on which there are deliveries on both sides. We would still use a single edge for a block with deliveries on only one side, e.g., facing a park or a parking lot, or a rural route with mailboxes on only one side of the road. A generalized graph in which there can be more than

one edge connecting a pair of nodes is called a *multigraph*. The entries of its adjacency matrix are the numbers of edges that join each pair of nodes. As an example, we give the multigraph for a very small delivery route, together with its adjacency matrix.

$$
\begin{bmatrix}
0 & 1 & 0 & 1 & 0 & 0 & 0 & 0 & 0 \\
1 & 0 & 1 & 0 & 2 & 0 & 0 & 0 & 0 \\
0 & 1 & 0 & 0 & 0 & 1 & 0 & 0 & 0 \\
1 & 0 & 0 & 0 & 2 & 0 & 0 & 0 & 0 \\
0 & 2 & 1 & 2 & 0 & 2 & 0 & 2 & 0 \\
0 & 0 & 1 & 0 & 2 & 0 & 0 & 0 & 1 \\
0 & 0 & 0 & 1 & 0 & 0 & 0 & 1 & 0 \\
0 & 0 & 0 & 0 & 2 & 0 & 1 & 0 & 1 \\
0 & 0 & 0 & 0 & 0 & 1 & 0 & 1 & 0
\end{bmatrix}
$$

Letter carriers often drive a vehicle from the post office to a starting point and proceed on foot from there. A good walking route would be a cycle in the multigraph—a walk whose starting and finishing nodes are the same, so that the carrier winds up back at the vehicle. The ideal route would be an Eulerian cycle, covering each street side exactly once and returning to the vehicle. There may not be such a route; where repeating some streets is necessary, the problem then is to find a suitable route of minimum length [Balakrishnan (1982)].

ination algorithm for solving a linear system (Section 3.4) and pivoting plus Gaussian elimination as an algorithm to find the inverse of a matrix (Section 3.5). In fact, much of your knowledge of mathematics consists of knowing how (and when) to perform certain algorithms to solve particular kinds of problems. Even solutions to problems in trigonometry or particular kinds of word problems in algebra (e.g., rate problems) can be algorithmic, once a pattern of solution emerges and becomes routine. You can even think of other activities in life as algorithms; for example, to explain to a young boy or girl how to use a pay phone, you would need to give a careful description of a sequence of steps to accomplish the task. How detailed the description should be would depend on the age and experience of the person, since you would need to know if you could rely on an understanding of certain terms ("quarter," "busy signal") and familiarity with certain operations (inserting a coin, pushing a button, etc.) without going into detail about them.

Similarly, the detail with which a mathematical algorithm should be described depends on the audience who will read and apply it. Sometimes, as with Gaussian elimination, the method of the algorithm can be conveyed effectively in an informal fashion, or by means of examples. Often, however, we want to give a precise formal description, as an aid in

➤ clarifying any imprecision in the informal description, so that we can be sure that we understand exactly how to perform the algorithm;

➤ proving that the algorithm in fact always does what it claims to do; and

➤ coding the algorithm as a computer program.

We now illustrate the formulation used in this book for describing algorithms formally.

Hodgson's Rule

An all-too-common problem that students face is sometimes having too much to do. In particular, you may be faced with deadlines for assignments in several classes but realize that you can't possibly finish all of the assignments by their respective due dates. What should you do?

You could take turns working on the different assignments, but that might result in all of them being late. You could try to minimize how late the latest assignment is. You could work intensively on the assignment that is due first, until you finish it or until it is late (whichever is sooner), then switch to the assignment due next. What you should do may depend on the penalty in each course for a late assignment.

Here we describe a strategy—an algorithm—to minimize the number of late assignments, called Hodgson's rule [Moore (1968)]. This algorithm has been applied in practice many times to situations of scheduling production in industry. For a particularly amusing account of an application to the manufacture of trailers, see Woolsey (1992).

We assume that you know the due dates for the assignments and know or can estimate the time that it will take to complete each one. Here is an informal description of the algorithm:

STEP 1:   Put the assignments in order of date due, left to right, from earliest to latest.

STEP 2:   Work out when the assignments would be started and finished if you did them one at a time from right to left. This is called *simulation* because you aren't actually doing it, only experimenting with how you would do it. Proceed until either

—you get through the whole list and no assignment is late, in which case stop; or

—you come to an assignment that would be late, in which case you continue to step 3.

STEP 3: Considering the first assignment that would be late and all assignments to the left of it, take the one with the longest processing time and remove it from the list (it is going to be late).

STEP 4: Return to step 2, and repeat steps 2–4 until done.

EXAMPLE 1   Illustration of Hodgson's Rule

Consider the list of assignments in the first row of Table 1, with the due dates (in days from now) in the second row and the times (in days) that they will take in the third row.

The assignments are already in order of date due, so we skip step 1 and proceed to step 2. We execute step 2, showing our calculations in the last row. That row shows what would happen if we did the first two assignments in order. The math assignment is done by the end of day 2, which is fine since it isn't due until day 3; but the economics assignment, which we don't start until after we finish math, gets finished on day 14, which is well past its deadline of day 5. Since

we have a late assignment, we continue to step 3. Since the economics assignment takes the longest time of the assignments up to and including itself, we remove it. Moving on to step 4, we return to step 2 and simulate once again, with the results shown in Table 2.

After that simulation, we now drop the chemistry assignment and simulate (repeating steps 2–4) again, with the results in Table 3. We get through the whole list with no late assignment, so we reach the alternative in step 2 that says to stop.

According to the algorithm, the best that we can do is complete four of the six assignments on time. The final line of Table 3 shows a way to complete the assignments for math, history, psychology, and art on time, with the assignments for economics and chemistry being late. This result may not be the only combination of four assignments that can be completed on time. (Can you find another one? See Exercise 10.) ▲

TABLE 2 Second Stage of the Sample Problem.

Assignment	Math	Hist.	Chem.	Psych.	Art
Due	3	9	14	36	48
Time	2	5	10	15	15
Total time	2	7	17		

TABLE 3 Final Result of Applying Hodgson's Rule to the Sample Problem.

Assignment	Math	Hist.	Psych.	Art
Due	3	9	36	48
Time	2	5	15	15
Total time	2	7	22	37

TABLE 1 Sample Problem for Hodgson's Rule.

Assignment	Math	Econ.	Hist.	Chem.	Psych.	Art
Due	3	5	9	14	36	48
Time	2	12	5	10	15	15
Total time	2	14				

What if you wanted to implement this algorithm on a computer? You would have to describe it in the code of a computer language, which is a big jump from an informal description. A useful intermediate step is to express it in pseudocode, as is done for algorithms in earlier chapters. Styles of pseudocode vary; the rules that we use are handily summarized in Appendix 2.

We describe Hodgson's rule in a pseudocode in Algorithm 1. The leftward arrow ($\leftarrow$) indicates that the value of the expression on the right is assigned to the variable on the left. Algorithm 1 assumes that the assignments have been given numbers and that they have already been arranged in order of date due, so that step 1 of Hodgson's rule has been done (and hence is not shown).

The best way to understand this description is to "desk-check" it: Work through it with the data from the example, and observe how the algorithm implements your informal understanding.

Components of Algorithm Description

Every algorithm takes one or more inputs and produces an output, which we want to be sure is the solution to our problem. In our formulation of an algorithm, we explicitly list

Input(s) and *Output(s)*, in part as a way of introducing notation and identifying variables that are used in the algorithm.

Inputs: due dates, Due[i], and times required, Time[i], for n assignments
Output: vector OnTime indicating assignments completable on time
Preconditions: 0 < Due[1] ≤ . . . ≤ Due[n], Time[i] > 0 for all i
Postcondition: number of on-time assignments is as large as possible

```
for k from 1 to n
   OnTime[k] ← true {we begin by being optimistic!}
repeat
   TotalTime ← 0
   i ← 0 {i is the current assignment}
   repeat
      i ← i + 1
      ThisOneWouldBeLate ← false {again, we start out optimistically}
      if OnTime[i] then {i is still on the list}
         TotalTime ← TotalTime + Time[i]
         ThisOneWouldBeLate ← (TotalTime > Due[i])
         if ThisOneWouldBeLate then {we must find one to omit}
            find j with 1 ≤ j ≤ i, OnTime[j] true, and
               Time[j] = max {Time[1], . . . , Time[i]}
            {we pick longest one so far that is still on the list}
            OnTime[j] ← false {we delete it from the list}
         endif
      endif
   until (ThisOneWouldBeLate) or (i = n)
until (i = n)
```

ALGORITHM 1 Hodgson's rule for maximizing the number of on-time assignments.

Preconditions, facts that must be true about the input(s) in order for the algorithm to work properly. For example, the bisection algorithm to find a real root of a polynomial (Section 1.10) requires that the polynomial have a real root (there are ways to test in advance whether this is so). Our preconditions here are that the assignments are in order of date due (in the future) and that the times for completion are all positive. If either of these conditions is not met, we cannot be sure that the algorithm will give a correct answer. If some of the inputs are entered as negative numbers instead of positive ones, we might find ourselves having already completed all of the assignments last week! For the sake of correctness on all inputs, a computer program should check whether the inputs to the program satisfy the preconditions of its algorithm.

Postconditions, features that are supposed to be true about the output. That they are indeed true and that the algorithm terminates are what it means for the algorithm to be correct. The postconditions and termination must be proved, just as we would prove a theorem in mathematics. Explicitly writing out the postconditions reminds us of what the algorithm is *supposed* to do, which we can then compare with what it actually does, as described in the body of the algorithm.

The formal description of Hodgson's rule makes it easier to reason clearly and specifically about the algorithm, implement it as a computer program, and prove features about it—for example, that it terminates (which is not obvious) and that it maximizes the number of on-time assignments (which is far from obvious). We do not prove these features here, but you may want to remember Hodgson's rule the next time that you face deadlines.

Exercises for Section 5.1

---

1. Does the graph of Figure 4a have
   (a) an Eulerian path?
   (b) an Eulerian cycle?

2. How many components does Figure 4a have?

3. Write out an algorithm for doing your laundry, in a form that would be understandable to a six-year-old child.

4. Write an algorithm for you to write and send a letter to a relative.

5. In what respects does the following procedure fall short of being an algorithm for directions to a hospital?
   Step 1. Go down the hill to Church St. and turn left.
   Step 2. Proceed to Woodward Ave. and turn right.
   Step 3. Go a block or two and turn left.
   Step 4. Keep going until you see the hospital.

6. In what respects does the following procedure fall short of being an algorithm to pass an exam?
   Step 1. Take 10 deep breaths.
   Step 2. Get lots of sleep the night before the exam.
   Step 3. Outline the material.

Step 4. Read the material.

Step 5. Listen attentively to class lectures.

Step 6. Show up for the exam on time.

7. Apply Hodgson's algorithm to the situation in the following table.

TABLE 4 Situation for Exercise 7.

Assignment	1	2	3	4
Due	2	4	8	12
Time	2	2	3	6

8. Desk-check Algorithm 1 by applying it to the situation of Exercise 7. In particular, make a table of the "states" of the "machine" as the computation proceeds, detailing the changes in the values of all the relevant variables. The two rows of Table 4 give the values in the arrays for Due and Time. You should follow the format of Table 5, starting a new row every time the value of a variable or expression changes. The values in OnTime and ThisOneWouldBeLate will be either true or false (T or F).

TABLE 5 Format for Desk-Check in Exercise 7.

			OnTime				
i	TotalTime	ThisOneWouldBeLate	1	2	3	4	j

9. Apply Hodgson's rule to determine the minimum number of late assignments for assignments with due dates 37, 14, 17, 21, 34, and 10 days from now and corresponding process times of 6, 8, 4, 6, 19, and 10 days.

10. For the illustration in Example 1, Hodgson's rule guarantees that the best that you can do is complete four assignments on time.

    (a) How many combinations of four assignments are there altogether?

    (b) How many combinations of four assignments are there that do not include economics (which, no matter what you do, you can't finish on time)?

    (c) What combinations of four assignments can you finish on time?

11. What inspired Euler's research into Eulerian paths was actually a recreational problem. In Euler's time the people of Königsberg, Prussia (now Kaliningrad, Russia), would take Sunday walks, trying to cross each of several bridges exactly once and return home (see Figure 6). Model the situation using a graph, and determine whether—and if so, how—the Sunday strollers could realize their goal.

FIGURE 6 Map of the Pregel River flowing through Königsberg, with island and bridges. [From Euler's original paper of 1736, reproduced in Biggs et al. (1976).]

12. **(Computer Project)** A folk puzzle is to draw a figure such as Figure 7a with a single continuous stroke of a pencil, without lifting the pencil from the paper. Such figures are sometimes called *unicursal.*

    The Tshokwe people of northern Angola draw intricate unicursal figures, called *sona,* in the course of telling stories. An example is shown in Figure 7b; in this figure, the cycle goes *around* the dots (so the dots are not nodes).

    If you have access to a Macintosh computer and can download files by anonymous ftp, you can explore Tshokwe unicursal drawings and create ones of your own. To download the program, ftp to cs.beloit.edu.

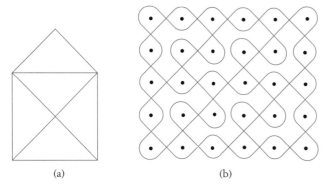

(a)   (b)

FIGURE 7 a. Can you draw this figure without lifting your pencil from the paper? b. The Tshokwe unicursal drawing called "Chased Chicken."

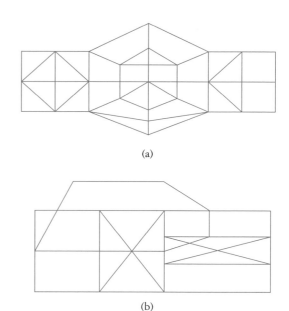

(a)

(b)

FIGURE 8 Graphs to try with Fleury's algorithm. [Figures courtesy of Darrah Chavey, Beloit College.]

Move up a directory level to the root directory /, and then move to the directory /Math-CS Dept./Paul Campbell/Public, where you will find the program Sona 1.4. This program runs under HyperCard and occupies about one megabyte of memory in addition to the memory HyperCard requires.

13. *Fleury's algorithm* is a simple algorithm for drawing an Eulerian path or cycle if one exists. If there is a node of odd degree, start there. Select and draw an edge incident to that node so that the undrawn edges remain connected, that is, any of them can be reached from any other. Move to the new node at the other end of the selected edge and repeat the selection step until either you arrive at an Eulerian path or you become blocked because there is no edge available that fulfills the condition—in which case there is no Eulerian path.

Apply Fleury's algorithm to the graphs in Figure 8, in which each crossing of lines is a node. Use a colored pencil or erasable pen so that you can easily see whether the undrawn edges are connected; to keep track, number the edges in the order in which you add them.

14. Formulate Fleury's algorithm of Exercise 13 in the algorithm format used in this text, using the control structures repeat . . . until, if . . . then, and if . . . then . . . else as needed. Be sure to specify any inputs, outputs, preconditions, and post-conditions.

15. Concerning Fleury's Algorithm of Exercise 13:
    (a) In what respect is it not deterministic? How could you fix it so that it is?
    (b) Give an argument that it is effective in producing a Eulerian path when one exists.
    (c) Give an argument that it is finite.

16. *Euclid's algorithm* is an ancient and efficient way to determine the greatest common divisor of two positive integers p and q—that is, the largest positive integer that divides both without remainder. Let g_1 be the larger of p and q, and let g_2 be the smaller. Let g_3 be the remainder when g_1 is divided by g_2; for example, $27 \bmod 4 \equiv 3$. A mathematical notation to express this relationship is $g_3 \equiv$

$g_1 \bmod g_2$, where mod comes from the word "modulo" and the equivalence sign $\equiv$ denotes the fact that the numbers on both of its sides have the same remainder when divided by g_2. A more familiar way to express this relationship is that g_3 equals g_2 plus some multiple of g_1, or $g_3 = g_2 + g_1 t_1$, just as $27 = 3 + 4(6)$.

Euclid's algorithm successively calculates $g_3 \equiv g_1 \bmod g_2$, then $g_4 = g_2 \bmod g_3$, then $g_5 = g_3 \bmod g_4$, and so forth, until it arrives at a g_k that is 0. Then the greatest common divisor of p and q is g_{k-1}. Here's an example: Let the two integers p and q be 25 and 85, respectively. Then we have

$$
\begin{aligned}
g_1 &= 85, \\
g_2 &= 25, \\
g_3 &\equiv 85 \bmod 25 \equiv 10, \quad 85 = 10 + 25(3), \\
g_4 &\equiv 25 \bmod 10 \equiv 5, \quad 25 = 5 + 10(2), \\
g_5 &\equiv 10 \bmod 5 \equiv 0, \quad 10 = 0 + 5(2),
\end{aligned}
$$

and we find the greatest common divisor of 85 and 25 to be 5. The algorithm can be generalized to find the greatest common polynomial divisor of two polynomials.

Apply Euclid's algorithm to find the greatest common divisor of 42 and 16.

17. Formulate Euclid's algorithm of Exercise 16 in pseudocode, in the format used in this text, using the control structures `repeat . . . until`, `if . . . then`, and `if . . . then . . . else` as needed. Be sure to specify any inputs, outputs, preconditions, and postconditions.

18. Concerning Euclid's Algorithm of Exercise 16:
    (a) Is it deterministic?
    (b) Show that it is effective. First show that the output of the algorithm must divide both of the original numbers. Then show by contradiction that there cannot be a larger integer that divides both of the original numbers, because such an integer would have to divide the output of the algorithm. (Hint: In both parts, use the fact that if an integer divides all the terms of an equation except possibly one, then it must divide that one too.)
    (c) Show that it is finite.

19. (**Computer Project**) Explore the capabilities of a computer algebra system available to you to represent and draw graphs.
    (a) The package Combinatorica extends Mathematica with functions for constructing graphs and other combinatorial objects, investigating these objects, and displaying them. The latest release is available by anonymous `ftp` from `ftp.cs.sunysb.eduftp.`; the package is documented in Skiena (1990). To use the functions, the package must first be loaded, via a command such as `《Combinatorica`; the loading can take several minutes.
    (b) In Maple V, the package `networks` provides many functions, including `addedge` (adds edge(s) to a graph), `addvertex` (adds node(s)), `draw` (draws a graph), `graph` (generates a graph from a specified set of nodes and edges), `incidence` (constructs the incidence matrix), `random` (n, m) (creates a random graph with n nodes and m edges). (Note: In Release 2 you must first enter the line `with (networks):` before calling these functions.)

SECTION 5.2 *Minimum Cost Spanning Cycles*

Meals on Wheels

The population of the United States is aging. The "baby boom" generation will begin to retire after the year 2000, and the number of Americans aged 85 and over will multiply sixfold by 2050. Most elderly people want to continue living at home, and many of them are able to if they have modest amounts of help. One service that many count on is "meals on wheels," the delivery to their homes of a hot meal once a day.

Suppose that you are the manager of the local organization providing meals on wheels. You would like to minimize the cost of delivery, which includes minimizing the miles the delivery vans travel.

A van delivering meals starts from the location of the kitchen, visits a certain set of homes, and returns to the kitchen. Figure 1 shows two different routes for a simplified problem in which the kitchen (K) and all three of the homes are located at the corners of the same city block.

This figure shows a simple situation, which isn't very hard to solve by trial and error. However, we will use this example to create the concepts we will need for larger real problems. Real delivery problems can be very much more complicated, and the order in which the meals are delivered can make a big difference in the cost of delivery.

No doubt, as manager of the local meals-on-wheels service, you have some ideas about how to design routes that are somewhat efficient. But how do you find a best possible route? It's not easy without a strategy, so you need to formalize your intuitions into an algorithm. Your algorithm must be general, so that it finds an optimal route no matter where the kitchen and the homes are located. To be sure that you have the best route, you need to *prove* that your algorithm gives an optimal route. Finally, the algorithm has to be efficient enough to produce new routes quickly, since you regularly receive last-minute cancellations and additions of clients.

We now make a brief detour to examine a situation in a completely different context, which turns out to lead to exactly the same mathematical problem. With these two applications as motivation, we will proceed to solve the common problem, examining variations and new approaches that are being actively researched today.

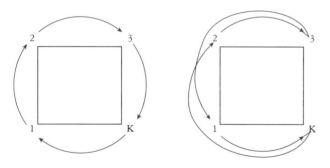

FIGURE 1 The kitchen and three homes for deliveries are all on the same square block. a. The driver delivers to homes in the order 1–2–3 and returns to the kitchen, all by making a single trip around the block, for a total distance of 4 block lengths. b. The driver delivers to homes in the order 2–3–1 and returns to the kitchen, for a total distance of 6 block lengths.

Machine Tools

A key ingredient in a nation's manufacturing ability and prosperity is its *machine-tool industry.* Workers use machine tools to make things: to cut and shape parts (the leg of a chair, the fender of a car); to make dies for metal or plastic objects (a metal wrench, the phone in your room); and to fasten parts together (weld car bodies, drill solder circuit boards).

A major consideration in using a machine tool is keeping costs down:

a. The work should make optimal use of materials. For example, a tool cutting circular parts from rectangular stock should waste as little material as possible. This kind of optimization task is a problem in geometry, and we do not pursue it here.

b. The tool should work as fast as possible, consistent with safety and accuracy. This aspect involves minimizing the movement of the *head* of the tool (e.g., the drill bit, the cutting edge, or the laser tube involved). Equivalently, if the head is fixed and the raw material moves past it (as in a supermarket scanner), we want to maximize the speed of the material's flow.

As our typical problem we take using a laser to drill identical holes at certain locations in a circuit board (see Figure 2). We suppose that the laser can drill the holes in any order. It starts at a fixed initial position, moves in some order to each hole location and drills there, and finally returns to the initial position. The time to do the actual drilling is the same for each hole; in fact, the total time for just the drilling is the same, no matter the order of visiting the holes. The only potential for time economy lies in

FIGURE 2 Diagram of a circuit board. The larger dots indicate locations where holes are to be drilled.

choosing the order for visiting the hole locations, so as to reduce the time for fast gross movement and then slower precise positioning of the laser.

Complete Graphs and Weighted Graphs

We may model each of these two problems, the meals-on-wheels problem and the drilling problem, by a graph. For the meals-on-wheels problem, we take the kitchen and homes to be nodes, with the edges being routes between homes and between the kitchen and homes. For the drilling problem, the nodes represent the drilling locations and the edges represent the routes that the laser takes in going from one location to another.

In the meals-on-wheels problem, we can go from any home directly to any other home; similarly, in the drilling problem, we can go directly from any drilling location to any other. We say that each of the graphs is *complete*.

D E F I N I T I O N

A *complete graph* is a graph in which every node is joined to every other node.

Figure 3 shows the resulting complete graph on the four nodes K, 1, 2, and 3. Notice that we have abstracted somewhat from the physical geography of the problem; for example, the edges (K, 2) and (1, 3) appear to cut through the city block diagonally, whereas in fact they are mathematical abstractions of the real physical routes (which do not cut through the block but go around it in the most efficient way).

Each edge of the meals-on-wheels problem graph has an associated cost—the cost of the most efficient trip between the two nodes (there's no need to con-

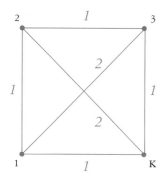

FIGURE 3 The complete graph on the four nodes of the simple meals-on-wheels delivery problem, with associated costs of each edge.

$$\begin{bmatrix} 0 & 1 & 2 & 1 \\ 1 & 0 & 1 & 2 \\ 2 & 1 & 0 & 1 \\ 1 & 2 & 1 & 0 \end{bmatrix}.$$

Spanning Cycles

The meals-delivery and drilling problems both become the mathematical optimization problem of finding a cycle that passes through every node of the graph exactly once and has minimum cost. We specify a route for the van or laser by listing all of the nodes in order of visit, beginning and ending with v_0. The resulting path is called a *Hamiltonian cycle* or *spanning cycle:*

spanning because it includes or "spans" all the nodes in the graph; and

cycle because following that path from any node eventually brings you back to that node without visiting any other node more than once.

sider out-of-the-way routes). Each edge of the drilling problem graph has an associated cost—the travel time of the laser head from the drilling location of one node to that of the other. The graph for each problem is a weighted graph.

D E F I N I T I O N

A *weighted graph* is a graph in which every edge has an associated number, called a *weight* or *cost*.

Let the initial location of the van or the laser be v_0, and let the locations to be visited be $v_1, \ldots, v_n$, so that there are n clients or holes. We designate the minimum cost for the trip from any location v_i to any location v_j by c_{ij} (if $i = j$, the cost is 0). (Recall that we are considering only a most efficient trip between two locations, not out-of-the-way routes.)

We can represent the entire problem by a matrix C of location-to-location costs. For the example of Figure 1 with the kitchen and homes all on the same block, the corresponding matrix, with the costs in terms of blocks traveled, is

---

D E F I N I T I O N

A cycle that contains every node of a graph is called a *spanning cycle* or *Hamiltonian cycle* of the graph.

The name "Hamiltonian" honors Sir William Rowan Hamilton (1805–1865), an Irish mathematician and astronomer who devised a game featuring the finding of cycles. However, the concept appears to have been first enunciated in mathematical terms by the Rev. Thomas Kirkman (1806–1895), an amateur mathematician.

Both of the two routes in Figure 1 are spanning cycles; the first is represented by K123K and the second by K231K. Figure 4 gives an example of a more complicated complete graph (just the nodes are shown, not all the edges), together with a spanning cycle.

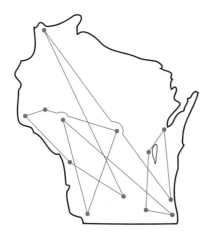

FIGURE 4 Nodes of a complete graph (most edges aren't shown) and a spanning cycle for it.

For each problem, we seek a *minimum cost spanning cycle* of a complete graph. We investigate this concept further next. Such "theoretical" concepts and models derive practical power from their generality, because generality is the key to wide-ranging applicability.

Brute Force

Our goal is to find an optimal ("best") route for the van or laser—a minimum cost spanning cycle. We say "an" optimal route, because several may be tied for best. One way to find a best route is to go through them all, keeping track of the best one so far. This algorithm—examining all solutions and picking a best one—is called *brute force* (see Algorithm 1).

This algorithm is not completely specified, be-

cause it does not describe how to generate all the spanning cycles, or in what order. For a situation as simple as Figure 1, however, it is easy to generate all of the spanning cycles and their costs by hand and then trace through the algorithm; Table 1 traces the execution of the algorithm for one particular ordering of the spanning cycles.

Does brute force always work? Yes, *if* we can finish the task. Because there are only finitely many cycles, one of them must be optimal; and our algorithm looks at all of them. Brute force is thorough, but there may be too many cycles to try them all. How many cycles are there? We start and finish at v_0. In between we are in the situation of Section 3 in Chapter 4, of choosing nodes without replacement; once we have visited a location, we do not want to return. There are n homes to visit or holes to be drilled; the different paths correspond to the $n!$ permutations of the n nodes $v_1, \ldots, v_n$. We can represent each possible path in a tree diagram, as in Figure 5. As we saw in Section 3 of Chapter 4, $n!$ can be an enormous number, even for relatively small n. For example, for $n = 50$, we have $50! \approx 3 \times 10^{64}$.

If there are too many cycles to enumerate, we need another line of attack. We can think of our problem as a *search* problem: We are searching, among all spanning cycles, for one with smallest length. Brute force amounts to what is called *sequential* or *linear* search, examining every spanning cycle, one at a time. There is no possibility of a shortcut, because we can't be sure that we've found a minimum-length spanning cycle until we've looked at all of them.

If we can't examine all of the spanning cycles with the resources that we have, then we must either

a.   apply more resources to the problem, or

Inputs: a weighted graph
       a spanning cycle InitialCycle for the graph
       a function Cost that calculates the cost of a spanning cycle
Output: a minimum cost spanning cycle for the given graph
Preconditions: none
Postcondition: BestCycleSoFar is a minimum cost spanning cycle

```
BestCycleSoFar ← InitialCycle
MinimumCostSoFar ← Cost(InitialCycle)
repeat
        generate a NewSpanningCycle for the graph
        if Cost(NewSpanning Cycle) < MinimumCostSoFar then
                    BestCycleSoFar ← NewSpanningCycle
        endif
until all spanning cycles have been examined
```

ALGORITHM 1 Brute-force algorithm to find a minimum cost spanning cycle.

TABLE 1 Finding a Minimum Cost Spanning Cycle for Figure 1.

InitialCycle	NewSpanningCycle	Cost	BestCycleSoFar	MinimumCostSoFar
K231K	undefined	undefined	K231K	6
K231K	K132K	6	K231K	6
K231K	K123K	4	K123K	4
K231K	K312K	6	K123K	4
K231K	K213K	6	K123K	4
K231K	K321K	4	K123K	4

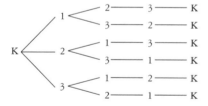

FIGURE 5 A tree, the paths through which represent the possible ways to complete a spanning cycle for the complete graph of Figure 3.

c. settle for the best solution that we can find with the time and resources available to us—a *near-optimal* solution, using a heuristic algorithm.

D E F I N I T I O N

A *heuristic algorithm,* or *heuristic,* for short, is a procedure that seeks solutions at manageable computational effort but offers no guarantees that its output is correct.

b. devise a more efficient algorithm, *using theory* to help develop it and to prove both that it works and that it is more efficient, or

The accompanying Spotlight gives details of a heuristic algorithm used to solve an actual instance of a meals-on-wheels problem. We will discuss heuristic algorithms further in Section 6 of this chapter.

Parallelizing

An example of devising a more efficient algorithm is to *parallelize* it: Modify it so that it can be run on a number of computers at the same time, each working on a different part of the problem. For example, if we had n computers, we could set the first computer to examine just the spanning cycles that go first from v_0 to v_1, the second to check those that go first to v_2, and so forth. Parallel algorithms are an exciting area of research in computer science, particularly since machines may now have thousands of processors. For our problem, however, even with n computers, each computer is still left with examining $(n-1)!$ spanning cycles, which isn't much of an improvement.

Random Search

For the minimum length spanning cycle problem, we could proceed by *random search,* generating at random and checking as many spanning cycles for we have time and resources.

At first that doesn't seem any better than just following the brute-force algorithm until we run out of time. However, with brute force we might not get beyond cycles that go first from v_0 to v_1, which might all be grossly inefficient. Generating spanning cycles at random gives a better "sample" of the universe of spanning cycles.

Can we find an optimal spanning cycle through random search? Yes, if we get lucky. If there is only a single spanning cycle with minimum cost, then we would expect to encounter it after generating about half of the spanning cycles. Whatever happens, we won't know for sure that the best that we can find is optimal unless we look at all of the spanning cycles.

For an intriguing scheme for code-breaking by combining parallel processing and random search, see Quisquate and Desmedt (1991).

Nearest-Neighbor Heuristic

An approach that can be applied in a wide variety of problem contexts is the greedy heuristic.

DEFINITION

The *greedy heuristic* approach is to build a solution by doing the best each stage, without looking ahead to see if this might be unwise in the long run.

For the meals-on-wheels and hole-drilling problems, the natural greedy heuristic is the *nearest-neighbor heuristic.*

DEFINITION

For a path in a weighted graph, the *nearest-neighbor heuristic* chooses as the next node, from those that are adjacent to the current node and not yet visited, one that can be reached along an edge of least weight.

According to this heuristic, we always go next to an unvisited building or undrilled hole that is *nearest* to the current one (if there is more than one, we go to any one of them). For the meals-on-wheels

A High-Math, Low-Tech Solution for Meals on Wheels

The average center providing meals on wheels is unlikely to have a computer, much less software to solve a minimum cost spanning cycle problem. What can mathematics do for such a center?

Exactly this problem was encountered by Prof. John J. Bartholdi III and his colleagues at the Georgia Institute of Technology, with the meals-on-wheels program of Senior Citizen Services, Inc., of Atlanta.

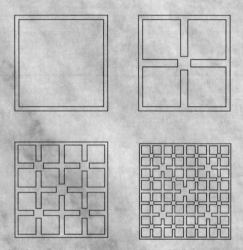

The first four iterations in constructing a curve to fill the unit square. {Figure courtesy of John J. Bartholdi III, Georgia Institute of Technology.}

They devised a heuristic that relies on the mathematics of *space-filling curves*. The figure above shows the first few iterations of a process that, if continued to the limit, results in a curving path that visits every point in the unit square. At every stage, each small square (minus a corner) is refined into a four-leaf clover of subsquares.

Every point (x, y) on the curve corresponds to a number θ between 0 and 1. Paradoxical as it may

seem, the curve establishes a correspondence between the points of a one-dimensional object (the line segment from 0 to 1) and those of a two-dimensional object (all the points in the unit square).

The idea of Bartholdi et al. was to superimpose the square and its curve on the map of the Atlanta area (see the figure below). They calculated once and for all a table of the θ-values for all street intersec-

Street map of Atlanta overlaid with an approximation of a space-filling curve. The highways joining near the top to form a "Y" are I-75 and I-85, and the horizontal highway near the bottom is I-20. {Figure courtesy of John J. Bartholdi, Georgia Institute of Technology.}

tions. Their heuristic calls for the delivery van to visit homes in the order in which the nearest intersection appears on the square-filling curve, in other words, in order of increasing value of θ. The final figure shows the *order* in which locations are visited, the actual route; the route itself is along city streets, pursuing a minimum-distance route from each location to the next.

Bartholdi et al. were able to show that this heuristic results in a route that is about 25% longer than the shortest route. A simple enhancement produces a route only about 15% longer than the shortest.

The space-filling curve heuristic also has the useful property that 1/kth of the length of the route con-

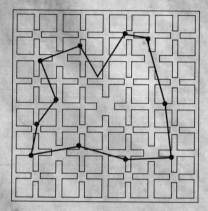

This heuristic route visits points in the order that they occur on the space-filling curve. {Figure courtesy of John J. Bartholdi, Georgia Institute of Technology.}

tains about 1/kth of the locations (assuming that they are distributed uniformly throughout the city). This means that you can partition the work among drivers by simply giving each driver the same number of the consecutive stops, thereby naturally balancing the distances traveled.

The system built by Bartholdi and colleagues consists of a standard street map, a precalculated table of θ-values for the intersections, and two card files. The supplies cost less than $50, the system requires no computer, and the savings in practice have been about 13% compared to previous performance [Bartholdi et al. (1983)].

The idea of using a space-filling curve was picked up by TRW Systems, one of the contractors for the now-defunct Strategic Defense Initiative. TRW's "delivery" problem was to target a space-based laser gun at thousands of incoming targets. Because the laser stays focused over great distance, one can imagine the targets as all appearing on a two-dimensional focal plane. The problem was to sequence the targets to minimize the time to reaim the gun between targets. The plan was to use a parallel computer with 64,000 processors (one for each target) and sequence the targets in the order along the space-filling curve. The heuristic had to be verifiable and analyzable (unlike, say, expert systems) and implementable on a computer that could be boosted into orbit, together with its power supply (a supercomputer is boostable into orbit, but its power supply is not!).

example of Figure 1, this heuristic would have us do the cycle K123K or else K321K.

Even such a simple heuristic requires some effort to describe carefully in pseudocode, as Algorithm 2 shows. The best approach to reading this formulation is to read the heading (*Input,* etc.) and then ignore the pseudocode itself and read just the comments in braces, which tell what the algorithm is supposed to do. (And the best way to develop an algorithm is to write the comments first.) In a later reading, you can see how the comments translate into the pseudocode itself.

Touring Millions of "Cities"

What kinds of problems give rise to having to find minimum cost spanning cycles on graphs with thousands—even millions—of nodes?

In making the printed circuit boards in computers and other electronic items, holes are drilled for electrical leads to connect components to be soldered onto the board. Often, multiple boards are fabricated as part of a large panel, which may require drilling thousands of holes. Since the drilling time is an important part of the cost, minimizing drill motion is desirable—which leads precisely to the minimum cost spanning cycle problem.

X-ray analysis of crystals (Exercises 6 and 7) typically involves 5,000 to 30,000 measurements, which are the nodes of the graph. The x-ray machine must move from one measurement position to another, and minimizing its travel time reduces the cost of the analysis.

Contemporary Approaches

How do you begin to solve a minimum cost spanning cycle problem with thousands of nodes? Even storing the data of the matrix of costs becomes a monumental problem!

First, there is a variety of algorithms; Exercise 3 examines another flavor of a greedy heuristic. Some problems have special properties that can be exploited, as in the wallpaper problem of Exercises 8 and 9.

For other problems, there may be special features in the data. For example, the nodes may correspond to points in the plane, and they may tend to occur in clusters. Then a heuristic that replaces each cluster by a single node can be very effective; after solving the reduced problem, the subproblems of touring each cluster can be managed. This is an example of the strategy of "divide and conquer": An original problem of 35 nodes occurring in 7 clusters of 5 nodes each is divided into 8 subproblems (1 of size 7 for joining the clusters together, and 7 of size 5, to tour each cluster). All the subproblems together can be solved much more efficiently than the original problem. For an approximate solution, two major heuristic approaches have arisen in the past 10 years.

Simulated annealing mimics the process of annealing a metal object, which strengthens it and reduces brittleness by heating it to a high temperature then slowly cooling it. In the algorithmic analog you begin with a suboptimal solution to your problem, randomly change it a little to get a new solution, and then decide whether to exchange your current solution for the new one. At first, you freely allow variations that are worse than your current solution, in hope of later improvements. As you proceed, you "lower the temperature" by reducing the probability of moving from your current solution to a worse one [Aarts (1992)].

Genetic algorithms follow the paradigm of biological reproduction, including mating, mutation, and natural selection ("survival of the fittest"). A genetic algorithm tries to improve an initial "population" of solutions by generating a random neighboring solutions ("mutation") and by combining ("mating") two solutions to yield a "progeny" solution. The algorithm proceeds by stages, mating and mutating the solutions of one generation, then selecting the best solutions to become the next generation [Cao and Ferris (1992)]. This technique has even been applied to "breeding" computer programs [Koza (1992)].

Input: a weighted graph with nodes $v_0, \ldots, v_n$
Output: a path in the graph, starting and ending at v_0
Precondition: the graph is connected
Postcondition: the path is a spanning cycle

```
path(0) ← v₀ {start the path at v₀}
CurrentNode ← v₀
for i from 1 to n
        Available(vᵢ) ← true {start with all nodes available}
endfor
for i ← 1 to n
        j ← 1 {look for first available node}
        while ((Available(vⱼ) = false) and (j < n))
                j ← j + 1 {skip unavailables}
        endwhile
        CheapestSoFar ← vⱼ {first available is tentative choice}
        for k ← 1 to n {check if a cheaper is available}
                if Available(vₖ) and
                   Weight(CurrentNode,vₖ) <
                        Weight(CurrentNode,CheapestSoFar) then
                                CheapestSoFar ← vₖ {go with the cheaper}
                endif
        endfor
        path(i) ← CheapestSoFar {add it to path}
        CurrentNode ← CheapestSoFar {move to new node}
        Available(CheapestSoFar) ← false {mark it unavailable}
endfor
path(n + 1) = v₀ {finish path at v₀}
```

HEURISTIC ALGORITHM 2 Nearest-neighbor heuristic for a minium cost spanning cycle.

The nearest-neighbor heuristic may or may not yield an optimal solution to a minimum cost spanning cycle problem. For the example of Figure 1, it does, as you can check. For the graph of Figure 6, starting at Beloit, it does not, as you can check in Exercise 1.

Does greed pay? Greedy heuristics are based on the hope and expectation that optimizing *locally* (at each step, we do as well as we can) will result also in *global* optimization (at the end, we have an optimal solution overall). Even if we don't get the absolute optimum, we may get a near-optimal solution, since the heuristic prevents us from doing really dumb things.

Traveling Salesperson Problem

The minimum cost spanning cycle problem is also traditionally called the *traveling salesperson problem* (TSP). A sales representative wants to visit each of a number of cities (each exactly once) and then return home, minimizing the mileage involved. Lawler et al. (1985, pp. 2–6) and Hobbs (1991, p. 266) give the history of the problem in this formulation.

At first glance, the TSP problem may not seem to have much practical application. It takes on more importance when you consider not just sales representatives but trucks making deliveries, housing inspectors checking complaints, clerks stocking and picking orders in warehouses, and visiting nurses making house calls. It is also a problem confronted and solved by millions of shoppers every day, as each decides on the order in which to visit stores.

In 1954 the largest problem that had been solved exactly had 49 cities, which meant finding the optimal route among the $48! \approx 1.2 \times 10^{61}$ possible routes [Dantzig et al. (1954)]. Currently, the largest TSP problem that has been solved exactly has 2,392 "cities" [Padberg and Rinaldi (1987)]. The improvement from 1954 to 1987 was due in part to computers becoming faster but much more so to faster and better algorithms.

Contemporary applications involve TSP problems with thousands and even *millions* of "cities." Since exact solution of problems this large is currently impossible, research in recent years has concentrated on finding near-optimal solutions, with highly useful results. In 1992, however, researchers showed that there is a limit to how close any efficient approximation algorithm can get to the least-cost route [Babai (1992); Cipra (1992)].

In other formulations, the minimum cost spanning cycle problem and generalizations of it have even wider application:

a. vehicle routing, including collection and delivery services by such organizations as
   —meals on wheels,
   —the postal service,
   —parcel carriers,
   —overnight express services,
   —food wholesalers, and
   —vending machine services (at colleges and elsewhere);

b. location of facilities;

c. design of circuit boards;

d. cutting wallpaper and other materials;

e. identifying the structure of crystals;

f. sequencing jobs on a machine; and

g. arranging data in clusters of related items.

The exercises explore some of these variations. The accompanying Spotlight discusses large TSP problems and methods to solve them that are currently being explored.

1. In the graph of Figure 6, the nodes are the Wisconsin cities of Beloit, Green Bay, Madison, and Milwaukee, with the edge weights the shortest highway distances between the pairs of cities.

   (a) Apply brute force, listing all of the possible spanning cycles. A spanning cycle can start anywhere, so you may as well start each one at Beloit. What is the minimum cost spanning cycle, and what is its cost?

   (b) Apply the nearest-neighbor heuristic to find a spanning cycle by starting at Beloit. What spanning cycle do you get, and what is its cost?

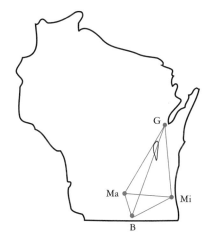

	Beloit (B)	Green Bay (G)	Madison (Ma)	Milwaukee (Mi)
Beloit		161	50	73
Green Bay			132	114
Madison				77

FIGURE 6 For this complete graph, the nearest-neighbor heuristic does *not* produce a minimum cost spanning cycle.

2. To the cities of Figure 6, add Eau Claire, which is 223 miles from Beloit, 193 miles from Green Bay, 176 miles from Madison, and 241 miles from Milwaukee.

   (a) How many spanning cycles are there?

   (b) Apply the nearest-neighbor heuristic to find a spanning cycle by starting at Beloit. What spanning cycle do you get, and what is its cost?

   (c) Can you find a better spanning cycle by using the nearest-neighbor heuristic but starting at a different city?

3. The nearest-neighbor heuristic is one way to apply the greedy heuristic to the minimum cost spanning cycle problem. Another greedy approach is to sort the edges by cost, then build a cycle by stages, and at each stage add in the "cheapest" unused edge that does not prematurely complete a cycle. In other words, we keep adding edges but make sure that we add no more than two edges at any one node. Apply this sorted-edges greedy heuristic to the graph of Figure 6. What spanning cycle do you get, and what is its cost? Do you arrive at the same spanning cycle as with the nearest-neighbor heuristic?

4. Apply the sorted-edges greedy heuristic of Exercise 3 to the graph of Exercise 2. What spanning cycle do you get, and what is its cost? Do you arrive at the same spanning cycle as with the nearest-neighbor heuristic?

5. A computer hard disk serving a multiprogramming environment (such as a time-sharing system) receives requests for disk access from different running processes, either to write to, or to retrieve data from, the disk. Often the processes make requests faster than the disk can service them, so a *queue* (a waiting line) builds up. One approach to servicing the queue is simply to process the requests on a first-come–first-served (FCFS) basis. This method has the advantage of being considered fair, but

it can be inefficient and result in very long waiting times for some or all users.

To see why FCFS can be inefficient, consider Figure 7, which shows a disk platter, the disk head, and the track locations of requests to be serviced. The head moves back and forth across the platter, while the platter rotates under the head. Although it is desirable to optimize total time, which includes both *seek time* for the head to move to the correct track and *rotational time* for the correct sector to rotate under the head, the crucial component is seek time, since on average it is 4 to 10 times as long as the rotational time. If four requests include the first and third for the outermost sector and the second and fourth for the innermost, FCFS would be grossly inefficient, with the disk head moving all the way back and forth across the surface three times.

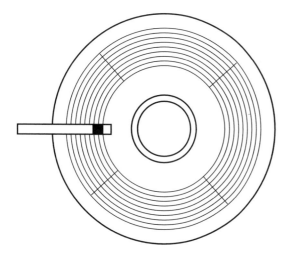

FIGURE 7 Diagram of a disk platter, tracks of the disk, and the disk head.

This situation should remind you of the minimum cost spanning cycle problem, with the difference that the disk head starts at a location to which it does not need to return. There is usually not time to do a brute-force calculation of the shortest route to service all the disk requests.

(a) Based on the ideas in this section, devise a disk-scheduling heuristic algorithm that in general is more efficient than FCFS, and describe it in the same style as the examples in the text.

(b) Are there any disadvantages to your algorithm?

6. Determining the detailed structure of a crystal involves taking a very large number (thousands) of x-rays, one for each triple (h, k, l) corresponding to a point in what is known as the reciprocal lattice of the crystal. Each coordinate of the triple is an integer in $0, \ldots, n$, and measurements must be taken for all combinations of triples of these integers. The dominant cost of taking measurements is *slewing time,* the time it takes for the motors to reposition the apparatus between measurements. There is a motor for each coordinate, and the time it takes to reposition is proportional to the difference in that coordinate between the present position and the next one. So, if the motors are positioned one at a time, the total time to reposition between $V = (h, k, l)$ and $V' = (h', k', l')$ is proportional to

$$C(V, V') = |h - h'| + |k - k'| + |l - l'|.$$

(a) We examine a simplified case. For $n = 1$, the possible values of h, k, and l are 0 and 1. By the multiplication principle of Section 1 of Chapter 4, there are eight triples, and we saw there how to list them in *lexical order.* List them and complete the matrix **C** of distances between pairs.

(b) Chemists have traditionally taken the measurements in lexical order, because it's a good way to ensure that you don't leave any out. How long does this take for the case of $n = 1$?

(c) Can you suggest an ordering that would take less time?

7. With improved apparatus, it should be possible to reposition the three motors of Exercise 6 simultaneously, so that the time to reposition all three becomes proportional to

$$C'(V, V') = \max\{|h - h'|, |k - k'|, |l - l'|\}.$$

(a) Complete the matrix $\mathbf{C}'$ of distances between the pairs.
(b) How long does it take to do the measurements in lexical order?
(c) Can you suggest an ordering that would take less time?

8. People who hang their own wallpaper are often disappointed by the large amount of paper that gets wasted. The wallpaper comes in long rolls, which the user cuts to fit the wall. Part of the waste is due to the desire to match the pattern of the wallpaper at the boundary of two sheets. Doing so involves taking into account the *drop* of the wallpaper, which is the distance that one of two identical sheets must be pushed upward (i.e., trimmed off) so that the pattern matches across the sheets.

Consider a wallpaper roll with the pattern of Figure 8. Notice that the pattern repeats itself every $r = 3$ units (lines), we have a drop of $d = 1$ unit, and it takes $l = 5$ units to cover the height of the wall. Number the sheets

ABCA
BCAB
CABC
ABCA
BCAB
CABC
ABCA
BCAB
CABC
ABCA
BCAB
CABC
ABCA
BCAB
CABC
ABCA
BCAB
CABC
ABCA
BCAB
CABC
ABCA

ABCA	BCAB	CABC
BCAB	CABC	ABCA
CABC	ABCA	BCAB
ABCA	BCAB	CABC
BCAB	CABC	ABCA

FIGURE 8 A wallpaper roll (in the leftmost column) and a wall (in the rightmost three columns) covered with three sheets cut from the roll, with the paper cut so that the ABC pattern continues across sheet boundaries.

of wallpaper from 1 to s, from left to right, here we have $s = 3$.

(a) Suppose you begin by cutting sheet 1 from the top of the roll. Notice that after you cut that sheet, you have CABC at the top of the roll. This is not what you want for the top of sheet 2 (you want BCAB), so you would have to waste the top two rows of patterns if you were to cut sheet 2 now. Verify that if you cut the sheets in the order 1–2–3, then a total of 4 units of wallpaper is wasted (apart from whatever remains on the roll).
(b) Verify that if you cut the sheets in the order 1–3–2, there is *no* waste.
(c) Construct a roll of wallpaper with $r = 4$ and $d = 3$.

9. Usually a roll of wallpaper is narrow compared to the expanse of wall to be covered, so the number s of sheets needed is many more than 3. In our example above, the amount of waste can be only 0, 1, or 2 units; in general, it can be only $0, 1, \ldots, r - 1$ units. After at most r sheets, we must repeat the pattern of the sequence of amounts of waste. In fact, the repetition takes place after exactly $n = r/\gcd(d, r)$ sheets, where $\gcd(d, r)$ is the greatest common divisor of d and r. (Euclid's algorithm in the Exercises of Section 1 finds the greatest common divisor of two positive integers.) Hence, for a large wall, we cycle through this sequence of amounts of waste. We are interested in the total waste in a cycle that starts from sheet 1, proceeds through the other $(n - 1)$ sheets, and returns to sheet 1.

This situation should remind you of the minimum cost spanning cycle problem! Take the sheets to be the nodes, let an edge between two nodes correspond to cutting one sheet right after the other, and let the edge weight be the waste involved in that cutting. The major difference from the problems that we have investigated so far is that here the weights are not necessarily symmetric: e.g., cutting sheet 1 right after sheet 2 generally involves a different amount of waste than cutting sheet 2 right after sheet 1.

(a) For the wallpaper that you constructed in part c of Exercise 8, give the matrix of weights for the sheets that you need to cover your walls.

(b) The matrix of part a of Exercise 8 has the remarkable property that each successive row is the same as the row above but rotated one position to the right. Such a matrix is called *circulant*. This special form has an exciting consequence: Although the nearest-neighbor heuristic applied to a general minimum cost spanning cycle problem does not necessarily yield an optimal solution, the nearest-neighbor spanning cycle for a circulant matrix *is* optimal [Garfinkel (1977), p. 749]. Apply the nearest-neighbor heuristic to the wallpaper that you devised in part c of Exercise 8, and demonstrate (by enumerating all spanning cycles) that it is optimal.

10. Suppose that a currency has positive integer denominations $\{d_1, \ldots, d_n\}$ with $d_1 < \cdots < d_n$.

(a) Write out the common greedy algorithm for making change, for an input N to be paid out.

(b) Exhibit a currency and a payout for which the algorithm fails.

(c) Find a precondition for the algorithm to succeed.

(d) Prove that with this precondition, the algorithm is correct—that is, write out a postcondition, and prove that it holds.

11. Apart from the greedy algorithm of Exercise 10, there are other algorithms for making change—for example, pay it all out in the smallest denomination (e.g., for U.S. currency, in pennies). Define the *bulk* of a payout to be the number of coins and bills used. Wouldn't it be nice if, no matter what the denominations of the currency are, the greedy algorithm always gave out the least bulk?

(a) Does the greedy algorithm always give out the least bulk?

(b) A currency is called *orderly* if for every payout value, the greedy algorithm generates the least possible bulk. Show that a necessary and sufficient condition for a currency to be orderly is that the difference between denominations does not get smaller as the denominations get larger, i.e.,

$$d_{j+2} - d_{j+1} \geq d_{j+1} - d_j, \qquad 1 \leq j \leq n - 2.$$

SECTION 5.3 *From Cycles to Trees*

Designing a Computer Network

In 1989 Wisconsin colleges received a grant to design and build from scratch the computer network Wiscnet, part of the worldwide Internet. A major concern was minimizing the cost of data communications lines between colleges. We can model the situation as a weighted complete graph, with colleges for the nodes, communications lines for the edges, and the cost of a communications line for the weight of an edge. The goal is to find a connected subgraph that includes all of the nodes and has the smallest total weight.

Earlier networks, such as the Bitnet network among college campuses, were not planned in advance but grew haphazardly as more colleges joined. A new institution would connect to the nearest network node, to minimize the cost of the communications line; so the network tended to grow according to a nearest-neighbor heuristic. Many local-area networks grow similarly as workstations are added. Each added campus or station is connected at the smallest cost for adding that particular node to the network already present. Such step-by-step economy, ("local optimization"), may not produce the smallest cost for the network as a whole ("global optimization"). If we can plan a network in advance, however, we can make sure to minimize the overall communications cost.

The Wiscnet problem differs in an important way from the spanning cycle problem of Section 5.2. The Wiscnet network sites do not have to be wired in a cycle; it's enough for any site to be able to reach any other site, through whatever number of intermediate sites. In fact, in the most efficient network, there shouldn't be any cycles; if there were, we could remove an edge and still be able to reach each node from any other one, just as a cut in a rubber band still leaves it in one piece. (However, when greater reliability is desired, or uninterrupted communication at all times is crucial, as in a life- or safety-critical application, we would deliberately build redundancy into the network.)

There is a special name for a connected graph with no cycles, like the graphs of the Bitnet and Wiscnet networks: a tree.

DEFINITION

A *tree* is a connected graph with no cycles. Figures 1 and 2 each show a graph of connections in a computer network, and each graph is a tree. Any complete graph on three or more nodes is not a tree (see Figure 3 in Section 5.2).

FIGURE 1 International lines of the EARN computer network, which take the form of a tree. [Donald L. Nash, The University of Texas System Office of Telecommunications Services.]

In this section we investigate trees and examine how well the nearest-neighbor heuristic serves to minimize their cost. We are interested in trees that span all of the nodes of a graph, and our problem is to find such a tree with minimum total weight of its edges. In other words, our focus has shifted, from the minimum cost spanning cycles of Section 5.2 to *minimum cost spanning trees* (MCST).

Trees have other uses. In Section 1 of Chapter 4, and in Section 2 of this chapter, we used trees in the form of tree diagrams to enumerate permutations. In the next section we will use trees to analyze strategies in games. The accompanying Spotlight notes still further uses of trees.

Parse Trees and Search Trees

Have you ever wondered how a computer interprets an arithmetic or algebraic expression, such as $250 * X + 30 * Y * Z$? The details depend on the computer language, but translators in all languages use *parse trees*. A parse tree represents the order in which to do a computation. Since addition, subtraction, multiplication, and division each operate on just two quantities at a time, an algebraic expression can be represented as a binary tree by putting an operation as a parent node and the operands as children of it. For example, the expression $250 * X + 30 * Y * Z$ can be represented as follows.

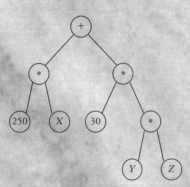

A parse tree for $250 * X + 30 * Y * Z$.

When values are supplied for the variables, the computer uses an algorithm to *traverse* (travel through) the tree, substituting the values for the variables and putting the values and operations together in the correct order.

Trees are commonly used for storing data that need to be retrieved quickly. If you were to search for a word in the dictionary by starting at the beginning with "a" and looking at each word until you found the one you were looking for, each search would, on average, involve looking at half of the words in the dictionary. Instead, you search based on a 26-way branching tree: You use the first letter of the word to indicate that you should look in the section of words that begin with that letter (some dictionaries even have thumb indexes to help you find those sections more easily), then you use the second letter of the word similarly, and so on.

A computer searching a file of thousands of payroll or medical records for your particular record, however, doesn't have thumb indexes to help it. Nor can it conveniently speed its search by skipping three-quarters of the way through the A's (as you would) in trying to locate "Jonathan M. Apple." In fact, there is no way for the computer to tell how long the A section is likely to be.

The computer *can* locate a record easily if there is an index of where each record is stored in the file, just as the directory of a computer diskette tells where on the diskette each file begins. Keeping an index reduces the record-location problem to index searching. The search can be made faster by storing the index in the form of a binary *search tree*.

The following figure shows how such a tree might be arranged. We use letters of the alphabet to stand for the index entries. The tree is binary-branching, with a parent node located in alphabetical order between its

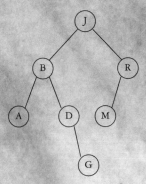

A tree for storing an index to records.

left child and its right child. Entries, called *keys,* are put into the tree in the order that they are received. For example, the entry J was received first and hence is at the root; similarly, B was received next and is the left child of J because it precedes J in alphabetical order. Note that D does not have a left child; a left child would have to be C, but no C entry has been received. Entries are made at new nodes in the tree by making comparisons with existing nodes, working down the tree. A search for an entry in the tree takes place in the same manner, until either the entry is found or it is apparent that it is not in the tree.

If the tree is built from n random entries, then inserting a node in the tree or searching for a particular entry takes on average about $2 \ln n$ comparisons. When n is small, say $n = 10$, not much time is saved over just searching through each entry ($2 \ln 10 \approx 4.6$ comparisons versus on average 5.5 comparisons); but for large n, say 100,000, the savings is remarkable ($2 \ln 100,000 \approx 23$ comparisons versus approximately 50,000 comparisons). In the worst case, however—if, say, every node of the tree has only a left child—then a search can require n comparisons. To avoid this possibility, computer scientists and mathematicians have devised more sophisticated techniques, both for "balancing" trees and for search problems in general.

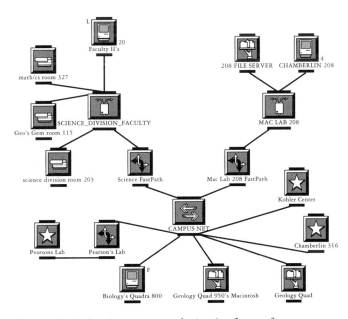

FIGURE 2 A local-area network, in the form of a tree.

Some Useful Facts About Trees

We note here some facts about trees that we will use later in this section.

Lemma 1. Removing an edge from a tree disconnects it.

Proof. By contradiction. Let a tree T be given. Consider any edge of the tree with incident nodes v_1 and v_2. Delete the edge (v_1, v_2) (but not the nodes themselves); we represent this deletion in Figure 3 by making a dashed line between v_1 and v_2. Call the resulting graph G, so that G is T minus the edge $v_1 v_2$.

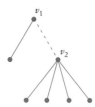

FIGURE 3 A tree with edge $v_1 v_2$ removed.

Suppose that G is connected. Then there is a path in G from v_1 to v_2. That path is also a path in T, where together with the edge $v_1 v_2$ it forms a cycle. But we have reached a contradiction of the fact that T is a tree, as a tree has no cycles.

▲ ▲ ▲

This lemma tells us that, in some sense, a tree can have as few edges as possible and still be connected. We can make this intuition quantitative, as the next two lemmas do. We will apply these lemmas in establishing facts about algorithms for a minimum cost spanning tree.

Lemma 2. Let a graph have v nodes and e edges. If the graph is connected, then

$$v \leq e + 1.$$

Proof. We proceed by *induction* on the number of edges e. We first note that the lemma holds for all connected graphs with zero edges. The only such graph has just one node; for it, we have $v = 1$ and $e = 0$, so $1 = v \leq e + 1 = 1$. (This may seem a strange beginning, but we have to start somewhere.)

Basis for the induction: $e = 0$

We then assume that

Inductive hypothesis: *The inequality holds for all connected graphs with n or fewer edges.*

Using that assumption, we prove that the lemma also holds for all connected graphs with $(n + 1)$ edges. With this "bootstrap" technique, we can go from 0 edges to 1 edge, from 1 edge to 2 edges, and so on, to whatever number of edges the graph has.

Suppose that we have a connected graph G with $e = (n + 1)$ edges. We temporarily take out one edge (anyone). After removing the edge, the remaining graph G', which has $v' = v$ nodes (the same number as G) and $e' = n$ edges (one fewer than G), may or may not be connected.

We analyze the situation to discover what must have been true of G *before* we took the edge out.

If G' is connected, then by our inductive hypothesis, we have $v' \leq e' + 1$; substituting the values we know for v' and e', we get $v \leq n + 1$. Further, we can say that $v \leq n + 1 \leq (n + 1) + 1 = e + 1$, which tells us that G satisfied the desired inequality *before* we took out the edge.

If G' is not connected, then it must consist of exactly two connected components, G_1 and G_2, each with n or fewer edges, and together having a total of v nodes. Our inductive hypothesis applies to both G_1 and G_2, so we have $v_1 \leq e_1 + 1$ and $v_2 \leq e_2 + 1$, where v_1 and e_1 are the numbers of nodes and edges of graph G_1, and similarly for v_2 and e_2. Add the two inequalities to get $v_1 + v_2 \leq e_1 + e_2 + 2$. Since $v_1 + v_2 = v$ and $e_1 + e_2 = n$, we have $v \leq n + 2 = (n + 1) + 1 = e + 1$. As in the other alternative, the desired inequality was satisfied by the graph *before* we took out the edge.

Hence every graph with $(n + 1)$ edges satisfies the inequality.

▲ ▲ ▲

Lemma 3. Let v be the number of nodes and e the number of edges of a tree. Then

$$v = e + 1.$$

Proof. We proceed again by induction on the number of nodes v. The result certainly holds for a tree with no edges, since it can have only one node.

Basis for the induction: $e = 0$

We then assume that

Inductive hypothesis: *The equality holds for all trees with n or fewer edges.*

Using that assumption, we prove that the lemma also holds for all trees with $(n + 1)$ edges.

Suppose that we have a tree T with $e = (n + 1)$ edges. We temporarily take out one edge (any one). After removing the edge, the remaining graph G', which has n edges and v nodes, definitely is not connected, by Lemma 1. It must consist of two connected components, G_1 and G_2. Neither of these components can have a cycle, as such a cycle would also be a cycle of T—which is a tree and can't have any cycles. Hence both of the components are also trees. G_1 and G_2 both have n or fewer edges, so the inductive hypothesis applies to them: $v_1 = e_1 + 1$ and $v_2 = e_2 + 1$. Adding gives $v_1 + v_2 = e_1 + e_2$. Since $v_1 + v_2 = v$ and $e_1 + e_2 = n$, we have $v = n + 2 = (n + 1) + 1 = e + 1$.

T satisfied the desired equality before we took out the edge. Hence every tree with $(n + 1)$ edges satisfies the inequality.

▲ ▲ ▲

Prim's Algorithm for a Minimum Cost Spanning Tree

The greedy heuristic of Section 5.2 suggests trying to build an MCST by starting with a fixed node and, one by one, adding nodes along least-cost edges. In the Wiscnet example, it makes sense to start from Madison (marked with an "M" in Figure 4a), which

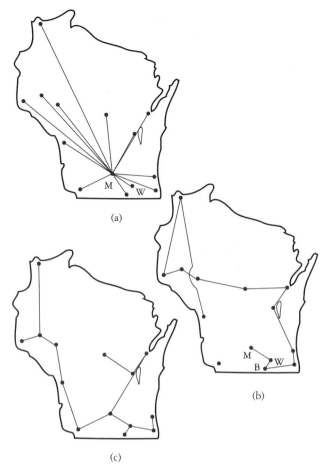

FIGURE 4 The results of three strategies for adding nodes, beginning at Madison (marked "M"). a. Add edges from Madison, getting what is called a "star" topology; this is a spanning tree but does not have least cost. b. Add a nearest neighbor to the current node. c. Add a nearest neighbor to the subgraph built so far.

a. We could add nodes and corresponding edges from Madison, getting a star–topology network—which is a tree, but maybe not the optimal tree (see Figure 4a). There's nothing very greedy about this approach, as we end up adding all the edges from Madison to everywhere.

b. We could move to the node just added (Whitewater) and, applying the nearest-neighbor version of the greedy heuristic, add a closest neighbor to that one (which would be Beloit, marked with a "B," via an edge from Whitewater) (see Figure 4b).

c. Following the principle of greediness most closely would dictate adding an edge with minimum cost—from any of the cities added so far (Madison or Whitewater), whichever has the least-cost edge—to some available (unvisited) city (once more, it would be Beloit via an edge from Whitewater). In other words, we add the nearest neighbor to the subgraph that we have built so far, taken as a whole (see Figure 4c). The main complication is to avoid adding an edge that completes a cycle, since we know that a cycle would be wasteful (and a tree cannot have any cycles).

is the connecting point to the rest of the Internet. From Madison we add an adjacent node corresponding to an edge with minimum cost—that would be to Whitewater (marked with a "W" in Figure 4a).

But then what do we do?

In all cases, we stop when there are no new nodes to add. Later in this section, we will examine still another fruitful approach, Kruskal's algorithm, which is also based on a greedy heuristic.

Adding a nearest neighbor to the subgraph built so far is the idea behind *Prim's algorithm* for a minimum cost spanning tree. We give an example, applying this algorithm to the miniaturized network problem in Figure 5; pseudocode for the procedure is in Algorithm 1.

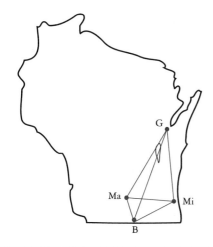

loit, Green Bay, Madison, and Milwaukee, shown in Figure 5 along with the relevant distances between pairs of cities. For readability, instead of the node labels $v_1, \ldots, v_4$, we use mnemonic abbreviations of the cities' names: B, G, Ma, and Mi. The weights of the six edges are given in Figure 5. We begin from Madison as the initial node. We present a trace of the execution of the algorithm in Table 1. ▲

	Beloit (B)	Green Bay (G)	Madison (Ma)	Milwaukee (Mi)
Beloit		161	50	73
Green Bay			132	114
Madison				77

FIGURE 5 Situation for Example 1.

E X A M P L E   1   Prim's Algorithm for a Minimum Cost Spanning Tree   We apply Prim's algorithm to determine a minimum cost spanning tree for the complete graph on the cities of Be-

We described this algorithm carefully in Algorithm 1 and stepped through it with an example. But do we know that the procedure *always* works? What could go wrong? There are six things that we want to be sure of: The procedure must

1. not "hang" (get stuck) because a step cannot be completed;

2. terminate (finish without going on forever);

3. output a graph;

Inputs: a weighted graph with nodes $v_1, \ldots, v_n$, and an InitialNode
Output: a graph with nodes $v_1, \ldots, v_n$
Preconditions: the original graph is connected, Tree is empty
Postcondition: the output graph is a minimum cost spanning tree

```
AvailableNodes ← {v₁, v₁, . . . , vₙ} − InitialNode
add InitialNode to the nodes of Tree
for i from 1 to n − 1
        choose an edge of smallest weight that has one node
            among AvailableNodes and the other not
        add that edge to Tree
        delete from AvailableNodes the new node incident to the new edge
        add that same node to the nodes of Tree
endfor
```

ALGORITHM 1 Prim's algorithm for a minimum cost spanning tree.

TABLE 1 Trace of the Execution of Prim's Algorithm. The Edge to Be Added Next Is Given in Bold Italic.

Initial node	Available nodes	i	Nodes of Tree	Edges with just one node in Tree	Edges in Tree
Ma					
	B, G, Mi				
			Ma		
		1			
				MaB (50), MaG (132), MaMi (77)	
					MaB
	G, Mi				
			B, Ma		
		2			
				BG (161), *BMi (73),* MaG (132), MaMi (77)	
					MaB, BMi
	G				
			B, Ma, Mi		
		3			
				BG (161), MaG (132), *MiG (114)*	
					MaB, BMi, MiG
	—				
			B, G, Ma, Mi		

4. output a *tree* (a connected graph, no cycles allowed);

5. output a *spanning* tree (all the nodes are included); and

6. output a *minimum cost* spanning tree.

How can we be sure? Only by *proving* in general that our procedure has these properties. The fact that the procedure may work on some small examples that we can examine by hand is not enough evidence.

One way that the algorithm could get stuck is if, in some iteration, either all nodes are available (not yet part of the subgraph) or else none of them are. They can't ever all be available, because we made the `InitialNode` unavailable before the loop and

nowhere in the algorithm do we change any unavailable nodes to being available. We begin with n nodes available, and each iteration deletes exactly one from the list; since we iterate exactly n times, we don't run out of available nodes until we exactly exhaust the supply on the last iteration.

There is another way to get stuck: if there is a node that we cannot reach from the nodes included so far, that is, no edges lead from them to it. But that in fact could happen (see Figure 6)! Our algorithm works only for connected graphs. We have discovered that what we want to claim about our algorithm requires a *precondition* (hypothesis) in order for the desired *postcondition* (conclusion) to be true.

Since it doesn't get stuck, the algorithm must terminate, because the loop is executed exactly $(n - 1)$ times. Thus, the first two properties hold.

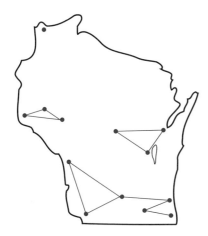

FIGURE 6 A graph that spans all the nodes but is not connected.

Since at each stage we add an edge incident to a node already in Tree and also add the new node at the other end of that edge, the output is a connected graph. Thus, the third and fourth properties hold.

We have designed the algorithm to ensure that we never add an edge that completes a cycle, because we never connect two nodes that are already in the tree that we are building (we always make a connection to one of the available nodes, the ones outside). Hence, the output is a tree. Since we add one node before the loop and exactly one node at each iteration, we add a total of n nodes—and that's all that there are—so the tree spans the original graph. That takes care of the fifth of the six properties.

Finally, we need to show that the spanning tree that we get has minimum cost. One way to show this is by verifying that our procedure's loop has an *invariant,* a feature that is true at all stages: just before we enter the loop, at the end of every iteration, and at the conclusion of the loop. The appropriate invariant is the following:

At every stage, the graph is contained in (could be expanded to) a minimum cost spanning tree.

Why would showing this clinch the matter? Because when the algorithm terminates, what we have is a spanning tree that is contained in a minimum cost spanning tree. Since the spanning tree is a subgraph of the minimum cost spanning tree, the total cost of the edges of the spanning tree must be less than or equal to the cost of the edges of the minimum cost spanning tree. The only way that this can happen is that our spanning tree must also have minimum cost. (In fact, the two trees must actually be identical, as Exercise 9 shows.)

The proof that the invariant holds at all stages is somewhat intricate, and we don't go into it here. Our main point is that *features of algorithms should not be taken for granted but must be proved.*

Prim's algorithm is a pleasant surprise. For a minimum cost spanning cycle, the nearest-neighbor greedy algorithm is not necessarily successful; here, for an MCST, Prim's greedy algorithm is *guaranteed* to be successful. We summarize this result in the following theorem.

Theorem 1. Prim's algorithm on a connected weighted graph yields a minimum cost spanning tree.

Kruskal's Algorithm

Another kind of greedy algorithm proceeds by first sorting the edges and then at each stage selecting the cheapest edge that "fits." That approach, applied to find a minimum cost spanning tree, is called *Krus-*

kal's algorithm. It builds a graph by adding edges that, *in the end,* will form a spanning tree. However, it is permissible to add an edge that is not connected to any of the edges added so far. The algorithm begins by preprocessing the edges, sorting them by weight from cheapest to most costly. The algorithm then loops exactly $(n - 1)$ times, adding each time a cheapest edge that won't create a cycle when added to the edges already included. Of course, as it adds an edge, it includes in the graph the nodes incident to the edge (see Algorithm 2).

EXAMPLE 2 Kruskal's Algorithm for a Minimum Cost Spanning Tree We apply Kruskal's algorithm to the same example of the four cities, Beloit, Green Bay, Madison, and Milwaukee, to which we applied Prim's algorithm earlier.

We first preprocess the six edges, listing them in order of cost:

BMa (50), BMi (73), MaMi (77), GMi (114), GMa (132), BG (161).

We trace the execution of the algorithm in Table 2. ▲

TABLE 2 Trace of the Execution of Kruskal's Algorithm. The Edge to Be Added Next Is Given in Bold Italic.

Available edges	i	Edges *in* Tree	Nodes *in* Tree
BMa (50), BMi (73), MaMi (77), GMi (114), GMa (132), BG (161)			
	1		
		BMa	
			B, Ma
BMi (73), MaMi (77), GMi (114), GMa (132), BG (161)			
	2		
		BMa, BMi	
			B, Ma, Mi
MaMi (77), ***GMi (114),*** GMa (132), BG (161)			
	3		
		BMa, BMi, GMi	
			B, Ma, Mi, G
MaMi (77), GMa (132), BG (161)			

Inputs: a weighted graph with n nodes
Output: a graph Tree
Preconditions: the original graph is connected, has n nodes
Postcondition: the output graph is a minimum cost spanning tree

```
make a list AvailableEdges of the edges, from cheapest to most costly
for i ← 1 to n − 1
    add to Tree a cheapest edge that won't create a cycle
        when added to already added edges
    add to Tree the nodes of that edge
    delete that edge from AvailableEdges
endfor
```

ALGORITHM 2 Kruskal's algorithm for a minimum cost spanning tree.

Kruskal's algorithm requires the same six facts to be proved as Prim's algorithm. The algorithm must

1. not "hang" (get stuck) because a step cannot be completed;

2. terminate (finish without going on forever);

3. output a graph;

4. output a *tree* (a connected graph, no cycles allowed);

5. output a *spanning* tree (all the nodes are included); and

6. output a *minimum cost* spanning tree.

We show that these properties do indeed hold.

1. The algorithm could get stuck if we run out of edges, which would be the case if we started with fewer than $(n - 1)$ edges. Could that happen? No; Lemma 2 tells us that a connected graph with n nodes must have at least $(n - 1)$ edges, so we can't run out. The algorithm could also get stuck if we got to a point where each of the remaining edges, if added to the tree, would force creation of a cycle. Then either we have incorporated all of the nodes of the original graph into Tree (which we can't do in fewer than $(n - 1)$ iterations, at which point we are done anyway), or else some node is not yet in Tree but no edge is incident to it (which would contradict the precondition that the original is connected).

2. The algorithm must terminate since the loop is executed a specific number $(n - 1)$ of times.

3. Because we make sure to include all the nodes incident to the edges that we add to Tree, the resulting structure is a graph.

4. Is the output a connected graph? The answer is not at all obvious, since at any stage we may have added an edge that was not connected to any of the edges added so far. Lemma 3, which says that for a tree we have $v = e + 1$, comes to the rescue. Suppose that the graph constructed consists of $k \geq 1$ disconnected components. Let the ith component have v_i nodes and e_i edges. Since we avoided ever adding any edge that would complete a cycle, *each component is a tree in its own right;* so we have $v_i = e_i + 1$. Adding all of the equations gives $\Sigma\, v_i = \Sigma\, e_i + k$. But $\Sigma\, e_i = n - 1$, the number of edges added; and $\Sigma\, v_i = n$, since there are n nodes altogether. Substituting, we have $n = (n - 1) + k$, so $k = 1$. So the graph we constructed has just one component and hence is connected.

Is the output a tree? Yes, because it is connected and we avoided ever including any edge that would complete a cycle.

5. Does the output span? Yes; since we have added $(n - 1)$ edges, we must have added at least n nodes to Tree, so all of the original nodes must be included.

6. Finally, is the output a minimum cost spanning tree? Amazingly, we have proved all of the facts so far without using the crucial feature of Kruskal's algorithm that it always puts into Tree the cheapest feasible edge. Certainly, that feature suggests that the spanning tree should have minimum cost. As was the case with the invariant in Prim's algorithm, however, the details of proving this final property of Kruskal's algorithm are too intricate for a first introduction to minimum cost spanning trees. We have proven the other features of the algorithm and realize that this property too requires proving, even though we do not develop the proof here.

Theorem 2. Kruskal's algorithm on a connected weighted graph yields a minimum cost spanning tree.

▲ ▲ ▲

We now have two algorithms for the same problem. Is one better than the other? This is the kind of question that we examine in the next chapter, where we relate the histories of Prim's and Kruskal's algorithms and compare them.

Exercises for Section 5.3

1. Use Prim's algorithm to find an MCST for the graph of Exercise 2 of Section 5.2, which adds the city of Eau Claire to the four cities of Figure 5.

2. Use Kruskal's algorithm to complete the task in Exercise 1.

3. Just as there may be more than one minimum cost spanning cycle for a graph, a graph may have more than one MCST.
    (a) Construct a weighted graph that has at least two minimum cost spanning cycles, and show what they are.
    (b) Can you convert the minimum cost spanning cycles into minimum cost spanning trees just by deleting a carefully chosen edge?
    (c) Construct a weighted graph that has at least two minimum cost spanning trees.

4. [Shier (1982)] Researchers in a number of fields need to be concerned about homogeneity, or how clustered or spread out things are. Chemists analyzing crystals calibrate x-ray diffraction equipment by using as a standard a disk on which particles of two metals are mixed thoroughly and spread evenly across the surface. Geologists look for nonhomogeneity as a clue to ore deposits, as in the concentration of copper from drill-hole assay samples

taken from different locations. Biologists may want to measure how clustered—as opposed to how spread out— plants or animals tend to be in a particular area.

We analyze a problem of this type and show how a minimum cost spanning tree can provide a measure of homogeneity. Consider the small example of Figure 7a, which represents a 4×4 sample grid in an area being studied (disk surface, ore field, habitat, or whatever). We make the grid into a graph by considering the gridpoints to be nodes and adding edges to connect adjacent (hori-

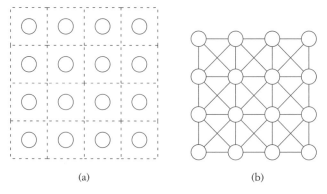

(a)   (b)

FIGURE 7 a. A 4×4 grid. b. The corresponding graph, with edges connecting nodes that are adjacent horizontally, vertically, or diagonally.

zontally, vertically, or diagonally) nodes, as shown in Figure 7b. Rank each of the nodes according to the value of the measured variable at that location, and define the edge costs as the difference in absolute value between the ranks of the joined nodes: For adjacent nodes v_i and v_j with ranks r_i and r_j, we set

$$\mathbf{C}_{ij} = |r_i - r_j|.$$

Why do we use ranks rather than the measurements themselves? We want a scale on which to assess homogeneity that does not depend on the scale used or the problem application area; measurements are affected by these factors, but ranks are not.

A cluster of nearby similar measurements gives small aggregate cost and leads to a small contribution to the total minimum cost. Thus, nonhomogeneity, corresponding to clustering, is signaled by a low value of the cost of a minimum cost spanning tree. If the values for the gridpoints are "randomly" (homogeneously) spread out over the grid, then we should expect large edge costs in our minimum cost spanning tree; so larger values of the cost of a minimum cost spanning tree correspond to situations that are more homogeneous.

How large is large? Statistical analysis shows that the total cost of a minimum cost spanning tree on an $n \times n$ grid has a normal distribution with mean and standard deviation respectively given by

$$\mu_n = 0.994(n^2 + 0.6115n + 1.652)^{1.984},$$

$$\sigma_n = 0.0763n^{2.989}.$$

For $n = 4$, the mean is 38.3 and the standard deviation is 4.8. An observed value more than two standard deviations below the mean, that is, less than 28.7, would be unlikely unless the data are nonhomogeneous.

(a) For the 4 × 4 grid of node ranks below, find a minimum cost spanning tree via Prim's algorithm.

1	5	16	9
8	4	12	7
3	11	10	15
14	13	2	6

(b) Does the cost of your minimum cost spanning tree suggest that the data are homogeneous or nonhomogeneous?

5. **(Computer Project)** If you have a Macintosh computer available, you may enjoy exploring the free Macintosh program, HandsOn, "a binary tree playground." This program can be downloaded from archives of Macintosh software, including `cs.beloit.edu` in directory `/MathDeptQuadra/Public/Macintosh/EducationalSoftware/ComputerScience`. The program will allow you to create a binary tree containing an alphabetically sorted list of words, create a parse tree for an arithmetic expression, and explore several ways of traversing the tree.

6. For the graph of Exercise 4, find a minimum cost spanning tree using Kruskal's algorithm instead. Do you get the same tree as in Exercise 4?

7. What is the greatest number of edges that a graph with v nodes can have?

8. If a graph with v nodes and e edges satisfies $v = e + 1$, does it have to be a tree?

9. Show that if a spanning tree is a subgraph of another tree on the same nodes, then the two are identical. (Hint: Proceed by contradiction, and apply one of the lemmas from this section.)

10. Is it true that if a spanning graph is a subgraph of another graph on the same nodes, then the two are identical?

SECTION 5.4 *Game Trees*

We have encountered trees in several contexts already:

➤ in Section 1 of Chapter 4, where we used a tree diagram to help in counting codons in DNA molecules;

➤ in Section 3 of this chapter, where we were concerned with building a minimum cost spanning tree.

A mathematical tree is a specialized kind of graph, but it is flexible enough to model a wide variety of situations. One such situation is a game of strategy, in which two players take turns making moves. Putting the game in a form of a tree allows us to attack such questions as

How long can the game last?

How many different playings of the game (played-out sequences of moves) are possible?

How many different positions are there in the game?

Can either player force a win?

How could we program a computer to play—and possibly win—the game?

We begin with a game played with matchsticks. Later we will see that strategies for this simple game are closely connected to strategies for checkers, chess, and other more complicated games.

The Game of Nim

The simple game that we investigate is called Nim. It can be played with matchsticks, poker chips, or any kind of counter. In the exercises, we explore variations in the rules; for now, we start with the simplest version.

We begin with 7 matchsticks on the table:

The two players take turns removing either 1 or 2 sticks. The player who removes the last stick loses. We refer to the two players as First Player and Second Player, according to who takes the first turn.

You should try playing this game a few times with someone else, taking turns being First Player and Second Player, and then a few times alone, playing the roles of both players. Is there a strategy that guarantees First Player a win, or one that guarantees Second Player a win?

A Tree for Nim

In Figure 1 we show how to begin to analyze the game by building a tree that keeps track of the sequence of moves. The tree may look peculiar to you,

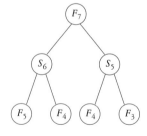

FIGURE 1 The first two rounds of Nim with 7 sticks, with notation for the subgame continuations.

because it grows *downward* from a root at the top. The reason for this mathematical convention is to leave as much room as possible on the page for the tree to grow, since we write from top to bottom, and the page is longer than it is wide.

The top node of a tree is called the *root* of the tree. The nodes in the tree are divided into levels; the *level* of a node is the number of nodes on the path from the root to that node, not including the node itself. Hence the root is at level 0, the two nodes below it in Figure 1 are at level 1, and so forth.

D E F I N I T I O N

The nodes incident to a particular node and below it in the tree are called its *children*. A node with no children is a *leaf*. A node that is not the child of any other node is the *root* of the tree. The *level* of a node in a tree is the number of nodes on the path from the root to that node, not including the node itself.

The nodes of the tree represent the decision points of the players, and we label each with the player to move (*F* for First, *S* for Second) at that point and the number of sticks remaining. For example, the root of the tree is labeled F_7, since First Player is to move and there are 7 sticks on the table.

At each move, the tree branches to the left if the player takes 1 stick and to the right if the player takes 2, and we mark the number of sticks on the tree branch (edge).

Each path through the tree from the root to a leaf represents the progress of one particular playing of the game. Figure 1 shows the first two rounds of the game, during which the tree is *binary-branching:* From each node, exactly two branches grow. Such a

tree has 2 branches that go to the first level, 4 that reach that the second, and 2^n branches that reach the nth level of the tree. This game tree branches binarily until it reaches a node where only 1 stick remains on the table. As long as at least 2 sticks remain, we follow the convention that the right-hand branch corresponds to taking two sticks, and the left-hand branch corresponds to taking one stick.

You may ask why the game has 7 sticks; it could have any number. In fact, you quickly realize in playing the game that after each move you are actually faced with playing a *subgame:* a game with fewer sticks, perhaps with the roles of First Player and Second Player interchanged (after the next move, they are changed back). We can represent a subgame by specifying whose move it is (the original First Player (*F*) or the Second Player (*S*)) and how many sticks remain on the table. So the game we started with is F_7; it branches into one of the two subgames S_6 or S_5. Note that the two games F_1 and S_1 both have the same number of sticks remaining; but in F_1, First Player has to move, necessarily picking up the last stick and losing, while in S_1, Second Player must move and lose.

Thus, we could complete the game tree of Figure 1 by specifying the subtrees that should be attached to the four leaves of the partial tree. There are only three different subgame continuations from those positions: F_5, F_4, and F_3. We show the trees of these subgames in Figure 2.

How Long Can a Game Last?

We show the entire game tree in Figure 3. The longest game is 7 moves long; it's the game in which each player takes 1 stick at each turn.

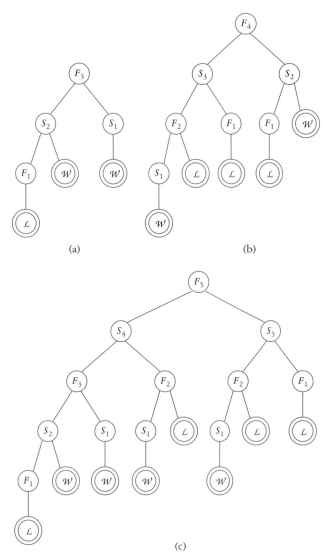

How Many Positions?

Counting the number of different positions is fairly easy in Nim. Each position corresponds to a subgame that can be reached from the original game F_7. The different possible subgames are

$$F_7, F_5, F_4, F_3, F_2 F_1, \qquad S_6, S_5, S_4, S_3, S_2, S_1,$$

giving a total of 12 positions. (Note that F_6 is not on our list because it is not reachable from F_7.) When we come to analyze strategies for the game, we can completely specify a strategy for F by saying what First Player should do in each of the subgames F_7, $\ldots$, F_1; and similarly for Second Player.

How Many Playings?

Counting the number of possible playings of the game is harder. The tree represents all possible playings, with each path through the tree representing the progress of a single playing. To find the total number of possible playings, we could count all of the paths through the tree. Alternatively, since each path ends with a leaf, we could count the number of leaves. Thus, in Figure 3, since there are 21 leaves, there are 21 playings of the game.

If we generalized the game to say 23 sticks, direct counting of the leaves would be infeasible (as would drawing the whole game tree in the first place). We want to find another method, a general method or formula, that can be used for bigger games of Nim.

Can we get a bound on the number of possible playings? The tree has height 7; if, along each branch, the tree branches binarily all the way out to that height, there would be $2^7 = 128$ leaves. Our

FIGURE 2 Different subgame continuations from the branches in Figure 1. "$\mathcal{W}$" (win) or "$\mathcal{L}$" (lose) is from the point of view of First Player. a. The continuation F_3. b. The continuation F_4. c. The continuation F_5.

D E F I N I T I O N

The *height* of a tree is the length of the longest possible path through the tree. The game tree for 7-stick Nim has height 7. A tree of height n has at least one node at level n but none at level $n + 1$.

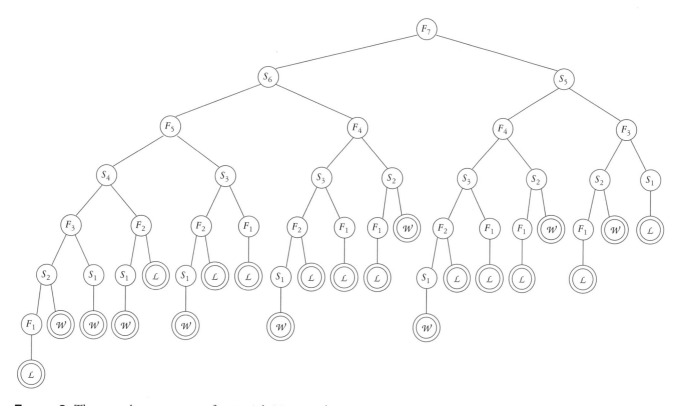

FIGURE 3 The complete game tree for 7-stick Nim. "$\mathcal{W}$" (win) or "$\mathcal{L}$" (lose) is from the point of view of First Player.

tree is not completely binary-branching, so it has fewer leaves. In fact, only that one playing lasts 7 moves; all the rest last 6 or fewer. So there can be at most $1 + 2^6 = 65$ playings, a better upper bound. We already know from Figure 1 that there are at least 4 different playings.

Can we determine the exact number without counting? Let

P_i be the number of playings of the game F_i

Our specific question concerns F_7, the game of Nim with 7 sticks, and the value of P_7; but we are interested in a general approach, for Nim games of any size. We observe that

$$P_7 = P_6 + P_5,$$

since the game with 7 sticks reduces to a game with 6 sticks or one with 5 sticks. In fact, we get the kind of generality we are looking for in

$$P_n = P_{n-1} + P_{n-2},$$

for $n \geq 3$. This is a second-order difference equation, similar to the difference equation for falling bodies in Example 6 of Section 4 of Chapter 1. We noted in Example 1 of Section 5 of Chapter 1 that a second-order difference equation requires two initial conditions for us to solve it. The initial conditions for our Nim game are

$$P_1 = 1 \quad \text{and} \quad P_2 = 2,$$

which we can obtain by counting the number of playings of F_1 and F_2.

From the difference equation and the initial conditions, we can easily calculate the successive values for P_n:

$$1, 2, 3, 5, 8, 13, 21, 34, 55, 89, 144, \ldots.$$

In particular, $P_7 = 21$; and for any relatively small number of sticks, we could easily write out this sequence as far as necessary. This sequence is called a *Fibonacci sequence,* after a twelfth-century Italian mathematician. In terms of differences, we have

$$\Delta P_n = P_{n+1} - P_n = P_{n-1}.$$

This is similar to, but not exactly the same as, the difference equations that we investigated of the form $\Delta a_n = c a_n$. The solutions to those difference equations were exponential functions; this difference equation has a solution that is one exponential function minus another. (We just exhibit the solution here, without showing how to obtain it.)

$$P_n = \frac{1}{\sqrt{5}} \left(\phi^{n+1} - \left(\frac{-1}{\phi} \right)^{n+1} \right)$$

$$= \frac{1}{\sqrt{5}} \left[\left(\frac{1 + \sqrt{5}}{2} \right)^{n+1} - \left(\frac{1 - \sqrt{5}}{2} \right)^{n+1} \right],$$

where ϕ is the Greek letter phi and the value of $\phi = (1 + \sqrt{5})/2 \approx 1.618$ is sometimes called the "golden ratio." Surprisingly, the square roots and fractions all balance each other out, so that the complicated expression always evaluates to an integer for integer $n \geq 1$. This expression allows us to calculate P_n handily even for very large values of n.

There is another approach to counting playings that also connects with material that you studied earlier in this book. We can record a playing by giving the sequence of subgames, as in the sequence $F_7 S_6 F_4 S_3 F_2 S_1$, in which Second Player loses. Since the players take turns, it suffices just to write the number of matches picked up at each turn, so that the same playing can be represented as 121111. If the total number of turns is odd, then First Player plays last and loses; if the number of turns is even, then First Player wins.

Each playing corresponds to an ordered sequence of 1s and 2s that totals 7. We have converted the problem of determining the number of playings into the more purely mathematical problem of counting how many such sequences there are.

The analysis that we began in trying to get a bound on the number of playings can be continued in the following vein. A playing can have only 0, 1, 2, or 3 turns in which one player or the other takes 2 sticks—it can't have any more, because 4 turns of taking 2 sticks would require 8 sticks, and we have only 7. In other words, the number of 2s in the sequence for the playing must be 0, 1, 2, or 3.

We make a table of the number of 2s, the number of 1s, and the total number of moves in the playing (see Table 1). The first row gives the possible numbers of moves that involve taking 2 sticks. Each 2-move takes up 2 sticks; the second row of the table is obtained by subtracting from 7 the total number of sticks involved in 2-moves. The third row is the

TABLE 1 Analyzing the Possible Games in F_7.

Number of 2s	0	1	2	3
Number of 1s	7	5	3	1
Length of sequences (number of moves)	7	6	5	4
Number of playings	1	6	10	4

sum of the first and second rows. We now show how the fourth row is obtained; notice that it sums to 21, which we know is the total number of playings of F_7.

Consider the rightmost column of the table. A playing with three 2s must have just a single 1, for a total of 4 moves; there are 4 such games:

$$2221, \quad 2212, \quad 2122, \quad 1222.$$

We may think of the sequence as consisting of four slots into which we distribute three 2s. Recall from Section 3 of Chapter 4 that the number of ways of choosing 3 things from 4 things is the number of combinations of 4 things taken 3 at a time, represented as $_4C_3$ or $\binom{4}{3}$.

Using this fact, and considering the other columns, we find that the total number of playings is, taking the columns from left to right,

$$\binom{7}{0} + \binom{6}{1} + \binom{5}{2} + \binom{4}{3} = 1 + 6 + 10 + 4 = 21.$$

If your calculator has a button marked "$_nC_r$", you can easily check these results.

How Do You Win?

You probably already have a pretty good idea from playing 7-stick Nim how to win that game. But how would you generalize what you know to 23-stick Nim? How would you describe your strategy, so that you could explain it to someone else, or program a computer to follow it?

What we can do is analyze smaller games and use them to give us an idea of strategy for larger games. In finding an optimal strategy, we have to assume not only that we play as well as possible but that our opponent does, too. We can't leave 2 sticks and hope that our opponent will elect to take both and thereby lose, or leave 3 and hope that the opponent will take only 1. In fact, if we do leave 2, we should assume that we are going to lose. In other words, S_2 is a losing subgame for First Player.

Suppose that we are First Player. Since we don't want to lose, we don't want to wind up at a leaf F_1, where we are forced to pick up the last stick. If we ever get to the subgames F_2 or F_3, we have a sure win: In F_2, take 1; in F_3, take 2.

What about larger games? First Player looks for moves that will lead us to psitions labeled W, while Second Player looks for moves to positions labeled L. We can use the subgame trees of Figure 2. In the game F_3 of Figure 2a, First Player is to move and can move to S_1, which is marked with a W. So First Player can win F_3. So mark the leaf for F_3 in our partial tree of Figure 1 with a W for "winning position" from the point of view of First Player (see Figure 4).

We continue with the idea. Since we must assume that our opponent plays as well as possible, we should plan on losing games S_2 and S_3. In F_4 of Figure 2b, from either S_3 or S_2 Second Player can choose a move that leads to an L. So label any instance of

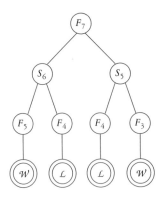

FIGURE 4 The first two rounds of Nim with 7 sticks, with each subgame marked as to whether First Player can win (W) or must lose (L) against best play.

F_4 in our partial tree of Figure 1 with an L for "losing position" from the point of view of First Player (see Figure 4).

Similarly, tracing up through F_5 of Figure 2c, we find that we should label it with a W.

In general, we can "back-propagate" the W's and L's as follows. We have to distinguish S nodes (where Second Player plays and we cannot control what branch is taken) from F nodes (where we can steer away from losing lines of play). Here is what to do:

1. If all of the children of an S node are labeled W, label that node W. No matter what Second Player does, First Player can win.

2. If all of the children of an F node are labeled L, label that node L also. Nothing First Player can do will lead to a win against best play by Second Player.

3. If any of the children of an S node are labeled L, label that node L also. We must assume that Second Player will choose a subgame in which First Player must lose against best play.

4. If any of the children of an F node are labeled W, label that node W. First Player can choose a line of play that leads to a win.

For example, at all locations in the tree, label S_1, F_2, F_3, S_4 and F_5 with W(First Player can win, no matter what Second Player does), but label F_1, S_2, S_3, F_4, S_5, and S_6 with L (First Player must lose against best play by Second Player).

We show the result of back-propagating the W's and L's through Figure 3 in Figure 5; to avoid clutter, we put the label for each node just below the node label. From the back-propagation and labeling, we find that from F_7 it doesn't matter what First Player does—we end up at either S_5 or S_6, both of which are losing positions for First Player. Hence F_7 is a losing game for First Player; label it with L. (So, if you are going to play a game with seven sticks, let your opponent go first!)

The back-propagation that we have described is actually a labeling algorithm. What do we need to be true for the labels to be useful in guiding our strategy? We need the labeling to be *complete* (every node of the tree gets labeled) and *consistent* (no node gets two different labels). These are postconditions that should be proved, which we do next.

Lemma 1. The W–L labeling algorithm for back-propagation on a finite tree, each of whose leaves is labeled either W or L but not both, terminates with a complete and consistent labeling of the tree.

Proof. We prove the lemma by induction on the height h of the tree. (Recall that the height of a tree is the same as the greatest level of any of its nodes.)

Basis for the induction: $h = 0$

A tree of height 0 has only one node, the root, which is also a leaf. As a leaf, by hypothesis of the

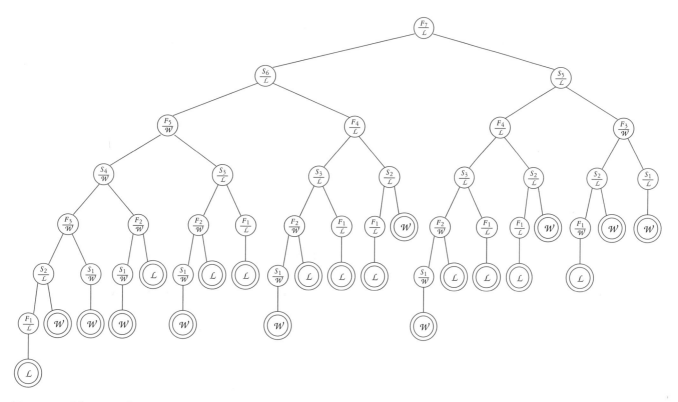

FIGURE 5 The complete game tree for 7-stick Nim, with labels propagated up from the leaves. "$\mathcal{W}$" (win) or "$\mathcal{L}$" (lose) is from the point of view of First Player.

lemma, the root is labeled either $\mathcal{W}$ or $\mathcal{L}$ but not both. Hence all nodes in the tree—all one of them— are labeled, and the labeling is consistent.

Inductive hypothesis: We assume that the lemma is true for all trees of height $h = n$.

We must show that the lemma holds for all trees of height $h = n + 1$. Let such a tree be given, and consider an arbitrary node v of level n.

If v is a leaf, then by hypothesis of the lemma, it is already labeled either $\mathcal{W}$ or $\mathcal{L}$ but not both.

If v is not a leaf, then it has one or more children at level $n + 1$; all of these children are leaves, because the tree has height $n + 1$. By the hypothesis of the lemma, each of them is already labeled either $\mathcal{W}$ or $\mathcal{L}$ but not both. The node v is either an S node or an F node.

If v is an S node, then the following two cases exhaust the possibilities:

1. All of its children are labeled $\mathcal{W}$, so that rule 1 of the labeling algorithm applies and labels v with a $\mathcal{W}$; or

2. At least one of its children is labeled $\mathcal{L}$, so

that rule 2 labels v with a $\mathcal{L}$. So v is labeled either $\mathcal{W}$ or $\mathcal{L}$ but not both.

If v is an F node, then the following two cases exhaust the possibilities:

1. All of its children are labeled $\mathcal{L}$, so that rule 3 labels v with a $\mathcal{W}$; or
2. At least one of its children is labeled $\mathcal{W}$, so that rule 4 labels v with a $\mathcal{W}$.

Whatever the situation of v, it is labeled either $\mathcal{W}$ or $\mathcal{L}$ but not both. Since v was an arbitrary node at level n, we can assert the same for any node at that level.

Having succeeded in labeling all of the nodes of level n consistently, we now strip off all of the nodes at level $n + 1$ (they are all leaves). The resulting subtree of the original tree has height n_0, and all of its leaves are labeled consistently. By the inductive hypothesis, the labeling algorithm labels the subtree completely and consistently. Now just reattach the deleted nodes at level $n + 1$, and we have a complete and consistent labeling of the original tree.

▲ ▲ ▲

Once we know that a game can be won, we still need to identify a strategy to win. But that's not hard from a labeled tree, since all we have to do is pick a child that is labeled $\mathcal{W}$. In Table 2 we show subgames and strategies for winning (where possible). As First Player, our only hope in games like F_7, F_4 is to give Second Player as many opportunities as possible to make a bad play and leave First Player with F_5, F_3, or F_2. The strategy of Table 2 is part of the strategy for Nim with any number of sticks.

TABLE 2 Strategy for 7-Stick Nim, for Either Player.

Sticks remaining	Take	Result
1	1	lose
2	1	win
3	2	win
4	1	loses against best play
5	1	can win
6	1	can win
7	1	loses against best play

Thus, while First Player will never face F_6 when starting from F_7, Second Player may; and First Player may face F_6 when the game starts with more than 7 sticks.

Even if First Player follows the strategy of Table 2, First Player is bound to lose F_7 against best play by Second Player. All Second Player has to do is follow the same strategy in the table. It isn't a fair game!

Like Prim's algorithm for a minimum cost spanning tree, an optimal strategy for a game is an algorithm that guarantees the best possible result under the circumstances and rules. The technique of eliminating ("pruning") from consideration losing lines of play is a general approach used in artificial intelligence in computer game-playing programs for chess, checkers, Go, and other games.

In Section 2 of Chapter 7, we will design the computer circuit for a special-purpose Nim machine that follows the strategy of Table 2 and is unbeatable at F_7 (provided it is allowed to move second!).

Tree Games

Nim is typical of a wide class of tree games. The following definition is from Beck (1969).

DEFINITION

A *tree game* is a game between two players

➤ who alternate clearly defined moves; and
➤ where each knows the moves of the other player; and which has
➤ a particular starting position;
➤ finitely many positions;
➤ a predetermined maximum number of moves, by which time the outcome is decided: a win for First Player, a win for Second Player, or a tie; and
➤ no involvement of chance.

Such games are called tree games because we can (potentially) enumerate all of the possibilities on each turn in the form of a game tree, just as we did for 7-stick Nim.

Card games, sports games, and the wealth of board games that use dice (from backgammon and parcheesi to the many trademarked games) all fail one or more of these criteria. But chess, checkers, Tic-Tac-Toe, Go, and Nim are all tree games.

DEFINITION

If one of the players can win, regardless of the play by the other, or if each player can prevent the other from winning, then we say that the game has a *natural outcome.*

The natural outcome may be a win for First Player (e.g., in F_6), a win for Second Player (e.g., in F_7), or a tie (in Tic-Tac-Toe). If a game has as its natural outcome a win for a particular player, then that player can *force* a win, no matter what the other player does.

Regarding chess and checkers, for example, there has long been speculation whether the natural outcome of the game is a tie or whether moving first may guarantee enough of an edge to make the natural outcome a win for First Player (few experts think that Second Player is the natural winner in these games).

Tree games have a remarkable property:

Theorem 1. Every tree game has a natural outcome.

Proof. We just generalize what we did to find a winning strategy for 7-stick Nim. In fact, the algorithm that we use to prove the theorem not only allows us to conclude that there is a natural outcome but in fact determines for us what it is.

Begin at the leaves of the game tree. Label a leaf with a W if First Player has won, an L if Second Player has won, and a T if the result is a tie.

Just as for Nim, we propagate the labeling back up the tree. Consider a node with one or more children. That node is a decision point for one of the players, who is about to make the last move of the game. Label the node

W if all its children are labeled W, because all lines of best play lead to a win for First Player;

W if First Player is to move and at least one child is labeled W, because First Player can elect that line of play;

L if all its children are labeled L, since all lines of best play lead to a loss for First Player;

L if Second Player is to move and at least one child node is labeled L, since best play dictates that Second Player choose such a child;

How to Dig a Better Mine

Open-pit mining involves digging a hole in the ground and carrying away thousands of tons of ore to process. Mining geologists help determine which areas underground contain the richest deposits of ore. The problem then becomes how to decide which areas are worth digging up. Specifically, mining companies want to dig a pit that is the most economical—yielding the most high-grade ore for the least digging—while also satisfying other constraints, such as the steepest slope that the side of the pit can safely sustain without falling in.

−3	−1	0	−1	−1
−5	1	1	0	1
−7	1	5	6	1
−5	6	20	12	1
−5	−4	0	−1	0

Blocks of ore in a vertical cross section of a mine, with lables indicating the profitability of removing them.

Ore deposits underground can be represented by a three-dimensional grid of numbers. Each number represents the profit corresponding to extracting the ore in a block that is a cube 10 meters on a side, i.e., the value of the ore minus the cost of removing the block. Geologists can determine the appropriate numbers by drilling test holes. The figure above shows blocks of ore in a vertical cross section of a mine, with labels indicating the hypothetical profitability of removing each. Negative numbers correspond to blocks that cost more to remove than their ore is worth.

The cost of removing a block, however, depends on which other blocks need to be removed to get at it. Here is where graph theory is useful. Let the blocks be nodes; in this graph, it is the nodes (not the edges) that have weights associated with them. For each block B, add an edge from it to any block that must be removed before you can get to B. Which blocks must be removed earlier depends on the maximum slope of the pit side. The relationship is not symmetric: If you have to remove A before B, you can't have to remove B before A.

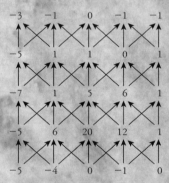

A directed graph showing which blocks must be removed to get at other blocks.

This fact is expressed by using directed edges, symbolized by arrows. The figure above shows the blocks as part of a directed graph, for a situation in which the maximum slope of the pit side can be 45°. Not all directed edges are shown, only the immediate ones.

For example, removing the very profitable block bearing the number 20 requires removing three blocks (with numbers 1, 5, and 6) in the level directly above it and five blocks in the level above those. Since we are looking at only one vertical face, you must realize that additional blocks would also need to be removed in front and behind this face.

A set of nodes in a directed graph is said to be *closed* if, whenever a node is in the set, then so are all the nodes to which its edges go. The mathematical translation of the mining problem is thus to find a closed set of nodes with the greatest total weight. These will correspond to the blocks to remove to realize the greatest profit.

There might be several hundred thousand blocks in a typical mine. As you can imagine, a graph of that size leads to an immense computing problem. The problem is solvable only because mathematicians have developed efficient algorithms for finding the optimum. Those algorithms execute in only a few minutes on a personal computer, so that a mining company can continually assess the profitability of its operations and adjust its planning.

$\mathcal{T}$ if First Player is to move, the node has no child labeled $\mathcal{W}$, and there is at least one child labeled $\mathcal{T}$;

$\mathcal{T}$ if Second Player is to move, the node has no child labeled L, and there is at least one child labeled $\mathcal{T}$.

This labeling algorithm terminates with a complete and consistent labeling of the tree (we could prove this by a slight extension of the induction argument in Lemma 1), including a label at the root of the tree. The root corresponds to the initial position of the game with First Player to move. The label there is the natural outcome of the game.

▲ ▲ ▲

The reason that the natural outcomes of games like chess, checkers, and Go have not been determined is that the game trees involved are so huge.

DEFINITIONS

If a tree game does not admit ties, we call it an *unfair* game, since one player or the other must be able to force a win under best play.

If, from any position in a tree game, exactly the same moves are available to either player (assuming that it's the player's turn), we say that the game is *impartial.*

Nim is impartial. Chess, checkers, and Go are not, because each player may move only the pieces of one particular color. Tic-Tac-Toe is impartial; although the players use different symbols, either may mark any empty square.

What makes Nim so fundamental is the following theorem, from Sprague (1936) and Grundy (1939).

Theorem 2. Every impartial unfair game is equivalent to a generalized Nim game with some number of sticks.

By "equivalent" we mean that there is a correspondence between the positions and the moves of the two games. By "generalized" we mean that the rule that a player may take 1 or 2 sticks is generalized so that the number of sticks that a player may leave comes from some specified list. The theorem means that if we know how to win (from a winning position) in Nim, we in effect know how to win (from a winning position) in *any* impartial game. We don't have to figure out from scratch a strategy for a new game, or work out its game tree; we just need to know how to translate the strategy into a Nim game, and how to translate the Nim strategy back into the new game. The details of the translation for a large number of games are the subject of the book by Berlekamp et al. (1982).

Games as Directed Graphs

No doubt you have noticed in our tree of F_7 in Figure 3 that the same subgame (i.e., S_3) appears at a number of nodes in the tree. You may also have noticed that the subtrees for the games F_3 and S_3 are the same except that the letters "F" and "S" and the results W and L are interchanged. Hence our tree representation of the game is considerably redundant.

A more compact representation can be achieved using a directed graph.

D E F I N I T I O N

A *directed graph,* or *digraph,* consists of a finite set of nodes and a finite set of ordered pairs of nodes, called *directed edges.*

We take as nodes the possible positions of the game, which for a 7-stick Nim game are just the possible numbers of sticks remaining, from 7 down to 0. We draw directed edges from each position to all of the positions that can result from a single move from that position. In other words, for two nodes u and v, we have a directed edge (u, v) from u to v if there is a move from u to v. For 7-stick Nim, we get the directed graph of Figure 6.

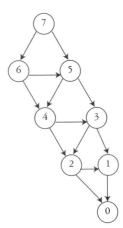

FIGURE 6 A directed-graph representation of 7-stick Nim.

We now want to label the nodes of the directed graph to indicate which player can win from each position. We label a node with N if the Next player—the one to move from that position—can win, and we label the node P if the *Previous* player can win. The result is the game graph for the game.

D E F I N I T I O N

The *game digraph* for a tree game is a directed graph whose nodes are the positions of the game, whose directed edges are the moves between positions, and each of whose nodes is labeled with who can win from that position (the player to move or the previous player).

The algorithm for labeling nodes is as follows:

Label all nodes with no successors with *P*;

Label a node *P* if all of its successors are labeled *N*;

Label a node *N* if one or more of its successors is labeled *P*.

This labeling algorithm will indeed terminate in a complete and consistent labeling for any tree game, a fact that could be proved by another application of mathematical induction. Applying it to Figure 6 gives Figure 7. The result is the game graph for 7-stick Nim—not for the version that we have considered so far, but for a version with the rules changed so that the player who takes the last stick wins.

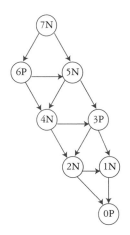

FIGURE 7 The game graph for normal-play 7-stick Nim: The player to move last wins.

The version of Nim described earlier is usually referred to as the *misère-play* version (pronounced

"mee-zare"), because the player who plays last (takes the last stick) loses. Under the *normal-play* convention, a player unable to move loses, so the player who takes the last stick wins.

The game graph for the misère-play 7-stick Nim can be obtained by taking the game graph for normal play, deleting ("pruning") all of the leaves (for 7-stick Nim, the only leaf is the 0 node), and applying the labeling algorithm to the remaining subtree. You should try doing this, and then compare your result with Figure 8. Notice that these labels are not just the reverse of those in the game graph for normal play in Figure 7.

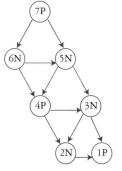

FIGURE 8 The game graph for misère-play 7-stick Nim: The player who can't move loses.

The accompanying Spotlight describes an industrial application of directed graphs.

1. Tchuka Ruma [Degrazia (1949)] is an Indonesian solitaire game (one that you play by yourself, with no opponent) played with holes and pebbles. There are five holes in a row, with two pebbles in each of the first four holes, while the last, called the *Ruma* (Indonesian for "house"), is empty:

hole number	4	3	2	1	R
number of pebbles	2	2	2	2	0

The object is to put all the pebbles into the Ruma. A move begins with picking up all the pebbles from any hole except the Ruma and "sowing" (placing) them, one per hole, into the next holes to the right. If you have pieces left over after dropping one into the last hole on the right (the Ruma), you put the next pebble into the hole at the far left and continue sowing from there to the right. If the last pebble lands in any hole other than the Ruma, you pick up all the pebbles in that hole and keep on sowing (you don't leave a pebble in that hole). If the last pebble lands in an empty hole, you lose. If the last pebble lands in the Ruma, you may choose any hole for your next move. You win if you get all the pebbles into the Ruma.

(a) Draw the complete game tree for Tchuka Ruma with $n = 4$ holes (apart from the Ruma) and $k = 1$ pebble per hole. Starting at the leaves of the tree, label each node of the tree as to whether it represents a position from which the player can win or not.

(b) Do the same for the game with $n = 4$ and $k = 2$. Mark the path through the tree that corresponds to winning; the path gives the strategy.

2. For the Tchuka Ruma game of Exercise 1 with $n = 4$ and $k = 2$:

(a) What is the maximum number of moves that a game can last?

(b) How many different possible arrangements are there for 4 pebbles distributed among the 4 holes, allowing 0 to 4 pebbles per hole?

(c) How many of these possible arrangements occur as positions in the course of playing the Tchuka Ruma game that starts with $n = 4$ and $k = 2$?

(d) How many different possible games are there? (Each path through the tree corresponds to a different game.)

3. Extend the strategy of Table 2 for 7-stick Nim to:

(a) Nim with 15 sticks;

(b) Nim with any number of sticks. (Hint: What is the pattern of the numbers of sticks that you leave your opponent?)

(c) Suppose that at some stage of the game, you are able to leave your opponent with $3n + 1$ sticks, for some n, i.e., a number that is equivalent to 1 modulo 3. Starting from such a position, if the opponent takes 1 stick, how many should you take? If the opponent takes 2 sticks, how many should you take? This description characterizes your strategy as an *answering strategy*, meaning that what you do depends on the immediate past move of your opponent. Explain why this particular strategy works.

4. For the normal-play convention:

(a) Reanalyze the strategy for 7-stick Nim;

(b) Extend your analysis to 15 sticks;

(c) Extend your analysis to any number of sticks.

5. One way to analyze Nim is to think in terms of *safe positions*. A position is safe for the player who has just moved to it if the subgame that the position represents has as its natural outcome a win for that player. In other words, the player whose turn it is to move cannot win against best play. The idea is to show that any move from a safe position leads to an unsafe position, and that from an unsafe position a player can move to a safe one.

    (a) What are the safe positions in Nim under misère play?

    (b) What are the safe positions in Nim under normal play?

    (c) What are the safe positions in Nim under normal play if a player may remove 1, 2, or 3 sticks?

6. The "classical" game of Nim, as described by Bouton (1901) (who gave it its name, from the German for "to take"), involves three rows (or heaps) of sticks (or objects of any kind). The number in each row is arbitrary. A move consists of a player selecting one of the piles and taking from it any number of the sticks, from one to the whole row. The one who takes the last stick wins (i.e., the normal-play convention).

    (a) Show that a player who first makes two rows equal, or first empties a row, can be forced to lose.

    (b) Starting with small numbers of sticks, determine what positions are safe in games with up to 7 sticks per row. (You will probably find it convenient to draw some trees!)

    (c) Safe positions are balanced in a certain sense. To discover what that balance is, write out some of the positions with the numbers of sticks in each row written in base 2 (i.e., binary notation; for example, writing the base that we are using as a subscript, we can say that the decimal number 6_{10} is the same as the binary number 110_2, which has a 0 in the 1's position, a 1 in the 2's position, and a 1 in the 4's position). Totaling for all three rows, how many 1s are there in the

4's positions? In the 2's positions? In the 1's positions? What is the general condition for a safe position?

    (d) Show that any move from any safe position leaves an unsafe position.

    (e) Show that from any unsafe position, it is possible to move to arrive at a safe position.

    (f) Does this characterization of safe positions work if the number of rows is different from 3?

7. Determine the safe positions and winning strategy for the misère form of Bouton's classical Nim of Exercise 6.

8. From your experience with Tic-Tac-Toe, you probably conclude that the natural outcome of that game is a tie.

    (a) What is the height of the game tree for Tic-Tac-Toe? (You can determine this without constructing the tree.)

    (b) Each playing of Tic-Tac-Toe is a path through the game tree. Give upper and lower bounds for the number of playings. (Many of these playings involve positions that are the same, but it is more difficult to calculate the number of different positions.)

    (c) The number of playings that must be investigated can be reduced by taking into consideration symmetries of the board. Show that there are essentially only three different opening moves.

    (d) Pursuing the essentially different responses by Second Player, show that after one move by each player there are only 12 essentially different positions. (The original game tree has 72 nodes at this point.)

    (e) Although it is not easy to show that a tie is the natural outcome of Tic-Tac-Toe, it is easy to show that "Second Player to win" is not. The key is to use the balancing idea from the strategy for multirow Nim, which here takes the form of an answering strategy. We proceed by contradiction and suppose that Second Player has a winning strategy. Pretend now that you are First Player and you know that strategy. As First Player, make your first move anywhere, then make your subsequent moves according to the winning strategy (the answering strategy for this game). Finish the argument by showing that you must arrive at a contradiction, so that the original supposition that Second Player has a winning strategy must be false.

9. Here is a game played on a 2 × 2 checkerboard. The game begins with a coin in the upper left square. Two players take turns moving the coin. There are two legal moves: one square down, or one square right. A player who moves off the board loses. Construct the game tree for this game, and determine its natural outcome.

10. Repeat Exercise 9, but for a 3 × 3 board.

11. Repeat Exercise 9, but for a 2 × 3 board. Can you see any connections between games of this type and Nim?

12. Verify by substitution that the expression given for P_n is indeed a solution to the difference equation $\Delta P_n = P_{n-1}$.

13. How many playings are there for the game
   (a) F_6?
   (b) S_8?

14. Analyze the following variations on Nim, for either normal play or misère play:
   (a) A player may take either 3 or 4 sticks, but if fewer than 3 remain, the player must take all of them.
   (b) A player may take no more sticks than the opponent did on the preceding move; the first player may take no more than half the sticks.
   (c) A player may take no more sticks than the square root of the number that the opponent

took on the previous move.
   (d) Start with an odd number of sticks. The winner is the player who possesses an odd number of sticks when all sticks have been taken.

15. The ancient Hawaiian game Konane is played on a black-and-white rectangular checkerboard. Play begins with a black piece on every black square and a white piece on every white square. The player who goes first removes a stone from a corner, the center, or a square next to the center. The other player removes an adjacent stone of the other color (it must the center stone if the first player took a stone next to the center stone). The players then play the pieces of the colors that they removed. A move consists of jumping (as in checkers)—horizontally or vertically, but not diagonally—and removing one or more stones of the other color. Multiple jumps are permitted provided the jumped stones lie in a straight line, each separated from the next by exactly one empty square. The first player unable to make a jump loses. Here we consider a tiny version of Konane, on a 4 × 4 board. (A traditional Konane board has from 8 to 20 rows and columns.)
   (a) Taking into account the symmetries of the board, how many distinct starting positions are there after removal of the first two stones?
   (b) Play Konane with a partner on a 4 × 4 board, with each of you getting a chance to play first multiple times. Record who wins (First Player or Second Player). Since the game cannot end in a tie, its natural outcome must be either a win for First Player or a win for Second Player. What do your data suggest? Does it make any difference which stones are removed at the start?

16. The game of Toads and Frogs is played on a one-dimensional board. First Player plays the toads, which move only to the right, and Second Player plays the frogs, which move only to the left. A toad or frog may either move over one square or, if next to an amphibian of the other kind, jump over to the square immediately be-

yond—provided in each case that the square to move to is empty. The first player unable to move loses.

    (a) Play this game with a partner, with each of you getting a chance to play first multiple times, starting from the position:

    (b) How many positions are there in this game? Give an upper bound.

    (c) Construct the complete game tree and the slightly simpler game graph for this game.

    (d) Label the leaves of the tree, and back-propagate the labels to determine the natural outcome of the game.

17. **(Computer Project)** If you have access to a computer running Unix, request a copy of the Gamesman's Toolkit [Wolfe (1994)] (e-mail addresses for requests are given in the References), read the simple directions for installation, and install it (if you have trouble, you may need to consult a local expert). Explore positions in Konane (see Exercise 15) or in Toads and Frogs (see Exercise 16).

SECTION 5.5   *Assignment Algorithms*

Bipartite Graphs

Suppose that we have

➤ companies X, Y, and Z, all with at least three job openings, and

➤ applicants Alicia, Bob, and Chris, each of whom applies to all three companies for a position.

How can we represent the outcome of whatever hirings take place? We can think of it as a kind of assignment. In this case, if each person can work for only one company, we can think of the jobs as being assigned to the people. While trees and directed graphs are useful for analyzing games, another specialized form of graph is a key way to represent assignments.

Figure 1a represents the outcome that company X hires Chris, company Y hires Alicia, company Z does not hire any of the three people, and Bob does not get an assignment with any of the three companies. The condition that a person can work for only one company is illustrated by there being no more than one edge incident to any of the nodes representing the people.

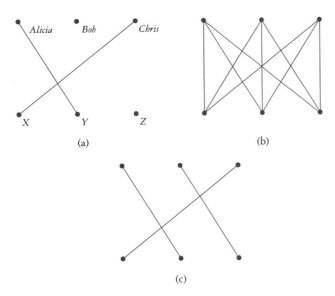

FIGURE 1 Examples of bipartite graphs, with the two subsets of nodes displayed in rows, so that each edge joins a node in the top row to a node in the bottom row. a. A bipartite graph. b. A complete bipartite graph. c. A matching.

DEFINITION

A *bipartite graph* is a graph whose nodes can be partitioned into two subsets, with no edge joining nodes that are in the same subset. A node in one of the subsets may be joined to all, some, or none of the nodes in the other. A bipartite graph is usually shown with the two subsets as top and bottom rows of nodes, as in Figure 1a, or with the two subsets as left and right columns of nodes.

Depending on the nature of the situation we are modeling, an assignment may be many to one (as in candidates to companies), one to one (if each company has only a single job), or many to many (as in students electing courses). Any assignment between two sets can be represented as a bipartite graph, and any such graph is a subgraph of a *complete* bipartite graph.

DEFINITIONS

If every node in one of the designated subsets of a bipartite graph is joined to every node in the other designated subset, then the graph is a *complete bipartite graph*. The graph of Figure 1b is a complete bipartite graph, while the graphs of Figure 1a and 1c are not.

An *assignment* between two sets S and T is any subset of $S \times T$, i.e., any set of ordered pairs whose first elements are from S and whose second elements are from T. An assignment can be represented as a subgraph of the complete bipartite graph with designated subsets S and T.

An assignment that is one-to-one is known as a matching.

DEFINITION

A *matching* in a bipartite graph is a set of edges with no endpoints in common. Figure 1c shows a matching.

Assignment Problems

We usually desire an assignment that is optimal in some sense while still meeting any necessary constraints. Here are some examples:

Each month the crews of an airline "bid" (apply) to work particular routes. The personnel and the routes

"Help Me, My Computer Is Down"

Most large companies have hundreds of employees using personal computers. What does an employee do when there's trouble with the computer? At Moody's Investors Services, a New York firm that evaluates the creditworthiness of companies, the employee calls a help desk.

Calls to the help desk include hardware and software problems that may range from being mildly annoying to preventing the employee from working. Some can be solved over the phone, but many require an on-site visit from a technician.

While problems vary in their urgency, technicians vary in their areas of expertise. The right kind of technician needs to be assigned to the problem, in a time frame that matches the urgency. Any scheduling algorithm, however, must also take into account how long a user has been waiting; a software upgrade should not be put off forever in favor of chronic urgent repairs.

Moody's Roger Stein, senior analyst in the Quantitative Analytics Group, uses a genetic algorithm to schedule the 30 technicians. Many feasible schedules are generated and evaluated for efficiency and for users' satisfaction. The best are used to produce a new generation of schedules, which in turn are evaluated for their fitness. The process is repeated until further generations produce no improvement.

The genetic algorithm software doesn't assign tasks to the technicians but suggests; any technician can change his or her recommended assignments. The software revises the current schedule every 10 minutes, based on updated information, technician availability, and technicians' preferences. The resulting schedules are not perfect; but they are good enough that management, technicians, and users are all pleased with the improved service.

form the two subsets of nodes. There are constraints, such as each plane needs a pilot. Assignments are made according to seniority to optimize satisfaction of longtime personnel.

Professional and college sports leagues need to devise season schedules. Pairs of teams are one kind of node and dates are the other. Depending on the sport, some pairs of teams may meet more than once (this

leads to a multigraph version of a bipartite graph) and others may not meet at all. In sports with frequent travel, like baseball, the sequencing of opponents for consecutive road games is done to minimize travel costs.

Every term students choose courses in the face of constraints (a student can't take two courses at the same hour, courses may have enrollment limits, etc.).

Optimizing the assignment by careful scheduling of course times is the subject of many algorithms and heuristics; Chavey et al. (1994) point to the literature and note factors that work against optimal scheduling.

In a similar way, we can also model assignment problems in which both parties have preferences:

A student applying to colleges has a preference ranking, and colleges accept students based on their own preferences for students. It would be ideal if each student went to as high-ranking (according to that student) a college as possible and each college enrolled the best class possible (according to the college's criteria). This is a "problem" that is "solved" every year, without any overall attempt at optimization.

A college must assign new students to dormitories, usually two to a room, with some attempt to meet preferences of the students (e.g., about whether the roommate smokes).

Each year graduates of medical schools are assigned to residencies at hospitals. In "The Match," as it is called, new doctors submit preferences to the assignment service, and hospitals submit preferences for particular doctors. Assignments are made according to an algorithm devised by hospital administrators. As we will see later, the algorithm has the very desirable feature that the assignment produced is *stable;* but it favors the hospitals' preferences over the doctors'.

The accompanying Spotlight discusses how genetic algorithms, which we described in Section 5.2,

are used in day-to-day solving of a changing assignment problem.

The Stable Marriage Problem

You can imagine (or may even have experienced) instability in an assignment of college roommates. Suppose that A and B are assigned as roommates, as are X and Y. However, A and X prefer to room together. If a switch is made, so that A and X room together and B and Y room together, everybody may be better off—or maybe not. Perhaps B and Y prefer their old roommates, or other people, to each other. Is it possible to find an assignment of roommates that would be stable, in the sense that no two people prefer each other to their assigned roommates?

Let us concentrate on a matching problem that, while artificial, captures the essential complexities of finding stable matchings.

The Marriage Problem

You have the honored position of being the matchmaker in a small, traditional ethnic community. Your task is to recommend suitable mates within the community for each of n men and women who wish to marry this year; in other words, you must match all of the n men with the n women. You base your recommendations on the preferences of the individuals, each of whom has given you a ranking of each of the individuals of the opposite sex, from 1 (most preferred) to n (least preferred), with no ties allowed. [Cultural notes: In some cultures you would instead use the preference rankings made up by the parents. Also, this problem is about arranging monogamous, heterosexual pairings; there are corresponding problems for homosexual and nonmonogamous pairings.]

We will draw serious conclusions from this whimsical fantasy.

Without some further goal or constraint, this isn't much of a problem: Just make any old assignment, as long as everybody gets matched up with somebody. The questions are: What are you trying to optimize? How do you take the preferences into account?

Well, you would be judged a very poor matchmaker if the matches you recommended didn't last! In particular, you want to avoid matchings that include an "unstable" situation, where a man and a woman who are married to other people would both rather be married to each other. Finding a stable matching, which is what you as a matchmaker want to do, is the *stable marriage problem.*

DEFINITIONS

A man–woman pair is a *blocking pair* for a matching if they are matched to others when both would rather be matched to each other. (Note, in particular, that a blocking pair for a matching cannot be matched in that matching.)

A matching is *stable* if there is no blocking pair for the matching (and is *unstable* if there is).

We now examine a particular situation to see examples of blocking and nonblocking pairs, and stable and unstable matchings. Table 1 shows preference rankings for a group of men and women. We use lowercase letters to stand for men's names and uppercase letters to stand for women's names. To make the discussion more personal, however, and es-

TABLE 1 Men's Preferences and Women's Preferences.

	Man				Woman			
	a	b	c	d	J	K	L	M
1st	J	M	M	K	a	d	a	c
2nd	K	L	L	J	c	b	d	d
3rd	L	J	J	L	b	c	b	a
4th	M	K	K	M	d	a	c	b

pecially for any classroom role-playing in simulating the algorithm, you may substitute names of your choice.

For the preference rankings of the table, the matching

$$a \,\&\, L, \qquad b \,\&\, J, \qquad c \,\&\, M, \qquad d \,\&\, K$$

is unstable, because **a & J** is a blocking pair. In fact, each is the other's first choice, but they are matched to others.

We can represent the situation of a blocking pair graphically. Figure 2a shows the preferences from Table 1 of man **a** (first column) and woman **J** (second column). The solid lines show actual matches in the

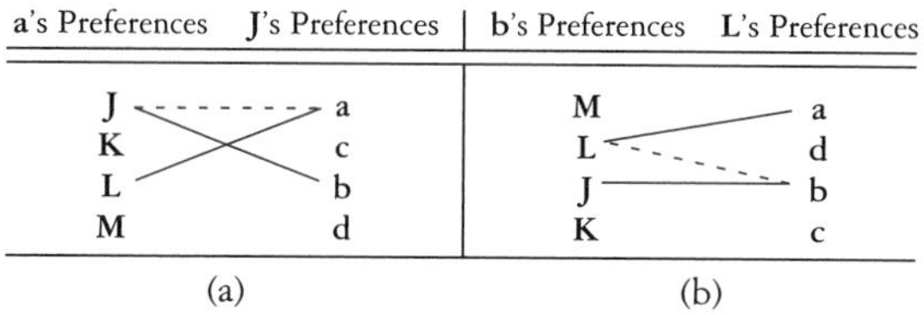

FIGURE 2 Graphical representations of the facts that: a. **a & J** is a blocking pair; b. **b & L** is not a blocking pair. The solid lines show actual matches in the matching, while the dashed line would match them to each other instead.

matching, while the dashed line shows what would happen if the man and woman in question were matched to each other instead. The fact that the dashed line lies above the two solid lines means that the man and woman are a blocking pair for the matching.

In Figure 2b we show what happens with a non-blocking pair, **b** & **L**. We set up the columns of preferences for **b** and for **L**; as before, we draw solid lines for the actual matchings of **b** and **L** and a dashed line between **b** and **L**. This time the dashed line lies between the solid lines. For a nonblocking pair, the dashed line will either lie between the solid lines or below them.

This is a case where a new stable matching can be made by matching **a** with **J**, matching up their former partners **b** and **L**, and leaving everyone else matched as they were, giving the matching:

$$\mathbf{a} \ \& \ \mathbf{J}, \quad \mathbf{b} \ \& \ \mathbf{L}, \quad \mathbf{c} \ \& \ \mathbf{M}, \quad \mathbf{d} \ \& \ \mathbf{K}$$

The major mathematical questions that arise for you as a matchmaker who wants to arrive at a stable matching are

How can you tell if a matching is stable?

Does a stable matching always exist?

If there is a stable matching, how can you find it?

These questions call for algorithms.

Is a Matching Stable?

To determine whether a matching is stable, we need to consider every possible blocking pair. There is no way around what amounts to a brute-force enumer-ation of all of the potential blocking pairs. How many are there? By the multiplication rule of Section 1 of Chapter 4, there are $n \times n = n^2$ pairs altogether. Of these, n are represented as matches in the matching and hence are not blocking pairs, leaving $(n^2 - n)$ pairs to check. To implement the checking in an algorithm, we can do an outside loop on the men, then for each man loop on all of the women with whom he is not matched. However, it is easier and not much less efficient simply to loop on all the women, rather than to check each time to see if the woman is the one to whom the man is already matched (see Algorithm 1).

We illustrate Algorithm 1 with the preferences of Table 1, for the unstable matching that we noted. We examine the men in alphabetical order, and the same for the women. We show the trace of the algorithm in Table 2. It ends so quickly only because the elements of the blocking pair, **a** and **J**, are both first in the ordering that we used (alphabetical). If they had been last, we would have gone through all of the possible combinations of men with women before finding out that the matching is unstable.

Is There a Stable Matching?

For some kinds of mathematical problems, it is possible to prove that a solution exists without showing how to find one. Such is the case with the theorem that a polynomial of odd degree whose coefficients are real numbers has a real root, and with the theorem that between two points where a polynomial changes sign there must be a root (you used these theorems when you found roots of polynomials in your study of algebra). An *existence proof* satisfies some of our curiosity and may be an important base for further theory; it certainly gives further motivation

```
Inputs: preference lists for each man and woman, and a
         matching
Output: prints a blocking pair or prints "match is stable"
Precondition: each list includes each potential partner once
Postcondition: output statement is correct
```

```
Stable ← true {no blocking pair found yet}
for each man m
       for each woman w
              if ((m prefers w to his current partner) and
                  (w prefers m to her current partner)) then
                     write "m and w are a blocking pair"
                     Stable ← false
                     exit loops {no point continuing}
              endif
       endfor
endfor
if Stable then
       write "match is stable"
endif
```

ALGORITHM 1 Algorithm to check if a matching is stable.

TABLE 2 Traces of the Algorithm for Checking if a Matching Is Stable, for the Unstable Matching a & L, b & J, c & M, and d & K.

Stable	man m	woman w	m perfers w to current partner?	w prefers m to current partner?
true				
true	a			
true	a	J		
true	a	J	true	
true	a	J	true	true
false				

to searching for an algorithm, since it guarantees that there is something to find.

Better than an existence proof is a *constructive proof,* one that uses an algorithm to produce a solution. We now give a constructive proof, via proving properties of an algorithm, that there is always a stable matching for any instance of the stable marriage problem.

The Gale–Shapley Algorithm for a Stable Matching

One way to find a stable matching, if there is one, or to be sure there isn't one is to proceed by means of our old friend, brute force: We "just" enumerate all of the possible matchings and check each one for stability. We've already examined the matter of checking a single matching; it is relatively simple and takes a number of steps roughly proportional to n^2.

The nemesis of brute force, however, is multitude. Just how many matchings are there? We can pair the first man with any of the n women, the second, with any of the remaining $(n - 1)$ women, and so forth: By Section 3 of Chapter 4, there are $n!$

matchings, a number that grows very fast. Even for $n = 10$, brute force is a poor approach to take, since $10! = 3,628,800$.

Gale and Shapley (1962) published the following algorithm for the stable marriage problem; unknown to them, a generalization of it had been used in the medical residents' matching since 1952.

The algorithm proceeds by rounds. Each round consists of two parts: Men make proposals of marriage, then women reject or (tentatively) accept. We will see that reversing these traditional roles can produce vastly different results.

Gale–Shapley Algorithm for Stable Matching

In the first round each man proposes to the woman whom he most prefers, even if someone else has already proposed to her. Then, from the proposals that she receives, each woman tentatively accepts the proposal from (becomes engaged to) the proposer whom she prefers the most; she rejects all the other proposals. A woman who does not receive any proposals waits for the next round.

In each subsequent round men who are currently engaged do nothing. Each man who is not engaged makes a new proposal, to the woman highest in his preference ranking who has not already rejected him, whether or not she is already engaged. In the women's part of the round, a woman accepts the proposal from the man highest in her ranking, rejecting all others and (if necessary) breaking her current engagement to become engaged to a man higher in her ranking. A woman who does not receive any proposals in this round waits for the next round.

As long as there are unengaged men at the end of a round, conduct another round.

We give a first-level pseudocode for a typical round of the Gale–Shapley algorithm in Table 3.

TABLE 3 First-level Pseudocode for a Typical Round of the Gale–Shapley Algorithm.

Inputs: preference lists for each man and woman
Output: a matching, consisting of a list of engaged pairs
Preconditions: each list includes each potential partner once; same number of men and women
Postcondition: a matching is produced that is stable

```
for each unengaged man
     send proposal to highest-ranked woman
        not yet proposed to
endfor
for each woman
     get engaged to most preferred man who has
        proposed, if any
endfor
```

We expand on this and give more detailed pseudocode in Algorithm 2.

Even the specification in Algorithm 2 is not completely satisfactory as a solution for our problem. The algorithm has three potential deficiencies, which need to be remedied by proofs:

1. It is not clear, without some argument, that the algorithm ever terminates.

2. It's a clever procedure, all right, but to solve our problem, we must be sure that it produces a matching and that the matching is stable.

3. The description is not completely deterministic. Especially for implementation on a computer, we should specify the order in which the men make proposals and the order in which the women act. We then need to wonder if the order is going to make any difference in the resulting matching; perhaps us-

Inputs: preference lists for each man and woman
Output: a matching, consisting of a list of engaged pairs
Preconditions: each list includes each potential partner once; same
         number of men and women
Postcondition: a matching is produced that is stable

```
for each man m
   engaged(m) ← false
endfor
for each woman w
   engaged(w) ← false
endfor
while there is man who is not engaged
   for each man, m
      if m is not yet engaged then {m springs into action}
         w ← highest on m's list to whom m has not yet proposed
         add w to m's list of women proposed to {m proposes to w}
   endfor
   for each woman w {engaged or not}
      if no proposers then {do nothing}
      else
         m ← highest among proposers to w {identify best suitor}
         if w is not engaged then {engage to best suitor}
            engaged(m) ← true
            engaged(w) ← true
            add (m,w) to list of engaged pairs
         else {w is already engaged}
            if w prefers m to current fiancé m' then
               {w dumps m' and engages to m}
               engaged(m') ← false
               delete (m', w) from list of engaged pairs
               engaged(m) ← true
               add (m, w) to list of engaged pairs
            else {w rejects m, so m remains unengaged}
            endif
         endif
      endif
   endfor
endwhile
write list of engaged pairs
```

ALGORITHM 2 The Gale–Shapley algorithm for a stable matching.

ing the "wrong" order could affect the possibility of getting a stable matching.

In other words, we must show that the Gale–Shapley procedure has the properties of an algorithm of being finite, effective, and deterministic. We settle these questions next.

Properties of the Gale–Shapley Algorithm

The Gale–Shapley algorithm terminates finitely. No man proposes to the same woman twice, so each man can make at most n proposals. Altogether, all of the men together can make no more than $n \times n = n^2$ proposals. In each round at least one proposal is made; so there can be at most n^2 rounds.

It produces a matching. When the algorithm terminates, all the men are engaged. Since each man is engaged to exactly one woman, and there are exactly as many women as men, we have a matching.

The matching is stable. If an arbitrary man prefers another woman to the one to whom he is matched, then he must have proposed to that other woman in some round. She must have rejected him because she preferred someone else. Hence, the man and woman in question cannot be a blocking pair. But since the man was arbitrary, and the woman was any woman whom he preferred over the one to whom he was matched, we have shown that there are no blocking pairs in the matching.

The matching produced is the same, no matter how we resolve the nondeterminism in the statement of the procedure. No man's proposal is affected in any way by what another man does. Likewise, no woman's choice is affected by what other women do. So, in each round it doesn't matter in what order the proposals are made or in what order the women make their choices.

No man could be better off in any other stable matching than he is in the Gale–Shapley one. Put another way: In the Gale–Shapley matching, each man is as well off as he could be in any stable matching. This doesn't sound so bad, though the proof is a bit complicated (and we omit it). Instead, assuming this property, we next consider and prove the corresponding downside.

Each woman winds up as badly off as she could possibly be in any stable matching. To prove this depressing result, we proceed by contradiction. Suppose that there is a stable non-Gale–Shapley matching under which **J** does worse than under Gale–Shapley (GS): She is matched in this other matching to **d**, whom she prefers even less than **b**, to whom GS matches her. Meanwhile, **b** is matched to **J** by the GS matching but to **M** by this non-GS matching. In Figure 3 we illustrate the GS matches with dashed lines and the non-GS matches with solid lines.

We have two cases:

1. **b** prefers **J** to **M**. Then we have the situation of Figure 3b. The dashed line in the figure depicts matching **b** to **J**. Because the dashed line lies above the two solid lines, **b** & **J** is a blocking pair for the non-GS matching and hence a contradiction to the supposition that the non-GS matching is stable.

2. **b** prefers **M** to **J**. Then we have the situation of Figure 3c. The non-GS matching matches **b** to someone (**M**) whom he prefers to his GS match (**J**); in other words, he is better off in the non-GS matching.

| b's Preferences | J's Preferences | b's Preferences | J's Preferences | b's Preferences | J's Preferences |

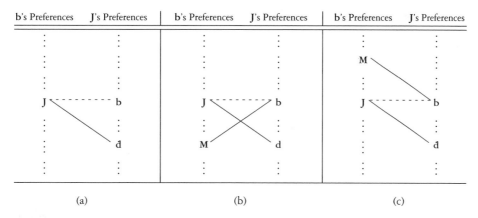

(a)   (b)   (c)

FIGURE 3 a. Under a hypothetical stable non-GS matching, **J** does worse (getting **d**) than under the GS matching (under which she gets **b**). b. The situation when **b** prefers **J** to **M**: **b** & **J** is a blocking pair. c. The situation when **b** prefers **M** to **J**: **b** is better off than in the GS matching.

This contradicts the fact that GS matches him to the best partner that he can have in any stable matching.

In either case, we reach a contradiction; so our supposition that there is a stable non-GS matching in which a woman does better than in the GS matching must be false.

What the last two properties show is that GS *is not a symmetrical algorithm.*

Among stable matchings, GS yields a matching that is the best possible for the group that does the proposing and the worst possible for the group that receives the proposals.

Of course, there are instances in which the same stable matching results when the roles of proposer and proposee are reversed; but in general, the two variations do not give the same matching.

The Roommates Problem

A natural generalization of the stable marriage problem is the *stable roommates problem.* We assume that each of n rooms is for two people and that we have $2n$ people. The people involved are not distinguished by sex (we allow for the possibilities of same-sex and opposite-sex roommates), but every one ranks every one else. Again, the goal is a stable matching.

The appropriate model is a complete graph on the $2n$ nodes corresponding to the individuals; it is not a bipartite graph, as we do not distinguish two kinds of nodes. The goal is to find a particular kind of subgraph.

You might think that, because the roommates problem eliminates a constraint of the stable marriage problem (that each pair must contain one member from each sex), solutions should be more plentiful and easier to find. However, our intuition leads us astray. An enormous difference between the two kinds of problems is that *there are roommates prob-*

Can We Make Matching Fair?

Interpreting the Gale–Shapley matchings back into the original context of a bipartite graph helps us to envision alternative approaches. There are two kinds of nodes: nodes for the men and nodes for the women. An edge between a man and a woman denotes that they have been matched, and a matching is a set of edges between men and women such that each man node is incident to one edge and each woman node is incident to one edge.

The complete bipartite graph would have an edge from each man to each woman. For the complete graph we could assign the man's preference ranking as a weight to an edge. What the GS algorithm does, with men as the proposers, is to generate a matching that, among all stable matchings, minimizes the sum of these weights (remember, the weight is 1 for the most preferred partner, up to n for the one least preferred).

Alternatively, we could use the women's preference rankings as weights. The GS algorithm with men as the proposers maximizes the sum of these weights (over all stable matchings) at the same time that it minimizes the sum of the men's rankings.

One way to take into account the preferences of *both* groups in a symmetric way is to label each edge with the sum of the man's and the woman's rankings. So, for example, if a man gives a particular woman a rank of 5, and she gives him a rank of 27, then we label

the edge between them with a weight of 32. Then we search for a stable matching that minimizes the sum of these combined weights; such a matching is called an *egalitarian stable matching.* Is there always an egalitarian stable matching? Yes. The proof is by means of a very complicated algorithm that constructs such a matching [Gusfield and Irving (1989), 128–133].

Another alternative is to change the system for evaluating potential partners from a ranking system to a rating system. For example, instead of ranking the potential partners from 1 (best) to n (worst), the participants could rate them on a scale from 1 (worst) to 10 (best), allowing any real number in between as a rating. The same approach as in egalitarian matchings can be used, this time maximizing the sum of the ratings by the men and the women over all stable matchings. Again, there is an algorithm for this problem, the *optimal stable matching* problem. Using ratings instead of rankings allows a person to express preferences more fully and hence may lead to matchings that offer greater satisfaction.

With a suitable assignment of weights (but based on rankings rather than ratings), the algorithm for an optimal stable matching can produce a stable matching under which the maximum number of men and women get their first choices (and among such matchings, the maximum number of remaining men and women get their second choices, and so on).

lems for which there is no stable matching (see Table 4). (The exercises investigate this and other examples.)

TABLE 4 Preferences in a Roommates Problem for Which There Is No Stable Matching.

	A	B	C	D
1st	C	A	B	any
2nd	B	C	A	any
3rd	D	D	D	any

Gusfield and Irving (1989, 165–183) give an efficient but very complicated algorithm to find a stable matching (or to determine that none exists) for the roommates problem.

Polygamy

The hospitals–residents problem can be thought of as a marriage problem in which polygamy is allowed by the hospitals. A resident can go to only one hospital, but a hospital may have more than one place available. In addition, not every person applies to (ranks) every hospital, nor does each hospital rank every student; we can consider that doctors or hospitals that do not appear on a preference list are unacceptable to the ranker.

There is a natural extension of GS to cover multiple partners. This is the algorithm used in The Match, with the hospitals in the proposing role. As in the single-partner situation, it can be shown that the stable matching produced favors the hospitals. Either a hospital fills all of its places with the best

residents that it could have, or else it has unfilled places that could not be filled under any stable matching. A resident is assigned to the worst hospital that the resident could have under any stable matching.

If the roles of residents and hospitals were reversed by having the residents as proposers, then each resident would be matched with the highest-ranked hospital that the resident could have in any stable match, and all of the hospitals would prefer any other stable matching.

Some remarkable results can be proved about what happens in any stable matching of residents and hospitals:

➤ A hospital receives the same number of residents;

➤ The same residents receive no assignment;

➤ Any undersubscribed hospital (one that does not get its full number of residents) receives the very same residents.

But what about students applying to colleges? The students seem to "propose" to the colleges for admission. The applications, however, can be thought of as just probes to see if a student is on a college's "acceptable" list. Presumably, all of the colleges to which a student applies are acceptable to the student. That the colleges are the proposers is clear when we realize that the students are the ones who can reject a college if a better offer comes along.

The accompanying Spotlight looks into the question of whether there is a way to make matching fair to both sides.

1. Apply the Gale–Shapley algorithm to find a stable matching for the preferences in Table 1, with the men taking the role of proposers.

2. Repeat Exercise 1, but with the women as proposers. Do you get the same matching as in Exercise 1?

3. Consider the following preferences:

	Man				Woman			
	a	b	c	d	J	K	L	M
1st	L	M	L	J	d	a	b	b
2nd	J	K	M	K	c	d	a	a
3rd	K	J	K	M	a	c	d	d
4th	M	L	J	L	b	b	c	c

Determine if the matching a & J, b & M, c & L, and d & K is stable.

4. Apply the Gale–Shapley algorithm to find a stable matching for the preferences in Exercise 3, with the men taking the role of proposers.

5. Repeat Exercise 4, but with the women as proposers.

6. Can the two versions of the Gale–Shapley algorithm, one with men proposing and the other with women proposing, ever produce the *same* matching for more than three men and three women? If so, give an example with at least four men and four women; if not, give an argument why not.

7. Here is another possible approach for an algorithm to find a stable matching: Start with any matching. Check if it is stable; if so, we are done. If it is not stable, select a blocking couple, match them to each other (thereby removing the block), and match their former partners to each other. Continue removing blocking couples until you arrive at a stable matching.

    (a) Apply this procedure to the preference rankings of Exercise 3, using the matching given there as the initial matching.

    (b) Comment on the procedure as an algorithm for the stable matching problem.

8. Is it possible for a man to get his last choice in the version of the Gale–Shapley algorithm in which men propose? Is it possible for two or more men to get their last choices?

9. Show that the situation of Figure 2 has no stable matching.

10. The handbook for The Match of residents and hospitals assures the medical students enrolling in the matching program that "You will be matched with your highest-ranked hospital that offers you a position" [National Resident Matching Program (1991)]. Discuss this claim.

11. It would be a great shame if the assignment of, say, students and colleges produced situations where, if two students accepted at different colleges changed places, they would both be happier (realize higher preferences) and the colleges would be at least no worse off (in terms of their preference rankings). Such an assignment would be unstable, in the sense that the individuals involved would have a motive to try to change the situation. Can this situation happen?

SECTION 5.6 *Heuristic Algorithms*

You may be surprised to learn that there are problems that no algorithm can solve. We give a few examples of historical significance, followed by examples of contemporary importance.

Unsolvable Mathematical Problems

Perhaps the most famous example of an unsolvable problem dates from before the time of Euclid (ca. 300 B.C.). The ancient Greeks believed that ratios of integers provided the basis for the reality around them, from lengths in geometry to musical scales. They were frustrated, however, at trying to find a way to express as a ratio of integers the length of the hypotenuse of a right triangle whose legs each measure 1. Eventually they were able to prove (and Euclid's *Elements* gives the proof) that there is no way to express $\sqrt{2}$ as a ratio of integers; today we say that $\sqrt{2}$ is irrational. Hence the search for an integer ratio, and for an algorithm to find it, had been in vain.

Similarly, mathematicians in the Middle Ages tried to find analogs of the quadratic formula for polynomials of higher degree. By the end of the sixteenth century, formulas for the zeros of cubic (third-degree) and quartic (fourth-degree) polynomials had been found. By the beginning of the nineteenth century, 250 years had elapsed without further substantial progress. Niels Henrik Abel (1802–1829) was able to show in 1826 that there are no formulas for the zeros of a general polynomial of degree five or higher. Of course, it is possible to find the zeros of a polynomial of any degree, to any desired degree of accuracy, by using approximation algorithms, such as the bisection algorithm of Section 10 of Chapter 1.

In 1900 David Hilbert (1862–1943), the leading mathematician at the turn of the century, set forth 23 problems that he felt would profit mathematicians to explore in the twentieth century. Hilbert's Tenth Problem asks for a general procedure to determine whether a *diophantine equation*—a polynomial equation with integer coefficients in several variables—has any solutions in integers. An example of such an equation is $x^2 - 3y^2 = 1$, which has the solutions $(\pm 7, \pm 4)$ (and others, too); a simple diophantine equation that has no solutions is $2x^2 + 3y^2 = 1$. The algorithm that Hilbert sought was not one to find integer solutions (if they exist) but one to tell whether or not there are any. The algorithm would have to be general—it would have to work on any diophantine equation, not just for special cases. In 1970 a Russian graduate student, Yuri Matiyashevich, proved that there is no such algorithm, meaning that Hilbert's Tenth Problem is unsolvable.

Turing Machines

Some remarkable examples of the nonexistence of algorithms involve the basic prototype model of a computer, the *Turing machine,* named after the pioneering computer scientist Alan Turing (1912–1954). The accompanying Spotlight tells a little about him.

Despite the simplicity of their description, Turing machines are capable of any calculation that modern general-purpose computers can do. Turing realized this, and he formulated the idea as what has become known as Turing's thesis.

Alan M. Turing (1912–1954)

Alan Turing showed early extraordinary ability in mathematics. In 1937, after graduating from Cambridge University, he published the paper that introduced what is now known as a Turing machine. The paper led to his invitation to do research at Princeton University, where he received a Ph.D. in mathematical logic in 1938 (on a completely different topic from Turing machines).

Alan Mathison Turing (1912–1954)

Turing returned to England, where he spent World War II working at Bletchley Park on decoding intercepted German and Japanese messages. In particular, he was involved with breaking the German Enigma cipher and he designed a machine to automate decoding.

After the war Turing worked in a British government laboratory on designing an electronic computer.

He also was interested in numerical analysis, and he published a paper showing that Gaussian elimination is not greatly affected by roundoff errors (i.e., by the fact that a computer cannot exactly represent most numbers but must round them off). Turing later worked for Manchester University and became a Fellow of the Royal Society. The most prestigious award of the Association for Computing Machinery, the main professional organization of computer scientists, is named after Turing. The award, a prize of $25,000, is given for technical contributions to computer science.

In the early 1950s Turing was arrested and convicted in Britain for homosexual activities; his sentence included hormone treatments. He died at age 41 in what was either a laboratory accident or (more likely) a deliberate suicide. *Breaking the Code,* a play by Hugh Whitemore (1987), dramatizes some of what must have been Turing's anguish.

DEFINITION

Turing's thesis proposes that anything that one could plausibly mean by an "effective procedure" can be accomplished by a Turing machine.

In other words, what can be computed is what can be computed by Turing machines. Indeed, all other formulations of what it could mean to compute have proven to be equivalent to computing with Turing machines. Thus, we can consider Turing machines as "model" computers.

DEFINITION

A *Turing machine* consists of a movable read/write head, a storage tape, and a control (or program). The tape is one-dimensional, infinite in both directions, and discretized into cells, over which the head moves and on which it reads and writes (see Figure 1).

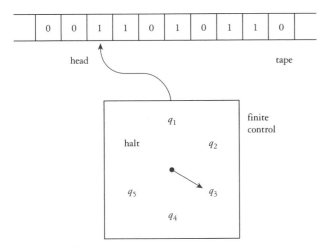

FIGURE 1 The components of a Turing machine. [Figure courtesy of Bryant A. Julstrom of St. Cloud State University.]

The head can read, write, and erase symbols from a finite alphabet. The simplest alphabet would have just a single character, which could be present or absent in each cell. Denoting its presence by 1 and its absence by 0, we may regard each cell of the tape as containing one of the two symbols, 1 or 0.

The control consists of a finite number of *states*, $q_1, \ldots, q_n$, and halt, each of which prescribes actions for the machine. Part of the action involves reading and writing, and part is to go to some other state. The ensuing state does not have to be the next one in the sequence of the q_i's (or in the clocklike order shown in Figure 1) but can be any state whatsoever. In general, the greater the number of states, the more the machine can do.

Each cycle of the machine consists of reading the symbol under the head of the tape, writing a symbol to that cell, moving the head one cell to the right or to the left, and entering a new state (which may be the same as the previous state). The symbol the machine writes, the direction it moves, and its next state are all determined by what it reads and by the current state. The machine has no way of knowing what symbols are on the tape to the left or right of where its head is unless it moves the tape to such a location and reads what's there. Furthermore, this machine has no memory and hence cannot recall anything that it read in the past.

Every Turing machine has a halt state; on entering that state, the machine stops. However, there is no requirement that the machine ever enter the halt state in the course of a computation—it could just keep on moving forever.

We regard the control as the program (or algorithm), the initial contents of the tape as its input, and whatever is written on the tape when the machine halts as the output. A particular Turing machine is not necessarily like a modern, general-purpose digital computer, which can run many different kinds of programs; a Turing machine runs just one program, as determined by its control. However, there are Turing machines that can perform any calculation that any Turing machine can do. These *universal* Turing machines have ability equivalent to any general-purpose digital computer.

One way to describe the control of a Turing machine is to exhibit a state table of the actions of the machine (what it writes, where it moves, what state it enters) for each of the combinations of each state with each possible symbol read. By the multiplication principle of Section 1 of Chapter 4, the number of entries in such a table will be twice the number of states, since there are two possible symbols (0 and 1). Alternatively, a Turing machine can be represented as a directed graph, known in this context as a transition diagram.

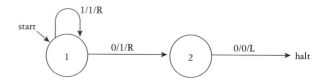

FIGURE 2 Transition diagram for a Turing machine that implements the function $f(n) = n + 1$. [Figure courtesy of Bryant A. Julstrom of St. Cloud State University.]

D E F I N I T I O N S

A *state table* is a representation in table form of the actions (program) of a Turing machine.

A *transition diagram* is a representation as a directed graph of the actions (program) of a Turing machine.

We now give an example of a simple Turing machine, describing it in terms of both a state table and a transition diagram.

E X A M P L E 1   A Turing Machine to Add 1
This machine implements the function $f(n) = n + 1$, for n a positive integer. The machine starts on a tape with n consecutive 1s; when the machine halts, there are $(n + 1)$ 1s on the tape. A precondition is that the machine starts with its head over one of the 1s (otherwise, we would need a more complicated machine, which could search for the block of 1s). The machine moves to the right until it goes past the last of the 1s, writes a 1 over a blank and moves right, moves left, and then halts. Figure 2 shows a state diagram for such a machine, in the form of a di-

rected graph, with the states as nodes and the transitions between them as directed edges. There are two states, indicated by circles, in addition to the halt state. The labeling over the arrows indicates what the machine does. For example, the label "0/1/R" over the arrow from state 1 to state 2 means that if the machine is in state 1 and reads a 0, then it writes a 1, moves to the right, and enters state 2. The alternative, "1/1/R," indicates that if the machine is in state 1 and reads a 1, it writes a 1, moves to the right, and stays in state 1 (this is our only use of a loop in a graph).

To make sense of the transition diagram, you should pretend to be the Turing machine and simulate what it does. You start at "start" with the head over one of the consecutive 1s and in state 1. You read the symbol under the head and take the corresponding path from state 1.

Notice that there is no arrow in the transition diagram from state 2 for the case when a 1 is read, as this case cannot occur, provided the precondition holds that the machine starts with its head over a 1.

The state table in Table 1 provides an equivalent description of the machine, with "h" denoting the halt state. The machine begins in state 1. There is no entry for being in state 2 and reading

TABLE 1 State Table for the Turing Machine of Figure 2.

	Symbol read					
	0			1		
State	Write	Move	Enter state	Write	Move	Enter state
1	1	R	2	1	R	1
2	0	L	h	—	—	—

a 1, as this situation should not occur. When such a situation occurs in a more extensive programming environment, we would have the computer enter an error state and halt. We do not need to list what the machine does in the halt state, as it just stops. ▲

Noncomputable Functions

Adding 1 is no great feat. How complicated a calculation is a particular Turing machine capable of? The number of states that it has gives a rough measure of how complicated it is (you could think of the states as analogous to brain cells). Another way that we can compare Turing machines is to take each one and feed it a blank tape. For each machine that halts, we count the number of 1s on the tape when it halts (we are not interested in machines that do not halt, even though they may produce an infinite number of 1s, as they are not useful for any practical purpose). Since a machine with more states ought to be able to write more 1s, a fair comparison would be among machines with the same number of states, which leads us to the *Busy Beaver Problem.*

A similar problem asks for the largest number $S(n)$ of tape moves that a halting Turing machine with n states can make, starting from a blank tape; $S(n)$ is called the *shift function.*

Why do we bring up these problems? Because of the surprising fact that both of these functions are *noncomputable*—there is no algorithm for either one. This means that there can be no computer program that, given n, evaluates $\Sigma(n)$ or evaluates $S(n)$. Just as was true for roots of polynomial equations of degree more than four, there is no formula that can be written for $\Sigma(n)$ or $S(n)$. The reason, though, is different; both of these functions *grow faster than any formula that can be written* with the usual arithmetic operations. In fact, no one knows the value of $\Sigma(5)$ (at least 4,098) or of $S(5)$ (at least 47,176,870) [Julstrom (1993)].

There is a similar negative answer to another famous problem, the *Halting Problem.*

The undecidability of this very basic question explains why there can be no test for whether a com-

Checking for Computer Viruses

A computer virus is a program that, when run on some input (e.g., a particular date of the year), alters the code of the operating system of the computer. The altered operating system may spread the virus by inserting the code for the virus into other programs, as well as perform other actions not ordinarily intended by the user (e.g., crash the machine, erase the hard disk, etc.). Some specialized computers, which store the operating system in read-only memory, are not susceptible to viruses, but most personal computers and mainframes are at risk.

Available antivirus utilities look for viruses that have already been spread or for unusual activity in the computer's operations. The ideal program would detect all possible viruses, including ones that have yet to be developed or spread.

Let us call a program *safe* if no matter what input it is given, the program does not alter the operating system. Suppose that there is an ideal virus-detection program Safe, which can determine for any program P whether P is safe on input I. (We will show that supposing that it exists leads to a contradiction.) We have

$$\text{Safe }(P, I) = \begin{cases} \text{true,} & \text{if P is safe on input I;} \\ \text{false,} & \text{otherwise.} \end{cases}$$

Notice that a program file, considered as text, can serve as data (input) to another program or even to itself; so we can consider situations like Safe (P, P), asking whether P is safe when we run it with a copy of itself as input.

Now we write a program Deceptive that, given a program P as input, makes a call to Safe as a subprogram and then does something devilish:

$$\text{Deceptive }(P) : \begin{cases} \text{alters the operating system, if} \\ \quad \text{Safe }(P, P) \text{ is true;} \\ \text{writes ``No virus found'' otherwise} \end{cases}$$

We ask the curious self-referential question, is Deceptive safe when run with a copy of itself as input?

- Let us first suppose so. Then, by the definition of Deceptive, the subprogram Safe must have returned the value "false" for the pair of arguments (Deceptive, Deceptive). So Safe *is incorrect.*

- Suppose, on the other hand, that Deceptive does alter the operating system when run on itself, i.e., it is not safe. This could happen either by 1) the "calling program" Deceptive doing the alteration or 2) by the subprogram Safe doing it. We examine each possibility:

1. Safe returned "true," so that Safe *is incorrect.*

2. Safe returned "false," in which case Safe is correct and Deceptive is unsafe on itself. Since the only two statements in Deceptive are a call to Safe and a write statement, the call to Safe must have altered the operating system: i.e., Safe *is itself unsafe.*

This clever argument by William F. Dowling (1989, 1990) shows that a general virus-checking program (algorithm) cannot be guaranteed to be safe. It also illustrates the value of computer science theory: Without knowing from a study of theory that a safe, general virus-checker is impossible, a programmer might waste time trying to write one.

puter program contains an infinite loop. This conclusion applies not just to Turing machines; by Turing's thesis (that everything computable is computable by a Turing machine) it applies to conventional computer languages, too, such as Basic, Pascal, and C. The accompanying Spotlight explains why there cannot be a program that detects all computer viruses without itself potentially acting like a virus.

Generalizing the Concept of Algorithm

Even if there is an algorithm to solve a particular problem, we may not be able to discover one or prove that it is effective, or it may take too long to execute to be practical.

When we come across a problem that we cannot solve, we often have to replace it by a "nearby" problem that we can solve. We are accustomed to doing this with approximation algorithms (e.g., for the root of a polynomial), since an approximation serves our purposes. In doing so, we give up on a guarantee that the approximation we find is an exact root, since it isn't. We give examples of two other kinds of problems in which we have to settle for solving a nearby problem instead: bin-packing and primality testing.

A Heuristic Algorithm for Bin-Packing

People moving their belongings from one place to another want to use as few car or truck trips as possible, and doing so depends on packing efficiently. Given the odd sizes of many belongings, it would

be hard to formulate an algorithm for packing them in the most efficient way. Bin-packing also arises, though, in the copying of computer files onto diskettes. If I want to copy 10 MB of files of varying sizes onto diskettes, each of which holds 1.44 MB (and no file is larger than that), am I likely to be able to get by with exactly 7 diskettes (because $7 \times 1.44 = 10.08 > 10$)? Possibly, but it's not very likely.

An important point is that I must copy all of a file onto the same diskette; I cannot split a file into pieces and put them on separate diskettes (though some sophisticated compression programs can do this). The worst I could do would be to put just one file on each diskette. What is the best I can do—the fewest number of diskettes that will suffice? How many I use will depend on how efficiently I use as much space as I can on each diskette.

With a large number of files, it would be too time-consuming to examine all possible combinations of ways to store the files onto diskettes. Another approach is the *first-fit heuristic:* Taking the files in the order presented, put each on the first diskette that has room for it. This is a heuristic algorithm; it is not guaranteed to find the optimal solution, though it may. Later in this section we will do an example of this heuristic algorithm and see that it does not do too badly. In fact, we will show that it cannot do too badly and that a simple improvement on it can be guaranteed to be even better.

DEFINITION

The *first-fit heuristic* for bin-packing: Take the items in the order presented, order the bins, and put each item into the first bin that has room for it.

"Packing" Rolls of Fabric

I work for Thomasville Upholstery in North Carolina, which makes custommade upholstered furniture. In early 1995 we opened a new automated warehouse with storage for more than 16,000 rolls of fabric.

Suppose that you order four chairs, a loveseat, and a sofa with matching upholstery. We estimate the yardage of material required, retrieve appropriate rolls, and cut the needed pieces. We want to get best usage from our rolls, discarding as little fabric as possible.

Author Baker with fabric rolls.

At the previous warehouse, all rolls of a particular material were piled together. Our pickers would look them over and pick out one or more, usually choosing larger rolls on top because they were easier to access. Cutters, who are paid according to speed, also favored going for "that nice big roll on top."

Now the pickers can no longer walk down a row and find what they want. Because of the time it takes to retrieve a roll (via radio-controlled forklift), we want to retrieve a roll as infrequently as necessary.

I had the task of designing the software to manage the new warehouse. I needed an algorithm to take pending orders and decide from which roll(s) to cut each order. I recognized that I faced a mathematical problem, similar to the bin-packing problem, except that we want to minimize not the number of rolls (bins)—the warehouse is huge!—but the total waste. We want to pack orders onto rolls so as to exhaust rolls as completely as possible.

Our problem is a version of the *knapsack problem:* You have a knapsack and items of various sizes, and you want a combination of items whose total size is as close as possible to the capacity of the knapsack. Like the bin-packing problem, the knapsack problem is NP-complete, meaning that there is no known algorithm to solve it that will execute in time polynomial in the size of the problem (the number of orders).

Our current strategy is a modification of the *worst-fit decreasing heuristic.* "Decreasing" means that we process pending orders from largest to smallest. "Worst-fit" means that we cut the largest order from the longest available roll. (Despite the name, it isn't the worst that we could do!) We keep cutting orders from that roll, repeatedly filling the largest remaining order, until we can no longer cut the next order. We then retrieve the next longest roll and cut, from largest to smallest, all of the consecutive orders that we can from that roll; and so on.

Devising the algorithm and designing the software were very challenging, as the work involved profitability of the company plus negotiating the interests of cutters, buyers, and inventory managers. Can we do better than our current strategy? No known algorithm is perfect for our problem, and we may need to adapt our strategy as we observe its performance.

—Nancy Baker, Programmer Analyst

EXAMPLE 2 A Heuristic Algorithm for Diskette-Copying For simplicity, we conceptualize the diskette-copying problem in the following way. We think of the diskettes as one-dimensional bins, all of the same size, which we take to be 1. The items (files) to be packed into these bins have lengths $x_1, \ldots, x_n$, between 0 and 1. We want to pack the items into as few bins as possible, without splitting an item into pieces and without any item poking out beyond the edge of a bin.

The first-fit heuristic calls for us to put x_1 into the first bin. For each succeeding x_i, we start again with the first bin and fit x_i into the first bin that has room for it or else—if there is no room in any of the bins used so far—we put it into a new bin. ▲

We mentioned that the first-fit heuristic is not too bad for this problem. In fact, we show the following:

Theorem 1. The first-fit heuristic requires at most twice as many bins as the optimal packing.

Proof. (This result may not sound terrific, but things could have have been worse!) The basic observation is that the first-fit heuristic can't leave two bins used but less than half full, because the heuristic would have placed the contents of the second into the first.

Every bin is at least half full, except possibly the last one. If every bin is at least half full, that's the same as saying that the contents occupy at least half the total capacity of the bins, which means that we could not have fit the contents into fewer than half as many bins. However, what if the last bin is less than half full? Suppose that first fit uses F bins when B_{opt} is the smallest number possible. Then the contents C satisfies $B_{opt} \geq C$, because all of the items fit into B_{opt} bins. With first fit, the first $(F - 1)$ bins are at least half full but don't include all the contents, so $C > (F - 1)/2$. Putting the two inequalities together gives $B_{opt} \geq C > (F - 1)/2$, or $B_{opt} > (F - 1)/2$, so $2B_{opt} > F - 1$, or $F < 2B_{opt} + 1$. Since F is an integer that is less than the integer $(2B_{opt} + 1)$, we must have $F \leq 2 B_{opt}$. So, whether or not the last bin is at least half full, the first-fit heuristic uses no more than twice as many bins as absolutely necessary.

▲ ▲ ▲

With a little extra effort, we can do even better. Instead of fitting the items in the order in which they are initially listed, sort them from largest to smallest before using the first-fit heuristic. This corresponds to the common strategy of packing the family car by putting the biggest items in first. Johnson et al. (1974) showed that the resulting heuristic, called *decreasing first-fit,* requires no more than $11B_{opt}/9 + 4$ bins, where B_{opt} is the minimum number of bins used by the best packing possible. We formulate this procedure in Heuristic Algorithm 1.

DEFINITION

Decreasing first-fit heuristic for bin-packing: Order the items from largest to smallest, order the bins, and starting with the largest item, put each item into the first bin that has room for it.

The accompanying Spotlight describes the roll that bin-packing plays in management of a computerized warehouse.

A Probabilistic Algorithm for Primality Testing

Large prime numbers are used in contemporary cryptography, so it is important to be able to determine if an integer of 100 or more digits is prime. The algorithm of trying every potential divisor would take far too long. Even more sophisticated mathematical algorithms for testing primality would still take too long. We need to relax our requirements on how strong a guarantee we are willing to accept. With a probabilistic algorithm that we describe shortly, we give up absolute mathematical certainty of the answer in exchange for quick computation and "almost certainty." We can make the probability of

---

Inputs: $x_1, \ldots, x_n$
Output: number k of bins of length 1 in which all the x_is fit
Preconditions: $0 \le x_1, \ldots, x_n \le 1$
Postconditions: $k \le 2 * B_{opt}$, where B_{opt} = minimum number needed; if
$\quad\quad x_1 \ge x_2 \ge \ldots \ge x_n$, then $k \le 11 * B_{opt}/9 + 4$

```
for i from 1 to n
        BinsUsed ← 0
        Bin ← 0 {denotes bin we try to fit xᵢ into}
        repeat
                Bin ← Bin + 1
                if xᵢ fits into Bin then
                        put it there
                endif
        until ((xᵢ is fitted into a bin) or (Bin = BinsUsed))
        if xᵢ doesn't fit into bins used so far then
                BinsUsed ← BinsUsed + 1 {put it into a new bin}
        endif
endfor
write (BinsUsed 'bins used')
```

HEURISTIC ALGORITHM 1 Bin-packing using decreasing first-fit.

getting a wrong answer as small as we like (though still nonzero). In proceeding this way, we relax the requirement on an algorithm that it be deterministic.

D E F I N I T I O N

A *probabilistic algorithm* for a decision problem is a procedure that produces an answer that has a prescribed probability of being correct.

E X A M P L E   3   Rabin's Probabilistic Algorithm for Primality-testing   In 1976 Michael O. Rabin published a famous probabilistic algorithm for testing whether a positive integer n is prime or not. For now we neglect the details and give you the big picture of how a probabilistic algorithm works. The basic ingredient in Rabin's approach involves making a test involving n on an integer a, with $1 \le a < n$, that is chosen at random. Rabin's test on a may report either that n is not prime (so n fails the test) or else that n may be prime (n passes the test, though we can't be sure that it is in fact prime).   ▲

D E F I N I T I O N

An integer a for which Rabin's test reports that n is not prime is a *witness* that n is not prime.

The following fact [from Rabin (1976)] about Rabin's test is the basis for the probabilistic algorithm.

Theorem 2.   If n is not prime, then at least three-fourths of the positive integers a, where $1 \le a < n$, are witnesses that n is not prime.

If n is not prime, then for at most one-fourth of the possible values for a, Rabin's test will report that n could be prime; for at least three-quarters of those values, n fails the test and a is a witness that n is not prime. Thus, if we take a value of a at random, there is at most a 25% chance that the test will say that n could be prime.

We can repeat the test with different random values for a. The probability that n will pass k tests with different random values for a when n is in fact not prime is less than $(1/4)^{-k} = 2^{-k}$ (the reason for this is that we can regard passing the test for different values of a as independent events to which the multiplication rule for probabilities of Section 4 of Chapter 8 applies). We can run the test as many times as we wish until we discover that n is not prime or until we reach a sufficiently small probability that a nonprime n would have passed that many tests.

We formulate this probabilistic algorithm in Algorithm 2. We go into the computational details of Rabin's test in Exercises 6, 7, 9, and 10.

Inputs: candidate n, probability tolerance ϵ
Output: "n is not prime" or "n is prime with probability $> 1 - \epsilon$"
Preconditions: n a positive integer, $\epsilon > 0$
Postconditions: none

```
Probability ← 1
repeat
      select a at random (without replacement) from 1, . . . , n
      apply Rabin's test using the value a
      if n passes the test then
            Probability ← Probability * 0.25
      else
            Prime ← false
      endif
until ((not Prime) or (Probability < ε))
if (not Prime) then
      write ('n is not prime')
else
      write ('n is prime with probability > 1 − ε')
endif
```

ALGORITHM 2 Probabilistic algorithm for primality, based on Rabin's test.

Exercises for Section 5.6

1. In Section 5.1 you encountered the following alleged algorithm for passing a true–false test:

Step 1. Bring a coin to the test.

Step 2. Flip the coin for each item on the test.

Step 3. If the coin lands heads, mark "true"; if it lands tails, mark "false."

Is this a probabilistic algorithm?

2. Here is a recipe for making a million dollars:

Step 1. Get four million quarters.

Step 2. Take the quarters to either Jackpot Junction or Oneida Bingo Hall & Casino.

Step 3. Play the slot machines until either you have eight million quarters or else you run out of quarters.

Step 4. If you run out of quarters, go back to step 1.

This approach certainly involves a great element of chance, but is it a probabilistic algorithm?

3. *Legendre's algorithm* provides an efficient way to calculate powers of a number, which Fermat's Little Theorem requires for its own efficiency. You probably discovered this technique yourself. To find, say, 3^{76}, it isn't necessary to multiply 3 by 3 a total of 75 times. Instead, the number of multiplications can be minimized by finding 3^2, multiplying that by itself to get 3^4, multiplying *that* by itself to get 3^8, and so on, working up by powers of 2. Of course, since 76 is not a power of 2, we need to do some further work as well.

(a) What powers of 3 in this sequence do we need to multiply together to get 3^{76}?

(b) Use Legendre's algorithm to calculate 11^7.

4. Formulate Legendre's algorithm of Exercise 3 in pseudocode, in the format used in this text. The algorithm should identify the needed powers, calculate them, and find a^n, for general positive integers a and n.

5. (**Computer Project**) Implement Legendre's algorithm of Exercise 3 in a computer language and on a computer of your choice. Is there any limit to the size of the integers that your program can handle?

6. The details of Rabin's test for primality begin with writing $(n - 1)$ as an odd number m times a power of 2:

$$n - 1 = 2^l m.$$

For example, for $n = 73$ we have $n - 1 = 72 = 2^3 \cdot 9$, so $l = 3$ and $m = 9$. The number m could be as small as 1 or as large as $(n - 1)$ itself (in which case $l = 0$). Having chosen a at random, calculate $a^m \pmod{n}$. If you get 1, report that n could be prime. Otherwise, square repeatedly and reduce mod n until you first get either

$-1 \pmod{n}$: report that n could be prime; or

$1 \pmod{n}$: report that n is not prime.

If you get to a^{n-1} without getting either (-1) or 1, report that n is not prime. [The explanation of why this works would involve more number theory than is appropriate for us to go into here, but you can find the details in Davenport (1992), 172–176.]

For example, let $n = 29$, so that $n - 1 = 28 = 2^2 \cdot 7$, so $l = 2$ and $m = 7$. Suppose that we choose a at random and get $a = 2$. Then we calculate

$$2^7 = 128 \equiv 12 \pmod{29},$$
$$12^2 = 144 \equiv -1 \pmod{29},$$
$$(-1)^2 \equiv 1 \pmod{29}.$$

Since we encounter $a - 1$ at the second step, we conclude that n could be prime.

(a) Apply Rabin's test to test whether $n = 5$ is prime, using $a = 3$.

(b) Repeat part (a), but with $a = 4$.

7. Apply Rabin's test of Exercise 6 to test $n = 6$, using

(a) $a = 3$

(b) $a = 4$

8. How many Rabin tests with random values of a are needed to determine whether a number is prime, if you want

(a) to be 99% sure of being correct?

(b) to be 99.99% sure of being correct?

9. Apply Rabin's probabilistic algorithm for primality of Exercise 6 to determine, with probability at least 99% of being correct, whether $n = 17$ is prime. (In these exercises we work with small numbers, which you already know to be prime or not, because of the great amount of calculation and the size of the numbers involved in testing numbers of even three digits.) For simplicity, use test numbers for a that you select at random from 00 through 15; you can produce such a random number by flipping a coin four times, with each flip determining a bit in the binary representation of the number. For example, letting a head correspond to a 1 and a tail to a 0, the sequence HHTT corresponds to the binary number 1100, which evaluates as

$$1 \cdot 2^3 + 1 \cdot 2^2 + 0 \cdot 2^1 + 0 \cdot 2^0 = 8 + 4 + 0 + 0$$
$$= 12.$$

If you get four tails, corresponding to 0, substitute 16 instead.

The test for $a = 12$ involves 12^{16}, a calculation that your calculator cannot do exactly. To avoid large numbers, use a modified version of Legendre's algorithm of Exercise 3, in which you immediately reduce mod 17 after evaluating a new power. For example, $12^2 = 144 \equiv 8 \bmod 17$, and so on.

10. Repeat Exercise 9, but for $n = 18$.

11. Pretend that you don't know whether 17 and 18 are prime.

    (a) From Exercise 9 you are at least 99% sure that 18 is not prime. Do your results allow you to be 100% sure?

    (b) From Exercise 9 you are at least 99% sure that 17 is prime. Do your results allow you to be 100% sure?

12. (**Computer Project**) Implement Rabin's probabilistic algorithm for primality testing in a computer language and on a computer of your choice. Is there any limit to the size of the integers that your program can test?

13. (**Computer Project**) Explore the capabilities of a computer algebra system available to you to determine whether an integer is prime or not. In particular, try to determine if 123,456,787 is prime.

    (a) In Mathematica 2.2, look into the built-in function `PrimeQ` and the function `ProvablePrime` in the package `NumberTheory'PrimeQ'`. (The function `ProvablePrime` prints a certificate that can be used to verify the result.)

    (b) In Maple V, the function `isprime(n)` returns false if n is shown to be composite and true otherwise (but the actual operation of the function depends on which release of the package that you are using). The function `isprime(n, t)` returns similar values, after doing t iterations of the Rabin primality test, where t must be no larger than 25.

14. Why is primality testing important today?

15. Fit into bins of size 10 items of size 4,7,9,1,1, 2,1,3,2, using

    (a) the first-fit heuristic

    (b) the decreasing first-fit heuristic

    (c) What is the least number of bins that will do?

How do the results of the previous parts of the problem compare with the guarantees for those heuristics, as given in the text?

16. (**Computer Project**) Implement either the first-fit or the decreasing first-fit heuristic for bin-packing in a computer language and on a computer of your choice, and explore how well the heuristic does on different sets of data.

17. The Turing machine in Figure 3 is started on a blank tape. What does this machine do? Will it ever halt? [Problem and figure courtesy of Bryant A. Julstrom, St. Cloud State University.]

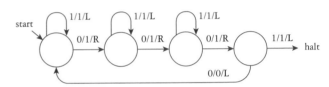

FIGURE 3 Turing machine for Exercise 17.

18. Design a Turing machine that will add two 1s to the number of consecutive 1s already on the tape. Describe the machine in the form of a transition diagram.

19. Design a Turing machine that will erase one 1 from the number of consecutive 1s already on the tape, leaving all remaining 1s consecutive. Describe the machine in the form of a transition diagram.

20. (**Computer Project**) Explore Turing machines using a computer program that simulates a Turing machine. A shareware program that is available for the Macintosh is Turing 1.0. (This program can be downloaded from many archives of Macintosh software, including `cs.beloit.edu` in directory `/MathDept-Quadra/Public/Macintosh/EducationalSoftware/ComputerScience/turing-machine.`)

Chapter Summary

Graph theory provides useful models for many kinds of common problems, from delivering meals to assigning roommates. A part from graphs in general, we have examined two special kinds of graphs, trees, and bipartite graphs. Graphs provide the data structure for a model of a situation, but solving a problem requires designing and analyzing algorithms and heuristics. For some problems, no algorithm exists; for others, optimal algorithms are easy to find (but may be difficult to prove); for still others, we must make do with heuristics. Algorithms themselves, and indeed any form of computation, can be modeled by Turing machines.

In Chapter 6 we compare the efficiency of different algorithms for the same problem; in Chapter 7 we build a machine to implement the winning strategy for Nim.

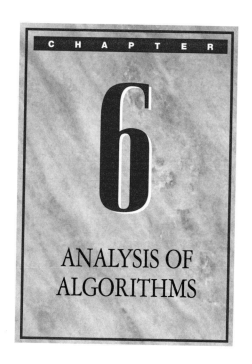

CHAPTER

6

ANALYSIS OF ALGORITHMS

Introduction

Often there is more than one way to get a job done. One important use of the combinatorics we learned in Chapter 4 is to help us evaluate how good an algorithm is—for example, how many steps it takes. Here is a simple example: Suppose you have a collection of 50 CDs and want to put them on a shelf in alphabetical order by album title. (Take time out to think about the process you would use to accomplish this before reading further.)

Here is how *not* to do the job: Make out all possible lists of the 50 album titles, with each list putting the titles in a different order. We know that each such list is a permutation, so there are 50!, or about 3×10^{64}, lists. Now look at each list to see

if the album names on it are in the desired order. When the right list is found, use it to put your actual CDs on the shelf. Have you stopped laughing? Of course, that is a ridiculous way to accomplish the job of ordering your CDs. But did you know that of the many, many shorter ways that people have used to order, or *sort,* a list, not all take the same number of steps?

There is a way to compare algorithms without relying on actually carrying out their steps; the technique depends on the concept of *time complexity*. Time complexity is a measure of how much work an algorithm must perform in order to carry out its various steps. By using time-complexity functions, we can often predict which of two algorithms would run faster without the overhead of coding and running them.

In the first two sections of this chapter we learn how to develop time-complexity functions relating the amount of work an algorithm does to n, the number of data items being processed. The variable n could be the number of nodes in a graph or the size of a data set to be put into alphabetical order. In Section 6.1 we develop the functions by using two basic principles to examine the algorithm. In Section 6.2 we use difference equations to describe the algorithm. This technique is very powerful, but it does require that we be able to solve those difference equations in order to get our functions.

Once you have some understanding of how time-complexity functions relate to specific algorithms, we will learn a pair of complementary techniques for comparing time-complexity functions. In Section 6.3 we use graphical methods in learning how to compare the functions for relatively small values of n. In Section 6.4 we learn about order of magnitude, a tool that tells us how functions behave for very large values of n.

SECTION 6.1 *Time-Complexity Functions*

Time complexity is a measure of how much work an algorithm requires. A time-complexity function for an algorithm is a rough attempt to measure how long it takes for the algorithm to solve the problem for which it was designed, to produce the output from the input(s). Time-complexity functions reflect the structure of the underlying algorithm, so different algorithms naturally have different time-complexity functions.

In Table 1 we consider some typical time-complexity functions. We assume that we have a computer that can carry out one million operations per second (a fast computer, but not the fastest available in the 1990s). The table entries give the time it would take for our computer to execute an algorithm of the given time complexity if the data set were of size n. Table 1 give us our inspiration for exploring time-complexity functions.

We notice that the functions in Table 1 are all increasing as n increases. This makes sense for time-complexity functions—you can't expect to process a larger data set in less time than it takes to process a smaller one.

In Table 1 we see that for functions like $\log_{10} n$ and monomials like n^3, the time to complete the algorithm is still within reason, seconds or minutes, when the data set contains 60 items. But for other functions, exponentials such as 2^n and the factorial function, $n!$, even a relatively small data set of 60 items would require *centuries* of processing.

It also seems that the functions have been listed so that the ones toward the bottom of Table 1 "grow faster" (from left to right as n increases) than those toward the top. We could speculate that there is a real difference among the functions listed and that multiplying any one of the functions by some constant, for example, multiplying n^3 by 589, would not make the resulting function grow faster than the next function, in this case n^5, at least not for very large values of n. Order of magnitude, which we ex-

TABLE 1 Time Needed to Execute an Algorithm of the Given Time Complexity on a Data Set of Size n Using a Computer Doing One Million Operations per Second.

Time-complexity functions	Size, n					
	10	20	30	40	50	60
$\log_{10} n$	0.0000010 seconds	0.0000013 seconds	0.0000015 seconds	0.0000016 seconds	0.0000017 seconds	0.0000018 seconds
n	0.00001 seconds	0.00002 seconds	0.00003 seconds	0.00004 seconds	0.00005 seconds	0.00006 seconds
n^2	0.0001 seconds	0.0004 seconds	0.0009 seconds	0.0016 seconds	0.0025 seconds	0.0036 seconds
n^3	0.001 seconds	0.008 seconds	0.027 seconds	0.064 seconds	0.125 seconds	0.216 seconds
n^5	0.1 seconds	3.2 seconds	24.3 seconds	1.7 minutes	5.2 minutes	13.0 minutes
2^n	0.001 seconds	1.0 seconds	17.9 minutes	12.7 days	35.7 years	366 centuries
3^n	.059 seconds	58 minutes	6.5 years	3855 centuries	2×10^8 centuries	1.3×10^{13} centuries
$n!$	3.6 seconds	771.5 centuries	8.4×10^{16} centuries	2.6×10^{32} centuries	9.6×10^{48} centuries	2.6×10^{66} centuries

Adapted from Garey and Johnson (1979), *Computers and Intractability: A Guide to the Theory of NP-Completeness,* New York: Freeman.

amine in Section 6.4, will confirm our suspicion that the differences among these functions are quite significant, more than can be overcome by multiplying by a constant.

Just Get a Faster Computer, or Won't That Help?

In Table 2 we see the effects of having faster computers on the size of a problem that can be run in one minute using algorithms of different time complexities. For example, line 4 of the table shows that if an algorithm has time complexity n^3 and can process a data set of size N_4 in one minute, then we see that if the same algorithm were run on a computer that were one million times faster, a data set of size $100N_4$ could be processed. Since we are considering one minute for calculations, the size of the problem that can be completed on a present computer is different for algorithms having different time-complexity functions in Table 2, hence the subscripts on the Ns.

The entries in Table 2 are quite dramatic, especially for the exponential and factorial functions.

TABLE 2 Growth of Size of Data Set Size That Could Be Processed in One Minute, for Algorithms Having Different Time Complexities and Three Different Computer Speeds.

Time-complexity functions	Present computer	Computer 100 times faster	Computer 1 million times faster
$\log_{10} n$	N_1	N_1^{100}	$N_1^{1,000,000}$
n	N_2	$100N_2$	$1,000,000N_2$
n^2	N_3	$10N_3$	$1000N_3$
n^3	N_4	$4.64N_4$	$100N_4$
n^5	N_5	$2.5N_5$	$15.8N_5$
2^n	N_6	$N_6 + 6.64$	$N_6 + 19.93$
3^n	N_7	$N_7 + 4.19$	$N_7 + 12.58$
$n!$	N_8	between N_8 and $N_8 + 3$*	between N_8 and $N_8 + 9$*

*The increase in the data set size in the case of an $n!$ algorithm depends on the size of N_8. In both columns, the larger increase corresponds to $N_8 = 2$, and for $N_8 \geq 100$ or $1,000,000$, depending on the column, there is no increase at all in data set size.
Adapted from Garey and Johnson (1979), *Computers and Intractability: A Guide to the Theory of NP-Completeness,* New York: Freeman.

As the saying goes, "A good (fast) algorithm on a slow computer is worth more than a bad (slow) algorithm on a fast computer."

How do these abstract time-complexity functions relate to the actual time complexity functions for some algorithms with which we are familiar? Prim's algorithm has time-complexity functions that behave like n^2. If tomorrow we could have computers that are 100 times faster than today's, then tomorrow we could apply Prim's algorithm to graphs that are 10 times larger than today's and do it in the same amount of time. For binary search, whose time-complexity function behaves like $\log n$, tomorrow's computer would allow us to search an enormously large data set, the size of today's raised to the power of 100. But what about the brute-force algorithm for the traveling salesperson problem, whose time-complexity function behaves like $n!$? Even if computers could be made one million times as fast as they are now, we could add at most nine nodes to the graph size and still do all the calculations. (You are probably wondering what the phrase "behaves like" means. We will learn its meaning in Section 6.4 when we learn about order of magnitude.)

In this section we develop time-complexity functions for some familiar algorithms. When we develop time-complexity functions, we focus on one or more elementary operations performed by the algorithm. Typical operations are arithmetic operations and comparisons. When computer implementation of an algorithm is considered, we could also consider assignment statements, which are data moves, and the overhead necessary for loop control. We can illustrate these possibilities in our first application: finding the scalar product (the output) of two vectors (the inputs).

Time Complexity: Scalar Product

E X A M P L E 1 Calculating the Scalar Product of Two Vectors by Hand We recall, from Section 2 of Chapter 3, the definition of the scalar product of two n-component vectors:

$$(x_1, x_2, x_3, \ldots, x_n) \cdot (y_1, y_2, y_3, \ldots, y_n)$$
$$= x_1 y_1 + x_2 y_2 + x_3 y_3 + \cdots + x_n y_n.$$

Counting the number of arithmetic operations is one way of expressing time complexity. What does a person do in calculating the scalar product of two n-component vectors? The person performs n multiplications and $n - 1$ additions, either by hand or using a calculator, for a total of $n + n - 1 = 2n - 1$ arithmetic operations. A time-complexity function, f, for finding a scalar product of two n-component vectors is $f(n) = 2n - 1$. ▲

Example 1 illustrates some important aspects of time complexity. First, since algorithms are designed to deal with input problems of all sizes, we describe the time complexity of an algorithm not by a number, but by a function of n, where n indicates the size of the problem being solved. Second, time-complexity functions count the number of occurrences of some "elementary" operations, not the actual time in seconds or some other time unit. Third, time-complexity functions often lump together counts of two or more different "elementary" operations, such as addition and multiplication, even though the actual time it takes either a person or a computer to carry out the different operations is not the same. This third aspect can introduce inaccuracies into the functions, and we will return to this issue later in this section.

It is important to recognize that we will be counting the *actual number of occurrences* of some elementary operation, such as an arithmetic operation or a comparison, in the execution of an algorithm. We will *not* be enumerating *total number of possibili-*

ties, as we did in Chapter 4, where we used the addition and multiplication principles. Since here we are counting repetitions of certain basic operations, we will develop some new guidelines for when to add and when to multiply.

E X A M P L E 2 Computer Calculation of a Scalar Product An algorithm for the computer calculation of the scalar product might look like this, where we use the notation `Vector1[i]` to denote the *i*th element of the first vector, namely x_i, and a similar notation for the second vector:

Input: n and two n-element vectors, Vector1
     and Vector2
Output: the scalar product of the vectors
Preconditions: n ≥ 0, n an integer
Postconditions: none

```
ScalarProduct ← 0        {Initialize "running
                            total"}
{One at a time, calculate the products and
 add them to "running total"}
For i from 1 to n
        ScalarProduct ← ScalarProduct +
        Vector1[i]·Vector2[i]
end for
```

When we analyze this algorithm, we find $2n$ arithmetic operations: one addition and one multiplication occurring in each of the *n* repetitions of the loop. Counting these arithmetic operations, we get a time-complexity function, *g*, for this algorithm as $g(n) = 2n$. We note that this function is not exactly the same as the $f(n) = 2n - 1$ we derived for hand calculation of the scalar product—the two expressions differ by a constant of 1. In Sections 6.3 and 6.4 we will focus on comparing two functions and learn which sorts of differences between functions are "significant" and

which can be ignored. (Do you think the difference between $2n$ and $2n - 1$ is significant?) ▲

How does a person decide what to count and what to ignore? In the computer algorithm for calculating a scalar product, there are assignment statements and loop control. There is 1 assignment statement initializing `ScalarProduct` to zero and *n* repetitions of the assignment to `ScalarProduct` that occurs within the loop, for a total of $n + 1$ assignment statements. In addition, there are assignments of values to the variable *i* and comparisons of the value of *i* to the limiting value, *n*. Why did our time-complexity function ignore all of this? This is a difficult question; it involves such technicalities of computer science as whether all arithmetic operations are equally time-consuming, how and where in the computer's architecture the loop control is actually being handled, and whether what appears to us as an assignment statement within the loop really involves accessing the computer's memory. We return to some of these issues in a bit more detail at the end of this section.

Since the algorithms we are analyzing do their work mainly by calculation and comparison, for simplicity, we will ignore both assignment statements and loop control in formulating our time-complexity functions. (We would be wrong to ignore assignment statements in algorithms whose main purpose is to rearrange data items, like sorting algorithms.) Further, we will typically count whichever basic operation appears more significant in our algorithms, either arithmetic operations or comparisons. In the exercises you will be told what to count.

Some mathematics computer packages (e.g., those for symbolic computations and graphing) include a "cost" feature that counts the number of some elementary operations needed to complete a calculation. These packages report several different operations separately: Multiplications, additions, comparisons, assignments, and subscript/loop index changes are typical operations considered. Some of the exercises suggest ways in which you could use a "cost" feature to help you understand time-complexity functions.

Time Complexity: Squaring a Matrix

E X A M P L E 3 Squaring a Matrix by Hand
Let us consider the number of arithmetic operations needed to square a matrix (see Chapter 3, Section 3). To gain some insight, we begin with the simplest cases. If the matrix is 1×1, then there is just 1 product to compute. If the matrix is 2×2, then there are $2 \times 2 = 4$ entries in the resulting matrix, each of which is a scalar product of two 2-element vectors; for example:

$$\begin{bmatrix} a & b \\ c & d \end{bmatrix}\begin{bmatrix} a & b \\ c & d \end{bmatrix} = \begin{bmatrix} a^2 + bc & ab + bd \\ ca + dc & cb + d^2 \end{bmatrix}$$
$$= \begin{bmatrix} row1 \cdot col1 & row1 \cdot col2 \\ row2 \cdot col1 & row2 \cdot col2 \end{bmatrix}.$$

To compute each scalar product, we need three arithmetic operations, two multiplications and one addition. Since we repeat the process of finding a scalar product 4 times, we have $4 \cdot 3 = 12$ arithmetic operations in all.

The number of entries in the square of an $n \times n$ matrix is $n \times n$, or n^2. Each of those entries is the result of a scalar product of two n-component vectors, which requires $2n - 1$ arithmetic operations when done by hand. Since we repeat the process of finding a scalar product n^2 times, we see that there are $n^2(2n - 1) = 2n^3 - n^2$ arithmetic operations needed to square an $n \times n$ matrix. Our time-complexity function for squaring an $n \times n$ matrix by hand is thus $f(n) = 2n^3 - n^2$. If we evaluate our function $f(n) = 2n^3 - n$ for $n = 1$ and $n = 2$, we get the same values we obtained informally at the beginning of this example: $f(1) = 2(1)^3 - (1)^2 = 2 - 1 = 1$; $f(2) = 2(2)^3 - (2)^2 = 16 - 4 = 12$. ▲

Many times we are presented with an algorithm presented as code or pseudocode. We need to be able to develop time-complexity functions directly from such presentations. As discussed earlier for Example 2, we will ignore assignment statements and loop control in our development.

E X A M P L E 4 Coded Algorithm for Squaring a Matrix Let us examine the coded algorithm, first seen in Section 3 of Chapter 3, for finding **S**, which is the square of an $n \times n$ matrix **M**. We develop a time-complexity function by examining this code, and then we compare it with the function we developed in Example 3, based on the hand computation of scalar products.

Input: n and an n × n square matrix M
Output: the n × n matrix S, the product of M
    × M
Preconditions: n > 0, n an integer
Postconditions: none

```
For i from 1 to n
      For j from 1 to n begin
            S[i, j] ← 0 {Initialize the
            "running total"}
            {One at a time, calculate
            products, add them to
            "running total"}
            For k from 1 to n
                  S[i, j] ← S[i, j]
                  + M[i, k]·M[k, j]
            end for
      end for
end for
```

We again count arithmetic operations. The

assignment statement $S[i, j] \leftarrow S[i, j] +$ $M[i, k] \cdot M[k, j]$ has one multiplication and one addition, for a total of two arithmetic operations. We need to know how many times each of these statements is repeated.

We will use "execution of a loop" to mean that the loop body is repeated the number of times specified in the loop control statement(s). Thus, the execution of the outer, middle, and inner loops in our example causes their respective loop bodies to be repeated n times, since i, j, and k begin at 1 and are incremented by 1s to a limit of n. Note the relationship of the loops; they are nested: The loop on i contains the loop on j, which in turn contains the loop on k.

We work from the innermost loop outward. The execution of the loop on k causes n repetitions of the assignment statement $S[i, j]$ $\leftarrow S[i, j] + M[i, k] \cdot M[k, j]$, which has 2 arithmetic operations, for a total of $2n$ arithmetic operations. One execution of the middle loop, on j, causes n repetitions of its loop body, which consists of the assignment statement $S[i, j] \leftarrow 0$, having no arithmetic operations, and the loop on k, having $2n$ arithmetic operations. The body of the loop on j thus has $2n$ arithmetic operations, which are repeated n times, so each execution of the loop on j results in $n(2n) = 2n^2$ arithmetic operations. One execution of the outermost loop, on i, causes its loop body, which is the loop on j, to execute n times. This means that one execution of the loop on i results in $n(2n^2) = 2n^3$ arithmetic operations. Thus, our time-complexity function is $f(n) = 2n^3$.

As with the comparison of the hand and computer computations of the scalar product, the hand and computer matrix multiplications do not give exactly the same time-complexity functions: $2n^3 - n^2$ by hand and $2n^3$ by computer. Again,

we wonder if the difference is significant, noting that the difference this time is a function of n, namely n^2, not just a constant. Stay tuned. ▲

The Nested Loops Principle

In the previous example we multiplied in a situation in which we had *nested loops*. This is possible because of the *nested loops principle*.

DEFINITION

The *nested loops principle* states that when two loops are nested, with the outer and inner loops causing their respective loop bodies to be repeated n and m times, then as a result of the nesting, the body of the inner loop will be repeated nm times.

The following algorithm shows the relationship of the two loops:

```
For i from 1 to n
        {this begins the body of outer loop,
         the loop on i}
         .
         .
         .
        For j from 1 to m
                {this begins the body of
                 inner loop, the loop on j}
                 .
                 .
                 .
                {this ends the body of inner
                 loop, the loop on j}
        end for
         .
         .
         .
        {this ends the body of outer loop,
         the loop on i}
end for
```

Matrix Multiplication in Computer Animation

Today there are many computer-generated animations, such as the ones in the movie *Jurassic Park* and in many television commercials. Computer animations require matrix multiplication and thus very many calculations.

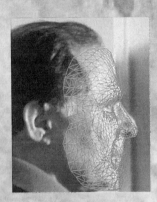

An animator creates a model of the basic object, for example, the dinosaur or a head, using a "skin" of many small, flat polygons. Data recorded for each polygon include where its corners are, what its color and texture are, how it reflects light, and so forth. Moving the object in space (rotation plus translation) means multiplying each 1×4 matrix describing a corner of a polygon by a 4×4 matrix describing the desired movement; that matrix multiplication requires 32 arithmetic operations.

In 1993 a state-of-the-art representation of a human consisted of 24,000 corners. Moving the entire shape required 768,000 operations. In order to fool the human eye into seeing motion, somewhere between 18 and 32 slightly different still scenes—called *frames*—must be played each second. For one second having 30 frames, there might be 23,040,000 calculations just to create motion. Calculations would also be needed to decide which parts of the body are now hidden and to position the body into the frame. Displaying color, texture, and reflective properties adds more steps.

The speed of the computer is crucial. Some very fast computers are constructed so that matrix multiplication is a basic operation, not requiring a coded algorithm, but existing on the same level as long division. Computer animation cannot be done without many, many calculations.

In finding time-complexity functions directly from code or pseudocode, we often use the nested loops principle. The crucial issues are the number of times we execute each loop and the number of steps in the innermost loop. Those values are multiplied together. As we saw in the *glass* and *black* boxes in Example 4 of Chapter 4, Section 2, we can get large numbers very quickly when we multiply.

Time Complexity: The Traveling Salesperson Problem (TSP)

In Section 3 of Chapter 5, we studied the traveling salesperson problem, or TSP, the problem of finding the minimal cost spanning cycle (visit each node exactly once, return to starting node when done) in a weighted graph. We now examine the time complexity for the two algorithms presented there.

Example 5 Brute-Force Algorithm for TSP

In the brute-force algorithm we generate all possible spanning cycles starting at a fixed node, determine the length of each cycle, and then find a minimal cycle. We found that for a minimal spanning cycle problem on a graph G having n nodes, there are $(n - 1)!/2$ different spanning cycles. To determine the weight of each cycle, we must add n weights found on the n edges of the cycle, at the cost of n additions. In rough pseudocode, we have the following situation:

Input: for each of the (n − 1)!/2 spanning
    cycles, the weights of all n edges in
    that cycle
Output: a list giving the CycleWeight of each
    of the (n − 1)!/2 cycles
Preconditions: n ≥ 0, n an integer
Postconditions: none

```
For i from first spanning cycle to the
  (n − 1)!/2th spanning cycle
        CycleWeight[i] ← 0
        For j from 1 to n     {One execution
          of this loop = n additions}
                CycleWeight[i] ←
                CycleWeight[i] + weight
                of jth edge of the ith
                cycle
        end for
end for
```

So, to find the weights associated with all the spanning cycles, we use the nested loops principle, to get a total of $n(n - 1)!/2 = n!/2$ additions.

Once that is done, the algorithm continues by identifying a cycle with minimum weight. Here is an algorithm for finding the Location of the minimum entry in a List of k items (we need to know where the minimum weight is on the list so we can find the corresponding cycle that has that weight):

Input: k and a List of k numbers
Output: the value of the Smallest item on
    List, and its Location
Preconditions: k ≥ 1, k an integer
Postcondition: if more than one item is smallest,
    the Location is that of the first
    one occurring on List

```
Smallest ← List[1]         {Assume that
first entry is Smallest}
Location ← 1
For i from 2 to k
        If List[i] < Smallest then  {New
        Smallest has been found}
                Smallest ← List[i]  {Record
                new Smallest and}
                Location ← i {its
                Location, so we can find
                it later}
        end if
end for
```

This "find-the-smallest" algorithm has a loop that executes $k - 1$ times. In it, we are not doing any arithmetic operations, but we are making comparisons. Thus we count the number of comparisons: There is one comparison in the If statement, and it is repeated once in each of the $k - 1$ repetitions of the loop, for a total of $k - 1$ comparisons. When we use this algorithm to find a minimal spanning cycle, $k = (n - 1)!/2$, so there are $((n - 1)!/2) - 1$ comparisons.

When we put the pieces together, we see that first we determine the weights of all the spanning cycles, at a cost of $n!/2$ additions, and then we locate a minimal spanning cycle, at a cost of $((n - 1)!/2) - 1$ comparisons. Our time complexity for the brute-force algorithm for TSP is thus given by the function $f(n) = (n!/2) + ((n - 1)!/2) - 1$. ▲

Separate Tasks Principle

It seems natural to add together the work required by the two successive parts of a task in order to get the work required to complete the whole task, as we just did in Example 5. This addition is an example of the second principle used to analyze time complexity of algorithms, the *separate tasks principle*.

DEFINITION

The *separate tasks principle* states that when both of two separate tasks, T_1 and T_2, are executed and they require w_1 and w_2 amounts of work, respectively, then the total amount of work done by the two tasks is $w_1 + w_2$.

In our use of the separate tasks principle, the two tasks were executed one after the other, and in fact, the second task could not be begun before the first one was completed. The separate tasks principle also applies to tasks that could be done independently and simultaneously, in the sense that the total work required is still the same. For example, if one person telephones the first half of a list of people and another person telephones the other half, the number of phone calls made does not depend on whether the first person works before, after, or simultaneously with the second person. However, if they are using two phones and working simultaneously, the same amount of work gets done in half the time. This reduction in time is the basis of parallel architecture computers, computers having more than one processor. Clearly, a time-complexity function for an algorithm takes on a very different meaning when more than one processor is available and when tasks can be run in parallel; time to complete the algorithm is no longer determined just by expressing the amount of work to be done. In this chapter we are assuming that just one processor will execute the algorithm.

More About TSP

E X A M P L E 6 Nearest-Neighbor Heuristic Algorithm for The TSP   To carry out the nearest-neighbor heuristic algorithm, we start at any one node and locate a minimum weight edge among the $n - 1$ edges connecting the starting node to the remaining $n - 1$ nodes. As in the discussion of the brute-force algorithm, we are trying to find a minimum, so we make one less comparison than the number of items on our list. So, to find the first edge using the nearest-neighbor algorithm, we carry out $n - 2$ comparisons. Once we have "traversed" the first edge, we are "at" the second node, and we must select from among $n - 2$ unvisited nodes for our third node. This selection can be done in $n - 3$ comparison steps. We summarize these results in Table 3.

The total number of comparisons is, by an extension of the separate tasks principle, the sum of all the comparisons listed in the second column, namely:

$$(n - 2) + (n - 3) + (n - 4) + \cdots + 2 + 1 + 0.$$

This is the sum of the first $n - 2$ integers. We recall from Section 2 of Chapter 1 that the

TABLE 3 Number of Comparisons Needed in Nearest-Neighbor Heuristic Algorithm.

Vertex to be visited	Number of comparisons needed
first (start)	none (random choice)
second	$n - 2$
third	$n - 3$
fourth	$n - 4$
$\vdots$	$\vdots$
$n - 2$th	$n - (n - 2) = 2$
$n - 1$th	$n - (n - 1) = 1$
nth	$n - n = 0$

sum of the first n integers has the formula $n(n + 1)/2$. So, by substituting $n - 2$ for n, we find our sum to be

$$\frac{(n - 2)(n - 1)}{2}.$$

Writing this formula as a polynomial gives us a time-complexity function for the nearest-neighbor algorithm, $f(n) = 0.5n^2 - 1.5n + 1$. ▲

In Table 4 we evaluate the time-complexity functions for the two algorithms for the TSP at a few values of n.

TABLE 4 Time Complexities for the TSP.

Number of vertices, n	Brute force, $(n!/2) + [(n - 1)!/2] - 1$	Nearest neighbor, $0.5n^2 - 1.5n + 1$
5	71	6
10	1,995,839	36
20	1.277×10^{18}	171
40	14.182×10^{47}	741

From experience we already know that carrying out the brute-force algorithm takes longer than carrying out the nearest-neighbor algorithm. Evaluating the time-complexity functions confirms our experience; the function values for nearest neighbor are consistently less than those for brute force. But Table 2 also shows dramatic differences between the time complexities of the two algorithms for larger values of n. Could we have predicted this dramatic difference without ever coding and running the algorithms or substituting specific values into the time-complexity functions? The answer is yes, and in Section 4 of this chapter we study the techniques mathematicians have developed for this purpose. In

the next section we learn how to use difference equations, which we studied in Chapter 1, to derive time-complexity functions.

Some General Considerations

Before we leave this section, we reconsider and expand upon some potentially important aspects of time complexity which we have minimized. One such aspect is the work needed to maintain loops. Typically, such maintenance involves adding an increment value to the counter and then comparing the counter to the limiting value. By ignoring loop control, we are potentially miscounting both additions and comparisons. However, many computers have special hardware that maintains loops without all the overhead of the usual adding and comparing, so our decision to ignore loop maintenance can be at least partially justified.

Another aspect that can be of crucial significance in a computer implementation is the time needed for retrieval and storage of data. This aspect can be involved every time a data item is mentioned in an assignment statement or any other type of statement and can even be affected by the order in which data are mentioned. Modern computer installations sometimes have the capability of recognizing the calculation of running totals, such as the variable `ScalarProduct` in Example 2, and can thus minimize the number of data moves. However, if the purpose of our algorithm is to rearrange data, then we should not ignore data moves. Our decision to ignore data moves is justified by the nature of the algorithms we are considering (their purpose is not data arrangement) and by a desire for simplicity.

A third aspect is the generation of all the cases. Examples include the $(n - 1)!/2$ different spanning cycles needed to work out the brute–force algorithm for TSP and the selection of the particular edges to be examined at any stage of the nearest-neighbor algorithm for TSP. We know from experience that just listing all the spanning cycles to work out the brute-force algorithm is a nontrivial, time-consuming task for a human. For some algorithms, the generation of all the cases can be the most time-consuming task performed.

A fourth aspect is the actual time required for carrying out the elementary arithmetic operations. For example, both by hand and in most computers, multiplication takes longer than addition; in some computers multiplication takes between 3 and 10 times as long as addition. We are ignoring this difference in lumping together all elementary arithmetic operations. In courses in numerical methods, such differences are important, and an expression such as $3n^2 + 5n + 23$ would be rewritten as $23 + n(5 + n(3))$. Both expressions require two additions, but the former requires four multiplications while the latter requires only two. A reduction in the number of multiplications needed for a calcula-

tion, often made at the expense of increasing the number of additions, has been major focus of concern for mathematicians for centuries. It was a prime reason for the invention of logarithm tables, whose use allowed the precalculator generations to convert multiplications to additions, and exponentiation to multiplication, which could then be converted to addition. Some of the exercises help highlight these ideas.

As you can see, the more we look into exactly what work is needed to execute an algorithm, whether it be done by human or by computer, the more tasks we see. It is typical in constructing mathematical models, such as our expressions of work needed, to simplify our examination by concentrating on a few "key" features. Thus, time-complexity functions are a *rough* estimate of the time cost of using an algorithm. This roughness not only makes our analysis simpler, but as we will learn in Section 4, it often does not affect our ability to compare the time complexities of two algorithms.

Exercises for Section 6.1

(Exercises referring to mathematics computer utilities are marked with a **T**. Features have been checked on Maple and may be possible on other utilities.)

1. Given an $n \times m$ matrix **A** and a scalar s, find the time-complexity function, in terms of both n and m, giving the number of arithmetic operations needed to compute s**A**. What is your function when $m = n$, when the matrix is square?

2. Given **A** and **B**, two $n \times m$ matrices, find the time-complexity function giving the number of arithmetic operations needed to compute **A** + **B**. Your function

should have two variables, n and m. What does your function become when $m = n$, when the matrix is square?

3. Suppose matrix **A** is $n \times m$ and matrix **B** is $m \times n$. Find time-complexity functions for the number of arithmetic operations needed to compute both **A** $\times$ **B** and **B** $\times$ **A**. Do both these computations require the same number of arithmetic operations? If not, assume that $n > m$, and see if the case with the bigger algebraic expression corresponds to the case in which you think you would have to do more work. Why should both your expressions become $2n^3 - n^2$ when $n = m$?

4. One method for determining the connectivity properties of a graph G having n nodes is to compute the matrix $\mathbf{R} = \mathbf{A} + \mathbf{A}^2 + \cdots + \mathbf{A}^{n-1}$, where $\mathbf{A}$ is the $n \times n$ adjacency matrix for G. Can you find an expression for the number of arithmetic operations needed to compute $\mathbf{R}$?

5. Suppose you wanted to write a time-complexity formula for the number of proposals made in carrying out the Gale–Shapley matching algorithm of Chapter 5 (Section 5) in the case of n men and an equal number of women. What difficulties or ambiguities might you encounter in writing such a formula? What assumption(s) might you decide to make in order to overcome those difficulties?

T 6. Some mathematics computer utilities have a "cost" function that reports the number of each type of operation that has been used to calculate a result. Here is how you would use that function in Maple to verify your results to Example 1 (the statements following the # signs are comments; Maple ignores them):

```
• with (linalg):      # Load in linear algebra
                        package
• readlib(cost):      # Load in the cost
                        function
• A := array(1..3,1..3, [[a,b,c], [d,e,f],
  [g,h,i]]);
                        [a b c]
                        [      ]
              A := [d e f]
                        [      ]
                        [g h i]
```

```
                # A is a square matrix with variables,
                  not numbers, as its entries
• B := scalarmul(A,s);    # Multiply matrix
                            A by scalar s
                        [sa   sb   sc]
                        [            ]
              B := [sd   se   sf]
                        [            ]
                        [sg   sh   si]
• cost(B);      # Find cost of computing B
    9 multiplications + 9 assignments + 9
    subscripts
```

Remember, your time-complexity function for Exercise 1 focuses only on arithmetic operations, but the utility you are using may, like Maple, also report counts for assignment statements and loop control or subscripts. Use the cost feature for several different values of n, the dimension of the matrix $\mathbf{A}$. Do the counts the utility gives fit into your time-complexity function?

T 7. Use the "cost" feature of a mathematics computer utility (see Exercise 6) to verify your results in Exercise 2: Use the cost feature for several different values of n and m, the dimensions of the matrices $\mathbf{A}$ and $\mathbf{B}$. Do the counts the utility gives fit into your time-complexity function?

T 8. Use the "cost" feature of a mathematics computer utility (see Exercise 6) to verify your results in Exercise 3. Use the cost feature for several different values of n and m, the dimensions of the matrices $\mathbf{A}$ and $\mathbf{B}$. Do the counts the utility gives fit into your time-complexity function?

SECTION 6.2 *Using Difference Equations to Find Time-Complexity Functions*

In Section 1 of this chapter we developed expressions for complexity of algorithms by directly counting the number of repetitions of such operations as additions, multiplications, and comparisons. The direct method is often very quick and useful. But there are other methods for developing complexity expressions, or functions. One of these is the use of difference equations, which we studied in Chapter 1. Difference equations are especially useful when it is easy to see how much extra work is required by an algorithm in one of two cases: when the size of a data set increases by 1, or when the size doubles.

Two Algorithms for Searching

We begin our discussion by focusing on the general problem of searching, looking for a desired (i.e., key) item in a collection of n items on a list. In any search, the process ends either when a match between the key and an item on the list is found or when the list is exhausted and thus there are no more items to use for comparison. We will examine two search algorithms: linear search (also known as sequential search) and binary search (also known as half-interval search). In linear search we compare the key with the items on the list, starting with the first item and proceeding one by one in their listed order. Linear search can be used with any list of items. By contrast, binary search requires that the items be in order before the search begins. In binary search the item in the middle of the list is compared with the key. If they do not match, then because the list is in order, we can determine which of the two halves of the list cannot possibly contain a match for the key, and we eliminate that half from further examination. The process then repeats by comparing the item in the middle of the remaining half of the list with the key.

These two search methods are often in evidence when people play the game "Guess the Number," a classic time-filler on long car trips. One person chooses a number (the key) between two values, say 1 and 100, and the other attempts to determine the chosen number. Most of us begin by asking, "Is it smaller than 50?" When the answer is no, then we ask, "Is it 50?" If the answer is no, we continue our search by focusing on the values 51 to 100, one-half of the original list. If we are told that the key number is smaller than 50, we continue our search using the other half of the original list, the values 1 to 49. This technique is the beginning of a binary search for the answer. However, when the list to be searched is quite short, for example 26 through 31, many of us use linear search. (Is it 26? Is it 27? . . .) Most people use a similar technique to look up a phone number, using something like binary search to locate the correct page in the directory, then using linear search to scan the page for the particular listing.

EXAMPLE 1   Binary Search in Action

Suppose we wish to locate the telephone number for Rodham Bader using binary search in a telephone book of 779 pages. We determine that the middle page is $(1 + 779)/2 = 390$. On page 390 we find last names beginning with "L," so when we ask, "Is Rodham Bader to be found at a page before 390?" the answer is yes. Table 1 summarizes our progress through the phone book (we drop all fractions in these calculations).

Once we have located the page, we can do a linear search on that page. Clearly, we could have skipped some steps by starting our search in the front of the phone book. But it is important to

TABLE 1 Searching for Rodham Bader.

Pages to be searched from–to	Number of the middle page	Last names on that page	Half to use to continue
1–779	390	L's	front
1–389	195	E's	front
1–194	97	C's	front
1–96	48	Be's	front
1–47	24	A's	back
25–48	36	Bar's	front
25–35	30	Bai–Bak's	front
25–29	27	A's	back
28–29	28	A's–Bada's	back
29–29	29	Bada–Bai's	this page

note that even with this inefficiency, we located the correct page in 10 steps. If the phone book had twice as many pages, 1558, one step would suffice to tell us which 779-page half to use to continue. Thus, doubling the number of pages adds but one step to the process. In our 779-page phone book, no binary search would require more than 10 steps, because $2^{10} = 1024 > 779$. Can you see how the 2^{10} is related to the binary search method? ▲

For many algorithms the amount of work to do depends on the scenario: The best case for a search would be finding the item on the first comparison; the worst case would be completing an entire search and not finding the item at all; and the average case would be finding it about halfway through an entire search. Typically, we don't spend much time analyzing best-case scenarios, and often it is more difficult to analyze average-case than worst-case scenarios. We will thus focus on the worst-case scenario, which for a search means completing an entire search process and not finding the key item. We will be attempting to find a time-complexity function $w(n)$, which describes the amount of work done in the worst case if n items are on the list. We will find the function w by developing, and then solving, a difference equation plus initial condition for it. We use comparisons as the work unit to be counted.

A Time-Complexity Function for Linear Search

In order to write a difference equation plus initial condition for the number of comparisons in a linear search, we need to focus on how the number of comparisons changes when the list grows from $n - 1$ to n items. We know that the number of comparisons needed for $n - 1$ items is given by the function w,

specifically, $w(n - 1)$. When the list has one more item, and keeping in mind that we have a worst-case scenario, we will need one more comparison in the case of n items than we did in the case of $n - 1$ items. These considerations give us the difference equation

$$w(n) = w(n - 1) + 1.$$

The initial condition will be $w(1) = 1$; only one comparison is needed when only one item is on the list. We solved this particular equation in Section 5 of Chapter 1. The function w is given by the formula $w(n) = n$.

The formula for the time complexity for linear search is not very surprising. If we need to compare the key item with every one of the n items on a list, then we expect to use n comparisons in the process. The purpose of the preceding example was to introduce difference equations for analysis of algorithmic complexity. The next investigation is more interesting; it is not obvious in advance what formula $w(n)$ will have for binary search.

A Time-Complexity Function for Binary Search

The crucial element in the analysis of binary search is recognizing that the significant change in the number of comparisons needed in the worst case is not related to the addition of one more item to the list, but is related to the doubling of the size of the list. When the list doubles in size, for example, from $n/2$ to n items, just one comparison puts us back into the previous situation of $n/2$ items, since half of the list is eliminated at each step. Thus, we have the difference equation

$$w(n) = w\left(\frac{n}{2}\right) + 1, \qquad (1)$$

and as in the linear search case, our initial condition is $w(1) = 1$. This difference equation does not appear similar to any of the ones we studied in Chapter 1. But with a change of variables, we can transform Eq. (1) into one that we can solve.

We want $n = 2^m$, so we define $m = \log_2 n$. This definition of m also gives us $n/2 = 2^m/2 = 2^{m-1}$. We substitute for n and $n/2$ in Eq. (1), getting

$$w(2^m) = w(2^{m-1}) + 1. \qquad (2)$$

We wish to express $w(2^m)$ in terms of m and $w(1)$. Here is our game plan: We begin by finding $w(2^m)$ in terms of $w(2^{m-2})$, then in terms of $w(2^{m-3})$ and so on; finally, we see a pattern by which we express $w(2^m)$ using $w(2^{m-k})$. Then we substitute for k so that $w(2^{m-k})$ becomes $w(2^0) = w(1)$, our initial condition.

Now to carry out the game plan. Substituting $m - 1$ for m into Eq. (2), we get $w(2^{m-1}) = w(2^{m-2}) + 1$, which we use to replace the $w(2^{m-1})$ in that same equation, resulting in

$$w(2^m) = (w(2^{m-2}) + 1) + 1 = w(2^{m-2}) + 2.$$

Continuing in this manner, we substitute $m - 2$ for m into Eq. (2). This gives $w(2^{m-2}) = w(2^{m-3}) + 1$, which we use to replace the $w(2^{m-2})$-term, resulting in

$$w(2^m) = (w(2^{m-3}) + 1) + 2 = w(2^{m-3}) + 3.$$

A pattern is developing here, which we now express in terms of m and k as

$$w(2^m) = w(2^{m-k}) + k.$$

Since our initial condition is given for $w(1)$, we want $2^{m-k} = 1$, so $m - k = 0$, or $k = m$ is the substitution we need. And since $w(1) = 1$, we get

$$w(2^m) = w(2^{m-m}) + m = w(1) + m = 1 + m.$$

But $m = \log_2 n$, so when we reverse our substitution of variables, we get this time-complexity function for binary search:

$$w(n) = 1 + \log_2 n.$$

We are acting here as if we only needed to consider values of n that are powers of 2, resulting in values for $m = \log_2 n$ that are nonnegative integers. The result we just obtained would still be valid if we added the requirement of rounding up to the next-higher integer when $w(n)$ is a fractional value. In Example 1 we used binary search to locate a page in a 779-page phone book. Since $\log_2 779$ is approximately 9.6, $w(779)$ would be $1 + 9.6 = 10.6$, rounded up to give 11. (In Example 1 we had only 10 steps because we ignored the last question, namely, "Is Rodham Bader's phone number on this page?")

Comparing the Two Algorithms

Most of us instinctively recognize that binary search is faster than linear search, as evidenced by how we play "Guess the Number." The advantage of binary search is made clearer by evaluating the two time-complexity functions for several values of n, the number of items in the list, as shown in Table 2.

The numbers are dramatic, confirming our instincts about the time complexity of these two search methods. We are tempted to ask, "Why would any-

A Legendary Problem—The Towers of Hanoi

An old Asian legend is said to describe a set of 3 tall towers, or poles, and a set of 64 rings of different diameters. The rings start out stacked on one of the towers, in size order with the largest on the bottom. The towers are tended by some monks, whose job it is to move the set of rings from the starting tower to one of the other towers, moving exactly one ring every day. The monks must obey one rule: A larger ring may never rest on a smaller one. They may use the third tower for temporary storage of rings. The legend predicts that when the monks finish their job, the world will end.

We do not know exactly when the monks started their task, but we can look at an algorithm they could follow and calculate the time it requires. Here is the idea behind the algorithm: If the monks could move the top 63 rings from the starting tower to the temporary tower, then they could move the bottom, largest ring from the starting tower to the destination tower and, finally, they could move the 63 rings from the temporary tower to the destination tower.

Let's see how that idea works with fewer rings. Here's how to move two rings from Tower A to Tower C: Move top ring from A to B (completes the first move of all but bottom ring); move biggest ring from A to C (the "middle" of the algorithm); move top ring from B to C (completes the second move of all but bottom ring). Now, if there are three rings on Tower A going to Tower C, we first move the top two rings to Tower B using the algorithm for two rings, but with the roles of some of the towers interchanged, then we move the bottom ring from A to C, and finally we move the top two rings from B to C, again using a variation of the algorithm for two rings.

How many moves does it take to reposition the three-ring tower? It takes the number of moves required for two rings—twice, once to Tower B and once from it—plus one move for the bottom ring. Since it takes 3 moves for a 2-ring tower, there are $2 \cdot 3 + 1 = 7$ moves for a 3-ring tower. That is, using a difference equation, $w(3) = 2w(2) + 1$. For the 64 rings of the old Asian legend, $w(64) = 2w(63) + 1$. Of course, $w(0) = 0$ is an initial condition. In Exercise 5 you are asked to verify that $w(n) = 2^n - 1$ is a solution of that difference equation plus initial condition. That means that $w(64) = 2^{64} - 1$, which is approximately $1.84 \cdot 10^{19}$ days, or about $5.05 \cdot 10^{14}$ centuries, a long, long time.

TABLE 2 Comparing Time Complexities of Linear and Binary Search.

# of items n	Linear search $w(n) = n$	Binary search $w(n) = 1 + \log_2 n$
2	2	2
4	4	3
8	8	4
16	16	5
32	32	6
64	64	7
128	128	8
256	256	9
512	512	10
⋮	⋮	⋮
32,768	32,758	16

one ever do a linear search?" The answer lies in the prerequisite for binary search—that the list of items be in order. For many types of data items we can get an ordered list, although there is a time cost to do so (see Examples 17 and 18 of Section 6.4.) But some kinds of items cannot easily be sorted into a list; for example, photographs of a farm or a collection of fingerprints. And in some kinds of computer data structures, access to data items is restricted, and so even if the items are in order, we cannot proceed directly to the "middle" one. For these reasons, linear search is still used.

What about less dramatic situations for which we have no instinct but do have time-complexity functions? In Section 6.3 we look at ways to choose between algorithms for specific intervals of n. In Section 6.4 we will study order of magnitude, a formal mathematical way of comparing functions. Using order-of-magnitude ideas, we can compare functions without needing to evaluate the functions for specific values of n.

Exercises for Section 6.2

1. Given an $n \times n$ matrix **A** and a scalar s, develop a difference equation and initial condition for the time-complexity function for computing s**A**. You need to ask yourself this question: How many more arithmetic operations are needed to find s**A** when **A** is $n \times n$ than are needed when **A** is $(n - 1) \times (n - 1)$? Use $n = 1$ to set up an initial condition.

2. Given **A** and **B**, two $n \times n$ matrices, develop a difference equation and initial condition for the time-complexity function for computing **A** + **B**. You need to ask yourself this question: How many more arithmetic operations are needed to find **A** + **B** when both are $n \times n$ than are needed when both are $(n - 1) \times (n - 1)$? As an initial condition, take the case for $n = 1$.

3. Determine if your time-complexity function for the square, $n \times n$ matrix from Exercise 1 of Section 6.1 satisfies your difference equation and initial condition from Exercise 1 here.

4. Determine if your time-complexity function for the square, $n \times n$ matrix from Exercise 2 of Section 6.1 satisfies the difference equation and initial condition in Exercise 2 here.

5. (Requires programming background) This Pascal procedure writes a list of the most efficient set of moves for the Towers of Hanoi problem with n rings of differing diameters. (See Spotlight: A Legendary Problem—The Towers of Hanoi.)

```
Procedure Tower (n: nonnegint; start,
  finish, other: towernames);
    If n > 0 then begin
          Tower (n − 1, start, other,
            finish);
          Writeln ("Move top ring from",
            start, "to", finish);
          Tower (n − 1, other, finish,
            start)
    end;
```

Write a difference equation and initial condition (for $n = 0$) for the **number of writeln's** executed by the above procedure. Determine if $w(n) = 2^n − 1$ satisfies your difference equation and initial condition. (Hint: It should.)

6. (Requires programming background) Here is Pascal code for an insertion sort.

```
Procedure Insertsort(n: 1..maxint; var
  data_array: array_type);
var
    k: 1..maxint;
procedure Insert(k: 1..maxint; var
  data_array: array_type);
```

```
{Code omitted. The already-sorted k − 1
  items are examined in reverse order,
  starting with the (k − 1)st, to see
  where the new item, currently in
  position k, should be inserted. As soon
  as one of the presorted k − 1 items is
  found that is smaller than the new item,
  we know where the new item goes, so the
  comparisons end.}
begin {Insertsort}
  if n > 1
    then begin
      Insertsort(n − 1, data_array);
      Insert(n, data_array)
    end
end;
```

In the worst case, the submodule `Insert` examines all $k − 1$ previously sorted items in order to position the new item; there are $k − 1$ comparisons. In the best case there is only 1 comparison. For each of these cases, write a difference equation and initial condition (for $n = 1$) for the number of comparisons needed to complete an insertion sort of n items. Also, describe the original order of the data items in each of these cases.

SECTION 6.3 *Comparing Time-Complexity Functions*

In the previous two sections we developed time-complexity functions for several algorithms in terms of n, the number of data items to be processed. One of the primary reasons we are interested in time complexity is to get guidelines about which of several possible algorithms to select when there is a choice. For many algorithms, development of time-complexity functions is not particularly difficult. There are algorithms for which such functions are not so easy to find, but we will not concern ourselves

with that type of algorithm here, except to mention that time-complexity estimates for such algorithms are typically obtained heuristically, by coding the algorithm and running the code for several well-chosen values of n. If we have time-complexity functions for the algorithms, we can compare the functions directly. In this section and the next one, we focus on the comparison process, and frequently the time-complexity functions we use do not come from specific algorithms. Rather, the functions we employ have been chosen for their usefulness in illustrating the mathematical points to be made.

Characteristics of Time-Complexity Functions

We need to be specific about the type of functions we are discussing. Since our functions were developed for an n that is the size of a data set, n is always a positive integer, that is, the domain of any time-complexity function is the positive integers. Similarly, the amount of work needed—the value of the function—is positive. Most of the functions we discuss would remain meaningful and positive if we were to substitute a real number such as 7.5 or π for n; however, except for displaying graphs of some functions, we can think of our functions as sequences defined on the positive integers, not on the real numbers.

When we use an algorithm, we can't do a bigger problem in less time than it takes us to do a smaller one. That is, if $n > m$, then $f(n) \geq f(m)$. We say that the functions we are considering are nondecreasing. Not only are time-complexity functions nondecreasing, but they grow without bound, or approach infinity. We can see this from Table 4 of Section 6.1, where the work required by either the brute-force or the nearest-neighbor algorithm grows larger as n gets larger. This unbounded growth is another feature of all of the time-complexity functions we will examine, and it seems natural: There is no one amount of work that effectively suffices for all problems no matter how large.

Because time-complexity functions are nondecreasing and unbounded, there may seem to be no particular reason to choose between two algorithms with different time-complexity functions. However, another look at the time-complexity functions in Table 4 of Section 6.1 shows that, for all values of $n > 3$ (there is no real TSP for a graph of fewer than 4 nodes), brute force takes more work than nearest neighbor. Thus, if the amount of work required were the only consideration, nearest neighbor would be preferred to brute force.

Comparing Two Time-Complexity Functions: Part I

The comparison of two time-complexity functions is not always straightforward. Often there are several intervals of n-values; on some of the intervals one algorithm is faster, and on the other intervals the other algorithm is faster. Two examples of such behavior will help us understand the issues and techniques we need.

EXAMPLE 1 Comparing $5n^2$ with $105n$

We begin by comparing $5n^2$ with $105n$ for a set of increasing values of n, as shown in Table 1. Both of these functions are nondecreasing and unbounded. Figure 1 contains a graph of these two functions for $0 \leq n \leq 30$.

Suppose that $5n^2$ and $105n$ are time-complexity functions for two algorithms, both of which provide a business with a solution to a frequent problem. For "small" values of n, both Table 1 and Figure 1 show that $5n^2 < 105n$. But for "large" n, both Table 1 and Figure 1 show that $5n^2 > 105n$. For "small" n the $5n^2$-algorithm is faster and thus cheaper; for "large" n the $105n$-algorithm is faster and cheaper. The graph indicates that somewhere near $n = 20$ the relative costs of the two algorithms change. In fact, by equating the two functions (the two functions are

TABLE 1 Comparing $5n^2$ with $105n$.

n	$5n^2$	$105n$
1	5	105
3	45	315
9	405	945
27	3645	2835
81	32,805	8505
241	290,405	25,515

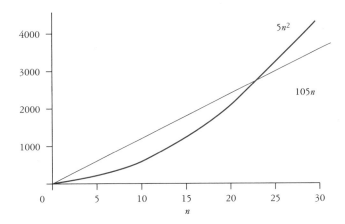

FIGURE 1 Graph of $5n^2$ and $105n$.

equal when their graphs cross), we get $5n^2 = 105n$, which has solutions $n = 0$ and $n = 21$; we noticed the crossing at $n = 21$. ▲

Some Interesting Questions

Example 1 raises two sets of questions:

1. How can we find the crucial values of n at which the costs of the two algorithms are the same? What is the relationship of the two algorithms for the intervals between crucial values?

2. What happens as n grows large, that is, when we process many data items? If we must choose just one algorithm to process all our data sets, and if our values of n can be anywhere, is there an algebraic method that tells us which algorithm to choose?

The first set of questions is important to us. Knowing the crucial values of n, the relationship of the algorithms in the intervals between crucial values, and the size of a typical data set can be important in deciding which algorithm should be used. In Example 2 we examine a second pair of algorithms,

learning more about how we can find crucial values and the relationship of the functions on the intervals.

The second set of questions is also important. A typical pair of time-complexity functions has a small number of crucial values; for the infinite interval, $n >$ the largest crucial value, the relationship of the two functions is always the same. If we need to choose just one algorithm, we often opt for the one with the lower cost over that infinite interval. In our previous example, we would choose the $105n$-algorithm, ignoring the fact that the $5n^2$-algorithm is faster for $n < 21$. Can we tell algebraically that for large n we should choose the linear $105n$ and not the quadratic $5n^2$? Will any linear function be preferable to any quadratic for large n? In Section 6.4 we take up these questions.

Comparing Two Time-Complexity Functions: A More Complicated Example

E X A M P L E   2   Finding Crucial Values and Function Relationships on Intervals   Again suppose that we need to choose between two algorithms, but this time let's say that algorithm A has time complexity of $n^3 + 20n^2 + 8000n + 10000$ and algorithm B has time complexity of $200n^2 + 300n + 70000$. We could use one of three techniques to help us analyze the relationship between the algorithms: numeric substitution, graphical analysis, and algebraic manipulation.

Numeric substitution simply means evaluating each time-complexity function for some value(s) of n. We evaluate by hand, with a calculator, or with a mathematics computer package. We would first determine the size of a data set, for example $n = 35$, and then calculate that algorithm A's complexity function would equal 357,375 while algorithm B's would equal

Solving Equations

Finding the value(s) for x that satisfy an equation like $x^2 + 2x - 1 = x + 2$ is one of the oldest mathematical problems. Today many calculators solve a wide variety of equations, even some that were considered an impossible challenge until recently.

Throughout most of the history of mathematics, formulas were sought that would give, in a finite number of steps, exact roots of an equation involving $+, -, \times, \div$, and $\sqrt{}$. An example is the quadratic formula, discovered around 2000 B.C.:

$$x = \frac{-b \pm \sqrt{b^2 - 4ac}}{2a}.$$

Solutions of this type to third- and fourth-degree equations were found by Tartaglia and Ferrari during the Renaissance. It's a colorful story, full of secrecy, betrayal, and a public contest of skill, well told by Eves (see the Recommended Readings). In the first half of the nineteenth century, Ruffini, Abel, and Galois proved that for fifth-degree or higher polynomial equations, there could be no such formula as the quadratic

formula to solve them. And nonpolynomial equations such as $\log(x) = x^2 - 3$ were mostly hopeless.

This seemed like a stone wall, but modern calculators can give you a very accurate solution to these "impossible equations." The theory used (called numerical analysis) jumps over the stone wall by not

Nicolo Tartaglia (1500–1557) Italian mathematician. Credited with the discovery of the cubic equation.

looking for exact answers via formulas that can be written down in a finite number of uses of $+, -, \times, \div$, and $\sqrt{}$. Algorithms are used, which give more and more accuracy the more steps you apply (like the bisection method of Chapter 1). This unlocks a very rich world of applications. It

would be interesting to know why exactness via a finite number of uses of $+, -, \times, \div$, and $\sqrt{}$ was considered so important for so long.

can know in advance the size of the data set to be processed. The hybrid would begin by testing n and then choosing the algorithm that is more efficient for that value. A more complex technique for "breeding" two or more algorithms is the development of "genetic" algorithms—see Chapter Exercise 17.

Using Graphs of the Ratio of Two Functions

There is another way to compare the behavior of two functions, namely by looking at their ratio. For example, we can take the two functions we just compared and graph their ratio:

$$\frac{n^3 + 20n^2 + 8000n + 10000}{200n^2 + 300n + 70000}.$$

What should we expect to see in the graph of the ratio? If the two functions are equal, the ratio will equal 1. If the numerator function is larger than the denominator function, the ratio will be larger than 1. If the denominator function is the larger one, the ratio will be smaller than 1. These features can be seen in the graph of this ratio in Figure 5.

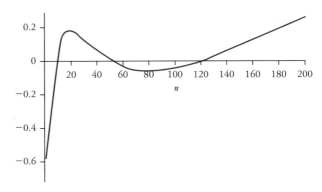

FIGURE 6 Graph of $(n^3 + 20n^2 + 8000n + 10000)/(200n^2 + 300n + 70000) - 1$.

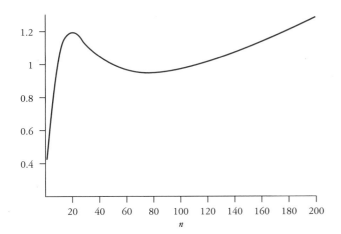

FIGURE 5 Graph of the ratio $(n^3 + 20n^2 + 8000n + 10000)/(200n^2 + 300n + 70000)$.

As we did when graphing the difference of two functions, when we want to locate precisely the value(s) of n at which the two functions are equal, we may need to look at related graphs. Two techniques are available:

1. Graph the ratio minus 1, as in Figure 6, putting all the critical values of n onto the horizontal axis; or

2. Graph both the ratio and the constant 1, as in Figure 7, and determine where the two graphs cross.

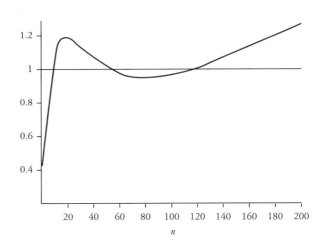

FIGURE 7 Graph of $(n^3 + 20n^2 + 8000n + 10000)/(200n^2 + 300n + 70000)$ and the constant $c = 1$.

The same zooming techniques we used before could be used with these ratio graphs to locate the critical values where the two functions in the ratio are equal.

As we end this section, we recall our second set of questions: What happens for large n? Will any quadratic, such as our $5n^2$, eventually be bigger than any linear function, such as our $105n$? Will any cubic, such as our $n^3 + 20n^2 + 8000n + 10000$, eventually be bigger than any quadratic, such as our $200n^2 + 300n + 70000$? The concept of order of magnitude, which we develop in the next section, provides us with just the algebraic technique we need to compare the behavior of functions for large values of n without relying on either numeric substitution or graphing.

Exercises for Section 6.3

(References to graphing calculators or to mathematics computer utilities are marked **T**.)

In each of the exercises you are given a pair of functions of n, which we can think of as time efficiencies of two algorithms, A and B, for the same problem. Your job is to find the crucial value(s) of n (or close approximations) at which the algorithms are equally efficient, list the intervals between these crucial values, and tell which algorithm is preferred for each interval. Since we are thinking of our functions as time-complexity functions, we can safely ignore any crucial values less than $n = 2$, and we can describe the intervals with integral endpoints.

There are several methods for finding the crucial values and the relationships on the intervals they determine:

1. Set the first function equal to the second, and try to solve the equation for the crucial value of n by factoring or (**T**) using a "solve" command.

T 2. Graph the difference of two functions, and find the desired values from the graph (this may require plotting several successive graphs, "zooming in" on the neighborhood where the graph crosses the horizontal axis).

3. Use the bisection method from Chapter 1 to solve the equation you get in method 1.

4. Use a combination of techniques.

Note that these problems do not require you to find the exact crucial values. Unless otherwise stated, there is only one crucial value to find in each exercise.

1. A: $100n^2$ and B: $4n^3$

2. A: $15n$ and B: $5n^2$

3. A: $3n^2 + 10n$ and B: $100n + 408$

4. A: n^3 and B: $64n$

5. A: 7^n and B: n^7

6. A: $n!$ and B: n^2

7. A: $5000n^3 + 800n^2$ and B: 4^n

8. A: $450n^3 + 60n^2 + 200n + 500$ and B: $n!$

9. A: $50n$ and B: $700 \log_2 n$

10. A: $500n \log_2 n$ and B: n^2

11. A: $n^3 + 25n^2 + 20000n$ and B: $265n^2 + 2100n + 420000$ (more than one crucial value)

12. A: $300n^2 + 50n + 50000$ and B: $3n^3 + 15n^2 + 7100n + 5000$ (more than one crucial value)

13. A: $300n^2 + 900n + 600000$ and B: $n^3 + 45n^2 + 21000n + 99500$ (more than one crucial value)

14. A: $n^3 + 50n^2 + 19000n + 5000$ and B: $300n^2 + 475n + 410000$ (more than one crucial value)

SECTION 6.4 *Order of Magnitude and Big-O Notation*

The concept of order of magnitude helps us express something about the growth rates of different functions. If we have two functions h and g, we want to be able to know when h is growing no faster than g. In Section 6.1 Table 1, we considered the following time-complexity functions: $\log(n)$, n, $n \log(n)$, n^2, n^3, 2^n, 3^n, and $n!$. We noticed that these functions are all increasing as n increases and that they seem to be listed so that the functions at the end of the list are "growing faster" than those at the beginning.

Even for n as small as 20, these functions vary greatly in size. So we are not surprised that when we graph these functions, as in Figure 1, we need two graphs with the horizontal and vertical scales adjusted for the particular functions in each. (The places where these graphs cross, our focus in Section 6.3, are not clearly shown in Figure 1, but we are focusing on the unbounded interval in this section, not the crossings.) There seems to be real differences among these functions. These differences are incorporated into the definition of order of magnitude.

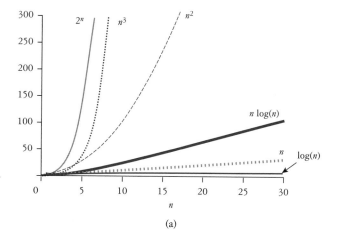

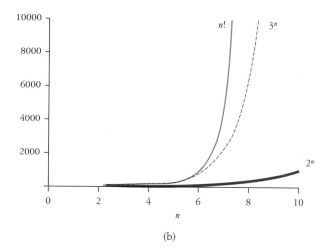

FIGURE 1 Growth rates of some common functions: (a) graph of $\log(n)$, n, $n \log(n)$, n^2, n^3, and 2^n; (b) graph of 2^n, 3^n, and $n!$.

Order of Magnitude and Constants

DEFINITION

Suppose we have a function g having non-negative real values. We define a corresponding set of functions, which we call $O(g)$ and read "*order of magnitude* of g" or "big O of g," as follows: Let h be any function having non-negative real values. Then h is in the set $O(g)$ if there is some finite constant c for which $h(n) \le cg(n)$ when n is large.

Expressing our intuitive ideas about the functions $\log(n)$, n, $n \log(n)$, n^2, n^3, 2^n, 3^n, and $n!$ in terms

of this definition, we would say that each of these functions is of a "higher" order of magnitude than the ones listed before it. Graphically, each function on the list seems to "curve upward faster" than the ones earlier on the list. In this section we will show that our intuition is correct. But first, let's see how this definition supports our notions that multiplying by a constant does not change order of magnitude.

Theorem 1. If k is any positive constant and f is a real-valued function, then $O(f) = O(kf)$, and we say that f and kf have the same order of magnitude. Informally, we say that when we are dealing with order of magnitude, constants "don't matter."

Proof. To prove this theorem, we need to show that the set $O(f)$ is contained in the set $O(kf)$ and also that the set $O(kf)$ is contained in the set $O(f)$. For this proof we assume that $k \geq 1$. If $k < 1$, the proof would hold for the constant $j = 1/k$, and the roles of the two functions would be reversed.

To show that the set $O(f)$ is contained in the set $O(kf)$, we need to find a constant c so that $f(n) \leq ckf(n)$ when n is large. Since $k > 0$, $f(n) \leq kf(n)$ for all values of n, and $c = 1$ works just fine. So we have shown that the set $O(f)$ is contained in the set $O(kf)$.

To show that the set $O(kf)$ is contained in the set $O(f)$, we again need to find a constant c, this time so that $kf(n) \leq cf(n)$ when n is large. Since $k > 0$, $kf(n) \geq f(n)$ for all values of n, but this relationship is "backwards" from the one we wish to have. c can be any positive constant; we choose c to be k. Using this value for c, the relationship becomes $kf(n) \leq kf(n)$, which is true for all values of n. So we have shown that the set of $O(f)$ is contained in the set $O(kf)$.

Thus we have shown that the sets $O(f)$ and $O(kf)$ are equal, so the functions f and kf have the same order of magnitude.

▲ ▲ ▲

Theorem 1 tells us that the functions $105n$ and n have the same order of magnitude. Also, $0.25n$ and n have the same order of magnitude. Similarly, n^2, $0.47n^2$ and $5n^2$ all have the same order of magnitude. Mathematicians, noting that the coefficient of n "doesn't matter," speak of the "order of magnitude" of n, the "order of magnitude" of n^2, and so on, using coefficients of 1. In general, for a constant c, we refer to $O(f)$ and not $O(cf)$, although they are the same set.

Polynomial Functions and "Big O"

EXAMPLE 1 **Comparing $O(n)$ and $O(n^2)$** Is n in $O(n^2)$? Is n^2 in $O(n)$? How can we tell if two functions that differ by more than a multiplicative constant are of the same order of magnitude?

First we look for a constant c so that $n \leq cn^2$ for large values of n. But we know that $n \leq n^2$ for $n \geq 1$, so $c = 1$ will suffice. Thus we have n in $O(n^2)$.

Our instinct is that n^2 is not in $O(n)$. Perhaps this instinct comes from looking at Figure 1. Since n has a straight line as a graph while n^2 has a graph that "curves upward," no amount of tilting a straight line (the equivalent of multiplying n by a constant) can overpower the effects of the upward curve. But how do we prove this?

In the definition of order of magnitude, we find "... h is in the set $O(g)$ if there is some finite constant c for which $h(n) \leq cg(n)$ when n is large." In other words, we can say that "for large values of n, $h(n)/g(n) \leq c$." (We have simply divided the inequality by $g(n)$, which does not change the direction of the inequality since g is positive.) In practice, we look at the behavior of the ratio of the two functions as n gets large, as n approaches infinity. If the ratio remains finite, then h is in $O(g)$; if the ratio grows without bound— approaches infinity—then h is not in $O(g)$.

We need to look at the ratio n^2/n as n grows large. This ratio reduces to just plain n, which grows without bound as n grows large. So there is no constant c for which $n^2 \leq cn$ for large values of n, and so n^2 is not in $O(n)$.

Figure 2 shows the relationship of $O(n)$ and $O(n^2)$. $O(n)$ is a proper subset of $O(n^2)$ since n is in $O(n^2)$ but n^2 is not in $O(n)$. ▲

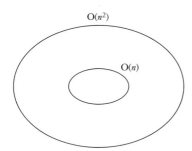

FIGURE 2 Relationship of the sets $O(n)$ and $O(n^2)$.

Although the definition of $O(g)$ gives us the set of all real-valued functions that grow no faster than the function g, we tend to think of $O(g)$ as the set of all real-valued functions that have the same growth rate as g because we are usually looking for the slowest growth rate that describes a function. In a similar spirit we would say that none of our relatives has more than, for example, 5 children, when it is equally true that none of them has more than 47 children. Just as we get "better" information from the 5 than the 47, we get better information by placing a function in the smallest big-O set to which it belongs. The rest of this section is mainly concerned with determining the relationship between two big-O sets.

E X A M P L E   2   Comparing the Sets $O(n^p)$ and $O(n^q)$ We now establish the relationship of the sets $O(n^p)$ and $O(n^q)$ for any real p and q. Suppose that $p < q$. We look first at the ratio n^p/n^q to determine if n^p is in $O(n^q)$; remember, the O set "belongs" to the function in the denominator.

The ratio simplifies to n^{p-q}. Since $p < q$, $p - q$ is negative, and thus n^{p-q} approaches zero as n approaches infinity. So n^p is in $O(n^q)$. We next look at the ratio n^q/n^p. This ratio simplifies to n^{q-p}. Since $p < q$, $q - p$ is positive, so n^{q-p} becomes infinite as n approaches infinity. Thus n^q is *not* in $O(n^p)$. ▲

In Example 2 we proved an important result about big O:

Theorem 2.   If $p < q$, then $O(n^p)$ is a proper subset of $O(n^q)$.

From Theorem 2 we can get a nesting of orders of magnitude, for example, $O(n^{1/2}) \subset O(n^{4/7}) \subset O(n) \subset O(n^2) \subset O(n^3) \subset O(n^4) \subset O(n^5) \subset O(n^{13}) \subset O(n^{267})$. Another way this same idea is often stated is to present functions in a list of increasing order of magnitude, for example, $n^{1/2}$, $n^{4/7}$, n, n^2, n^3, n^4, n^5, n^{13}, n^{267}. Theorem 2 tells us that some of the time-complexity functions in the list at the start of this section were in fact listed in increasing order of magnitude. (Which ones?)

E X A M P L E   3   Polynomial Functions   Next, we expand our knowledge of order of magnitude from monomials to polynomials, such as the time-complexity functions for algorithms A and B from Example 2 of Section 6.3. How does $n^3 + 20n^2 + 8000n + 10000$ compare with $200n^2 + 300n + 70000$? We first show that each polynomial belongs to big O corresponding to its highest power. The ratio $(n^3 + 20n^2 + 8000n + 10000)/n^3$ simplifies to $1 + (20/n) + (8000/n^2) + (10000/n^3)$. As n approaches infinity, this expression approaches 1 since the three fractions all approach zero. Thus $n^3 + 20n^2 + 8000n + 10000$ is in $O(n^3)$. A similar calculation would show us that $200n^2 + 300n + 70000$ is in $O(n^2)$. So, using Theorem 2, we can say that $200n^2 + 300n + 70000$ is of a smaller order of magnitude than $n^3 + 20n^2 + 8000n + 10000$. ▲

We state general forms of the results of the previous two examples:

Theorem 3. Given a polynomial M of degree p, M is in $O(n^p)$.

Theorem 4. Given two polynomials M_1 and M_2 of degrees p_1 and p_2, respectively, if $p_1 < p_2$, M_1 has a lower order of magnitude than M_2, and $O(M_1)$ is a proper subset of $O(M_2)$. If $p_1 = p_2$, then $O(M_1) = O(M_2)$.

The following theorem completes our results concerning polynomials. Its proof is found in Exercise 5a.

Theorem 5. $O(0)$ consists of just the function that is identically zero. $O(1)$ consists of the constant functions $f(n) = c$. $O(0)$ is a proper subset of $O(1)$.

Using Graphs of Ratios to Learn About O

At the end of Section 6.3 we have several graphs of the ratio

$$\frac{n^3 + 20n^2 + 8000n + 10000}{200n^2 + 300n + 70000}.$$

Figure 3 looks again at a graph of this ratio, but this time we are not looking for the values of n at which the numerator and denominator are equal. Rather, we want to use the graph to give us insight into what happens to the ratio when n gets very large.

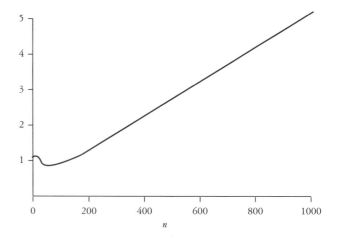

FIGURE 3 Graph of the ratio $(n^3 + 20n^2 + 8000n + 10000)/(200n^2 + 300n + 70000)$ for large n.

For large values of n, the graph of the ratio $(n^3 + 20n^2 + 8000n + 10000)/(200n^2 + 300n + 70000)$ appears to be a straight line. It certainly is increasing; it is not a horizontal line, as would be the case if the ratio were some finite constant. So from the graph we would guess that the numerator, a cubic, is not in the set $O(n^2)$, which is the order of magnitude of the denominator.

In Figure 4 we graph the ratio $(200n^2 + 8000n + 10)/(20n^2 + 300n + 70000)$ of two functions that Theorem 4 tells us are of the same order magnitude. We expect the graph to be a horizontal line for large values of n, and that is exactly what we see. Do you know why the horizontal line seems to be at height 10? In general, the two functions in a ratio are of the same order of magnitude if, for large n, the graph of the ratio is a straight line corresponding to a nonzero constant.

It is important that the height of the horizontal line not be zero, since the definition of order of magnitude specifically requires the constant c to be positive, $c > 0$. Here's why. Instead of graphing the ratio $(n^3 + 20n^2 + 8000n + 10000)/(200n^2 +$

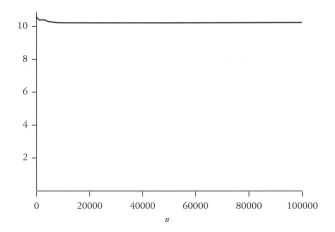

FIGURE 4 Graph of the ratio $(200n^2 + 8000n + 10)/(20n^2 + 300n + 70000)$.

$300n + 70000)$ as in Figure 3, we graph the inverse ratio, namely $(200n^2 + 300n + 70000)/(n^3 + 20n^2 + 8000n + 10000)$, in Figure 5. Figure 5 shows that as n grows large the ratio seems to be approaching a straight line at height 0. But we know that the denominator function is of a higher order of magnitude than the numerator, both from Theorem

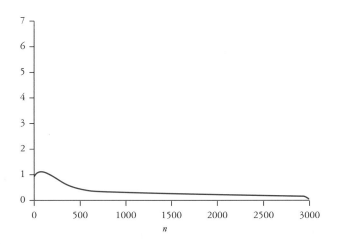

FIGURE 5 Graph of the ratio $(200n^2 + 300n + 70000)/(n^3 + 20n^2 + 8000n + 10000)$.

4 and from our intuition about the graph in Figure 3. So Figure 5 must be telling us the same information: If the ratio becomes a straight line corresponding to a constant $c = 0$, then the denominator function is of a higher order of magnitude than the numerator function. That is why $c > 0$ in the definition.

Looking at graphs is not the same as constructing proofs. But graphs do give us insight as to what statements might be true.

"Big O" for Nonpolynomial Functions

We recall that this discussion of big O and order of magnitude was motivated by our desire to compare time complexities of algorithms. In our examination of time complexities, we encountered functions that were not polynomials. For example, we had the time-complexity function $(n - 1)[(n - 1)!/2 + 1]$ for the brute-force algorithm for the TSP and the function $1 + \log_2 n$ for binary search. We will state some further theorems about order of magnitude here without proofs but with some suggestive graphs; proofs of some of these theorems are given as exercises.

Theorem 6. For any $a, b > 1$, $O(\log_a n) = O(\log_b n)$. Thus, we usually write simply $O(\log n)$.

In Figure 6 the graph of the ratio $\log_2 n/\log_{10} n$ appears to be a horizontal line corresponding to a nonzero constant.

Theorem 7. For $p > 0$, $\log n$ is in $O(n^p)$, but n^p is never in $O(\log n)$. Or equivalently, $O(\log n)$ is a proper subset of $O(n^p)$.

Theorem 7 tells us that in our list of functions at the start of this section, it was correct to list

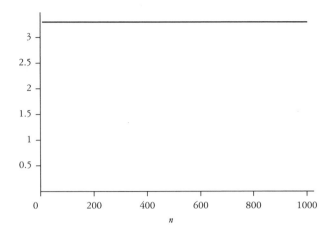

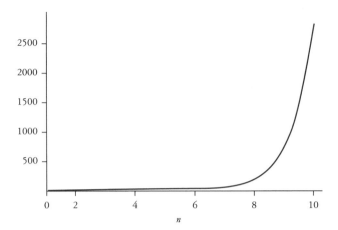

FIGURE 6 Graph of the ratio $\log_2 n/\log_{10} n$.

FIGURE 7 Graph of the ratio $7^n/n^5$.

$\log_{10} n$ first, since $\log_{10} n$ has a lower order of magnitude than n.

Theorem 8. $O(1)$ is a proper subset of $O(\log n)$.

In Exercise 5b you are asked to prove Theorem 8. Theorem 8 cautions us that while $\log n$ grows slowly, it does in fact grow and is thus of a higher order of magnitude than a constant.

Theorem 9. For $p > 0$ and $b > 1$, $O(n^p)$ is a proper subset of $O(b^n)$.

Theorem 9 confirms what we see in Figure 1, namely that polynomial functions of n, like n^3, are of lower orders of magnitude than exponential functions, such as 2^n. For example, in Figure 7 the graph of the ratio $7^n/n^5$ grows larger as n gets larger, suggesting that Theorem 9 is indeed true.

Theorem 10. For $a > b > 1$, $O(b^n)$ is a proper subset of $O(a^n)$.

You are asked to give an intuitive proof of this theorem in Exercise 5c. Theorem 10 confirms the

placement of 2^n before 3^n on our list. A graph of the ratio $2^n/3^n$, as shown in Figure 8, supports the statement of Theorem 10. Recall that if the graph of the ratio approaches a horizontal line of height 0, the denominator function has a higher order of magnitude than the numerator.

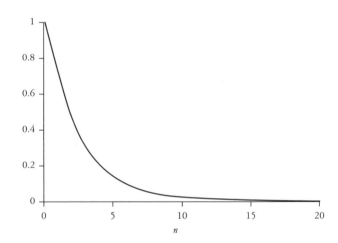

FIGURE 8 Graph of ratio $2^n/3^n$.

Theorem 11. For $b > 1$, $O(b^n)$ is a proper subset of $O(n!)$.

Theorem 11 confirms the placement of $n!$ at the end of our list, since it has the highest order of magnitude of all the functions listed. A graph of the ratio $n!/7^n$, as shown in Figure 9, supports the statement of Theorem 11. You are asked to give an intuitive proof of this theorem in Exercise 5d.

The final theorem of this section gives us rules for finding O for sums and products of the "basic" functions we have been discussing.

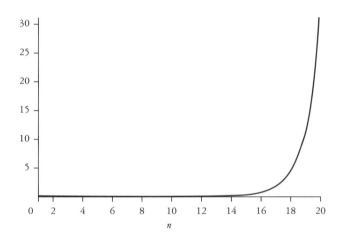

FIGURE 9 Graph of the ratio $n!/7^n$.

Theorem 12. Given two functions f_1 in $O(g_1)$ and f_2 in $O(g_2)$, then

a. $f_1 \cdot f_2$ is in $O(g_1 \cdot g_2)$;

b. $f_1 + f_2$ is in $O(g_1 + g_2)$ provided both f_1 and f_2 have only positive values;

c. if $O(g_1)$ is a proper subset of $O(g_2)$ and f is any positive function, then $O(f \cdot g_1)$ is a proper subset of $O(f \cdot g_2)$.

E X A M P L E 4 The Term of the Highest Order-of-Magnitude Rules Theorem 12b tells us that we need only look at the term with the highest order of magnitude in determining the order of magnitude for a multiterm function. For example, $7n! + 5n^3$ is in $O(n! + n^3)$. But since n^3 is in $O(n!)$, $O(n! + n^3)$ is the same set as $O(n!)$, so we say that $7n! + 5n^3$ is in $O(n!)$. In general, a sum of several functions belongs to the big O of the individual function (term) with the highest order of magnitude. ▲

E X A M P L E 5 In O, We Can "Ignore" Some Terms in a Function In Section 6.1 Example 2 ended with a question as to whether the difference between $2n$ and $2n + 1$ is significant. Similarly, Example 4 of that section ended with a similar question regarding $2n^3 - n^2$ and $2n^3$. From Theorem 12b we now know that these differences are not significant when considering order of magnitude. ▲

Some further examples of the theorems of this section are in order.

E X A M P L E 6 Results of Using Some Theorems in This Section

a. $\log_2 n$ has a smaller order of magnitude than $n^{0.5}$ (Theorem 7).

b. 2^n has a larger order of magnitude than n^{235} (Theorem 9).

c. n^{235} and 2^n both have smaller orders of magnitude than $n!$ (Theorems 9 and 11).

d. $n \log(n)$ has a larger order of magnitude than n but a smaller order of magnitude than n^2. (Theorems 7, 8 and 12c, applied twice. For $n \log(n)$

versus n, think of n as $n \cdot 1$; for $n \log(n)$ versus n^2, think of n^2 as $n \cdot n$.)

e.   These functions are listed in increasing order of magnitude:

any constant, $\log(n)$, n, $n \log(n)$, n^2, n^{1776}, n^{1994}, 2^n, 15^n, 43^n, $n!$. ▲

Using "Big O" to Compare Time-Complexity Functions

E X A M P L E   7   The TSP Revisited   We recall that the time-complexity function for the brute-force algorithm is $(n - 1)[(n - 1)!/2 + 1]$, and the corresponding function for the nearest-neighbor heuristic algorithm is $0.5n^2 - 1.5n + 1$. The time complexity for brute force is O($n!$) and that for nearest neighbor is O(n^2). Since we know that O(n^2) is a proper subset of O($n!$), we know that the nearest-neighbor algorithm is faster to execute than the brute-force algorithm. Using order of magnitude makes comparing the time complexity of these two algorithms very easy. A look back at Table 2 in Section 6.1 reminds us why we use heuristic algorithms for the TSP, since we gain so little by using a faster computer with an O($n!$) algorithm. We can judge heuristic algorithms by how close their results are to an optimal solution. This means that we may need to balance two often-conflicting considerations: time complexity and closeness to optimality. See the Spotlight "Very Complex Problems" for more about problems whose only known solutions have nonpolynomial order of magnitude. ▲

We can use order of magnitude to compare "ugly" functions quickly, like the two time-complexity functions for the TSP algorithms, with-out graphing them or needing to locate crucial values. But we pay for the ease of comparison by losing certain details. One such detail is that for some finite intervals the relative costs of two algorithms, for example, A and B from Example 2 of Section 6.3 can be different than on the infinite interval, $n >$ the biggest crucial value. Another detail is the constant, which "doesn't matter" but in fact can make a difference.

E X A M P L E   8   Order of Magnitude Is Not the Whole Story   In determining big O, we first concentrate on the one term in the sum of terms which has the highest order of magnitude, and then we further ignore the coefficient of that fastest-growing term. Both of these practices hide differences among functions with the same order of magnitude. Using order of magnitude, we would classify $w(n) = 3n^2 + 200n$ as "the same as" $f(n) = n^2 + 5n$. But if these functions represent the work needed for two different algorithms to complete the same job, then clearly the algorithm corresponding to the first function would take at least three times longer than the other algorithm. ▲

In general, techniques and algorithms that "divide and conquer" are faster than those that are linear. Many algorithms make use of this idea by cutting the original problem into two half-size problems. For example, in a binary search, each time we make a comparison and do not find a match, we eliminate "half" of the remaining items, either those before or those after the present item. Often some variation of the word "binary" is in the name of the algorithm, as in binary search and the bisection method for finding roots, but sometimes it is not, as in the sorting techniques merge sort and quick sort.

Very Complex Problems

Mathematicians define the complexity of a problem as the time complexity of the fastest solution algorithm for it. For example, finding an item on an ordered list is $O(\log n)$ because although sequential search is $O(n)$, binary search is $O(\log n)$. Mathematicians generally consider problems to be "nice" if they have solution algorithms whose complexities are polynomial, belong to $O(n^p)$, for some $p > 0$. "Nice" problems are said to belong to the class P (for polynomial).

At the other end of the scale are problems that can be stated quite simply and for which mathematicians have proven that there can never be any solution algorithm at all, no matter how complex we allow that algorithm to be. An example of such a problem is that of safely checking for computer viruses. (See Chapter 5 Spotlight "Checking Computer Viruses.")

In the middle are many interesting problems for which there are solution algorithms, but the known algorithms are not polynomial on real, finite computers. Some of those algorithms could run in "polynomial time," but only if we could run them on computers with unlimited numbers of processors and thus could perform an unlimited number of simultaneous operations. Such problems belong to the class called NP. (The N stands for "nondeterministic," referring to the unlimited nature of the computer.) A major unsolved question is whether the P and NP classes are the same.

The traveling salesperson problem, TSP, is one of many NP problems. The brute-force algorithm produces an optimal solution, but it has the nonpolynomial time complexity $O(n!)$ when we think of it as being run on a typical, limited computer. The nearest-neighbor heuristic algorithm has complexity $O(n^2)$, but it does not provide a *solution* to the TSP, only an *approximate solution.* Much work has been put into the TSP, but no one knows whether there is a (yet undiscovered) solution algorithm for it which has polynomial complexity on a computer with a limited number of processors, that is, whether it is in the class P.

An interesting result in the research in this field of mathematics has grouped some NP problems and shown that if any problem in that group can be solved with a polynomial time algorithm, then all of them can be solved with a polynomial algorithm. And should one of the problems in this group be "cracked" in polynomial time, that algorithm would show the way for us to find "nice" algorithms for all other problems in the group. Thus, a breakthrough on one of these special NP problems not only improves our understanding of that problem but is a breakthrough for all the others as well. This is a surprising and marvelous result, considering that, superficially, the problems in this group do not appear at all similar.

The order of magnitude for divide-and-conquer algorithms usually involves log n.

There are dozens of sorting techniques, algorithms for putting data into order. We recall that the binary search, which is O(log n), requires sorted data as a prerequisite, while the linear search can be used with data in any order but is O(n). Thus time complexity of sorting is a matter of some interest. Here we informally present two sort algorithms and their complexities. The algorithms are insertion sort and binary tree sort.

EXAMPLE 9 Insertion Sort In the *insertion sort* we are building an ordered list from data items we receive one at a time; we put each item into its proper relationship to the other items already received before we accept the next item. Each of the n items that arrives must be compared with the items on the list, starting at one end of the list and proceeding sequentially, until the location for the new item is found. Then the new item is positioned in the list and the process repeated. The search for the proper location is a sequential, or linear, search, which is O(n), and there are n repetitions of this process, so insertion sort is O(n^2). There are other considerations, for example, the computer storage technique, also known as the "data structure." The data structure (linked list) that allows for the easiest insertion of the new item dictates a linear search for proper positioning; a binary search would be possible if the data were stored in an array, but then the positioning would require "moving the data down to make room for the new item." ▲

EXAMPLE 10 Binary Tree Sort In *binary tree sort* we first locate the proper position for the new item and then put that item there. For example, if we want to add "Campbell" to the bi-

nary tree in Figure 10a, we would start at the top of the tree and compare "Campbell" with "Giordano." The tree is constructed so that on the left of any entry are all the items coming before that entry, and similarly those coming after it are on the right. So "Campbell" goes to the left of "Giordano." The same reasoning would put "Campbell" to the right of "Bumcrot" and to the left of "Gallian." In Figure 10b "Campbell" has been added to the tree. The location process is like a binary search, with complexity O(log n), provided that the tree is "nice," not lopsided, too heavy on either side. At each stage in the process we eliminate about half of the remaining items from further consideration. Since we are positioning n new items, in a "nice" tree we have the complexity of O(n log n). But if the data come to us already in order, then the straightforward construction of a "binary tree" gives us a very "un-nice" tree indeed, and we wind up doing some-

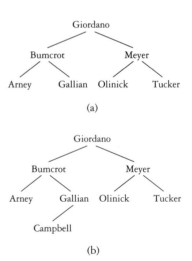

(a)

(b)

FIGURE 10 Adding an item to a binary tree: (a) tree before adding "Campbell"; (b) tree after adding "Campbell."

thing closely resembling an insertion sort, which is much slower, $O(n^2)$. ▲

E X A M P L E   1 1   Prim's and Kruskal's Algorithms   Not only does Kruskal's algorithm for finding a minimal cost spanning tree require more computer memory than Prim's (See the Chapter 4 Spotlight, "Combinatorics Tells Which Algorithm Is Better"), it also requires more work. In Prim's there are $n - 1$ repetitions of the process of finding which edge to use to connect a new node to the growing tree. Finding that new edge is an $O(n)$ process. Using the nested loops principle, we find that Prim's algorithm is an $O(n^2)$ algorithm. The sorting required by Kruskal's algorithm could be performed by one of the faster sorting algorithms, which are $O(k \log k)$ for

a list of k items. Since we have $n(n - 1)/2$ edges to sort, $k = n(n - 1)/2$, or about n^2, making the sorting process one of $O(n^2 \log n^2)$, which is of higher order of magnitude than the $O(n^2)$ we found for Prim's. ▲

The message of this chapter is that analysis of algorithms, and the related data structures, is a field in which mathematics can shed insight. Initially we are happy to have a solution algorithm to a problem for which we had none. But stopping with the first algorithm we get can be a more time-consuming approach than continuing to look for better, faster algorithms. Since algorithms are the heart of the computer programs that perform so many tasks for us in the modern world, it is clear that efficient algorithms are important.

Exercises for Section 6.4

1. For each pair of functions of n, tell which has the bigger order of magnitude.
   - (a) $100n^2$ and $4n^3$
   - (b) $15n$ and $5n^2$
   - (c) $3n^2 + 2n$ and $10n + 408$
   - (d) n^3 and $64n$
   - (e) 5^n and 7^n

2. For each pair of functions of n, tell which has the bigger order of magnitude.
   - (a) 7^n and n^7
   - (b) $n!$ and n^2
   - (c) $4 \log_2 n$ and $n/4$
   - (d) 6^n and $n!$
   - (e) $n^3 + 2^n$ and $n^2 + 3^n$

3. For each of these functions of n, give the order of magnitude in big-O notation.
   - (a) $0.5n + 7n^2 + 13n^{1/2} - 90 + 2n^3$
   - (b) $3n! + 375n^{100}$
   - (c) $7n \log n + n^2$
   - (d) $4n + 3n \log n$

4. For each of these functions of n, give the order of magnitude in big-O notation.
   - (a) $2n^{1/2} + 5 \log n$
   - (b) $7^n + n^7$
   - (c) $9n + 7n^{1/3} + 135 \log n$
   - (d) $13n! + 13^n$

5. Try to construct proofs or intuitive arguments for some of these unproved theorems about order of magni-

tude. Examine the suggested ratios either by using a calculator or mathematics computer package, or by reasoning from what you know about limits, or by using both methods.

(a) O(0) is a proper subset of O(1). Look at the limits as n approaches infinity of the two ratios $1/0$ and $0/1$.

(b) O(1) is a proper subset of O(log n). Look at the limits as n approaches infinity of the two ratios $1/\log n$ and $\log n/1$. Remember that $\log n$ is an increasing function of n, that is, for $n > m$, $\log n > \log m$.

(c) For $a > b > 1$, O(b^n) is a proper subset of O(a^n). Look at the ratio b^n/a^n. It is the same as the fraction b/a, raised to the power n. What happens to that ratio when n grows large?

(d) For $b > 1$, O(b^n) is a proper subset of O($n!$). Look at the limits as n approaches infinity of the two ratios $b^n/n!$ and $n!/b^n$. Both b^n and $n!$ have n factors. So in this case we need to look at the fact that all the factors in b^n are equal to b, but in $n!$ the factors are different. What happens to the "newer" factors in each product as n approaches infinity?

Chapter Summary

Human beings never seem content with just one way to solve a problem, so we develop many ways. When we are faced with several algorithms, it is natural to compare them on various dimensions. The dimension of time complexity is meaningful to us because it can tell us how large a problem we can expect to solve, using existing technology and within a given time period. In this chapter we first focused on finding time-complexity functions for algorithms. Then we compared algorithms by comparing their time-complexity functions. Perhaps the most surprising part of this study is the realization that "a good algorithm on a slower computer is better than a bad algorithm on a faster one."

Chapter 6 Exercises

(Exercises referring to graphing calculators or mathematics computer utilities are marked **T**. Features have been checked on Maple and may be possible on other utilities.)

For Exercises 1–4, follow the instructions to the Exercises in Section 6.3. Then give the order of magnitude of each of the two functions, and explain how the order of magnitude can be used to predict the relative behaviors of the functions on the unbounded interval.

1. A: $4 \log_2 n$ and B: $n/4$

2. A: 6^n and B: $n!$

3. A: $n!$ and B: $720n^5$

4. A: $4 \log n$ and B: $n^{0.5}$

5. In Section 6.1 we developed the two different time-complexity functions, $2n^3 - n^2$ and $2n^3$, for the number of arithmetic operations in an algorithm for calculating S, the matrix that is the square an $n \times n$ matrix **M**. The first function was for hand calculation, the second for pseudocode. Here is another pseudocoded algorithm for finding S:

```
For i from 1 to n
        For j from 1 to n
                {Initialize the total to
                 first product}
                S[i,j] ←M[i,1]·M[1,j]
                {One at a time, calculate
                 the remaining products
                 and add them to the
                 total}
                For k from 2 to n
                        S[i,j] = S[i,j] +
                        M[i,k]·M[k,j]
                end for
        end for
end for
```

Develop a time-complexity function for the number of arithmetic operations in this algorithm, and compare it to the previous formulas. Do all formulas have the same order of magnitude? Is one of the pseudocoded algorithms faster?

6. We can evaluate an $n \times n$ determinant using expansion by minors, which are $n - 1 \times n - 1$ determinants. The value of the determinant is a sum of n terms, each of which is the product of a minor, which of course is just a scalar, times another scalar. Write a difference equation for the time-complexity function for evaluating an $n \times n$ determinant in terms of the evaluation of an $n - 1 \times n - 1$ determinant. For this difference equation, the initial condition would be $w(1) = 0$, since there is no work required to find a 1×1 determinant. Solving this difference equation is very difficult,

beyond the scope of this text. However, we can evaluate $w(n)$ for $n = 2, 3, 4,$ and 5. What order of magnitude would you guess corresponds to the function w?

T 7. Some mathematics utilities can solve some kinds of difference equations (e.g., the `resolve` feature in Maple). Using this feature, try to solve the difference equation and initial condition from Exercise 6. If your utility can solve the difference equation, it may give you a solution in terms of another function, called Gamma. The Gamma function can also show up as the solution to the difference equation $f(n) = nf(n - 1), f(0) = 1$. Do you know what familiar function, having a very high order of magnitude, is a solution to that difference equation?

T 8. If the mathematics utility available to you has the "cost" feature and can evaluate determinants, use the "cost" feature of the package (see Exercise 6 of Section 6.1) to verify the time complexities you found in Exercise 6 for evaluating a determinant. Find the actual cost for $n \times n$ determinants where n is larger than 5. Does your guess as to the order of magnitude for evaluating a determinant remain the same as it was in Exercise 6?

9. In Section 4 of Chapter 3 we learned about Gaussian elimination as a method to solve a system of n equations in n variables. It involves operating on a matrix, **A**, of n rows and $n + 1$ columns, as described in the following pseudocode:

```
For i from 1 to n
        For k from 1 to n, but not k = i
                Calculate the value
                 -A(k,i)/A(i,i), call this
                 value the "multiplier."
                Create a temporary row by
                 multiplying each element
                 of row i by the
                 "multiplier."
                Add the temporary row to row
                 k, component by component.
        end for
end for
```

Write a formula for $w(n)$, the number of arithmetic operations performed in a Gauss–Jordan elimination for n linear equations in n variables.

T 10. If the mathematics computer utility available to you has the "cost" feature and can do Gaussian elimination, use the "cost" feature of the package (see Exercise 6 of Section 6.1) to verify the time complexity you found in Exercise 9 for Gaussian elimination. Find the actual cost for several systems of equations of various sizes. Do the numbers of arithmetic operations actually performed fit into the function you wrote in Exercise 9?

11. Using the results of Exercises 6–8 and 9–10, compare two techniques for solving a system of n equations in n variables: Gaussian elimination and Cramer's rule. The latter expresses the value of each variable as the quotient of two $n \times n$ determinants; the denominator determinant is the same for all n variables, but each variable has a unique determinant in the numerator.

12. Sometimes we are not as interested in the precise value of $n!$ but in approximating $n!$. We might want to know, for example, the number of digits in $n!$ for some particular value of n. In 1730 James Stirling published his famous formula for approximating $n!$:

$$n! \sim (2\pi)^{1/2} n^{n+1/2} e^{-n}.$$

The $\sim$ indicates that the ratio of the two sides of the formula tends to the value 1 as n approaches infinity; the two sides of the formula have the same order of magnitude. Compare the values of $n!$ and of Stirling's formula for each of these values of n, and for each determine the percentage error of Stirling's formula based on the correct value of $n!$: $n = 5, 10, 15, 20$, and 25. (An example: 1! is 1, but Stirling's formula gives 0.9221 for an error of $(1 - 0.9221)/1 = 0.0779/1$, or 8%.) Do you see any trend in the error values?

T 13. Use a mathematics computer package to graph the percentage error function for Stirling's formula (Ex-

ercise 12). That function is $100(n! - (2\pi)^{1/2} n^{n+1/2} e^{-n})/n!$. Can you find how large n must be for the percentage error to be less than 0.1%?

14. Stirling did not have electronic calculators at his disposal, but he did have tables of logs from which he could find the log of a number and the inverse log of a log, i.e., the number whose log is known. Compare how many calculations and table referrals are necessary to find $n!$, both as it is defined and as it is approximated by Stirling's formula. Assume that logs are used to simplify both exponentiation and multiplication. Which method takes less work? Can you find big Os for the number of arithmetic operations plus table lookups in each of these two calculation methods? Recall the techniques for calculating with logs (logs do not produce "short cuts" for $+$ and $-$):

$x \cdot y$ is found by taking the inverse log of $(\log x + \log y)$;

x^p is found by taking the inverse log of $(p \cdot \log x)$.

15. (**Programming Exercise**) This exercise assumes that accurate execution times can be obtained for programs or parts of programs. If you are using a standalone desktop computer and can access the internal clock, then such times can be obtained. In some networked installations, accurate execution times can also be obtained. Compare the execution times for squaring an $n \times n$ matrix of integers smaller than 10 with those for squaring an $n \times n$ matrix whose entries are real values having five nonzero decimal places. Use these values of n: 10, 20, 30. Try to learn how your computer does integer and real arithmetic. Are there really more "basic" operations for one type of arithmetic than for the other?

16. (**Writing Project**) Look up several sorting algorithms in a data structures and algorithms text (see Recommended readings). For each one, describe the average and worst cases and find the big O for those cases. Demonstration software for experimenting with sorting algorithms may also be useful (Meyer in Recommended readings). Can you explain why there is no one "best" sort algorithm?

17. (**Writing Project**) Recently, the search for more efficient algorithms has involved "genetic algorithms," in which several solutions for the same problem are pitted against one another and the best of them are "allowed to breed." That is, parts of two solutions are switched, forming a new solution, much as an offspring in the world of living things is formed from parts of its parents. The parent and offspring algorithms then compete for survival. Read about "genetic algorithms" (see the Recommended Readings by Holland), and write an explanation for your classmates.

CHAPTER

7

LOGIC AND THE
DESIGN OF
"INTELLIGENT"
DEVICES

Introduction

In previous chapters we examined many algorithms. We saw that algorithms can solve problems or carry out tasks by specifying a set of instructions or rules for what action to take at each step, based on the previous actions taken and the data at hand. We know that people can follow, or execute, algorithms; but we are also aware that in today's world, electronic devices also execute algorithms.

In fact, there are two types of electronic devices that execute algorithms: *programmable devices,* such as computers and some calculators, and *dedicated devices.* Programmable devices are capable of carrying out many different algorithms, even algorithms that have yet to be developed. A dedicated device is built

to carry out a specific algorithm, such as warning us that the key is in the ignition if we seem to be leaving the car without removing our keys from the ignition.

In this chapter we will build some dedicated devices. Some will carry out simple algorithms like the "key-in-ignition" warning buzzer. A more complex device will embody the winning strategy we developed in Section 4 of Chapter 5 for the game of Nim. Our devices will be primitive, but in building them we learn the basics of mathematical logic and circuit design needed to build both dedicated circuits and programmable computers.

We begin this chapter by studying basic logic because we need to know how to combine elementary elements into more complex ones. For example, as drivers we are happy to have the key-in-ignition message when we are about to leave the car and have forgotten to take our keys, but certainly we do not want the message to appear while we are driving. So the dedicated devices in our cars typically test for three elementary conditions to determine when to alert us about the keys:

1. the key—in the ignition;

2. the engine—not running;

3. the driver's door—open.

Only when all three of these conditions hold will the message, usually a buzzer, occur. As we learn about mathematical logic, we will learn about the circuits that embody that logic.

After our introduction to logic and circuits in Section 7.1, we will construct some dedicated devices in Section 7.2. In Section 7.3 we will take a brief excursion into quantifiers and mathematical proofs.

SECTION 7.1 *Logic Operations and Gates*

The theory of logic that we introduce here was pioneered by Leibniz, Boole, and others in the 1800s. They hoped to turn the process of human thought into something mechanical. Their dreams have not been realized; we do not see logic symbols and truth tables in our local newspaper! However, the foundations of mathematics have been clarified by logic, which is one reason we study it.

Another reason we study logic is that it is precisely the mathematics needed to create computer chips, which can execute algorithms. It remains to be seen how much of human thinking can be done by algorithms. The many researchers pursuing this question today are the modern heirs of Leibniz and Boole.

Some Basic Definitions

In logic we start with *statements*, often called *propositions*, which have a truth value, that is, they are either true or false. Simple statements have one subject and one predicate. Typical simple statements are as follows:

The key is in the ignition.

The engine is running.

The driver's door is open.

Many simple English sentences are not statements, because they do not have truth values, although these sentences have important uses in human communication. For example, these sentences are not statements:

Is the engine running? (A question does not have a truth value.)

It is open. (The word "it" has not been specified; a sentence with an unspecified subject is an *open sentence*.)

Simple statements can be combined into *compound statements,* those having the more complex grammatical structures typical of human communications and interactions. Three fundamental logic operations are the basis of these combinations: *and, or,* and *not.* We examine these three first. Other logic operations are also useful; we will explore them later. For convenience, we often use a capital letter to represent a simple statement.

EXAMPLE 1 We associate simple English statements with some capital letters:

$$W = \text{“Ten is divisible by two”};$$
$$X = \text{“Ten is divisible by five”};$$
$$Y = \text{“Ten is divisible by three”};$$
$$Z = \text{“Ten is a prime number.”}$$

We know that both W and X are true, while Y and Z are both false. ▲

In the classical logic of Leibniz and Boole, there are no "maybes," no "shades of gray." A statement is either true or false, never somewhere in between those two absolutes. (Recently, some mathematicians have begun studying "fuzzy logic," in which "maybes" are possible; see Chapter Exercise 23.) We see the two-valued nature of classical logic in examining the logic operation *not.* Suppose we wish to convey "Ten is not a prime number," or equivalently, "It is not true that ten is a prime number." This statement is the negation of the simple statement Z. Since we believe Z to be false, we believe its negation to be true. Similarly, if we negate a true statement, such as W, the result is a false statement.

We have just defined *not* informally. Mathematicians define logic operations using truth tables. In a truth table each individual statement becomes a

column in the table, and the rows list all the possibilities for truth values of those statements. There is a column in the truth table for each compound statement. In a definition truth table, there may be just one such column, while in other truth tables there may be many, as we will see. To complete the truth table, we give the truth values for the compound statements, the result of the logic operations. Many mathematical logic utilities, available in some computer packages and on some sophisticated calculators, are capable of generating truth tables.

DEFINITION

The *negation* of a statement changes its truth value. We symbolize the negation of a statement by putting the symbol "~" to the left of the statement being negated. The truth table for negation is

P	$\sim P$
T	F
F	T

We say "not P" or "the negation of P" for $\sim P$.

Negation is a unary operation; it "operates" on one logic statement, producing a different but related statement. Both *and* and *or* are binary operations, that is, they "operate" on two logic statements to produce a different but related new statement. Using the multiplication principle from Section 1 of Chapter 4, we see that we have two cases to define for *not,* since *not* operates on just one statement, which can be either true or false. For *and* and *or,* since we need two statements, each of which independently can be either true or false, the multiplication principle tells us that we have four cases. The number of cases translates into the number of rows in the truth table definitions. When there are n simple statements in a truth table, that table must have 2^n rows.

In English we could write, "Ten is divisible by two and ten is divisible by five." Equivalently, we could write, "Ten is divisible by both two and five." These English sentences are formed by putting W and X together using the word "and."

DEFINITIONS

The binary operation *and* is symbolized by putting "∧" between the statements being joined. The truth table for *and* is

P	Q	P ∧ Q
T	T	T
T	F	F
F	T	F
F	F	F

The compound logic statement $P \wedge Q$ is called a *conjunction* and is read "P and Q" or "the conjunction of P and Q." A conjunction is true if both of its constituent statements are true; otherwise it is false.

EXAMPLE 2 **Some Conjunctions** The compound statement "Ten is divisible by two and ten is divisible by five" would be symbolized as $W \wedge X$. Since we have agreed that both W and X are true, the combination $W \wedge X$ is also true. However, since Z is false, the statement $W \wedge Z$ is false. An English translation of $W \wedge Z$ would be "Ten is divisible by two and ten is a prime." ▲

Finally, we examine the statement "Either ten is divisible by two or ten is divisible by three." This compound statement uses the word "or" to link the simple statements we have symbolized using W and Y.

DEFINITIONS

The binary operation *or* is symbolized by putting "∨" between the statements being joined. The truth table for *or* is

P	Q	P ∨ Q
T	T	T
T	F	T
F	T	T
F	F	F

The compound logic statement $P \vee Q$ is called a *disjunction* and is read "P or Q" or "the disjunction of P and Q." A disjunction is true if one or more of the constituent statements is true and is false if both of them are false.

EXAMPLE 3 **Some Disjunctions** The compound statement "Either ten is divisible by two or ten is divisible by three" would be symbolized as $W \vee Y$. We would agree that since W is true, the entire statement is true, regardless of the truth value of Y (which happens to be false). We would also declare $W \vee X$ ("Ten is divisible by two or ten is divisible by five") to be true since both of the constituents are true, but $Z \vee Y$ ("Ten is divisible by three or ten is a prime number") would be false since both of the constituent statements are false. ▲

The disjunction is sometimes called the *inclusive or,* because we define it to be true if both of the constituent statements are true. In typical, non-mathematical usage, people often use a different meaning: People saying "or" typically mean "either one or the other, but not both," which mathematicians call *exclusive or.* When you decide to order either coffee or tea at a restaurant, you are using the *exclusive or.* In our discussion of the addition princi-

"Why Does It Have To Be (Wrong or Right)?"

In the title of their 1986 hit song, Restless Heart poses a question that lies at the center of a recent development in mathematical logic. Classical logic obeys the *law of the excluded middle,* first stated by Aristotle (384–322 B.C.): There is no truth value other than true (1) or false (0). Since the statement "*a* is an element of set *S*" would be either true or false, in classical set theory an element is either in a set or not. As Restless Heart's song title reminds us, there are every-day situations in which the middle ground is where we want to be.

Lotfi Zadeh

In 1965 Lotfi Zadeh published a groundbreaking paper, "Fuzzy Sets." In the logic governing fuzzy sets, a statement like "*a* is an element of the set *S*" can have a truth value anywhere between 0 and 1. The in-between values correspond to "shades of gray."

For example, in evaluating the statement "52° F is a cold temperature," we could say that it is 0.2 true, meaning that 52° F belongs to the set "cold temperature" with truth value 0.2. We might also say that "52° F is a cool temperature" has truth value 0.8. Using fuzzy logic, a controller for a fan blowing warm air into a room would use the 52° by combining 0.2 of the "cold temperature" rule and 0.8 of the "cool temperature" rule to determine the fan speed.

"Fuzzy logic" has been used in applications as diverse as washing machines, which determine how long the wash cycle should be, and automated subway car controllers, which regulate train speed. Fuzzy logic devices often outperform the standard control devices; they sometimes even outperform humans.

ple, in Section 1 of Chapter 4, we referred to "mutually exclusive" options and noted that the word "or" is often used in a situation calling for that principle. The form of "or" in which options are mutually exclusive is the *exclusive or.* We will return to further discussion of the *exclusive or* after we concentrate on the plain vanilla (*inclusive*) *or.*

For some purposes, it is unnecessary to distinguish between the two *or*s. If someone makes a comment like, "The Yankees will win this game or I'm a monkey's uncle," the person surely intends for us to take the entire compound statement to be true.

Further, we are expected to treat the second of the simple statements as false. Thus, we must conclude that the first of the simple statements is true, no matter which interpretation we use for "or."

Gates for the Basic Operations

Let us return to the problem of designing dedicated devices. In our example of the buzzer that alerts us that the key is in the ignition, we found three constituent conditions that needed to be tested. In representing a typical electrical device, we use an input wire as the equivalent of a simple statement, such as "Driver's door is open." The input wire is said to be "1" when the statement is true and "0" when the statement is false. In practice, the 1 and 0 correspond to two voltage settings. Standard symbols called *gates* are used in diagrams to show the operations being performed on the voltages on the wires. The diagrams are usually called *circuit* diagrams. The word "circuit" refers to a closed path, like a cycle in a graph (Chapter 5), because in order for any electrical device to work, the wires and components must form a closed circuit. However, it is typical to present circuit diagrams as if they just have inputs and outputs, and to ignore the closure of the circuit, assuming it to occur off the diagram somewhere. The gates for the logic operations we have studied thus far—*not, and,* and *or*—are shown in Figure 1.

The *not* gate

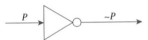

The *and* gate

The *or* gate

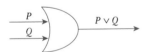

FIGURE 1 Gates for *not, and,* and *or.*

Each line, or edge, of the gates in Figure 1 has a direction indicated by an arrowhead. This is to indicate direction within the circuit. When the arrowheads are omitted in complicated circuit diagrams, we read the circuit from left to right.

EXAMPLE 4 **A Truth Table and the Corresponding Circuit** We can produce more complicated statements involving more than two letters or more than one operator, using parentheses in the expression to group items, thus specifying the order of operations. Suppose A = "I have a delicious dinner," B = "I get a good night's sleep," and C = "I am happy." Then the sentence "Either I have a delicious dinner and get a good night's sleep or I am not happy" would be symbolized as $(A \wedge B) \vee \sim C$. We will construct a truth table for this compound statement; since there are three letters in the compound statement, we will need $2^3 = 8$ lines in the table. The first three columns are the individual letters, the rightmost column is the final statement, and the two intermediate columns are parts of the final statement.

A	B	C	$A \wedge B$	$\sim C$	$(A \wedge B) \vee \sim C$
T	T	T	T	F	T
T	T	F	T	T	T
T	F	T	F	F	F
T	F	F	F	T	T
F	T	T	F	F	F
F	T	F	F	T	T
F	F	T	F	F	F
F	F	F	F	T	T

The hierarchy of logic operations places *not* above *and* and *and* above *or*. Thus, we did not need to put parentheses around $\sim C$, although we could have. We also did not actually need the set of parentheses we do have in the $(A \wedge B)$ portion of the statement. Typically, we use parentheses in logic statements for clarity, not relying exclusively on the hierarchy.

We can also think of A, B, and C as input wires to a circuit whose output is the combination $(A \wedge B) \vee \sim C$. Each of the input wires would have the value "1" if the corresponding statement is true and "0" if the statement is false. The output from the circuit will be a "1" if the compound statement is true and "0" if it is false. The *signal table,* showing these 1s and 0s, is merely a transformation of the truth table, showing Ts and Fs, as we see from the following table:

A	B	C	A ∧ B	~C	(A ∧ B) ∨ ~C
1	1	1	1	0	1
1	1	0	1	1	1
1	0	1	0	0	0
1	0	0	0	1	1
0	1	1	0	0	0
0	1	0	0	1	1
0	0	1	0	0	0
0	0	0	0	1	1

In further discussions we will understand that truth and signal tables are interchangeable for our purposes, and so we will not present both of them.

Note that the pattern of the 1s and 0s in the columns for the statements *A*, *B*, and *C* is similar to the pattern we used in Example 10 from Section 1 of Chapter 4. The rows here are the same as the rows there but in exactly the reverse order.

Figure 2 is a circuit diagram for (*A* ∧ *B*) ∨ ~*C*. We can think of this circuit as a testing device: Given the truth values of the input statements, the device tells us if the compound statement is true or false. Each of the columns in the truth table for (*A* ∧ *B*) ∨ ~*C* corresponds to the output of a gate in this diagram. ▲

In a circuit diagram the difference between an actual connection, such as a split, and an "accidental" crossing of lines, forced by a planar drawing of a nonplanar circuit, is shown by placing a solid dot at an actual connection. Where lines cross and there is no dot, the wires cross over and under one another.

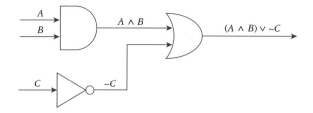

FIGURE 2 Circuit diagram for (*A* ∧ *B*) ∨ ~*C*.

Tautologies and Contradictions

Some statements always have the same truth value; we have special names for such statements.

DEFINITIONS

A *tautology* is a statement that is always true.

A *contradiction* is a statement that is always false.

Tautologies and contradictions are so important that many mathematical logic utilities have special functions that evaluate a compound statement to determine if it is one of these two special types.

EXAMPLE 5 **A Tautology and a Contradiction** We use a truth table to show that *A* ∨ ~*A* is a tautology and *A* ∧ ~*A* is a contradiction.

A	~A	A ∨ ~A	A ∧ ~A
T	F	T	F
F	T	T	F

Intuitively, we understand that no statement can be both true and false, so we believe that *A* ∧ ~*A* is a contradiction. Similarly, any statement is either true or false, and A ∨ ~*A* is thus a tautology. ▲

EXAMPLE 6 Looks Aren't Everything

Sometimes logic expressions look similar but do not have the same meaning. Other times logic expressions look different but do have the same meaning. A truth table for $\sim A \vee \sim B$, $\sim(A \wedge B)$, and $\sim(A \vee B)$ will illustrate these points.

A	B	$\sim A$	$\sim B$	$\sim A \vee \sim B$	$A \wedge B$	$\sim(A \wedge B)$	$A \vee B$	$\sim(A \vee B)$
T	T	F	F	F	T	F	T	F
T	F	F	T	T	F	T	T	F
F	T	T	F	T	F	T	T	F
F	F	T	T	T	F	T	F	T

If we assign A the meaning "Alice is talented," and B the meaning "Bijan is hard-working," then $\sim A$ has the meaning "Alice is not talented," and $\sim B$ has the meaning "Bijan is not hard-working." The compound statement $\sim A \vee \sim B$ would translate as "Either Alice is not talented or Bijan is not hard-working," while $\sim(A \wedge B)$ would translate as "It is not true that Alice is talented and Bijan is hard-working," and $\sim(A \vee B)$ would translate as "It is not true that Alice is talented or Bijan is hard-working."

The truth table tells us that the sentences $\sim A \vee \sim B$, "Either Alice is not talented or Bijan is not hard-working," and $\sim(A \wedge B)$, "It is not true that Alice is talented and Bijan is hard-working," have the same logical meaning, but the sentence $\sim(A \vee B)$, "It is not true that Alice is talented or Bijan is hard-working," does not share that meaning. When we define more binary operations, we will explore some other English sentences that correspond to these logical expressions. ▲

We note that the expressions $\sim A \vee \sim B$ and $\sim(A \wedge B)$ have the same truth table, but the expressions $\sim A \vee \sim B$ and $\sim(A \vee B)$ have different truth tables even though both expressions in the second pair involve disjunctions while in the first pair we have one conjunction and one disjunction. If we had a symbol with the meaning "has the same truth table as," we could create more tautologies by formulating statements of the form "This expression has the same truth table as that expression."

DEFINITION

The symbol "↔" is used for "has the same truth table as," often expressed in mathematics as "is logically equivalent to," "if and only if," or "iff." There is no gate for ↔. The truth table is

P	Q	$P \leftrightarrow Q$
T	T	T
T	F	F
F	T	F
F	F	T

EXAMPLE 7 Relationship of Equivalence and Tautology

Based on Example 6, we can state that $(\sim A \wedge \sim B) \leftrightarrow \sim(A \vee B)$ is a tautology because the truth tables for the left and right halves of the equivalence are the same. ▲

Some tautologies occur frequently and are usually listed in texts on logic. We follow that tradition in Table 1, using the convention that "T" will stand for any true statement or tautology and "F" will stand for any false statement or contradiction.

One way to prove that a statement is a tautology is to construct its truth table and show that all rows under the statement have the value true. In effect, we proved one of the DeMorgan's tautologies in Example 6 and both of the inverse tautologies in Ex-

TABLE 1 Frequently Occurring Tautologies.	
Commutativity:	$(P \wedge Q) \leftrightarrow (Q \wedge P)$
	$(P \vee Q) \leftrightarrow (Q \vee P)$
Associativity:	$[P \wedge (Q \wedge R)] \leftrightarrow [(P \wedge Q) \wedge R]$
	$[P \vee (Q \vee R)] \leftrightarrow [(P \vee Q) \vee R]$
Distributivity:	$[P \vee (Q \wedge R)] \leftrightarrow [(P \vee Q) \wedge (P \vee R)]$
	$[P \wedge (Q \vee R)] \leftrightarrow [(P \wedge Q) \vee (P \wedge R)]$
Identity:	$(P \vee F) \leftrightarrow P$
	$(P \wedge T) \leftrightarrow P$
Inverse:	$(P \vee \sim P) \leftrightarrow T$
	$(P \wedge \sim P) \leftrightarrow F$
DeMorgan:	$\sim(P \wedge Q) \leftrightarrow (\sim P \vee \sim Q)$
	$\sim(P \vee Q) \leftrightarrow (\sim P \wedge \sim Q)$

DEFINITIONS

We define *nand, nor,* and *exclusive or,* which we will represent symbolically as |, ↓, and ⊕, respectively, in the truth table in Table 2.

TABLE 2 Definitions for *nand, nor,* and *exclusive or.*

P	Q	$P \mid Q$	$P \downarrow Q$	$P \oplus Q$
T	T	F	F	F
T	F	T	F	T
F	T	T	F	T
F	F	T	T	F

ample 5. In those examples the column with the actual tautology was omitted, but the two equivalent halves of the tautology were given, and the truth values in both those columns match, row for row. The proofs of the other tautologies are left as exercises for the reader.

Mathematical logic utilities often include a "simplify" command. Such a command uses tautologies such as the ones in Table 1 to rewrite a compound logic expression with fewer symbols. For example, using the inverse and identity tautologies, we can rewrite $Q \wedge (P \vee \sim P)$ as $Q \wedge T$, which simplifies to just Q. If we asked a utility to simplify $Q \wedge (P \vee \sim P)$, it would respond with Q. One of the essential steps in designing a circuit is to simplify the logical expression it embodies, so that the circuit can be made with a minimum number of gates. Chapter Exercise 24 refers to minimization algorithms.

More Operations and Their Gates

As we mentioned earlier, other binary operations in logic are useful.

We have previously discussed the meaning of *exclusive or.* The *nand* is the negation of an *and,* and the *nor* is the negation of an *or.*

There is no English word with the same meaning as the logical *nand.* The statement $\sim(A \wedge B)$, from Example 6, is a *nand* in structure, since it is the negation of an *and.* It was translated as "It is not true that Alice is talented and Bijan is hard-working." Using a DeMorgan tautology, $\sim(A \wedge B)$ becomes $\sim A \vee \sim B$, which translates as "Either Alice is not talented or Bijan is not hard-working."

We usually find "nor" as the second half of a pair, the first half of which is "neither." The $\sim(A \vee B)$ from Example 6 has the structure of a *nor.* It was translated there as "It is not true that Alice is talented or Bijan is hard-working." Another translation comes from the *nor* structure: "Neither is Alice talented nor is Bijan hard-working." And using a DeMorgan tautology, we arrive at $\sim A \wedge \sim B$, which we might translate as "Alice is not talented and Bijan is not hard-working." All of these translations correspond to the same logic expression, but we may find some of them awkward in English. It is important to realize that *none* of the sentences corresponding to $\sim(A \wedge B)$ is a correct translation for $\sim(A \vee B)$.

The gates for these three operations are shown in Figure 3. Note that the gate symbol for *nand* is just an *and* gate with the addition of the small circle from the *not* gate. The gate for *nor* is constructed analogously.

The *nand* gate

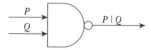

The *nor* gate

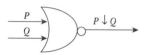

The *exclusive or* gate

FIGURE 3 Gates for *nand, nor,* and *exclusive or.*

We have defined five gates that can be used to convert two binary input signals into a binary output signal. These five are symbolized as $\wedge$, $\vee$, $|$, $\downarrow$, and $\oplus$, and they correspond to aspects of how we speak in English. Thus they hold some intrinsic interest for us. When we define a binary operation, we specify a column of four truth values, as we did in columns 3, 4, and 5 of Table 2. Since each of those four values could be either T or F, by the multiplication principle from Chapter 4 we have $2^4 = 16$ possible different binary operations in logic. Five of these possibilities are the abovementioned gates. Many of the

remaining 11 possibilities, but not all of them, might strike us as arbitrary definitions having no particular meaning in terms of how we think or speak.

We presented $\wedge$, $\vee$ and $\sim$ (a unary operation) before the three additional binary operations because the three others can be represented, albeit awkwardly, in terms of $\wedge$, $\vee$ and $\sim$. Further, $\wedge$, $\vee$ and $\sim$ can each be defined in terms of either $|$ or $\downarrow$. Here are some examples of such representations. We will explore equivalent expressions further in some of the exercises.

E X A M P L E 8 Logically Equivalent Expressions Both the expressions $(A \vee B) \wedge \sim(A \wedge B)$ and $(A \wedge \sim B) \vee (\sim A \wedge B)$ are equivalent to $A \oplus B$. By using the special symbol for *exclusive or,* we convey the meaning quickly and the expression is shorter. ▲

E X A M P L E 9 Complete Boolean Operators—*nand* and *nor* In the following truth table we express $\sim A$ in terms of *nand* as $A|A$ and in terms of *nor* as $A \downarrow A$.

A	$\sim A$	$A|A$	$A \downarrow A$
T	F	F	F
F	T	T	T

Once we can express $\sim A$ using a *nand*, we can see that we can easily use just *nand*s and create the equivalent of an *and*. For example, $A \wedge B \leftrightarrow (A|B) | (A|B)$. Then we can use De-Morgan's tautologies and use just *nand*s to write the equivalent of an *or*. Similarly, we could write the equivalent of *not* and *or*, and thus of *and*, using only *nor*s. Both *nand* and *nor* are called *complete* because any Boolean expression can be represented using either just *nand*s or just *nor*s (see Exercises 9 and 10 at the end of this section). No other binary operation—no other two-input gate—has this property, which is why the *nand*

and *nor* gates are often preferred for building actual circuits. ▲

In the next section we will use the gates introduced here to build some dedicated devices, one of which plays the game of Nim.

$$\boxed{\textbf{E x e r c i s e s \quad f o r \quad S e c t i o n \quad 7 . 1}}$$

All of these exercises are intended to be worked by hand; a mathematical logic utility may be useful to check on the truth tables, but such a utility is not necessary for solving any of these problems.

1. Translate the logic expressions into English, using these definitions of the simple statements: C = "I am a college student"; H = "My parents dislike my hair style"; V = "I voted in the last Presidential election"; G = "I play the guitar."

    (a) $C \lor \sim H$
    (b) $G \land V$
    (c) $\sim(C \land V)$
    (d) $(V \lor H) \land G$
    (e) $\sim G \land C$

2. Translate the logic expressions into English, using these definitions of simple statements: S = "I like to sleep late"; P = "I eat pizza every day"; M = "Math is my most interesting course"; R = "I read a lot."

    (a) $\sim S \land M$
    (b) $P \lor R$
    (c) $\sim(S \lor R)$
    (d) $(\sim P \land S) \lor M$
    (e) $M \lor \sim R$

3. Using the definitions in Exercise 1, write the symbolic equivalents of these English sentences:

    (a) I am a college student who does not play the guitar.
    (b) Either I am a college student or my parents like my hair style.
    (c) I am a guitar-playing college student who did not vote in the last Presidential election.

    (d) Either I voted in the last Presidential election or I am a college student whose parents dislike my hair style.
    (e) It is not true that I am a guitar-playing college student.

4. Using the definitions in Exercise 2, write the symbolic equivalents of these English sentences:

    (a) I eat pizza every day and like to sleep late.
    (b) Math is not my most interesting course, but I read a lot.
    (c) It is not true that either I read a lot or math is my most interesting course.
    (d) Either I read a lot or I like to sleep late, and I eat pizza every day.
    (e) Math is my most interesting course and I like to read a lot, and I do not eat pizza every day.

5. Use truth tables to determine if the two expressions are logically equivalent:

    (a) $A \land \sim B$ and $\sim(\sim A \land B)$
    (b) $M \lor \sim N$ and $\sim(\sim M \land N)$
    (c) $P \land (\sim Q \land R)$ and $(P \land \sim Q) \land R$

6. Use truth tables to determine if the two expressions are logically equivalent:

    (a) $\sim(M \land \sim N)$ and $\sim N \lor \sim M$
    (b) $\sim(A \lor B)$ and $\sim A \land \sim B$
    (c) $\sim P \land (Q \lor R)$ and $(\sim P \land Q) \lor (\sim P \land R)$

7. Use truth tables to determine if each of these statements is a tautology, a contradiction, or neither:
   (a) ~(P ∧ ~Q) ↔ (~P ∨ Q)
   (b) [A ∨ (B ∧ C)] ↔ [(A ∨ B) ∧ C]

8. Use truth tables to determine if each of these statements is a tautology, a contradiction, or neither:
   (a) [(M ∧ ~M) ∨ N] ↔ [(M ∨ ~M) ∧ ~N]
   (b) [P ∧ (~Q ∧ (R ∨ ~R))] ↔ [(P ∧ ~Q)]

9. For each of these four expressions, find an equivalent one using only *nand*s:
   (a) ~A
   (b) A ∧ B
   (c) A ∨ B (Hint: Use a DeMorgan tautology.)
   (d) A ⊕ B (Hint: A ⊕ B ↔ ~(~(A ∧ ~B) ∧ ~(~A ∧ B)). Verify and use the equivalence.)

10. For each of these four expressions, find an equivalent one using only *nor*s:
    (a) ~A
    (b) A ∨ B
    (c) A ∧ B (Hint: Use a DeMorgan tautology.)
    (d) A ⊕ B (Hint: A ⊕ B ↔ ~(~(A ∨ B) ∨ ~(~A ∨ ~B)). Verify and use the equivalence.)

11. Construct circuits corresponding to these logic expressions:
    (a) (P ∧ Q) ∨ (P ∧ (~Q ∨ R))
    (b) (~P ∧ ~Q ∧ ~R) ∨ ~(~P ∨ ~Q ∨ ~R)

12. Construct circuits corresponding to these logic expressions:
    (a) (P ∧ ~Q ∧ R) ∨ (~P ∧ ~Q ∧ R) ∨ (P ∧ Q ∧ ~R)
    (b) ~(P ∧ Q) ∨ (R ∧ (~P ∨ Q))

13. Find logic expressions corresponding to these circuits:

(a)
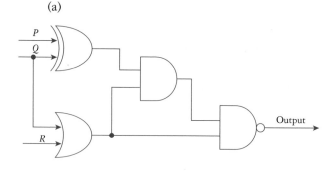

(b)
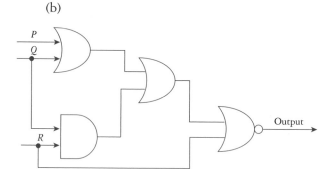

14. Find logic expressions corresponding to these circuits:

(a)
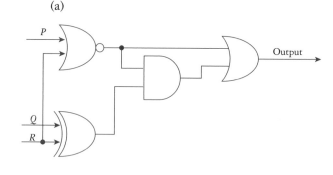

(b)
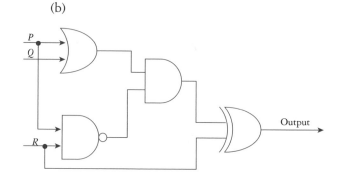

Circuits, Boolean Functions, and Dedicated Devices

TABLE 1 Defining the Key-In-Ignition Warning Buzzer.

Input variables			Output variable
K	E	D	B
T	T	T	F
T	T	F	F
T	F	T	F
T	F	F	T
F	T	T	F
F	T	F	F
F	F	T	F
F	F	F	F

A Key-in-Ignition Warning Buzzer

We begin our construction of dedicated devices with the "key is in the ignition" buzzer mentioned in the introduction. Our analysis of the behavior of a typical warning buzzer suggests that three aspects of the car's status, to which we assign variable names, determine the buzzer's behavior:

K = The key is in the ignition;

E = The engine is running;

D = The driver's door is closed.

We could have defined these variables in other ways; for example, giving the opposite meaning to E, namely that the engine is *not* running. What is important is that we have identified the inputs whose values are used to determine whether or not the buzzer buzzes. Now we need to specify exactly which combinations of these inputs cause the buzzing. So we need another variable:

B = The buzzer buzzes.

Table 1 is a definition table. It shows all possible truth value combinations for the inputs and defines how the output variable, B, is supposed to behave.

The T/F values under the input variables are arranged in the standard way for a truth table having three variables. The fourth variable, B, is really a function of the first three, $B = f(K, E, D)$. B is a Boolean function, and K, E, and D are Boolean variables.

DEFINITIONS

A *Boolean variable* is a variable that can have only two values, usually thought of as T = "true" and F = "false."

A *Boolean function* is a function that has only two values in its range, again T and F. A Boolean function can be a function of one, two, or any other number of variables.

We need to be able to express our function B as a Boolean expression. Then we will be able to create a circuit corresponding to that expression.

Before we get to the expression, we need to focus on the T/F values for the variable B. How were they chosen? They were assigned by our understanding of how the buzzer actually works. In our experience, just one combination of the three input variables sets the buzzer off:

1. The key is in the ignition.

2. The car is not running.

3. The driver's door is open.

In terms of our input variables, this combination corresponds to K being true and both E and D being false. In the truth table, therefore, the line with that particular definition of the input variables also has B set as T, meaning that the buzzer buzzes. In all other cases, the buzzer should not buzz, so there we set $B = $ F.

An important part of designing a dedicated device is the kind of thought needed to create the definition table in Table 1. In order to create that table, we needed to identify the input variables and clearly specify which combinations of the inputs should cause the output to be true and which should cause it to be false. This analysis is critical and cannot be done by a machine—what machine could tell us how *we* want the buzzer to behave?

Now we learn how to translate the table definition of our Boolean function into a Boolean expression. The problem we are about to solve is the reverse of the problems we encountered in the previous section. There we had a Boolean function, which we called a logical expression, and constructed a circuit to embody it, or we had a circuit and found a Boolean function (logical expression) that the circuit evaluated. Here we have, as yet, neither circuit nor Boolean function. We need to find the Boolean function corresponding to B in order to build the circuit that has B as its output.

We know that the Boolean function B has the input variables K, E, and D. This is similar to knowing that y is a function of the variable x but not knowing whether the expression for that function is $4x^2$, or $\sin(x)$, or $3x + 7$, and so on. In solving a difference equation plus initial condition, as we did in Chapters 1 and 6, our goal is to find a formula for the function, f or w, in terms of the variable, x or n. Here we also seek a formula, but the method we use is different.

We begin by noting that only line 4 of Table 1 represents a combination of inputs that correspond to a T as the value of the Boolean function B. The other combinations listed correspond to a value of F. For that line of Table 1 where $B = T$, we will create a Boolean expression that equals T when the input variables have the values in that line but equals F for any other line. Our expression will be a conjunction, an "*and*ing," involving the input variables. What conjunction of the input variables would be true if the actual values of those variables are the ones found on line 4? Here is one that works: $K \wedge {\sim}E \wedge {\sim}D$. Each input variable whose value on line 4 is T shows up in the conjunction as itself. And each input variable whose value on line 4 is F shows up negated in the conjunction. Since on line 4 both E and D are false, negating them gives the value true. "*And*ing" the three trues together results in the output true, which is what we desire for the function B.

Before we decide that the Boolean expression $K \wedge {\sim}E \wedge {\sim}D$ will work to define B, we need to check that it never has the value true on any of the seven other lines of Table 1. Pick any other line, for example, line 6. On that line K has the value F, so our expression, a conjunction, will also have the value F. This same argument can show that our expression has the value F on lines 5, 7, and 8, since K is false on those lines also. What about a line where K is true, for example, lines 1, 2, and 3? On each of those lines one (or both) of the variables E and D is true. But we are negating the value of both of those variables before we form the conjunction, so there would be a false value as input to the conjunction, making the output also false. Our expression is carefully designed to give us an output T for exactly one line of the definition table for our function B.

It is a relatively easy job to draw the circuit that

embodies the Boolean function B; that circuit is shown in Figure 1.

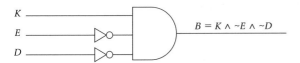

FIGURE 1 Circuit for a key-in-ignition warning buzzer.

The commutativity and associativity tautologies in Table 1, Section 1, allow us to expand our *and* and *or* gates to gates having more than two inputs. An expanded *and* gate, such as in Figure 1, will produce an output of T only in the case that all inputs are T. An expanded *or* gate, such as we will see in Figure 2, will produce an output of F only in the case that all inputs are F. Expanded gates can be physically constructed from several basic gates of the corresponding type.

The approach we just used can work for more complicated circuits: those in which there is more than one line of input for which we desire that the output be T, and those with more than one output. Before we look at the more complicated cases, we will make a slight change in our notation for Boolean expressions.

An Alternative Notation

Here is an alternative notation for the logic operations:

Our original notation	Alternative notation
$A \wedge B$	AB
$A \vee B$	$A + B$
$\sim A$	A'

Using this alternative notation, our expression $B = K \wedge \sim E \wedge \sim D$ becomes

$$B = KE'D'.$$

We will use this alternative notation for the remaining portion of this section. It is more compact and easier to type. Note that the juxtaposition we are using means "and," and the "$+$" means "or," not their usual meanings in arithmetic. This is important to remember because sometimes people refer to an expression like $KE'D'$ as a "product" and an expression like $K + D$ as a "sum."

We will also switch at this point to using "1" instead of "T" and "0" instead of "F" as the values for our Boolean variables and functions. The 1 and 0 are the typical values used in circuitry design. An added bonus to this change will become apparent when we develop a Nim-playing device and need binary numeration.

Building Circuits for Passive Restraint Seatbelts

While I was writing this chapter, I purchased a 1994 Saturn. So naturally, when I noticed that the passive restraint (automatically engaging) seatbelt on the driver's side and the one on the passenger's side seemed to be obeying different rules, the mathematician part of my brain started working.

The hard part of the analysis was to determine the inputs. The outputs are clearer to describe—they are the signals that tell the seatbelt holders over the driver's and passenger's doors to remain where they are or to move to the alternative position. So the first step is easy, define an output variable C (change), with these meanings:

C = 1   means change the position of the holder, regardless of which position it is now in;

C = 0   means do not change the position; preserve the status quo.

Determining what prompted the holders to change position required that I sit in the car and try various things, changing the status of parts of the car and noting how the holders responded. As we noted above, this is the critical part of creating (or recreating) a circuit, and also a very interesting part. The holders on the 1994 Saturn respond to various combinations of key, door, and belt statuses. The key status is the most subtle, involving the several possible positions of the key. The true picture, therefore, would require four or more input variables. For the sake of simplicity I have reduced these subtleties to just two key statuses: "in" and "out." This simplification does suffice to explain the differing behaviors of the driver's and passenger's door. Here are the variables I used for the driver's door:

K = Key status: 1 = in   0 = out

D = Door status: 1 = shut   0 = open

B = Belt status: 1 = back,   0 = forward,
                         engaged   open

Table 2 gives the definition of C as a Boolean function of the variables K, D, and B for the driver's side of the car.

Line 1 of Table 2 tells us that if the key is in the ignition with the engine running, and the driver's door is shut, and the seat belt is engaged, then there should be no change in the status of the seat belt holder. By contrast, line 3 tells us that if the key is in the ignition with the engine running, and the

TABLE 2 Function Table for Driver's Side Passive Restraint Seatbelt.

Input variables			Output variable
K	D	B	C
1	1	1	0
1	1	0	1
1	0	1	1
1	0	0	0
0	1	1	1
0	1	0	0
0	0	1	1
0	0	0	0

driver's door is open, and the seat belt is engaged, then there should be a change in the status of the seat belt holder. A reason for the value given to C on line 3 might be that a driver might want to momentarily exit the car while it was running, for example, to check the engine timing, which cannot be checked when the engine is not running. You can probably think of reasons why all the values for C have been set as in Table 2.

For the driver's seat belt, we want to find the Boolean function $C = f(K, D, B)$. The first thing that we might notice is that, in contrast to the buzzer problem, the output variable in this case has several 1s (same as Ts) in its definition column. Well, when do the holders change position? The answer is that they do so if either the status is as described on line 2, or the status is as described on line 3, or on line 5, or on line 7. What we need to do, therefore, is describe each line where there is a 1 under the C, in the same way we did earlier, and then "*or* them all together."

Recall the conjunction we formed for line 4 in the buzzer analysis. Each input variable whose value on the line is 1, or T, shows up in the conjunction as itself. Each input variable whose value on the line is 0, or F, shows up negated in the conjunction. Using this technique, we get KDB' as the conjunction for line 2, since on line 2 both K and D are 1 and B is 0. Similarly, we get $KD'B$ as the conjunction for line 3. For lines 5 and 7 we get, respectively, $K'DB$ and $K'D'B$.

We want to form a disjunction of these four conjunctions, to "*or* them all together." When we do that, we get $KDB' + KD'B + K'DB + K'D'B$. How do we know that this is a correct Boolean expression for C? Since the expression is a disjunction, it will have the value 1 whenever any one of its individual terms is a 1. The individual terms are each designed to have the value 1 for exactly 1 of the lines in Table 2. So $C = KDB' + KD'B + K'DB + K'D'B$ will have the value 1 exactly when the status of the car corresponds to one of the four lines we used to make up the four conjunctions. That is precisely what we want.

In Figure 2 we see a circuit that corresponds to our Boolean expression for C, namely $C = KDB' + KD'B + K'DB + K'D'B$. In Figure 2 the gates for *and* and *or* have been expanded to accept more than

two input lines. We have "split" the three input lines, thus feeding each input (or its negation) into each "and" gate.

You may remember that the reason this analysis took place is because the driver's and passenger's passive restraint seatbelts in the author's new car do not behave by the same rules. Table 3 contains the definition table for the passenger's side. To avoid confusion, the output variable in this case is P, having the same meaning as the C for the driver's side. Also, we use the variable R instead of D for the door, since the driver's door is distinctly different from the passenger's door.

TABLE 3 Function Table for Front Passenger's Passive Restraint Seatbelt.

Input variables			Output variable
K	R	B	P
1	1	1	0
1	1	0	1
1	0	1	1
1	0	0	0
0	1	1	0
0	1	0	0
0	0	1	1
0	0	0	0

There are only three 1s in the definition of P. (The difference between Tables 2 and 3 is on line 5; can you explain this design choice?) The Boolean function we get from Table 3 is $P = KRB' + KR'B + K'R'B$. A circuit corresponding to our Boolean expression for P is shown in Figure 3.

The format of the Boolean expressions we derived in the seatbelt analysis is a disjunction of conjunctions. We will sometimes refer to this form by its formal name, *disjunctive normal form.* Even though

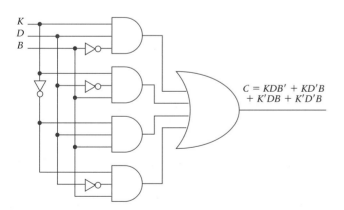

$$C = KDB' + KD'B + K'DB + K'D'B$$

FIGURE 2 A circuit for the driver's seatbelt.

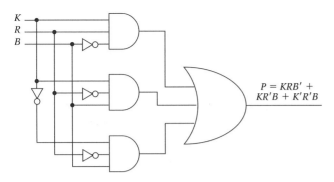

FIGURE 3 A circuit for the passenger's seatbelt.

$$P = KRB' + KR'B + K'R'B$$

no arithmetic is involved, the disjunctive normal form is sometimes called "sums of products."

Building a Nim-Playing Device: Getting Started

We will now build a dedicated device that can execute an algorithm to win one particular version of the game of Nim. Our version of Nim begins with 7 sticks. Each player in turn must remove either 1 or 2 sticks; the player who removes the last stick loses. Our Nim-playing device will be the "player" who moves second. This is the version of Nim we analyzed using a game tree in Section 4 of Chapter 5. The result of the game tree analysis is summarized in Table 4.

TABLE 4 Strategy for Second Player in 7-Stick Nim; Taking Last Stick = Loss.

Possible number of sticks on the board	Number of sticks the device should remove
6	2
5	1
3	2
2	1

The "possibilities" that are missing in Table 4 are "possibilities" that never occur. For example, the device's first move occurs after the human has removed either 1 or 2 of the original 7 sticks, so the device does not need a response to finding 7 sticks on the board. Since the device's strategy forces the person to face 4 sticks on his or her second move, the device never faces 4 sticks either. The missing "possibilities" of 1 and 0 sticks are similarly explained.

Building a Nim Machine: Expressing Inputs and Outputs as Binary Signals

We begin by thinking of the number of sticks on the board and the number of sticks the computer should remove, respectively, as the inputs and outputs to a circuit. We seek circuits to link up those inputs to those outputs, to embody the winning strategy, or algorithm.

The inputs to the Nim-playing device could be any number from the left column of Table 4, and the output is either a "1" or a "2." The circuits we constructed earlier had inputs that could be "1" (T) or "0" (F). We can imagine our circuit as having four input wires, corresponding to such statements as, "The number of sticks on the table is 5," and "The number of sticks on the table is 3."

But there is another point of view, one that requires fewer input wires in this case and is easily generalizable to other situations. We can convert each of the possible numbers of sticks to its binary form. Since we need three bits to express the Arabic number 6, we will pad all our binary inputs to three bits with leading 0s if necessary. Similarly, all the outputs are expressed using two bits. In Table 5 we rewrite our strategy table, Table 4, using padded binary form.

TABLE 5 Strategy for Nim in Binary Form (with decimal equivalents).

Possible number of sticks on the board		Number of sticks the device should remove	
110	(6)	10	(2)
101	(5)	01	(1)
011	(3)	10	(2)
010	(2)	01	(1)

In Table 5 the input and output to our desired circuit are presented in the form of *binary strings,* rows of 1s and 0s concatenated ("glued") together. We need to convert the strings to individual "bits" (binary digits) in the inputs and outputs. This has been done in Table 6. Reading from left to right, we denote the three input bits by I_2 through I_0 and denote the two output bits as O_1 and O_0. The subscripts used to denote the elements in the inputs and outputs correspond to the power of 2 represented by that element of the string. For example, I_1 corresponds to the 2^1 bit of the number of sticks on the table.

TABLE 6 Binary Strategy for Nim with Digits Labeled.

Possible number of sticks on the board				Number of sticks the device should remove		
I_2	I_1	I_0	(decimal)	O_1	O_0	(decimal)
1	1	0	(6)	1	0	(2)
1	0	1	(5)	0	1	(1)
0	1	1	(3)	1	0	(2)
0	1	0	(2)	0	1	(1)

Recall that a truth table for three input variables would have $2^3 = 8$ lines. Thus, if we expand Table 6 to include all 8 possible lines that a table with

three inputs could have, we get Table 7. On those lines of Table 7 which are not real possibilities facing our device, we have arbitrarily entered 0 as the number of sticks for the device to remove. These lines are sometimes called "don't cares" and are the subject Chapter Exercise 26.

TABLE 7 Expansion of Binary Strategy for 7-Stick Nim.

Possible number of sticks on the board				Number of sticks the device should remove			Line from Table 6
I_2	I_1	I_0	(decimal)	O_1	O_0	(decimal)	
1	1	1	(7)	0	0	(0)	no
1	1	0	(6)	1	0	(2)	yes
1	0	1	(5)	0	1	(1)	yes
1	0	0	(4)	0	0	(0)	no
0	1	1	(3)	1	0	(2)	yes
0	1	0	(2)	0	1	(1)	yes
0	0	1	(1)	0	0	(0)	no
0	0	0	(0)	0	0	(0)	no

Building the Nim-Playing Device: Outputs as Boolean Functions of Inputs

The eight lines of Table 7 are fixed in the sense that any circuit having three inputs would have output possibilities for all eight of those input possibilities. There are two output columns in Table 7. Each of those output columns defines a Boolean function of three variables.

We will build our dedicated device by dividing it into two subproblems, solving each separately, and then putting the answers together. The first subproblem will be to build a circuit that computes one digit, O_1, of the answer from the inputs I_2, I_1, and I_0. The second subproblem will be to build a circuit that computes the other digit, O_0, from the inputs I_2, I_1, and I_0.

Figure 4 shows where the individual circuits will

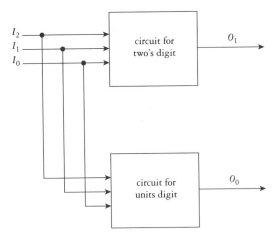

FIGURE 4 Putting together circuits for O_1 and O_0.

fit in the overall solution. Each of the three wires carrying the values of the variables I_2, I_1, and I_0 is "split" so that its signal will go to each of the separate circuits we design. The answers we desire appear on the two wires, O_1 and O_0, emerging from the two separate circuits.

To express the Boolean functions O_1 and O_0 in terms of the variables I_2, I_1, and I_0, we use the same techniques as we did for the seatbelts. For each line of Table 7 where $O_1 = 1$, we create a Boolean expression that has the value 1 for the inputs in that line but has the value 0 for any other line. (For this part of our work, we ignore the O_0 column.) Two lines in the table represent combinations of inputs that correspond to a 1 as the value of the Boolean function O_1. The conjunction we form for each line has each input variable negated if the value of the variable is 0 and not negated if the value of the variable is 1. The conjunctions that correspond to the line 110 (6 sticks on the board) is $I_2 I_1 I_0'$. Similarly, the line 011 (3 sticks on the board) corresponds to the conjunction $I_2' I_1 I_0$. So the disjunctive normal

form expression for O_1 is given by $O_1 = I_2 I_1 I_0' + I_2' I_1 I_0$. Figure 5 shows a circuit for O_1.

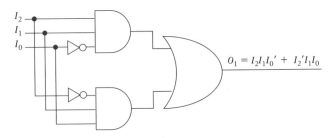

FIGURE 5 Circuit to compute O_1, the two's digit of device's move in Nim.

Now we do a similar calculation for the units digit, the Boolean function O_0. The two lines for which O_0 must be 1 are 101 (5 sticks on the board) and 010 (2 sticks on the board), corresponding, respectively, to the conjunctions $I_2 I_1' I_0$ and $I_2' I_1 I_0'$. Thus, $O_0 = I_2 I_1' I_0 + I_2' I_1 I_0'$. Figure 6 contains the circuit that produces O_0.

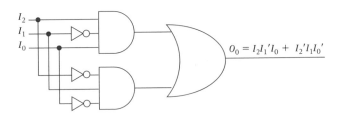

FIGURE 6 Circuit to compute O_0, the units digit of device's move in Nim.

When we put the two circuits from Figures 5 and 6 into the master circuit of Figure 4, we have completed the construction of our Nim-playing device. Now let's review the process by which we built our dedicated device and see how that process also relates to the architecture of programmable computers.

Building Circuits: Dedicated Devices and Programmable Computers

The process of building a dedicated device has several steps. We could follow this process to build a device to carry out any algorithm:

1. State the real-world outcomes (seatbelt holder changes position; number of sticks to pick up) desired for the various possible real-world situations (status of key/engine, door, holder; number of sticks on the table).

2. Describe an algorithm for getting the desired outcomes. (In the seatbelt example, the first two steps were done by the author's examination of how her car works; in our Nim example, it was done in Section 4 of Chapter 5.)

3. Identify the outcomes as output variables and the situations as input variables. Define the meaning of 1 and 0 for each of the input and output variables.

4. Set up a table with all possible combinations of values for the input variables. Express each output in terms of the inputs—for each line in the table, decide if the output variable should be a 0 or a 1 in that particular case.

5. Write each output as a Boolean function of the inputs in disjunctive normal form: Write a conjunction corresponding to each line on which the output is a 1, and then form a disjunction of those conjunctions.

6. Construct the logic circuits that embody these Boolean functions. (In actual practice, the Boolean functions, originally in disjunctive normal form, would be simplified, often using an algorithm in a mathematics logic utility, before constructing the circuit.)

Our Nim device is an example of a dedicated device, sometimes called a dedicated computer. Dedicated devices are ubiquitous in the modern world, showing up in such diverse applications as warning devices in automobiles and security systems protecting our homes. Since these devices are dedicated, the circuitry that causes a display of the message "Porch door open" cannot be transported into a car and display the message "Your headlights are on."

Although dedicated devices are often called "computers," they are not programmable. In programmable computers, the hardware—the dedicated circuitry—performs very general functions. Some circuits do arithmetic, some circuits serve as memory, and so on. The memory circuits hold binary strings that can be interpreted in various ways: Some strings are the data to use; some are the addresses of the places in memory to find the appropriate data. Other strings, called instructions, tell the computer which of its many dedicated circuits is to operate on which data. A program is a sequence of these instructions. Programmable computers have internal clocks that are used, in part, to make sure that the right instruction and data are used at each step of the program sequence (see Chapter Exercise 25). The flexibility and power of the programmable computer comes from its ability to follow a program stored in its memory and to accept different programs for accomplishing different objectives. The design of a programmable computer is quite complex; it is studied in courses in electronics and computer architecture.

Daniel Leibholz, Designer of Computer Chips

Daniel Leibholz is a principal engineer at Digital Semiconductor, a division of Digital Equipment Corporation. His group designs the circuitry for computer chips, working from a very general description of what the chip should do. Rochelle Meyer talked with Dan about his work.

Rochelle: What kind of math do you use in your work?

Dan: Of course, you can't design circuits without Boolean logic. But we also use algebra, statistics, calculus, and discrete math. To make a physical layout of the circuit, we also use geometry.

Rochelle: How does your group actually produce a chip design?

Dan: Two closely linked activities are needed to design a microprocessor chip. Activity one is specifying the architecture—

Daniel Leibholz

the performance and behavior of the chip. This activity includes decisions about the flow of information on the chip and making sure that the chip gives the right answers. Speed is a paramount constraint in chip design. A second constraint is cost. Computer architects balance the desired performance characteristics and the constraints, and then define the behavior of the chip in general terms.

Activity two is implementing the design in actual circuitry, building the circuit from the transistors up.

In order to create a chip, the circuit designers need to know all about gates and basic circuits; for example, differences in circuit speed. The architects give the designers a general design and not exact specifications, so the designers have the flexibility to come up with new and better types of circuits.

Rochelle: How closely do the architects collaborate with the circuit designers?

Dan: The architects and designers are involved in an iterative process: the circuit designers try to find the best way to implement an architectural idea, and

Computer chip

the architects try to redefine the chip behavior so that it can make use of the highest performing circuits but still get the right answers. Together they try to make the high-priority functions of the chip work fast and efficiently. The architects and circuit designers share in the task of making sure that the chip works reliably and gets the right answers.

Rochelle: Can you tell us more about the nitty-gritty of circuit design?

Dan: Circuit designers use several methodologies which differ in their degree of customizing. The choice among them is often made based on the priority of a given circuit function and the performance constraints of the chip. To create the most exotic circuits, the designers start from scratch, taking advantage of the characteristics of the underlying transistor technology. Creating full-custom circuits involves- much more than Boolean reduction; you need to understand not just the logic but the technology, and which circuits go faster than others. You often get the best results from these circuits, but they take the most time to design.

Sometimes the designers "semi-custom" a circuit by connecting predesigned logic blocks that perform basic functions. For the most routine circuits, a computer program is written which expresses the logic the circuit is to perform. This program is fed as data into an automated "circuit synthesis" tool, another program, which produces a "wire list" of the necessary components and their connections.

Rochelle: Doesn't that mean that computers are being used to design other computers?

Dan: We definitely need computers in order to design computers. But we also need people. The circuit produced by the synthesis tool can only be as good as the designers who wrote that tool. In addition, we need people to define and describe all of the constraints to the tool, and to interpret whether the results from the tool are usable.

Hand calculators are an interesting category since many of them are a collection of dedicated devices requiring us to direct every step of a calculation. Many of the more sophisticated calculators are programmable. Programmable calculators allow us to enter a series of instructions; for example, "Add the first number to the product of the next three numbers, divide the result by a fifth number, then subtract a sixth number, and show me the square root of that result." (Of course, the instructions are not given to the calculator in this form.) The calculator stores these instructions, and carries them out with each set of six numbers you enter, until you instruct the calculator do perform some other task.

More About Boolean Functions

The unary and binary logic operations we studied in the first section of this chapter can also be viewed as Boolean functions: Negation can be viewed as a Boolean function N of one variable: $N(1) = 0$ and $N(0) = 1$. Conjunction can be viewed as a Boolean function C of two variables: $C(1, 1) = 1; C(1, 0) = 0; C(0, 1) = 0;$ and $C(0, 0) = 0$.

How many different Boolean functions of a fixed number of variables could there be? When there are 4 variables, a Boolean function can be specified by a truth table column of 16 entries, each entry a 0 or a 1. Using the multiplication principle from Chapter 4, we find that there are 2^{16} possible such columns. But where did the 16 come from? That was based on the number of lines in the truth table, which is the same as the number of input variables, 4. Our $16 = 2^4$. So the number of possible Boolean functions for 4 inputs is 2^{2^4}.

In general, if we have n inputs, we have 2^{2^n} possible Boolean functions. Each of those functions corresponds to both a logic expression and a circuit. We can determine the logic expression as we did for the Nim device by looking at the lines of the truth table for which the function must be a 1. The circuit can be built from that expression.

Another Device: A Seven-Element Display

Our Nim-playing device is very primitive. For example, in order to make sense of its output, we would need to incorporate signal lights to show us the values of O_1 and O_0. Then we would have to convert the binary number shown by the lights to its decimal equivalent. This might not seem to be too much trouble when our Nim device can give us only two possible answers, but what about the results we get from our hand calculators? Surely we do not want to use hand calculators with outputs that are merely a string of lights that indicate the binary equivalent of the answer. In fact, the display of a calculator is precisely what we want to construct. Each digit in that display has seven elements, each of which is a light in the shape of a line segment. There are three horizontal segments and four vertical ones. We name these seven elements in Figure 7.

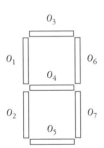

FIGURE 7 A seven-element display.

In Table 8 we present the relationship between the four-bit padded strings for the digits 0 through 9 and the on–off status of the output elements O_1 through O_7. (We use 1 for on and 0 for off.) The usual order of the rows in such tables has been reversed here, since we are only interested in the "bottom" rows. The subscripts on the input variables have the same meanings here as in the Nim device.

TABLE 8 Outputs for a Seven–Element Display.

Binary string, $I_3I_2I_1I_0$	(decimal value)	O_1	O_2	O_3	O_4	O_5	O_6	O_7
0000	(0)	1	1	1	0	1	1	1
0001	(1)	0	0	0	0	0	1	1
0010	(2)	0	1	1	1	1	1	0
0011	(3)	0	0	1	1	1	1	1
0100	(4)	1	0	0	1	0	1	1
0101	(5)	1	0	1	1	1	0	1
0110	(6)	1	1	0	1	1	0	1
0111	(7)	0	0	1	0	0	1	1
1000	(8)	1	1	1	1	1	1	1
1001	(9)	1	0	1	1	0	1	1
1010								
1011								
1100		From here to the bottom of the table, all outputs are 0, since these strings do not correspond to single-digit decimal numbers.							
1101								
1110								
1111								

From Table 8 we can construct a Boolean expression and logic circuit for any of the seven output segments. Then in the same way as we did for the Nim device, we can create the complete circuit for

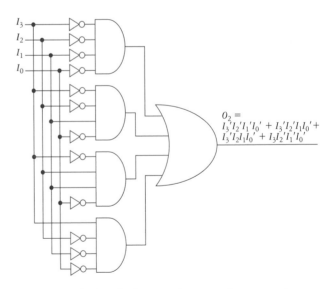

$$O_2 = I_3'I_2'I_1'I_0' + I_3'I_2'I_1I_0' + I_3'I_2I_1I_0' + I_3I_2'I_1'I_0'$$

FIGURE 8 A circuit for one element of a seven-element display.

all the outputs. Here we will focus on O_2, leaving the other outputs as exercises. The expression we desire is

$$O_2 = I_3'I_2'I_1'I_0' + I_3'I_2'I_1I_0' + I_3'I_2I_1I_0' + I_3I_2'I_1'I_0'.$$

We note that the four conjunctions making up the expression for O_2 correspond to precisely those decimal digits (namely, 0, 2, 6, and 8) for which we want element O_2 to be turned on. The circuit that corresponds to O_2 is given in Figure 8.

<hr>

Exercises for Section 7.2

1. (a) For the table below, write the Boolean expression in disjunctive normal form for the output.
   (b) Draw the circuit for part (a).

p	q	r	output
1	1	1	1
1	1	0	1
1	0	1	0
1	0	0	0
0	1	1	0
0	1	0	1
0	0	1	0
0	0	0	1

2. (a) For each of the two outputs in the table below, determine a Boolean expression in disjunctive normal form for the outputs.
   (b) Draw a circuit with two outputs for part (a).

p	q	output #1	output #2
1	1	1	1
1	0	0	0
0	1	0	1
0	0	1	1

3. Explain how you can tell from a definition table how many *or* and how many *and* gates you'll need for the circuit corresponding to the disjunctive normal form Boolean function it defines. You should be able to do this before you work out the disjunctive normal form.

4. Suppose a table defining a Boolean expression E has just a single line where there is a 0. Show how to fit the table with a Boolean function that is simpler than the one you get if you use disjunctive normal form.

5. A lawn sprinkler system goes on whenever the day of the month is even, provided that the rain gauge is dry. There is a wire (W) from the rain gauge which carries a 1 if it is wet. There is another wire (E) which carries a 1 if the day is even. We need to hook these wires up with gates so that the circuit has a single output wire (S) which leads to the sprinkler. The sprinkler will go on if the output is 1.
   (a) Make a table showing all combinations of 0 and 1 for the inputs E and W, showing what the value of S should be in each case.
   (b) Show how to hook up the wires and appropriate gates in a circuit.

$$\underline{\quad E \quad}$$
$$\qquad\qquad\qquad ? \quad \underline{\quad S \quad}$$
$$\underline{\quad W \quad}$$

6. In the previous problem, suppose the sprinkler system is upgraded to have manual control, which allows you to turn the sprinkler on whether or not the day is even or the rain gauge is dry. This manual override is a

button hooked up to a wire (*M*) that will carry a 1 if the button is pushed and 0 otherwise.

   (a)   Make a new table showing all combinations of 0 and 1 for the three inputs.

   (b)   Redesign the circuit to deal with this third input.

In Exercises 7–12, follow the techniques used in the text to construct circuits for these outputs of the seven-element display.

7. O_1 8. O_3 9. O_4

10. O_5 11. O_6 12. O_7

In Exercises 13–16, you are to construct Nim-playing devices for some variations of the game analyzed in the text. In all these variations, a person plays a dedicated device, and there is a winning strategy for the device. In order to construct these dedicated devices, you will need to analyze the game, as you did in Section 4 of Chapter 5. In Exercise 13 you only need 2 input lines; in the others you need 3.

13.   In this variation of Nim, the player who picks up the last stick *loses.* We start with 4 sticks on the board, and the person (not the device) moves first.

14.   In this variation of Nim, the player who picks up the last stick *loses.* We start with 6 sticks on the board, and the device moves first.

15.   In this variation of Nim, the player who picks up the last stick *wins.* We start with 5 sticks on the board, and the person moves first.

16.   In this variation of Nim, the player who picks up the last stick *wins.* We start with 4 sticks on the board, and the device moves first.

17.   (**Writing project**) In the title of this chapter, the word "intelligent" is in quotation marks, signifying that perhaps the devices we are building are just mimicking intelligence, not displaying it. The British mathematician Alan Turing proposed a test for determining whether a device (computer) has intelligence: If a person is communicating, having a "conversation," with something or somebody hidden in a room, and the responses from the room have all the characteristics of human intelligence, then whatever is in the room has intelligence. Read about tests of computer intelligence using Turing's ideas, and write a report, including not just the results of the tests, but your own ideas about whether computers might ever display "intelligence."

SECTION 7.3 *Quantifiers and Counterexamples*

Quantifiers

Many sentences, both in mathematics and in other parts of the real world, include words like "all," "some," or "none." Examples of such sentences are as follows:

1.   All math majors are intelligent.

2.   Some of the things politicians say make me angry.

3.   None of my friends likes jazz as much as I do.

4.   Every matrix has an inverse.

5.   Some systems of linear equations can't be solved.

Sentences like these can be expressed as statements in logic by using quantifiers. There are two quantifiers: the universal quantifier, meaning "for all," and the existential quantifier, meaning "there exists (at least one)." Using them, plus negation, we will be able to create logic statements corresponding to sentences like those above. Note: We use the term "sentence" to refer to an English sentence, and the term "statement" to refer to a logical expression.

DEFINITION

The *universal quantifier,* meaning "for all," is symbolized by ∀.

EXAMPLE 1 Using ∀ We symbolize the open sentence "*x* is intelligent" as *B*(*x*). (Recall that an open sentence in one in which the subject is unspecified.) Then, using the universal quantifier, we can write the quantified statement (∀*x*) *B*(*x*), which would have the meaning "For all *x*, *x* is intelligent." This sentence is related to, but not the same as, "All math majors are intelligent." ▲

We recall that an open sentence, like "*x* is intelligent," or "It is green," by itself has no truth value, since the "*x*" or the "it" has not been defined. The necessary defining is done when we specify a domain of definition, the set of items under scrutiny.

DEFINITION

The *domain of definition, D,* of a quantified statement is the set of replacement items that can be substituted for the variable in an open sentence.

EXAMPLE 1 Using ∀ (continued) The natural domain of definition for the sentence "All math majors are intelligent" is the set of all math majors, so *D* = {all math majors}. The link between that domain and the open sentence *B*(*x*) is

provided by the quantifier (∀*x*). In our example, the quantified statement (∀*x*) *B*(*x*) now has a truth value, assuming for the moment that the terms "math major" and "intelligent" have the same meaning for all of us. ▲

We note that the *subject* of the original English sentence, "math majors," corresponds to the domain of definition, while the *predicate,* "are intelligent," corresponds to the open sentence. This is the typical correspondence between the English sentence and the quantified logic statement and thus is very useful in translating between them.

EXAMPLE 2 Choice of Domain Is Critical
The truth value of a quantified statement can change when we change *D*. If we define *D* = {all matrices} and *I*(*x*) = "*x* has an inverse," then the quantified statement (∀*x*) *I*(*x*) would correspond to the English sentence "Every matrix has an inverse." This quantified statement is false. However, if we redefine *D* = {square matrices with nonzero determinant}, then (∀*x*) *I*(*x*) becomes a true statement. Can you write the associated English statement? ▲

DEFINITION

The *existential quantifier,* meaning "there exists," "at least one," or "some," is symbolized by ∃.

EXAMPLE 3 Using ∃ If *H*(*x*) = "*x* is a history book" and *D* = {books in my local library}, the (∃*x*) *H*(*x*) corresponds to the English sentence "There is a history book in my local library." This statement would be true as long as the local library has at least one book about history. The statement (∃*x*) *H*(*x*) would also be true if the library has 17 books about history or even if it contained nothing but history books. ▲

EXAMPLE 4 ∃ **and the Word "Some"**   The existential quantifier is often used where casual language has the word "some." At the start of this section, we had the statement "Some of the things politicians say make me angry." To translate this into a quantified logic statement, we need to find the subject and predicate of the English sentence. The domain corresponds to the subject of the English sentence "The things politicians say," so D = {the things politicians say}. The predicate in the original statement, the basis of the open sentence, is "make me angry," so our open sentence is "x make(s) me angry," which we can symbolize as $A(x)$. Thus, our original statement, "Some of the things politicians say make me angry," is translated as $(\exists x)\, A(x)$.

Our definitions of the domain and of $A(x)$ also allow us to translate the symbolic statement $(\exists x)\, A(x)$ as, "There is at least one thing said by a politician that makes me angry." Typically, the logic statement allows for several equivalent English sentences, and we usually write the English sentence that is least awkward. ▲

In Table 1 we summarize some typical relationships between logic statements and the correspond-

ing English sentences. In the table, we use the ellipsis, ". . ." to stand for the key words from $S(x)$ and realize that the resulting English sentences are quite rough-hewn.

Quantifiers as "Gates"

In Section 7.1, we learned the gate symbols used for negation and for the binary logic operations. In Section 7.2 we broadened our concept of the *and* and *or* gates to more than two inputs. Here we develop extensions of those two gates as ways of thinking about quantified statements. We begin with the universal quantifier.

EXAMPLE 5 ∀ **as a "Big" *and***   We set D = {the authors of this text}, define $M(x) = x$ is a male, and look at $(\forall x)\, M(x)$. The domain is a finite set of manageable size (eight). We can imagine someone writing eight statements of the form "Michael Olinick is a male," "Alan Tucker is a male," "Rochelle Meyer is a male," and so on, and then noting that the combination "Michael Olinick is a male *and* Alan Tucker is a male *and* Rochelle Meyer is a male *and* . . ." is equivalent to the statement $(\forall x)\, M(x)$. If $M(x)$ is true for every element x of the domain, then the quantified statement $(\forall x)\, M(x)$ is true. If $M(x)$ is false for any one element of the domain, then the quantified statement $(\forall x)\, M(x)$ is false. The universal quantifier is like a "big" *and*. ▲

Example 5 is a specific case of a general principle. Suppose we have a statement $(\forall x)\, S(x)$ with a domain of definition $D = \{x_1, x_2, x_3, \ldots, x_n\}$. The quantified statement $(\forall x)\, S(x)$ can be regarded as an extension of the logical *and* from a binary operation to a "big" *and* of some number, n, of statements. For each ele-

TABLE 1   Typical Relationships Between Logic Statements and English Sentences.

Logic statement	English sentence
$(\forall x)\, S(x)$	All x . . . Every x . . .
$(\forall x)\, {\sim} S(x)$	All x are not . . . No x . . .
$(\exists x)\, S(x)$	Some x . . . There is an x so that . . .
$(\exists x)\, {\sim} S(x)$	Some x are not . . . There is an x for which . . . is false.

ment x_i of the domain of definition, there will be one statement $S(x_i)$. We have the equivalence

$$(\forall x)\, S(x) \leftrightarrow S(x_1) \wedge S(x_2) \wedge S(x_3) \wedge S(x_4) \wedge \cdots \wedge$$
$$S(x_i) \wedge \cdots \wedge S(x_n), \quad \text{for } x_i \text{ in } D.$$

A "gate" for $(\forall x)\, S(x)$ is shown in Figure 1; it is an *and* gate with n inputs.

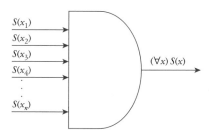

FIGURE 1 A "gate" representing the universal quantifier, $(\forall x)\, S(x)$.

As we did with the universal quantifier, we look for an analogy between the existential quantifier and the basic logic gates.

E X A M P L E   6 $\exists$ **as a "Big"** *or*   We let $D =$ {all positive multiples of 6 less than 20}, define $O(x) =$ "x is odd," and consider $(\exists x)\, O(x)$. The domain consists of 6, 12, and 18. So $(\exists x)\, O(x)$ means that *either* 6 is odd *or* 12 is odd *or* 18 is odd. We see that $(\exists x)\, O(x)$ is true if $O(x)$ is true for at least one of the elements in D and false if no x makes $O(x)$ true. The existential quantifier can be viewed as a "big" *or,* since it is true as long as any one of the constituent statements is true and is false only in the case that all the constituent statements are false. ▲

In general, given the statement $S(x)$, and $D =$ $\{x_1, x_2, x_3, \ldots, x_n\}$, we have the equivalence

$$(\exists x)\, S(x) \leftrightarrow S(x_1) \vee S(x_2) \vee S(x_3) \vee S(x_4) \vee \cdots \vee$$
$$S(x_i) \vee \cdots \vee S(x_n), \quad \text{for } x_n \text{ in } D.$$

A "gate" for the existential quantifier is shown in Figure 2 as an *or* with n inputs.

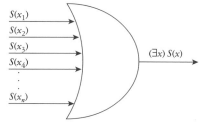

FIGURE 2 A "gate" representing the existential quantifier, $(\exists x)\, S(x)$.

Quantified statements can have domains of definition that are infinite sets (see the Spotlight "Beyond Finite Sets" in Chapter 4). We could still picture those statements as corresponding to "gates," but such gates would need infinitely many inputs and thus could not actually be constructed.

Negating Quantified Statements

Some quantified statements are false. In what follows, we will use our interpretation of quantified statements as "big" gates to help us write true English sentences expressing the negations of the false quantified statements.

E X A M P L E   7   These quantified statements are false:

1. $D =$ {the rational numbers}, $P(x) = x$ is a positive number, $(\forall x)\, P(x)$. In English, "All rational numbers are positive." Since the English sentence is false, we know that "It is not true that all rational numbers are positive" is true, but it is an awkward sentence.

2. $D = \{\text{multiples of } 4\}$, $P(x) = x$ is prime, $(\exists x)\, P(x)$. In English, "There exists some multiple of 4 which is prime." Again, the English sentence is false, so we know that "It is not true that there exists some multiple of 4 which is prime" is true. But we seek a more direct way of saying the same thing. ▲

We restate the ideas from the previous example in more general terms. If, for a given domain of definition and open sentence $S(x)$, the quantified statement $(\forall x)\, S(x)$ is false, then what quantified statement involving $S(x)$ might be true? One answer is that $\sim[(\forall x)\, S(x)]$ is true. But the translation of that quantified statement into English would be the somewhat awkward statement "It is not the case that for all x, $S(x)$." Similarly, $\sim[(\exists x)\, S(x)]$ would correspond to "It is not the case that there is an x for which $S(x)$." Can we express $\sim[(\forall x)\, S(x)]$ and $\sim[(\exists x)\, S(x)]$ in some other way as logic statements? In particular, we seek a quantified statement whose corresponding English sentences will be less awkward.

We recall that $(\forall x)\, S(x) \leftrightarrow S(x_1) \wedge S(x_2) \wedge S(x_3) \wedge \cdots \wedge S(x_n)$. If we think of $\sim[(\forall x)\, S(x)]$ as negating a "big" *and*, we get $\sim[S(x_1) \wedge S(x_2) \wedge S(x_3) \wedge \cdots \wedge S(x_n)]$. We can remove the outer parentheses—and simultaneously the need for awkward English—by applying an extended form of the DeMorgan tautology:

$$\sim[S(x_1) \wedge S(x_2) \wedge S(x_3) \wedge \cdots \wedge S(x_n)]$$
$$\leftrightarrow [\sim S(x_1) \vee \sim S(x_2) \vee \sim S(x_3) \vee \cdots \vee \sim S(x_n)].$$

The left-hand side of the equivalence corresponds to "It is not the case that for all x, $S(x)$." The right side of this equivalence is a "big" *or* and thus translates as "There is at least one x that makes $S(x)$ false." Another way of stating the same condition is

"There is at least one x that makes $\sim S(x)$ true." So we arrive at Equivalence 1:

Equivalence 1. $\sim[(\forall x)\, S(x)] \leftrightarrow (\exists x)\, \sim S(x)$.

E X A M P L E 8 Using Equivalence 1 Instead of the awkward "It is not true that all rational numbers are positive," from Example 7, we can write, "There is a rational number that is not positive." ▲

We also need to consider the negation of a statement involving an existential quantifier such as $(\exists x)\, S(x)$, which we recall is equivalent to $S(x_1) \vee S(x_2) \vee S(x_3) \vee \cdots \vee S(x_4)$. The negation is $\sim[(\exists x)\, S(x)]$, which in English is "It is not true that there is an x for which $S(x)$," and which is equivalent to $\sim[S(x_1) \vee S(x_2) \vee S(x_3) \vee \cdots \vee S(x_n)]$. And as we did before, we look at an extended DeMorgan's tautology, namely

$$\sim[S(x_1) \vee S(x_2) \vee S(x_3) \vee \cdots \vee S(x_n)]$$
$$\leftrightarrow [\sim S(x_1) \wedge \sim S(x_2) \wedge \sim S(x_3) \wedge \cdots \wedge \sim S(x_n)].$$

The right-hand side of the equivalence is a "big" *and*, which we recognize as a universally quantified statement. Thus we have Equivalence 2:

Equivalence 2. $\sim[(\exists x)\, S(x)] \leftrightarrow (\forall x)\, \sim S(x)$.

E X A M P L E 9 Using Equivalence 2 Instead of the awkward "It is not true that there exists some multiple of 4 which is prime," from Example 7, we can write, "Every multiple of 4 is not prime," or better yet, "No multiple of 4 is prime." ▲

Here is an informal summary of the effects of negating quantifiers:

~all = some do not;

~some = none.

Proposed Theorems and Counterexamples

Many theorems in mathematics take the form of universally quantified statements:

All primes bigger than 2 are odd numbers.

All isosceles triangles have base angles equal.

All solutions to linear programming problems occur at corner points.

Sometimes proposed theorems also take the same form:

All primes are odd numbers.

All quadratic equations have two real-valued roots.

All matrices have inverses.

You may suspect that the proposed theorems are not true. How would you show that the statement "All primes are odd numbers" is a false statement? We recognize that the statement can be translated into a universally quantified statement having a $D = \{$all primes$\}$ and an open sentence of $O(x) =$ "x is an odd number." Since we believe $(\forall x)\, O(x)$ is false, we believe that $\sim[(\forall x)\, O(x)]$ is true, which by Equivalence 1 means that we believe that $(\exists x)\sim O(x)$ is true. This translates into "There is a prime that is not an odd number." So if we can produce such a prime, we have shown that $(\exists x)\sim O(x)$ is true, which in turn tells us that $\sim[(\forall x)\, O(x)]$ is true, which tells us that $(\forall x)\, O(x)$ is false, that is, our proposed theorem, "All primes are odd numbers," is a false statement. But we know a prime that is not an odd number, namely 2, so by the chain of reasoning we just constructed, we know that the statement "All primes

are odd numbers" is a false statement. The number 2 in this discussion is called a counterexample to the proposed theorem.

DEFINITION

A *counterexample* is an element of the domain of definition of a universally quantified statement which makes the open sentence false.

We note that all we need is one counterexample to show that a proposed universally quantified theorem is false. If we now turn to the other two proposed theorems, both of which happen to be false, we can try to find a counterexample for each. Both of those proposed theorems have many counterexamples, but producing just one will suffice to demonstrate the falsehood of the proposal. Can you find counterexamples?

Sometimes mathematical theorems appear as existentially quantified statements; for example:

There is at least one real root among the roots of any cubic equation.

Another example is

For any weighted graph, there exists a spanning tree that is minimal.

Now let us examine a proposed theorem of this form; for example,

There is at least one number that has 10 as a factor but does not have 5 as a factor.

Suppose we suspect that this is a false statement. We have $D = \{$numbers having 10 as a factor$\}$, and the open sentence is $F(x) =$ "x does not have 5 as a factor." Our proposed theorem has the form $(\exists x)\, F(x)$, which we believe is false, so we believe that $\sim[(\exists x)\, F(x)]$ is true, which means that we believe that $(\forall x)\, \sim F(x)$ is true. So in order to show that the original proposal is false, we must show that "Every number that has 10 as a factor does have 5 as a factor" is a true statement. In this example, a

counterexample will not work to prove the original statement false. We need a proof showing that *every* element of D has a particular property. Such proofs are often more complex logically than finding counterexamples, and we will not discuss them further here. The technique needed to prove a quantified statement false depends strongly on the quantifier.

Exercises for Section 7.3

1. For each domain of definition, D, open sentence, and quantified expression, write a corresponding English sentence. What truth value would you assign in each quantified statement?
    (a) $D = \{$the counting numbers$\}$, $P(x) = $ "x is a positive number," $(\forall x)\, P(x)$.
    (b) $D = \{$all primes$\}$, $O(x) = $ "x is odd," $(\forall x)\, O(x)$.
    (c) $D = \{$all counting numbers$\}$, $N(x) = $ "x is negative," $(\exists x)\, N(x)$.
    (d) $D = \{$all integers$\}$, $N(x) = $ "x is negative," $(\exists x)\, N(x)$.

2. For each domain of definition, D, open sentence, and quantified expression, write a corresponding English sentence. What truth value would you assign to each quantified statement?
    (a) $D = \{$roots of quadratic equations$\}$, $P(x) = $ "x is a positive number," $(\forall x)\, P(x)$.
    (b) $D = \{$all integers$\}$, $R(x) = $ "x is a rational number," $(\forall x)\, R(x)$.
    (c) $D = \{$all rational numbers$\}$, $I(x) = $ "x is an integer," $(\exists x)\, I(x)$.
    (d) $D = \{$all possible SAT scores$\}$, $I(x) = $ "x is an integer," $(\exists x)\, \sim I(x)$.

3. For each sentence, state a domain of definition and open sentence, and then write the sentence as a quantified statement.

    (a) On a circle of radius r, every point is a distance r from the center.
    (b) The relationship $c^2 = a^2 + b^2$ is true for some triangles.
    (c) Some square matrices do not have inverses.
    (d) "All is fish that comes to net." (John Heywood)
    (e) "Somebody's always throwing bricks." (Vachel Lindsay)
    (f) "Something there is that doesn't love a wall." (Robert Frost)
    (g) "All the world's a stage." (William Shakespeare)
    (h) "The deepest feeling always shows itself in silence." (Marianne Moore)

4. For each sentence, state a domain of definition and open sentence, and then write the sentence as a quantified statement.
    (a) Every spanning tree obtained from Prim's algorithm is minimal.
    (b) Some spanning cycles obtained from the Nearest Neighbor heuristic algorithm are minimal.
    (c) All orders for adding a set of n vectors give the same resultant vector.
    (d) "Some come to take their ease." (William Shakespeare)
    (e) "Every day, itsa getting closer." (Buddy Holly)

(f) "Sometimes a cigar is just a cigar." (attributed to Sigmund Freud)

(g) "All religions, laws, moral and political systems are but necessary means to preserve social order." (Ch'en Tu-hsiu)

(h) "Wit has truth in it." (Dorothy Parker)

In problems 5–8, the proposed theorem is false; write each proposed theorem as a quantified statement. Then write the negation of the quantified statement and the corresponding English sentence. Finally indicate whether you would search for a counterexample or need a proof to show that the original proposed theorem is false.

5. Every straight line has a y-intercept.

6. Every quadratic equation has a graph which intersects the x-axis.

7. There is a straight line which intersects neither the x-axis nor the y-axis.

8. There is a point on a circle whose distance from the center of the circle is twice the radius.

Chapter Summary

Mathematical logic gives us techniques for combining and evaluating expressions consisting of Boolean variables, variables that can only have two values, usually called "true" and "false." Among the many uses of mathematical logic, the one emphasized here is the construction of circuitry, or dedicated devices, which carry out algorithms. The essential steps in the construction of such devices are the identification of the necessary inputs and desired outputs, and the specification of an algorithm that produces those desired outputs from those necessary inputs. Once these steps are complete, a table is used to define the outputs in terms of the inputs; from the definition table we get a Boolean expression from which we can build the circuitry for our device. Since there are many Boolean expressions that are equivalent, deciding which one to use for a particular circuit can make important differences in properties of the resulting circuit, such as the number of required gates and whether it can be constructed on a single plane.

Other important uses of mathematical logic are in quantified statements and as the basis for all proofs in mathematics. Thus we could say that, in some sense, all mathematics relies on mathematical logic.

Chapter 7 Exercises

All of these exercises are intended to be worked by hand; a mathematics computer package may be useful to check on the truth tables, but such a package is not necessary for solving any of these problems.

1. Using truth tables, verify the commutativity and associativity tautologies from Section 7.1.

2. Using truth tables, verify the distributivity and DeMorgan tautologies from Section 7.1.

3. Use the tautologies from Section 7.1 to determine if the pairs of expressions in Exercise 5 of Section 7.1 are logically equivalent.

4. Use the tautologies from Section 7.1 to determine if the pairs of expressions in Exercise 6 of Section 7.1 are logically equivalent.

5. The following diagrams show two circuits that are useful in computers. One of these circuits is an *adder* and does just what its name implies. The other circuit is called a *comparator* and is used to determine if two binary strings are identical. Determine which circuit performs which function, and explain how the gates accomplish that function. Hint: What would a 1 or a 0 output mean in terms of the inputs?

Circuit 1:

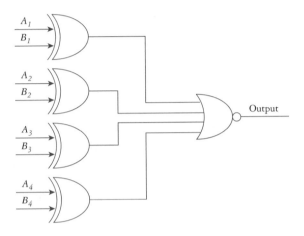

Circuit 2:

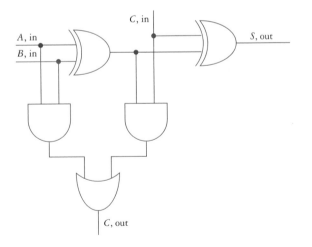

6. The next diagrams contain two circuits, both with two inputs and two outputs. For each circuit determine the logic expression for Out #1 and Out #2. Do these circuits serve the same purpose? How might these circuits be useful in obtaining planar layouts for larger circuits? What drawbacks do these circuits present?

Circuit 1:

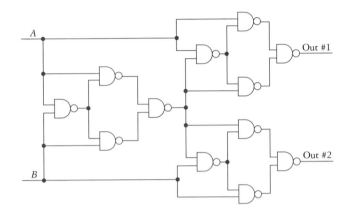

Circuit 2:

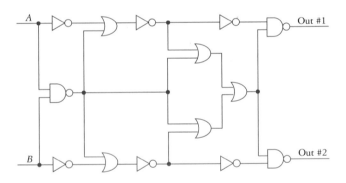

In problems 7–14, you are asked to construct a circuit which performs a specific task. These circuit-design problems require you to set up a table showing all inputs with all possible combinations of values, and then making one column for each output and setting the outputs as described in the problem. Write each output as a Boolean functions, like we did in building our dedicated circuits, and then draw the circuit.

7. A system of security alarms for the outside of a house has monitors that detect whether human beings are moving about on the grounds. There is a wire (H) which carries a 1 if any monitor senses a human being. In addition, there is a wire (D) hooked up to a light meter which carries a 1 if it is daytime. Finally, there is an output wire (A) which sounds the alarm if it carries a 1. We want to create a circuit which will sound the alarm when humans are detected at night.

8. The system described in the previous exercise has some disadvantages. Name one and redesign the system (you may use additional inputs, for example).

9. An espresso coffee vending machine allows the customer to have three items added to the coffee: cream (C), lemon peel (L), anise flavoring (A). For each there is a button the customer can push if he wants that additive. The customer can push as many buttons as he/she likes. But if he/she pushes more than one, a sign lights up asking for an additional dime to be inserted. Imagine that each button is hooked up to a wire which will carry current if the button is pushed and carry no current otherwise. There will also be a wire leading to the sign which will light up the sign if the wire carries current (S = 1) and not otherwise.

10. Draw a circuit for squaring a 2-bit number. Notice that the answer may need as many as 4 bits (for example, (11)*(11) = 1001) so you will need 4 output lines. If the answer has fewer than 4 bits, pad it with leading zeros (e.g., replace 1 by 0001). [If you work out the multiplications correctly you will notice that one digit in the answer is always 0. In the circuit this can be represented with a wire which never carries current and is, therefore, unaffected by the inputs. Represent this wire as a line, unconnected to the inputs for the number to be squared, having a break in it.]

11. A decoder is a circuit with n input lines and 2^n output lines. The settings on the n input lines, taken together as one n-bit binary number, determine which of the 2^n output lines shall be set to 1. The other output lines are set to 0. Construct a decoder for $n = 3$. Hint:

If the input lines show the pattern 101, then output line 5 should be set to 1.

12. An encoder is a circuit with 2^n input lines, only one of which is set at 1, the rest of which are set to 0, and n output lines. The output lines are set so that, taken together as one n-bit binary number, they correspond to the binary equivalent of which input line is a 1. Construct an encoder for $n = 3$. Hint: If input line 2 is the unique input line set to 1, then the output lines should be set to the pattern 010.

13. A demultiplexer is a circuit which has 1 data input line and 2^n data output lines and n control lines. The control lines should be thought of as input lines. The settings on the n control lines, taken together as one n-bit binary number, determine which of the 2^n output lines shall receive signal from the input line. Construct a demultiplexer for $n = 3$. Hint: If the three control lines show the pattern 011, then output line 3 receives the data from the input line.

14. A multiplexer is a circuit which has 2^n data input lines and 1 data output line and n control lines. The control lines should be thought of as input lines. The settings on the n control lines, taken together as one n-bit binary number, determine which of the 2^n input lines should be fed directly into the output line. Construct a multiplexer for $n = 3$. (Hint: If the three control lines show the pattern 110, then the data from input line 6 are sent to the output line.)

15. For each sentence state a domain of definition and open sentence, and then write the sentence as a quantified statement. In these problems you may need to negate quantifiers.
    (a) "We cannot all be masters." (William Shakespeare)
    (b) "There is no royal road to geometry." (Euclid)
    (c) "On the day of victory no one is tired." (Arab proverb)
    (d) "Nature never wears a mean appearance." (Ralph Waldo Emerson)
    (e) "Nobody who has not been in the interior of a family can say what the difficulties of any

individual of that family might be."
(Jane Austen)

(f) "We don't like flowers that do not wilt."
(Marianne Moore)

16. For each sentence state a domain of definition and open sentence, and then write the sentence as a quantified statement. In these problems you may need to negate quantifiers.

(a) "For there was never yet philosopher that could endure the toothache patiently."
(William Shakespeare)

(b) "Never in the field of human conflict was so much owed by so many to so few."
(Winston Churchill)

(c) "For another, better thing than a fight required of duty exists not for a warrior."
(Bhagavad Gita)

(d) "Nothing is useless or low."
(Henry Wadsworth Longfellow)

(e) and (f) "I joked about every prominent man in my lifetime, but I never met one I didn't like." (Will Rogers) Treat this quotation as two separate quantified statements.

In Exercises 17–24, write each proposed theorem as a quantified statement. Then decide if the proposed theorem is true or false. If the proposed theorem is false, write the negation of the quantified statement and the corresponding English sentence, and then indicate whether you would search for a counterexample or need a proof to show that the proposed theorem is false.

17. There is a quadratic equation with real coefficients having $2 + 3i$ and 7 as its roots.

18. Every system of two equations in two variables has a unique point (x, y) as its solution.

19. There are systems of linear equations which can't be solved.

20. Every square matrix has a determinant.

21. Every matching produced by the Gale–Shapley algorithm is stable.

22. There is a polynomial-time algorithm that always gives us a minimal spanning cycle for the TSP.

23. (Writing project) Read about "fuzzy logic," and write a report on it for your classmates. A good article on this subject is the Recommended Readings by Kosko and Isaka.

24. (Writing project) Read about minimization algorithms such as Karnaugh maps or the Quine-McKlusky method in a discrete mathematics text (Gersting; Maurer and Ralston; Ross and Wright in the Recommended Readings), and write a description of them, with examples, for your classmates.

25. (Writing project) Read the chapter about microarchitecture in the Recommended Readings by Kruse, and write a report for your class based on it.

26. (Project) Suppose we have two input signals, A and B, and need to construct a circuit that is defined by $F(A, B)$ in the following table. Since two of the four lines in the table correspond to situations in which we don't care about the value for F, we have four possibilities (why 4?) for a definition of F, which we show in the table as P_1, P_2, P_3, and P_4.

A	B	$F(A, B)$	P_1	P_2	P_3	P_4
1	1	don't care	0	0	1	1
1	0	1	1	1	1	1
0	1	don't care	0	1	0	1
0	0	0	0	0	0	0

Of these four possibilities, the one setting both "don't cares" to 0 is probably the worst, since the formula it gives requires a *negation* and an *and* gate. Two of the remaining possibilities require just one gate, and one possibility requires no gates at all! (You need to recognize the pattern in the column, not rely on disjunctive normal form to verify this.) This example suggests checking out the various "don't care" possibilities before physically making circuits.

In the Nim circuits we designed in Section 7.2, there were "don't cares." Can you change some "don't care" settings from 0 to 1 in the definitions of O_1 or O_0 and thereby change the resulting Boolean expression into one whose corresponding circuit has fewer gates? (Be careful, not all 0s are "don't cares.") For the disjunctive normal forms we found and for those you derive, it may also be possible to reduce the necessary number of gates by using tautologies from Section 7.1.

In general, checking out all the "don't cares" is very time-consuming. No algorithm exists for generating and checking the possibilities efficiently. If there are many don't cares, say 10 of them, then there are $2^{10} = 1024$ different ways to set the don't cares, and we would have to examine each of them. A time-complexity function for the examination of n don't-care lines would be $O(2^n)$. If this nonpolynomial situation reminds you of the brute–force algorithm for the traveling salesperson problem, you are right on track. Checking all the possibilities is, in general, not an efficient way to find an optimal solution. (See the spotlight "Very Complex Problems" in Chapter 6.)

CHANCE

Introduction

Many of the applications you have studied in earlier chapters of this book or in previous encounters with mathematics are characterized by the fact that having sufficient information about the current status of a system lets you make correct and exact predictions about its status at future times. If, for example, you know the initial position of a robot, the angle along which it is moving, and its speed, then you can calculate the robot's position at any future time. If you know the cost of a new automobile, the monthly interest rate, and the period of your loan, then you can compute how much remains to be paid back at the end of any future month. If you know the production characteristics of three oil refineries your company owns and the demands for heating oil, diesel fuel, and gasoline, then you can determine how much petroleum to send to each refinery.

Many interesting problems society must attempt to solve are of a distinctly different nature: Some amount of uncertainty about the future remains even after all the relevant knowledge has been obtained and thoroughly studied. We cannot say with absolute certainty, for instance, whether it will rain tomorrow, and we can give only the most tentative predictions about the weather a month from now. Similarly, we are unable to determine precisely the lifetime of the first child born in the twenty-first century or even that child's gender. We are likely never to be able to arrive at an exact value for the percentage of criminal defendants that juries correctly convict.

These situations all involve components of *chance* or *uncertainty* or *probability*. People have wrestled with problems of this type throughout all periods of human civilization. In the past century, we have made great progress in discovering how to incorporate chance events into our analysis in a rigorous and productive manner. As one writer phrases it,

> Probability and statistics transformed our ideas of nature, mind, and society. These transformations have been profound and wide-ranging, changing the structure of power as well as of

knowledge. These transformations have shaped modern bureaucracy as well as the modern sciences.[1]

In this chapter we shall discuss the role of chance in modeling real-world phenomena, introduce some of the basic concepts of probability, and show how they are used to solve important problems of human concern.

---

[1] Gerd Gigerenzer et al. (1989), *The Empire of Chance: How Probability Changed Science and Everyday Life,* Cambridge, England: Cambridge Univ. Press.

SECTION 8.1 *The Need for Probability Models*

The principal structure for applying mathematics to the study of real-world problems is a *mathematical model.* In this section we'll look at two familiar models, provide a more formal definition of a model, and then discuss the distinction between deterministic models and probabilistic models. Our familiar models deal with falling apples and rising populations.

In the seventeenth century, Isaac Newton (1642–1727) used the newly forged tools of calculus along with a handful of simplifying physical assumptions to investigate the motions of the Earth, its moon, the other planets, and the sun. Although the focus of Newton's study was the heavenly bodies, his approach could also be applied to the analysis of more familiar situations.

We can predict, for example, that if an apple initially F feet above the ground is released, then it

will hit the ground a certain number T of seconds later. The numbers F and T are related by the equation

$$F = 16T^2.$$

From this equation we can make certain predictions. If we observe that an apple takes 2 seconds to fall to the Earth, the equation tells us that it fell from a height of 64 feet. We can also work backwards, if desired, to compute the time it takes the object to fall from a known height.

In the classical model of exponential growth, one investigates the assumption that a population increases at a rate proportional to its numbers. If it takes D hours for such a population to double in size, then it can be shown that the population P at any time t is given by the equation

$$P = P_0 2^{t/D},$$

where P_0 is the initial size of the population.

What Is a Model?

These two examples—falling bodies and rising populations—are examples of *deterministic models.*

By a *model,* we mean a representation or imitation of some important features of a real-world object or system in a simpler, more easily understood and manipulated medium. A road map, for example, is a model of a geographic region. A wind-up toy car is a model of a real automobile. Architects often construct scaled-down models of proposed buildings. Children build model towns out of building blocks. These are all examples of *physical models.*

A *mathematical model* is a representation in the language of mathematics of the essential features of a "real" system. A mathematical model of a complex phenomenon or situation has many of the advantages and limitations of other types of models. Some factors in the situation will be omitted, while others are stressed. Mathematics is a precise and unambiguous language. In order to use mathematics, we must first clarify to ourselves just what are the underlying assumptions we are making about the real-world problem. The mathematical model forces us to organize our thoughts in a more systematic way.

Once we have formulated the model, we can use mathematical tools to derive new conclusions that may have escaped us had we proceeded with a more intuitive approach. Not only will these conclusions shed light on the assumptions we have originally made about the system under study, but they will also suggest further experiments and observations that will lead to a more complete understanding.

The assumption behind a deterministic model of a process that changes over time is that the entire future behavior is exactly and explicitly determined by the present status of the system and the forces acting on it. In other words, if we know everything about the system at a particular moment, then we can, at least in principle, predict its behavior at every future instant.

This belief was the driving force that led to the very fruitful development of the physical sciences during the last three centuries. Much of our understanding of the behavior of the physical systems we observe in daily life comes from deterministic models. Physical scientists and engineers developed powerful analytic techniques to analyze more and more complex deterministic models. The availability of these techniques and the predictive success of these models in the physical sciences motivated many thinkers to employ similar models in the study of social and biological systems.

Deterministic Versus Probabilistic Models

In many real-life instances, deterministic models may not be appropriate. Our knowledge about the current status of a particular system may be too imprecise to stipulate exactly the levels of each of the critical variables or the impact of every force impinging on it. We may be comfortable only in asserting that a given number lies within some range or in giving an estimate of the relative likelihood that a certain force will affect the system in one of several different ways. As our knowledge of the system increases, we may obtain sharper estimates, but some uncertainty may always remain. In such cases, we do not attempt to make deterministic predictions of the state of the system at a particular point in the future; instead, we offer probabilistic judgments: "There is a 60% chance of rain tomorrow"; "Odds makers give the Chicago Bulls a 3–2 chance of winning the basketball championship"; "About 35% of the people think the president is doing a satisfactory job."

It might be the case that even with full knowledge of the present state of the system and the forces acting on it, we may still believe that in principle the outcome is still subject to some unpredictable, chance process. As Pierre-Simon de Laplace (1749–1827), the great French mathematician, wrote, "The most important questions of life are, for the most part, really only questions of probability. Strictly speaking, one may even say that nearly all our knowledge is problematical; and in the small number of things which we are able to know with certainty, even in the mathematical sciences themselves, induction and analogy, the principal means of discovering truth, are based on probabilities, so that the entire system of human knowledge is connected with this theory."

Let's examine some real-life situations in which chance seems to play an important role:

Six Problems

1. *Women and witchcraft:* Of 21 men and 68 women facing witchcraft charges in a seventeenth-century New England town, only 2 of the men were convicted while 14 of the women were found guilty. Were men and women judged by different standards?

2. *Screening for AIDS:* Although blood tests designed to identify the presence of certain antibodies may have high levels of sensitivity and selectivity, when such tests are given to a large population where the incidence of the particular antibody is low, most "positives" are false positives. Why is this so? What are the implications of mandatory tests for the AIDS virus or for drugs?

3. *Reliability of safety devices:* A radiation meter at a nuclear power plant is supposed to activate a loud warning alarm if the reactor reaches an unacceptable level of radiation leakage. The alarm sometimes rings when the radiation level is far below the danger level; there is suspicion that it may fail to go off even when there is real danger. How reliable is this safety system? Can we increase the reliability by installing several additional meters?

4. *Stolen cars:* Crime statistics show that in your neighborhood, one of every five persons is a victim of car theft each year. What are your chances of avoiding a theft over a three-year period? How long should a typical resident expect to wait before his or her car is stolen? How many years would you estimate it would take for *everyone* in your neighborhood to be a victim?

5. *Job interviews:* You've just landed interviews with three prospective employers. Each one has three different positions open: Fair, Good, and Excellent, with annual salaries of $25,000, $30,000, and $35,000, respectively. Your college's placement office estimates that at each firm your chances of being offered a Fair job are 4 in 10, while the prospects for Good and Excellent positions are 3 in 10 and 2 in 10, respectively. There is a 1 in 10 chance that you won't be given an offer. Suppose each company requires you to accept or reject a job offer at the conclusion of the interview. What strategy should you follow? Should you, for example, reject a Good offer from the first company and gamble that one of the two remaining offers will be better?

6. *Lotteries:* In the Virginia state lottery, six numbers between 1 and 44 are randomly selected. If you hold a lottery ticket with these six numbers highlighted, then you will win the grand prize. In February 1992, a group of Australian investors tried to

win the Virginia lottery, with an estimated jackpot of $27 million, by purchasing every single possible combination of six numbers. How much would it cost the investors to do this? Was it worth the risk?

Some Common Features of These Problems

Each of these problems involves **uncertainty** or **chance**. If 20% of the cars on your block are stolen each year, then it's possible that everyone will be a victim within five years. But it's also possible that some individuals will never have a car stolen, while others will lose several vehicles to thieves. With good luck, you might never have to report a stolen car; with bad luck, you might become a familiar face at the police precinct house.

If your blood tests positive for the AIDS virus, that might be an instance of a true positive or a false positive. We can't say for sure, because the test is not perfectly reliable. We might be able to estimate a probability or likelihood of the chances that it is a true positive.

We also face the issue of false positives and false negatives in the nuclear reactor example. A false positive means that a warning is issued when none is necessary; a false negative is a situation where the warning device showed no trouble but there was a real danger present. In this situation, a false negative may pose far graver danger than a false positive. We might then want to analyze the situation to find ways to lower the chances of one. If several copies of the warning device are in operation, then many patterns of successes and failures of the individual devices could occur, each happening with a different likelihood. Even if every warning device is silent, for example, we still can't be absolutely certain that the reactor is operating safely.

In the witchcraft example, we're interested in determining if the historical data justify a conclusion that gender bias was present. An alternative explanation is that whether or not a given individual was convicted was a matter of chance; the numbers we see may be consistent with a random selection from the pool of men and women.

In considering playing a weekly lottery, we must balance off the overwhelming chances that a single ticket will lose and we'll be out the $1 we paid for it against the very unlikely event that we'll pick the lucky combination of numbers and win millions of dollars. We might get some of the lucky numbers, but not all of them, and win one of several smaller cash prizes. How do we achieve this balancing act? We could consider buying a single ticket week after week after week and then figure out the average amount we win (or loss) on each ticket. This average is called the *expected value* of the lottery. If in 50 successive weeks, we won $25 once, $15 another time, and lost on the remaining 48 tickets, then our net gain would be $25 + $15 − $50 = −$10, so the average value would be −$10/50, or −$0.20, per week. We could, of course, increase our chances of winning in a single week by purchasing more than one ticket. What if we tried, as the Australian gamblers did, to buy all possible combinations? How many tickets would that be? Would it be worth the effort?

In the scenario we've described involving job interviews, you might try to develop a strategy ahead of time that will guide you when the moment comes to accept or reject a job offer. There's no doubt about what to do if you're offered an Excellent position, but for any other offer there are risks associated with either decision. If you turn down a Good or Fair job from one employer, you might not receive an offer from any of the subsequent companies. On the other hand, if you take one of these offers, you pass up the chance of the Excellent post with its substantial added yearly income. What strategy maximizes your chances of getting a Good or Excellent job? What strategy yields the largest expected salary? Are these strategies the same? How should each be modified if the probability estimates are changed or the salary figures are altered?

In order to obtain more precise formulations of these questions and related problems so that we can eventually find more exact solutions, we need to develop some of the tools of probability.

Common Ingredients

We can find some more specific ingredients that are common to probability problems:

1. a fixed set of well-defined possible outcomes;

2. uncertainty as to which outcome will actually occur;

3. some knowledge of the relative likelihood of each possible outcome; and

4. measurable penalties/rewards attached to each outcome.

Let's illustrate these with the job interview example. The possible *outcomes* of each interview are as follows: No offer, Fair offer, Good offer, Excellent offer. Exactly one of these events will happen, but we don't know which one; hence there is *uncertainty*. We have estimated the relative likelihood of each of these possible outcomes:

Outcome	Relative chances
No offer	1 in 10
Fair offer	4 in 10
Good offer	3 in 10
Excellent offer	2 in 10

Finally, we have associated with each outcome a penalty or reward; in this case, the starting annual salary that goes with each offer:

Outcome	Reward
No offer	$0
Fair offer	$25,000
Good offer	$30,000
Excellent offer	$35,000

These four ingredients—outcomes, uncertainty, relative likelihood of each outcome, and risk/reward—will normally be present in all situations involving chance outcomes. We will need to learn ways to assign *probabilities* or numerical measures of the relatively likelihood of the possible outcomes in a consistent manner that reflects the underlying reality of the situation. We also need to examine ways in which the probability of a more complex event can be determined if we know the probabilities of the simpler events that make it up. For example, what is the probability that at least one of the three job interviews will result in an Excellent offer? What is the probability that all three employers will offer Fair jobs? In this chapter we will develop the tools to answer such questions.

Why a Probabilistic Approach May Provide Better Models

There are several important objections to the use of deterministic models in the social and life sciences. First, some philosophers argue that deterministic models of social systems must necessarily assume that human beings have no free will; few people are willing to accept this view of humans. Social problems deal with individuals or groups of individuals. We can never completely predict the exact future behavior of any person in a specific situation, no matter how well we understand the person or the social context.

Probability and the Law

Questions involving probabilities are increasingly finding their way into the courtroom. The guilt or innocence of criminal defendants may often be determined by a jury that, in the absence of eye witnesses, must weigh testimony about DNA "fingerprinting," similarity of hair fibers, or matches of carpet threads. Television's broadcast of the 1995 O. J. Simpson murder trial brought such issues into the living rooms of millions of Americans.

A central question about the admissibility in a trial of a piece of evidence is its relevance; relevance is defined by the *Federal Rules of Evidence* in terms of probability: Evidence is relevant if it has a "tendency to make the existence of any fact that is of consequence to the determination of the action more probable or less probable than it would without the evidence."

A 1968 California case, *People* v. *Collins,* illustrates how probability can be used (and misused) in a criminal trial. Witnesses reported seeing a blond white woman with a ponytail running with a bearded, mustachioed black man from an alley in a Los Angeles suburb where an elderly person had just been mugged. The couple drove off in a partly yellow car.

Police subsequently arrested Janet and Malcolm Collins. They owned a partly yellow Lincoln, she usually wore her blond hair in a ponytail, and he was black. Although clean-shaven when arrested, Collins showed traces of recently having a beard and mustache.

At the trial, the prosecutor argued that he had "mathematical proof" of the Collins's guilt. He assigned the following "conservative probabilities" to the characteristics noted by witnesses:

man with mustache	1/4
woman with ponytail	1/10
woman with blond hair	1/3
black man with beard	1/10
interracial couple in car	1/1000
partly yellow car	1/10

The prosecutor then argued that the product of these probabilities was 1/12,000,000, and thus the chances that there was another couple in the Los Angeles area having all these characteristics was less than one in a billion.

The jury convicted the couple, but the California Supreme Court reversed the conviction on appeal, citing several major misuses of the probability-based argument. The Collins case has been widely debated in legal circles with debates reaching the pages of prestigious law journals.

An analysis of the Collins case by a noted Constitutional scholar (and undergraduate mathematics major!) can be found in Laurence Tribe (1971), "Trial by mathematics: Precision and ritual in the legal process," *Harvard Law Review* 84:1329–1391, while a more general survey of the use of probabilities as evidence is provided by Mary W. Gray (1983), "Statistics and the law," *Mathematics Magazine* 56:67–81.

A second objection arises from the discovery of Werner Heisenberg (1901–1976) in the early part of the twentieth century that purely deterministic models are insufficient even to study *physical* processes. Heisenberg showed that it is impossible, even in theory, to know the exact state of a physical system. We can conclude from models only the *average behavior* of a group of atoms, but we cannot assert with certainty anything about the future course of an individual atom. This observation has fundamentally affected the physical sciences and led to the introduction of probabilistic models.

Nobel Laureate Richard Feynman emphasized the new probabilistic viewpoint of physics in this excerpt from his famous book, *Lectures on Physics:*

> We would like to emphasize a very important difference between classical and quantum mechanics. We have been talking about the probability that an electron will arrive in a given circumstance. We have implied that in our experimental arrangement (or even in the best possible one) it would be impossible to predict exactly what would happen. We can only predict the odds! This would mean, if it were

true, that physics has given up on the problem of trying to predict exactly what will happen in a definite circumstance. Yes! Physics *has* given up. We do not know how to predict what would happen in a given circumstance, and we believe that it is impossible—that the only thing that can be predicted is the probability of different events. It must be recognized then that this is a retrenchment in our earlier ideal of understanding nature. It may be a backward step, but no one has seen a way to avoid it. . . . We suspect very strongly that it is something that will be with us forever . . . that this is the way nature really is.[2]

We see then that the probabilistic viewpoint will come to dominate quantitative models and mathematical thinking about physical, scientific, technological, and social problems. In this chapter we will explore some of the elementary concepts of probability.

---

[2]Richard P. Feynman, Robert B. Leighton, and Matthew Sands (1965), *The Feynman Lectures on Physics,* Vol. 3, Reading, MA: Addison-Wesley.

Exercises for Section 8.1

1. Using the model $F = 16T^2$, determine the initial height of an apple if it hits the ground in
   (a) 1 second   (c) 4 seconds
   (b) 3 seconds

2. Show that the time of fall of the apple can be written in terms of the initial height by solving the equation $F = 16T^2$ for T in terms of F. Determine how long it will take the apple to reach the ground if its initial height is

   (a) 64 feet   (c) 400 feet
   (b) 100 feet

3. Using the model $P = P_0 2^{t/D}$, find the population after 120 years if the initial population is 100,000 and the doubling time is 30 years.

4. Determine the population after 80 years if the doubling time is 40 years and the initial population is 1,000,000.

5. Suppose the initial population is 400 and the population after 30 years is 3200. Determine the doubling time.

6. Discuss three additional examples of physical models.

7. Discuss three examples of deterministic situations and three probabilistic situations that occur in everyday life.

8. Classify each of the following behaviors as deterministic or probabilistic:
   (a) The color of light displayed by the traffic light nearest your home at a particular moment
   (b) The amount of traffic at any instant passing through the intersection where the traffic light in (a) is located
   (c) The percentage of voters favoring a particular presidential candidate at a given moment
   (d) The time and date of the next complete solar eclipse
   (e) The grades you will receive in your courses at the end of this term

9. Before you make a formal study of probability, write down what you think reasonable answers would be for each of the six problems concerning
   (a) women and witchcraft
   (b) screening for AIDS
   (c) reliability of safety devices
   (d) stolen cars
   (e) job interviews
   (f) lotteries

10. (**Writing Project**) Historically "the founding document of mathematical probability" is said to be a letter in July 1654 from Blaise Pascal to Pierre de Fermat concerning a gambling problem. Even today many probability problems are cast in the form of gambling questions. Investigate several popular casino games (e.g., blackjack, keno, roulette, slot machines). What are the rules? What variations do casinos offer to decrease your chances of winning? What types of lotteries does your state offer? How do you determine the odds of winning these lotteries?

11. (**Writing Project**) Some decisions we make in everyday life can also be viewed as gambles. When your family buys fire insurance on its house, there is a need to balance the certainty of the insurance premium against the relatively small risk of a catastrophic loss. Over a 40-year period, the 20 families on my block paid out more than $250,000 in fire insurance premiums, yet none of them ever had a house burn down. Most purchased their homes for about $25,000 and saw them appreciate to about $60,000. Write a short essay discussing the insurance as a gamble. Does it make sense for you to buy fire insurance? health insurance? life insurance? Does it make economic sense for companies to sell such insurance? What considerations should you make in determining if the cost of an insurance policy is fair?

SECTION 8.2 *What Is Probability?*

Probabilistic Statements

"Don't worry about smoking cigarettes. You'll either get cancer or you won't. So there's a fifty–fifty chance that cigarettes won't harm you at all."

"I always carry a bomb with me when I fly, because the chances of there being a bomb on the plane are 1 in 100,000, but the chances of two bombs are 1 in a billion."

"There are three ways of rolling a sum of 6 with a pair of dice: 1 and 5, 2 and 4, 3 and 3. There are also three ways of rolling a sum of 7: 1 and 6, 2 and 5, 3 and 4. We can conclude that the chances of rolling a 6 and rolling a 7 are the same."

Each of these three quotations is a *probabilistic statement.* It is an assertion about the *chances* that something will happen. People make probabilistic statements about situations where one of a designated and known set of outcomes will occur but we cannot assert with certainty precisely which one it will be. If you flip a fair coin, it will land either heads or tails. If you smoke cigarettes, either you will develop lung cancer or you will not. If you roll a pair of dice, then you will obtain a sum that is one of the 11 numbers 2, 3, 4, . . . , 12.

Many probabilistic statements are *qualitative* in nature:

"There's hardly any chance of rain tomorrow."

"Scientists believe there will very likely be a major earthquake in California in the 1990s."

"This medication will almost certainly lower your blood pressure."

Qualitative statements are often helpful in making decisions: If I believe it's very unlikely to rain, then I'll probably leave my umbrella at home. But qualitative statements are often too imprecise. If a physician assures you that a risky operation "will probably help you," does this give you enough information to make an informed decision? One of the goals of probability theory is to develop tools that enable us to make more precise, *quantitative* statements about the relative likelihood of outcomes. Each of the three quotations at the beginning of this section is a quantitative probabilistic statement (although not necessarily a valid one!).

There are two major interpretations of probability: the *frequency interpretation* and the *subjective judgment interpretation.* In the frequency interpretation, probabilities are conceived as approximations to long-run relative frequencies. In the subjective judgment interpretation, probabilities are viewed as expressing the opinion of some person regarding the degree of certainty that a particular outcome will occur. Our treatment of probability will emphasize the frequency interpretation. A *relative frequency* is a proportion (or fraction) that measures how frequently some particular outcome occurs in a sequence of observations or experiments. Consider some situation that can be repeated under identical conditions over and over again. Some examples are tossing a coin, rolling a pair of dice, picking a letter at random from some long text, and noting the gender of newborn babies. Each such repetition is called a *trial.* Let A be a possible result of such a trial; for example, the result A might be the coin lands tails, the dice sum to 8, the letter is an "e," the baby is a girl. If the result A happens m times in n trials, then the fraction m/n is the relative frequency of A in the n trials.

In the frequency interpretation, the probability of a result A is the estimated relative frequency of A in a very large number of trials. If we toss a fair coin a large number of times, we expect that the relative frequency of obtaining heads to be about one-half. For any particular value of n, we would not expect to get exactly the same number of heads and tails, but if n is very large, then we would expect to find that close to half the time we get tails. If $n(A)$ denotes the number of times the result A occurs in n trials, then the probability of A is estimated by the fraction $n(A)/n$. If a particular result necessarily always occurs, then its relative frequency would be 1. If a particular result can never occur, then its relative frequency would be 0. Thus, the relative frequency interpretation of probability always gives a number between 0 and 1. Impossible outcomes are assigned probability 0, and "sure things" are assigned probability 1.

EXAMPLE 1 Baseball A baseball player's batting average is the ratio of hits to times at bat. The batting average is the relative frequency of successes (obtaining a hit) among official trips to the plate. We can interpret a player's batting average as the probability that he will get a hit in a randomly selected time at bat. For example, the legendary Ty Cobb—considered by many baseball experts to be the greatest athlete in the history of the game—compiled the following statistics in more than two decades of play:

Times at bat	11,429
Hits	4191
Singles	3052
Doubles	724
Triples	297
Homeruns	118
Strikeouts	357

We can estimate the probability that Ty Cobb got a hit in any one time at bat as his batting average, $4191/11{,}429 = .367$. The probability that he hit a homerun in a randomly chosen appearance is the relative frequency of homeruns, $118/11{,}329 = .010$ (these numbers have been rounded off to three decimal places). ▲

Fundamental Definitions of Probability

Although the frequency interpretation of probability is often quite useful, mathematics requires us to have

a more rigorous and precise definition of probability if we are going to develop a useful theory. Our first definition provides such an approach. It is based on the fundamental mathematical concept of *sets*. (You may wish to review some of the material on set theory in the appendix.)

DEFINITION

Let E be a set with a finite number of elements. A *probability measure on E* is defined to be a real-valued function Pr whose domain consists of all subsets of E and which satisfies three rules:

1. $Pr(E) = 1$;

2. $Pr(A) \geq 0$ for every subset A of E; and

3. $Pr(A \cup B) = Pr(A) + Pr(B)$ for every pair of disjoint subsets A and B of E.

Note that the first two conditions of a probability measure are consistent with the observations we made about relative frequencies. The third condition is also a very natural one. The set $A \cup B$ contains all those elements belonging to A or to B or to both. If A and B are disjoint sets that correspond to two different outcomes, then whenever one of them occurs, the other cannot have occurred. Thus, the number of times A or B occurs is simply the number of times A occurs plus the number of times B occurs. For relative frequencies, we have

$$\frac{n(A \cup B)}{n} = \frac{n(A)}{n} + \frac{n(B)}{n} \quad \text{if } A \text{ and } B \text{ are disjoint.}$$

Recall also the presentation and discussion of the addition principle in Section 4.1.

We can also interpret the third condition in our definition in a geometric way (see Figure 1). Imagine that we select a point from set U. The probability of selecting a point from a subset S is the relative area of S: the area of S divided by the area of U. Thus, if A and B are disjoint, the area of $A \cup B$ is the sum of the areas of A and B.

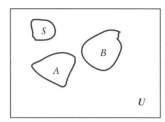

FIGURE 1 area($A \cup B$)/area(U) = area(A)/area(U) + area(B)/area(U)   if A and B are disjoint.

DEFINITION

A *probability space* is a set E, called the *sample space*, together with a probability measure Pr.

Examining several examples will help us understand these definitions.

EXAMPLE 2   Coin Flips   Let E be the set of possible outcomes in an experiment consisting of flipping a coin and noting which side faces up when the coin lands. Then E has two elements: h for "heads" and t for "tails." In this case the set $E = \{h, t\}$ has four subsets: E, $\{h\}$, $\{t\}$, and the empty set $\varnothing$.

The assumption that the coin is fair, that is, there is as much chance of heads as tails, would be reflected by the probability measure:

$$\Pr(\varnothing) = 0, \qquad \Pr(E) = 1,$$
$$\Pr(\{h\}) = \Pr(\{t\}) = \tfrac{1}{2}. \ \blacktriangle$$

EXAMPLE 3   If we have reason to believe that the coin of Example 2 has been weighted so that heads appears three times as often as tails, then our probability measure would become

$$\Pr(\varnothing) = 0, \qquad \Pr(E) = 1, \qquad \Pr(\{h\}) = \tfrac{3}{4},$$
$$\Pr(\{t\}) = \tfrac{1}{4}. \ \blacktriangle$$

Notice that Examples 2 and 3 are two different sample spaces with the same underlying set.

EXAMPLE 4   We can define a probability space without regard to any particular real-world setting. Suppose we let the set $E = \{b, g\}$ and define a probability measure by the assignments

$$\Pr(\varnothing) = 0, \qquad \Pr(E) = 1, \qquad \Pr(\{b\}) = \tfrac{7}{9},$$
$$\Pr(\{g\}) = \tfrac{2}{9}.$$

This gives a probability measure satisfying our definition.

In this case we can interpret this probability space as a model for the experiment of choosing a sock at random from a drawer of hosiery that contains exactly 7 blue socks and 2 green ones. Then the probability of selecting a blue sock would be 7/9, with a probability of 2/9 of picking a green one. $\blacktriangle$

EXAMPLE 5   The Working Class   In a study of the labor market, researchers constructed three categories of workers: Unskilled, Skilled, and Professional. An experiment consists of selecting a worker at random for the purposes of an in-depth interview. We could then let our set be $E = \{$Unskilled, Skilled, Professional$\}$. Our probability measure Pr must then assign numbers to eight subsets:

E,

$\emptyset$,

{Unskilled},

{Skilled},

{Professional},

{Unskilled, Skilled},

{Unskilled, Professional},

{Skilled, Professional},

which we will denote, in the obvious way, as

E, $\emptyset$, $\{U\}$, $\{S\}$, $\{P\}$, $\{U, S\}$, $\{U, P\}$, $\{S, P\}$.

If we think there are equal numbers of workers in all three categories, then an appropriate probability measure would be

set A	E	$\emptyset$	$\{U\}$	$\{S\}$	$\{P\}$	$\{U, S\}$	$\{U, P\}$	$\{S, P\}$
$\Pr(A)$	1	0	1/3	1/3	1/3	2/3	2/3	2/3

Suppose, however, that studies show that 40% of workers are Unskilled, 50% are Skilled, and 10% are Professional. In this case our probability measure would look like

set A	E	$\emptyset$	$\{U\}$	$\{S\}$	$\{P\}$	$\{U, S\}$	$\{U, P\}$	$\{S, P\}$
$\Pr(A)$	1	0	.4	.5	.1	.9	.5	.6

We can make an extremely important observation based on these examples. Once the probability measures are assigned for the one-element subsets of E, the probability measures for all other subsets are determined. This is a consequence of property 3 of a probability measure: *The measure of*

the union of two disjoint sets is the sum of the measures of the two sets. Taking our last probability measure as an example,

$$\Pr(\{U, S\}) = \Pr(\{U\} \cup \{S\}) = \Pr(\{U\}) + \Pr(\{S\})$$
$$= .4 + .5 = .9. \ \blacktriangle$$

Let's call the one-element subsets of E *elementary events,* and we'll call the probability measure assigned to a particular elementary event the *weight* of that event. For Example 5, the weight of P would be .1 since $\Pr(\{P\}) = .1$.

We have then the following important result:

Theorem 1. The probability measure of an arbitrary subset A of E is the sum of the weights of the elements of A.

Next we'll list some of the elementary laws of probability, which flow easily from the definition of a probability measure. They are gathered together in the form of a theorem, whose proof is mostly left as an exercise.

Theorem 2. If Pr is a probability measure on a finite set E, then the following statements are true for all subsets $A, B, A_1, A_2, \ldots, A_k$ of E:

1. If A is a subset of B, then $\Pr(A) \leq \Pr(B)$.

2. $\Pr(\emptyset) = 0$.

3. $\Pr(A \cup B) = \Pr(A) + \Pr(B) - \Pr(A \cap B)$.

4. $\Pr(A') = 1 - \Pr(A)$, where A' is the complement of A; that is, $A' = E - A$.

5. $\Pr(B) = \Pr(A \cap B) + \Pr(A' \cap B)$.

6. If $A_1, A_2, \ldots, A_k$ are mutually disjoint subsets of E, then

$$\Pr(A_1 \cup A_2 \cup \cdots \cup A_k)$$
$$= \Pr(A_1) + \Pr(A_2) + \cdots + \Pr(A_k)$$
$$= 1 - \Pr(A_1' \cap A_2' \cap \cdots \cap A_k').$$

7. $\Pr(A \cap B) = \Pr(B \cap A)$

Proof. To establish part 5, note that we may write the set B as the union of two disjoint subsets

$$B = (A \cap B) \cup (A' \cap B), \qquad (1)$$

and hence by the definition of a probability measure,

$$\Pr(B) = \Pr(A \cap B) + \Pr(A' \cap B). \qquad (2)$$

To show that part 4 is true, rewrite this equation as

$$\Pr(A' \cap B) = \Pr(B) - \Pr(A \cap B), \qquad (3)$$

and use the fact that the union $A \cup B$ may be written as the union of two disjoint sets

$$A \cup B = A \cup (A' \cap B), \qquad (4)$$

we have

$$\Pr(A \cup B) = \Pr(A) + \Pr(A' \cap B)$$
$$= \Pr(A) + \Pr(B) - \Pr(A \cap B), \qquad (5)$$

which shows the validity of part 4.

The remaining parts of the theorem can be proved in a similar fashion. We leave these as exercises.

▲ ▲ ▲

Let's illustrate property (3) geometrically (see Figure 2). The total area in the union $A \cup B$ is the sum of the areas of A and B *minus* the common area in the intersection $A \cap B$, because that area

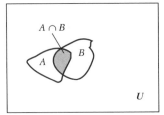

FIGURE 2 $\Pr(A \cup B) = \Pr(A) + \Pr(B) - \Pr(A \cap B).$

is counted twice, once as part of A and once as part of B.

The Equiprobable Situation

One especially important case of a sample space occurs when each of the elementary events has the same weight. This would occur in any situation where we had reason to believe that all elementary events were equally likely to occur. When we flip a fair coin, we believe that heads and tails have the same chance of being the outcome. In fact, we often define "fair coin" to meant just that; that is, $\Pr(\text{heads}) = \Pr(\text{tails})$. Since the two probabilities are equal and they add up to 1, each must be one-half.

In any case where the elementary events are equally likely to occur, we make use of the *equiprobable measure*. In an experiment with n distinct outcomes, the equiprobable measure assigns probability $1/n$ to each of these outcomes. In rolling a standard die, each of the six possible outcomes is given probability $1/6$, unless we have evidence that the die has been "fixed" so that some outcomes are more likely and others are less likely.

Calculations of probabilities of subsets are generally easier to do in sample spaces having equiprobable measure. The reason for this is easy to see. In any sample space, the probability of a particular subset A is the sum of the weights of the elements that make up A. In an equiprobable case, each of the elements has the same weight, $1/n$, where n is the number of elements in the underlying set E. Thus, in the equiprobable case, the probability measure of a subset becomes the proportion of elements of the underlying set that actually belong to the subset. We'll state this formally as a theorem:

Theorem 3. For the equiprobable measure,

$$\Pr(A) = \frac{\text{number of elements in } A}{\text{number of elements in } E}.$$

The computation of probabilities in the equiprobable situation reduces to the problem of counting the number of elements in a set. We need to ask two questions:

How many different outcomes (elementary events) are there?

For how many of these outcomes is the desired condition true?

EXAMPLE 6 Arrivals Bruce, Priscilla, and Janine are students in the same class. If all patterns of their arrival are equally likely, what is the probability that Priscilla arrives before Janine?
Solution: List all the possible arrival patterns. In this case, there are six:

1.	Bruce	Priscilla	Janine
2.	Bruce	Janine	Priscilla
3.	Priscilla	Janine	Bruce
4.	Priscilla	Bruce	Janine
5.	Janine	Priscilla	Bruce
6.	Janine	Bruce	Priscilla

Then the event {A = Priscilla arrives before Janine} occurs exactly for patterns 1, 3, and 4. Hence $\Pr(A) = 3/6 = 1/2$. ▲

EXAMPLE 7 Fair Coin A fair coin is tossed four times. What is the probability of getting two heads and two tails?
Solution. First we determine the possible patterns of heads and tails that could conceivably occur. There are 16 possible patterns. Using H to represent heads and T for tails, we list them all:

HHHH	HTHH	THHH	TTHH
HHHT	HTHT	THHT	TTHT
HHTH	HTTH	THTH	TTTH
HHTT	HTTT	THTT	TTTT

Since the coin is fair, each of the 16 possibilities is as likely to occur as any other outcome. We can then make use of the equiprobable measure. List all the ways in which we see exactly two heads and two tails. There are six possible patterns:

HHTT HTHT HTTH THHT THTH TTHH

Thus the probability of obtaining two heads and two tails in four tosses of fair coin is

$$\frac{\begin{array}{c}\text{number of ways to get}\\ \text{2 heads and 2 tails}\end{array}}{\begin{array}{c}\text{number of ways for}\\ \text{experiment to end}\end{array}} = \frac{6}{16} = \frac{3}{8}. \text{▲}$$

In Examples 6 and 7, we listed every single possible outcome of the experiment and then counted every possible outcome in which the desired condition is true. Such an approach is always *logically possible* in an equiprobable situation, but it has severe practical problems. For example, suppose we ask for the probability that in 40 flips of a fair coin, we will get 20

Probability and Secret Codes

Probability plays an important role in *cryptology*, the making and breaking of secret codes. Keeping messages indecipherable to enemies but easily translated by allies has long been a concern to diplomats and generals. The great Roman soldier and statesman Julius Caesar enciphered messages by replacing each letter by the letter three places later in the Latin alphabet.

Concerns about security and guarantees of privacy have increased today as financial reports, medical records, and other sensitive information are passed from computer to computer along networks where messages are easily intercepted.

An important characteristic of written language is that the individual letters do not appear with equal frequency. In English, for example, the letter "e" accounts for about 12% of all characters in ordinary text, while "t" makes up about 8.5% and "j" occurs far less than 1% on the average. A simple cipher, such as the one used by Caesar (but applied to the English alphabet), which replaces CODES by FRGHV, is easily broken by a *frequency analysis* of the letters in the enciphered message. The most frequently appearing letter is likely to represent an E, the next most common letter a T, and so on.

Modern cryptographic schemes use techniques to ensure that each character in a coded message appears with equal probability. A theoretically unbreakable cipher is the "one-time pad." The pad is a long list of *random numbers*, each between 1 and 26. Such a pad might begin 19, 7, 12, 1, 3, 8, ... To encipher a word such as "eleven," you replace the first letter by the one 19 spaces later in the alphabet, the second letter by the one 7 spaces later, and so on. Thus, ELEVEN becomes XSQWHV. Note that although "E" appears 3 times in the "plaintext" ELEVEN, it is replaced by three different letters in the "ciphertext": XSQWHV.

The Soviet Union employed the one-time pad throughout World War II and for many years afterward. In July 1995 the U.S. National Security Agency revealed that it had broken the Soviet code in 1944. Russian agents had used some of the sequences of random numbers repeatedly, causing patterns to appear in their messages that the codebreakers were able to exploit.

heads and 20 tails. It's futile to attempt to list all of the possible patterns of 40 heads and tails. There are more than a *trillion* such patterns. (More exactly, we know from Chapter 4 that there will be 2^{40} such patterns. Since $2^{10} = 1024$, which is about 1000, we can make the approximation

$$2^{40} = 2^{10 \times 4} = (2^{10})^4 \sim (1000)^4 = (10^3)^4 = 10^{12}.$$

Of these one trillion plus patterns that can result from flipping a coin 40 times, how many of the patterns contain exactly 20 heads and 20 tails? Each such pattern has precisely 20 specific flips that resulted in heads. The total number of such patterns will be the number of ways we can designate 20 particular flips in a sequence of 40 as heads. We learned in Chapter 4 that this will be the number of ways to choose 20 items from a collection of 40; that is, the binomial coefficient $_nC_k = \binom{n}{k}$, with $n = 40$ and $k = 20$.

The number represented by $\binom{n}{k}$ can be calculated by hand for small values of n and k. For larger values it is quicker and more accurate to use computer software. With the help of such a computer program, we evaluated "40 choose 20" to be about 1.378465×10^{11}. Thus, the probability of obtaining exactly 20 heads (and therefore 20 tails) on 40 flips of a fair coin is

$$\frac{\binom{40}{20}}{2^{40}},$$

which is approximately equal to .125. (*Caution:* If we flip a fair coin 40 times, we expect to get about 20 heads and 20 tails, but the probability that we get exactly an even split is only about 1/8.)

EXAMPLE 8 Who Is Earlier? Suppose Priscilla and Janine are students in a class of 70. If all patterns of their arrival are equally likely, what is the probability that Priscilla arrives before Janine?

Solution: There are 70! different possible patterns of arrival. This number is greater than 10^{100}, so we have no real possibility of listing them all. We can arrive at a correct answer, however, by "counting without counting." Consider any pattern of arrival in which Priscilla comes before Janine. Keeping everyone else's position the same, interchange Priscilla and Janine. This gives a pattern with Janine arriving before Priscilla. We then see that there is a one-to-one correspondence between the patterns with Priscilla before Janine and the patterns with Janine before Priscilla. Now in every pattern, either Priscilla is ahead of Janine or Janine comes in before Priscilla. Thus, the entire set of 70! patterns breaks down into two mutually exclusive sets of equal size. The probability that Priscilla arrives before Janine is 1/2. ▲

To understand this argument about a one-to-one correspondence between the patterns with Priscilla before Janine and the patterns with Janine before Priscilla, suppose the total class size is 4 students instead of 70. Let the other two students be Matt and Steve. Then the 4! = 24 patterns of delivery can be listed in two columns:

Priscilla, Janine, Matt, Steve	Janine, Priscilla, Matt, Steve
Priscilla, Janine, Steve, Matt	Janine, Priscilla, Steve, Matt
Priscilla, Matt, Janine, Steve	Janine, Matt, Priscilla, Steve
Priscilla, Matt, Steve, Janine	Janine, Matt, Steve, Priscilla
Priscilla, Steve, Janine, Matt	Janine, Steve, Priscilla, Matt
Priscilla, Steve, Matt, Janine	Janine, Steve, Matt, Priscilla
Matt, Priscilla, Janine, Steve	Matt, Janine, Priscilla, Steve
Matt, Priscilla, Steve, Janine	Matt, Janine, Steve, Priscilla
Steve, Priscilla, Janine, Matt	Steve, Janine, Priscilla, Matt
Steve, Priscilla, Matt, Janine	Steve, Janine, Matt, Priscilla
Matt, Steve, Priscilla, Janine	Matt, Steve, Janine, Priscilla
Steve, Matt, Priscilla, Janine	Steve, Matt, Janine, Priscilla

Note that each row in the right column is obtained by interchanging *Priscilla* and *Janine* in the listing for that row in the left column of the table. The left-hand side of the table shows all of the arrival patterns in which Priscilla precedes Janine. Thus, the right-hand side contains all of the patterns in which Janine arrives before Priscilla. The table exhibits the desired one-to-one correspondence.

Exercises for Section 8.2

1. Suppose automobile license plates consist of either a single letter followed by three digits or a pair of letters followed by two digits. How many different license plates are possible?

2. In a round-robin basketball tournament, each of 16 teams will play each other once. How many games will be played? In a single-elimination tournament, like the NCAA competition, a team can't advance to a subsequent round unless it wins the current game. How many games are played in such a tournament?

3. A small New England college divides its curriculum into four areas. Certain courses in each division have been designated as "foundation courses." The divisions and the respective number of foundation courses are

Humanities	20
Foreign language	16
Social sciences	8
Natural sciences	17

Each student must take a foundation course in at least three of the four divisions. In how many different ways can a student satisfy this requirement?

4. You decide to invest a large inheritance in a portfolio of eight stocks and four bonds. Your broker recommends 12 different stocks and 7 bonds as good investments. In how many ways can you create the portfolio?

5. Use the data in Example 1 to estimate the probability that Ty Cobb
   (a) hit a double in any time a bat
   (b) struck out

6. The career statistics for Babe Ruth are as follows:

Times at bat	8399
Hits	2873
Singles	1517
Doubles	506
Triples	136
Homeruns	714
Strikeouts	1330

Estimate the probability that in any one time at bat, Babe Ruth
   (a) had a hit
   (b) hit a home run
   (c) struck out

7. Find the probability of obtaining three heads and two tails in five tosses of a fair coin.

8. What is the probability that in 10 tosses of a fair coin, you will get 5 heads and 5 tails?

9. Consider the experiment of rolling a pair of fair dice, Die 1 and Die 2.
   (a) Use the multiplication principle of Section 4.1 to show that there are 36 equally likely outcomes.
   (b) Determine the sum of the two die for each of these outcomes.
   (c) Show that there are precisely five ways of rolling a sum of 8.
   (d) Find the probability of getting a sum equal to 7.
   (e) Find the probability of getting a sum larger than 7.
   (f) What are the least likely sums and most likely sums?

10. On a cold winter morning, the probability of my station wagon starting is 3/4, and the probability of my pickup truck starting is 5/8. What is the probability that at least one of the vehicles will start?

11. Thirty percent of students in first-semester calculus at a western engineering school fail the course or drop out before the end of the term.
   (a) What is the probability that a student selected at random at the beginning of the term will pass the course?
   (b) What is the probability that a particular student, Henry Brewster, will pass the course?
   (c) Why are the answers to (a) and (b) not the same?

12. At a large midwestern university, 25 men and 30 women belong to the mathematics club. An executive committee of five members will be selected at random from the membership. What is the probability that the committee will consist of three men and two women? What is the probability that the committee will have more females than males?

13. A recent study of U.S. Army personnel on active duty yielded the following figures:

	Men	Women
Commissioned officers	80,091	8,921
Enlisted personnel	606,956	67,077

Using these data, determine:
   (a) the total number of men, women, commissioned officers, and enlisted personnel
   (b) The probability that a randomly chosen Army member is
      (i) a woman
      (ii) a male commissioned officer
      (iii) an enlisted person
   (c) the probability that a randomly chosen female Army member is an officer
   (d) the probability that a randomly chosen officer is a woman

14. Linda is 31 years old, single, outspoken, and very bright. She majored in philosophy. As a student, Linda was deeply concerned with issues of discrimination and social justice, and she participated in antinuclear demonstrations. Use your judgment to rank the following statements by their probabilities:

A: Linda is a bank teller.

B: Linda is active in the feminist movement.

C: Linda is a bank teller and is active in the feminist movement.

Most people rank C as a more likely statement than A; that is, they assert Pr(C) > Pr(B). But C is a subset of A (and also of B) so, however you rank the three statements, C should be the least likely. How do you explain this apparent contradiction?

Criticize the following response from someone who asserts that C is more likely than A: "It's not very likely that Linda is a bank teller in the first place. But if she is a bank teller, then she is very likely to be active in the feminist movement. She's much more likely to be an active feminist bank teller than a nonfeminist bank teller."

Amos Tversky and Daniel Kahneman, who originally formulated this problem, found that 85% to 90% respondents rated C as more likely than A. They labeled this response the *conjunction fallacy*.

15. (**Writing Project**) Psychologists have found many interesting results about how people interpret probabilistic statements. See Exercise 14 for one instance. As another, the two statements, "You have a 95% chance of surviving this heart operation" and "You have a 5% chance of dying during this heart operation," are logically equivalent, but patients (and their physicians) often assess them differently. They may be more willing to have the operation if they hear the first statement than the second. Prepare a report on the unconscious biases people may have in assessing probabilities or the inconsistent ways in which they may assign likelihoods of uncertain events. A good starting reference is D. Kahneman, P. Slovic, and A. Tversky (eds.) (1982), *Judgment Under Uncertainty: Heuristics and Biases,* Cambridge, England: Cambridge Univ. Press. (This book also includes Tversky's and Kahneman's discussion of the fallacy introduced in Exercise 14.)

SECTION 8.3 *Conditional Probability*

Definition and Elementary Examples

What is the probability that in three flips of a fair coin we will have more heads than tails? For three flips of a coin, we have eight possible sequences of heads and tails:

E: HHH, HHT, HTH, THH, HTT, THT, TTH, TTT.

For a fair coin, each of these eight sequences is equally likely. Let B be the result of obtaining more heads than tails. Then B corresponds to the set of outcomes

B: {HHH, HHT, HTH, THH}.

Thus $\Pr(B)$, the probability of more heads than tails, is $4/8 = 1/2$. If we are given no information about the results of any flip, then we would estimate the probability that heads outnumber tails to be $1/2$.

Suppose, however, that we are told that the first flip was heads and then we are asked to find the probability that more heads than tails resulted. The additional information about the outcome of the first flip may cause us to revise our estimate. If we let A represent the event "First coin is heads," then A corresponds to the set of outcomes

A: {HHH, HHT, HTH, HTT}.

Each of these four sequences is equally likely. In three of these four sequences, we obtain more heads than tails. Thus, we would estimate that the probability of more heads than tails in three flips of a fair coin, *given* that the first flip is heads, is $3/4$. The additional information has enabled us to reduce the set of possible outcomes from E to A and so has changed our estimate of the probability that heads outnumber tails, from $1/2$ to $3/4$.

The situation we have just described occurs quite often in applications. Our estimate of the probability of an outcome may be *conditioned,* or affected, by knowledge that some other related event has occurred. We formalize the concept of conditional probability in our next definition.

DEFINITION

Let Pr be a probability measure defined on a set E. If A and B are any two subsets of E with $\Pr(A) > 0$, then the *conditional probability of B given A,* denoted $\Pr(B \mid A)$, is defined by

$$\Pr(B \mid A) = \frac{\Pr(B \cap A)}{\Pr(A)}.$$

To illustrate the definition, consider our coin-tossing problem. In this case, the event $B \cap A$ is "more heads than tails and first coin is a head," which corresponds to the set of three outcomes

$$B \cap A: \quad \{\text{HHH, HHT, HTH}\}.$$

Thus we have

$$\Pr(B \mid A) = \frac{\frac{3}{8}}{\frac{4}{8}} = \frac{3}{4}.$$

The initial probability estimate $\Pr(B)$ is often called an *a priori estimate,* and the estimate $\Pr(B \mid A)$ made after learning that A has occurred is called an *a posteriori estimate.*

E X A M P L E 1 Used Cars Your friendly local used-car sales agent offers you the following deal: a guaranteed $500 off the negotiated price or a chance to draw blindfolded an envelope from a bowl. The bowl contains 10 envelopes, of which 6 are blue and 4 are red. Three of the red envelopes and one of the blue envelopes each contain a $1,000 bill. The others contain blank pieces of paper. You get to keep the contents of the envelope you randomly select. If you decide to choose an envelope,

a. What is the probability that you win $1,000?

b. What is the probability that you win $1,000 if you choose a blue envelope?

c. What is the probability that you choose a blue envelope given that you won $1,000?

Solution

a. Let B be the outcome of winning $1,000. Of the 10 envelopes, 4 contain $1,000 bills; thus $\Pr(B) = 4/10$. The car dealer's offer amounts to giving you a choice of the certainty of $500 and a gamble with a 40% chance of winning $1,000 and a 60% chance of winning nothing.

b. Let A be the outcome of choosing a blue envelope. Since the bowl contains 10 envelopes of which 6 are blue, $\Pr(A) = 6/10$. The probability that you win $1,000 given that you choose a blue envelope is

$$\Pr(B \mid A) = \frac{\Pr(B \cap A)}{\Pr(A)}.$$

The outcome $B \cap A$ corresponds to the one blue envelope containing $1000. Hence

$$\Pr(B \mid A) = \frac{\Pr(B \cap A)}{\Pr(A)} = \frac{\frac{1}{10}}{\frac{6}{10}} = \frac{1}{6}.$$

You might make such a calculation if the dealer proposed an alternative gamble: He will remove all the red envelopes before you draw from the bowl.

c. We seek $\Pr(A \mid B)$. Using the definition of conditional probability and the results of (a) and (b), we calculate

$$\Pr(A \mid B) = \frac{\Pr(A \cap B)}{\Pr(B)} = \frac{\Pr(B \cap A)}{\Pr(B)} = \frac{\frac{1}{10}}{\frac{4}{10}} = \frac{1}{4}. \; \blacktriangle$$

Tree Diagrams and Conditional Probability

Tree diagrams provide a convenient geometric way to represent situations involving conditional probabilities. Suppose we consider a two-step process. At the first step, we determine whether or not the event A occurred. At the second step, we consider the occurrence of event B. We represent this process as a tree diagram:

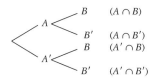

The conditional probability equation $\Pr(C \cap D) = \Pr(D \mid C)\Pr(C)$ can be shown in the diagram by writing the probabilities along corresponding branches:

Pr(B|A) B   (A ∩ B)
Pr(A)   A Pr(B'|A)
           B'   (A ∩ B')
        Pr(B|A') B   (A' ∩ B)
Pr(A') A'
           B'   (A' ∩ B')
        Pr(B'|A')

To find the probability of reaching the end of branch at the right of the tree diagram, we multiply together the probabilities on the branches that give the unique path from the tree's root on the left to the particular branch end. Thus, for example, $\Pr(A' \cap B) = \Pr(B \mid A')\Pr(A')$.

To illustrate this process, we represent Example 1 as a tree diagram. Stage 1 is the determination of the color of the envelope, and stage 2 is finding out how much money is in the envelope. An appropriate tree diagram incorporating the probabilities given in Example 1 looks like this:

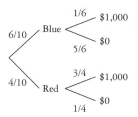

The 1/6, for example, represents the conditional probability of finding $1,000 given that you chose a blue envelope. The conditional probability of finding $1,000 given that you selected a red envelope is 3/4, since 3 of the 4 red envelopes have $1,000 bills in them. The probability of winning $1,000 can then be calculated by adding up the probabilities of each path from the root to a $1,000 node, because these paths are mutually exclusive. Thus

$$\Pr(\text{you win } \$1{,}000) = (\tfrac{1}{6})(\tfrac{6}{10}) + (\tfrac{3}{4})(\tfrac{4}{10}) = \tfrac{1}{10} + \tfrac{3}{10} = \tfrac{4}{10}.$$

Bayes' Theorem

We can rewrite the equation that defines conditional probability as

$$\Pr(B \cap A) = \Pr(B \mid A)\Pr(A). \qquad (1)$$

This form of the equation is useful in computing $\Pr(A \mid B)$ in certain instances when the probability $\Pr(B \mid A)$ is given. This is true since

$$\Pr(A \mid B) = \frac{\Pr(A \cap B)}{\Pr(B)} = \frac{\Pr(B \cap A)}{\Pr(B)}$$
$$= \frac{\Pr(B \mid A)\Pr(A)}{\Pr(B)}.$$

EXAMPLE 2 Homework? A multiple-choice exam has four proposed answers for each question, only one of which is correct. A student who has done her homework is certain to identify the correct answer. If a student skips her homework, she chooses an answer at random. Suppose that two-thirds of the class has done the homework. In grading the exams, the teacher observes that Jennifer has the right answer to the first problem. What is the probability that Jennifer did the homework?

Solution: Let A be the event "Jennifer has done the homework" and B denote the event "Jennifer has the correct answer." The information given in the problem translates into three probability statements:

$$\Pr(A) = \tfrac{2}{3}, \qquad \Pr(B \mid A) = 1, \qquad \Pr(B \mid A') = \tfrac{1}{4}.$$

We seek the conditional probability $\Pr(A \mid B)$. By definition of conditional probability, we can compute this probability if we know $\Pr(A \cap B)$ and $\Pr(B)$.

We'll work first on $\Pr(B)$. To find this probability, we break the event B into the union of two disjoint events,

$$B = (B \cap A) \cup (B \cap A').$$

Now use the fact that the probability of the union of two disjoint sets is the sum of their probabilities:

$$\Pr(B) = \Pr(B \cap A) + \Pr(B \cap A').$$

Going one step further, using Eq. (1) twice, we have

$$\Pr(B \cap A) = \Pr(B \mid A)\Pr(A)$$

and

$$\Pr(B \cap A') = \Pr(B \mid A')\Pr(A').$$

Finally, we'll put this together with the definition of conditional probability to obtain

$$
\begin{aligned}
\Pr(A \mid B) &= \frac{\Pr(A \cap B)}{\Pr(B)} \\
&= \frac{\Pr(B \mid A)\Pr(A)}{\Pr(B \mid A)\Pr(A) + \Pr(B \mid A')\Pr(A')} \\
&= \frac{(1)(\tfrac{2}{3})}{(1)(\tfrac{2}{3}) + (\tfrac{1}{4})(\tfrac{1}{3})} = \frac{8}{9}. \quad \blacktriangle
\end{aligned}
\tag{2}
$$

The type of calculation used to solve Example 2 occurs in a great many applications. The general rule that underlies it is called *Bayes' theorem,* named after the Reverend Thomas Bayes (1702–1761). To see how we might generalize Eq. (2), let $A_1 = A$ and $A_2 = A'$. Then A_1 and A_2 are disjoint sets whose union is E. Equation (2) then takes the form

$$
\Pr(A_1 \mid B) = \frac{\Pr(B \mid A_1)\Pr(A_1)}{\Pr(B \mid A_1)\Pr(A_1) + \Pr(B \mid A_2)\Pr(A_2)}.
\tag{3}
$$

Bayes' theorem generalizes Eq. (3) from two disjoint subsets of E to a partition of E into k nonoverlapping subsets. Let's state this famous theorem formally:

Theorem 1. Let Pr be a probability measure defined on a set E, and suppose B is a subset of E with $\Pr(B) > 0$. If $A_1, A_2, \ldots, A_k$ is any collection of mutually disjoint subsets of E whose union is all of E, and each A_i has positive probability, then

$$Pr(A_j \mid B) = \frac{Pr(B \mid A_j)Pr(A_j)}{\displaystyle\sum_{i=1}^{k} Pr(B \mid A_i)Pr(A_i)}$$

$$\text{for each } j = 1, 2, \ldots, k.$$

Proof. First note

$$Pr(A_j \mid B) = \frac{Pr(A_j \cap B)}{Pr(B)} = \frac{Pr(B \cap A_j)}{Pr(B)}$$
$$= \frac{Pr(B \mid A_j)Pr(A_j)}{Pr(B)}.$$

Then write

$$B = (B \cap E) = B \cap (A_1 \cup A_2 \cup \cdots \cup A_k)$$
$$= (B \cap A_1) \cup (B \cap A_2) \cup \cdots \cup (B \cap A_k),$$

so that

$$Pr(B) = \sum_{i=1}^{k} Pr(B \cap A_i).$$

Note then that $Pr(B \cap A_i) = Pr(B \mid A_i)Pr(A_i)$, so that

$$Pr(B) = \sum_{i=1}^{k} Pr(B \cap A_i) = \sum_{i=1}^{k} Pr(B \mid A_i)Pr(A_i).$$

▲ ▲ ▲

E X A M P L E 3 Economic Advice The president of the United States often seeks recommendations from the Council of Economic Advisers. Suppose there are three advisers (Perlstadt, Kramer, and Oppenheim) with different theories about the economy. The president is considering the adoption of a new policy on wage and price controls and is concerned about the impact of this policy on the unemployment rate. Each adviser gives the president a personal prediction about the impact, the predictions being probabilities that the unemployment rate will lessen, remain unchanged, or increase. These are summarized in the following table:

	Probabilities of changes in unemployment rate		
	Decrease (D)	Remain Same (S)	Increase (I)
Perlstadt	.1	.1	.8
Kramer	.6	.2	.2
Oppenheim	.2	.6	.2

From previous experience with these advisers, the president has formed a priori estimates about the chances that each adviser has the correct theory of the economy. We indicate these by

$$Pr(\text{Perlstadt right}) = \tfrac{1}{6},$$

$$Pr(\text{Kramer right}) = \tfrac{1}{3},$$

$$Pr(\text{Oppenheim right}) = \tfrac{1}{2}.$$

Suppose the president adopts the proposed policy. One year later, the unemployment rate has increased. How should the president readjust his estimates about the correctness of his advisers' theories?

Solution: We are looking for Pr(Perlstadt right | I), Pr(Kramer right | I) and Pr(Oppenheim right | I).

First we compute the probability of an increase in the unemployment rate:

$$\begin{aligned}
Pr(I) &= Pr(I \mid \text{Perlstadt right})Pr(\text{Perlstadt right}) \\
&\quad + Pr(I \mid \text{Kramer right})Pr(\text{Kramer right}) \\
&\quad + Pr(I \mid \text{Oppenheim right})Pr(\text{Oppenheim right}) \\
&= (\tfrac{8}{10})(\tfrac{1}{6}) + (\tfrac{2}{10})(\tfrac{1}{3}) + (\tfrac{2}{10})(\tfrac{1}{2}) = \tfrac{3}{10}.
\end{aligned}$$

Now we use Bayes' theorem:

$$\begin{aligned}
&\Pr(\text{Perlstadt right} \mid I) \\
&= \frac{\Pr(I \mid \text{Perlstadt right})\Pr(\text{Perlstadt right})}{\Pr(I)} \\
&= \frac{\frac{8}{10}\frac{1}{6}}{\frac{3}{10}} = \frac{4}{9},
\end{aligned}$$

$$\begin{aligned}
&\Pr(\text{Kramer right} \mid I) \\
&= \frac{\Pr(I \mid \text{Kramer right})\Pr(\text{Kramer right})}{\Pr(I)} \\
&= \frac{\frac{2}{10}\frac{1}{3}}{\frac{3}{10}} = \frac{2}{9},
\end{aligned}$$

$$\begin{aligned}
&\Pr(\text{Oppenheim right} \mid I) \\
&= \frac{\Pr(I \mid \text{Oppenheim right})\Pr(\text{Oppenheim right})}{\Pr(I)} \\
&= \frac{\frac{2}{10}\frac{1}{2}}{\frac{3}{10}} = \frac{3}{9}.
\end{aligned}$$

Note that Perlstadt's theory seemed the least correct before the policy was adopted, but it appears to be the most correct one a year later. ▲

We will now examine in more detail three applications that involve applying conditional probability to the examination of contemporary issues. We begin with an extended discussion of an important public policy issue: mass screening for the AIDS virus.

Example 4: Screening for AIDS

The AIDS Disease

The *acquired immune deficiency syndrome* (AIDS) is a terrifying, contagious disease that many observers believe has already reached epidemic proportions in the United States. In October 1986, the National Academy of Sciences said that the AIDS problem required "perhaps the most wide-ranging and intensive efforts ever made against an infectious dis-ease. . . . If the spread of the virus is not checked, the present epidemic could become a catastrophe."[3]

AIDS is invariably a fatal illness; AIDS victims rarely live more than 5 years. By September 1994, AIDS had infected about 16 million people worldwide and killed more than 2 million. Dr. Ward Cates of the Federal Center for Disease Control warned that

> Anyone who has the least ability to look into the future can already see the potential for this disease being much worse than anything mankind has seen before.[4]

AIDS is transmitted from one person to another primarily through sexual relations, intravenous drug use, and contact with blood. To halt the disease's spread, we need first to identify the carriers of the AIDS virus so that they may take proper precautions to avoid infecting others. Scientists have been searching for an effective method to identify AIDS virus carriers. Such methods have been found for other diseases; for example, chest x-rays are routinely given now to detect tuberculosis.

Blood Tests

One technique is a blood test (ELISA) that detects the presence of certain antibodies that the body produces in reaction to the AIDS virus. While such a test may be extremely accurate, it is subject to two possible types of errors. First, it may fail to detect the illness in someone who actually suffers from it; this is called a *false negative*. Second, it may "detect"

---

[3]*The New York Times* (Oct. 30, 1986).
[4]James I. Slaff and John K. Brubaker, (1985), *The Aids Epidemic,* New York: Warner Books, p. 4.

AIDS in a person who really doesn't have the disease, a *false positive*.

Suppose the blood test is refined to the point where it correctly identifies AIDS in 95% of the population that has the disease; thus 5% of those who have AIDS get a "false negative" result from the test. Furthermore, suppose that 99% of those tested who are free of the AIDS virus also receive a negative test result. This means that 1% of AIDS-free individuals will receive a "false positive" test result.

Many people have argued that the government should fund a mass screening program to detect AIDS carriers. Suppose such a program is undertaken and all individuals have this blood test. The actual number of AIDS virus carriers in the United States is unknown, but it is estimated that approximately 1 person per 1000 is so afflicted. If your blood is tested and the result is positive, how worried should you be? What are the chances that you actually have AIDS?

An Informal Solution

We're assuming that the ELISA test (enzyme-linked immunosorbent assay) correctly points to the presence of AIDS virus in 95% of those who actually have the virus and that it incorrectly identifies the presence of the virus in only 1% of those who really don't have it. We've also assumed that approximately 1 in every 1000 persons in the population actually carries the AIDS virus.

Suppose then we give the ELISA blood-screening test to 100,000 randomly chosen people. We would expect that about 100 of these people have the AIDS virus. About 95 of these 100 will receive positive test results. Of the 99,900 people who do not carry the AIDS virus, we expect that 1% of them, or 999, will receive a false positive. There are 95 + 999 = 1094 persons with positive tests, of which only 95

truly have the AIDS virus. Thus, among those who have positive tests, the probability of having the AIDS virus is

$$\frac{95}{1094} = .087,$$

which is less than 9%. We would conclude that more than 90% of those who receive a positive test for the AIDS virus are actually free of the virus! The vast majority of people who are told they have a positive test for the virus are needlessly frightened. What can we do to prevent this unnecessary panic? There are several approaches we might take.

Improving the Procedure

Sensitivity

First, try to improve the accuracy of the test. For an illness like AIDS, which can be spread easily, we definitely do not want our screening procedure to result in many false negatives. By the **sensitivity** of a screening test or procedure, we mean the probability that a person who has the disease has a positive test result. So we might concentrate our efforts in increasing the 95% accuracy figure.

Suppose we were incredibly successful at this effort and developed a screening test that had a 100% sensitivity. Then, in our numerical example, the number of people who receive positive tests increases from 95 + 999 to 100 + 999 = 1099. The conditional probability that a person carries the virus given that the test is positive now becomes

$$\frac{100}{1099} = .091,$$

which is just over 9%. We have not improved the probability very much.

We've just seen that efforts to improve a screening procedure that already has a high sensitivity by increasing that sensitivity are not likely to increase dramatically the chances that a positively tested individual carries the AIDS virus.

The other problem with boosting sensitivity is that it often increases the probability of false positives. The screening technique may be based on a "threshold" value for some substance. If the measure is above the threshold value, then we call the test result positive. One way to improve the sensitivity of the test is to lower the threshold value. Then more people who actually carry the virus will receive positive tests. Unfortunately, so will more people who don't have the disease.

To illustrate the drawbacks of this approach, suppose the sensitivity is raised to 100% at the cost of raising the percentage of positive tests among those free of the disease from 1% to 2%. Then of the 100,000 people tested, all 100 of the virus carriers are identified as positive. Of the 99,900 virus-free people, 1998 will test as positive. The probability of having the virus given a positive test is

$$\frac{100}{(100 + 1998)} = \frac{100}{2098} = .05,$$

which is worse than the 8% probability we had originally.

Specificity

As a second attempt to improve the situation, we might bend our efforts to reduce the error rate of false positives. The **specificity** of a screening test is a measure of how well the test performs in this aspect. The specificity is the probability of a virus-free person having a negative test. The more specific a test is, the smaller the proportion of false positives is.

In our original numerical example, the specificity was 99%, so that the probability of a positive test for a virus-free person was .01. Suppose we could reduce this probability by a factor of three. Then the number of people with false positives would drop from 999 to 333. The probability that a person has the virus given a positive test would then be

$$\frac{95}{(95 + 333)} = .22,$$

which is a significant improvement.

In many screening procedures, increased specificity comes only at the cost of decreased sensitivity. If we raise the threshold for a positive test, then fewer people without the virus will test positive. Hence the specificity will go up. But then more people who actually have the virus will also test negative, and we will fail to diagnose their illness. In our continuing example, suppose that effort which reduced the number of false positives by a factor of three also lowers the sensitivity from 95% to 90% of the population. Then the proportion of those with positive tests who have the disease becomes

$$\frac{90}{(90 + 333)} = .21,$$

which is not very different from the 22% figure we obtained above.

If our primary goal is to raise the probability that a positively tested person actually has the disease, then our analysis so far shows that we should concentrate our efforts on improving the specificity of the screening procedure even at the cost of lowering the test's sensitivity. We should realize, however, that the pursuit of this primary goal may mean a greater failure to detect the virus in those who have it at an early stage, where effective treatment may one day be possible.

Incidence

A third important factor enters into these calculations. This is the relative **incidence** of the virus among the population being tested. In our example, we assumed that about 1 person per 1000 actually had the virus. If the actual proportion is as high as 1 in 100, then the probability that a person with a positive test has the virus will change. Of the 100,000 people in the study, about 1000 will have the virus. Among these 1000, we expect that 950 (95%) will test positive. There will be about 99,000 people without the virus. One percent of them, or 990, will test positive. Thus, there will be about $1000 + 990 = 1990$ positive tests. The proportion of virus carriers among the positive tests will be

$$\frac{1000}{1990} = .503,$$

or just over one-half.

If our virus is present in 1 out of every 10 people, then the test will be positive for about 9500 of the 10,000 people with the virus. The test will also be positive for about 900 of the 90,000 who don't carry the virus. In this case, the probability that a positively tested person has the virus jumps to

$$\frac{9500}{(9500 + 900)} = .91,$$

so that less than 10% of those with positive tests are actually free of the virus.

These last scenarios indicate that certain aspects of medical screening may be most effective when applied to a population that is believed to have a much higher rate of incidence of the targeted disease than the general population. In March 1986, U.S. health officials recommended that Americans at "high risk" of contracting AIDS should undergo periodic blood tests to determine if they have become infected with the virus. Those seen as having an increased risk of infection include

➢ homosexual and bisexual men,
➢ present or past intravenous drug abusers,
➢ people with signs of symptoms compatible with AIDS,
➢ people born in Haiti and certain central African nations,
➢ male or female prostitutes,
➢ sexual partners of infected people or of people at increased risk,
➢ hemophiliacs who have received blood-clotting-factor products,
➢ newborn children of infected mothers.

A More Formal Solution

In this section we use the basic probability rules to derive an expression for the conditional probability that a person has the AIDS virus given that the test result was positive.

Suppose

Pr(AIDS) = Probability that randomly chosen person has the AIDS virus = a,

so that

Pr(No AIDS) = Pr(AIDS′) = 1 − Pr(AIDS) = 1 − a.

Also suppose

$\text{Pr(Positive | AIDS)}$ = Probability that test indicates presence of AIDS virus given that person actually has AIDS virus ("true positive") $= p$,

so that

$\text{Pr(Negative | AIDS)}$ = Probability that test fails to show presence of AIDS virus given that person actually has AIDS virus ("false negative") $= 1 - p$.

Finally, suppose

$\text{Pr(Positive | AIDS}')$ = Probability that test indicates presence of AIDS virus given that person does not have AIDS virus ("false positive") $= q$,

so that

$\text{Pr(Negative | AIDS}')$ = Probability that test fails to show presence of AIDS virus given that person does not have AIDS virus ("true negative") $= 1 - q$.

Thus,

$$\begin{aligned}\text{Pr(Positive test)} &= \text{Pr(Positive} \cap \text{AIDS)} \\ &\quad + \text{Pr(Positive} \cap \text{AIDS}') \\ &= \text{Pr(Positive | AIDS)Pr(AIDS)} \\ &\quad + \text{Pr(Positive | AIDS}')\text{Pr(AIDS}') \\ &= pa + q(1 - a).\end{aligned}$$

We want $Pr(\text{AIDS | Positive})$. Now,

$$\begin{aligned}\text{Pr(Positive} \cap \text{AIDS)} &= \text{Pr(AIDS} \cap \text{Positive)} \\ &= \text{Pr(AIDS | Positive)Pr(Positive)}.\end{aligned}$$

Hence

$$\begin{aligned}\text{Pr(AIDS | Positive)} &= \frac{\text{Pr(Positive} \cap \text{AIDS)}}{\text{Pr(Positive)}} \\ &= \frac{pa}{pa + q(1 - a)}.\end{aligned}$$

In our initial example, $p = .95$, $q = .01$, and $a = .001$.

Example 5: Lie Detector Tests

If the police unjustly accuse you of a crime you didn't commit, should you volunteer to take a "lie detector" test to prove your innocence?

The *polygraph* is a psychophysiological device that records physiological variables such as heart and respiration rates, blood pressure, and galvanic skin response, which are under autonomic control. Advocates of polygraph use claim that when these variables are measured while a subject is interrogated, telling lies makes a characteristic pattern of responses that a trained examiner can detect.

If we consider the use of a polygraph as a diagnostic tool, then we are seeking to find the liars in a particular population. In this setting, sensitivity would be the conditional probability that the polygraph asserts the subject lied given that the subject did in fact lie. Specificity would be the conditional probability that the polygraph indicates the subject told the truth given that the subject did speak the truth.

In a recent study, three physicians examined a number of experimental studies about polygraphs.[5] They found an average sensitivity of .76, with a low mark of .64 and a high mark of .82. The average value of specificity in these studies was .63, ranging from a low of .5 to a high of .82.

---

[5]"Predictive power of the polygraph: Can the 'lie detector' really detect lies?" *The Lancet* (Mar. 8, 1986).

Getting Truthful Answers to Sensitive Questions

Political pollsters, public opinion surveyors, social scientists, and others often want to measure accurately the percentage of people who hold a particular belief or engage regularly in some specific behavior. Their starting point is to obtain truthful answers to their questions from a randomly chosen subset of the group.

But people are often skeptical about promises of confidentiality and may be unwilling to respond honestly to an interviewer they do not know.

In 1965 Stanley L. Warner devised a procedure using elementary concepts of probability to overcome this unwillingness. Warner's *randomized response* technique asks people to select randomly one of two proposed questions without telling the interviewer which question they have answered. One of the questions is about a sensitive or potentially embarrassing topic; the other is innocuous. Respondents are free to reply honestly since only they know which question they answered.

Here's an illustration. The innocuous question is, "Is the third digit of your social security number odd?" The sensitive question is, "Do you use an illegal drug at least once a month?" We then ask respondents to flip a coin. If they get heads, they answer the first question; otherwise they answer the second.

Suppose we follow this procedure with 200 respondents and receive 64 "yes" responses. Since the probability of a head is 1/2, we expect that about 100 people answered the first question. Since the third digit of a social security number is equally likely to be odd or even, of the 100 who reply to the first question, about 1/2, or 50, should answer "yes." Thus, about 64 − 50, or 14, "yes" answers came from the 100 people who answered the sensitive question. Hence we would estimate that 14/100, or about 14%, of the group uses an illegal drug at least once a month.

Mathematics at Work: Jennifer Ahner

Walt Meyer: Tell me something about the work you do, Jenny.

Jennifer Ahner: I do actuarial work for Insurance Services Office, Inc., a nonprofit company in the insurance field. We serve as a data analysis and information bureau that sends reports and data to insurance companies all around the country. I work in the personal automobile division.

Jennifer Ahner

Walt Meyer: Is there a lot of math in that?

Jennifer Ahner: Yes, I use math to analyze data, select trends, and make projections. In addition, to advance as an actuary you have to pass professional exams in calculus, probability, statistics, and insurance concepts.

Walt Meyer: Did you study much math?

Jennifer Ahner: As a student at Adelphi, I majored in math. Some of it is essential, like probability for example, but all of it provided a valuable background. Besides using math on my job, a couple of times a year I'll also try to use it to win at blackjack in Atlantic City.

Walt Meyer: For someone entering your line of work, what courses would you recommend in college?

Jennifer Ahner: Mathematics is essential, not just for the specific skills but for the way of thinking. Computer, writing, and communication skills and economics courses are also helpful.

Walt Meyer: Do you like your work?

Jennifer Ahner: Yes! I really like the people I work with, and my organization provides an essential service for insurance companies. The company is generous in providing study time for the exams. There are raises for passing each test and room for advancement. I work in the World Trade Center and find New York City an exciting place.

Using the average values of .76 for sensitivity and .63 for specificity, they constructed a table showing the expected results if a polygraph test was administered to a population of 1000 people, 5% of whom were actually lying:

Polygraph result	Liars	Truth tellers	Total
Positive (read as lying)	38 true positives	351 false positives	389 total positives
Negative (read as truthful)	12 false negatives	599 true negatives	651 total negatives
Total	50	950	1000

From this table, we can easily calculate the predictive value of a positive test. Recall that this is the probability that a person with a positive polygraph test is lying. This would be

$$\frac{\text{true positives}}{\text{total positives}} = \frac{38}{389} = .098,$$

or just under 10%.

Example 6: A Witness Identification Problem

Recently there has been an increasing realization of the importance of probabilistic reasoning in judicial situations. Witnesses, for example, do not recall perfectly the details of an event months, days, or even minutes after it occurred. In his book *Innumeracy: Mathematical Illiteracy and Its Consequences,* John Allen Paulos presents a modification of a problem originally formulated by Daniel Kahneman and Amos Tversky:[6]

---

[6]Daniel Kahneman and Amos Tversky (1972), "On prediction and judgment," *ORI Research Monographs* 12(4).

A man is downtown, he's mugged, and he claims the mugger was a black man. However, when the scene is reenacted many times under comparable lighting conditions by a court investigating the case, the victim correctly identifies the race of the assailant only about 80% of the time. What is the probability his assailant was indeed black?

Most people give .8 as the answer to this question. But a careful analysis based on reasonable assumptions shows that the true answer may be far different. Suppose that in this city, about 90% of the population is white and 10% is black. Suppose further that these percentages also hold for those people downtown at the time of the mugging and that neither race is more likely to mug someone. We assume finally that the victim is equally likely to make wrong identifications in both directions, white for black and black for white.

Then our assumptions and the information we have been given can be represented in a two-step tree diagram. The first stage corresponds to the random selection of a person from the city to be the assailant. Since whites and blacks have the same rates of mugging, the associated probabilities of selecting a white man or a black man are simply their relative proportions in the population. The second stage shows the victim's identifications.

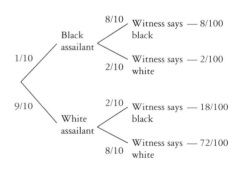

From the diagram, we see that the probability that the witness will identify his assailant as black is $8/100 + 18/100 = 26/100$. But a good chunk of this probability, $18/100$ in particular, is due to incorrect identifications! The proportion of correct identifications is $8/26$, or about $.31$.

To write a solution up more rigorously, we seek the conditional probability

Pr(Assailant is black | Witness says he is black),

which by the definition of conditional probability is

$$\frac{\text{Pr(Assailant is black } \textbf{and}\text{ Witness says he is black)}}{\text{Pr(Witness says he is black)}}$$

$$= \frac{\left(\frac{8}{100}\right)}{\left(\frac{26}{100}\right)} = \frac{8}{26}.$$

Under our assumptions, therefore, there is only a little more than 30% chance that the assailant really was black. In the exercises, you will explore how this number changes if we vary the assumptions.

Exercises for Section 8.3

1. A poll of 40 juniors and 45 sophomores found 21 sophomores preferred diet soda and 47 people favored regular soda. What is the probability that a person randomly chosen from the poll prefers diet soda? What is the probability that a randomly chosen person is a junior who favors regular soda? Suppose a person is chosen at random and we learn that she prefers diet soda; what's the probability that she is a junior?

2. Suppose the sensitivity of a screening process is $.98$ and the specificity is $.92$. If the process is used to detect a condition present is 5% of the population, determine the probability of having the condition given a positive test.

3. A multiple-choice exam has five suggested answers to each question, only one of which is correct. A student who has done her homework is certain to identify the correct answer. If a student skips her homework, she chooses an answer at random. Suppose that one-third of the class has done the homework. While grading, the teacher observes that Anne has the right answer to the first problem. What is the probability that she did the homework?

4. The Internal Revenue Service uses computers to identify suspicious tax returns that warrant further auditing. Approximately 2% of all returns are spotted by the computer for auditing. Of those returns which are audited, about 85% turn out to be ones in which the taxpayer owes the IRS more money. Of the returns that are not audited, experts estimate that perhaps one in five are cases where taxpayers have underpaid. Suppose a tax return is randomly selected. What is the probability that it represents a case of underpayment? If it is an underpayment, what is the probability that the computer will spot it for an audit?

5. At an Atlantic City gambling casino, you are dealt 2 cards from a well-shuffled standard deck of 52 cards.
   (a) What is the probability that the first card is red?
   (b) What is the probability that the second card is red given that the first card is black?

(c) What is the probability that the second card is red given that the first card is red?

(d) What is the probability that the second card is red?

6. Weather records in Vermont indicate that during winter months, it snows on 42% of Saturdays and 48% of Sundays. On 30% of the weekends, it snows on both days. It's a Saturday morning in February and you look out your Vermont College dorm window and see that is snowing. What's your estimate of the probability that it will snow tomorrow?

7. The admissions office of a college divides the United States into three geographic regions:

A: New England;

B: outside of New England but east of the Mississippi River;

C: west of the Mississippi.

In a recent graduating class, 40% of the students came from a state in region A, 35% from region B, and the rest from region C. Two percent of students from region A graduated with highest honors, while the comparable figures from regions B and C were 4% and 5%, respectively.

(a) Which region produced the largest number of highest-honors graduates?

(b) What is the probability that a randomly chosen senior was a highest-honors student from New England?

(c) What is the probability that a randomly chosen senior came from outside New England and didn't achieve highest honors?

(d) Each year a graduating senior is chosen at random to be photographed for the alumni magazine cover. Given that this student achieved highest honors, what was the probability that he or she came from region B?

8. Two dice are rolled and the sum is 7. What is the probability that at least one of the die is a 2? What is the probability that both are 2s?

9. Two dice are rolled and they show different faces. What is the probability that one of them shows at 4?

10. Three dice are rolled and they show three different faces. What is the probability that one of them shows a 4?

11. Generalize Exercises 9 and 10 for four dice, five dice, and six dice.

12. In Tom Stoppard's play *Rosencrantz and Guildenstern Are Dead,* a coin is flipped repeatedly and keeps showing heads. Suppose 1 coin in every 10 million is misengraved and is given two heads. Suppose you flip a coin 10 times and obtain heads each time. What is the probability that you have a coin with two heads?

13. An automotive manufacture produces the Neptune model car at factories in Detroit and Houston. Sixty percent of the Neptunes are made in Detroit and the remainder in Houston. The Detroit-built cars average 4 serious defect in every 1000 vehicles; the defect rate for Houston is 9 in 1000. The cars are shipped from both factories to dealerships all over the country. The service manager for a car dealer in Seattle reports that 200 Neptunes were brought in for service on serious defects. Estimate the number of these Neptunes that came off the Detroit assembly line.

14. **(Writing Project)** The most celebrated murder trial in recent history was that of O. J. Simpson, who was charged with brutal slayings of Nicole Brown Simpson and Ronald Goldman. Since there were apparently no eyewitnesses, all of the evidence was circumstantial. There were many arguments between the prosecutors and the defense attorneys on the admissibility of evidence. Some of these debates involved important ideas in probability. One stormy issue—which involved the conditional probability—was the defense's attempt to exclude evidence of O.J.'s prior abuse of his wife Nicole. Investigate the arguments by both sides and the decision handed down by Judge Lance Ito. How clearly did the mass media present the probability ideas? An excellent starting reference is Jon F. Merz and Jonathan P. Caulkins (1995), "Propensity to abuse—Propensity to murder?" *Chance* 8, no. 2:14.

SECTION 8.4 *Independent Events*

Knowledge about some aspects of the outcome of an experiment on a sample space can influence the estimate of the probability of other aspects of the outcome. This influence is measured using conditional probability. If you're told that the result of rolling a fair die was an even number, then you would revise your estimate of the probability of a 4 from 1/6 to 1/3. Similarly, you would change your guess on the probability of a 5 from 1/6 to 0.

Sometimes, however, the extra knowledge does not influence the estimate of probabilities. Imagine an experiment that consists of flipping a coin and rolling a die. The outcome of the experiment consists of two observations: a head or tail for the coin, and the number of spots (1, 2, 3, 4, 5, or 6) on the die. How the die rolls is in no way affected by how the coin flips. The answer to the question "What is the probability that the die shows a 3?" will be the same whether or not you know how the coin landed. More exactly, the conditional probability that the die shows a 3 given that the coin shows a Head is the same as the probability that the die shows a 3 given no information at all about the coin. A probabilist would say that the coin flip and the die roll are independent of each other. The formal definition looks like this:

DEFINITION

Let Pr be a probability measure on a set E. If A and B are subsets of E with $\Pr(A) > 0$ and $\Pr(B) > 0$, then A and B are *independent events* if $\Pr(B \mid A) = \Pr(B)$.

If A and B are independent events, then

$$\Pr(A \mid B) = \frac{\Pr(B \mid A)\Pr(A)}{\Pr(B)} = \frac{\Pr(B)\Pr(A)}{\Pr(B)} = \Pr(A),$$

which shows that independence is a *symmetric* relationship; that is, if A and B are independent, then B and A are also independent.

The definition also gives the very important multiplication rule for independent events.

Theorem 1. Suppose $\Pr(A) > 0$ and $\Pr(B) > 0$. Then A and B are independent if and only if

$$\Pr(A \cap B) = \Pr(A)\Pr(B).$$

Proof. Since

$$\Pr(A \cap B) = \Pr(B \cap A) = \Pr(B \mid A)\Pr(A),$$

we have $\Pr(A \cap B) = \Pr(A)\Pr(B)$ if and only if $\Pr(B/A) = \Pr(B)$.

▲ ▲ ▲

E X A M P L E 1 Rolling a Die Roll a fair die.

a. Let A be the event that the number displayed is less than 5, and let B be the event that the number is odd. Then $A = \{1, 2, 3, 4\}$, $B = \{1, 3, 5\}$, and $A \cap B = \{1, 3\}$. Here we have

$$\Pr(A)\Pr(B) = \tfrac{4}{6}\tfrac{3}{6} = \tfrac{1}{3}, \quad \text{while} \quad \Pr(A \cap B) = \tfrac{2}{6} = \tfrac{1}{3},$$

so A and B are independent.

b. If A is the event that the number displayed is less than 4, then $A = \{1, 2, 3\}$ and $A \cap B$ is still $\{1, 3\}$. In this case

$$\Pr(A)\Pr(B) = \tfrac{3}{6}\tfrac{3}{6} = \tfrac{1}{4} \neq \Pr(A \cap B),$$

so A and B are not independent. ▲

Our definition of independent events makes explicit use of the idea of conditional probability. Hence it is restricted to events with positive probabilities. But the multiplicative rule $\Pr(A \cap B) = \Pr(A)\Pr(B)$ may hold even for events with zero probabilities. For this reason, some probability theorists *define* two events to be independent exactly when the multiplicative rule is valid.

Warning: There's a common mistake students often make in thinking about independent events. The independence of two events is not determined strictly from the intrinsic nature of the two events. Independence is also a function of the probability measure that has been assigned to the original set of outcomes. Two events may be independent under one probability measure but not independent under another. Consider carefully the next two examples.

EXAMPLE 2 Rolling a Pyramid A pyramid is a solid figure with four triangular faces. Suppose the faces are labeled with the letters a, b, c, and d. Roll the pyramid and observe which triangle faces the ground when the pyramid comes to rest. The set E of outcomes may be denoted by $E = \{a, b, c, d\}$. Let A be the subset $\{a, c\}$ and B be the subset $\{b, c\}$. Then $A \cap B = \{c\}$. If Pr is the equiprobable measure on E, then $\Pr(A \cap B) = 1/4$, while $\Pr(A)\Pr(B) = (2/4)(2/4) = 1/4$. Thus A and B are independent events in this sample space. ▲

EXAMPLE 3 Consider the same situation as in Example 2, except that the probability measure is defined by assigning a, b, c, and d weights of .4, .4, .1, and .1, respectively. Then $\Pr(A \cap B) = .1$, while $\Pr(A)\Pr(B) = (.5)(.5) = .25$. The events A and B are not independent in this sample space. ▲

We can make use of the multiplicative rule to extend the idea of independence to a collection of events. Three events A, B, and C will be called *mutually independent events* if each pair of events is independent and if, in addition,

$$\Pr(A \cap B \cap C) = \Pr(A)\Pr(B)\Pr(C).$$

More generally, a collection of events A_1, A_2, . . . , A_n in a sample space is mutually independent if the probability of the intersection of any k distinct events in the set is equal to the product of the probabilities of events, where $k = 2, 3, 4, \ldots, n$.

Bernoulli Trials

Independence is an important concept in the discussion of situations in which the same experiment is repeated under identical conditions a number of times. Suppose, for example, that a fair coin is tossed three times. It is reasonable to assume that successive tosses of the coin do not influence each other—the coin "has no memory" of how it landed on earlier tosses. In other words, the sequence of outcomes is a mutually independent set. Let H_i be the subset corresponding to obtaining a head on the ith toss, for $i = 1, 2, 3$. The probability of obtaining heads on all three tosses is $\Pr(H_1 \cap H_2 \cap H_3)$. By the assumption of independence, this is equal to $\Pr(H_1)\Pr(H_2)\Pr(H_3) = (1/2)(1/2)(1/2) = 1/8$.

EXAMPLE 4 Weighted Coin Suppose the coin has been weighted so the probability of a head on a single toss is 1/3. If the coin is flipped three times, what is the probability of getting exactly one head? If A is the subset corresponding to getting precisely one head in three tosses, then we can write A as the union of three mutually disjoint subsets:

$$A = (H_1 \cap T_2 \cap T_3) \cup (T_1 \cap H_2 \cap T_3)$$
$$\cup (T_1 \cap T_2 \cap H_3),$$

where T_i indicates a tail on toss i. By the assumption of independence, we have

$$\Pr(A) = \Pr(H_1)\Pr(T_2)\Pr(T_3) + \Pr(T_1)\Pr(H_2)\Pr(T_3)$$
$$+ \Pr(T_1)\Pr(T_2)\Pr(H_3) = (\tfrac{1}{3})(\tfrac{2}{3})(\tfrac{2}{3}) + (\tfrac{2}{3})(\tfrac{1}{3})(\tfrac{2}{3})$$
$$+ (\tfrac{2}{3})(\tfrac{2}{3})(\tfrac{1}{3}) = \tfrac{12}{27}. \quad \blacktriangle$$

As a significant generalization of this example, consider an experiment with precisely two outcomes with associated probabilities p and q, where $q = 1 - p$. Call the outcome with probability p a "success" and the other outcome a "failure." Repeat this experiment a number of times in such a manner that the outcomes of any one experiment in no way affect the outcomes in any other experiment; that is, assume the sequence of outcomes forms a mutually independent set. Let A_i represent the outcome of a success on the ith trial of the experiment and B_i the outcome of a failure on the ith trial. Then $\Pr(A_i) = p$ and $\Pr(B_i) = q$ for each i.

Suppose the experiment is repeated four times. The probability that there are successes on the first and fourth trials and failures on the second and third trials is given by

$$\Pr(A_1 \cap B_2 \cap B_3 \cap A_4),$$

which by independence is equal to

$$\Pr(A_1)\Pr(B_2)\Pr(B_3)\Pr(A_4) = pqqp = p^2q^2 = p^2(1 - p)^2.$$

It should be clear that any other prescribed sequence of two successes and two failures in four trials will also have probability $p^2(1 - p)^2$. In general, if the experiment is repeated n times, then the proba-

bility of obtaining a prescribed sequence of exactly k successes and $n - k$ failures will be $p^k q^{n-k} = p^k(1 - p)^{n-k}$.

A related question concerns the probability of obtaining exactly k successes in n trials. This probability will be $p^k q^{n-k}$ multiplied by the number of distinct ways one can prescribe a sequence of k successes and $n - k$ failures. This number is equal to $\binom{n}{k}$. Thus, the probability of precisely k successes in n trials is

$$\binom{n}{k}p^k(1 - p)^{n-k}.$$

This formula, which occurs in many applications, defines the *binomial probability distribution* over the possible number of successes. A sequence of repetitions of the same experiment in which the outcomes of successive trials are independent is called a sequence of *Bernoulli trials*.

EXAMPLE 5 Floppy Disks The probability that an individual computer floppy disk is errorfree is .99. The probability that in a box of 7 disks, exactly 4 will be free of errors is

$$\binom{7}{4}.99^4(1 - .99)^{7-4} = 35(.99)^4(.01)^3,$$

which is approximately .0000336. In the language of Bernoulli trials, each trial is the selection of a disk from a box of disks. We are repeating the trial 7 times, with the probability of a "success" (selecting an errorfree disk) being .99. Also, we are assuming that the outcomes are independent events. $\blacktriangle$

How to Win the Lottery

"The lucky numbers this week are 8 . . . 11 . . . 13 . . . 15 . . . 19 . . . and . . . 20, and the jackpot is $27 million!" So announced the directors of the Virginia lottery in February 1992.

The Virginia lottery follows a common format. You pay $1 for a ticket and select a combination of six numbers from 1 to 44. At the close of ticket sales, six numbers are selected at random. If the numbers you picked match the numbers drawn, you win the lottery. Since there are $_{44}C_6$ = 7,059,052 possible combinations, your chances of winning big are pretty small if you buy only one ticket.

You can *guarantee* that you hold a winning ticket, however, if you buy every possible combination. That's exactly what a group of Australian investors tried to do in 1992. State lottery officials reported that the Australians succeeded in purchasing about five million different combinations before time ran out. Luckily for them, one of the combinations they did buy was the winning set of numbers.

Assuming you had time to purchase every possible combination of numbers, would it be a wise move?

Getting $27 million back on a $7 million investment seems to be a great opportunity. Unfortunately, you don't receive all the money right away. The prize is paid over a 20-year period. You receive $1.35 million each year. Federal and state income taxes reduce that amount to about $850,000.

Is there a better alternative investment? If you have $7 million at your disposal, you should be able to command a high rate of interest. Suppose you buy a 20-year municipal bond that pays 10% interest per year. Since the revenue from such bonds is tax-free, you have a clear $700,000 per year for 20 years. As an added bonus, you still have the $7 million when the bond matures.

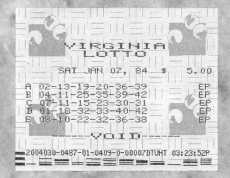

A final thought for those who still want to buy all the possible combinations: What if someone else also holds the winning combination and you have to split the jackpot?

E X A M P L E   6   Fast Food   The manager of Burger Buddy's restaurant has determined that 90% of the customers order a hamburger-related sandwich for lunch (hamburger, cheeseburger, Big Buddy, Little Buddy, etc.) and 10% order fish sandwiches. Anticipating 50 customers for today's lunch hour, the manager has 5 Swimwiches (fish sandwiches) prepared in advance. Find the probability that

a.   exactly 5 Swimwiches are ordered (supply equals demand),

b.   fewer than 5 Swimwiches are ordered (supply exceeds demand), and

c.   more than 5 Swimwiches are ordered (demand exceeds supply).

Solution: We model the customer orders as a sequence of Bernoulli trials with "success" if the customer orders a Swimwich. We assume the orders are independent. We then have $n = 50$ trials with p, the probability of success, equal to $1/10$. Let S denote the number of successes.

a.   We must find the probability of exactly 5 successes in 50 trials. This is given by

$$Pr(S = k) = \binom{n}{k} p^k (1 - p)^{n-k},$$

with $n = 50$, $k = 5$, and $p = 1/10$. Thus, the probability that supply equals demand is

$$\binom{50}{5}\left(\frac{1}{10}\right)^5\left(\frac{9}{10}\right)^{45} = 2{,}118{,}760\left(\frac{9^{45}}{10^{50}}\right) \approx .185.$$

b.   Supply will exceed demand if $S < 5$; that is, if $S = 0, 1, 2, 3,$ or 4:

$$Pr(S < 5) = Pr(S = 0) + Pr(S = 1) + Pr(S = 2) \\ + Pr(S = 3) + Pr(S = 4)$$

$$= \binom{50}{0}(\tfrac{1}{10})^0(\tfrac{9}{10})^{50} + \binom{50}{1}(\tfrac{1}{10})^1(\tfrac{9}{10})^{49}$$

$$+ \binom{50}{2}(\tfrac{1}{10})^2(\tfrac{9}{10})^{48} + \binom{50}{3}(\tfrac{1}{10})^3(\tfrac{9}{10})^{47}$$

$$+ \binom{50}{4}(\tfrac{1}{10})^4(\tfrac{9}{10})^{46}$$

$$= (\tfrac{1}{10})^{50}\left[\binom{50}{0}9^{50} + \binom{50}{1}9^{49}\right.$$

$$\left. + \binom{50}{2}9^{48} + \binom{50}{3}9^{47} + \binom{50}{4}9^{46}\right],$$

which is approximately $.431$.

c.   Demand exceeds supply if $S > 5$; that is, $S = 6, 7, 8, \ldots, 50$. We can then compute $Pr(S > 5)$ as a sum of 45 terms:

$$\sum_{k=6}^{50}\binom{50}{k}(\tfrac{1}{10})^k(\tfrac{9}{10})^{50-k}.$$

An alternative approach that involves less calculation can be used. Recall that the probabilities of an event and its complement sum to 1. Thus

$$Pr(S > 5) = 1 - Pr(S \le 5)$$
$$= 1 - [Pr(S = 5) + Pr(S < 5)]$$
$$= 1 - Pr(S = 5) - Pr(S < 5).$$

Using the results of parts (a) and (b), we estimate that the likelihood that more fish sandwiches will be ordered than were prepared is approximately $1 - .185 - .431 = .384.$ ▲

In a Bernoulli trials situation, our experiment consists of n independent repetitions of the same process. For each such experiment, we determine the number S of successes. In our next section, we will examine many other situations when a number, called a *random variable,* is attached to each outcome of an experiment. For Bernoulli trials, we have a relatively simple formula that determines the probability for each possible value of *S*. Such a formula is called, in general, a *probability distribution.* Because the probability distribution for *S* involves a binomial coefficient, it is called the *binomial probability distribution.*

1. A monkey sits at a typewriter with keys for the 26 letters of the alphabet and a space bar. If the monkey hits three keys at random, what is the probability that it spells ape? What is the probability that the monkey will strike the pattern consonant–vowel–consonant?

2. If you choose three people at random and ask about the month in which they were born, what is the probability that at least two of them were born in the same month? Suppose you choose seven people; what is the probability? What is the probability if you ask 13 people?

3. Approximately 1 of every 4000 female sterilization surgeries fails, while male sterilization procedures are 99.9% effective. What is the probability that a couple chosen at random, both of whom have had sterilization surgery, can conceive a child?

4. Three keys are taken at random from a set of 16, one of which will open Room 314 of Warner Hall. There is a 2/5 chance that this room is unlocked. What is the probability that the room can be entered without breaking in or returning for more keys?

5. Two metal detectors designed for airport security checks are on the market. The first costs $20,000 and will detect weapons with a probability of .999. The second has a lower detection probability of .99, but costs only $10,000. It's possible to install two of the cheaper devices in a way in which they will act independently of each other; if this is done, then a passenger must step through each of the devices in turn. Is it better to buy the two $10,000 detectors or one $20,000 device?

6. Fly-By-Night Computer Sales offers bargain prices on computer systems. Unfortunately, not all the equipment it ships works. About 20% of the printers, 15% of the central processing units, and 12% of the monitors that consumers receive are faulty. You're thinking about purchasing a system from this company. Determine the probability that
    (a)   the monitor and printer will be defective
    (b)   all three components won't work
    (c)   all three components will work
    (d)   at least one component will fail

7. Babe Ruth hit 714 homeruns in 8399 times at bat. Estimate the probability that in a single game in which he batted four times, Babe would hit four homeruns.

8. Hank Aaron holds the major league record for most career homeruns: 755 in 12,364 times at bat. Estimate the probability that Hank slugged three or more homers in a single game if he batted
    (a)   four times
    (b)   five times

9. A typical student stereo system is arranged as follows:

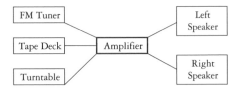

Given the way the system has been set up, we will be able to hear some music provided

at least one of the initial devices (FM tuner, tape deck, turntable) works, **and**

the amplifier works, **and**

at least one of the speakers works.

If both speakers work, then we can produce stereophonic sound. Each of the six components has its own *reliability;* that is, the probability that it will function. The reliabilities are as follows:

Component	Reliability
FM tuner	.98
Tape deck	.993
Turntable	.95
Left speaker	.97
Right speaker	.97
Amplifier	.96

Determine the probability that
(a) all components work
(b) we can hear some kind of music out of the system
(c) we can hear some stereo music
(d) the system fails
(e) the tape deck, amplifier, and left speaker work
(f) only the tape deck, amplifier, and left speaker work

10. Consider the experiment of flipping two fair coins. Let A be the event "head on first coin," B the event "head on second coin," and C the event "same result on both coins." Show that
(a) A and B are independent

(b) A and C are independent
(c) B and C are independent
(d) A, B, and C are not mutually independent events

11. A die is weighted so that the six faces have the following probabilities of showing:

1: $\frac{2}{16}$ 2: $\frac{5}{16}$ 3: $\frac{1}{16}$ 4: $\frac{6}{16}$ 5: $\frac{1}{16}$ 6: $\frac{1}{16}$

Let A be the event "die shows 1 or 4," B the event "die shows 1, 2, or 5," and C the event "die shows 1, 2, or 3." Show that $\Pr(A \cap B \cap C) = \Pr(A)\Pr(B)\Pr(C)$ but that A, B, and C are not mutually independent.

12. If A and B are mutually exclusive (disjoint) events each having positive probability, then show that A and B cannot be independent.

13. (**Writing Project**) Prepare an essay addressing the following questions about independence:

a. Many people have an intuitive sense that two events are independent if "they have nothing to do with each other." They might argue that two disjoint events "have nothing to do with each other" since they have no elementary events in common. This reasoning suggests that A and B are independent if they are disjoint, but this contradicts the result of Exercise 12. In what sense does independence mean "having nothing to do with each other"?

b. The most natural extension of the definition of independence for two events to independence of three events is simply the condition $\Pr(A \cap B \cap C) = \Pr(A)\Pr(B)\Pr(C)$, but our definition requires more. What is gained or lost by adding the extra restrictions? Keep in mind the results of Exercises 10 and 11.

SECTION 8.5 *Expected Values*

Assessing a Gamble

A "friend" offers you the following gamble. She will provide a coin, which you get to flip. If the coin lands as heads, she will pay you $3. If the coin lands as tails, you must pay her $2. If the coin is fair, that is, equally likely to land heads as tails, then you would enthusiastically participate in the gamble. You would win about as many flips as you would lose; the payoff for winning is greater than the penalty for losing. The gamble is stacked in your favor. Can we measure how favorable it is to you? Would the gamble still be favorable if the coin has been weighted so that heads appear only with probability .45? What if heads only come up 40% of the time?

One natural way to respond to such questions is to consider what our average profit per flip would be if we spent the evening gambling in this pleasant manner. We'll start our analysis with the fair coin. Suppose we flip it 100 times. Then we would expect to see heads about 50 times and tails about 50 times. Thus we would receive a total of about $(50)(\$3) = \$150$ and pay out a total of about $(50)(\$2) = \$100$. Our total profit would be about $\$150 - \$100 = \$50$. This would amount to an average profit of about $\$50/100$, or 50 cents per flip.

If we flip the coin 200 times, then our estimated net profit would be $(100)(\$3) - (100)(\$2) = \$100$, but the average profit per flip would remain the same: 50 cents.

If we flip the coin N times, then our estimated net profit would be $(N/2)(\$3) - (N/2)(\$2) = \$N/2$. The average profit per flip would be $(\$N/2)/N$, which is still 50 cents. We can also write this calculation as

average profit per flip
$$= \frac{(N/2)(\$3) - (N/2)(\$2)}{N} = (\$3)\left(\frac{1}{2}\right) + (-\$2)\left(\frac{1}{2}\right)$$
$$= (\$3)(\text{probability of heads}) + (-\$2)(\text{probability of tails}).$$

The average profit per flip is called the *expected value* of the gamble. It is a measure of what you would expect as your average profit if you were to repeat the gamble a large number of times. For this example, the expected value is 50 cents. Note that on any particular flip of the coin, the result is either a gain of $3 or a loss of $2. There is no single flip on which you gain 50 cents. But if you were to spend the evening playing this game, you would expect over the "long haul" to average a half-dollar gain per flip.

Random Variables

Many situations have a structure similar to the gamble just described, where a number is associated to each possible chance outcome of an experiment. The possible outcomes of the coin flip experiment are a head or a tail. Associated with a head is a payoff of 3 to you; associated with a tail is a payout by you of 2. The term "random variable" is used whenever we have numerical values tied to the outcomes of a chance experiment.

DEFINITION

A *random variable* is a numerical-valued function defined on a sample space.

Some other typical random variables are the number of heads in three flips of a coin, the sum of the spots on two dice that are rolled, and the score on the final exam of a randomly chosen student in a course. In each case there is a rule that uniquely assigns a number X to each elementary event of the sample space.

E X A M P L E 1 Three Coin Flips Consider the experiment of tossing a fair coin three times. An elementary event for this sample space is a three-term sequence whose elements are heads or tails. There are eight equally likely elementary events:

TTT, HTT, THT, TTH, HHT, HTH, THH, HHH.

Let X represent the number of heads. Then X is a random variable that takes on one of the values 0, 1, 2, or 3 for each repetition of the experiment. The random variable X has the value 1 if exactly one of the three flips resulted in a head. There are three distinct elementary events in which this occurs: HTT, THT, TTH. Thus the probability that $X = 1$ is 3/8. The following table summarizes the possible values for X, the elementary events that yield these values, and the probability that X takes on each of its possible values:

Values z of X	Outcomes	$Pr(X = z)$
$z = 0$	TTT	1/8
$z = 1$	HTT, THT, TTH	3/8
$z = 2$	HHT, HTH, THH	3/8
$z = 3$	HHH	1/8 ▲

E X A M P L E 2 Course Enrollments Registration figures at a large southern college showed the following percentages of students enrolled for different numbers of courses in a particular semester

Number of courses	Percentage
5	5%
4	75%
3	7%
2	12%
1	1%

Tuition charges at this college are $10,000 if the student is enrolled in 4 or 5 courses, $8,000 for 3 courses, and $5,000 for 1 or 2 courses.

Suppose we select a student at random and let the random variable X be that student's tuition charge for the semester. Then the possible values for X are $10,000 or $8,000 or $5,000. Then

$$Pr(X = \$8,000) = Pr(\text{Student is enrolled in } 3 \text{ classes})$$
$$= .07.$$

We also have

$$Pr(X = \$10,000) = Pr(\text{Student is enrolled in } 4 \text{ classes or } 5 \text{ classes})$$
$$= .75 + .05 = .80.$$

Finally,

$$Pr(X = \$5,000) = Pr(\text{Student is enrolled in } 1 \text{ or } 2 \text{ classes})$$
$$= .12 + .01 = .13. \blacktriangle$$

Distribution of a Random Value

The essential characteristics of a random variable are the values it can take on and the corresponding probabilities with which it assumes these values. We can describe a random variable X by a *probability density function* f, where for each real number z, $f(z)$ is the probability that X takes on value z, that is, $f(z) = Pr(X = z)$.

E X A M P L E 3 Spots on a Die If we roll a fair die and let the random variable X be the number of spots on the top face, then X takes on the values 1, 2, 3, 4, 5, or 6 each with probability 1/6. The probability density function for the random variable X is

$$f(z) = \begin{cases} \frac{1}{6} & \text{if } z = 1, 2, 3, 4, 5, \text{ or } 6, \\ 0 & \text{otherwise.} \end{cases}$$

Figure 1 shows the graph of this density function. ▲

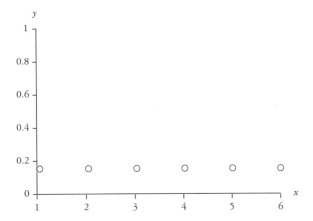

FIGURE 1

E X A M P L E 4 Number of Heads The probability density function for the random variable X, which is the number of heads in three flips of a fair coin (see Example 1), is

$$f(z) = \begin{cases} \frac{1}{8} & \text{if } z = 0, \\ \frac{3}{8} & \text{if } z = 1, \\ \frac{3}{8} & \text{if } z = 2, \\ \frac{1}{8} & \text{if } z = 3, \\ 0 & \text{otherwise.} \end{cases}$$

We can also completely describe a random variable by a *cumulative distribution function F,* which measures the probability that the random variable has values that do not exceed various levels. To be more precise, for each real number z,

$F(z)$ is the probability that X has a value less than or equal to z; that is, $F(z) = \Pr(X \leq z)$. ▲

E X A M P L E 5 The cumulative distribution function for the number of heads in three flips of a fair coin is given by

$$F(z) = \begin{cases} 0 & \text{if } z < 0, \\ \frac{1}{8} & \text{if } 0 \leq z < 1, \\ \frac{4}{8} & \text{if } 1 \leq z < 2, \\ \frac{7}{8} & \text{if } 2 \leq z < 3, \\ 1 & \text{if } z \geq 3. \end{cases}$$ ▲

Expected Value for a Two-Outcome Experiment

If an experiment or gamble has two possible numerical outcomes or payoffs: a, which occurs with probability p, and b, which occurs with probability $1 - p$, then the expected value EV is defined to be

$$\text{EV} = ap + b(1 - p).$$

In our example, $a = 3$, $b = -2$, and $p = 1/2$, so that EV $= 1/2$. If the coin is weighted so that p, the probability of a head, is .45, then

$$\text{EV} = 3(.45) + (-2)(.55) = 1.35 - 1.10 = .25.$$

For such a gamble, the expected value per flip is 25 cents. The gamble is still favorable to you but not as advantageous as it would be with a fair coin.

If the coin is so unbalanced that heads appear only 40% of the time, then $p = .4$ and the expected value is

$$\text{EV} = 3(.4) + (-2)(.6) = 1.2 - 1.2 = 0.$$

The expected value per flip here is 0. You should expect to "break even" in the long run, on such a gamble. We would call such a gamble a fair bet.

If the probability p of a head drops below .4, then

the expected value will become negative. You would expect a net loss, on the average, per flip.

Another way to think about the expected value is as a *weighted average* of the two outcomes a and b, where each outcome is weighted by the likelihood of its occurrence. If the two outcomes are equally likely, then the probability of each is 1/2, and the expected value becomes

$$a\left(\frac{1}{2}\right) + b\left(\frac{1}{2}\right) = \frac{a + b}{2},$$

which is the usual average of the two numbers a and b.

Expected Value for a Multiple-Outcome Experiment

Suppose an experiment has many different possible numerical outcomes; in other words, we have a random variable on a sample space that can take on many different values. The expected value is still defined as the sum of the products of each numerical outcome multiplied by the probability that the particular outcome occurs. The expected value is the weighted sum of the possible outcomes, where the weights are the relative likelihoods.

More formally, if X is a random variable that takes on distinct possible values $x_1, x_2, \ldots, x_k$ with respective probabilities $p_1, p_2, \ldots, p_k$, then the expected value of X is defined by

$$\mathrm{EV}(X) = x_1 p_1 + x_2 p_2 + \cdots + x_k p_k.$$

EXAMPLE 6 Summer Job Prospects Suppose your aunt is the director of the personnel office at a large university. She agrees to hire you for a summer job but isn't sure now exactly what

position she can give you. She offers the following estimates:

Position	Summer salary	Probability of getting the job
Custodian	$1,200	.4
Dishwasher	$1,350	.3
Painter	$1,425	.2
Programmer	$2,200	.1

Your expected summer salary would be

$$(1200)(.4) + (1350)(.3) + (1425)(.2) + (2200)(.1)$$
$$= 480 + 405 + 285 + 220 = 1390$$

dollars. ▲

EXAMPLE 7 Expected Number of Heads
We flip a fair coin three times and record the number of heads. There are four possibilities:

0 heads if outcome is TTT;

1 head if outcome is HTT, THT, or TTH;

2 heads if outcome is HHT, HTH, or THH;

3 heads if outcome is HHH,

so the expected number of heads is

$$0(\tfrac{1}{8}) + 1(\tfrac{3}{8}) + 2(\tfrac{3}{8}) + 3(\tfrac{1}{8}) = 1.5. \ ▲$$

EXAMPLE 8 Expected Sum of Dice We'll calculate the expected value of the sum when rolling a pair of fair dice (see Exercise 7 of Section 8.2). Here we have an experiment with 11 different numerical outcomes: 2, 3, 4, . . . , 12. Our random variable X is the sum of the spots on the two dice. The expected value for the sum is

$$2 \Pr(\text{sum is } 2) + 3 \Pr(\text{sum is } 3) + \cdots + 12 \Pr(\text{sum is } 12)$$

$$= 2\left(\frac{1}{36}\right) + 3\left(\frac{2}{36}\right) + 4\left(\frac{3}{36}\right) + 5\left(\frac{4}{36}\right) + 6\left(\frac{5}{36}\right) + 7\left(\frac{6}{36}\right)$$

$$+ 8\left(\frac{5}{36}\right) + 9\left(\frac{4}{36}\right) + 10\left(\frac{3}{36}\right) + 11\left(\frac{2}{36}\right) + 12\left(\frac{1}{36}\right)$$

$$= \frac{2 + 6 + 12 + 20 + 30 + 42 + 40 + 36 + 30 + 22 + 12}{36}$$

$$= \frac{252}{36} = 7. \ \blacktriangle$$

Finally, let's note that we can compute the expected value of a random variable X on a sample space in an alternative way. Recall our theorem that the probability measure of an arbitrary subset of E is the sum of the weights of the elements of that subset. In evaluating the expected value of a random variable X, we multiply each possible value x_i of X by the probability $\Pr(X = x_i)$. Let A_i be the subset of E on which X takes on value x_i. Then e belongs to A_i exactly if $X(e) = x_i$. Thus $\Pr(X = x_i) = \Pr(A_i) = $ sum of weights of elements of A_i. Thus we see the following result.

Theorem 1. If the elements of the underlying set E are $e_1, e_2, \ldots, e_m$, with respective weights $w_1, w_2, \ldots, w_m$, then the expected value of X is also given by

$$\text{EV}(X) = X(e_1)w_1 + X(e_2)w_2 + \cdots + X(e_m)w_m$$

$$= \sum_{i=1}^{m} X(e_i)w_i,$$

where $X(e_i)$ is, of course, the value of the random variable X on the element e_i.

Properties of Expected Value

Recall that if f and g are real-valued functions defined on a set A and c is any real number, then we can define new functions, $f + g$ (the sum of f and g) and cf (the scalar multiple of f by c), by the rules

$$(f + g)(x) = f(x) + g(x),$$

where the addition on the right-hand side is the sum of real numbers, and

$$(cf)(x) = cf(x),$$

where the multiplication on the right is the product of the real numbers c and $f(x)$.

Now if X and Y are random variables defined on some sample space and c is any real number, we can use these rules to define two new random variables $X + Y$ and cX. We will see that there is a very simple relationship between the expected values of these random variables and the expected values of X and Y.

If X has the values $x_1, x_2, \ldots, x_k$, with respective probabilities $p_1, p_2, \ldots, p_k$, then cX has the values $cx_1, cx_2, \ldots, cx_k$, with respective probabilities $p_1, p_2, \ldots, p_k$. Thus, the expected value of cX is

$$\text{EV}(cX) = \sum_{i=1}^{k} (cx_i)p_i = cx_1 p_1 + cx_2 p_2 + \cdots + cx_k p_k$$

$$= c(x_1 p_1 + x_2 p_2 + \cdots + x_k p_k)$$

$$= c \sum_{i=1}^{k} x_i p_i = c\text{EV}(X).$$

To determine the expected value of $X + Y$, we will make use of our alternative equation for the expected value:

$$\text{EV}(X + Y) = (X + Y)(e_1)w_1 + (X + Y)(e_2)w_2$$
$$+ \cdots + (X + Y)(e_m)w_m,$$

which, by definition of the sum of two random variables,

$$
\begin{aligned}
&= [X(e_1) + Y(e_1)]w_1 + [X(e_2) + Y(e_2)]w_2 \\
&\quad + \cdots + [X(e_m) + Y(e_m)]w_m \\
&= X(e_1)w_1 + X(e_2)w_2 \\
&\quad + \cdots + X(e_m)w_m + Y(e_1)w_1 + Y(e_2)w_2 \\
&\quad + \cdots + Y(e_m)w_m \\
&= EV(X) + EV(Y).
\end{aligned}
$$

Thus we obtain the very important result that expected value satisfies the two properties of *linearity;* it preserves sums and scalar multiples. We summarize these properties in the statement of the following theorem.

Theorem 2. If X and Y are random variables defined on the same probability space and c is any constant, then

a. $EV(cX) = cEV(X)$,

b. $EV(X + Y) = EV(X) + EV(Y)$.

Expected Value for an Experiment with Infinitely Many Outcomes

In many instances, we can determine the expected value for a random variable defined on a discrete sample space with infinitely many outcomes. The basic idea is the same: The expected value is the weighted average of the possible values of the random variable, each value being weighted by the probability of its occurrence.

E X A M P L E 9 We have a random variable X whose possible values are the positive integers $1, 2, 3, \ldots$, and the probability that X takes on the value k is p_k where each p_k is nonnegative and the sum of the p_ks is 1. Then the expected value is defined to be

$$
\begin{aligned}
EV &= 1\,\Pr(X = 1) + 2\,\Pr(X = 2) + 3\,\Pr(X = 3) \\
&\quad + \cdots = 1p_1 + 2p_2 + 3p_3 + 4p_4 + \cdots
\end{aligned}
$$

if this infinite sum converges. (You may wish to review the discussion of convergence of infinite sums in Section 1.9.) ▲

As a simple example, suppose $\Pr(X = k) = p_k = 1/k(2^k)$. Then

$$
\begin{aligned}
EV(X) &= 1\,\frac{1}{1(2^1)} + 2\,\frac{1}{2(2^2)} + 3\,\frac{1}{3(2^3)} \\
&\quad + \cdots = 1\left(\frac{1}{2}\right) + \left(\frac{1}{2}\right)^2 + \left(\frac{1}{2}\right)^3 + \cdots .
\end{aligned}
$$

This is a geometric series with initial term $a = 1$ and common ratio r equal to $1/2$. The series sums to

$$
\frac{a}{1 - r} = \frac{1}{1 - \frac{1}{2}} = \frac{1}{\frac{1}{2}} = 2,
$$

and so the expected value is 2.

E X A M P L E 10 Job Applicants Consider an employer interviewing job applicants until she finds one with a college degree. We assume that 10% of the applicants met this qualification and that the job seekers are scheduled in random order for interviews. Thus, the probability that a particular applicant is a college graduate is 1/10. We pose three questions about this process:

1. What is the probability that the employer will interview more than five applicants?

2. What is the probability that the employer interviews exactly six applicants?

3. How many applicants should she expect to interview before finding one with a college degree? ▲

The answers to the first two questions are relatively easy; the third question requires a bit more complicated analysis.

The employer would interview more than five applicants only when all of the first five were not college graduates. Since each candidate has a probability of 9/10 of not holding a college degree and they arrive independently of each other, the probability of five successive applicants without college degrees is $(9/10)^5$, or about .59.

The probability that the sixth person interviewed is the first one with a college degree is the probability of five successive nongraduates followed by a graduate. Assuming independence again, this probability is

$$(\tfrac{9}{10})(\tfrac{9}{10})(\tfrac{9}{10})(\tfrac{9}{10})(\tfrac{9}{10})(\tfrac{1}{10}) = (\tfrac{9}{10})^5(\tfrac{1}{10}).$$

To determine the expected number of interviews, let X be the random variable whose value is the number of applicants. Thus $X = n$ if the nth person interviewed is a college graduate but the first $n - 1$ are not. It's easy to see that the probability of this occurring is given by

$$\Pr(X = n) = (\tfrac{9}{10})^{n-1}(\tfrac{1}{10}).$$

The expected number of interviews is

$$EV(X) = 1\,\Pr(X = 1) + 2\,\Pr(X = 2) \\ + 3\,\Pr(X = 3) + 4\,\Pr(X = 4) + \ldots,$$

or

$$1(\tfrac{9}{10})^0(\tfrac{1}{10}) + 2(\tfrac{9}{10})^1(\tfrac{1}{10}) + 3(\tfrac{9}{10})^2(\tfrac{1}{10}) + 4(\tfrac{9}{10})^3(\tfrac{1}{10}) + \ldots.$$

We can more easily sum this series if we rewrite the sum as follows:

$$\tfrac{1}{10}$$
$$+ (\tfrac{9}{10})^1(\tfrac{1}{10}) + (\tfrac{9}{10})^1(\tfrac{1}{10})$$
$$+ (\tfrac{9}{10})^2(\tfrac{1}{10}) + (\tfrac{9}{10})^2(\tfrac{1}{10}) + (\tfrac{9}{10})^2(\tfrac{1}{10})$$
$$+ (\tfrac{9}{10})^3(\tfrac{1}{10}) + (\tfrac{9}{10})^3(\tfrac{1}{10}) + (\tfrac{9}{10})^3(\tfrac{1}{10}) + (\tfrac{9}{10})^3(\tfrac{1}{10})$$
$$+ \ldots.$$

Let's evaluate the sum by adding down the columns. Each column is a geometric series with common ratio $r = 9/10$ but with different initial terms. The first column has initial term 1/10, the second column has $(9/10)^1(1/10)$, the third column has $(9/10)^2(1/10)$, etc. In general, the kth column has initial term $(9/10)^{k-1}(1/10)$. Thus, the sum of the kth column is

$$\frac{a}{1 - r} = \frac{(\tfrac{9}{10})^{k-1}(\tfrac{1}{10})}{1 - (\tfrac{9}{10})} = \frac{(\tfrac{9}{10})^{k-1}(\tfrac{1}{10})}{\tfrac{1}{10}} = (\tfrac{9}{10})^{k-1}.$$

The first few column sums are

$$1, \tfrac{9}{10}, (\tfrac{9}{10})^2, (\tfrac{9}{10})^3, \ldots,$$

and so the expected value is

$$1 + (\tfrac{9}{10}) + (\tfrac{9}{10})^2 + (\tfrac{9}{10})^3 + \ldots,$$

which is a geometric series with initial term 1 and common ratio $r = 9/10$. Thus, it adds up to

$$\frac{1}{1 - (\tfrac{9}{10})} = \frac{1}{(\tfrac{1}{10})} = 10,$$

so the employer should, on the average, expect to interview 10 candidates.

This argument is easily modified for other probabilities. If, for example, the probability that a particular applicant is a college graduate is 2/5, then the expected number of interviews is 5/2. Replacing 1/10 by p in the argument yields the result that the expected number of interviews is $1/p$ if the probability that a randomly chosen applicant is a graduate is p. The following theorem states the result in a slightly more general setting.

Theorem 3. Suppose the probability that a particularly desired outcome—call it a *success*—occurs in a single trial of an experiment is p. Then the expected number of trials until a first success is $1/p$.

Some interesting paradoxes arise if the infinite series does not converge to any finite value. Perhaps the most famous is the St. Petersburg paradox. It concerns a game in which you toss a coin until it lands tails. If this happens for the first time on the kth toss, you win $\$2^k$. Thus you win $\$2$ if your first toss results in a tail; you win $\$4$ if you flip a head and then a tail; $\$8$ if the sequence is head, head, tail; and so forth. How much money are you willing to pay to enter this game? As we have seen, the usual answer to this question is the expected value of your payoff. In order to win $\$2^k$, you must flip $(k - 1)$ heads and then a tail. The probability of such an event is $(1/2)^{k-1}(1/2)$, or $(1/2)^k$. The standard calculation of the expected value leads to the sum

$$2(\tfrac{1}{2})^1 + 4(\tfrac{1}{2})^2 + 8(\tfrac{1}{2})^3 + 16(\tfrac{1}{2})^4 + \cdots + 2^k(\tfrac{1}{2})^k,$$

in which each term is equal to 1. Thus the sum would be infinite! Since you have an unlimited expected payoff, you should be willing to pay any fee, no matter how large, to enter this game. The paradox arises because most people would not pay very much to play.

If, for example, you pay $\$100$ as an entrance fee and flip a tail on the first toss, you actually lose $\$98$. This will happen about half the time. If you flip 5 consecutive heads and then a tail, you're paid $\$2^6 = \$64$, so your net loss is $-\$36$. To make any money at all, you have to flip at the start at least 6 consec-

utive heads; the probability of doing this is only 1/64.

Expected Value When Sampling Without Replacement

In many of the situations we have examined in this section, we are dealing with repetitions of an experiment in which the probability of a particular outcome is the same on every trial. For example, each job applicant was assumed to have a probability of .1 of being a college graduate. As another example, consider selecting a card from a standard deck of playing cards. Repeat this experiment until you obtain the ace of spades. After each selection, you *replace* the card and shuffle the deck. Since each of the 52 cards is equally likely to be chosen, the probability of selecting the ace of spades on any trial is 1/52. By Theorem 2, the expected number of trials is 52. This example illustrates a situation called *sampling with replacement*.

In other applications, you may be faced with *sampling without replacement:* You select one object from a larger population, record its value, and *discard* it, and then you select another object from the larger population. In sampling without replacement, there is no possibility of selecting the same object twice; the size of the larger population decreases by one after each selection. In sampling with replacement, the size of the larger population remains fixed during the entire selection process; we might choose some objects several times.

As an example of sampling without replacement, suppose that you're trying to guess the correct answer to a multiple-choice question. Four suggested answers are given. Only one of them is correct. You may make repeated guesses. After each guess, you're told if you are right or wrong. How many guesses would you expect to make?

Let's assume that the game will end when you guess the correct answer. We'll also assume that you won't repeat a guess that you've been told is a wrong one. This means that you will certainly guess the correct answer by the fourth try. You might get it as early as the first try. The expected number of guesses is

1× (probability you guess on first try) +

2× (probability you guess on second try) +

3× (probability you guess on third try) +

4× (probability you guess on fourth try).

The probability that you guess correctly on the first try is simply 1/4, since there are four equally likely choices but only one correct one.

In order for you to guess correctly on the second try, your first guess must be wrong but your second guess must be right. The probability that the first guess is wrong is 3/4. After making the first guess, you're told that it is incorrect. That means that one of the remaining three answers must be the right one. So on the second try, you are guessing at random from a collection of three objects; the probability of a success is 1/3. Thus, the probability of guessing correctly on the second try is

$$\text{Probability(Incorrect, Correct)} = (\tfrac{3}{4})(\tfrac{1}{3}) = \tfrac{1}{4}.$$

In order for you to guess correctly on the third try, your first two guesses must be wrong but your third guess must be right. We're looking for

$$\text{Probability(Incorrect, Incorrect, Correct)} = (\tfrac{3}{4})(\tfrac{2}{3})(\tfrac{1}{2}) = \tfrac{1}{4}.$$

The last factor (1/2) comes from the consideration that your first two incorrect guesses have narrowed the set of possibly correct answers down to $4 - 2 =$

2 and that you will pick one of these two at random.

Finally, the only way not to guess correctly until the fourth try is to have three incorrect guesses followed by the right answer:

$$\text{Probability(Incorrect, Incorrect, Incorrect, Correct)} = (\tfrac{3}{4})(\tfrac{2}{3})(\tfrac{1}{2})1 = \tfrac{1}{4}.$$

Thus, the expected number of guesses is

$$1(\tfrac{1}{4}) + 2(\tfrac{1}{4}) + 3(\tfrac{1}{4}) + 4(\tfrac{1}{4}) = \tfrac{10}{4} = 2.5.$$

If you suffer from severe short-term memory loss and promptly forget which answer you just guessed, then you can repeatedly select an incorrect answer you picked on an earlier guess. You would then be dealing with a sampling with replacement and, by Theorem 2, the expected number of guesses would be $1/(1/4) = 4$.

Expected Values for Continuous Distributions

We have seen that in the case of a finite sample space or an infinite discrete sample space, it is possible to define a notion of average value or expected value of a random variable X by computing a weighted average of the values taken on by X, where the weights are simply the probabilities or relative frequency of these values.

We can also develop an analogous concept of expected value for a *continuous distribution*. Continuous distributions are used when we have a random variable that can take on all possible numbers in an interval. In the case of a finite sample space or a discrete infinite sample space, we can think of a total probability of 1 which is parceled out by daubing each element with a certain amount of probability (its weight). In the continuous case, we think of the total probability "smeared" over the interval of values. The probability can be smeared evenly or unevenly. If all parts of the interval are evenly smeared with probability, then we have the *uniform distribution*. In many instances, however, the probability is not evenly spread. A continuous distribution is often described by its cumulative distribution function F, where $F(z) = \Pr(X \leq z)$.

EXAMPLE 11 Suppose we select a number k at random in the interval $[0, 2]$ and let X be the square of that number. Then X can also take on any value between 0 and 4. Thus

$$F(\tfrac{16}{9}) = \Pr(X \leq \tfrac{16}{9}) = \Pr(k^2 \leq \tfrac{16}{9}) = \Pr(k \leq \tfrac{4}{3}). \ \blacktriangle$$

A full treatment of continuous distributions requires the tools of calculus, particularly the idea of a definite integral. You'll enjoy exploring this aspect of probability once you've studied some calculus. Here we will treat only one simple situation. Suppose we are dealing with the *uniform distribution* on the closed unit interval $[0, 1]$. The uniform distribution for a continuous distribution is analogous to the equiprobable distribution for a finite sample space. The cumulative distribution function for the uniform distribution is $F(z) = z$. This implies that the likelihood that a random chosen number lies in any particular subinterval A of $[0, 1]$ is equal to the length of A. Thus, the probability that a number lies between 0.25 and 0.60 is $0.60 - 0.25 = 0.35$.

To determine the expected value of a random variable associated with the uniform distribution on $[0, 1]$, suppose our random variable is a function f that assigns some real number $f(x)$ to each x in the interval. See Figure 2. We can approximate this random variable by partitioning $[0, 1]$ into a large number of nonoverlapping subintervals,

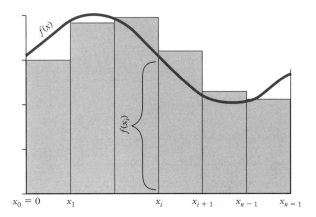

FIGURE 2

$$[x_0, x_1] \cup [x_1, x_2] \cup [x_2, x_3] \cup \cdots \cup [x_{n-1}, x_n],$$

and considering the new (approximating) random variable X whose value on $[x_i, x_{i+1}]$ is $f(x_i)$. Then X takes on only a finite number of values $f(x_0)$, $f(x_1)$, $\ldots$, $f(x_{n-1})$. The probability that X takes on the value $f(x_i)$ is the probability that we selected a number in the interval $[x_i, x_{i+1}]$. For the uniform distribution, this probability is the length of the subinterval: $x_{i+1} - x_i$. Thus the expected value of X is given by

$$\sum_{i=0}^{n-1} f(x_i)\Pr(X = f(x_i)) = \sum_{i=0}^{n-1} f(x_i)(x_{i+1} - x_i).$$

Note that each term, $f(x_i)(x_{i+1} - x_i)$, in this sum is the area of a rectangle of height $f(x_i)$ and width $(x_{i+1} - x_i)$. This area is approximately the area bounded by the graph of f: the x-axis and the vertical lines at $x = x_i$ and $x = x_{i+1}$. If n is very large, then the graphs of X and f are very similar so that X provides an excellent approximation to f. The area under the graphs of X and f are virtually identical. Since the expected value of X is the area under the graph of X, it is reasonable to define the expected

China's Family-Size Policy

Using elementary probability theory often helps us analyze important public-policy decisions. A carefully crafted mathematical model may yield some surprising results that are contrary to our intuition.

China's 1995 population of 1.2 billion people is the world's largest and it continues to increase. In an effort to curb population growth, the Chinese government has attempted to enforce a "one-child" policy, limiting each family to only one offspring. Although many follow the policy as a matter of choice, *The New York Times** reports that "Coercion still underlies the one-child policy, and the rationing of the right to become pregnant remains a source of tension and bitterness. . . . Many peasants grumble that the policy . . . should not be applied to them until they give birth to a son." Because of the importance many Chinese families attach to having a son, women are often forced to have abortions if tests show the fetus is a female. There have also been reports of the infanticide and abandonment of infant girls.

What are the implications of limiting families to one *son*, rather than to one *child?* What would be the ratio of births of girls to boys? What would be the average number of children in a family?

Since there is no restriction on the number of girls, we can expect millions of families with two girls and a boy, and perhaps thousands of families with six or more girls before a son is born. Our intuition may lead us to conjecture that the average family would have three, four, or perhaps more children and that there will be many times more girls than boys.

A careful probabilistic analysis, however, shows that the expected family size is two children and that about as many girls as boys will be born. Thus the more lenient policy would not affect the girl/boy ratio nor lead to a large average family size.

*Nicholas D. Kristof, "More in China willingly rear just one child," The New York Times, May 9, 1990, A1.

value of f to be the area under the graph of f. More precisely, it is the area bounded by the graph of f: the x-axis and the vertical lines at $x = 0$ and $x = 1$.

For example, if our random variable is the constant function $f(x) = 1$ for all x, then it should certainly be the case that the expected value of f is 1. The area in question for the constant function $f(x) = 1$ is the area of the square with vertices $(0, 0)$, $(1, 0)$, $(1, 1)$, and $(0, 1)$, as Figure 3 shows. Since this square has sides of length 1, its area is also equal to 1.

random from the unit interval $[0, 1]$. What does your intuition suggest the answer should be? A reasonable response might be as follows: Since all the numbers between 0 and 1 are equally likely to be chosen, if we repeat the experiment many times, then about half the time the numbers will be less than $1/2$ and half the time they will be greater than $1/2$; thus the expected value should be $1/2$. If we examine the relevant area for the function $f(x) = x$, as shown in Figure 4, it turns out to be the area of the triangle with vertices at $(0, 0)$, $(1, 0)$, and $(1, 1)$. This triangle does indeed have area $1/2$.

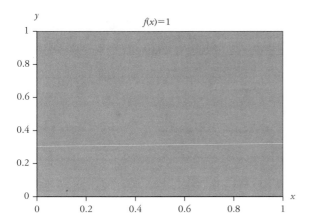

FIGURE 3

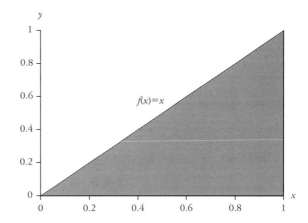

FIGURE 4

This example, although clearly a trivial one, illustrates an important principle of mathematics. In developing a definition for a general concept, test to make sure that it yields the correct value for any special case where you already know the answer.

Let's look at a more interesting example. Suppose our random variable is the function $f(x) = x$. This is the random variable that corresponds to selecting a point from our sample space $[0, 1]$ and noting its value. The expected value of f would then represent the average value of a number selected at

As a third example, consider the function $f(x)$ defined by the rule

$$f(x) = \begin{cases} 2x & \text{if } x \le \frac{1}{2} \\ 1 & \text{if } x > \frac{1}{2} \end{cases}.$$

Here the area is made up of a triangle with vertices at $(0, 0)$, $(1/2, 0)$, and $(1/2, 1)$ and a rectangle with vertices at $(1/2, 0)$, $(1, 0)$, $(1, 1)$, and $(1/2, 1)$ which have areas of $1/4$ and $1/2$, respectively (see Figure 5). The total area is $1/4 + 1/2 = 3/4$ and is the expected value of f.

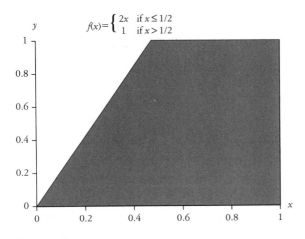

$$f(x)=\begin{cases} 2x & \text{if } x \le 1/2 \\ 1 & \text{if } x > 1/2 \end{cases}$$

FIGURE 5

For our final example, consider the random variable that results if we select a number at random from $[0, 1]$ and compute its square. The function f in this case is given by $f(x) = x^2$. The corresponding

area is shown in Figure 6. It is difficult to calculate this area with only the tools of geometry. When you study calculus, you'll discover that it's relatively easy to show that this area has value $1/3$.

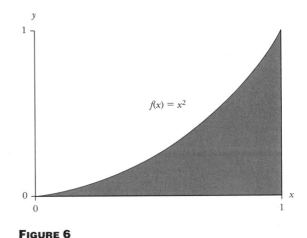

$f(x) = x^2$

FIGURE 6

Exercises for Section 8.5

1. Determine the probability density function and cumulative distribution function for the random variable in Example 2.

2. If a coin is flipped 8 times, find the expected number of heads assuming
   (a) the coin is fair
   (b) the coin is weighted so that it comes up heads with probability $3/5$

3. My automobile insurance company charges $500 a year for an accident policy. It claims that about 8% of its clients file a claim each year and that the average claim is $1200. What is the company's expected gain per policy?

4. A pair of fair dice is tossed with the following payoffs (losses) determined:

   2: win $0   3: win $5   4: lose $3
   5: win $3   6: lose $5   7: win $7
   8: lose $3   9: lose $2   10: win $2
   11: win $5   12: win $10

   What is the expected value of rolling the dice?

5. In the game of "craps" a player rolls a pair of dice. If the sum of the numbers shown is 7 or 11, she wins. If it is 2, 3, or 12, she loses. If it is any other sum, she must continue rolling the dice until either she repeats the same sum (in which case she wins) or she rolls a 7 (in which

case she loses). Suppose the outcome of each round is a win or a loss of $5. What is the probability that our player wins a round? What is the expected value of the player's winnings in shooting craps?

6. A roulette wheel has the numbers 0, 1, 2, . . . , 36 marked on 37 equally spaced slots. The numbers from 1 to 36 are evenly divided between red and black. A player may bet on either color. If a player bets on red and a red number turns up after the wheel is spun, he receives twice his stake. If a black number turns up, he loses his stake. If 0 turns up, then the wheel is spun again until it stops on a red or a black. If this is red, the player receives only his original stake, and if it is black, he loses his stake. If a player bets $1 on red with each spin, what is the expected value of his winnings?

7. (Expected-value considerations applied to decision-making) A construction firm is considering bidding on a contract to build one of two new schools. The firm may submit a bid to construct a high school. It estimates that there would be a $50,000 profit on this project, but that it would cost $1,000 to prepare the plans that must be submitted with the bid. The other possibility is a bid on an elementary school. The firm has built several elementary schools in the recent past and estimates that the preparation costs for a bid would be only $500 while the potential profit is $40,000. In estimating profit, all costs—including that of the bid—have been considered. The construction company has enough time to prepare only one bid. Past experience leads the company to estimate that it has one chance in five of winning the high school contract and one chance in four for the elementary school. The firm's goal is to maximize expected profit. Which bid should the company prepare and submit?

8. National safety experts estimate that one in every four Americans is injured each year in a household accident. Calculate the expected number of years before every family suffers an accident.

9. In a large office building, five elevators are arranged in a row so that the distance between the center of one elevator and the center of an adjacent elevator is six steps. Suppose each elevator is equally likely to arrive next.

(a) If you stand in front of elevator 2, what is the expected number of steps to reach the next available elevator?
(b) Where should you stand to minimize the expected number of steps?
(c) How would your answer to (b) change if there were six elevators? an odd number of elevators? an even number of elevators?

10. A survey of U.S. high schools determined that one-half of the students drop out or fail mathematics each year. What is the expected number of years of mathematics study for a high school student selected at random?

Exercises 11–17: Assuming the uniform distribution on [0,1], sketch the graph of each of the following random variables, and determine its expected value by determining the area bounded by the graph.

11. $f(x) = \begin{cases} 2x & \text{if } 0 \le x \le 1/2 \\ -2x + 2 & \text{if } 1/2 < x \le 1 \end{cases}$

12. $f(x) = \begin{cases} 3x & \text{if } 0 \le x \le 1/3 \\ -3/2x + 3/2 & \text{if } 1/3 < x \le 1 \end{cases}$

13. $f(x) = \begin{cases} nx & \text{if } 0 \le x \le 1/n \\ -(n/(n-1))x + n/(n-1) & \text{if } 1/n < x \le 1 \end{cases}$
where n is an integer greater than 1

14. $f(x) = \begin{cases} 3x & \text{if } 0 \le x \le 1/3 \\ 1 & \text{if } 1/3 < x \le 2/3 \\ -3x + 3 & \text{if } 2/3 < x \le 1 \end{cases}$

15. $f(x) = \begin{cases} 4x & \text{if } 0 \le x \le 1/4 \\ 1 & \text{if } 1/4 < x \le 3/4 \\ -4x + 4 & \text{if } 3/4 < x \le 1 \end{cases}$

16. $f(x) = \begin{cases} 2 & \text{if } 0 \le x \le 1/2 \\ 3 & \text{if } 1/2 < x \le 1 \end{cases}$

17. $f(x) = c$ for all x in $[0, 1]$ where c is a constant

18. (Writing Project) The expected value (or average value) of a random variable is a covenient single number to summarize what may be a complicated probability distribution. We often read about the average income in a country or the average cost of living in a particular city.

By itself, however, the expected value can mask important differences. Show that the random variables X and Y described by

$$\Pr(X = a) = \begin{cases} .1 & \text{if } a = 100{,}000, \\ .9 & \text{if } a = 0; \end{cases}$$

$$\Pr(Y = a) = \begin{cases} 1 & \text{if } a = 10{,}000, \\ 0 & \text{otherwise} \end{cases}$$

have the same expected value. If X and Y give the probability distributions for the incomes in two different communities, then both have the same average income but vastly different economic conditions.

A second number, the *variance,* is often computed for a random variable as a measure of how far the distribution varies from its expected value. Investigate the definition, properties, uses, and limitations of the variance measure.

An excellent reference on this topic and, indeed, on all aspects of probability is J. Laurie Snell, *Introduction to Probability,* New York: Random House, 1988. Snell presents in-depth discussions of discrete and continuous probability and includes a number of interesting historical notes.

SECTION 8.6 *Some Probabilistic Models*

In this section we use the probability tools developed in this chapter to examine in some detail three of the situations described in Section 8.1: The witchcraft problem; The stolen cars problem; and the reliability of warning devices problem.

The Witchcraft Problem

In seventeenth-century America, a good many people were accused of witchcraft. Some received formal trials; most of the accused were found not guilty. Both men and women faced witchcraft charges, but not in equal numbers. Historians wish to know if there was a sex bias in these trials: Were the women more likely to be convicted than the men?

Some Numerical Evidence

For a fascinating study of witchcraft, see *Entertaining Satan: Witchcraft and the Culture of Early New En-* gland, by John Putnam Demos, published by Oxford University Press in 1982. The following table contains some data Professor Demos collected about witchcraft trials in New England between 1630 and 1690. Of the 89 cases brought to trial, there were 68 accused women and 21 accused men. Fourteen of the women and two of the men were convicted. We arrange the data in a table with variables Gender and Convict.

Gender	Convict No	Yes	Total
Male	19	2	21
Female	54	14	68
Total	73	16	89

Examine these data for a moment. Were women more likely than men to be *accused* of witchcraft? Were women more likely than men to be *convicted?* Are the answers "obvious" from the data? We will examine possible bias in conviction rates.

One way to get a better understanding of the witchcraft problem is to convert the *absolute frequen-*

cies into *relative frequencies* or *percentages*. By the "absolute frequency" of an event, we mean the *number* of times it occurred. By the *"relative frequency,"* we mean the *proportion* of times this event happened among all relevant observations. There were 2 convicted males among the 89 observed; thus the convicted men make up 2.2% of the total cases. We might examine the relative frequency of convicted men among all men in the study. Of the 21 men charged with witchcraft, only 2 were convicted. Thus the absolute frequency of convictions is 2. This gives a relative frequency or percentage of (2/21) × 100 or 9.5%. Similarly, the percentage of women acquitted (54 of 68) is (54/68) × 100 or 79.4%. The following table shows the percentages for all categories.

| | Convict | | |
Gender	No	Yes	Total
Male	90.5%	9.5%	100%
Female	79.4%	20.6%	100%
Total	82.0%	18.0%	100%

Does the presentation of the data in this format make it seem more likely that there was bias against women? Or does it look less likely that there was sex discrimination?

Perhaps you focused on the "Yes" column. Here we observe that the rate of conviction for women (20.6%) is more than twice as great as the conviction rate (9.5%) for men. It does *appear* from this view that women suffered a disproportionately high rate of conviction. But if we examine the "No" column, we may reach a different conclusion. Here we note that the percentage of males found innocent (90.5%) is only 11 percentage points higher than the figure (79.4%) for women. Moreover, the overall acquittal rate for the entire accused group is 82%, just slightly more than the rate for the women. Perhaps the data indicate that men and women were treated equally?

Is the difference between the conviction rate for men and that for women large enough to be significant? After all, three times as many women were accused as men. Perhaps this pattern of conviction is about what we should expect to see. On the other hand, seven times as many women as men were convicted. Maybe this pattern did not occur by chance. How can we resolve these questions? Is there a simple way to determine whether or not a real relationship between gender and the outcome of the accusations exists?

As a first approach to this problem, let's see what patterns might occur by chance. Then we can compare these to the historical data. We want to test a *hypothesis:* Women were more likely than men to be convicted of witchcraft. If chance patterns as extreme or more extreme than the historical data are relatively rare, then we may conclude that our hypothesis is correct. If such patterns are reasonably common, then we have no grounds for accepting the hypothesis as an explanation for the observed historical data.

Simulating a Chance Process

What do we mean by the "chance process"? There are a total of 89 people in our sample population of accused witches. Of this number, 21 were male and 68 were female. We also know that 16 were convicted. If there was no gender bias, then each of the 89 people had the same chance to be selected as "guilty." By a *chance process* here, we mean one that will select 16 of the 89 people at random so that each accused person has the same probability of selection as every other accused person. We will note the sex of each person selected and keep track of the total number of men and women chosen.

There are many ways to simulate this chance process. We could take 89 index cards and label 21 with the word "Male" and 68 with the word "Female." Shuffle the cards together and then deal out 16. Alternatively, you could put 21 green jelly beans and 68 red ones into a bag, shake the bag several times, and then reach in and select—without looking—16 of the candies. If you don't have access to a computer, we suggest that you try one of these experiments.

A computer simulation of this chance process is easy to create. The computer simulation has many advantages. It is easy to repeat the experiment many hundreds, or even thousands, of times in a few moments. So it is much faster (and less fattening) than using jelly beans. The output of a typical run of this program is

There were 3 men and 13 women selected.

There's not very much we can conclude from a single run of the program. To get a better sense of what typical patterns of choices result from this simulation, we need to run the program a larger number of times. In the following table we show the results of two different sets of repetitions of the program. Each set shows the frequencies that were seen in one thousand runs of the program.

Results of two thousand repetitions of WITCHES program

Number of women chosen	6	7	8	9	10	11	12	13	14	15	16
Frequency (Sample 1)	0	3	11	47	90	189	206	219	152	70	13
Frequency (Sample 2)	1	3	15	31	108	159	244	225	140	62	12

In the total of 2000 simulations of choosing 16 people from the group of 89, we never wound up with fewer than 6 women. We did obtain a total of six instances in the 2000 where exactly 7 women and 9 men were picked. It is *possible* then for purely chance drawings with no gender bias to result in

cases where more men than women are selected. The simulation leads us to believe, however, that such cases are extremely rare. If the historical pool of 89 accused witches did contain only 21 men and if the convicted 16 had a majority of male members, we might reasonably conclude that there was gender bias in the conviction process. But what about the historical data we have that showed 14 women and 2 men in the convicted group?

Note that in the first sample of 1000 simulations of the 16 draws, we had 235 (152 + 70 + 13) cases when 14 or more women were selected. In the second sample we saw 214 such cases. From our simulations it would be reasonable to conclude that if individual men and women had equal chances of being selected as one of the 16 "guilty" parties, we would expect to see in slightly more than 20% of the cases situations when 14, 15, or 16 were women. Since results as extreme, or more extreme, than the ones recorded in the historical data would appear by chance roughly one in five times—a not very rare proportion—there is no sufficient evidence that gender bias existed.

Using Probabilistic Reasoning

Can we obtain a more precise estimate of the probability that 14 or more women would be chosen? Yes. Probability theory gives us the tool to calculate this number exactly. The outcome of obtaining at least 14 women in 16 can be broken up into three mutually exclusive outcomes: exactly 14 women, exactly 15 women, and exactly 16 women. Thus

$$
\begin{aligned}
\Pr(14 \text{ or more women}) = {} & \Pr(\text{exactly 14 women}) \\
& + \Pr(\text{exactly 15 women}) \\
& + \Pr(\text{exactly 16 women}).
\end{aligned}
$$

Let's consider, then, how to calculate the probability of the first of these outcomes: exactly 14 women (and hence exactly 2 men). We are selecting 16 objects, without replacement, from a collection of 89 objects each of which has the same chance of being picked. We know that the sampling is *without replacement* since 16 different objects must be selected. Thus

Pr(exactly 14 women) =
$$\frac{\text{number of ways 14 women and 2 men can be selected}}{\text{number of ways 16 people can be selected from 89}},$$

which, using the multiplication principle of Chapter 4 is equal to

$$\frac{\left(\begin{array}{c}\text{number of ways 14 women}\\\text{can be selected from 68}\end{array}\right)\left(\begin{array}{c}\text{number of ways 2 men}\\\text{can be selected from 21}\end{array}\right)}{\text{number of ways 16 people can be selected from 89}}.$$

Each of the three numbers involves counting the number of ways of selecting k objects from a collection of n objects; that is, $_nC_k = \binom{n}{k}$. Thus Pr(exactly 14 women and 2 men were convicted) is given by

$$\frac{\binom{68}{14}\binom{21}{2}}{\binom{89}{16}}.$$

The probability of exactly 15 women being selected is a similar fraction. The denominator is the same, but the numerator would be

$$\binom{68}{15}\binom{21}{1}.$$

Can you see now what is the probability that all 16 chosen people are women? The probability that 14 or more women will be picked is the sum

$$\frac{\binom{68}{14}\binom{21}{2} + \binom{68}{15}\binom{21}{1} + \binom{68}{16}\binom{21}{0}}{\binom{89}{16}}.$$

Although in principle we could find the value of this fraction with pencil and paper calculations, the necessary computations for evaluating some of these binomial coefficients are so arduous that we turn to the computer for help. The computer calculates the value of this fraction to be .2077, so we see again that the probability that at least 14 women will be chosen—if gender plays no role in the selection—is slightly more than .20. Note that our simulation results are quite close to this value.

The Stolen Cars Problem

The Questions

In our problem, thieves make off with 20% of all the cars in the neighborhood each year. We can easily pose three questions of interest to us:

1. What are the chances that my car will be stolen this year?

2. How many years can I expect to pass before thieves get my car?

3. How long will it take before everybody in the neighborhood suffers from auto theft?

Some Simplifying Assumptions

We'll see that it's not difficult to answer these questions if we make a few simplifying assumptions about the situation. We'll assume that each person's car is as likely to be stolen as any other person's car. We'll also assume that successive car thefts are independent of each other. We're asserting here that you and your next-door neighbor each have the same chance (1 in 5) of being the next victim of automobile theft. We're also saying that the chances that the next car stolen belongs to a particular person are not dependent on the previous history of car thefts.

In the real world, these assumptions are probably not strictly valid. Whether a thief decides to go after your car rather than your neighbor's may depend on such factors as the age, make, or model of the cars, whether the car is on the street on in a garage, whether the owner is home or not, and so on. Also, if your car is stolen one year, you may take extra security measures that will deter thieves in subsequent years. The likelihood of your being a victim again may go down. On the other hand, since the thieves were successful against you last time, they may be more likely to return to the "scene of the crime" and try again.

The Answers

The answers to our first two questions are easily obtained from our previous work. For our basic model, we're asserting that there is a probability of .2 that any particular car will be stolen in a given year. Thus, you have a 20% chance of having your automobile stolen this year and an 80% chance of completing a full year with uninterrupted possession of your car.

If the thefts are independent of each other, then the probability of two successive years without suffering a theft is

$$(.8)(.8) = .64,$$

so there is a probability of .36 of having at least one car stolen in two years.

The likelihood of going three years in a row without a car theft is

$$(.8)(.8)(.8) = .512,$$

or a little more than one-half.

From Theorem 3 in our section on expected value, we see that, on average, you should expect five years to go by before your car is stolen. You should compute the probability that your car has not been stolen during a five-year period; the answer is $.8^5 = .32768$.

The one problem we have not yet solved is how long it should take on the average before everyone in the neighborhood experiences a stolen car. A first approach to this problem is to do some simulations. You could simply wait and observe the car thefts in the neighborhood for several years. You might even join (or organize) a car theft ring and keep careful statistics on your operations. Such methods of collecting experimental insight into the problem might have severe consequences for you.

A cheaper *physical simulation* is available. Take an ordinary six-sided die (that's one of the members of a pair of dice). Roll it repeatedly until each of the numbers one through five has been observed. Keep track of the number of rolls until you've obtained all five numbers; simply ignore any of the rolls that results in a six.

In this approach and our subsequent analyses, we'll assume for simplicity that there are exactly five different families in our neighborhood, each with one automobile, and that one car theft occurs each year. Also, to simplify our discussion, we'll assume that each stolen car is recovered or replaced by an identical model.

A third alternative is to write a computer program to simulate the process. When such a program was run one thousand times, we found that the average number of thefts that occurred before all five families were affected was 11.59. The following table gives a more detailed report. It shows how many times we observed 5 thefts, 6 thefts, ... 39 thefts.

Thefts	Number of times	Thefts	Number of times
5	39	22	13
6	76	23	7
7	116	24	9
8	103	25	5
9	93	26	2
10	99	27	2
11	71	28	5
12	63	29	2
13	56	30	0
14	48	31	1
15	45	32	2
16	33	33	2
17	33	34	1
18	22	35	0
19	17	36	0
20	18	37	0
21	14	38	2
		39	1

Note that in just under 4% of the trials (39 of 1000), we found all 5 cars stolen in the first 5 thefts. In the most extreme cases, we had 39 thefts before getting a complete set of cars. If you examine the table, you'll see that there were 11 ($2 + 0 + 1 + 2 + 2 + 1 + 0 + 0 + 0 + 2 + 1$) cases involving more than 28 thefts; this amounts to just over 1% of the trials. Examining the rows of the table for 5 to 19 thefts shows a total of 914 cases. Thus, in 90% of the trials, at most 19 thefts were required.

Let's see if we can attack the problem analytically. It is easy to find the expected number of thefts to obtain the first car. It will be one. Now let's jump ahead a little bit. Suppose that we have already had two different cars stolen. We wish to determine the expected number of thefts to get a third car different from the first two. There are three cars we haven't

seen yet. Let's call it a *success* if a theft results in one of these three cars. For any randomly chosen theft, the probability of a success is 3/5. As we saw earlier in our section on expected value (Theorem 2), the expected number of trials for a success is 5/3.

What we have just done in finding the expected number of thefts to find the third car can be imitated for each of the other stages as well. If we have three different cars and are searching for the fourth, then either of the two remaining cars will do. There is a probability of 2/5 that any single theft will yield a new car. The expected number of thefts for the fourth car is 5/2. This reasoning holds equally well for other numbers of cars. We summarize the results in the following table.

Car	Probability of success	Expected number of thefts to get this car
First	1	1
Second	4/5	5/4
Third	3/5	5/3
Fourth	2/5	5/2
Fifth	1/5	5

Putting this information together with the definition of expected value, we see that the expected number of thefts to get all 5 cars is

$$1 + \left(\tfrac{5}{4}\right) + \left(\tfrac{5}{3}\right) + \left(\tfrac{5}{2}\right) + 5 = 11\tfrac{5}{12} = 11.417.$$

Since there is one theft per year for any group of five cars, we conclude that on the average we would expect a period of about 11 1/2 years before everyone in the neighborhood is the victim of an automobile theft.

The Reliability of Warning Devices Problems

Our newspapers and television broadcasts often present us with graphic images of the consequences of technological failures: the Challenger explosion, the Three Mile Island nuclear plant accident, the release of poisonous gases from a chemical plant in India, the midair collision of two passenger airplanes, to name but a few.

The Promise and Perils of Warning Devices

In an effort to prevent such disasters, engineers often install warning devices inside of complex technological artifacts. The warning devices should alert the human workers to take emergency action.

Consider, for example, a nuclear reactor plant generating electricity for home and commercial uses. A warning device should give a clear alarm if it detects an unacceptably high level of radiation. The human response would then be to shut down the plant and begin a careful inspection of the reactor.

The United States and Russia both maintain systems to warn if a massive missile strike has been launched by the "other side." Some military experts have argued that there is insufficient time to alert the American or Russian president and wait for a decision; the only "reasonable" response is an automatic launching of a retaliatory strike. Thus they call for an automation of the system; when the warning bell rings, the system responds without human intervention.

In both of these cases—and a host of similarly structured situations—there are serious consequences if the warning device does not function properly. The warning device could fail by not sounding its warning when there is excessive radiation or there has been an enemy missile strike, or by falsely issuing a warning when there is no danger.

Here are some other examples of real-world situations in which warning devices play an important role:

1.   home smoke detectors;

2.   devices on board commercial aircraft that alert pilots if another plane is on a collision course;

3.   National Weather Service declarations of hurricane or tornado alerts and warnings;

4.   radar detectors on board trains which warn of obstacles in the path of these high-speed vehicles.

As in the case of our HIV-virus screening problem, we are faced with a mechanism that may give false positives or false negatives. One solution calls for installing several copies of the warning device in the system. If the social and economic costs of a malfunctioning warning device are high *and* the cost of multiple warning devices is low, then using several devices could be effective *if* we can show that this duplication increases *reliability*.

The Two-Device Situation

Let's begin with a simple example. Suppose that a single warning device functions properly 90% of the time. Then, the device will alert us about 90 out of 100 times when a true danger is present. Similarly, it will falsely signal in only about 10 of 100 times when there is no true danger. Will installing additional warning devices enable us to improve both these probabilities? We would ideally like to raise the probability of acting appropriately if there is a true danger and lower the probability of acting inappropriately if there is no danger.

Suppose we install two copies, A and B, of this device. We'll assume that the two devices operate independently of each other. Then, in the case of true danger, the probability that both A and B alert us is $(.9)(.9) = .81$. The following table summarizes the possible cases.

The Monte Carlo Method

Large-scale computer simulations of complicated deterministic and probabilistic models are widely used today. They help us gain better understanding of the behavior of systems whose dynamics we understand well enough to generate some governing equations whose solutions we cannot compute directly.

One of the first serious attempts to apply computer simulation of probabilities to study complex phenomena arose during the development of the atomic bomb in World War II.

Physicists working at the Los Alamos Scientific Laboratory needed to discover how far neutrons would travel through different materials. They understood the basic laws governing the neutron's movement and had gathered a great deal of appropriate data. They knew the average distance a neutron would travel before it collided with an atomic nucleus, the probabilities that the neutron would be absorbed by the nucleus or would bounce off, and the amount of energy a neutron was likely to lose after a collision. What they lacked was equations whose solutions could be calculated straightforwardly to answer their questions.

Two of the mathematicians, John von Neumann and Stanislaus Ulam, proposed solving the problem by modeling it by chance devices on a computer. von Neumann chose the code name "Monte Carlo" for this secret project because it suggested the randomness of casino gambling devices. The name has stuck, and such methods of simulation are still known as *Monte Carlo methods*.

The problem of generating "random numbers" on a computer remains. The goal is to produce a very large stream of numbers in a rapid, replicable manner in such a way that each number has an equally likely chance of appearing next but with no discernible patterns in the stream. The methods used are completely deterministic, but good methods produce sequences of numbers that behave as if they were truly random. In attempts to achieve the goal, many insights into the concept of "randomness" have been obtained but numerous questions still persist. The generation of "random numbers" remains an interesting field of current research.

	True danger	No real danger
A warns, B warns	(.9)(.9) = .81	(.1)(.1) = .01
A warns, B silent	(.9)(.1) = .09	(.1)(.9) = .09
A silent, B warns	(.1)(.9) = .09	(.9)(.1) = .09
A silent, B silent	(.1)(.1) = .01	(.9)(.9) = .81

What should our policy be? There are basically two policy decision rules we could follow:

1. Act as if there is a real danger only when both A and B sound the signal.

If there is a true danger, we will act appropriately 81% of the time. If there is no real danger, then we will take inappropriate action only 1% of the time.

2. Act as if there is a real danger if at least one of the devices sounds the signal.

If there is a true danger, the probability that at least one of the devices works correctly is .99 (that is, .81 + .09 + .09). If there is no real danger, then we will take unnecessary actions 19% of the time.

Notice that the first policy decreases the likelihood of correctly identifying a real danger while increasing the chances of correctly ignoring a false danger signal. The second policy has the reverse effect. It makes it more likely that we will recognize an actual danger, but it also boosts the chances that we will assume a dangerous situation when one does not exist.

The Three-Device Situation

Going from one warning device to two hasn't solved our problem. We have not yet found a way to improve both probabilities simultaneously. But consider what would happen if we installed three devices (A, B, and C), each with the same individual reliability and operating independently of the others.

	True danger	No real danger
A warns, B warns, C warns	(.9)(.9)(.9) = .729	.001
A warns, B warns, C silent	(.9)(.9)(.1) = .081	.009
A warns, B silent, C warns	(.9)(.1)(.9) = .081	.009
A warns, B silent, C silent	(.9)(.1)(.1) = .009	.081
A silent, B warns, C warns	(.1)(.9)(.9) = .081	.009
A silent, B warns, C silent	(.1)(.9)(.1) = .009	.081
A silent, B silent, C warns	(.1)(.1)(.9) = .009	.081
A silent, B silent, C silent	(.1)(.1)(.1) = .001	.729

Here we can formulate three decision-making policies:

1. Assume there is danger only if all three devices sound the alarm.

Where this is a real danger, this will be detected with probability .729. In the cases where there is no real danger, we will falsely believe there is danger with a probability of only .001. Policy 1 does lower the probability of acting inappropriately if there is no real danger, but it does not raise the probability of getting a warning when there is a true danger.

2. Assume there is danger if at least one device sounds.

Under this decision policy, we will make the right decision in 999 of 1000 cases when there is a true danger, but the probability of our assuming there is a risk when no danger is present jumps to .271. This policy does better than the single-warning device if there is a danger but much worse if there is no danger.

3. Assume there is a danger if a majority of the devices (2 or 3) sound a warning.

When there is a true danger, this policy recognizes it with probability .971 (that is, .729 + .081 + .081 + .081). When no danger is present, we falsely assume it is with probability .028.

The *majority-rule* policy achieves our goal. It improves upon the chances of acting correctly in both the true danger and false danger situations. Compared with having a single-warning device, using a two-out-of-three-or-better system boosts the chances of making the correct judgment in a true danger situation from .9 to .971. It also decreases the probability of making the wrong judgment when no danger is present from .1 to .028.

Exercises for Section 8.6

1. Would you conclude that gender bias was involved in the witchcraft situation if three men had been convicted? five men?

2. There were 60 finalists for a national fellowship program awarding 35 prizes. Of the finalists, 15 were women and 45 were men. Only 5 of the women received prizes, while 30 of the men were winners. Is there sufficient evidence here to conclude that gender played a role in the selection of the winners?

3. During an 11-year period, 870 people from one Texas county were summoned as potential grand jurors. Mexican-Americans made up 79% of the county's population, but only 339 Mexican-Americans (39%) were selected for grand-jury duty. What calculations would you make to determine if the small number of Mexican-Americans was due to chance? The U.S. Supreme Court dealt with this problem in the 1977 case *Castaneda* v. *Partida* (430 U.S. 482, 51 L. Ed. 2d 498). The Court accepted arguments based on probability models that grand-jury selection was not free of discrimination against Mexican-Americans.

4. Your favorite fast-food restaurant, Cramps, has announced a new giveaway. Inside each of its hamburgers is a plastic letter: C, R, A, M, P, or S. The first person who collects all six letters will win a prize of $1000. If equal numbers of each of the letters are inserted at random into the sandwiches, how many hamburgers would you expect to buy before obtaining all six letters? If a substantial number of people each purchase this many burgers and no one has discovered a full set of letters, what conclusions can you draw about the distribution of the letters?

5. Your intensive, beginning Russian class has 15 students. To keep students on their toes, your instructor rapidly fires questions randomly at students. Determine each of the following:
   (a) the probability that you will be asked the first two questions
   (b) the probability that after seven questions you still have not been asked one
   (c) the expected number of questions asked before everyone in class has been asked at least one
   (d) the expected number of questions you will have to answer in a 50-minute class if questions are asked every 20 seconds

6. On March 30, 1981, a young college dropout, John W. Hinckley, Jr., attempted to assassinate President Ronald Reagan. He wounded Reagan, his press secretary, and two security men. At his trial in 1982, Hinckley pleaded not guilty by reason of insanity. Among the 18 physicians who testified was Dr. Daniel R. Weinberger. He told the court that when persons diagnosed as schizophrenic are given CAT scans, the scans show brain atrophy in 30% of the cases while only 2% of the scans on normal people display atrophy. Hinckley's defense lawyer sought to introduce Hinckley's CAT scan as evidence. The defense argued that since Hinckley's scan showed

atrophy it was more likely that he suffered from mental illness. The incidence of schizophrenia in the United States is approximately 1.5%. Suppose a person is chosen at random and given a CAT scan. Let S be the event "the person is schizophrenic" and A the event "the scan shows brain atrophy."

    (a)   Show that $\Pr(A \mid S) = 0.30$ and $\Pr(A \mid S') = 0.02$.

    (b)   Show that the a priori estimate of schizophrenia is 0.015.

    (c)   Use Bayes' theorem to show that the a posteriori estimate $\Pr(S \mid A)$ is 0.186.

    (d)   How much would the result of (c) convince you, as a juror, of Hinckley's likely insanity?

    (e)   Since Hinckley was not a randomly chosen person (he did in fact shoot the President), does it make sense to use .015 as the a priori estimate of his being schizophrenic?

    (f)   Let p be the a priori estimate of Hinckley's schizophrenia before the CAT-scan evidence is introduced. Determine the a posteriori probability as a function of p. [If $p = .1$, for example, then $\Pr(S \mid A) = .63$.]

For additional information, see A. Barnett, I. Greenberg, and R. Machol, "Hinckley and the chemical bath," *Interfaces,* 14(1984):48–52.

7. Analyze the two- and three-device situations if a single-warning device functions properly 95% of the time.

8. A warning device will sound its alarm with probability .99 when there is danger present but will falsely ring 3% of the time when there is no danger. How does the reliability change if we adopt a majority-rule strategy based on three such devices?

9. Returning to our original situation with a single device operating correctly 90% of the time, determine the improvement in reliability if we install 5 of these devices and make our decision using a majority rule.

10. Is majority rule the optimal strategy to follow if we have five identical devices?

11. Is there an optimal strategy if we have four identical devices?

12. A radio talk show host's program is heard in New York, New Jersey, and Connecticut. The number of calls he receives from each of the states is proportional to the population of the states. The estimated populations, in millions are 18.5 (New York), 8.1 (New Jersey), and 3.6 (Connecticut).

    Determine

    (a)   the probability that a randomly selected call is from New York

    (b)   the probability that the first two calls of the day are from New Jersey

    (c)   the probability that neither of the first two calls is from Connecticut

    (d)   the expected number of calls until all three states are represented if the calls are accepted at random

13. (**Writing Project**) Computer simulations of probabilistic models begin with the generation of random numbers. To simulate the flip of a fair coin, for example, the computer selects a random number between 0 and 1 and declares the result a Head if the number lies between 0 and 1/2 and a Tail otherwise. But computers are inherently deterministic devices; they cannot perform "random acts." Investigate algorithms that have been developed to generate sequences of real numbers that appear to be random. Possible references include Averill M. Law and W. David Kelton (1982), *Simulation Modeling and Analysis,* New York: McGraw-Hill, and Paul Bratley, Bennett L. Fox, and Linus E. Schrage (1983), *A Guide to Simulation,* New York: Springer-Verlag.

SECTION 8.7 *Decision Theory*

What Is Decision Theory?

Decision theory, which is also called *decision analysis,* is a collection of mathematical models and tools developed in the twentieth century to assist people in choosing among alternative actions in complex situations. In many cases the options we have are clearly defined and we also know the collection of consequences that may eventually occur, but we are uncertain exactly which of these outcomes will happen. Between the moment we act and the time the process concludes, there may be other forces affecting the action, forces over which we have no control. We may, however, have some information about the relative likelihood of each outcome flowing from each action. This may sound a little vague and abstract, so let's look at a simple example or two.

It's a warm, humid morning and you're about to leave your apartment to walk down to the campus six blocks away. You're worried that it might rain, so you consider taking your raincoat. Although your raincoat is great at protecting you from downpours and keeping you nice and dry, it's rather bulky and ugly and you'd prefer not to take it along if you don't really need it. You have two possible actions: take your raincoat with you or leave your raincoat at home. There is an unknown "state of nature": It may or may not rain. There are four possible outcomes, depending on your action and the state of nature:

A: Take raincoat, it rains

B: Take raincoat, it doesn't rain

C: Leave raincoat, it rains

D: Leave raincoat, it doesn't rain

You will obviously be happier with some outcomes than with others. The worst possibility is that you decide to leave the raincoat in your apartment and it pours as you walk to campus, drenching you to the skin, soaking your clothes, and perhaps ruining your books. The happiest eventuality occurs when you leave the raincoat behind and no rain comes down. Then you are both dry and free of the burden of lugging the coat. The other two outcomes (when you take the raincoat) are somewhere in between; once you've gone to the bother of wearing the coat, you'd probably have a happier sense of vindication if it did rain than if it didn't.

In his provocative book, *The Foundations of Statistics* (second edition, New York: Dover, 1972), which laid the foundations of modern decision theory, Leonard J. Savage presents another intriguing example. Your spouse has just broken five good eggs into a bowl when you come in and volunteer to finish making the omelet. A sixth egg, which for some reason must be either used for the omelet or wasted altogether, lies unbroken beside the bowl. You must decide what to do with this unbroken egg.

You have three possible courses of action: break the egg into the bowl containing the other five; break it into a saucer for inspection; or throw it away without inspection. Depending on the state of the egg, each of these acts will have some consequence of concern to you. We illustrate the situation in a table.

	States of nature	
	Egg is good	*Egg is rotten*
Break into bowl	Six-egg omelet	No omelet, and five good eggs destroyed
Break into saucer	Six-egg omelet, and a saucer to wash	Five-egg omelet, and a saucer to wash
Throw away	Five-egg omelet, and one good egg destroyed	Five-egg omelet

In both of these examples where you have to make a decision, there is an unknown state of nature. To make progress toward a rational basis for deciding what course of action to take, you need some knowledge of the probabilities associated with the possible states of nature. You also need to know how you assess the values of the possible outcomes.

Modern decision analysis helps illuminate many complicated situations, where a very large number of possible actions and a multitude of different states of nature may exist, where decision makers must balance the conflicting demands of maximizing expectation and minimizing risk, where decisions must be taken at various points in a multistep process. But one guiding principle is always present: Determine as best you can the probabilities at every stage for the various outcomes as well as the value, or *utility,* of these outcomes; once you've done this, calculate the expected utility of each course of action and consider the strategy that yields the largest expected utility. This is a familiar theme in our study of probability theory. We have random variables—utilities you may receive under various courses of action—whose expected value we wish to compute. After using their probability distributions to compute the expected values, we select the option that has the greatest expected value.

We'll illustrate this process with an example from business.

A Manufacturing Problem

The top executives at a manufacturing company are hopeful that demand for the product the firm produces will increase. The factory is currently operating at full capacity with regular employees working 40 hours a week. The executives are considering meeting the anticipated new demand either by putting workers on overtime shifts or by acquiring new production equipment. Market analysts determine that there is a 60% chance of a 15% increase in demand for the product, but they warn that the economy may deteriorate, as there is a 40% chance that the demand will actually decrease by 5%. The accountants are asked to determine the net gain to the company under the proposed actions and also to construct figures reflecting the situation if the company maintains the current levels of equipment and personnel hours.

We summarize the information in the following table, which shows each action, each state of nature, the probability of each of these states, and the net value to the company for each action under each state.

	States of nature	
Action	5% drop (probability = .4)	15% increase (probability = .6)
Maintain current levels	$300,000	$340,000
Go to overtime	300,000	420,000
Add new equipment	260,000	440,000

Decision analysts often find it useful to present the information in the form of a *decision tree,* which is quite similar to the tree diagrams we have already seen. The decision tree for this manufacturing problem looks like the following illustration:

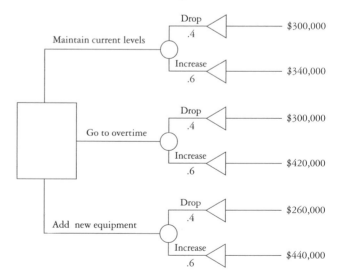

A decision tree is read from left to right in the same manner as other tree diagrams. The first component on the tree is a square *decision node,* marking the initial decision awaiting us. From this node lead decision branches, one for each possible action that could be taken at the decision node. If the ultimate outcome of the action depends on the state of nature, then the decision branch leads to a circular *uncertainty node.* From each uncertainty node we find *outcome branches,* each marked with its probability of occurrence, leading to triangular *terminal nodes.* At the end of each terminal node is a *consequence branch* showing the value of that occurrence. Decision trees can become very complex, modeling situations where a sequence of decisions needs to be made.

We can use either the table or the decision tree to calculate the expected value of each course of action:

expected value(current levels)
= ($300,000)(.4) + ($340,000)(.6) = $324,000,

expected value(overtime)
= ($300,000)(.4) + ($420,000)(.6) = $372,000,

expected value(add new equipment)
= ($260,000)(.4) + ($440,000)(.6) = $368,000.

The action with the highest expected value ($372,000) is to have the employees work overtime.

David and Goliath

The famous story of the battle between David and Goliath is recounted in the Old Testament (1 Samuel 17). The armies of the Israelites and the Philistines face each other, each on a mountain side with a valley between them, prepared to battle.

From the camp of the Philistines strides forth Goliath, a fearsome warrior. He stands 10 feet tall, wears a coat of armor weighing 150 pounds, and carries a spear with a shaft "like a weaver's beam." Goliath challenges the Israelites to decide the outcome of the battle by a fight between two men. Goliath will take on any person from the ranks of Israel. They will fight to the death, and the nation of the loser will becomes servants to the nation of the winner.

The men of Israel are "dismayed and greatly afraid" of this proposal. No one will venture forth to take on Goliath, who repeats his challenge twice morning and evening for 40 days. At this point David arrives on the scene, bringing provisions for three of his older brothers, who are members of the Israelite army. David is very interested in Goliath's proposition and inquires among the soldiers about the rewards that might come from defeating Goliath. He learns that "the man who kills Goliath, the king will enrich with great riches, and will give him his daughter, and make his father's house free in Israel."

David faces a decision. Should he fight Goliath? The outcomes of a decision to fight have been graphically portrayed: wealth, a princess bride, and great prestige if he wins; death if he loses.

Let p be the probability that David will beat Goliath, and so $1 - p$ is the probability that Goliath will kill David. To use decision theory, we need to attach numerical values to the possible outcomes. One standard approach in such a situation is to assign the value 0 to the worst possible outcome and the value 1 to the best outcome. Then we determine intermediate values, usually called *utilities,* to all the other outcomes. In this case the worst outcome for David is death and the best (riches, bride, and all those other goodies) is killing Goliath. Only one intermediate outcome is relevant here: If David decides not to fight, then he avoids the risk of death but forsakes the joys of victory, but at least he lives to try to get his fame and fortune in another way on another day. Let's denote by u the utility to David of this outcome; it is some number between 0 and 1 yet to be determined.

Now we can compute the expected utility to David of each course of action. The choice to fight has payoffs 1 with probability p and 0 with probability $1 - p$. The expected value is $1p + 0(1 - p) = p$. On the other hand, the expected value of the option don't fight is $u \times 1 = u$. Our decision-theoretic analysis of the situation leads to this advice: David should fight if and only if $p > u$. If David believes, for example, that he has a .75 chance of beating Goliath and he rates the utility of declining the giant's challenge at .67, then decision theorists would advise David to enter the fight. The decision tree for this analysis follows.

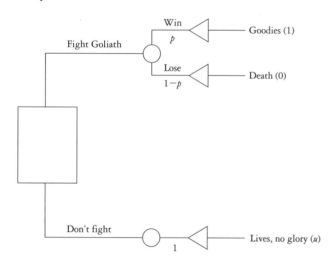

Douglas's Decision

In the Spring of 1992, James Douglas needed to make a decision. Douglas was Vermont's Secretary of State, an elected position he had held for several terms. Douglas had served as a member of the state House of Representatives and as an administrative assistant to the governor. He was a well-known Re-

publican in Vermont and was a popular incumbent who had always won election and reelection by wide margins. His present term of office expired at the end of 1992. If he decided to seek another term as Secretary of State, he had an extremely strong chance of being returned again to that office by the voters.

Many colleagues had been urging Jim to run for a higher office, however, such as the governorship or a spot in Congress. There was an election for one of Vermont's two U.S. Senate seats scheduled for November 1992. The incumbent, Patrick Leahy, had already announced that he would seek reelection for a fourth term. Leahy, a Democrat, was also well known and admired in the state. Six years earlier, the last time he faced the voters, Leahy was returned to office by a wide margin.

If Douglas challenged Leahy in the election, he would have to give up the possibility of also running for another term as Secretary of State. The prize, a seat in the Senate in Washington, was a big one. It would be tough to get, but the electorate was showing increasing signs of frustration with incumbents, and the voters might be ripe for a change.

We'll display Douglas's decision problem in the form of a decision tree.

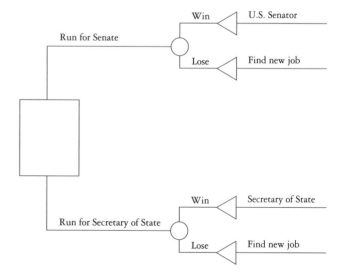

To apply decision theory to this problem, we need first to determine the probability estimates for Douglas's chances of winning the Senate election and winning the Secretary of State contest. In this setting the interpretation of "probability" as *relative frequency* may seem to make little sense. We cannot conduct an experiment of running an election between Leahy and Douglas a large number of times, keeping track of the relative number of times Douglas wins. The interpretation of "probability" in this context is often described as *subjective probability,* the degree of belief one has in the likelihood of the outcome. The subjective probability might be based on public opinion polls of the voters' preferences before the election or on some political analysis of similar races in Vermont and other states. In any event, decision theory rests on the assumption that it is possible consistently to assign probabilities to the various possible outcomes of an action so that, for any two events A and B, $\Pr(A) > \Pr(B)$ exactly when the decision maker believes that A is more likely to occur than B and $\Pr(A) = \Pr(B)$ exactly when the decision makes believes they are equally likely to occur.

The second step is to give utilities to the various consequences. We denote by $U(A)$ the utility of outcome A. Then we require that utilities be assigned in a consistent way so that, for any two outcomes A and B, we have $U(A) > U(B)$ exactly when the decision maker prefers A to B and that $U(A) = U(B)$ exactly when he has no preference between the two outcomes.

As we noted earlier, a standard approach is to make the utility of the worst outcome 0 and the utility of the most preferred outcome 1 and then to come up with intermediate values for the other outcomes. In our election example, we presume that of the four possible consequences, Douglas would most prefer being elected to the U.S. Senate and that his least desired alternative is entering and losing the election as Secretary of State. In both cases, where he loses the election, he needs to find a new job, but the ignominy of losing a reelection bid for his old office would be greater than failing to upset a popular incumbent.

How do decision theorists determine utilities? One common procedure is to develop a "reference gamble." We start with the best and worst possible outcomes of the decision we would make. We want to develop a utility for some intermediate outcomes A. Now consider a gamble where the possible outcomes are "Best" and "Worst." You get to choose between taking the gamble and getting outcome A for certain. Which would you take? Your answer would probably be, "It all depends on the chances of getting Best as the result of the gamble. If the probability of obtaining Best is very high, I would select the gamble, but if there's not much hope of winning Best, I'd stick with A as a sure thing."

Denote a gamble that has probability p of outcome Best and $1 - p$ of outcome Worst by

$$[p \text{ Best}, (1 - p) \text{ Worst}].$$

For values of p close to 1, you would choose the gamble over the certainty of A. For values of p close to 0, you would pick A. Hence if we start with a relatively large value for p and gradually lower it, we would eventually find some value $p^\#$ so that you would feel indifferent between choosing A and taking the gamble

$$[p^\# \text{ Best}, (1 - p^\#) \text{ Worst}].$$

The decision theorist would then assign $p^\#$ as the utility to you of outcome A.

For a more detailed discussion of this procedure and many other topics and applications in decision analysis, see Robert D. Behn and James W. Vaupel

(1982), *Quick Analysis for Busy Decision Makers*. New York: Basic Books.

Sensitivity Analysis

We seldom know the exact probabilities of the various states of nature. Usually we have some approximate values, and we may have the option of obtaining more precise figures for some additional cost. Whether it's worth the price to hone our estimates more sharply is a concern that decision analysts study under the topic of "value of information."

In a complex decision setting, there may be many different probabilities to assess. Some may be more important than others in the sense that small changes in the values of these probabilities may cause great changes in the relative orderings of the various expected values and hence may change our optimal strategies. Studying the effect of small changes in selected values in a problem on the proposed solution is called *sensitivity analysis*.

We'll illustrate a very simple case of sensitivity analysis for our manufacturing problem. In that problem we assumed that the probability of a 15% increase in demand for a product was .6. Suppose that the probability of this increase in demand is p, so that the likelihood of a 5% drop is $1 - p$. Then expected values of the possible courses of action (in tens of thousands of dollars) become

$$\text{expected value(current levels)} = (30)(1 - p) + (34)(p)$$
$$= 30 + 4p,$$

$$\text{expected value(overtime)} = (30)(1 - p) + (42)(p)$$
$$= 30 + 12p,$$

$$\text{expected value(add new equipment)} =$$
$$(26)(1 - p) + (44)(p) = 26 + 18p.$$

Comparing expected values, we see that the overtime option is always better than maintaining current lev-

els (since $30 + 12p > 30 + 4p$, for all p between 0 and 1). The action to add new equipment is superior to that of overtime for all p such that

$$26 + 18p > 30 + 12p$$

or

$$p > \tfrac{2}{3}.$$

Thus, if we think that p is substantially below .67 (our current estimate was .6), we should elect the overtime option; but if we believe that it is much above this level, then we should opt for the new equipment choice.

The Job Interview Problem

Recall the job interview problem that we posed at the beginning of this chapter: You obtained interviews with three prospective employers. Each has different positions open: Fair, Good, and Excellent, with annual salaries of $25,000, $30,000, and $40,000, respectively. You estimate that your chances of being offered a Fair job are 4 in 10 while the prospects for Good and Excellent positions are 3 in 10 and 2 in 10, respectively. There is a 1 in 10 chance that you won't be given an offer. Each company requires you to accept or reject a job offer at the conclusion of the interview. What strategy should you follow? Should you reject, for example, a Good offer from the first company and gamble that one of the two remaining offers will be better?

If all of your job prospects are tied up in these three interviews, then you face decision points at the conclusion of the first two interviews: Do you take the job offered at that point or go on to the next interview? Some advice is easy to give, as it requires little analysis: If you are offered an Excellent job at any point, take it and cancel the remaining interviews. You can only do worse if you keep trying. At the other extreme, if an interview results in No Job, then you lose nothing by continuing on the next interview: You can only do better. The quandary occurs if interview 1 or 2 finishes with an offer of a Good or Fair job. If you accept one of these, then you give up the possibility of a more lucrative opportunity at a later stage. If you turn it down, then you run the risk that later interviews may finish with less desirable outcomes. Our analysis of the chances of various jobs suggests that we assume Pr(Fair job) = .4, Pr(Good job) = .3, Pr(Excellent job) = .2, and Pr(No job) = .1.

The decision analyst would recommend basing a decision on what to do at the first interview by considering the expected value of each course of action you might take. This is difficult to do in this problem because one course of action (go on to interview 2) has an uncertain outcome since we will have another decision to make at the end of that interview. This problem illustrates a common feature of complex decision making: The outcomes of immediate decisions cannot be evaluated until future decisions are made. There is a way out of this difficulty: Analyze the future decisions first. This is sometimes called a *future-first* approach or a *folding back* or *backward induction of dynamic programming.*

Let's consider the situation if you reach the third interview without accepting a position. Recall the possible outcomes (with their corresponding annual salaries) and their respective probabilities:

Outcome	Probability
Fair: $25,000	.4
Good: $30,000	.3
Excellent: $40,000	.2
None: $0	.1

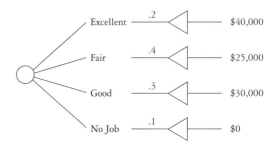

The expected value of the job from this employer is easily calculated as

$$(\$25,000)(.4) + (\$30,000)(.3) + (\$40,000)(.2) + (\$0)(.1) = \$27,000.$$

Knowing the expected value of the third interview, we can work backward to decide on the best action to take at the conclusion of the *second* interview. We already know that we'll accept an Excellent offer and that we will go on to interview 3 if we get no offer. If we're offered a Fair job, then we must choose between taking that job (expected value = $25,000) and trying our luck on the third interview (expected value = $27,000). Since the latter choice has a larger expected value, that's the one we should take. On the other hand, if employer 2 offers a Good job, then that has a higher expected value ($30,000 vs. $27,000) so we should accept and cancel the third interview.

To summarize our optimal strategy at Interview 2: Accept a Good or Excellent job, reject a Fair one.

What is the expected value of the second interview under this strategy?

Outcome of interview 2	Value	Probability
Fair: Go to interview 3	$27,000	.4
Good: Take it	$30,000	.3
Excellent: Take it	$40,000	.2
None: Go to interview 3	$27,000	.1

The expected value is

$$(\$27,000)(.4) + (\$30,000)(.3) + (\$40,000)(.2)$$
$$+ (\$27,000)(.1) = \$30,500.$$

We back up now to the first interview. If we're presented with a Fair job, then we face the choice of an expected value of $25,000 if we take it or an expectation of $30,500 if we don't. We go for the higher figure and reject the Fair offer. Should the offer be for a Good job, then our alternatives are $30,000 for that job compared with an expected value of $30,500. To maximize our expected value, we should also turn down the Good position at this stage.

Thus, our best strategy at the first interview is to take the job only if it is an Excellent one; otherwise, move on to interview 2.

Outcome of interview 1	Value	Probability
Fair: Go to interview 2	$30,500	.4
Good: Go to interview 2	$30,500	.3
Excellent: Take it	$40,000	.2
None: Go to interview 2	$30,500	.1

Our optimal overall strategy for the interview situation is now clear: At the first interview, accept only an Excellent offer; otherwise go on to the second interview. At the second interview, accept a Good or an Excellent offer; otherwise, go on to the third interview, where you accept any job that is offered.

The expected value associated with this strategy is

$$(\$30,500)(.4) + (\$30,500)(.3) + (\$40,000)(.2)$$
$$+ (\$30,500)(.1) = \$32,400.$$

We have introduced only the most elementary aspects of decision theory in this section. It should be apparent, however, that even these simple ideas can be used productively to analyze complicated situations and to help lead us to prudent choices among alternative strategies. We conclude with an assessment by Simon French, a noted British decision theorist:

When it was first adopted, . . . decision theory provided such an idealized simplified view of decision making that it was difficult to believe that the theory had anything to offer practice in the complex, ill defined world of real decision making. Gradually it became apparent that it has. It can offer structure: structure in which to explore and analyze a problem; structure in which the decision makers can articulate their views and communicate with each other; structure in which parts of a problem can be analyzed independently of other parts: and, in general, structure in which to organize thought.

1. Suppose Douglas estimates the probability of winning the Senate election and that of the Secretary of State race to be .42 and .88, respectively. He rates the utility of the Secretary of State position to be .45 and the utility of finding a new job after failing to unseat Senator Leahy to be .20. On the basis of maximizing expected utility, what should be his decision?

2. North Carolina Attorney General Rufus Edmisten was considering seeking the Democratic senatorial nomination to run against the incumbent Republican Jesse Helms. Edmisten faces a compound decision: If he won the primary election and got the nomination, should he resign his state office? There was public sentiment against politicians remaining in one office while seeking another office, so Edmisten's chances of defeating Helms were slighter if he didn't resign.

Suppose Edmisten estimated his probability of winning the primary to be .75 and that his chances of winning the general election against Helms were 0.4 if he resigned the attorney generalship and 0.3 if he didn't resign. Suppose further that Edmisten has the following utilities associated with the possible outcomes:

He becomes U.S. Senator	1
He wins primary, resigns AG post, loses senate election	0
He loses primary, remains AG but tarnished by primary defeat	.45
He wins primary, doesn't resign, loses senate election	.40

Draw a decision tree that reflects three strategy alternatives facing Edmisten:

A: Seek nomination and, if nominated, resign
B: Seek nomination and if nominated, do not resign

C: Do not seek nomination

Which alternative maximizes Edmisten's expected utility?

For more information about this problem and alternative ways of forming Edmisten's decision tree, see Robert D. Behn and James W. Vaupel (1982), *Quick Analysis for Busy Decision Makers,* New York: Basic Books, from which this exercise was adapted.

3. For each of the following strategies in the job interview problem, determine
   (a) The probabilities of each of the following outcomes if you follow that strategy:
      —You get a Fair Job
      —You get a Good Job
      —You get an Excellent Job
      —You get no job
   (b) The *expected value* of the salary you would receive

Strategy 1: Accept an Excellent offer at interview 1; otherwise, continue to interview 2.
Accept an Excellent or Good offer at interview 2; otherwise continue to interview 3.
Accept any offer at interview 3.

Strategy 2: Accept an Excellent or Good offer at interview 1; otherwise continue to interview 2.
Accept an Excellent or Good offer at Interview 2; otherwise continue to interview 3.
Accept any job offer at interview 3.

Strategy 3: Accept an Excellent or Good offer at interview 1; otherwise continue to interview 2.
Accept any job offer at interview 2; otherwise continue to interview 3.
Accept any job offer at interview 3.

Strategy 4: Make up your own strategy and investigate it.

4. Rework part (b) of the Exercise 3 if the salary of an Excellent job is $33,000 rather than $35,000.

5. Determine the optimal strategy in the job interview problem if the salary for an Excellent job is $33,000 but all the other data remains the same.

6. Consider Strategy A ("accept only Excellent jobs at interviews 1 and 2") and Strategy B ("accept the first job offered").
   (a) Verify that Strategy A has a higher expected salary than Strategy B if an Excellent job pays $35,000, but that Strategy B has a higher expected salary than Strategy A if an Excellent job pays $33,000, all other factors being equal.
   (b) Determine the salary for an Excellent job which would produce equal expected salaries for strategies A and B.

7. For the original job interview problem, suppose the salaries for Excellent, Good, and Fair jobs remain the same ($35,000, $30,000, and $25,000, respectively) but the chances of obtaining each of these jobs change to:

Excellent	1 in 6
Good	1 in 4
Fair	1 in 3
No Job	1 in 4

Determine the optimal strategy which maximizes the expected salary.

8. A fledgling manufacturer has just signed a contract that will allow him to market his toasters in Korea, where he expects to sell 1500 units a month. He also hopes to negotiate a similar deal with China; if successful, he will be able to sell an additional 5500 toasters monthly. To obtain these contracts, the manufacturer is willing to sell the toasters at a price that will earn a profit of $1 on each toaster sold.

The manufacturer must now start producing the toasters for Korea before the Chinese make a final decision. He has to decide between leasing a small-capacity machine and a large-capacity one. The smaller unit can produce 2000 toasters a month and costs $12,000 a year to lease. The larger machine can make up to 10,000 toasters each month; its lease costs $40,000 per year.

(a) Show that the manufacturer's net profit for a year if he leases the smaller machine and gets the Chinese contract is

$$\$2000 \times 12 - 12,000 = \$12,000$$

(b) Show that the net yearly profit if he leases the smaller machine and doesn't get the contract from China is $6,000.
(c) Show that the manufacturer loses $22,000 if he leases the larger machine but gets a negative decision from the Chinese.
(d) Determine the yearly profit if the larger machine is leased and the Chinese approve the contract.
(e) Draw a decision tree for the manufacturer's problem.
(f) Determine the expected profit if there is a probability of .5 that the China contract will be approved.
(g) If p is the probability that the Chinese offer the contract, find how large p must be to make leasing the larger machine have a higher expected profit than leasing the smaller machine.
(h) Before making his decision, the manufacturer is approached by a industrial espionage agent who offers to sell him advance knowledge of the Chinese decision for $15,000. Draw a decision tree that includes the option of spending this $15,000. What decision maximizes the expected profit?

9. Gail Kalmowitz sued Brooklyn's Brookdale Hospital, claiming she suffered 22 years of near blindness because the hospital administered excessive oxygen to her

when she was born two months prematurely. While the jury was deliberating its decision, she was offered a $165,000 out-of-court settlement. Construct a decision tree that would help Ms. Kalmowitz decide whether to accept the settlement or gamble that the jury would award her substantially more. Suppose her lawyer advises her that there is a 50% chance the jury will decide the hospital was not responsible for her blindness and award her nothing, a 40% chance that it awards her $200,000, and about a 10% chance that the jury will assess the hospital one million dollars. Should she accept the settlement? Suppose her lawyer works on a contingency fee and collects a tenth of whatever money she collects. Should she gamble on the jury in this case? (See "Nearly blind student accepts $165,000, forfeiting $900,000 award from jury," *New York Times,* March 27, 1975. The lawyer's estimates are fictitious).

10. (Adapted from J. W. Smith, *Decision Analysis: A Bayesian Approach,* London: Chapman and Hall, 1988) The federal government offers a particular oil company the option to drill in an Atlantic coast offshore field or in a Pacific coast offshore field but not both. If the company strikes oil in the Atlantic field, then it expects to make a profit of $77 million; the corresponding figure for the larger Pacific field is $195 million. The probabilities that the Atlantic and Pacific fields contain oil are 0.4 and 0.2, respectively; these are independent events. The company is considering one of three decisions:

A: accept the option, pick a field, and start drilling.

B: decline the option to drill in either field.

C: Pick a field, pay for an investigation of it, and then choose option A or B after analyzing the results of the investigation. Note that the company could investigate

the Atlantic field and then decide to drill in the Pacific field.

Option C seems the most prudent but the cost of investigating a field is $6 million and the results are not definitive. If oil is present, then the investigation has a 0.8 probability of detecting it. If there is no oil, then the investigation has a 0.4 probability of a false positive.

Draw a decision tree and determine the action the company should take to maximize its expected profit if the cost of drilling a field is $31 million.

11. (**Writing Project**) The relative frequency interpretation of probability is a useful one but it is sometimes difficult to apply in a particular real-world situation where probabilities must be assessed. There may be no natural "experiment" that is reasonable or possible to repeat a large number of times so that we can count the relative number of times a certain event occurs. For example, what experiment would you do to estimate the probability that a catastrophic earthquake will strike San Francisco in the next decade? There are also situations that have already occurred but we are unsure of the outcome. Here the uncertainty lies within our heads, not in reality. For example, what is the probability that the Boston Red Sox won the 1946 World Series? A different approach to assigning numerical weights for events is necessary. *Subjective probability* is the term used for methods of assigning measures of uncertainty. Decision theory is built on the assumption that it is possible to assign subjective assessments of probabilities in a careful, coherent and useful way. Prepare a report on subjective probability. Two good references with which to begin your investigation are

Robert T. Clemen (1996), *Making Hard Decisions: An Introduction to Decision Analysis,* Belmont: Duxbury Press.

R. L. Winkler (1972). *Introduction to Bayesian Inference and Decision,* New York: Holt, Rinehart and Winston.

Chapter Summary

Probability theory is the branch of mathematics that analyzes the uncertainty of chance events. In this chapter, we introduced some of the fundamental ideas of probability theory in the context of a number of real-world applications.

We began by reviewing the concept of a *mathematical model* as a representation in the language of mathematics of the essential features of a "real" system. We needed to distinguish between *deterministic* models and *probabilistic* models. In a probabilistic model, we assert only that a given number lies within some range or estimate the relative likelihood that a certain event will take place.

Our approach to probability emphasized the *relative frequency* interpretation: the probability of an outcome is the relative fraction of times it occurs in a large number of repetitions of the same trial. We also saw that a *subjective judgment* interpretation may be more useful in some contexts.

The formal definition of a probability measure follows. A *probability space* is a set E, called the *sample space*, together with a probability measure, which is a real-valued function Pr whose domain consists of all subsets of E and which satisfies three rules: $Pr(E) = 1$, $Pr(A) \geq 0$ for every subset A of E, and $Pr(A \cup B) = Pr(A) + Pr(B)$ for every pair of disjoint subsets A and B.

Knowledge that one event has occurred may alter our estimate of the probability that another event has happened. We formalized this observation in the concept of conditional probability. We stressed the use of Bayes Theorems, which provides a means of estimating $Pr(A \mid B)$ from $Pr(B \mid A)$. We applied Bayes Theorem to examine the complexities of diagnostic tests where issues of sensitivity, specificity, and incidence affect the degree to which we are willing to accept a positive result of a diagnostic test as evidence that the underlying condition is actually present.

We also examined the important case of *independent events* with emphasis on a commonly occurring structure, Bernoulli trials, a sequence of repetitions of the same experiment where the outcomes on successive trials are independent. We derived the binomial probability distribution which gives the probability of precisely k successes in n trials.

Associated with many experiments are numerical payoffs attached to the various possible outcomes. The notion of a random variable, a numerical-valued function defined on a sample space, captures this idea. The most important property of a random variable is its expected value, the sum of the products of each numerical outcome multiplied by the probability that the particular outcome occurs. We extended the study of expected value from finite sample spaces to infinite discrete spaces and then to continuous distributions.

We then turned to a more detailed analysis of three probability questions initially posed in our first section and showed how we can apply the tools we developed to solve them.

In our final section, we introduced the rudiments of decision theory, the use of probabilistic models and dynamic programming to select the most optimal strategy in complex decision-making situations with uncertain outcomes.

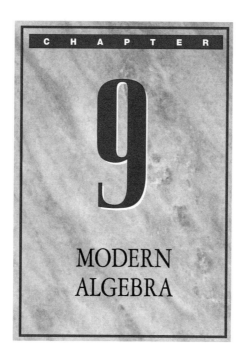

C H A P T E R

9

MODERN
ALGEBRA

Finite Symmetry Groups

Principles of symmetry permeate mathematics, science, engineering, architecture, and art. The symmetry a molecule possesses has a bearing on its chemical properties. The internal structure of crystalline solids consists of symmetrical arrays of atoms or ions. The symmetry an organism has gives a clue to its way of life and may influence its mating prospects (see the Spotlight). Many viruses have a high degree of symmetry. Symmetry is a guiding principle in theoretical physics. Finally, symmetry is intimately linked with our concept of beauty.

Look at the logos in Figures 1 and 2. What comes to mind? The figures in Figure 1 have rotational symmetry, whereas those in Figure 2 have both rotational and reflective symmetry. Notice that the four-sided figure in Figure 1 (third entry in first row) can be rotated by 0°, 90°, 180°, or 270° and coincide with its original position, whereas the four-sided figure in Figure 2 (third entry in the second row) can be rotated by 0°, 90°, 180°, or 270°, as well as reflected (flipped) across vertical, horizontal, or diagonal axes, and coincide with its original position.

The logos in Figures 1 and 2 raise interesting questions. What is the mathematical definition of symmetry? How many kinds of symmetry are there?

Symmetry's Role in Mate Selection

Elk antler
Male elks with the largest and most symmetrical racks of antlers have the largest harems. Elks that lose fights may lose antler symmetry—and females.

Barn swallow tail
Female barn swallows prefer a long-tailed male with a symmetrical wishbone pattern of feathers the same size and color on both sides of the tail.

Scorpion fly
A male scorpion fly with symmetrical wings can be detected not only by sight but by scent. For some reason, there is an association between wing symmetry and hormone signals. Even minute differences count.

Evolutionary biologists have gathered evidence from studies on subjects as diverse as zebra finches, scorpion flies, elk, and humans that creatures appraise the overall worthiness of a potential mate by looking for at least one classic benchmark of beauty: symmetry. The biologists postulate that the choosier partner in a pair—usually the female—seeks in a mate the maximum possible balance between the left and right halves of the body.

Reporting in a recent issue of the journal *Nature,* two researchers found that when they put a variety of colored leg bands on male zebra finches, the females vastly preferred males with symmetrically banded legs over those with bands of different colors on each leg.

Researchers who study elk have determined that males who possess the largest harems of females have the most symmetrical racks.

The researchers hypothesize that a symmetrical body demonstrates to females that the male's central operating systems were all in peak form during important phases of his growth and that the male possesses an immune system capable of resisting parasites. In the case of elk, a symmetrical rack indicates that the male has not lost part of its rack in a fight with another male.

{Adapted from Natalie Angier, "Why birds and bees, too, like good looks," The New York Times (Feb. 8, 1994): C1.}

FIGURE 1 Logos with rotational symmetry.

FIGURE 2 Logos with rotational and reflective symmetry.

Is there a mathematical theory of symmetry? In this section we provide answers to these questions.

DEFINITION

A *plane isometry* is a function from the plane to itself that preserves distances; that is, for any points p and q in the plane and for any plane isometry T, the distance from $T(p)$ to $T(q)$ is the same as the distance from p to q.

A consequence of preserving distance is that size and shape are also preserved. Thus, the image of a tri-

angle is a triangle congruent to the original; the image of a circle is a circle of the same size. Familiar examples of plane isometries are rotations, translations, and reflections. A plane is rotated by turning it by a specific angle about a point called the *center of rotation*.

DEFINITION

A *translation of the plane* is a function that carries all points the same distance in the same direction.

For example, if p and q are points in a plane and T is a translation, then the two vectors joining p to $T(p)$ and q to $T(q)$ have the same length and direction (see Figure 3). A vector of the form $\overrightarrow{pT(p)}$, where p is any point in the plane, is called the *translation vector of T*. (Intuitively, one may think of a translation T with translation vector $\overrightarrow{pT(p)}$ as sliding in the plane so that the point p moves along the vector $\overrightarrow{pT(p)}$ to the point $T(p)$; see Figure 3.) When the plane is coordinatized, a translation has the form $T(x, y) = (x + a, y + b)$.

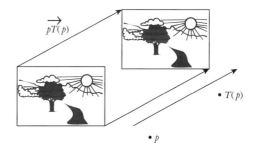

FIGURE 3 Translation taking p to $T(p)$.

DEFINITION

A *reflection across a line L* is the function F that leaves each point of L fixed and takes every point q not on L to the point $F(q)$ so that L is the perpendicular bisector of the line segment joining q and $F(q)$.

For example, the function $F(x, y) = (x, -y)$ is a reflection across the x-axis, and the function $F(x, y) = (y, x)$ is a reflection across the line $y = x$. Other examples are shown in Figure 4.

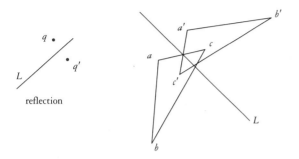

FIGURE 4 Reflections taking point x to point x'.

It is a remarkable fact that there remains only one other kind of plane symmetry.

DEFINITION

A *glide reflection* is a composition of a translation and a reflection across the line containing the translation vector. In symbols, if T is a translation and F is a reflection, then the two composition functions $(T \circ F)(q) = T(F(q))$ and $(F \circ T)(q) = F(T(q))$ are glide reflections.

Geometrically, $T \circ F$ can be thought of as "T following F." In Figure 5, the vector $q\overrightarrow{T}(q)$ gives the direction and the length of the translation and lies on the axis of reflection.

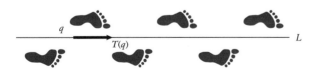

FIGURE 5 Footprints along a line have glide reflection symmetry.

It has been known for more than 150 years that the composition of any of the four types of symmetries we have thus far defined is just one of these four types again. For example, the composition of a rotation and a translation is another rotation (see Figure 6); the composition of two reflections across intersecting lines is a rotation (see Figure 7); the composition of two reflections across parallel lines is a translation (see Figure 8). Rotations and translations preserve orientation, whereas reflections and glide reflections switch orientation. (That is, reflections and glide reflections transform clockwise orientation to counterclockwise and right-handedness to left-handedness. See Figure 9.)

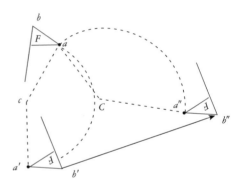

FIGURE 6 The composition of a rotation and a translation is a rotation. The centers of rotations are c and C'. Points a and b are carried to points a' and b' by a rotation. Points a' and b' are carried to points a'' and b'' by a translation.

DEFINITIONS

A *symmetry* of a figure S in a plane is an isometry T of the plane with the property that $T(S) = S$. The *symmetry group* of a plane figure is the set of all plane symmetries of the figure.

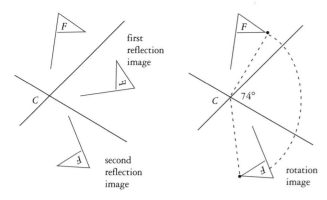

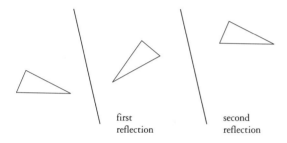

FIGURE 7 The composition of two reflections across intersecting lines is a rotation with center at c.

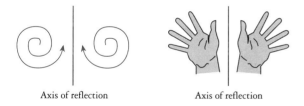

FIGURE 8 The composition of two reflections across parallel lines is a translation.

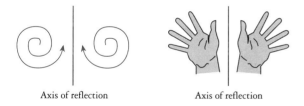

FIGURE 9 Reflected images reverse orientation.

Thus a symmetry of a figure S is a distance-preserving function with the property that the image of S is S itself. It follows from the definition of a symmetry that the composition of two symmetries is a symmetry. Two symmetries X and Y are called *inverses* if their composition is the identity function; that is, if $(X \circ Y)(p) = p$ for all points p in the

plane. The inverse of a rotation of θ degrees (positive angles are measured counterclockwise) about a point p is a rotation of $-\theta°$ (or $360° - \theta°$) about p; a reflection is its own inverse; the inverse of a translation or a glide reflection with translation vector $\mathbf{v}$ is the corresponding symmetry with translation vector $-\mathbf{v}$. Notice that the inverse of the four kinds of plane symmetries are themselves plane symmetries.

EXAMPLE 1 Symmetry Groups for Figure 1 Let us examine the symmetry groups of the logos in Figure 1. Consider the one shown here.[1]

Obviously, it has rotational symmetries of $0°$, $90°$, $180°$, and $270°$ and no others. (A rotation of $-90°$ is the same function as a rotation of $270°$; a rotation of $450°$ is the same function as a rotation of $90°$, and so on.) This figure is said to have *fourfold* rotational symmetry. Its symmetry group is called the *cyclic* group of order 4 (because the figure cycles around after four turns of $90°$ each). Chemists denote this group with C_4. Analogously, the symmetry groups of the other figures in Figure 1 are C_2, C_3, C_5, C_6, C_8, C_{16}, and C_{20}, in left-to-right order, beginning with the top row. ▲

[1]The symmetries of a figure are best studied with the use of a transparency with the figure and a piece of paper with the identical figure. The paper is held fixed while the superimposed transparency is manipulated.

EXAMPLE 2 Symmetry Groups for Figure 2

The logos in Figure 2 have both rotational symmetry and reflective symmetry. Notice that the following logo can be reflected about a horizontal axis, a vertical axis, and two diagonal axes and coincide with itself.

Let us denote these four reflection symmetries respectively by H, V, D, and D' and a rotation of θ degrees by R_θ. (D denotes the reflection that fixes the diagonal running from the upper left to the lower right; D' is the reflection that fixes the opposite diagonal.) Since symmetries are functions and function composition is done right to left, Table 1 is the function composition table for this group of symmetries. For instance, $H \circ R_{90}$ means a 90° rotation followed by a horizontal reflection. To verify that the entries in the table are correct, simply select the point p of the figure in the up-

per left corner for reference (any point but the center of rotation will do) and perform the relevant symmetries. For example, Figure 10 shows that $R_{90} \circ H = D'$. Similarly, one can show that $H \circ R_{90} = D$.

Does the fact that $R_{90} \circ H \neq H \circ R_{90}$ surprise you? It shouldn't. You certainly know that for 2×2 matrices $\mathbf{A}$ and $\mathbf{B}$, $\mathbf{AB}$ and $\mathbf{BA}$ need not be equal. Likewise, if $f(x) = 2x$ and $g(x) = x^2$, then $f \circ g \neq g \circ f$. Symmetry groups for which $XY = YX$ for all group elements X and Y are called *commutative* or *Abelian* (in honor of the mathematician Niels Abel). The group in Table 1 is called the *dihedral group of order 8* and is denoted by D_8. In general, a symmetry group consisting of n rotations and n reflections is called the *dihedral group of order 2n* and is denoted by D_{2n}. The cyclic groups C_n are Abelian, whereas the dihedral groups with at least six elements are non-Abelian. The symmetry groups of the figures in Figure 2 are D_4, D_{32}, D_{28}, D_8, D_{12}, D_{10}, D_8, and D_4. ▲

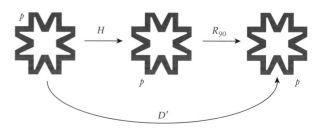

FIGURE 10 Verification that $R_{90} \circ H = D'$.

Notice that each logo in Figure 2 has an equal number of rotations and reflections (counting the 0° rotation). It can be proved that if a finite symmetry group contains at least one reflection, then it has an equal number of rotations and reflections. In fact, if

TABLE 1 Function Composition Table for D_8.

	R_0	R_{90}	R_{180}	R_{270}	H	V	D	D'
R_0	R_0	R_{90}	R_{180}	R_{270}	H	V	D	D'
R_{90}	R_{90}	R_{180}	R_{270}	R_0	D'	D	H	V
R_{180}	R_{180}	R_{270}	R_0	R_{90}	V	H	D'	D
R_{270}	R_{270}	R_0	R_{90}	R_{180}	D	D'	V	H
H	H	D	V	D'	R_0	R_{180}	R_{90}	R_{270}
V	V	D'	H	D	R_{180}	R_0	R_{270}	R_{90}
D	D	V	D'	H	R_{270}	R_{90}	R_0	R_{180}
D'	D'	H	D	V	R_{90}	R_{270}	R_{180}	R_0

$R_1, R_2, \cdots, R_n$ are the rotational symmetries from a group (the subscripts here do not denote the amount of rotation) and F is a reflective symmetry, then R_1F, $R_2F, \cdots, R_nF$ are the reflections as illustrated in Table 1. Remarkably, the cyclic groups and dihedral groups are the only finite plane symmetry groups possible. This fact has been attributed to Leonardo da Vinci (1452–1519), who studied symmetry in connection with designing buildings.

Strip Groups

What about infinite plane symmetry groups? There are two kinds: continuous and discrete. Examples of continuous plane symmetry groups include the symmetries of a circle, the symmetries of a line, and the symmetries of the plane itself. In the case of a circle, notice that there are rotational symmetries R_θ for arbitrarily small positive θ. For a line, there are translational symmetries with translation vectors of arbitrarily small magnitude. In contrast, a discrete symmetry group contains a rotational symmetry of smallest positive angle. Likewise, in a discrete symmetry group that has a nonzero translation, there is one with a translation vector of smallest positive magnitude. We will focus on discrete plane symmetry groups. Examples of figures with discrete symmetry groups include the graphs $y = \sin x$, $y = |\sin x|$, $|y| = |\sin x|$, and $y = \tan x$ (see Figure 11) and the plane periodic designs shown in Figures 12 and 13. The designs in Figure 12 are called *strip* patterns (or *one-dimensional patterns*), whereas those in Figure 13 are called wallpaper patterns (or *two-dimensional patterns*).

Strip patterns are defined as patterns whose translational symmetries vectors all have the form $n\mathbf{v}$, where n is an integer and $\mathbf{v}$ is a fixed vector.

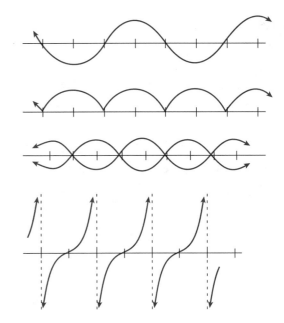

FIGURE 11 Graphs with discrete symmetry groups.

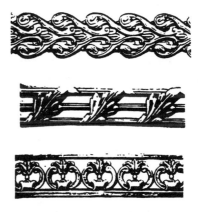

FIGURE 12 Strip patterns.

Wallpaper patterns are defined as plane patterns whose translational symmetries vectors have the form $m\mathbf{v} + n\mathbf{u}$, where $\mathbf{v}$ and $\mathbf{u}$ are independent fixed vectors (that is, $\mathbf{v}$ and $\mathbf{u}$ are not parallel) and m and n are integers. Notice that if a pattern has a glide

FIGURE 13 Wallpaper patterns.

reflection with glide vector **v**, then the pattern has a translational symmetry with vector 2**v** (see Figure 5).

Although the number of strip patterns and wallpaper patterns is unlimited, there is a severe limit on the number of symmetry groups that patterns in a plane can have. Indeed, there are only 7 kinds of symmetry groups for strip patterns and 17 kinds for wallpaper patterns. This fact has important practical consequences. For example, a wallpaper company can make every possible wallpaper pattern using only 17 computer programs for its machines. In three dimensions there are 230 symmetry groups. This permits mineralogists to classify the symmetries of crystals into a small number of families.

We examine the seven strip patterns first. Pattern I (Figure 14) consists of translations only. Letting t denote a translation to the right of one unit (that is, the distance between two consecutive Rs), t^{-1} a translation of one unit to the left, t^2 two units to the right, and so on (t^0 denotes a translation of 0 units to the right, which is no motion at all). Using

R   R   R   R

FIGURE 14 Pattern I.

Z to denote the integers, we may write the symmetry group of pattern I as

$$P_1 = \{t^n \mid n \in Z\}.$$

The group for pattern II (Figure 15), like that of pattern I, consists of powers (positive, negative, and zero) of a single symmetry. Letting g denote a glide reflection, we may write the symmetry group of pattern II as

$$P_2 = \{g^n \mid n \in Z\}.$$

Notice that g^2 is a translational symmetry and all translational symmetries of the pattern are powers of g^2.

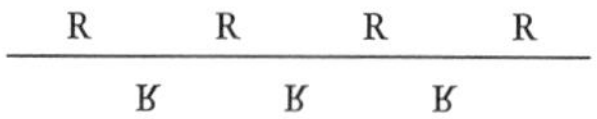

R   R   R   R
  Я   Я   Я

FIGURE 15 Pattern II.

The symmetry group of pattern III (Figure 16) consists of all strings of a translation t, its inverse t^{-1}, and a reflection v across a fixed vertical line. (There are infinitely many axes of reflective symmetry, including those midway between consecutive pairs of opposite facing Rs. Any one will do, but once it is selected it remains fixed for reference purposes.) Since even powers of v are the identity and odd powers of v are the same as v itself, the entire group is

$$P_3 = \{t^n \mid n \in Z\} \cup \{t^n v \mid n \in Z\}.$$

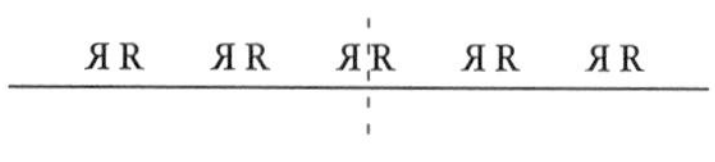

ЯR   ЯR   ЯR   ЯR   ЯR

FIGURE 16 Pattern III.

In Figure 17 we demonstrate that in P_3, $vt \neq tv$ by tracing the results of vt and tv upon the R in the

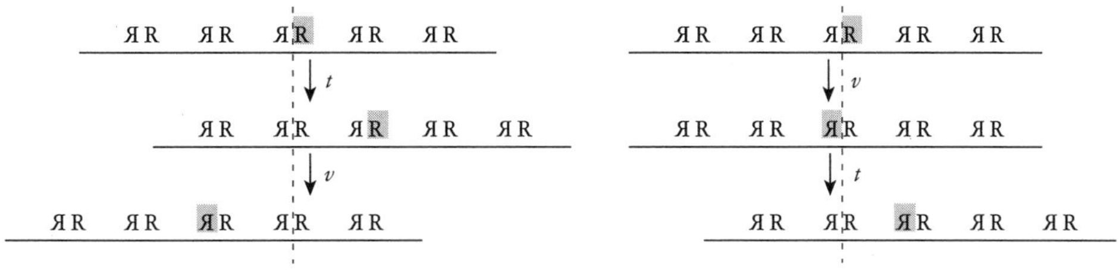

FIGURE 17 In P_3, $vt \neq tv$.

shaded box. Note that after applying t followed by v, the shaded R on the left side of Figure 17 lies to the left of the fixed vertical axis of reflection, whereas after applying v followed by t, the shaded R on the right side of Figure 17 lies to the right of the reflection axis, thereby proving that $vt \neq tv$. Hence, P_3 is non-Abelian (that is, not commutative). We can also see from the left side of Figure 17 that $vt = t^{-1}v$.

In pattern IV (Figure 18) the symmetry group consists of all strings of a translation t, its inverse t^{-1}, and a rotation r of $180°$ about a point p midway between consecutive Rs. This group, like P_3, is also non-Abelian. (Verify this yourself.)

$$P_4 = \{t^n | n \in Z\} \cup \{t^n r | n \in Z\}.$$

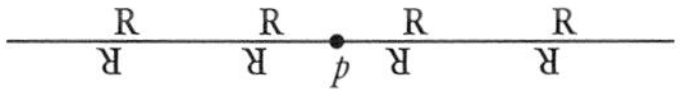

FIGURE 18 Pattern IV.

Can you see that every symmetry of pattern IV has the form t^n or $t^n r$? No? Well, recall that there are only four possible plane symmetries: translations, rotations, reflections, and glide reflections. Notice that t and its powers (positive, negative, and zero) account for all translations of pattern IV. Certainly, the only possible nonzero rotation for any strip pattern is $180°$. Also, for any strip pattern the only possible reflections are across the horizontal axis or across vertical axes. Observe that pattern IV has no such symmetries. We also note that pattern IV does not have a glide-reflective symmetry. So, every member of P_4 consists of strings of t terms and r terms. Finally, using the fact that $rt = t^{-1}r$, $rt^{-1} = tr$, and r^2 is the identity, we can take any string comprised of t terms and r terms and collect the t terms on the left. For example, $r^3 tr = r^2 rtr = rtr = (rt)r = (t^{-1}r)r = t^{-1}r^2 = t^{-1}$. As another example we have $rt^{-2}rt = (rt^{-1})t^{-1}rt = (tr)t^{-1}rt = t(rt^{-1})rt = t(tr)rt = t^2 r^2 t = t^2 t = t^3$.

The group for pattern V (Figure 19) is a non-Abelian group consisting of elements that are strings of a glide reflection g, its inverse g^{-1}, and a rotation r of $180°$ about the point p. Notice that pattern V has vertical reflection symmetry, but it is just gr. The rotation points, such as p, are midway between the vertical reflection axes.

$$P_5 = \{g^n | n \in Z\} \cup \{g^n r | n \in Z\}.$$

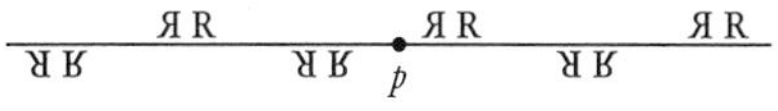

FIGURE 19 Pattern V.

The symmetry group of pattern VI (Figure 20) consists of strings of a translation t, its inverse t^{-1},

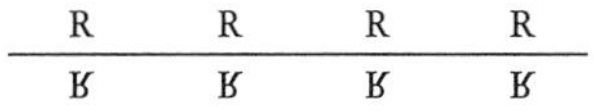

FIGURE 20 Pattern VI.

and a horizontal reflection h. Since $th = ht$, the symmetry group is Abelian (see Figure 21) and the elements are

$$P_6 = \{t^n | n \in Z\} \cup \{t^n h | n \in Z\}.$$

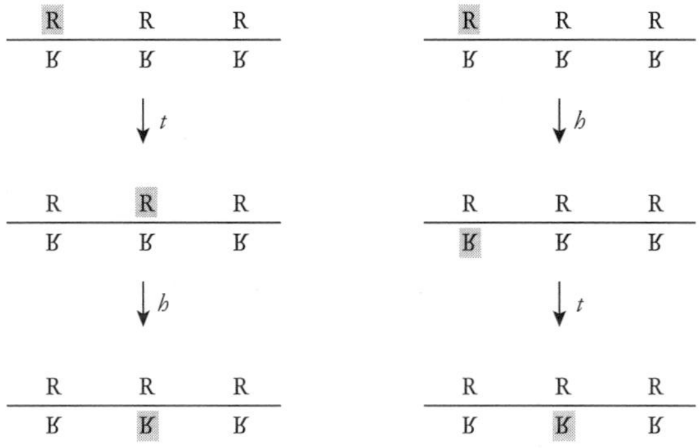

FIGURE 21 In P_6, $th = ht$.

(Note that since the final position of the shaded R on the left side and the shaded R on the right side are identical, $th = ht$.) Observe that pattern VI coincides with itself under a glide reflection also, but in this case the glide reflection is considered trivial since it is the product of t and h. (Conversely, a glide reflection is *nontrivial* if its translation component and reflection component are not elements of the symmetry group.)

The symmetry group of pattern VII (Figure 22) consists of strings of a translation t, its inverse t^{-1}, a horizontal reflection h, and a vertical reflection v. Just as in pattern III, it is non-Abelian since $tv \neq vt$. Note that the composition of h and v in either order is a $180°$ rotation.

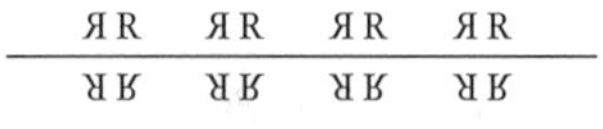

FIGURE 22 Pattern VII.

$$P_7 = \{t^n | n \in Z\} \cup \{t^n h | n \in Z\} \cup \{t^n v | n \in Z\}$$
$$\cup \{t^n hv | n \in Z\}.$$

Astute readers may be concerned that pattern VII has two kinds of vertical reflections; one kind splits the back-to-back Rs as shown in Figure 23, whereas the other is midway between consecutive

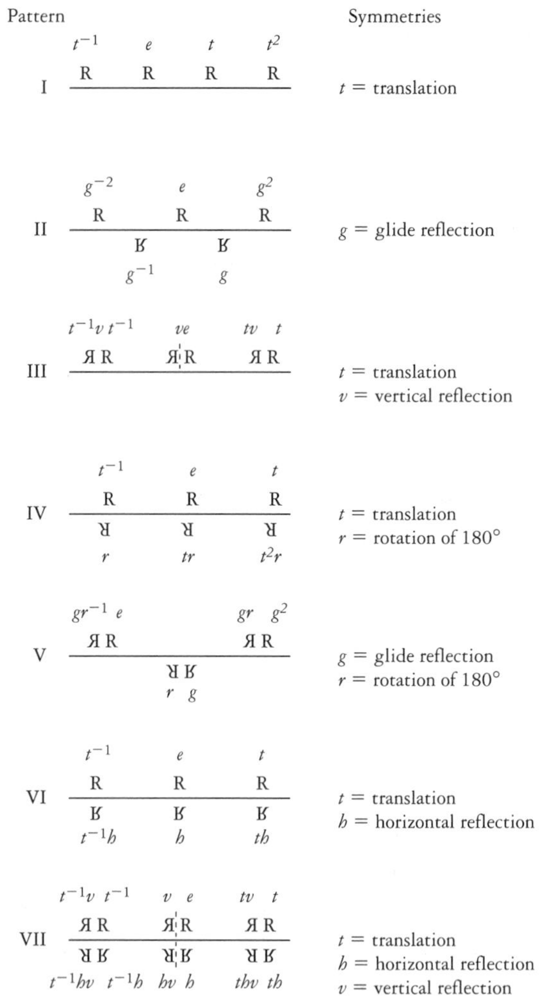

FIGURE 23 The seven strip patterns and their symmetries. Every R is labeled with the symmetry that would carry the one labeled with e to that R.

pairs of back-to-back Rs. However, the latter kind of reflection is accounted for in P_7 since it is tv.

The preceding discussion of the seven types of symmetry groups is summarized in Figure 23. Figure 24 provides an identification algorithm for the strip patterns. The symbol e is the identity function (that is, no motion at all). The label associated with each R in the figure indicates the effect that symmetry has on the R labeled with e. For example, in pattern VII the R in the lower left hand corner is labeled with $t^{-1}hv$ since it is the image of the R labeled with e after the composition of v, h, and t^{-1}.

In describing the seven strip groups, we have written all the elements in a specific form such as $t^n r^m$ or $t^n h^m v^k$. Implicitly, this means that the product of two elements of one of these forms must be expressible in the same form (the technical term for this property is *closure*). To find the expression for the

product of two elements written in one of these forms as another of the same form, we can always use the geometry to determine the result. For example, we know that every element of P_7 can be written in the form $t^n h^m v^k$. So, just for fun, let's determine the appropriate values for n, m, and k for the product of $t^{-1}hv$ and tv. That is, we wish to determine n, m, and k so that $x = (t^{-1}hv)(tv) = t^n h^m v^k$. We may do this simply by looking at the effect that x has on pattern VII. For convenience, we will pick out a particular R in the pattern and trace the action of x one step at a time starting with the v on the right, and following with t, v, h, and t^{-1} in that order. To distinguish this R, we enclose it in a shaded box. Also, we draw the fixed axis of the vertical reflection v as a dotted line segment. See Figure 25.

Now comparing the starting position of the shaded R with its final position on both the left and

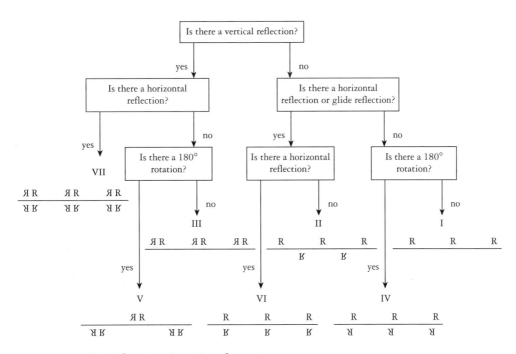

FIGURE 24 Identification algorithm for strip patterns.

$\mathbf{FIGURE\ 25}$ Geometric verification that $t^{-1}hvtv = t^{-2}h$ in P_7.

right sides of the figure, we see that $t^{-1}hvtv = t^{-2}h$. Exercise 32 suggests how one may arrive at the same result through purely algebraic manipulation.

The Crystallographic Groups

The seven strip groups catalog all discrete plane symmetry groups whose translation vectors are all multiples of just one translation vector. However, there are 17 additional kinds of discrete plane symmetry groups that arise from infinitely repeating designs in a plane. These groups are the symmetry groups of plane patterns whose translation vectors have the form $m\mathbf{u} + n\mathbf{v}$, where $\mathbf{u}$ and $\mathbf{v}$ are independent (nonparallel) vectors. These 17 groups were first studied by nineteenth-century crystallographers (that is, chemists who study crystals) and are called the *plane crystallographic* groups or *wallpaper* groups.

Our approach to the crystallographic groups will be geometric. Our goal is to enable the reader to determine which of the 17 plane symmetry groups

corresponds to a given periodic pattern. We begin with some examples.

The simplest of the 17 crystallographic groups contains translations only. In Figure 26 we give an illustration of a representative pattern for this group (imagine that the pattern is repeated to fill the entire plane). The crystallographic notation for it is $p1$.

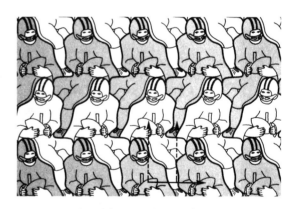

$\mathbf{FIGURE\ 26}$ Study of Regular Division of the Plane with Fish and Birds, 1938. Escher graphic with symmetry group $p1$.

The symmetry group of the pattern in Figure 27 (disregarding shading) contains translations and

$\mathbf{FIGURE\ 27}$ Escher-like tessellation by J. L. Teeters with symmetry group pg (disregarding shading). The solid arrow is a translation vector. The dashed arrows are glide-reflection vectors.

M. C. Escher

M. C. Escher was born on June 17, 1898, in The Netherlands. His artistic work prior to 1937 was dominated by the representation of visible reality, such as landscapes and buildings. Gradually, he became less and less interested in the visible world and became increasingly absorbed in an inventive approach to space. He studied the abstract space-tiling patterns used in the Moorish mosaic in Alhambra, Spain. He also studied the mathematician George  Polya's paper on the 17 plane crystallographic groups. Instead of the geometrical motifs used by the Moors and Polya, Escher preferred to use animals, plants, or people in his space-filling prints.

Escher was fond of incorporating various mathematical ideas into his works. Among these are infinity, Möbius bands, stellations, deformations, reflections, Platonic solids, spirals, and the hyperbolic plane. This latter idea was suggested to Escher by a figure in a paper by the geometer H. S. M. Coxeter.

Although Escher originals are now quite expensive, it was not until 1951 that he derived a significant portion of his income from his prints. Today, Escher is widely known and appreciated as a graphic artist. His graphics have appeared on postage stamps, bank notes, a candy box (in the shape of an icosahedron!), note cards, T-shirts, jigsaw puzzles, record album covers, and the covers of scores of scientific publications. His prints have been used to illustrate ideas in hundreds of scientific works. Despite his popularity among scientists, Escher has never been held in high esteem in traditional art circles. Escher died on March 27, 1973, in Holland.

glide reflections. (Try locating an axis for a glide reflection for this pattern.) This group has no (nonzero) rotational or reflective symmetry. The crystallographic notation for it is *pg* (the "*g*" stands for "glide reflection"). Figure 28 (disregard shading) has translational symmetry and threefold rotational symmetry (that is, the figure can be rotated 120° about certain points and brought into coincidence with itself). The notation used by crystallographers for this group is *p*3 (the 3 indicates that the smallest positive rotational symmetry is 360/3 = 120°).

Representative patterns for all 17 plane crystallographic groups using a triangle motif are given in Figure 29.

Identification of Plane Periodic Patterns

To decide to which of the 17 classes any particular plane periodic pattern belongs, we may use the algorithm in Figure 30. This is done by determining the smallest positive rotational symmetry and whether or not the pattern has reflection symmetry or nontrivial glide-reflection symmetry.

FIGURE 28 Study of Regular Division of the Plane with Human Figures, 1938. Escher graphic with symmetry *p*3 (disregarding shading).

E X A M P L E 3 Symmetry Groups for Figure 27 Consider the pattern in Figure 27. First we observe that there is no nontrivial rotational symmetry; next we check for a reflection (there is none); lastly, we see that there is a glide reflection. Thus, the algorithm tells us that the symmetry group is *pg*. ▲

E X A M P L E 4 Symmetry Groups for Figure 31 Consider the two patterns in Figure 31 with a hockey-stick motif. Referring to the algorithm in Figure 30, we first check for the smallest nontrivial rotational symmetry. This is 120° in both cases. Then we observe that both have reflectional and nontrivial glide-reflectional symmetry. Now, according to the algorithm, these patterns must be of type *p*3*m*1 or *p*31*m* (the presence of the *m* indicates that the pattern has reflective symmetry—the "*m*" is for "mirror"). But notice that the pattern on the left has all its centers of rotational

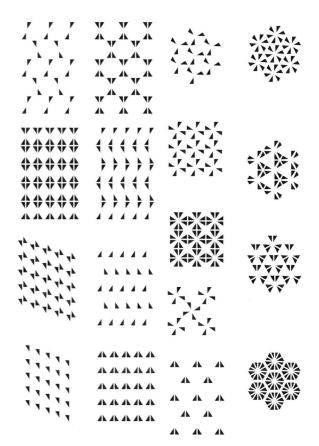

FIGURE 29 The 17 plane periodic patterns formed with a triangle motif.

symmetry on reflection axes, whereas the pattern on the right does not. Thus, the left pattern is *p*3*m*1, and the right pattern is *p*31*m*. ▲

Notice that the algorithm reveals that the only possible *n*-fold rotational symmetries for a repeating pattern are when *n* = 1, 2, 3, 4, and 6 (an *n-fold* rotation is a rotation of 360°/*n*). This fact is commonly called the *crystallographic restriction*. A geometric proof of it was first given more than 100 years ago by the English mathematician Peter Barlow.

The symmetry groups for three-dimensional ob-

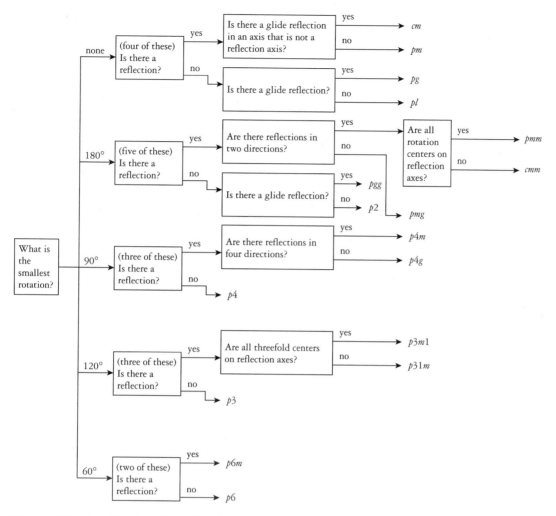

FIGURE 30 Identification algorithm for plane periodic patterns.

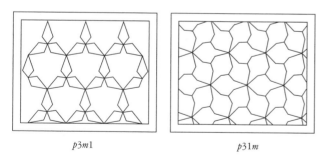

| p3m1 | p31m |

FIGURE 31 Patterns with a hockey-stick motif.

jects are more numerous than the groups for two dimensions. Although there are infinitely many three-dimensional objects, there are essentially only five types of finite groups of rotational symmetry: the cyclic groups C_n; dihedral groups D_{2n}; the rotations of a regular tetrahedron; the rotations of a cube; and the rotations of a regular dodecahedron (Figure 32 shows a tetrahedron and a dodecahedron). Later

Repeating Patterns in Hyperbolic Geometry

Among the artist Escher's most visually interesting and intellectually intriguing designs are those based in hyperbolic geometry as shown here. Hyperbolic geometry differs from Euclidean geometry (the geometry taught in high school) with regard to the existence of a line parallel to a given line passing through a given point. In Euclidean geometry there is exactly one such line, whereas in hyperbolic geometry there are infinitely many.

Douglas Dunham of the University of Minnesota, Duluth, has written computer programs incorporating hyperbolic geometry and group theory that can produce repeating patterns like those of Escher. Although there are only 17 symmetry groups for repeating patterns in Euclidean geometry, there are infinitely many symmetry groups for repeating patterns in hyperbolic geometry. Dunham used group theory to provide the set of instructions to the computer for replicating the initial image in the desired pattern.

Tetrahedron

Dodecahedron

FIGURE 32 Tetrahedron and dodecahedron.

in this chapter we will show how one may count the number of symmetries of a solid. The classification of the three-dimensional groups of repeated patterns (often called *space groups*) was done in the late nineteenth century by crystallographers. There are 230 such groups.

As one might expect, the crystallographic groups are fundamentally important in the study of crystals.

In fact, a crystal is defined as a rigid body in which the component particles are arranged in a pattern that repeats in three directions (the repetition is caused by the chemical bonding). A grain of salt and a grain of sugar are two examples of common crystals. In crystalline materials the motif units are atoms, ions, ionic groups, clusters of ions, or molecules (see Figure 33).

The two-dimensional crystallographic groups we have examined are important to mineralogists. In fact, mineralogists determine the internal structure of crystals by studying two-dimensional x-ray projections of the atomic make-up of the crystals. The symmetry present in the projections reveals the internal symmetry of the crystals themselves. Commonly occurring symmetry patterns are D_8, the symmetry group of a square, and D_{12}, the symmetry group of a regular hexagon (see Figure 34).

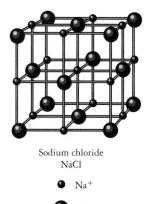

Sodium chloride
NaCl

● Na⁺

● Cl⁻

FIGURE 33 Arrangement of atoms in a crystal.

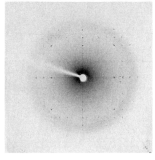

FIGURE 34 X-ray diffraction photos revealing D_8 symmetry patterns in crystals.

Perhaps it is fitting to conclude this section by recounting two episodes in the history of science where an understanding of symmetry groups was crucial to a great discovery. In 1912 Max von Laue, a young German physicist, hypothesized that a narrow beam of x-rays directed onto a crystal with a photographic film behind it would be deflected (the technical term is "diffracted") by the unit cell (made up of atoms or ions) and show up on the film as spots (see Figure 34). Shortly thereafter, two British scientists, Sir William Henry Bragg and his 22-year-old son, William Lawrence Bragg, who was a student, noted that von Laue's diffraction spots together with the known information about crystallographic space groups could be used to calculate the shape of the internal array of atoms. This discovery marked the birth of modern mineralogy. From the first crystal structures deduced by the Braggs to the present, x-ray diffraction has been the means by which the internal structures of crystals have been determined. von Laue was awarded the Nobel Prize in physics in 1914, and the Braggs were jointly awarded the Nobel Prize in physics in 1915. (William Lawrence Bragg is the youngest person ever to receive a Nobel prize.)

Our second episode took place in the early 1950s when a small number of scientists were attempting to learn the structure of the DNA molecule—the basic genetic material. One of these was a graduate student named Francis Crick; another was an x-ray crystallographer, Rosalind Franklin. On one occasion, Crick was shown one of Franklin's research reports and an x-ray diffraction photograph of DNA.

At this point, we let Horace Judson,[2] our source, continue the story.

> Crick saw in Franklin's words and numbers something just as important, indeed eventually as visualizable. There was drama, too: Crick's insight began with an extraordinary coincidence. Crystallographers distinguish 230 different space groups, of which the face-centered monoclinic cell with its curious properties of symmetry is only one—though in biological substances a fairly common one. The principal experimental subject of Crick's dissertation, however, was the x-ray diffraction of the crystals of a protein that was of exactly the same

[2]H. Judson (1979), *The Eighth Day of Creation*, New York: Simon and Schuster, pp. 165–166.

space group as DNA. So Crick saw at once the symmetry that neither Franklin nor Wilkins had comprehended, that Perutz, for that matter, hadn't noticed, that had escaped the theoretical crystallographer in Wilkins' lab, Alexander Stokes—namely, that the molecule of DNA, rotated a half turn, came back to congruence with itself. The structure was dyadic, one half matching the other half in reverse.

This was a crucial fact. Shortly thereafter, James Watson and Crick built an accurate model of DNA. In 1962 Watson, Crick, and Maurice Wilkins received the Nobel Prize in medicine and physiology for their discovery. The opinion has been expressed that, had Franklin correctly recognized the symmetry of the DNA molecule, she might have been the one to unravel the mystery and receive the Nobel prize.

Exercises for Section 9.1

1. With pictures and words, describe each symmetry in D_6 (the set of symmetries of an equilateral triangle).

2. Write out a complete multiplication table for D_6.

3. Is D_6 Abelian?

4. Describe in pictures or words the elements of D_{10} (symmetries of a regular pentagon).

5. For $n \geq 3$, describe the elements of D_{2n}. (Hint: You will need to consider two cases—n even and n odd.) How many elements does D_{2n} have?

6. If r_1, r_2, and r_3 represent rotations from D_{2n} and f_1, f_2, f_3 represent reflections from D_{2n}, determine whether $r_1 r_2 f_1 r_3 f_2 f_3 r_3$ is a rotation or a reflection.

7. Find elements of A, B, and C in D_8 such that $AB = BC$ but $A \neq C$. (Thus, "cross" cancellation is not valid.)

8. For reference purposes, label the n vertices of a regular n-gon clockwise 1 through n. Explain what the following diagram proves about the group D_{2n}.

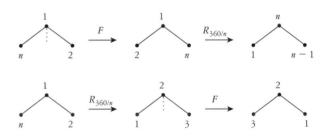

9. Describe the symmetries of a nonsquare rectangle. Construct the corresponding multiplication table.

10. Describe the symmetries of a parallelogram that is neither a rectangle nor a rhombus. Describe the symmetries of a rhombus (a quadrilateral with four sides of equal length) that is not a rectangle.

11. Describe the symmetries of a noncircular ellipse. Do the same for a hyperbola.

12. Consider an infinitely long strip of equally spaced Hs:

$$\cdots \text{H H H H} \cdots$$

Describe the symmetries of this strip. Is the group of symmetries of the strip Abelian?

13. For each of the snowflakes in the following figure, find the symmetry group and locate the axes of reflective symmetry (ignore imperfections).

14. Find the plane symmetry group for the Canadian hendecagonal one-dollar coin shown below. Disregard the printing and scene on the coin.

15. Determine the plane symmetry groups for the hub-caps shown here.

16. Determine the plane symmetry groups for the hub-caps shown here.

17. Show that a translation (in a plane) composed with a translation is a translation.

18. Show that a plane isometry preserves angles. (That is, if T is a plane isometry and A is an angle, then $T(A)$ and A are congruent.) Hint: Form a triangle and use side-side-side.

19. Show that a plane isometry is completely determined by the image of three noncollinear points.

20. Suppose that a plane isometry leaves three noncollinear points fixed. Which isometry is it?

21. Suppose that a plane isometry fixes exactly one point. What type of isometry is it?

22. Which of the letters shown here have exactly one reflective symmetry? Which have two reflection symmetries? Which have a nonzero rotational symmetry?

ABCDEFGHIJKLMNOPQRSTUVWXYZ

23. Determine the symmetry group for each pattern in Figure 13.

24. In the strip group P_3, write vt in the form t^n or $t^n v$.

25. In the strip group P_4, write $rt^{-2}rt^3r$ in the form t^n or $t^n r$.

26. Determine the symmetry group of each of the following strip patterns.

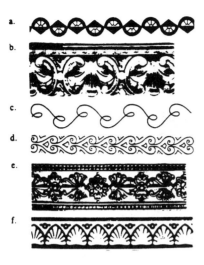

a.
b.
c.
d.
e.
f.

27. Determine the plane symmetry group for each of the patterns shown in Figure 29.

28. In the strip group P_7, write t^2hvtv in the form t^n, t^nh, t^nv, or t^nhv.

29. In the strip group P_7, write $t^{-3}vthv$ in the form t^n, t^nh, t^nv, or t^nhv.

30. In the strip group P_7, show that $hv = vh$ and $th = ht$.

31. In the strip group P_7, show that $vtv = t^{-1}$.

32. Use the results of Exercises 30 and 31 to do Exercises 28 and 29 through symbol manipulation only (i.e., without referring to the pattern).

SECTION 9.2 *Abstract Groups*

Although group theory provides the proper framework in which to study symmetry, symmetry groups are just one of many concrete realizations of the notion of an abstract group. In this section we introduce the abstract group concept and illustrate it with many other concrete realizations.

Definition and Examples of Groups

The term "group" was coined about 160 years ago to describe certain sets of functions defined on finite sets. As is the case with most fundamental concepts in mathematics, the modern definition of a group that follows is the result of a long evolutionary process. Although this definition was given in 1882, it did not gain universal acceptance until this century.

> **DEFINITION**
>
> Let G be a set. A *binary operation* on G is a function that assigns to each ordered pair of elements of G an element of G.

A binary operation on a set G, then, is simply a method (or formula) by which an ordered pair from G combines to yield a new member of G. The most familiar binary operations are the ordinary addition, subtraction, and multiplication of integers. The division of integers is not a binary operation on the integers because an integer divided by an integer need not be an integer.

DEFINITION

Let G be a nonempty set together with a binary operation that assigns to each ordered pair (a, b) of elements of G an element in G, which we denote ab. (When a specific binary operation such as addition is used, the notation $a + b$ is used to denote the element assigned to the ordered pair (a, b).) We say G is a *group* under this operation if the following three properties are satisfied:

1. *Associativity.* The operation is associative; that is, $(ab)c = a(bc)$ for all a, b, c in G.

2. *Identity.* There is an element e (called the *identity*) in G, such that $ae = ea = a$ for all a in G.

3. *Inverses.* For each element a in G, there is an element b in G (called an *inverse* of a) such that $ab = ba = e$.

In words, then, a group is a set together with an associative operation and an identity such that every element has an inverse and every pair of elements can be combined without going outside the set. This latter condition is called *closure.* Be sure to verify closure when testing for a group (see Example 5).

As was the case with symmetry groups, if a group has the property that $ab = ba$ for every pair of elements a and b, we say the group is *Abelian.* A group is *non-Abelian* if there is some pair of elements a and b for which $ab \neq ba$. When encountering a particular group for the first time, one should determine whether or not it is Abelian.

Students may wonder about the motivation for our choice of the axioms for a group. Well, associativity, the existence of an identity, and inverses are precisely the properties needed to solve equations. Consider $2x = 4$. Here are *all* the steps involved in solving for x:

$$2x = 4,$$
$$2^{-1}(2x) = 2^{-1} \cdot 4 \quad \text{(existence of inverse)},$$
$$(2^{-1} \cdot 2)x = 2 \quad \text{(associativity)},$$
$$(1)x = 2 \quad \text{(definition of inverse)},$$
$$x = 2 \quad \text{(definition of identity)}.$$

To solve an equation such as $2x + 5 = 4$, we must use the fact that the set of real numbers is a group under addition and under multiplication (excluding 0 for multiplication).

Now that we have the formal definition of a group, our first job is to build a good stock of examples. Of course, we already have the finite plane symmetry groups Z_n and D_{2n}, as well as the strip groups and crystallographic groups. The following are nongeometric examples.

EXAMPLE 1 Additive Groups The set of integers Z (so denoted because the German word for integers is "Zahlen"), the set of rational numbers Q (for "quotient"; rational numbers have the form a/b, where a and b are integers and $b \neq 0$), and the set of real numbers R are all groups under ordinary addition. In each case, the identity is 0 and the inverse of a is $-a$. ▲

EXAMPLE 2 Failure of Inverses The set of integers under ordinary multiplication *is not* a group. Property (3) fails. For example, there is no *integer b* such that $5b = 1$. ▲

EXAMPLE 3 A Two Element Group The set $\{1, -1\}$ is a group under multiplication. Note that -1 is its own inverse. ▲

EXAMPLE 4 The Positive Rationals The set Q^+ of positive rationals *is* a group under ordinary multiplication. The inverse of any a is $1/a = a^{-1}$. ▲

A 15,000-Page Proof!

Danny Gorenstein

John Thompson

Michael Aschbacher

Three key figures in the classification of finite simple groups

The largest coordinated effort ever undertaken to discover and prove a single theorem was the so-called classification of finite simple groups. Among all finite groups there are certain ones called simple groups that serve as the building blocks for all finite groups, much like the chemical elements make up the component pieces of all chemical compounds. A complete list of the finite simple groups is for group theory what the periodic table is for chemistry.

From the time simple groups were first defined in 1830 by Galois until the mid-1950s, many infinite families of finite simple groups were found but there was no organized attempt to discover *all* of them and to prove that there were no more to be discovered.

Beginning in the late 1950s and ending in the early 1980s, an informal "team" of well over 100 mathematicians cooperated on an effort to create a list of all simple groups and to prove the list was complete.

The smallest non-Abelian simple group consists of the 60 rotational symmetries of the icosahedron.

After hundreds of thousands of person-hours and hundreds of journal articles, the project was complete. The proof runs approximately 15,000 pages. Many people rate this classification among the greatest achievements in the history of mathematics.

EXAMPLE 5 **Failure of Closure** The set S of negative real numbers together with 1 under multiplication satisfies the three properties given in the definition of a group but is *not* a group since multiplication is not a binary operation on S. Indeed, $(-2)(-2) = 4$, so S is not closed under multiplication. ▲

EXAMPLE 6 **Matrices under Addition** The set of all 2×2 matrices with real entries is a group under componentwise addition. That is,

$$\begin{bmatrix} a_1 & b_1 \\ c_1 & d_1 \end{bmatrix} + \begin{bmatrix} a_2 & b_2 \\ c_2 & d_2 \end{bmatrix} = \begin{bmatrix} a_1 + a_2 & b_1 + b_2 \\ c_1 + c_2 & d_1 + d_2 \end{bmatrix}.$$

The identity is $\begin{bmatrix} 0 & 0 \\ 0 & 0 \end{bmatrix}$ and the inverse of $\begin{bmatrix} a & b \\ c & d \end{bmatrix}$ is $\begin{bmatrix} -a & -b \\ -c & -d \end{bmatrix}$. ▲

Before presenting our next two examples, we first introduce two important binary operations on the set $\{0, 1, \ldots, n - 1\}$. They are an abstraction of a method of counting that you often use. For example, if it is now September, what month will it be 25 months later? Of course, the answer is October, but the interesting fact is that you didn't arrive at the answer by starting with September and counting off 25 months. Instead, without even thinking about it, you simply observed that $25 = 2 \cdot 12 + 1$, and you added 1 month to September. Similarly, if it is now Wednesday, you know that in 23 days it will be Friday. This time, you arrived at your answer by noting that $23 = 7 \cdot 3 + 2$, so you added 2 days to Wednesday instead of counting off 23 days. If your electricity is off for 26 hours, you must advance your clock 2 hours since $26 = 2 \cdot 12 + 2$. Surprisingly, this simple idea has numerous important applications in mathematics and computer sci-

ence. You will see a few of them in this section. The following notation is convenient. If a and b are integers and n is a positive integer, we often write $a = b \bmod n$ whenever n divides $a - b$. (This is shorthand for $a \bmod n = b \bmod n$). Thus,

$$3 \bmod 2 = 1 \text{ since } 3 = 1 \cdot 2 + 1;$$
$$6 \bmod 2 = 0 \text{ since } 6 = 2 \cdot 3 + 0;$$
$$4 \bmod 3 = 1 \text{ since } 4 = 1 \cdot 3 + 1;$$
$$15 \bmod 3 = 0 \text{ since } 15 = 5 \cdot 3 + 0;$$
$$12 \bmod 10 = 2 \text{ since } 12 = 1 \cdot 10 + 2;$$
$$37 \bmod 10 = 7 \text{ since } 37 = 3 \cdot 10 + 7;$$
$$98 \bmod 85 = 13 \text{ since } 98 = 1 \cdot 85 + 13;$$
$$342 \bmod 85 = 2 \text{ since } 342 = 4 \cdot 85 + 2;$$
$$62 \bmod 85 = 62 \text{ since } 62 = 0 \cdot 85 + 62.$$

In our applications, we will use addition and multiplication mod n. When you wish to compute $ab \bmod n$ or $(a + b) \bmod n$ and a or b is greater than n, it is easier to "mod first." By this we mean

$$(ab) \bmod n = ((a \bmod n)(b \bmod n)) \bmod n.$$

Similarly,

$$(a + b) \bmod n = ((a \bmod n) + (b \bmod n)) \bmod n.$$

Here are some examples.

$$(17 + 23) \bmod 10 = ((17 \bmod 10) + (23 \bmod 10)) \bmod 10$$
$$= (7 \bmod 10 + 3 \bmod 10) \bmod 10$$
$$= 10 \bmod 10 = 0$$

$$(17 \cdot 23) \bmod 10 = ((17 \bmod 10)(23 \bmod 10)) \bmod 10$$
$$= (7 \cdot 3) \bmod 10 = 21 \bmod 10 = 1$$

$$(22 \cdot 19) \bmod 8 = ((22 \bmod 8)(19 \bmod 8)) \bmod 8$$
$$= (6 \cdot 3) \bmod 8 = 18 \bmod 8 = 2$$

EXAMPLE 7   Group of Integers Modulo *n*

The set $Z_n = \{0, 1, \ldots, n - 1\}$ for $n \geq 1$ *is* a group under addition modulo *n*. For any *j* in Z_n, the inverse of *j* is $n - j$. For example, in Z_7, 4 is the inverse of 3 since $3 + 4 = 7 = 0$ mod 7. This group is usually referred to as the *group of integers modulo n*. ▲

As we have seen, the real numbers, the 2×2 matrices with real entries, and the integers modulo *n* are all groups under the appropriate addition. But what about these same sets under multiplication? In each case the existence of some elements that do not have inverses prevents the set from being a group under the usual multiplication. However, we can form a group in each case by simply throwing out the rascals. The next three examples illustrate this.

EXAMPLE 8   Group of Units Modulo *n*   An

integer *a* has a multiplicative inverse modulo *n* if and only if *a* and *n* are relatively prime (*a* and *n* are relatively prime if their greatest common divisor is 1). So, for each $n > 1$, we define $U(n)$ to be the set of all positive integers less than *n* and relatively prime to *n*. Then $U(n)$ is a group under multiplication modulo *n* and is called the *group of units modulo n*.

For $n = 10$, we have $U(10) = \{1, 3, 7, 9\}$. The multiplication table for $U(10)$ is

mod 10	1	3	7	9
1	1	3	7	9
3	3	9	1	7
7	7	1	9	3
9	9	7	3	1

For $n = 12$, we have $U(12) = \{1, 5, 7, 11\}$. The multiplication table for $U(12)$ is

mod 12	1	5	7	11
1	1	5	7	11
5	5	1	11	7
7	7	11	1	5
11	11	7	5	1	▲

EXAMPLE 9   The set $R^\#$ of nonzero real

numbers is a group under ordinary multiplication. The identity is 1. The inverse of *a* is $1/a$. ▲

EXAMPLE 10   General Linear Group over *R*   The set

$$GL(2, R) = \left\{ \begin{bmatrix} a & b \\ c & d \end{bmatrix} \middle| a, b, c, d \in R, ad - bc \neq 0 \right\}$$

of 2×2 matrices with real entries and nonzero determinant is a non-Abelian group under matrix multiplication.

The first step in verifying that this set is a group is to show that the product of two matrices with nonzero determinant also has nonzero determinant. This follows from the fact that for any pair of 2×2 matrices **A** and **B**, $\det(\mathbf{AB}) = (\det \mathbf{A})(\det \mathbf{B})$ (see Theorem 4 of Section 6 of Chapter 3).

Associativity can be verified by direct (but cumbersome) calculations. The identity is $\begin{bmatrix} 1 & 0 \\ 0 & 1 \end{bmatrix}$; the inverse of $\begin{bmatrix} a & b \\ c & d \end{bmatrix}$ is

$$\begin{bmatrix} \dfrac{d}{ad - bc} & \dfrac{-b}{ad - bc} \\ \dfrac{-c}{ad - bc} & \dfrac{a}{ad - bc} \end{bmatrix}$$

(the denominators in the terms of this matrix explain the requirement that $ad - bc \neq 0$). This very important group is called the *general linear group* of 2×2 matrices over *R*. ▲

EXAMPLE 11 Failure of Inverses   The set of 2 × 2 matrices with real number entries is *not* a group under matrix multiplication. Inverses do not exist when the determinant is zero (see Theorem 6 of Section 6 of Chapter 3). ▲

EXAMPLE 12 Failure of Inverses   The set {0, 1, 2, 3} is *not* a group under multiplication modulo 4. From the multiplication table we see that although 1 and 3 have inverses, the elements 0 and 2 do not since the identity 1 does not appear in the columns headed by 0 and 2.

mod 4	0	1	2	3
0	0	0	0	0
1	0	1	2	3
2	0	2	0	2
3	0	3	2	1	▲

EXAMPLE 13 Failure of Associativity
The set of integers under subtraction is *not* a group, since the operation is not associative. For instance, $(7 - 5) - 3 \neq 7 - (5 - 3)$. ▲

With the examples given this far as a guide, it is wise for the reader to pause here and think of his or her own examples. Study actively! Don't just read along and be spoon-fed by the book.

EXAMPLE 14 The Special Linear Groups
The sets of all 2 × 2 matrices with determinant 1 with entries from Q, R, or Z_p (p a prime) are non-Abelian groups under matrix multiplication. These groups are called the *special linear groups* of 2 × 2 matrices over Q, R, or Z_p denoted by SL(2, Q), SL(2, R) and SL(2, Z_p), respectively. For

any special linear group, the formula given in Example 10 for the inverse of $\begin{bmatrix} a & b \\ c & d \end{bmatrix}$ simplifies to $\begin{bmatrix} d & -b \\ -c & a \end{bmatrix}$. When the matrix entries are from Z_p, we use modulo p arithmetic to compute determinants, matrix products, and inverses. To illustrate the case SL(2, Z_5), consider the element $A = \begin{bmatrix} 3 & 4 \\ 4 & 4 \end{bmatrix}$. Then det $A = 3 \cdot 4 - 4 \cdot 4 = -4 = 1 \bmod 5$ so that $A \in$ SL(2, Z_5) and the inverse of A is $\begin{bmatrix} 4 & -4 \\ -4 & 3 \end{bmatrix} = \begin{bmatrix} 4 & 1 \\ 1 & 3 \end{bmatrix}$. Note that $\begin{bmatrix} 3 & 4 \\ 4 & 4 \end{bmatrix}\begin{bmatrix} 4 & 1 \\ 1 & 3 \end{bmatrix} = \begin{bmatrix} 1 & 0 \\ 0 & 1 \end{bmatrix}$ when the arithmetic is done modulo 5. ▲

Example 10 is a special case of the following more general construction.

EXAMPLE 15 General Linear Group over *F*   Let F be any of Q, R, or Z_p (p a prime). The set GL(2, F) of all 2 × 2 matrices with non-zero determinant and entries from F is a non-Abelian group under matrix multiplication. As in Example 14 when F is Z_p, modulo p arithmetic is used to calculate determinants, the matrix products, and inverses. The formula given in Example 10 for the inverse of $\begin{bmatrix} a & b \\ c & d \end{bmatrix}$ remains valid for elements from GL(2, Z_p) provided we interpret division by $ad - bc$ as multiplication by the inverse of $ad - bc$ modulo p. For example, in GL(2, Z_7), consider $\begin{bmatrix} 4 & 5 \\ 6 & 3 \end{bmatrix}$. Then the determinant $ad - bc$ is $12 - 30 = -18 = 3 \bmod 7$, and the inverse of 3 is 5 (since $3 \cdot 5 = 1 \bmod 7$). So, using the

Fermat's Last Theorem

More than 350 years ago, French mathematician Pierre de Fermat wrote in the margin of a book that he had proved that the equation $x^n + y^n = z^n$ has no positive integer solutions when $n > 2$ but did not have room in the margin for the proof. (When $n = 2$, there are infinitely many solutions, including $3^2 + 4^2 = 5^2$ and $5^2 + 12^2 = 13^2$.) Since then, many distinguished mathematicians have tried in vain to discover a proof of Fermat's assertion. In 1993 Andrew Wiles of Princeton University electrified the mathematics community by announcing that he had proved Fermat's theorem after seven years of effort. His proof, which ran 200 pages, relied heavily on group theory. Because of Wiles' solid reputation and the fact that his approach was based on deep results that had already shed much light on the problem, many experts in the field believed that Wiles had succeeded where so many others had failed. Wiles' achievement was reported in newspapers and maga-

zines around the world. *The New York Times* ran a front-page story on it and one TV network announced it on the evening news. Wiles even made *People* magazine's list of the 25 most intriguing people of 1993! In San Francisco a group of mathematicians rented a 1200-seat movie theater and sold tickets for $5.00 each for public lectures on the proof. Scalpers received as much as $25.00 a ticket for the sold-out event.

The bubble soon burst when experts had an opportunity to scrutinize Wiles' manuscript. By December Wiles released a statement that he was working to resolve a gap in the proof. In September of 1994, a paper by Richard Taylor, a former student of Wiles, and Wiles circumvented the gap in the original proof. Mathematicians who have seen the paper have not found any errors. One mathematician was quoted as saying "the exuberance is back." In 1996 Wiles received the prestigious Wolf Prize for his "spectacular contributions to number theory."

facts that $-5 = 2 \bmod 7$ and $-6 = 1 \bmod 7$, we see that the inverse of $\begin{bmatrix} 4 & 5 \\ 6 & 3 \end{bmatrix}$ is

$$\begin{bmatrix} (3 \cdot 5) \bmod 7 & (2 \cdot 5) \bmod 7 \\ (1 \cdot 5) \bmod 7 & (4 \cdot 5) \bmod 7 \end{bmatrix} = \begin{bmatrix} 1 & 3 \\ 5 & 6 \end{bmatrix}.$$

(The reader should check that $\begin{bmatrix} 4 & 5 \\ 6 & 3 \end{bmatrix}\begin{bmatrix} 1 & 3 \\ 5 & 6 \end{bmatrix} = \begin{bmatrix} 1 & 0 \\ 0 & 1 \end{bmatrix}$ in $GL(2, Z_7)$). ▲

EXAMPLE 16 Nonzero Elements of Z_n
The set $\{1, \dots, n - 1\}$ *is* a group under multiplication modulo n if and only if n is prime. When n is prime, the set $\{1, \dots, n - 1\}$ is simply $U(n)$ (see Example 8). When n is not prime, we may write $n = ab$, where a and b belong to $\{1, \dots, n - 1\}$. But then this set is not closed since $ab = n = 0 \bmod n$. ▲

Our last example introduces a family of groups

that is important in computer science and data communication.

EXAMPLE 17 Binary Strings For a fixed, positive integer n, let Z_2^n denote the set of strings of the form $a_1 a_2 \cdots a_n$, where each a_i is 0 or 1. We add two strings $a_1 a_2 \cdots a_n$ and $b_1 b_2 \cdots b_n$ componentwise modulo 2. That is,

$$a_1 a_2 \cdots a_n + b_1 b_2 \cdots b_n = c_1 c_2 \cdots c_n,$$

where $c_i = (a_i + b_i) \bmod 2$. For instance, $11000111 + 01110110 = 10110001$; $00111011 + 01100101 = 01011110$; and $10011100 + 10011100 = 00000000$. Members of Z_2^n are called *binary strings of length n.* Notice that the string $00 \cdots 0$ is the identity and every element is its own inverse. Closure and associativity follow from the corresponding properties of Z_2 itself. ▲

Table 1 summarizes many of the specific groups

				TABLE 1		
Group	Operation	Identity	Form of element		Inverse	Abelian
Z	Addition	0	k		$-k$	Yes
Q^+	Multiplication	1	m/n $m, n \geq 0$		n/m	Yes
Z_n	Addition mod n	0	k		$n - k$	Yes
$R^{\#}$	Multiplication	1	x		$1/x$	Yes
$GL(2, R)$	Matrix multiplication	$\begin{bmatrix} 1 & 0 \\ 0 & 1 \end{bmatrix}$	$\begin{bmatrix} a & b \\ c & d \end{bmatrix}$ $ad - bc \neq 0$		$\begin{bmatrix} \dfrac{d}{ad - bc} & \dfrac{-b}{ad - bc} \\ \dfrac{-c}{ad - bc} & \dfrac{a}{ad - bc} \end{bmatrix}$	No
$U(n)$	Multiplication mod n	1	k $\mathrm{GCD}(k, n) = 1$		solution to $kx = 1 \bmod n$	Yes
$SL(2, R)$	Matrix multiplication	$\begin{bmatrix} 1 & 0 \\ 0 & 1 \end{bmatrix}$	$\begin{bmatrix} a & b \\ c & d \end{bmatrix}$ $ad - bc = 1$		$\begin{bmatrix} d & -b \\ -c & a \end{bmatrix}$	No
D_{2n}	Composition	R_0	R_α, F		$R_{360 - \alpha}, F$	No
Z_2^n	Componentwise addition mod 2	$00 \cdots 0$	$a_1 a_2 \cdots a_n$ $a_i = 0$ or 1		$a_1 a_2 \cdots a_n$	Yes

that we have presented thus far. The symbol F denotes a reflection; R_α denotes a rotation of α degrees.

As the preceding examples demonstrate, the notion of a group is indeed very broad. The goal of the axiomatic approach is to find a set of properties that are general enough to permit many diverse examples that possess these properties and that are specific enough to allow one to deduce many interesting consequences of these properties. All one knows or needs to know is that these operations, whatever they may be, have certain properties. We then seek to deduce consequences of these properties. It must be remembered, however, that when a specific group is being discussed, a specific operation must be given (at least implicitly).

Elementary Properties of Groups

Now that we have seen many diverse examples of groups, we wish to deduce some properties they share. The definition itself raises some fundamental questions. Every group has an identity. Could a group have more than one? Every group element has *an* inverse. Could an element have more than one? The examples suggest not. But examples can only suggest. One cannot *prove every* group has a unique identity by looking at examples, because each example inherently has properties that may not be shared by all groups. We are forced to restrict ourselves to the properties that all groups must have. That is, we must view groups as abstract entities rather than argue by example. The next three theorems illustrate the abstract approach.

Theorem 1. In a group G, there is only one identity element.

Proof. Suppose both e and e' are identities of G. Then, by definition,

1. $ae = a$ for all a in G, and

2. $e'a = a$ for all a in G.

The choice of $a = e'$ in (1) and $a = e$ in (2) yields $e'e = e'$ and $e'e = e$. Thus, e and e' are both equal to $e'e$ and so are equal to each other.

▲ ▲ ▲

Because of this theorem we may unambiguously speak of "the identity" of a group and denote it by "e" (because the German word for identity is "Einheit").

Theorem 2. In a group G, the right and left cancellation laws hold; that is, $ba = ca$ implies $b = c$, and $ab = ac$ implies $b = c$.

Proof. Suppose $ba = ca$. Let a' be an inverse of a. Then, multiplying on the right by a' gives $(ba)a' = (ca)a'$. Associativity yields $b(aa') = c(aa')$. Then, since $aa' = e$, we have $be = ce$ and, therefore, $b = c$ as desired. Similarly, one can prove $ab = ac$ implies $b = c$ by multiplying by a' on the left.

▲ ▲ ▲

Be careful using Theorem 2, the cancellation theorem. Because groups need not be commutative, you can cancel only on the right or on the left. In particular, it is not always true that $ab = ca$ implies $b = c$.

A consequence of the cancellation property is the uniqueness of inverses.

Theorem 3. For each element a in a group G, there is a unique element b in G such that $ab = ba = e$.

Proof. Suppose b and c are both inverses of a. Then $ab = e$ and $ac = e$, so that $ab = ac$. Now, by Theorem 2, we can cancel a to obtain $b = c$.

Notation for Groups

As was the case with the identity element, it is reasonable, in view of Theorem 3, to speak of "the inverse" of an element g of a group; and, in fact, we may unambiguously denote it by g^{-1}. This notation is suggested by that used for ordinary real numbers under multiplication. Similarly, when n is a positive integer, g^n is used to denote the product

$$\underbrace{gg \cdots g}_{n \text{ factors}}$$

We define $g^0 = e$; when n is negative, we define $g^n = (g^{-1})^{|n|}$ (for example, $g^{-3} = (g^{-1})^3$). With this notation, the familiar laws of exponents hold for groups; that is, for all integers m and n and any group element g, we have $g^m g^n = g^{m+n}$. Although the way one manipulates the group expressions $g^m g^n$ and $(g^m)^n$ coincides with the laws of exponents for real numbers, the laws of exponents fail to hold for expressions involving two group elements. Thus, for groups in general, $(ab)^n \neq a^n b^n$ (commutativity is needed for equality to hold in all cases). For example, in the dihedral group shown in Table 1 of Section 1 of this chapter, $(R_{90}H)^2 \neq R_{90}^2 H^2$.

Also, one must be careful with this notation when dealing with a specific group whose binary operation is addition and is denoted by $+$. The definition and group properties expressed in multiplicative notation must be translated to additive notation. For example, in additive notation the inverse of g is written as $-g$. Likewise, for example, g^3 means $g + g + g$ and is usually written as $3g$, whereas g^{-3} means $(-g) + (-g) + (-g)$ and is written as $-3g$. When this notation is used, do not interpret ng as combining n and g under the group operation; n may not even be an element of the group! Unlike the case for real numbers, in an abstract group, we do not permit noninteger exponents such as $g^{1/2}$. Table 2 shows the common notation and corresponding terminology for groups under multiplication and groups under addition. As is the case for real numbers, we use $a - b$ as an abbreviation for $a + (-b)$.

TABLE 2			
Multiplicative group		*Additive group*	
$a \cdot b$ or ab	Multiplication	$a + b$	Addition
e or 1	Identity or one	0	Zero
a^{-1}	Multiplicative inverse of a	$-a$	Additive inverse of a
a^n	Power of a	na	Multiple of a
ab^{-1}	Quotient	$a - b$	Difference

Mathematicians prefer to use the multiplicative notation when dealing with groups in general, reserving the additive notation for specific instances where the operation is traditionally addition, such as the case with the integers and the integers modulo n.

Because of the associative property, we may unambiguously write the expression abc, for this can

Math Practitioner: Patricia Hersh

After finishing her junior year at Harvard in 1994, where she was majoring in mathematics and computer science, Patricia Hersh accepted a summer research position at the Center for Communications Research (CCR) at Princeton, New Jersey.

Patricia Hersh

Joe: What do mathematicians at the CCR work on?

Tricia: They work on cryptologic problems to provide secure communications for the United States government and businesses.

Joe: Were you required to have a security clearance?

Tricia: Yes. I took a lie detector test and they made a background check.

Joe: What kind of mathematics is used at the CCR?

Tricia: They employ a mixture of coding theory, number theory, group theory, statistics, and computer science to analyze various methods for encoding and decoding data.

Joe: Was it a good experience for you?

Tricia: Definitely. It was inspiring to see how the math I had learned in my classes could be used in beautiful and surprising applications. I appreciated the emphasis not on competition, but on everyone contributing what they could to solving compelling problems.

Joe: Were there other undergraduate math students there with you?

Tricia: Yes, I was one of four that summer. The National Security Agency also has a summer intern program for undergraduate math majors. They had 18 interns in 1994.

reasonably be interpreted as only $(ab)c$ or $a(bc)$, which are equal. In fact, by using repeated application of the associative property, one can prove a general associative property that essentially means balanced pairs of parentheses can be inserted or deleted at will without affecting the value of a product involving any number of group elements. Thus,

$$a^2(bcdb^2) = a^2b(cd)b^2 = (a^2b)(cd)b^2 = a(abcdb)b,$$

and so on.

Applications to Cryptography

In many situations there is a desire for security against the unauthorized interpretation of coded data. Among the most obvious situations are banking, military, and diplomatic transmissions. Premium television services such as Home Box Office (HBO), Showtime, and Disney also need to protect their television signals to local cable operators and satellite dish subscribers from being received free by dish owners.

We conclude this section with two applications of the material presented here to *cryptography*—the science of sending and deciphering secret messages.

Data Security

Because computers are built from two-state electronic components, it is natural to represent information as binary strings (see Example 17). That is, as elements of the group Z_2^n. The fact that in Z_2^n

$$a_1 a_2 \cdots a_n + b_1 b_2 \cdots b_n = 00 \cdots 0$$

if and only if the sequences are identical is the basis for a data-security system used by HBO to protect its television signals.

Beginning in 1984, HBO scrambled its signal. To unscramble the signal, a cable system operator or dish owner who pays a monthly fee has to have a "key." A password, which is changed monthly, is transmitted along with the scrambled signal. The technical term for the scrambling process is *encryption*. Although HBO uses binary strings of length 56, we will illustrate the method with strings of length 8. Let us say that the password for this month is p. Each authorized user of the service is assigned a binary string uniquely associated with him or her. Let us call these the *keys* and label them $k_1, k_2, \ldots$ (the keys do not change from month to month). HBO transmits the password p and the sequences $k_1 + p, k_2 + p, \ldots$ (that is, one sequence for each authorized user). A microprocessor in each subscriber's decoding box adds its key, say k_i, to each of the sequences. That is, it calculates $k_i + (k_1 + p), k_i + (k_2 + p), \ldots$. As it does so, the microprocessor compares each of these calculated sequences with the correct password p. When one of the sequences matches p, the microprocessor will unscramble the signal.

Notice that the correct password p will be produced precisely when k_i is added to $k_i + p$, since $k_i + (k_i + p) = (k_i + k_i) + p = 00 \cdots 0 + p = p$ and $k_i + (k_j + p) \neq p$ when $k_j \neq k_i$. If a subscriber with key k_i fails to pay the monthly bill, HBO can terminate the subscriber's service by not transmitting the sequence $k_i + p$ the next month.

To illustrate, let us say that the password for this month is $p = 10101100$ and your key is $k = 00111101$. One of the sequences transmitted by HBO is $k + p$: $00111101 + 10101100 = 10010001$. Your decoder box adds your key $k = 00111101$ to all the sequences received. Eventually, it finds the sequence obtained by adding the password to your key (namely, $p + k = 10010001$) and calculates $00111101 + 10010001 = 10101100$ to obtain the password p. This password then permits the decoder to unscramble the TV signal.

One might suspect that a computer hacker could find the password by simply trying a large number of possible keys until one "unlocks" the password. However, with sequences of length 56, there are 2^{56} possible keys, even though perhaps only a few million are used by HBO to unlock its monthly password and the number 2^{56} is so large (it exceeds 72 quadrillion) that even if one tries one billion possible keys, the chance of finding one that works is essentially 0.

Cryptography

In the mid-1970s, Ron Rivest, Adi Shamir, and Len Adleman devised an ingenious method that permits each person that is to receive a secret message to publicly tell how to scramble messages sent to him or her. And even though the method used to scramble the message is known publicly, only the person to whom it is intended will be able to unscramble the message. The idea is based on the facts that there exist methods for finding in a matter of seconds very large prime numbers (say about 100 digits long) and for multiplying them, but even the fastest computer would require years to factor integers with 200 digits using known algorithms.

So, the person who is to receive the message finds a pair of large primes p and q and chooses an integer r with $1 < r < (p-1)(q-1)$ so that r is relatively prime to $(p-1)(q-1)$ (that is, r and $(p-1) \cdot (q-1)$ have no common divisors other than 1). This person calculates $n = pq$ and announces that a message M is to be sent to him or her publicly as M^r mod n. Although r, n, and M^r are available to everyone, only the person who knows how to factor n as pq will be able to decipher the message.

The algorithm described here is called the *RSA public key encryption scheme* in honor of Rivest, Shamir, and Adleman, who developed the method.

Receiver

1. Pick very large primes p and q and compute $n = pq$.

2. Compute the least common multiple of $p-1$ and $q-1$; let us call it m.

3. Pick r so that it has no divisors in common with m other than 1 (any such r will do).

4. Find s so that $rs = 1$ modulo m. (To find s, raise r to successive powers until you reach 1 modulo m. The previous power of r mod m is s. For example, if $r = 3$ and $m = 16$, we have 3^4 mod $16 = 1$, so that $s = 3^3$ mod $16 = 11$.)

5. Publicly announce n and r, but keep s secret.

Sender

1. Convert the message to a string of digits.

2. Break up the message into uniformly sized blocks of digits appending 0s in the last block if necessary; call them $M_1, M_2, \ldots, M_k$. For example, for a string such as 2105092315, we could use $M_1 = 2105$, $M_2 = 0923$, and $M_3 = 1500$.

3. Check to see that the greatest common divisor of each M_i and n is 1. If not, n can be factored and the code is broken. (In practice, the primes p and q are so large that they exceed all M_i, so this step may be omitted.)

4. Calculate and send $R_i = M_i^r$ mod n.

Receiver

1. For each received message R_i, calculate R_i^s mod n.

2. Convert the string of digits back to a string of characters.

To present a simple example that nevertheless illustrates the essential features of the method, say we wish to send the message "IBM". We will send letters one at a time by converting the message to digits by replacing A by 1, B by 2, . . . , and Z by 26. (In practice, the messages are not sent one letter at a time. Rather, the entire message is converted to decimal form with A represented by 01, B by 02, . . . , and the space by 00. The message is then broken up into blocks of uniform size and the blocks are sent. See step 2 under Sender above.) So, the message "IBM" becomes 9213. The person to whom the message is to be sent has picked two primes p and q, say $p = 5$ and $q = 17$ (in actual practice, p and q would have a hundred or so digits), and a number r that has no divisors in common with the least common multiple m of $p - 1 = 4$ and $q -$

Leonard Adleman

Leonard Adleman grew up in San Francisco. He did not have any great ambitions for himself and, in fact, never even thought about becoming a mathematician. Upon enrolling at the University of California at Berkeley, Adleman at first declared he was going to be a chemist and then changed his mind and said he would be a doctor. Finally, he settled on a mathematics major. "I had gone through a zillion things and finally the only thing that was left where I could get out in a reasonable time was mathematics," he said.

Leonard Adleman

Adleman graduated in 1968, "wondering what I wanted to with my life." He took a job as a computer programmer at the Bank of America and soon after applied to medical school. Then, he said, "I decided maybe I should be a physicist." Once again, Adleman lost interest. "I didn't like doing experiments, I liked thinking about things," he said. Later, he returned to Berkeley, with the aim of earning a Ph.D. in computer science. "I thought that getting a Ph.D. in computer science would at least further my career," he said.

But while in graduate school, something else happened to Adleman. He finally understood the true nature and compelling beauty of mathematics. He discovered, he said, that mathematics "is less related to accounting than it is to philosophy. . . . People think of mathematics as some kind of practical art," Adleman said. But, he added, "the point when you become a mathematician is where you somehow see through this and see the beauty and power of mathematics."

Adleman received his Ph.D. in 1976 and immediately landed a job at MIT. There he met Rivest and Shamir, who were trying to invent an unbreakable public key system. They shared their excitement about the idea with Adleman, who greeted it with a polite yawn, thinking it impractical and not very interesting. Nevertheless, Adleman agreed to try to break the codes Rivest and Shamir proposed. Rivest and Shamir invented 42 coding systems, and each time Adleman broke the code. Finally, on their 43rd attempt, they hit upon what is now called the RSA scheme.

Asked what it is like to simply sit and think for six months, Adleman responded: "That's what a mathematician always does. Mathematicians are trained and inclined to sit and think. A mathematician can sit and think intensely about a problem for 12 hours a day, 6 months straight, with perhaps just a pencil and paper." The only prop he needs, he said, is a blackboard to stare at.

{*Adapted from an article by Gina Kolata,* The New York Times *(Dec. 13, 1994).*}

1 = 16 (in this case, $m = 16$) other than 1, say $r = 3$, and published $n = pq = 85$ and r in a public directory. The receiver computes m, the least common multiple of $p - 1$ and $q - 1$, and must find a number s so that $r \cdot s = 1 \bmod m$ (this is where the knowledge of p and q is necessary). That is, $3 \cdot s = 1 \bmod 16$. This number is 11. (The number s is unique since it is the inverse of 3 in $U(16)$.) We consult this directory to find n and r, then send the "scrambled" numbers $9^3 \bmod 85$, $2^3 \bmod 85$, and $13^3 \bmod 85$ rather than 9, 2, and 13, and the receiver will unscramble them. Thus we send

$$9^3 \bmod 85 = 49,$$

$$2^3 \bmod 85 = 8,$$

and

$$13^3 \bmod 85 = 72.$$

Now the receiver must take the numbers he or she receives, 49, 8, and 72, and convert them back to 9, 2, and 13 by calculating $49^{11} \bmod 85$, $8^{11} \bmod 85$, and $72^{11} \bmod 85$, respectively.

The calculation of $49^{11} \bmod 85$ can be simplified as follows:[1]

---

[1]To determine $49^2 \bmod 85$ with a calculator, enter 49×49 to obtain 2401, then divide 2401 by 85 to obtain 28.247058. Finally, enter $2401 - 28 \times 85$ to obtain 21.

$49 \bmod 85 = 49;$

$49^2 \bmod 85 = 2401 \bmod 85 = 21;$

$49^4 \bmod 85 = 49^2 \cdot 49^2 \bmod 85 = 21 \cdot 21 \bmod 85$
$\qquad = 441 \bmod 85 = 16 \bmod 85;$

$49^8 \bmod 85 = 49^4 \cdot 49^4 \bmod 85 = 16 \cdot 16 \bmod 85 = 1.$

So,

$49^{11} \bmod 85 = (49^8 \bmod 85)(49^2 \bmod 85)(49 \bmod 85)$
$\qquad = (1 \cdot 21 \cdot 49) \bmod 85$
$\qquad = 1029 \bmod 85$
$\qquad = 9 \bmod 85.$

Thus, the receiver has correctly determined the code for "I". The calculations for $8^{11} \bmod 85$ and $72^{11} \bmod 85$ are left as exercises for the reader. Notice that without knowing how pq factors, one cannot find the least common multiple of $p - 1$ and $q - 1$ (in our case, 16) and therefore the s that is needed to determine the intended message.

Why does the RSA method work? It succeeds because of a basic property of the group $U(n)$ and the choice of r. It so happens that the number m has the property that for each x having no common divisors with n except 1 (that is, any member of $U(n)$), we have $x^m = 1 \bmod n$. So, because each message M_i has no common divisors with n except 1, and r was chosen so that $rs = 1 + mt$ for some t, we have, modulo n,

$$R_i^s = (M_i^r)^s = M_i^{rs} = M_i^{1+mt} = M_i(M_i^m)^t = M_i 1^t = M_i.$$

1. Give two reasons why the set of odd integers under addition is not a group.

2. For the set of nonzero real numbers, is division associative?

3. Show that $\begin{bmatrix} 2 & 2 \\ 1 & 1 \end{bmatrix}$ does not have a multiplicative inverse.

4. Show that the group GL(2, R) of Example 10 is non-Abelian, by exhibiting a pair of matrices **A** and **B** in GL(2, R) such that $\mathbf{AB} \neq \mathbf{BA}$.

5. Explain why division is not a binary operation on the set of real numbers.

6. Give an example of group elements a and b with the property that $a^{-1}ba \neq b$.

7. Translate each of the following multiplicative expressions to its additive counterpart:
   (a) a^2b^3
   (b) $a^{-2}(b^{-1}c)^2$
   (c) $(ab^2)^{-3}c^2 = e$

8. For any elements a and b from a group and any integer n, prove that $(a^{-1}ba)^n = a^{-1}b^na$.

9. For each part below, explain how modular arithmetic can be used to answer the question.
   (a) If a clock indicates that it is now 4 p.m., what will it indicate in 75 hours?
   (b) In the game of Monopoly, if a piece is now on the third space after the "pass and go" space, what space will the piece be on after advancing 87 spaces? (A Monopoly board has 40 spaces.)
   (c) If the five-digit odometer of an automobile reads 97,000 miles now, what will it read in 12,000 miles?

10. Find the inverse of the element $\begin{bmatrix} 2 & 6 \\ 3 & 5 \end{bmatrix}$ in GL(2, Z_{11}).

11. Let a, b, and c be elements of a group. Solve the equation $axb = c$ for x. Solve $a^{-1}xa = c$ for x.

12. Show that the set $\{5, 15, 25, 35\}$ is a group under multiplication modulo 40.

13. In a group, given that $abc = e$, show that $bca = e$. Find an example that shows that $abc = e$ does not imply $bac = e$.

14. In a group, given that $(ab)^3 = e$, show that $(ba)^3 = e$. Is the same true if 3 is replaced by any integer?

15. Let n be a fixed positive integer. Show that the set of all complex numbers x such that $x^n = 1$ is a group under multiplication. Is the same true if n can vary with x?

16. Suppose an instructor assigned every other odd exercise as homework. What modular arithmetic equation do these numbers satisfy? Use this equation to decide if 51 or 65 is assigned.

17. Is the binary operation defined by the following table associative? Is it commutative?

	a	b	c	d
a	a	b	c	d
b	b	a	d	c
c	c	d	a	b
d	a	d	b	c

18. A teacher intended to give a typist a list of nine integers that form a group under multiplication modulo 91. Instead, one of the nine integers was inadvertently left out so that the list appeared as 1, 9, 16, 22, 53, 74, 79, 81. Which integer was left out? (This really happened!)

19. (Law of exponents for Abelian groups) Let a and b be elements of an Abelian group, and let n be any integer. Show that $(ab)^n = a^nb^n$. Is this also true for non-Abelian groups?

20. (Socks–shoes property) In a group, prove that $(ab)^{-1} = b^{-1}a^{-1}$. Find an example that shows it is possible to have $(ab)^{-2} \neq b^{-2}a^{-2}$. Find a non-Abelian example that shows it is possible to have $(ab)^{-1} = a^{-1}b^{-1}$ for some distinct nonidentity elements a and b. Draw an analogy between the statement $(ab)^{-1} = b^{-1}a^{-1}$ and the act of putting on and taking off your socks and shoes.

21. Prove that a group G is Abelian if and only if $(ab)^{-1} = a^{-1}b^{-1}$ for all a and b in G.

22. In a group, prove $(a^{-1})^{-1} = a$ for all a.

23. Let F denote a reflection in D_{20}. If R_α denotes a rotation of α degrees, express the element $(R_{36}F)^{-1}$ as a product without using negative exponents.

24. If $a_1, a_2, \ldots, a_n$ belong to a group, what is the inverse of $a_1 a_2 \cdots a_n$?

25. The integers 5 and 15 are among a collection of 12 integers that form a group under multiplication modulo 56. List all 12.

26. Give an example of a group with 105 elements. Give two examples of groups with 42 elements.

27. Let a and b belong to a group. Show that $a^3 = e$ if and only if $(b^{-1}ab)^3 = e$. Is the same true if 3 is replaced by any other integer?

28. Show that a group with exactly three elements must be Abelian.

29. Show that in a group table every element of the group appears exactly once in each row and in each column.

30. Suppose the table below is a group table. Fill in the blank entries.

	e	a	b	c	d
e	e	—	—	—	—
a	—	b	—	—	e
b	—	c	d	e	—
c	—	d	—	a	b
d	—	—	—	—	—

31. Prove that if $(ab)^2 = a^2b^2$ in a group G, then $ab = ba$.

32. Use the RSA scheme with $n = 85$ and $r = 3$ to determine the numbers sent for the message VIP.

33. Use the RSA scheme with $p = 5$, $q = 17$, and $r = 3$ to decode the received numbers 52 and 72.

34. In the RSA scheme with $p = 5$, $q = 17$, and $r = 5$, determine the value of s.

35. Explain what goes wrong in the RSA scheme if one tries using $p = 5$, $q = 17$, and $r = 4$.

SECTION 9.3 *Coding Theory*

Coding of Information

Did you know that bar code readers can detect certain kinds of errors that can occur in the printing of the code or reading of the code? Have you heard about compact disc players that can correct errors caused by dust and dirt? Have you seen a fax machine with an "error correction mode"? In this section you will learn a few of the many methods that are used to detect or correct errors in data transfer. The methods are applications of groups. We also illustrate a way data can be coded to reduce transmission time and storage space.

A *code* is a set of symbols that represents information. Codes existed thousands of years ago: hieroglyphics, the Greek alphabet, and Roman numerals are three examples. Many codes have been developed for a particular application: musical scores, the Morse code, and the "genetic code" used to describe the makeup of DNA. About 40 years ago an MIT graduate student named David Huffman invented a code that is now used in computers, high-definition television, modems, and even VCR Plus+ devices that automatically program a VCR. In recent decades coding schemes have been invented that do more than simply represent data. For example, errors in data from space probes, signals from compact discs, and transmissions from modems, fax machines, and high-definition television satellites can be corrected. In this section we examine some of the methods used to code information.

Decimal Codes

Let's begin with a coding scheme you encounter daily that uses the group Z_{10}. Most products sold in supermarkets have an identification number coded with bars that are read by optical scanners. In Figure 1 the identification number is 021000658978.

0 21000 65897 8   **FIGURE 1**   Bar code.

This code is called the Universal Product Code (UPC). In most cases the product is assigned a 12-digit number. The first six digits identify the manufacturer, the next five the product, and the last digit is extra information for detecting errors. It is called a *check digit.* (For many items, the twelfth digit is not printed but is always bar coded.) In Figure 1 the check digit is 8.

To explain how the check digit is calculated, it is convenient to introduce the scalar product notation for two k-tuples:

$$(a_1, a_2, \ldots, a_k) \cdot (w_1, w_2, \ldots, w_k) = a_1 w_1 + a_2 w_2 + \cdots + a_k w_k.$$

Although the digits $a_1, a_2, \ldots, a_{11}$ for a UPC number have no mathematical function, the last digit is chosen so that

$$(a_1, a_2, \ldots, a_{12}) \cdot (3, 1, 3, 1, \ldots, 3, 1) = 0 \bmod 10.$$

In particular, the check digit is the integer between 0 and 9 that when added to $3a_1 + a_2 + 3a_3 + \cdots + 3a_{11}$ results in a sum that ends with a zero. (Put another way, the check digit is the additive inverse of $3a_1 + a_2 + \cdots + 3a_{11}$ in the group Z_{10}.) To verify that the number in Figure 1 satisfies the above condition, we calculate

$$\begin{aligned} &0 \cdot 3 + 2 \cdot 1 + 1 \cdot 3 + 0 \cdot 1 + 0 \cdot 3 + 0 \cdot 1 + 6 \cdot 3 \\ &+ 5 \cdot 1 + 8 \cdot 3 + 9 \cdot 1 + 7 \cdot 3 + 8 \cdot 1 \\ &= 90 = 0 \bmod 10. \end{aligned}$$

The fixed k-tuple used in the calculation of check digits is called the *weighting vector.* Thus the UPC weighting vector is $(3, 1, 3, 1, \ldots, 3, 1)$.

Now, suppose a single error is made in entering the number in Figure 1 into a computer. Say, for instance, that 021000958978 is entered (notice that the seventh digit is incorrect). Then, the computer calculates

$$0 \cdot 3 + 2 \cdot 1 + 1 \cdot 3 + 0 \cdot 1 + 0 \cdot 3 + 0 \cdot 1 + 9 \cdot 3$$
$$+ 5 \cdot 1 + 8 \cdot 3 + 9 \cdot 1 + 7 \cdot 3 + 8 \cdot 1$$
$$= 99 = 9 \bmod 10.$$

Since the result is not 0, the entered number cannot be correct. We don't know what the error is or in which position it is. Nor do we know that only one error was made. But we do know that the number entered is in error. In such cases, the computer would request that the number be reentered.

But why should a weighting vector be used at all? One could detect all single-digit errors in a string $a_1 a_2 \cdots a_k$ by simply choosing a check digit a_{k+1} so that $a_1 + a_2 + \cdots + a_k + a_{k+1}$ is divisible by 10. For surely any error in exactly one position will result in a sum that does not end with a 0. Yes, but the second most common human error, the transposition of adjacent digits, is never detected by simply summing the digits, whereas nearly all such errors are detected when a weighting vector is employed. To illustrate, let us say that the identification number given in Figure 1 is entered as 021000658798. Notice that the last two digits preceding the check digit have been transposed. Calculating the scalar product, we obtain 94, which is not 0 mod 10, so we have detected an error. In fact, the only undetected transposition errors of adjacent digits a and b are those where $|a - b| = 5$. To verify this, we observe that a transposition error of the form

$$a_1 a_2 \cdots a_i a_{i+1} \cdots a_{12} \rightarrow a_1 a_2 \cdots a_{i+1} a_i \cdots a_{12}$$

is undetected if and only if

$$(a_1, a_2, \ldots, a_{i+1}, a_i, \ldots, a_{12}) \cdot (3, 1, 3, 1, \ldots, 3, 1)$$
$$= 0 \bmod 10.$$

Since $a_1 a_2 \cdots a_{12}$ satisfies $(a_1, a_2, \ldots, a_{12}) \cdot (3, 1, 3, 1, \ldots, 3, 1) = 0 \bmod 10$, the previous equation is equivalent to

$$(a_1, a_2, \ldots, a_{i+1}, a_i, \ldots, a_{12}) \cdot (3, 1, 3, 1, \ldots, 3, 1)$$
$$- (a_1, a_2, \ldots, a_i, a_{i+1}, \ldots, a_{12})$$
$$\cdot (3, 1, 3, 1, \ldots, 3, 1) = 0 \bmod 10.$$

This expression reduces to $\pm 2(a_{i+1} - a_i) = 0 \bmod 10$. But since a_i and a_{i+1} are between 0 and 9, we see that $a_{i+1} - a_i$ must be ± 5. Although the UPC scheme detects all single errors, it does not detect all instances in which two or more errors occur.

Banks in the United States employ a different method to append a check digit to their identification number printed on checks. Identification numbers printed on bank checks (see Figure 2) consist of an eight-digit number $a_1 a_2 \cdots a_8$ and a check digit a_9 so that

$$(a_1, a_2, \ldots, a_9) \cdot (7, 3, 9, 7, 3, 9, 7, 3, 9) = 0 \bmod 10.$$

As was the case for the UPC scheme, this method detects all single-digit errors and all errors involving

FIGURE 2 Bank check with identification number 07100001 and check digit 3.

the transposition of adjacent digits *a* and *b* except when $|a - b| = 5$. But it also detects most errors of the form $\cdots abc \cdots \rightarrow \cdots cba \cdots$, whereas the UPC method detects no errors of this form.

That the numbers 1, 3, 7, and 9 used in the weighting vectors for the UPC scheme and the bank scheme comprise the group $U(10)$ is no coincidence. Indeed, it is not hard to prove that an identification number $a_1 a_2 \cdots a_k$ that satisfies

$$(a_1, a_2, \ldots, a_k) \cdot (w_1, w_2, \ldots, w_k) = 0 \bmod 10$$

will detect all single digit errors if and only if each w_i belongs to the group $U(10)$.

In some schemes modular arithmetic is applied directly to the identification number to obtain the check digit. (That is, no weighting vector is employed.) For example, the check digit used by the United States Postal Service on money orders is simply the serial number of the money order modulo 9. Thus, the money order with the serial number 6244256379 (see Figure 3) has the check digit 3 since $6244256379 = 3 \bmod 9$. If the number 62442563793 were incorrectly entered into a computer (programmed to calculate the check digit) as, say, 62492563793 (error in position 4), the machine would calculate the check as 8, whereas the entered check digit is 3. Thus the error is detected.[2]

Airline companies, the United Parcel Service, and Avis and National rental car companies use the modulo 7 value of identification numbers to assign check digits. Thus, the airline identification number 00121410700093 (see Figure 4) has appended the

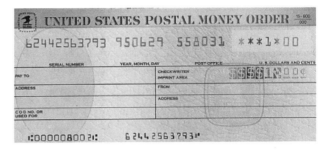

FIGURE 3 Money order with serial number 6244256379 and check digit 3.

FIGURE 4 Airline ticket with identification number 00121410700093 and check digit 4.

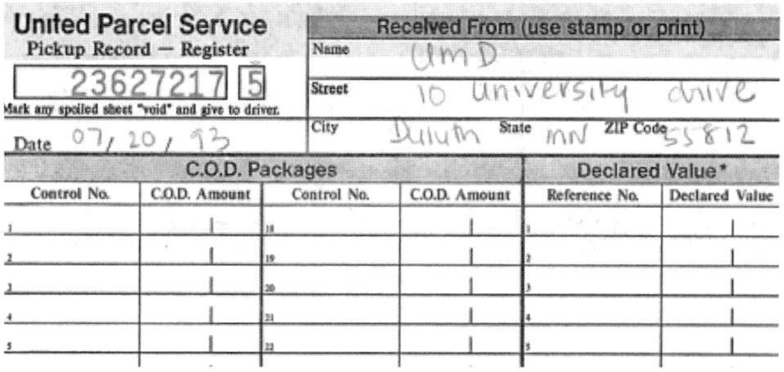

check digit 4 because $00121410700093 = 4 \bmod 7$. Similarly, the UPS pickup record number 23627217, shown in Figure 5, has appended a check digit 5.

FIGURE 5 UPS Pickup Record number 23627217 with check digit 5.

---

[2]It is a property of division by 9 that, for any positive integer $a_1 a_2 \cdots a_k$, it is true that $a_1 a_2 \cdots a_k \bmod 9 = (a_1 + a_2 + \cdots + a_k) \bmod 9$. For example, $6244256379 \bmod 9 = (6 + 2 + 4 + 4 + 2 + 5 + 6 + 3 + 7 + 9) \bmod 9 = 48 \bmod 9 = 3$.

The money order, airline, and UPS schemes do not detect all single-digit errors nor all transposition errors. In particular, for the money order scheme, any substitution of b for a among the noncheck digits or the transposition of adjacent digits a and b among the noncheck digits will not be detected when $|a - b| = 9$. For the airline and UPS schemes, any substitution of b for a among the noncheck digits or the transposition of adjacent digits a and b among the noncheck digits will not be detected when $|a - b| = 7$.

It is possible to detect 100% of single-digit errors and 100% of transposition errors involving adjacent digits with a single check digit. To see how to create such a scheme, we use a function σ defined as

$$\sigma(0) = 1, \qquad \sigma(5) = 8,$$
$$\sigma(1) = 5, \qquad \sigma(6) = 3,$$
$$\sigma(2) = 7, \qquad \sigma(7) = 0,$$
$$\sigma(3) = 6, \qquad \sigma(8) = 9,$$
$$\sigma(4) = 2, \qquad \sigma(9) = 4$$

and a group operation $*$ defined by the group table given in Table 1. Now for any string of digits

TABLE 1

*	0	1	2	3	4	5	6	7	8	9
0	0	1	2	3	4	5	6	7	8	9
1	1	2	3	4	0	6	7	8	9	5
2	2	3	4	0	1	7	8	9	5	6
3	3	4	0	1	2	8	9	5	6	7
4	4	0	1	2	3	9	5	6	7	8
5	5	9	8	7	6	0	4	3	2	1
6	6	5	9	8	7	1	0	4	3	2
7	7	6	5	9	8	2	1	0	4	3
8	8	7	6	5	9	3	2	1	0	4
9	9	8	7	6	5	4	3	2	1	0

$a_1 a_2 \cdots a_{n-1}$, we append the check digit a_n so that $\sigma(a_1) * \sigma^2(a_{n-2}) * \cdots * \sigma^{n-1}(a_{n-1}) * a_n = 0$. (Here $\sigma^2(x) = \sigma(\sigma(x))$, $\sigma^3(x) = \sigma^2(\sigma(x))$, etc.) For example, say we have the number 3971 and wish to append a check digit using this method. We first calculate

$$\sigma(3) = 6,$$
$$\sigma^2(9) = \sigma(\sigma(9)) = \sigma(4) = 2,$$
$$\sigma^3(7) = \sigma^2(\sigma(7)) = \sigma^2(0) = \sigma(\sigma(0)) = \sigma(1) = 5,$$
$$\sigma^4(1) = \sigma^3(\sigma(1)) = \sigma^3(5) = \sigma^2(\sigma(5)) = \sigma^2(8) = \sigma(\sigma(8)) = \sigma(9) = 4.$$

Then we use Table 1 to compute $6 * 2 * 5 * 4 = (6 * 2) * 5 * 4 = 9 * 5 * 4 = (9 * 5) * 4 = 4 * 4 = 3$. Finally, we choose the check digit a_5 as 2 because from Table 1 we see that $3 * 2 = 0$.

At this point you should be bursting with questions: What is special about σ? Why is the operation $*$ used? Does this scheme really do all that is claimed? Well, there are two things special about σ. First, σ is a permutation on the set $\{0, 1, \ldots, 9\}$. This means that σ is a function from $\{0, 1, \ldots, 9\}$ to itself with the property that $\sigma(b) \neq \sigma(c)$ when $b \neq c$. And second, $b * \sigma(c) \neq c * \sigma(b)$ when $b \neq c$ (as can be checked by a case-by-case analysis). The operation $*$ is used for two reasons. First, we need a group operation on the digits 0 through 9 so that we can do computations and simplify equations (for which associativity, an identity, and inverses are needed). Second, the operation must be noncommutative since it can be proved that there is no commutative group of order 10 for which there is a permutation σ with the property that $b * \sigma(c) \neq c * \sigma(b)$ for distinct b and c. (In fact, the multiplication table for $*$ is simply a numerical representation of D_{10}, the dihedral group of order 10. The digits 0 through 4 are the rotations, whereas 5 through 9 are the reflections.)

To see why all the properties we have stated are

necessary and that the scheme does detect 100% of all single-digit errors and 100% of all transpositions involving adjacent digits, we will examine two special cases that illustrate the general arguments. Let us say we have a number $a_1a_2a_3a_4$ satisfying $\sigma(a_1) * \sigma^2(a_2) * \sigma^3(a_3) * a_4 = 0$ and that we make an error in position 2 so that $a_1a_2a_3a_4$ becomes $a_1a_2'a_3a_4$, where $a_2 \neq a_2'$. For this error to go undetected, we must also have $\sigma(a_1) * \sigma^2(a_2') * \sigma^3(a_3) * a_4 = 0$. Thus,

$$\sigma(a_1) * \sigma^2(a_2) * \sigma^3(a_3) * a_4 = \sigma(a_1) * \sigma^2(a_2') * \sigma^3(a_3) * a_4.$$

Using cancellation (here we are using the fact that D_{10} is a group), we reduce this to

$$\sigma^2(a_2) = \sigma^2(a_2').$$

Now we use the fact that σ^2 is a permutation (the composition of permutations is a permutation) to conclude that $a_2 = a_2'$. Since this contradicts our assumption that $a_2 \neq a_2'$, there are no undetected single-digit errors.

Now suppose we have a transposition error of the form $a_1a_2a_3a_4 \rightarrow a_1a_3a_2a_4$, where $a_2 \neq a_3$. For this error to go undetected, we must have

$$\sigma(a_1) * \sigma^2(a_2) * \sigma^3(a_3) * a_4 = \sigma(a_1) * \sigma^2(a_3) * \sigma^3(a_2) * a_4.$$

Using cancellation once again, we reduce the preceding to

$$\sigma^2(a_2) * \sigma^3(a_3) = \sigma^2(a_3) * \sigma^3(a_2).$$

Finally, letting $b = \sigma^2(a_2)$ and $c = \sigma^2(a_3)$, we reduce the last equation to $b * \sigma(c) = c * \sigma(b)$. But σ was chosen precisely so that this equality cannot occur when $b \neq c$. In our situation, the equality cannot occur when $\sigma^2(a_2) \neq \sigma^2(a_3)$ and, since σ^2 is a per-

mutation, the equality cannot occur when $a_2 \neq a_3$. Thus all transpositions of adjacent digits are detected.

EXAMPLE 1   German Banknotes   The scheme involving σ and D_{10} is used to append a check digit to German banknotes. Each banknote has a 10-character serial number comprised of letters and numbers. The letters are assigned numerical values as follows:

A	D	G	K	L	N	S	U	Y	Z
0	1	2	3	4	5	6	7	8	9

Upon converting letters to digits, the resulting number $a_1a_2 \cdots a_{10}$ is assigned the check digit a_{11} so that

$$\sigma(a_1) * \sigma^2(a_2) * \cdots * \sigma^{10}(a_{10}) * a_{11} = 0.$$

To simplify the procedure, here is the table of values for $\sigma, \sigma^2, \ldots, \sigma^{10}$.

	0	1	2	3	4	5	6	7	8	9
σ	1	5	7	6	2	8	3	0	9	4
σ^2	5	8	0	3	7	9	6	1	4	2
σ^3	8	9	1	6	0	4	3	5	2	7
σ^4	9	4	5	3	1	2	6	8	7	0
σ^5	4	2	8	6	5	7	3	9	0	1
σ^6	2	7	9	3	8	0	6	4	1	5
σ^7	7	0	4	6	9	1	3	2	5	8
σ^8	0	1	2	3	4	5	6	7	8	9
σ^9	1	5	7	6	2	8	3	0	9	4
σ^{10}	5	8	0	3	7	9	6	1	4	2

The banknote (featuring the mathematician Carl Gauss) in the following illustration has serial number AU3630934N and check digit 7. To verify that 7 is the appropriate check digit, we observe that

$$\sigma(0) * \sigma^2(7) * \sigma^3(3) * \sigma^4(6) * \sigma^5(3) * \sigma^6(0) * \sigma^7(9)$$
$$* \sigma^8(3) * \sigma^9(4) * \sigma^{10}(5) * 7 = 1 * 1 * 6 * 6 * 6$$
$$* 2 * 8 * 3 * 2 * 9 * 7 = 0,$$

▲

In view of the complicated nature of the scheme just discussed, it is natural to wonder if there is a scheme that uses modular arithmetic to detect 100% of single-digit errors and 100% of transposition errors involving adjacent digits. Such schemes do exist, but in every instance there is some drawback. The ISBN scheme used on books (see Exercise 10) is one such scheme that uses modulo 11. The drawback for using this method is that some numbers require the check "digit" to be 10, which is not a single character. In these cases the character X is assigned as the check "digit." Using an X could be avoided by simply not issuing numbers that require a check digit of 10. But this is also a drawback.

Error Correction

What about error correction? Suppose you have a number such as 73245018 and you would like to be sure that even if a single mistake were made in entering this number into a computer, the computer would nevertheless be able to determine the correct number. (Think of it. You could make a mistake in dialing a telephone number but still get the correct phone to ring!) This is possible. But at least two check digits are required. One of the check digits determines the magnitude of any single-digit error, while the other check digit locates the position of the error. With these two pieces of information you can fix the error. To illustrate the idea, let us say that we have 8-digit identification numbers $a_1 a_2 \cdots a_8$. We assign two check digits a_9 and a_{10} so that

$$a_1 + a_2 + \cdots + a_9 + a_{10} = 0 \bmod 11$$

and

$$(a_1, a_2, \ldots, a_9, a_{10}) \cdot (1, 2, 3, \ldots, 10) = 0 \bmod 11$$

are satisfied.

E X A M P L E 2 Error Correction Let's try an example. Say our number before appending the check digits is 73245018. Then a_9 and a_{10} are chosen to satisfy

$$7 + 3 + 2 + 4 + 5 + 0 + 1 + 8 + a_9$$
$$+ a_{10} = 0 \bmod 11 \tag{1}$$

and

$$7 \cdot 1 + 3 \cdot 2 + 2 \cdot 3 + 4 \cdot 4 + 5 \cdot 5 + 0 \cdot 6$$
$$+ 1 \cdot 7 + 8 \cdot 8 + a_9 \cdot 9 + a_{10} \cdot 10$$
$$= 0 \bmod 11. \tag{2}$$

Since $7 + 3 + 2 + 4 + 5 + 0 + 1 + 8 = 30 = 8 \bmod 11$, Eq. (1) reduces to

$$8 + a_9 + a_{10} = 0 \bmod 11. \tag{1'}$$

Likewise, since $7 \cdot 1 + 3 \cdot 2 + 2 \cdot 3 + 4 \cdot 4 + 5 \cdot 5 + 0 \cdot 6 + 1 \cdot 7 + 8 \cdot 8 = 10 \bmod 11$, Eq. (2) reduces to

$$10 + 9a_9 + 10a_{10} = 0 \bmod 11. \tag{2'}$$

How do we solve Eqs. $(1')$ and $(2')$ simultaneously? The same way two linear equations are usually solved, by elimination. Thus, multiplying Eq. $(1')$ by 10 and observing that $80 = 3$ mod 11, we have

$$3 + 10a_9 + 10a_{10} = 0 \text{ mod } 11. \qquad (1'')$$

Subtracting Eq. $(2')$ from Eq. $(1'')$ yields $-7 + a_9 = 0$ mod 11. Thus $a_9 = 7$. Now substituting $a_9 = 7$ into Eq. $(1')$, we obtain $a_{10} = 7$ as well. So, the number is encoded as 7324501877. Now let us suppose that this number is erroneously entered into a computer programmed with our encoding scheme as 7824501877 (an error in position 2). The computer calculates $7 + 8 + 2 + 4 + 5 + 0 + 1 + 8 + 7 + 7 = 49 = 5$ mod 11. This means that some digit is 5 too large (assuming only one error has been made). But which digit? Say the error is in position i. Then the second scalar product has the form $a_1 \cdot 1 + a_2 \cdot 2 + \cdots + (a_i + 5)i + a_{i+1} \cdot (i + 1) + \cdots + a_{10} \cdot 10 = (a_1, a_2, \ldots, a_{10}) \cdot (1, 2, \ldots, 10) + 5i = 0 + 5i$ mod 11. So, since $5i$ mod 11 = $(7, 8, 2, 4, 5, 0, 1, 8, 7, 7) \cdot (1, 2, 3, 4, 5, 6, 7, 8, 9, 10) = 10$ mod 11, we see that $i = 2$. Our conclusion: The digit in position 2 is 5 too large. We have successfully corrected the error. ▲

Although our error was deliberately selected to make the arithmetic simple, the method works for all single-digit errors. Here is an illustration where the arithmetic is a bit tricky.

E X A M P L E 3 Error Correction Say the error is 7324501877 → 7324201877 (error in the fifth position.) Then, since $7 + 3 + 2 + 4 + 2 + 0 + 1 + 8 + 7 + 7 = 8$ mod 11, we

know that one digit is 8 too much. (In fact, in this case, the digit in the fifth position is 3 too little rather than 8 too much but since $-3 = 8$ mod 11, we will still arrive at the correct digit assuming only one error has been made.) Also, since $7 \cdot 1 + 3 \cdot 2 + 2 \cdot 3 + 4 \cdot 4 + 2 \cdot 5 + 0 \cdot 6 + 1 \cdot 7 + 8 \cdot 8 + 8 \cdot 9 + 7 \cdot 10 = 7$ mod 11. We know the value of 7 is the result of an error of 8 too much in some position i. That is, $8i = 7$ mod 11. This time the value of i is not apparent. Not to worry. As long as the modulus is a prime p, an equation of the form $ax = b$ mod p $(a \neq 0, b \neq 0)$ can always be uniquely solved for x. (Since a and b belong to the group $U(p)$, $x = a^{-1}b$ in $U(p)$.) In our case we merely observe that because $7 \cdot 8 = 1$ mod 11, we have $7(8i) = 7 \cdot 7$ mod 11 or $i = 5$ mod 11. Thus our analysis tells us that the digit 2 in the fifth position is 8 "too much." So, $2 - 8 = -6 = 5$ mod 11 = 5 must be the correct digit for the fifth position. And indeed it is. ▲

Binary Linear Codes

Because computers are built from two-state electronic components, it is natural in computer applications to represent information as strings of 0s and 1s. (For example, each cell on a floppy disk can be polarized by electric current into positive or negative states, which can be denoted by 0 and 1 respectively.) Over the past 40 years, highly sophisticated encoding and decoding schemes for error detection and error correction have been developed for such information.

The idea behind error-correction schemes is simple and one you often use. To illustrate, suppose you are reading the employment section of a newspaper

Neil Sloane

In the middle of Neil Sloane's office, which is in the center of AT&T Bell Laboratories, which in turn is at the heart of the Information Age, there sits a tidy little pyramid of shiny steel balls stacked up like oranges at a neighborhood grocery. Sloane has been pondering different ways to pile up balls of one kind or another for most of his professional life. Along the way he has become one of the world's leading researchers in the field of sphere packing, a field that has become indispensable to modern communications. Without it we might not have modems or compact discs or satellite photos of Neptune. "Computers would still exist," says Sloane. "But they wouldn't be able to talk to one another."

To exchange information rapidly and correctly, machines must code it. As it turns out, designing a code is a lot like packing spheres: Both involve cramming things together into the tightest possible arrangement. Sloane, fittingly, is also one of the world's leading coding theorists, not least because he has studied the shiny steel balls on his desk so intently.

Here's how a code might work. Imagine, for example, that you want to transmit a child's drawing that used every one of the 64 colors found in a jumbo box of Crayola crayons. For transmission, you could code each of those colors as a number—say, the integers from 1 to 64. Then you could divide the image into many small

Neil Sloane

units, or pixels, and assign a code to each one based on the color it contains. The transmission would then be a steady stream of those numbers, one for each pixel.

In digital systems, however, all those numbers would have to be represented as strings of 0s and 1s. Because there are 64 possible combinations of 0s and 1s in a 6-digit string, you could handle the entire Crayola palette with 64 different 6-digit "code words." For example, 000000 could represent the first color, 000001 the next color, 000010 the next, and so on.

But in a noisy signal, two different code words might look practically the same. A bit of noise, for example, might shift a spike of current to the wrong place, so that 001000 looks like 000100. The receiver might then wrongly color someone's eyes. An efficient way to keep the colors straight in spite of noise is to add four extra digits to the six-digit code words. The receiver, programmed to know the 64 permissible combinations, could now spot any other combination as an error introduced by noise and it would automatically correct the error to the "nearest" permissible color.

In fact, says Sloane, if any of those 10 digits were wrong, you could still figure out what the right crayon was.

{Source: Adapted from an article by David Berreby, Discover, October 1990.}

Jessie MacWilliams

Jessie MacWilliams

One of the important contributors to coding theory was Jessie MacWilliams. Born in 1917 in England, she received a B.A. in 1938 and an M.A. degree in 1939 from Cambridge University and came to America to study at Johns Hopkins. After one year at Johns Hopkins, she went to Harvard, where she studied a second year. In 1955, with three children aged 13, 11, and 9, MacWilliams became a programmer at Bell Labs, where she learned about coding theory. Although she made a major discovery about codes while a programmer, she could not obtain a promotion to a math research position with-

out a Ph.D. degree. She completed some of the requirements for the Ph.D. while working full-time at Bell Labs and looking after her family. She then returned to Harvard for one year (1961–1962) and finished her degree. Interestingly, both MacWilliams and her daughter Ann were studying mathematics at Harvard at the same time. MacWilliams returned to Bell Labs, where she remained until her retirement in 1983. While at Bell Labs she made many contributions to the subject of error-correcting codes, including her book: *The Theory of Error-Correcting Codes,* written jointly with Neil Sloane—a book that is still a leader in the field. One of her results of great theoretical importance is known as the "the MacWilliams identity." She died on May 27, 1990, at the age of 73.

and you see the phrase "must have a minimum of bive years experience." Instantly you detect an error since "bive" is not a word in the English language. Moreover, you are fairly confident that the intended word is "five." Why so? Because "five" is a word and it makes the phrase sensible. In other phrases, words such as "bike" or "give" might be sensible alternatives to "bive." Using the extra information provided by the context, we are often able to infer the intended meaning when errors occur. In essence, error-correcting coding schemes put messages in a "context" that often permits the inference of the correct message from an incorrect one.

As a simple example let us say that our set of messages consists of the 16 possible 4-tuples of 0s and 1s. (See the left column of Table 2.) We encode these messages with the aid of the diagram in Figure 6. Begin by placing the four message digits in the four overlapping regions I, II, III, and IV, with the digit in position 1 in region I, the digit in position 2 in region II, and so on.

For regions V, VI, and VII, assign 0 or 1 so that the total number of 1s in each circle is even (see Figure 7).

Now suppose the message 1001, which we have encoded as 1001101, is received as 0001101 (an er-

TABLE 2

Message	Encoder **G**	Code word
0000	→	0000000
0001	→	0001011
0010	→	0010111
0100	→	0100101
1000	→	1000110
1100	→	1100011
1010	→	1010001
1001	→	1001101
0110	→	0110010
0101	→	0101110
0011	→	0011100
1110	→	1110100
1101	→	1101000
1011	→	1011010
0111	→	0111001
1111	→	1111111

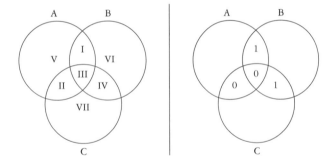

FIGURE 6 Diagram of message 1001.

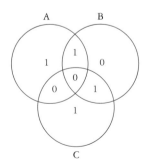

FIGURE 7 Diagram of encoded message 1001101. The sum of the digits in each circle is even.

ror in the first position). How would we know an error was made? We place each digit from the received message in its appropriate region as in Figure 8.

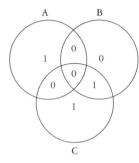

FIGURE 8 Diagram of received message 0001101. The sum of the digits in each circle is not even.

Noting that each of the circles A and B in Figure 8 has an odd number of 1s we instantly realize that something is wrong since the intended message had an even number of 1s in each circle. How do we correct the error? Since circles A and B have the wrong parity (an integer has even parity if it is an even integer and odd parity if it is odd) and C does not, the source of the error is the position of the diagram in circles A and B but not in circle C. See Figure 9.

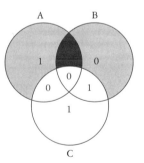

FIGURE 9 Circles A and B but not C have wrong parity.

That is, region I. Here we also see the value of using only 0s and 1s. Since the 0 in region I is incorrect, we know 1 is correct.

In practice, matrices rather that diagrams are used to encode and decode strings of 0s and 1s. Here is an example that shows how our diagram encoding scheme can be duplicated using a matrix. For each of the 16 possible 4-tuples of 0s and 1s in the left column of Table 2, we multiply on the right by the matrix **G** below and do all arithmetic modulo 2. The result is shown in the right column of Table 2.

$$
\mathbf{G} = \begin{bmatrix} 1 & 0 & 0 & 0 & 1 & 1 & 0 \\ 0 & 1 & 0 & 0 & 1 & 0 & 1 \\ 0 & 0 & 1 & 0 & 1 & 1 & 1 \\ 0 & 0 & 0 & 1 & 0 & 1 & 1 \end{bmatrix}
$$

Now, how is decoding done when a matrix is used to encode the messages? Say, for instance, that our message 1000, which has been encoded as $\mathbf{u} = 1000110$, is received as $\mathbf{r} = 0000110$. We simply compare $\mathbf{r}$ with each of the 16 possible correct messages in Table 2 and decode it as the one that agrees with $\mathbf{r}$ in the most positions. In this case, $\mathbf{r}$ will be decoded as $\mathbf{u}$ since it differs from $\mathbf{u}$ in only one position, whereas it differs from all others in the table in at least two positions. If there are two or more code words that agree with r in the most positions, we do not decode. This method is called *nearest neighbor* decoding. Assuming that errors occur independently, this means decoding each received message as the one it most likely represents.

To illustrate why the diagram method and the nearest neighbor method give the same results, let us look at the sixth position (a similar analysis applies to all positions). The diagram method yields 0 in region VI if the sum of the digits in regions I, III, and IV is even and 1 if the sum is odd. Thus the value put in position 6 of the code word is the sum of the digits in positions 1, 3, and 4 modulo 2. On the other hand, the sixth component of the code word obtained by multiplying a four-digit message

on the right by the matrix **G** is the scalar product of the message and (1, 0, 1, 1) mod 2. This scalar product is also the sum of the digits in positions 1, 3, and 4 modulo 2.

We now formalize the ideas we have introduced so far.

D E F I N I T I O N

An (n, k) *binary linear code* C is a set of 2^k n-dimensional vectors with components from Z_2 with the property that if $\mathbf{u}$ and $\mathbf{v}$ belong to C, then $\mathbf{u} + \mathbf{v}$ belongs to C. Members of C are called *code words*.

Notice that an (n, k) binary linear code is simply an Abelian group of n-tuples whose components are added modulo 2. You should think of an (n, k) binary linear code as a set of n-tuples each of which is composed of two parts: the message part, consisting of k digits; and the redundancy part, consisting of the remaining $n - k$ digits. Where there is no possibility of confusion, it is customary to denote an n-tuple $(a_1, a_2, \cdots, a_n)$ more simply as $a_1 a_2, \cdots, a_n$, as we did in Table 2.

E X A M P L E 4   A (7, 3) Binary Linear Code
The set

{0000000, 0010111, 0101011, 1001101, 1100110, 1011010, 0111100, 1110001}

is a (7, 3) binary code. To verify this, one must show that this set of 2^3 7-tuples is closed under addition modulo 2. ▲

E X A M P L E 5   A (4, 2) Binary Linear Code
The set {0000, 0101, 1010, 1111} is a (4, 2) binary code. Here it must be verified that the set of 2^2 4-tuples is closed under addition modulo 2. ▲

Once we have a binary linear code, how can we tell if it will correct errors? To facilitate this discussion, we introduce some terminology and notation.

D E F I N I T I O N

The *Hamming distance* between two vectors with the same number of components is the number of components in which they differ. The *Hamming weight* of a vector is the number of nonzero components of the vector. The *Hamming weight* of a binary linear code is the minimum weight of any nonzero vector in the code.

We will use $d(\mathbf{u}, \mathbf{v})$ to denote the Hamming distance between the vectors $\mathbf{u}$ and $\mathbf{v}$, and $wt(\mathbf{u})$ for the Hamming weight of the vector $\mathbf{u}$.

E X A M P L E 6   Distances and Weights Let $\mathbf{s} = 0010111$, $\mathbf{t} = 0101011$, $\mathbf{u} = 1001101$, and $\mathbf{v} = 1101101$. Then, $d(\mathbf{s}, \mathbf{t}) = 4$, $d(\mathbf{s}, \mathbf{u}) = 4$, $d(\mathbf{s}, \mathbf{v}) = 5$, $d(\mathbf{u}, \mathbf{v}) = 1$; and $wt(\mathbf{s}) = 4$, $wt(\mathbf{t}) = 4$, $wt(\mathbf{u}) = 4$, $wt(\mathbf{v}) = 5$. ▲

The Hamming distance and Hamming weight have the following important properties (see Exercise 29). (Note that for binary vectors $\mathbf{u}$ and $\mathbf{v}$, the vectors $\mathbf{u} - \mathbf{v}$ and $\mathbf{u} + \mathbf{v}$ are the same.)

Theorem 1.   For any vectors $\mathbf{u}$, $\mathbf{v}$, and $\mathbf{w}$, $d(\mathbf{u}, \mathbf{v}) \leq d(\mathbf{u}, \mathbf{w}) + d(\mathbf{w}, \mathbf{v})$ and $d(\mathbf{u}, \mathbf{v}) = wt(\mathbf{u} - \mathbf{v})$.

With the preceding definition and theorem, we can now explain why the codes given in Table 2 and Example 4 will correct any single error, but why the code in Example 5 will not.

Theorem 2.   If the Hamming weight of a binary linear code is at least $2t + 1$, then the code can correct any t or fewer errors, or the same code can detect any $2t$ or fewer errors.

Proof.   We will use nearest neighbor decoding; that is, we will decode any received vector $\mathbf{r}$ as the code word $\mathbf{r}'$ such that the Hamming distance $d(\mathbf{r}, \mathbf{r}')$ is a minimum. (If there is more than one such $\mathbf{r}'$, we do not decode. Typically a retransmission is requested.) Now, suppose that the intended code word $\mathbf{u}$ is received as the vector $\mathbf{r}$ and that at most t errors were made in transmission. We will show that among all code words, the intended code word $\mathbf{u}$ is the one closest to $\mathbf{r}$. That is, $\mathbf{r}' = \mathbf{u}$. By definition of the distance between two words, $d(\mathbf{r}, \mathbf{u}) \leq t$. If $\mathbf{w}$ is any code word other than $\mathbf{u}$, then $\mathbf{w} - \mathbf{u}$ is a nonzero code word (here we are using the facts that in binary $\mathbf{w} - \mathbf{u} = \mathbf{w} + \mathbf{u}$ and binary linear codes are closed under addition). Thus, because of our assumption that the weight of the code is at least $2t + 1$, we have

$$2t + 1 \leq wt(\mathbf{w} - \mathbf{u}) = d(\mathbf{w}, \mathbf{u}) \leq d(\mathbf{w}, \mathbf{r}) + d(\mathbf{r}, \mathbf{u}) \leq d(\mathbf{w}, \mathbf{r}) + t,$$

and it follows that $t + 1 \leq d(\mathbf{w}, \mathbf{r})$. So, for any code word $\mathbf{w} \neq \mathbf{u}$, we have $d(\mathbf{r}, \mathbf{u}) \leq t < t + 1 \leq d(\mathbf{w}, \mathbf{r})$, and therefore the code word closest to the received vector $\mathbf{r}$ is $\mathbf{u}$, and $\mathbf{r}$ is thus correctly decoded as $\mathbf{u}$.

To show that the code can detect up to $2t$ errors, we suppose a transmitted code word $\mathbf{u}$ is received as the vector $\mathbf{r}$ and that at least one error, but no more than $2t$ errors, was made in transmission. Because only code words are transmitted, an error will be detected whenever a received word is not a code word. But $\mathbf{r}$ cannot be a code word, since $d(\mathbf{r}, \mathbf{u}) \leq 2t$, whereas we know that $\mathbf{u}$ is a code word and the minimum distance between distinct code words is at least $2t + 1$.

▲ ▲ ▲

Theorem 2 is often misinterpreted as meaning that a binary linear code with Hamming weight $2t + 1$ can correct any t errors *and* detect any $2t$ or fewer errors simultaneously. This is not the case. The user must choose one or the other role for the code before decoding. Consider, for example, the Hamming $(7, 4)$ code given in Table 2. By inspection, the Hamming weight of the code is $3 = 2 \cdot 1 + 1$, so we may elect either to correct any single error or to detect any one or two errors. To understand why we can't do both at the same time, consider the received word 0001010 and the code given in Table 2. In this situation, either we can decode 0001010 as its nearest neighbor or we can opt not to decode at all. Certainly, we cannot do both! Of course, because 0001010 is not a code word, we know at least one error occurred. If we knew that only one error occurred, we could decode 0001010 correctly as its nearest neighbor since only one code word differs from the received word in one position. On the other hand, if we knew that two errors occurred, we could settle for error detection and not decode. But since there is no way to know which of these two possibilities occurred, there is no way we can know which of the two options to take. So, we must decide on one or the other in advance.

EXAMPLE 7   Triple Error Detection   Since the Hamming weight of the linear code given in Example 4 is 4, it will correct any single error or it will detect any triple error (here $2t = 3$). ▲

It is natural to wonder how the matrix G used to produce the code in Table 2 was chosen. Better yet, in general, how can one find a matrix G that carries a set of k-tuples to a set of n-tuples in such a way that for any k-tuple v, the vector vG will agree with v in the first k components and build in enough redundancy in the last $n - k$ components to be able to correct an error? Such a matrix is a $k \times n$ matrix of the form

$$\begin{bmatrix} 1 & 0 & \cdots & 0 & a_{11} & \cdots & a_{1n-k} \\ 0 & 1 & \cdots & 0 & \vdots & & \vdots \\ \vdots & \vdots & & \vdots & & & \\ 0 & 0 & \cdots & 1 & a_{k1} & \cdots & a_{kn-k} \end{bmatrix} = [\mathbf{I}_k | \mathbf{A}],$$

where the a_{ij}'s are 0s and 1s and the rows of $\mathbf{A}$ are nonzero and distinct. A matrix of this form is called the *standard generator matrix* for the resulting code. The presence of the $k \times k$ identity matrix guarantees that vG agrees with v in the first k components, whereas the requirements that the rows of $\mathbf{A}$ are nonzero and distinct are sufficient to ensure that we can correct a single error. That the set of vectors of the form vG is a closed set (which is necessary to have a binary linear code) follows from the fact that for any vectors $\mathbf{u}$ and $\mathbf{v}$ we have $\mathbf{u}G + \mathbf{v}G = (\mathbf{u} + \mathbf{v})G$.

EXAMPLE 8   Standard Generator Matrix
From the set of messages

$$\{000, 001, 010, 100, 110, 101, 011, 111\},$$

we may construct a $(6, 3)$ binary linear code with the standard generator matrix

$$\mathbf{G} = \begin{bmatrix} 1 & 0 & 0 & 1 & 1 & 0 \\ 0 & 1 & 0 & 1 & 0 & 1 \\ 0 & 0 & 1 & 1 & 1 & 1 \end{bmatrix}.$$

The resulting code words are given in Table 3.
Since the minimum weight of any nonzero code word in Table 3 is 3, this code can be used to correct any single error or detect any double error (but not both). ▲

TABLE 3

Message	Encoder **G**	Code word
000	→	000000
001	→	001111
010	→	010101
100	→	100110
110	→	110011
101	→	101001
011	→	011010
111	→	111100

Parity-Check Matrix Decoding

At this point we can conveniently encode messages with a standard generator matrix and use the nearest neighbor method to decode them. But when there is a large number of code words, the nearest neighbor method is too time-consuming to use. In this section we provide an alternative method for decoding, which, in the case where at most one error per code word occurred, is a fairly simple method for decoding. (In the unlikely event that two errors occurred in the same code word, our decoding method fails.)

To describe this method, suppose that we have binary linear code C given by the standard generator matrix $\mathbf{G} = [\mathbf{I}_k | \mathbf{A}]$, where $\mathbf{I}_k$ represents the $k \times k$ identity matrix and $\mathbf{A}$ is the $k \times (n - k)$ matrix obtained from $\mathbf{G}$ by deleting the first k columns of $\mathbf{G}$. Then, the $n \times (n - k)$ matrix

$$\mathbf{H} = \left[\frac{\mathbf{A}}{\mathbf{I}_{n-k}}\right],$$

where $\mathbf{I}_{n-k}$ is the $(n - k) \times (n - k)$ identity matrix, is called the *parity-check matrix* for C. (Notice that

$$\mathbf{GH} = [\mathbf{I}_k | \mathbf{A}]\left[\frac{\mathbf{A}}{\mathbf{I}_{n-k}}\right] = \mathbf{A} + \mathbf{A} = [0] = \text{the zero}$$

matrix.)

The decoding procedure is as follows:

1. For any received word **r**, compute **rH**.

2. If **rH** is the zero vector, assume that no error was made.

3. If **rH** is the ith row of **H** assume an error was made in the ith component of **r**.

4. If **rH** does not fit into either of category (2) or category (3), we know that at least two errors occurred in transmission and we do not decode.

EXAMPLE 9 A Parity-Check Matrix Consider the Hamming (7, 4) code given in Table 2. The generator matrix is

$$\mathbf{G} = \begin{bmatrix} 1 & 0 & 0 & 0 & 1 & 1 & 0 \\ 0 & 1 & 0 & 0 & 1 & 0 & 1 \\ 0 & 0 & 1 & 0 & 1 & 1 & 1 \\ 0 & 0 & 0 & 1 & 0 & 1 & 1 \end{bmatrix},$$

and the corresponding parity-check matrix is

$$\mathbf{H} = \begin{bmatrix} 1 & 1 & 0 \\ 1 & 0 & 1 \\ 1 & 1 & 1 \\ 0 & 1 & 1 \\ 1 & 0 & 0 \\ 0 & 1 & 0 \\ 0 & 0 & 1 \end{bmatrix}.$$

By the way **H** was defined, the parity-check matrix method correctly decodes any received word in which no error has been made. But it will also allow us to correct any single error. For example, let us say that the message 1000110 has been received as **r** = 1001110 (an error in the fourth position). We compute **rH** = 011 and see that this matches the fourth row of **H** and no

David Huffman

Large networks of IBM computers use it. So do high-definition televisions, modems, and a popular electronic device that takes the brain work out of programming a videocassette recorder. All these digital wonders rely on the results of a 40-year-old term paper by an MIT graduate student—a data compression scheme known as Huffman encoding.

In 1951 David Huffman and his classmates in an electrical engineering graduate course on information theory were given the choice of a term paper or a final exam. For the term paper, Huffman's professor had assigned what at first appeared to be a simple problem. Students were asked to find the most efficient method of representing numbers, letters, or other symbols using binary code. Huffman worked on the problem for

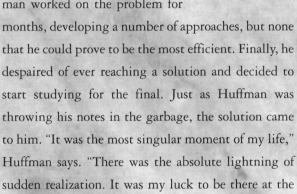
David Huffman

months, developing a number of approaches, but none that he could prove to be the most efficient. Finally, he despaired of ever reaching a solution and decided to start studying for the final. Just as Huffman was throwing his notes in the garbage, the solution came to him. "It was the most singular moment of my life," Huffman says. "There was the absolute lightning of sudden realization. It was my luck to be there at the right time and also not have my professor discourage me by telling me that other good people had struggled with the problem." he says. When presented with his student's discovery. Huffman recalls, his professor exclaimed: "Is that all there is to it!"

"The Huffman code is one of the fundamental ideas that people in computer science and data communications are using all the time," says Donald Knuth of Stanford University. Although others have used Huffman code to help make millions of dollars, Huffman's main compensation was dispensation from the final exam. He never tried to patent an invention from his work and experiences only a twinge of regret at not having used his creation to make himself rich. "If I had the best of both worlds, I would have had recognition as a scientist, and I would have gotten monetary rewards," he says. "I guess I got one and not the other."

But Huffman has received other compensation. A few years ago an acquaintance told him that he had noticed that a reference to the code was spelled with a lowercase "H." Remarked his friend to Huffman, "David, I guess your name has finally entered the language."

{Adapted from an article appearing in Scientific American *(Sept. 1991) by Gary Stix.}*

The Ubiquitous Reed–Solomon Codes

One of the mathematical ideas underlying the current error-correcting techniques for everything from computer hard disk drives to CD players was first introduced in 1960 by Irving Reed and Gustave Solomon. Reed–Solomon codes made possible the stunning pictures of the outer planets sent back by the space probes Voyager I and II. They make it possible to scratch a compact disc and still enjoy the music.

Irving Reed and Gustave Solomon monitor the encounter of Voyager II with Neptune at the Jet Propulsion Laboratory in 1989.

"When you talk about CD players and digital audio tape and now digital television, and various other digital imaging systems that are coming—all of those need Reed–Solomon [codes] as an integral part of the system," says Robert McEliece, a coding theorist at Caltech.

Why? Because digital information consists of 0s and 1s and a physical device that may occasionally confuse the two. Voyager II, for example, was transmitting data at incredibly low power—barely a whisper—over billions of miles. Error-correcting codes are a kind of safety net—mathematical insurance against the vagaries of an imperfect material world.

In 1960 the theory of error-correcting codes was only about a decade old. Through the 1950s, a number of researchers began experimenting with a variety of error-correcting codes. But the Reed–Solomon paper. McEliece says, "hit the jackpot." "In hindsight it seems obvious," Reed recently said. However, he added, "coding theory was not a subject when we published the paper." The two authors knew they had a nice result; they didn't know what impact the paper would have.

Three decades later, the impact is clear. The vast array of applications, both current and pending, has settled the questions of the practicality and significance of Reed–Solomon codes. Billions of dollars in modern technology depend on ideas that stem from Reed and Solomon's original work.

other row. Thus the fourth component of **r** is assumed to be incorrect. On the other hand, if more than one error is made, the parity-check matrix method can give the wrong code word. For example, if the message 1000110 is received as **r** $= 0001110$ (mistakes in positions 1 and 4), we have **rH** $= 101$ and we incorrectly assume that a single error in position 2 has been made. ▲

Theorem 3. For a binary linear code, parity-check matrix decoding will correct any single error if the rows of the parity-check matrix are nonzero and distinct.

Although we have only considered binary linear codes, in some applications the code words are n-tuples whose components are from Z_p, where p is a prime. Such a code is called linear if it is an Abelian group under addition modulo p.

Data Compression

An (n, k) binary code is a fixed-length code. In a fixed-length code, each code word is represented by the same number of digits (or symbols). In contrast, the Morse code (see Figure 10), designed for the telegraph, is a *variable-length* code, that is, a code in which the number of symbols for each code word may vary.

FIGURE 10 Morse code.

Notice that in the Morse code the letters that occur most frequently have the shortest coding whereas the letters that occur the least frequently have the longest coding. By assigning the code in this manner, telegrams could convey more information per line than would be the case for fixed-length codes or a randomly assigned variable-length coding of the letters. The process of encoding data so that the most frequently occurring data are represented by the fewest symbols is an example of *data compression*. Figure 11 shows a typical frequency distribution for letters in English-language text material.

	A	B	C	D	E	F	G	H	I	J	K	L	M
Percentage:	8	1.5	3	4	13	2	1.5	6	6.5	0.5	0.5	3.5	3

	N	O	P	Q	R	S	T	U	V	W	X	Y	Z
Percentage:	7	8	2	0.25	6.5	6	9	3	1	1.5	0.5	2	.25

FIGURE 11 A widely used frequency table for letters in normal English language.

Let us illustrate the principles of data compression with a simple example. Biologists are able to describe genes by specifying sequences comprised of the four letters A, T, G, and C, which represent the four nucleotides adenine, thymine, guanine, and cytosine, respectively. One way to encode a sequence such as AAACAGTAAC in fixed-length binary form would be to encode the letters as

$$A \rightarrow 00 \quad C \rightarrow 01 \; T \rightarrow 10 \; G \rightarrow 11.$$

The corresponding binary code for the sequence AAACAGTAAC is then

$$00000001001110000001.$$

On the other hand, if we knew from experience that the hierarchy of occurrence of the letters is A, C, T, and G (that is, A occurs most frequently, C second most frequently, and so on) and that A occurs much more frequently than T and G together, the most efficient binary encoding would be

$$A \to 0 \quad C \to 10 \quad T \to 110 \quad G \to 111.$$

For this encoding scheme, the sequence AAACAG-TAAC is encoded as

$$0001001111100010.$$

Notice that this binary sequence has 20% fewer digits than our previous sequence in which each letter was assigned a fixed length of 2 (16 digits versus 20 digits). However, to realize this savings, we have made decoding more difficult. For the binary sequence using the fixed length of two symbols per character, we decode the sequence by taking the digits two at a time in succession and converting them to the corresponding letters. For the compressed coding, we can decode by examining the digits in groups of three.

EXAMPLE 10 Decode 0001001111100010
Consider the compressed binary sequence 0001001111100010. Look at the first 3 digits: 000. Since our code words have one, two, or three digits and neither 00 nor 000 is a code word, the sequence 000 can only represent the *three* code words 0, 0, and 0. Now look at the next three digits: 100. Again, since neither 1 nor 100 is a code word, the sequence 100 represents the *two* code words 10 and 0. The next three digits, 111, can only represent the code word 111 since the other three code words all contain at least one 0. Next consider the sequence 110. Since neither 1 nor 11 is a code word, the sequence 110 can only

represent 110 itself. Continuing in this fashion, we can decode the entire sequence to obtain AAACAGTAAC. ▲

The following observation can simplify the decoding process for compressed sequences. Note that 0 occurs only at the end of a code word. Thus each time you see a 0, it is the end of the code word. Also, because the code words 0, 10, and 110 end in a 0, the only circumstances under which there are three consecutive 1s is when the code word is 111. So, to quickly decode a compressed binary sequence using our coding scheme, insert a comma after every 0 and after every three consecutive 1s. The digits between the commas are code words.

EXAMPLE 11 Encoding and Decoding
Code AGAACTAATTGACA and decode the result. Recall: $A \to 0$, $C \to 10$, $T \to 110$, and $G \to 111$. So,

$$\begin{array}{c} AGAACTAATTGACA \\ \to 0111001011000110110111100. \end{array}$$

To decode the encoded sequence, we insert commas after every 0 and after every occurrence of 111 and convert to letters:

$$\begin{array}{cccccccccccc} 0, & 111, & 0, & 0, & 10, & 110, & 0, & 0, & 110, & 110, & 111, & 0, & 10, & 0 \\ A, & G, & A, & A, & C, & T, & A, & A, & T, & T, & G, & A, & C, & A. \end{array}$$ ▲

Modern data compression schemes were first invented in the 1950s. They are now routinely used by modems and fax machines for data transmissions and by computers for data storage. In many cases data-compression results in a savings of up to 50% on telephone charges or storage space.

In this section, we have presented coding theory in its simplest form. A more sophisticated treatment involves advanced algebraic concepts called "rings" and "fields." In some instances, two error-correcting codes are employed. The European Space Agency space probe Giotto, which encountered Halley's comet, had two error-correcting codes built into its electronics. One code checked for independently oc-curring errors, and another—a so-called Reed–Solomon code (see the Spotlight)—checked for bursts of errors. Giotto achieved an error detection rate of .999999. Reed–Solomon codes are used on compact discs. They can correct thousands of consecutive errors.

1. Determine the check digit for a money order with identification number 7234541780.

2. Suppose a money order with the identification number and check digit 21720421168 is erroneously copied as 27750421168. Will the check digit detect the error? Explain your reasoning.

3. A transposition error involving distinct adjacent digits is one of the form $\ldots ab \ldots \rightarrow \ldots ba \ldots$, with $a \neq b$. Prove that the money order check digit scheme will not detect such errors unless the check digit itself is transposed.

4. Suppose that in one of the noncheck positions of a money order number, the digit 0 is substituted for the digit 9, or vice versa. Prove that this error will not be detected by the check digit. Prove that all other errors involving a single digit are detected.

5. Determine the check digit for the United Parcel Service (UPS) identification number 873345672.

6. Show that a substitution of a digit a_i' for the digit a_i ($a_i' \neq a_i$) in a noncheck position of a UPS number is detected if and only if $|a_i - a_i'| \neq 7$.

7. Show that a transposition of adjacent digits a and b not involving the check digit in a UPS number is detected if and only if $|a - b| \neq 7$.

8. Use the UPC scheme to determine the check digit for the number 07312400508.

9. Use the bank scheme to determine the check digit for the number 09190204.

10. Every book is assigned a 10-digit identification number (called ISBN) that satisfies the condition $(a_1, a_2, \ldots, a_{10}) \cdot (10, 9, \ldots, 1) = 0 \bmod 11$. Verify the check digit for the ISBN assigned to this book.

11. Determine the check digit for the Avis rental car with identification number 540047.

12. Determine the check digit for the airline ticket number 30860422052.

13. Determine the check digit for the ISBN number 0-669-19493.

14. For some products such as soft drink cans and magazines, an 8-digit UPC number called Version E is used instead of the 12-digit number. The method of calculating the eighth digit, which is the check digit, depends on the value of the seventh digit. Use the fact that the check digit a_8 for a UPC Version E identification number $a_1a_2a_3a_4a_5a_6a_7$, where a_7 is 0, 1, or 2, is chosen so that $a_1 + a_2 + 3a_3 + 3a_4 + a_5 + 3a_6 + a_7 + a_8 = 0 \bmod 10$ to determine the check digit for the following Version E numbers:

   (a)   0121690   (c)   0760022
   (b)   0274551   (d)   0496580

15. Use the fact that the check digit a_8 for a UPC Version E identification number $a_1a_2a_3a_4a_5a_6a_7$, where a_7 is 4, is chosen so that $a_1 + a_2 + 3a_3 + a_4 + 3a_5 + 3a_6 + a_8 = 0$ mod 10 to determine the check digit for the following numbers:

(a) 0754704 (c) 0724444

(b) 0774714

16. Explain why the bank scheme will detect the error $751 \ldots \to 157 \ldots$ but the UPC scheme will not.

17. The invalid ISBN 0-669-03925-4 is the result of a transposition of two adjacent digits not involving the check digit. Determine the correct ISBN.

18. (IBM check digit method) Major credit cards, banks in Germany, many libraries in the United States, and South Dakota driver's license number employ the following check digit method. Define a function σ from $\{0,1,2,3,4,5,6,7,8,9\}$ to itself by $\sigma(0) = 0$, $\sigma(1) = 2$, $\sigma(2) = 4$, $\sigma(3) = 6$, $\sigma(4) = 8$, $\sigma(5) = 1$, $\sigma(6) = 3$, $\sigma(7) = 5$, $\sigma(8) = 7$, $\sigma(9) = 9$. To a number $a_1a_2\cdots a_k$, where k is odd, append a check digit c so that

$$\sigma(a_1) + a_2 + \sigma(a_3) + \cdots + \sigma(a_k) + c = 0 \text{ mod } 10.$$

[If k is even, σ is applied to the digits with even subscripts instead of the ones with the odd subscripts.] For example, for 64387 we have the check digit 4 because $(3 + 4 + 6 + 8 + 5)$ mod $10 = 26$ mod $10 = 6$ mod 10. Calculate the check digit for the number 3125600196431. Prove that this method detects all single-digit errors. Determine which transposition errors involving adjacent digits go undetected by this method.

19. Many identification number schemes utilize both numbers and letters. In these cases the "check" can be a numeral or a letter. One of the most common places of these systems is the so-called Code 39. For the purpose of computing the check character, the letter A is assigned the value of 10, B the value of 11, and so on. For a character string of length n, the check character is the number or letter that results in the scalar product of the identification number and the weighting vector $(n, \ldots,$

3, 2, 1) being 0 modulo 36. For example, the check character of the number 2705A0086164ZE is N since

$$(2, 7, 0, 5, 10, 0, 0, 8, 6, 1, 6, 4, 35, 14)$$
$$\cdot(15, 14, 13, 12, 11, 10, 9, 8, 7, 6, 5, 4, 3, 2)$$
$$= 589 = 13 \text{ mod } 36 \text{ and } 13 + 23$$
$$= 0 \text{ mod } 36.$$

Use the Code 39 scheme to determine the check character for the number 2105A0055186ZA.

20. Determine the check digit for the German banknote with the serial number DA6819403G (see Example 1).

21. Use the method illustrated in Example 2 to append two check digits to the number 73445860.

22. Suppose that an eight-digit number has two check digits appended using the method given in Example 2 and it is incorrectly transcribed as 4302511568. If exactly one digit is incorrect, determine the correct number.

23. Suppose the check digit a_{10} of ISBN numbers were chosen so that $a_1 + 2a_2 + 3a_3 + 4a_4 + 5a_5 + 6a_6 + 7a_7 + 8a_8 + 9a_9 + 10a_{10} = 0$ mod 11 instead of the way described in the chapter. How would this compare with the actual check digit?

24. Suppose the check digit a_9 for bank checks were chosen to be the last digit of $3a_1 + 7a_2 + a_3 + 3a_4 + 7a_5 + a_6 + 3a_7 + 7a_8$ instead of the way described in the chapter. How would this compare with the actual check digit?

25. Use the diagram method to verify the code word in Table 2 for the message 0101.

26. Referring to Table 2, use the nearest neighbor method to decode the received words 0000110 and 1110101.

27. Find the Hamming distance between the following pairs of vectors:

$$\{1101, 0111\}, \{0110, 1100\}, \{11101, 00111\}.$$

28. How many code words are there in an (8, 5) binary linear code?

29. For any n-dimensional vectors $\mathbf{u}$, $\mathbf{v}$, $\mathbf{w}$, prove that the Hamming distance has the following properties:
   (a) $d(\mathbf{u}, \mathbf{v}) = wt(\mathbf{u} - \mathbf{v})$
   (b) $d(\mathbf{u}, \mathbf{v}) = d(\mathbf{v}, \mathbf{u})$ (symmetry)
   (c) $d(\mathbf{u}, \mathbf{v}) = 0$ if and only if $\mathbf{u} = \mathbf{v}$
   (d) $d(\mathbf{u}, \mathbf{v}) \leq d(\mathbf{u}, \mathbf{w}) + d(\mathbf{w}, \mathbf{v})$ (triangle inequality)
   (e) $d(\mathbf{u}, \mathbf{v}) = d(\mathbf{u} + \mathbf{w}, \mathbf{v} + \mathbf{w})$ (translation invariance)

30. Determine the (6, 3) binary linear code with generator matrix

$$\mathbf{G} = \begin{bmatrix} 1 & 0 & 0 & 0 & 1 & 1 \\ 0 & 1 & 0 & 1 & 0 & 1 \\ 0 & 0 & 1 & 1 & 1 & 0 \end{bmatrix}.$$

31. Show that for binary vectors, $wt(\mathbf{u} + \mathbf{v}) \geq wt(\mathbf{u}) - wt(\mathbf{v})$ and equality occurs if and only if the ith component of $\mathbf{u}$ is 1 whenever the ith component of $\mathbf{v}$ is 1.

32. Let C be a binary linear code. Show that the code words of even weight form a subcode of C. (A *subcode* of a code is a subset that is closed under addition modulo 2.)

33. Use the diagram method to decode the received messages 0111011 and 0100110.

34. Extend the code words listed in Table 2 to eight digits by appending a 0 to words of even weight and a 1 to words of odd weight. What is the error-detecting and error-correcting capability of the new code?

35. The *Caesar cipher* encrypts messages by replacing each letter of the alphabet with the letter shown beneath it:

ABCDEFGHIJKLMNOPQRSTUVWXYZ
DEFGHIJKLMNOPQRSTUVWXYZABC.

Use the Caesar cipher to encrypt the message RETREAT. Determine the intended message corresponding to the encripted message DWWDFN. Describe the rule for the Caesar cipher using mod arithmetic.

36. Explain why no (6, 3) binary linear code can detect all possible double errors.

37. A (4, 2) *ternary code* is formed by starting with all possible pairs of 0s, 1s, and 2s and appending two extra digits that are also 0s, 1s, or 2s by multiplying a pair $a_1 a_2$ on the right by a 2×4 matrix $\mathbf{G}$ with entries consisting of 0s, 1s, and 2s and using modulo 3 arithmetic.

Form a ternary code by appending to each message $a_1 a_2$ the check digits $c_1 c_2$ using:

$$\mathbf{G} = \begin{bmatrix} 1 & 0 & 1 & 1 \\ 0 & 1 & 2 & 1 \end{bmatrix}.$$

For example, 12 is encoded as $12\mathbf{G} = 1220$. List the elements of this code.

38. Use the ternary code in the previous exercise and the nearest neighbor method to decode the received word 1201.

39. A (6, 4) ternary code is formed by starting with all possible 4-tuples of 0s, 1s, and 2s and appending two extra digits that are also 0s, 1s, and 2s. How many code words are there in a (6, 4) ternary code? How many possible received words are there in a (6, 4) ternary code?

40. Discuss the relative merits of a (7, 3) binary linear code with weight 4 and a (4, 2) ternary linear code with weight 3.

41. Let

$C = \{0000000, 1110100, 0111010, 0011101,$
$1001110, 0100111, 1010011, 1101001\}.$

What is the error-correcting capability of C? What is the error-detecting capability of C?

42. Suppose the parity-check matrix of a binary linear code is

$$H = \begin{bmatrix} 1 & 0 \\ 0 & 1 \\ 1 & 1 \\ 1 & 0 \\ 0 & 1 \end{bmatrix}.$$

Can the code correct any single error?

43. Consider the binary linear code

$$C = \{00000, 10011, 01010, 11001, 00101, 10110, \\ 01111, 11100\}.$$

Use nearest neighbor decoding to decode 11101.

44. Construct a (6, 3) binary linear code with generator matrix

$$G = \begin{bmatrix} 1 & 0 & 0 & 1 & 1 & 0 \\ 0 & 1 & 0 & 0 & 1 & 1 \\ 0 & 0 & 1 & 1 & 0 & 1 \end{bmatrix}.$$

Decode each of the received words,

$$001001, \ 011000, \ 000110, \ 100001,$$

using the nearest neighbor method and the parity-check matrix method.

45. Suppose the minimum weight of any nonzero code word in a binary linear code is 6. Discuss the options for error correction and error detection.

46. If the parity-check matrix for a binary linear code is

$$H = \begin{bmatrix} 1 & 1 & 0 \\ 0 & 1 & 1 \\ 1 & 0 & 1 \\ 1 & 0 & 0 \\ 0 & 1 & 0 \\ 0 & 0 & 1 \end{bmatrix},$$

will the code correct any single error? Why?

47. Suppose we code a five-symbol set $\{A, B, C, D, E\}$ into binary form as follows:

$$A \to 0, B \to 10, C \to 110, D \to 1110, E \to 1111.$$

Convert the sequence *AEAADBAABCB* into binary code. Determine the sequence of symbols represented by the binary code 01000110100011111110.

48. Devise a variable-length binary coding scheme for a six-symbol set $\{A, B, C, D, E, F\}$. Assume that A is the most frequently occurring symbol, B is the second most frequently occurring symbol, and so on.

SECTION 9.4 *Permutation Groups*

It often happens in science and engineering that one has a set of objects that are to be arranged to meet certain specifications. In these situations it may be important to know how many such arrangements are possible. For example, chemists are interested in molecules that can be viewed as six carbon atoms arranged in a hexagon with any combinations of the radicals *COOH* or *OH* attached at each carbon atom as illustrated in Figure 1.

Mathematicians call an arrangement of a finite set of objects a *permutation*. In this section we will view a permutation of the objects from a finite set A as a function from A to itself and study groups of permutations. In the early and mid-nineteenth century, groups of permutations were the only groups

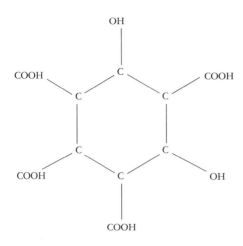

FIGURE 1 A molecule.

mathematicians investigated. It was not until around 1850 that the notion of an abstract group was introduced, and it took another quarter century before the idea firmly took hold.

DEFINITION

A *permutation* of a finite set A is a function α from A to A with the property that $\alpha(i) \neq \alpha(j)$ whenever $i \neq j$. A *permutation group of a set A* is a set of permutations of A that forms a group under function composition.

Since A is finite, it is customary, as well as convenient, to take A to be a set of the form $\{1, 2, 3, \ldots, n\}$ for some positive integer n. Unlike in calculus, where most functions are defined on infinite sets and are given by formulas, in algebra, permutations of finite sets are usually given by an explicit listing of each element of the domain and its corresponding functional value. When the sets that are permuted are fairly small, listing the elements and the values they are assigned is reasonable. It follows

from the definition of a permutation that each element of A appears exactly once as a functional value. For example, we define a permutation α of the set $\{1, 2, 3, 4\}$ by specifying

$$\alpha(1) = 2, \quad \alpha(2) = 3, \quad \alpha(3) = 1, \quad \alpha(4) = 4.$$

A more convenient way to express this correspondence is to write α in array form as

$$\alpha = \begin{bmatrix} 1 & 2 & 3 & 4 \\ 2 & 3 & 1 & 4 \end{bmatrix}.$$

Here $\alpha(j)$ is placed directly below j for each j. Similarly, the permutation β of the set $\{1, 2, 3, 4, 5, 6\}$ given by

$$\beta(1) = 5, \quad \beta(2) = 3, \beta(3) = 1, \beta(4) = 6,$$
$$\beta(5) = 2, \beta(6) = 4$$

is expressed in array form as

$$\beta = \begin{bmatrix} 1 & 2 & 3 & 4 & 5 & 6 \\ 5 & 3 & 1 & 6 & 2 & 4 \end{bmatrix}.$$

Notice that the bottom row of an array is simply an arrangement of the integers in the top row. Thus, our definition of permutation is consistent with the one given in Chapter 4.

Composition of permutations expressed in array notation is carried out from right to left by going from top to bottom, then top to bottom (although our notation for permutations is the same as the notation for matrices, the composition of two permutations expressed in this notation is not matrix multiplication). For example, letting

$$\sigma = \begin{bmatrix} 1 & 2 & 3 & 4 & 5 \\ 2 & 4 & 3 & 5 & 1 \end{bmatrix}$$

and

$$\gamma = \begin{bmatrix} 1 & 2 & 3 & 4 & 5 \\ 5 & 4 & 1 & 2 & 3 \end{bmatrix},$$

we have

$$\gamma\sigma = \begin{bmatrix} 1 & 2 & 3 & 4 & 5 \\ 5 & 4 & 1 & 2 & 3 \end{bmatrix} \begin{bmatrix} 1 & 2 & 3 & 4 & 5 \\ 2 & 4 & 3 & 5 & 1 \end{bmatrix}$$
$$= \begin{bmatrix} 1 & 2 & 3 & 4 & 5 \\ 4 & 2 & 1 & 3 & 5 \end{bmatrix}.$$

In the array for $\gamma\sigma$ we have 4 under 1, since $(\gamma\sigma)(1) = \gamma(\sigma(1)) = \gamma(2) = 4$, so $\gamma\sigma$ sends 1 to 4. The remainder of the bottom row of $\gamma\sigma$ is obtained in a similar fashion.

We are now ready to give some examples of permutation groups. Throughout this section we will use ϵ to denote the identity permutation.

E X A M P L E 1 Symmetric Group S_3 Let S_3 denote the set of all permutations of $\{1, 2, 3\}$. Then S_3, under function composition, is a group with six elements. The six elements are

$$\epsilon = \begin{bmatrix} 1 & 2 & 3 \\ 1 & 2 & 3 \end{bmatrix}, \qquad \alpha = \begin{bmatrix} 1 & 2 & 3 \\ 2 & 3 & 1 \end{bmatrix},$$
$$\alpha^2 = \begin{bmatrix} 1 & 2 & 3 \\ 3 & 1 & 2 \end{bmatrix},$$

$$\beta = \begin{bmatrix} 1 & 2 & 3 \\ 1 & 3 & 2 \end{bmatrix}, \qquad \alpha\beta = \begin{bmatrix} 1 & 2 & 3 \\ 2 & 1 & 3 \end{bmatrix},$$
$$\alpha^2\beta = \begin{bmatrix} 1 & 2 & 3 \\ 3 & 2 & 1 \end{bmatrix}.$$

Note that $\beta\alpha = \begin{bmatrix} 1 & 2 & 3 \\ 3 & 2 & 1 \end{bmatrix} \neq \alpha\beta$, so that S_3 is non-Abelian. ▲

Example 1 can be generalized as follows.

E X A M P L E 2 Symmetric Group S_n Let $A = \{1, 2, \ldots, n\}$. The set of all permutations of A is called the *symmetric group of degree n* and is denoted by S_n. Elements of S_n have the form

$$\alpha = \begin{bmatrix} 1 & 2 & \cdots & n \\ \alpha(1) & \alpha(2) & \cdots & \alpha(n) \end{bmatrix}.$$

It is easy to compute the number of elements in S_n. There are n choices of $\alpha(1)$. Once $\alpha(1)$ has been determined, there are $n - 1$ possibilities for $\alpha(2)$ (since α is one-to-one, we must have $\alpha(1) \neq \alpha(2)$). After choosing $\alpha(2)$, there are exactly $n - 2$ possibilities for $\alpha(3)$. Continuing along in this fashion, we see that S_n must have $n(n - 1) \ldots 3 \cdot 2 \cdot 1 = n!$ elements. We leave it to the reader to prove that S_n is non-Abelian when $n \geq 3$ (see Example 1 for a hint). ▲

E X A M P L E 3 Symmetries of a Square As a third example, we associate each symmetry in D_8 (see Table 7 in Section 1 of this chapter) with the permutation of the locations of each of the four corners of a square. For example, if we label the four corner positions as in the following figure and keep these labels fixed for reference, we may describe a 90° rotation by the permutation

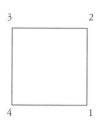

$$\rho = \begin{bmatrix} 1 & 2 & 3 & 4 \\ 2 & 3 & 4 & 1 \end{bmatrix}.$$

A reflection across a horizontal axis yields

$$\phi = \begin{bmatrix} 1 & 2 & 3 & 4 \\ 2 & 1 & 4 & 3 \end{bmatrix}.$$

Since $D_8 = \{\epsilon, \rho, \rho^2, \rho^3, \phi, \rho\phi, \rho^2\phi, \rho^3\phi\}$, we see that the entire group consist of combinations of ρ and ϕ. ▲

Permutation groups have many applications in mathematics, computer science, physics, and chemistry. Permutation groups also naturally arise when objects are being rearranged such as card shuffling and Rubik's cube. Before illustrating a few of these applications, we introduce some definitions and notation.

DEFINITIONS

Let G be a group of permutations of set $\{1, 2, \ldots, n\}$. For each i from 1 to n, let $\mathrm{orb}_G(i) = \{\phi(i) | \phi \in G\}$. The subset $\mathrm{orb}_G(i)$ of $\{1, 2, \ldots, n\}$ is called the *orbit of i under G*.

Let G be a group of permutations of $\{1, 2, \ldots, n\}$. For each i from 1 to n, let $\mathrm{stab}_G(i) = \{\phi \in G | \phi(i) = i\}$. We call $\mathrm{stab}_G(i)$ the *stabilizer of i in G*.

The student should verify that $\mathrm{stab}_G(i)$ is itself a group under the operation of composition.

EXAMPLE 4 Orbits and Stabilizers Let G be the group consisting of the six permutations of the set $\{1, 2, 3, 4, 5, 6, 7, 8\}$ listed below.

$$\beta_1 = \begin{bmatrix} 1 & 2 & 3 & 4 & 5 & 6 & 7 & 8 \\ 1 & 2 & 3 & 4 & 5 & 6 & 7 & 8 \end{bmatrix}$$

$$\beta_2 = \begin{bmatrix} 1 & 2 & 3 & 4 & 5 & 6 & 7 & 8 \\ 3 & 1 & 2 & 6 & 4 & 5 & 8 & 7 \end{bmatrix}$$

$$\beta_3 = \begin{bmatrix} 1 & 2 & 3 & 4 & 5 & 6 & 7 & 8 \\ 3 & 1 & 2 & 6 & 4 & 5 & 7 & 8 \end{bmatrix}$$

$$\beta_4 = \begin{bmatrix} 1 & 2 & 3 & 4 & 5 & 6 & 7 & 8 \\ 2 & 3 & 1 & 5 & 6 & 4 & 7 & 8 \end{bmatrix}$$

$$\beta_5 = \begin{bmatrix} 1 & 2 & 3 & 4 & 5 & 6 & 7 & 8 \\ 2 & 3 & 1 & 5 & 6 & 4 & 8 & 7 \end{bmatrix}$$

$$\beta_6 = \begin{bmatrix} 1 & 2 & 3 & 4 & 5 & 6 & 7 & 8 \\ 1 & 2 & 3 & 4 & 5 & 6 & 8 & 7 \end{bmatrix}$$

Then,

$\mathrm{orb}_G(1) = \{1, 3, 2\}$, $\mathrm{stab}_G(1) = \{\beta_1, \beta_6\}$,

$\mathrm{orb}_G(2) = \{2, 1, 3\}$, $\mathrm{stab}_G(2) = \{\beta_1, \beta_6\}$,

$\mathrm{orb}_G(3) = \{3, 2, 1\}$, $\mathrm{stab}_G(3) = \{\beta_1, \beta_6\}$,

$\mathrm{orb}_G(4) = \{4, 6, 5\}$, $\mathrm{stab}_G(4) = \{\beta_1, \beta_6\}$,

$\mathrm{orb}_G(5) = \{5, 4, 6\}$, $\mathrm{stab}_G(5) = \{\beta_1, \beta_6\}$,

$\mathrm{orb}_G(6) = \{6, 5, 4\}$, $\mathrm{stab}_G(6) = \{\beta_1, \beta_6\}$,

$\mathrm{orb}_G(7) = \{7, 8\}$, $\mathrm{stab}_G(7) = \{\beta_1, \beta_3, \beta_4\}$,

$\mathrm{orb}_G(8) = \{8, 7\}$, $\mathrm{stab}_G(8) = \{\beta_1, \beta_3, \beta_4\}$. ▲

The preceding example also illustrates the following theorem. For any finite set A, we use $|A|$ to denote the number of elements in A. $|A|$ is called the *order* of A.

Buckyballs: A New Form of Carbon

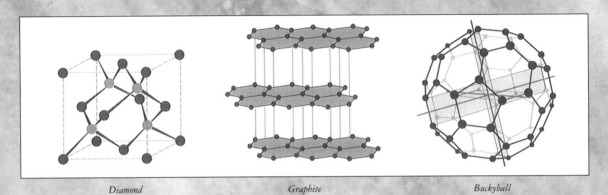

Diamond   Graphite   Buckyball

Until recently, it was believed that carbon occurred in only two principal forms: diamond and graphite. In 1985 chemists Robert Carl and Richard Smalley of Rice University stirred tremendous excitement in the scientific community when they created a new form of carbon by using a laser beam to vaporize graphite. The structure of the new molecule is composed of 60 carbon atoms arranged in the shape of a soccer ball! Because the shape of the new molecule reminded them of the dome structures built by the architect R. Buckminster Fuller, Carl and Smalley named their discovery *buckminsterfullerenes,* or "buckyballs." Buckyballs are the roundest, most symmetrical large molecules known.

Within 8 years of their discovery, buckyballs were the subject of more than 1500 research papers. *Science* magazine chose the buckyball as its "molecule of the year" in 1991. Group theory has been particularly useful in illuminating their structure and properties since the absorption spectrum of a molecule depends on its symmetries and chemists classify various molecular states according to their symmetry properties.

Theorem 1. Let G be a finite group of permutations on the set $\{1, 2, \ldots, n\}$. Then, for any i from 1 to n, $|G| = |\text{orb}_G(i)||\text{stab}_G(i)|$.

It cannot be overemphasized that Theorem 1 is a *counting* theorem. It enables one to determine the number of elements in a group by counting the number of elements in two smaller sets. To see how Theorem 1 is useful, we will determine the order of the group of rotational symmetries of a cube and the order of the group of rotational symmetries of a soccer ball. That is, we wish to find the number of essentially different ways that we can take a cube or a soccer ball in a certain location in space and physically rotate it in such a way that, after it has been rotated, it still occupies the same location in space.

EXAMPLE 5 Rotations of a Cube Let G be the group of rotational symmetries of a cube. (It may not be obvious that the composition of two rotational symmetries is a rotational symmetry, but this is indeed the case. Try experimenting with a cube.) We will determine $|G|$ using Theorem 1. But first let us see how many rotations we can find by looking at a cube (it is best to have one in hand for this discussion). Certainly, we can rotate a cube 90°, 180°, or 270° about the axis joining the centers of the top and bottom of the cube (see Figure 2a) and likewise for the axes joining the centers of the front and back sides and the centers of left and right sides. These gives us nine nonidentity rotations. Four more axes of rotational symmetry are the diagonals joining opposite vertices (see Figure 2b). For each of these axes there are 120° and 240° rotations. Finally, there are six axes of 180° rotational symmetry joining the midpoints of opposite edges (see Figure 2c). So, counting the identity, we have found 24 rotational symmetries. Could there be more? Perhaps, but it seems as though we have them all. Now let's see what Theorem 1 tells us. Label the six faces of the cube 1 through 6. (Call the top 1.) Since any rotational symmetry of the cube must carry each face of the cube to exactly one other face of the cube and different rotations induce different permutations of the faces, G can be viewed as a group of permutations on the set $\{1, 2, 3, 4, 5, 6\}$. Clearly, any face can be taken to the position held by face 1 by a suitable rotation; therefore $|\text{orb}_G(1)| = 6$. Next, we consider $\text{stab}_G(1)$. Here, we are asking for all rotations of a cube that leave face number 1 where it

is. Surely, there are only four such motions—rotations of 0°, 90°, 180°, 270°—about the line perpendicular to face number 1 and passing through its center (see Figure 2a). Thus, by Theorem 1, $|G| = |\text{orb}_G(1)||\text{stab}_G(1)| = 6 \cdot 4 = 24$. ▲

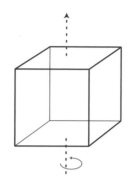

FIGURE 2a Rotation about axis joining centers of the top and bottom.

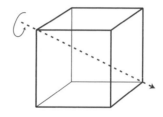

FIGURE 2b Rotation about a diagonal joining opposite vertices.

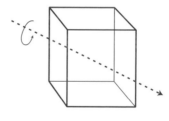

FIGURE 2c Rotation about axis joining midpoint of opposite edges.

Notice that by using Theorem 1 instead of just looking for as many rotations as we can find, we need far less observational skill and we are sure we have counted all rotations. Example 5 illustrates the power of abstraction.

EXAMPLE 6 Rotations of a Soccer Ball A soccer ball has 20 faces that are regular hexagons

George Pólya

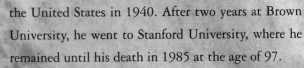

In 1924 the mathematician George Pólya published a paper in a crystallography journal in which he classified the plane symmetry groups and provided a full-page illustration of the corresponding 17 periodic patterns. B. G. Escher, a geologist, sent a copy of the paper to his artist brother M. C. Escher, who used Pólya's black and white geometric patterns as a guide for making his own interlocking colored patterns featuring birds, reptiles, and fish.

George Pólya was born in Budapest, Hungary, on December 13, 1887. He received a teaching certificate from the University of Budapest in languages before turning to philosophy, mathematics, and physics.

In 1912 he was awarded a Ph.D. in mathematics. Horrified by Hitler and World War II, Pólya came to

George Pólya

the United States in 1940. After two years at Brown University, he went to Stanford University, where he remained until his death in 1985 at the age of 97.

Pólya contributed to many branches of mathematics, and his collected papers fill four large volumes. One of his well-known discoveries is a counting technique that generalizes Theorem 2 of this section. Pólya is also famous for his books on problem solving and his teaching. The Society for Industrial and Applied Mathematics, the London Mathematical Society, and the Mathematical Association of America each have prizes named after Pólya.

Pólya taught courses and lectured around the country into his 90s. He never learned to drive a car and took his first plane trip at age 75. He was married for 67 years and had no children.

---

and 12 faces that are regular pentagons (see Figure 3). (The technical term for this solid is *truncated icosahedron*.) To determine the number of rotational symmetries of a soccer ball using Theorem 1, we may chose our set S to be the 20 hexagons or the 12 pentagons. Let us say that S is the set of 12 pentagons. By carefully experiment-

ing with a soccer ball, you can verify that any pentagon can be carried to any other pentagon by some rotation so that the orbit of the group of rotational symmetries is S. Also, there are five rotations that fix (stabilize) any particular pentagon. Thus, by Theorem 1 there are $12 \cdot 5 = 60$ rotational symmetries. ▲

FIGURE 3 Soccer ball.

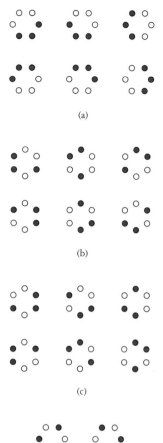

(a)

(b)

(c)

(d)

FIGURE 4 Twenty possible designs with three black beads and three white beads.

To appreciate the utility of Theorem 1, try to count the number of rotational symmetries of a soccer ball without using it.

Permutation groups naturally arise in many situations involving symmetrical designs or arrangements. Consider, for example, the task of coloring the six vertices of a regular hexagon so that three are black and three are white. Figures 4a–d show the 20 possibilities (6 vertices chosen 3 at a time). However, if these designs appeared on one side of hexagonal ceramic tiles, it would be nonsensical to count the designs shown in Figure 4a as different since all six designs shown there can be obtained from one of them by rotating. (A manufacturer would only make one of the six.) In this case we say the designs in Figure 4a are *equivalent* under the group of rotations of the hexagon. Similarly the designs in Figure 4b are equivalent under the group of rotations, as are the designs in Figures 4c and 4d. And since no design from any one of the four figures can be obtained from a design from a different figure by rotation we see that the designs within each figure are equivalent to each other but inequivalent to any design in another figure. In contrast, the designs in Figures 4b and 4c are equivalent under the dihedral group D_{12} since the designs in Figure 4b can be reflected across a vertical axis to give the designs in Figure 4c. For example, when arranging three black beads and three white beads to form a necklace, one could consider the designs shown in Figures 4b and 4c to be equivalent.

In general, we say two designs (arrangements) A and B are *equivalent under a group G* of permutations if there is an element ϕ of G such that $\phi(A) = B$. That is, two designs are equivalent under G (the set being permuted is the set of all possible designs or arrangements) if they are in the same orbit under G (see Exercise 20). It follows then that the number of inequivalent designs under G is simply the number of orbits of G since elements in the same orbit are the same design whereas elements from different orbits are different designs.

Notice that the designs in Figures 4a–d divide

up into four orbits under C_6, the group of six rotations, but only three orbits under the group D_{12} since the designs in Figures 4b and 4c form a single orbit under D_{12}. Thus one could obtain all 20 tile designs from just 4 tiles (one from each orbit under C_6), but we could obtain all 20 necklaces with just 3 of them (one from each orbit under D_{12}).

Although the problems we have just posed are simple enough to solve by observation, more complicated ones require a more sophisticated approach. Such an approach was provided by the English mathematician William Burnside in his classic book *Theory of Groups of Finite Order* published in 1911. But first, one more definition.

FIGURE 5 Tile designs fixed by 120° and 240° rotations.

TABLE 1

Element	Number of designs fixed by element
Identity	20
Rotation of 60°	0
Rotation of 120°	2
Rotation of 180°	0
Rotation of 240°	2
Rotation of 300°	0

DEFINITION

For any group G of permutations on a set S and any ϕ in G, we let fix(ϕ) = $\{i \in S | \phi(i) = i\}$. This set is called the *elements fixed by* ϕ (or more simply, the "fix of ϕ").

Theorem 2. If G is a finite group of permutations on a set S, then the number of orbits of G on S is

$$\frac{1}{|G|} \sum_{\phi \in G} |\text{ fix}(\phi)|.$$

In the case of counting hexagonal tiles with 3 black vertices and 3 white vertices, the set of objects being permuted is the 20 possible designs, whereas the group of permutations is the group C_6 of 6 rotational symmetries of a hexagon. Obviously, the identity fixes all 20 designs. We see from Figure 4 that rotations of 60°, 180°, or 300° fix none of the 20 designs. Finally, Figure 5 shows fix(ϕ) for the rotations of 120° and 240°. These data are collected in Table 1.

So, applying Burnside's theorem (Theorem 2) we obtain that the number of orbits under the group of rotations is

$$\tfrac{1}{6}(20 + 0 + 2 + 0 + 2 + 0) = 4.$$

Now let's use Burnside's theorem to count the number of necklace arrangements consisting of three black beads and three white beads. (For the purpose of analysis, we may arrange the beads in the shape of a regular hexagon.) For this problem, two arrangements are equivalent if they are in the same orbit under D_{12} since a manufacturer would not distinguish between a particular bead arrangement and any of its rotations or reflections. (A necklace with black and white beads can be flipped, whereas a tile with a design on one side cannot.) Figure 6 shows the arrangements fixed by a reflection across a diagonal. Table 2 summarizes the information needed to apply Burnside's theorem. (Note that Table 1 is a subset of Table 2.)

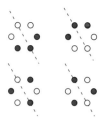

Figure 6 Tile designs fixed by a reflection across a diagonal.

TABLE 2

Type of element	Number of elements of this type	Number of arrangements fixed by type of element
Identity	1	20
Rotation of 60°	1	0
Rotation of 120°	1	2
Rotation of 180°	1	0
Rotation of 240°	1	2
Rotation of 300°	1	0
Reflection across diagonal	3	4
Reflection across side bisector	3	0

So, there are

$$\tfrac{1}{12}(1 \cdot 20 + 1 \cdot 0 + 1 \cdot 2 + 1 \cdot 0 + 1 \cdot 2 + 3 \cdot 4 + 3 \cdot 0)$$
$$= 3$$

inequivalent ways to string three black beads and three white beads on a necklace.

Of course, you are wondering, who, besides mathematicians, is interested in counting problems like the ones we have discussed above. Well, chemists are. One kind of molecule chemists study has six carbon atoms arranged in the shape of a hexagon and at each carbon atom there is one of three radicals (NH^3, $COOH$, or OH). So, let's use Burnside's theorem to count the number of such molecules. Clearly, the symmetry group is D_{12}. At each of the 6 carbon atoms (vertices) of the hexagon we can place

any of the 3 radicals (with repetition) so that the total number of arrangements is 3^6. By definition, the identity fixes all 3^6 of these arrangements. An arrangement fixed by a 60° or a 300° rotation ($-60°$) must have the same radical at each vertex. Thus, there are three fixed arrangements for each of these two rotations (one for each radical). In order for a 120° or a 240° rotation to fix an arrangement, the hexagon must have the same radical at every other vertex. So, the fixed arrangements have the form

This yields 3^2 fixed arrangements for the 120° rotation and the same number for the 240° rotation. The arrangements fixed by the 180° have the form

There are 3^3 such arrangements. Next consider a reflection across a diagonal joining two vertices. Fixed arrangements of this type of symmetry have the form

So, there are 3^4 fixed arrangements of this form for each of the three diagonal reflections. Finally, the

fixed arrangements for a reflection across a side bisector have the form

There are 3^3 of these for each of the three side bisectors. Now applying Burnside's theorem, we have

$$\tfrac{1}{12}(1 \cdot 3^6 + 1 \cdot 3 + 2 \cdot 3^2 + 1 \cdot 3^3 + 3 \cdot 3^4 + 3 \cdot 3^3) = 91.$$

Thus there are 91 possible molecules with six carbon atoms arranged in a hexagon with one of three kinds of radicals at each carbon atom. Surely it would be difficult to determine this number without Burnside's theorem.

Exercises for Section 9.4

1. Let

$$\alpha = \begin{bmatrix} 1 & 2 & 3 & 4 & 5 & 6 \\ 2 & 1 & 3 & 5 & 4 & 6 \end{bmatrix} \text{ and}$$

$$\beta = \begin{bmatrix} 1 & 2 & 3 & 4 & 5 & 6 \\ 6 & 1 & 2 & 4 & 3 & 5 \end{bmatrix}.$$

Compute each of the following:
   (a) α^{-1} (b) $\alpha\beta$ (c) $\beta\alpha$ (d) α^{100} (e) β^5

2. Represent the symmetry group of an equilateral triangle as a group of permutations of its vertices (see Example 3).

3. Verify the assertion made in Example 3 that $D_8 = \{\epsilon, \rho, \rho^2, \rho^3, \phi, \rho\phi, \rho^2\phi, \rho^3\phi\}$.

4. In Example 4 determine which of the βs is the same as $\beta_2\beta_4$. Which β is the same as β_5^{-1}?

5. In Example 3 determine
   (a) the stabilizer of 1 and the orbit of 1
   (b) the stabilizer of 2 and the orbit of 2

6. Let

$$G = \left\{ \begin{bmatrix} 1 & 2 & 3 & 4 & 5 & 6 \\ 1 & 2 & 3 & 4 & 5 & 6 \end{bmatrix}, \begin{bmatrix} 1 & 2 & 3 & 4 & 5 & 6 \\ 2 & 1 & 4 & 3 & 5 & 6 \end{bmatrix}, \right.$$

$$\begin{bmatrix} 1 & 2 & 3 & 4 & 5 & 6 \\ 2 & 3 & 4 & 1 & 6 & 5 \end{bmatrix}, \begin{bmatrix} 1 & 2 & 3 & 4 & 5 & 6 \\ 3 & 4 & 1 & 2 & 5 & 6 \end{bmatrix},$$

$$\begin{bmatrix} 1 & 2 & 3 & 4 & 5 & 6 \\ 4 & 1 & 2 & 3 & 6 & 5 \end{bmatrix}, \begin{bmatrix} 1 & 2 & 3 & 4 & 5 & 6 \\ 3 & 2 & 1 & 4 & 6 & 5 \end{bmatrix},$$

$$\left. \begin{bmatrix} 1 & 2 & 3 & 4 & 5 & 6 \\ 4 & 3 & 2 & 1 & 5 & 6 \end{bmatrix}, \begin{bmatrix} 1 & 2 & 3 & 4 & 5 & 6 \\ 1 & 4 & 3 & 2 & 6 & 5 \end{bmatrix} \right\}.$$

   (a) Find the stabilizer of 1 and the orbit of 1.
   (b) Find the stabilizer of 3 and the orbit of 3.
   (c) Find the stabilizer of 5 and the orbit of 5.

7. In S_n, find $|\mathrm{stab}_G(1)|$.

8. If G is a permutation group on $\{1, 2, \ldots, n\}$ and $1 \le i \le n$, prove that $\mathrm{stab}_G(i)$ is itself a group under composition.

9. Suppose

$$\alpha\beta = \begin{bmatrix} 1 & 2 & 3 & 4 & 5 \\ 2 & 5 & 4 & 1 & 3 \end{bmatrix}, \quad \beta\alpha = \begin{bmatrix} 1 & 2 & 3 & 4 & 5 \\ 2 & 4 & 1 & 5 & 3 \end{bmatrix}$$

and $\alpha(1) = 3$. Find α and β.

10. Suppose the vertices of a regular octagon are labeled consecutively counterclockwise 1 through 8. Similar to what was done in Example 3, write a 135° rotation in permutation form. Write the reflection across the axis joining vertex 1 and vertex 5 as a permutation.

11. A card-shuffling machine always rearranges cards in the same way relative to the order in which they were given to it. Suppose that all the hearts arranged in order from ace to king were put into the machine, and then the shuffled cards were put into the machine again to be shuffled. If the cards emerged in the order 6, 10, A, Q, 9, K, J, 7, 4, 8, 3, 2, 5, in what order were they arranged after the first shuffle?

12. Prove that S_n is non-Abelian for all $n \geq 3$.

13. Determine the number of rotation symmetries in the solid shown here. (The figure to the right of the solid is an "unfolded" version of the solid.)

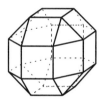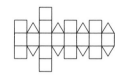

14. Determine the number of rotation symmetries in the solid shown here. (The figure to the right of the solid is an "unfolded" version of the solid.)

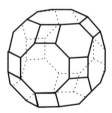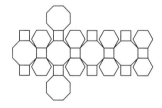

15. Calculate the number of elements in the group of rotations of a regular octahedron (a solid with eight congruent triangles as faces).

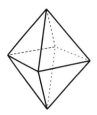

16. Calculate the number of elements in the group of rotations of a regular dodecahedron (a solid with 12 congruent pentagons as faces).

17. Calculate the number of elements in the group of rotations of a regular icosahedron (a solid with 20 congruent triangles as faces).

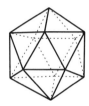

18. A soccer ball has 20 faces that are regular hexagons and 12 faces that are regular pentagons. Use the orbit-stabilizer theorem (Theorem 1) to explain why a soccer ball cannot have a 60° rotational symmetry about a line through the centers of two opposite hexagonal faces.

19. A volleyball has six identical faces of the following shape:

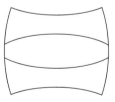

How many rotational symmetries does it have? Is the symmetry group Abelian?

20. Let G be a group of permutations of a set. Prove that $i \in \text{orb}_G(j)$ if and only if $j \in \text{orb}_G(i)$.

21. Let G be a group of permutations of a set S. Prove that if i and j belong to S, then $\mathrm{orb}_G(i) = \mathrm{orb}_G(j)$ or $\mathrm{orb}_G(i) \cap \mathrm{orb}_G(j) = \emptyset$. (Thus, two orbits are identical or disjoint.)

22. Determine the number of indistinguishable ways there are to color the four corners of a square with two colors. (It is permissible to use just a single color on all four corners.)

23. Determine the number of distinguishable ways the edges of a square can be colored with six colors with no restriction placed on the number of times a color can be used.

24. Determine the number of ways there are to color the vertices of an equilateral triangle with five colors so that at least two colors are used.

25. Determine the number of different necklaces that can be made using 13 white beads and 3 black beads.

26. Determine the number of distinguishable ways the edges of a square can be colored with six colors so that no color is used on more than one edge.

Chapter Summary

Although group theory is a cornerstone of theoretical mathematics, we have focused on some applications of group theory, such as symmetry, coding, and counting. Symmetry is pervasive in science, nature, technology, and art. Symmetry groups provide a systematic method for classifying and analyzing symmetry. Groups also are used to append extra digits to numbers for the purpose of error detection and error correction. Error-detection schemes are used on such items as retail products, credit cards, bank checks, airline tickets, and books. Error-correction schemes are used by space probes, satellites, compact discs, fax machines, and modems. Group theory can be used to solve sophisticated counting problems concerning the arrangements of objects. Such problems arise in science, engineering, computer science, and mathematics.

Chapter 9 Exercises

1. Prove that if G is a group with the property that the square of every element is the identity, then G is Abelian.

2. Show that in a group that has exactly four elements, $a^3 \neq e$ for all $a \neq e$.

3. Show that in a binary linear code, either all the code words end with 0 or exactly half end with 0. What about the other components?

4. Find all code words of the $(7, 4)$ binary linear code whose generator matrix is

$$G = \begin{bmatrix} 1 & 0 & 0 & 0 & 1 & 1 & 1 \\ 0 & 1 & 0 & 0 & 1 & 0 & 1 \\ 0 & 0 & 1 & 0 & 1 & 1 & 0 \\ 0 & 0 & 0 & 1 & 0 & 1 & 1 \end{bmatrix}.$$

Find the parity-check matrix of this code. Will this code correct any single error?

5. How many elements β in S_5 have the property that $\beta(1) = 3$?

6. Explain why it is impossible to find α and β in S_5 so that αβ =

$$\begin{bmatrix} 1 & 2 & 3 & 4 & 5 \\ 2 & 3 & 4 & 5 & 1 \end{bmatrix} \text{ and } \beta\alpha = \begin{bmatrix} 1 & 2 & 3 & 4 & 5 \\ 2 & 1 & 3 & 5 & 4 \end{bmatrix}.$$

7. Let G be a finite group and a an element of G. Show that there exist distinct integers i and j so that $a^i = a^j$.

8. Let G be a group, a an element of G, and $a \neq e$. If $a^7 = e$, show that $a \neq a^{-1}$. Is the same true if 7 is replaced by any odd integer? Give an example to show that if 7 is replaced by any even integer, it is possible that $a = a^{-1}$.

9. Determine the strip group corresponding to each of the following graphs (see Figure 11 of Section 9.1):
   (a) $y = \sin x$ (c) $|y| = |\sin x|$
   (b) $y = |\sin x|$ (d) $y = \tan x$

10. Determine the symmetry group of the tessellation of the plane exemplified by the brickwork shown here:

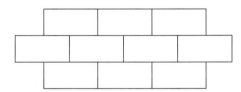

11. The state of Utah appends a ninth digit a_9 to its eight-digit driver's license number $a_1 a_2 \cdots a_8$ so that

$$(a_1, a_2, \ldots, a_8, a_9) \cdot (9, 8, 7, 6, 5, 4, 3, 2, 1) = 0 \bmod 10.$$

   (a) If the first eight digits of a Utah license number are 14910573, what is the ninth digit?
   (b) Suppose a legitimate Utah license number 149105767 is miscopied as 149105267. How would you know a mistake was made? Is there any way you could determine the correct num-

ber? Suppose you know the error was in the seventh position; could you correct the mistake?
   (c) If a legitimate Utah number 149105767 was miscopied as 199105767, would you be able to tell a mistake was made? Explain.
   (d) Explain why any transposition error involving adjacent digits of a Utah number would be detected.

12. The Canadian province of Quebec uses the weighting vector (12, 11, 10, ..., 2, 1) and modulo 10 arithmetic to append a check digit to the driver's license numbers. Criticize this method. Describe all single-digit errors that this scheme does not detect. How does the transposition of two adjacent digits of a number affect the check digit of a number?

13. The group D_8 acts as a group of permutations of the points making up the squares shown below. (The axes of symmetry are drawn for reference purposes.) For each square, locate the points in the orbit of the indicated point under D_8. In each case, determine the stabilizer of the indicated point.

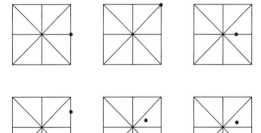

14. Calculate the number of elements in the group of rotations of a regular tetrahedron (see Figure 32 in Section 9.1).

15. Determine the symmetry group of the outer shell of the cross-section of the HIV virus shown below.

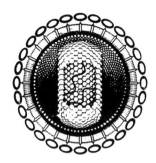

16. Determine which of the strip groups is the symmetry group of each of the following strip patterns.

(a) $\cdots$ D D D D $\cdots$ (e) $\cdots$ N N N N $\cdots$
(b) $\cdots$ V Λ V Λ $\cdots$ (f) $\cdots$ H H H H $\cdots$
(c) $\cdots$ L L L L $\cdots$ (g) $\cdots$ L Γ L Γ $\cdots$
(d) $\cdots$ V V V V $\cdots$

Writing Projects

17. Visit a large parking lot and take note of the symmetry groups you find on wheels and hupcaps. Ignore things such as names, logos, bolts, and valve stems. Focus on the symmetry pattern. Prepare a report on the symmetry groups you discover.

18. Look at the yellow pages of the phone directory for a midsize city and observe the logos used by various companies and businesses. Prepare a report of the symmetry groups of the logos.

19. Prepare a report on identification numbers used in your location. Possibilities for investigation include driver's license numbers in your state, student ID numbers at your school, and bar-coded numbers used by your school library, and city library. Determine whether a check digit is employed and how it is calculated. Include samples. The article "The mathematics of identification numbers," *The College Mathematics Journal,* 22:194–202 (1991) by J. A. Gallian, has information that will help you.

20. Prepare a report on the driver's license coding schemes that use a check digit as part of the number. The article "Assigning driver's license numbers," *Mathematics Magazine,* 64:13–22 (1992) by J. A. Gallian, has the information you will need.

21. Imagine that you are employed by a small company that does not use identification numbers and bar codes for its employees or products and that your boss has asked you to prepare a report discussing the various methods and to make a recommendation. Chapter 9 of the book *For All Practical Purposes,* 3rd ed., Freeman, NY, 1994, has the information you will need.

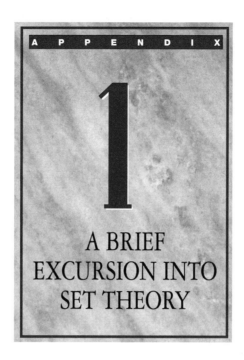

1

A BRIEF EXCURSION INTO SET THEORY

We do not formally define "set," but we all understand that a set is a collection of things we call "elements." For example, suppose the possible majors at Alright College consist of geology, psychology, and business. We can write these majors as a set: A = {geology, psychology, business}. The order in which we list the elements is not important in defining a set. Thus, we also have A = {business, geology, psychology}. A set can have no elements; we call such a set "the empty set," and use the symbol $\varnothing$ to denote it.

Combining Two Sets

Several basic operations occur in set theory; we begin by defining two binary operations:

DEFINITIONS

The *union* of two sets A and B is a set made up of all the elements of A together with all the elements of B, with the understanding that if some element, x, is in both A and B, then x is listed just once, not twice, in the union. We symbolize the union with $\cup$ and write $A \cup B$.

The *intersection* of two sets A and B is a set of just those elements of A which are also elements of B (or those elements of B which are also elements of A). We use the symbol $\cap$ for intersection, and we write $A \cap B$.

EXAMPLE 1 Unions and Intersections

Suppose the possible majors at Alright College

form the set A = {geology, psychology, business}. Similarly, we could have as possible majors at Biggish College and Correct College the respective sets B = {mathematics, geology, business} and C = {mathematics, biology, music}. Then the unions of possible majors at different pairs of colleges are

$A \cup B$ = {geology, psychology, business, mathematics},

$B \cup C$ = {mathematics, geology, business, biology, music},

$A \cup C$ = {geology, psychology, business, mathematics, biology, music}.

Using these same sets, we can also ask whether there are any majors that are possible at two of the colleges, majors the two colleges have in common. To answer this, we use the intersection operation:

$$A \cap B = \{\text{geology, business}\},$$

$$B \cap C = \{\text{mathematics}\},$$

$$A \cap C = \{\ \} \text{ or } \varnothing. \ \blacktriangle$$

It is often important to know whether or not the intersection of sets is empty, and additional vocabulary reflects this importance.

DEFINITIONS

Two sets S and T are called *disjoint* if S and T have no elements in common, that is, if $S \cap T = \varnothing$.

A collection C of sets is said to be *pairwise disjoint* if every pair of sets of C is disjoint.

EXAMPLE 2 Disjoint and Pairwise-Disjoint Sets   In Example 1 the sets A and C are disjoint. However, the collection of the three sets A,

B, and C is not pairwise disjoint, because $A \cap B \neq \varnothing$. An important counting principle says that the number of elements in the union of a collection of pairwise-disjoint sets is the sum of the numbers of elements in the individual sets of the collection (see Chapter 4, Section 1). ▲

If you have studied mathematical logic (for example, in Chapter 7 of this text) you will feel comfortable with these somewhat more formal definitions of union and intersection:

DEFINITIONS

$A \cup B$ = {elements, x, for which x is in A *or* x is in B}.

$A \cap B$ = {elements, x, for which x is in A *and* x is in B}.

Complement of a Set

The close relationship between mathematical logic and set theory might suggest that there should be a set theory equivalent of the logical *not*, negation. How would you define "not A" for our set A? Since A = {geology, psychology, business}, we would probably agree that none of the three possible majors listed in A could be in the set "not A." But what should be in "not A"? Is "physics" an element of "not A"? Could "not A" have "peanut butter and jelly sandwich" as an element? That last question opens up a whole area for exploration, namely, what should we consider as legitimate candidates for "not A"? We need to define a world that contains all the possible elements in our discussion. In this discussion, we probably would consider possible majors as elements of our world of interest, but not sandwiches, people, or cruise ships. In practice, the set of all elements of interest is usually stated as part of any set theory problem. Once we know all the possible elements, we can define the "negation" of a set.

DEFINITIONS

A *universal set,* denoted U, is the set of all the elements in a particular set theory discussion.

The *complement* of a set A is written as A' or $\sim A$ and has as its elements all the elements of U that are *not* elements of A.

EXAMPLE 3 Universal Set and Complement Suppose U = {geology, psychology, business, mathematics, biology, music, history, journalism}. Then using the definitions from Example 1, A' is {mathematics, biology, music, history, journalism} and B' = {psychology, biology, music, history, journalism}. But, if U = {geology, psychology, business, mathematics, biology, music}, then A' = {mathematics, biology, music} and B' = {psychology, biology, music}. ▲

Pictorial Representations of Sets

When we are dealing with just a few sets, it is often useful to accompany the discussion with *Venn diagrams*—pictures that show us the relationships among the sets. Figure 1 shows a Venn diagram with three sets, A, B, and C. The numbered regions of the Venn diagram in Figure 1 correspond to sets that we can describe using intersection and complement, as shown in Table 1.

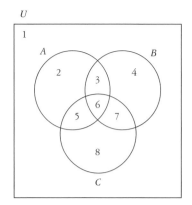

FIGURE 1 Venn diagram with three sets and numbered regions.

TABLE 1 Set Descriptions of the Eight Regions of a Three-Set Venn Diagram.

Region #	Set corresponding to that region
1	$A' \cap B' \cap C'$
2	$A \cap B' \cap C'$
3	$A \cap B \cap C'$
4	$A' \cap B \cap C'$
5	$A \cap B' \cap C$
6	$A \cap B \cap C$
7	$A' \cap B \cap C$
8	$A' \cap B' \cap C$

We can also begin with a set name and then locate the regions of the Venn diagram which correspond to it, as shown in Table 2.

TABLE 2 Describing Sets as Regions in a Venn Diagram.

Set	Regions making up that set
A	2, 3, 5, and 6
B	3, 4, 6, and 7
C	5, 6, 7, and 8
U	1, 2, 3, 4, 5, 6, 7, and 8
$A \cup B$	2, 3, 4, 5, 6, and 7
$A \cup C$	2, 3, 5, 6, 7, and 8
$B \cup C$	3, 4, 5, 6, 7, and 8

APPENDIX

2

PSEUDOCODE

Describing the Steps in an Algorithm

Mathematics involves more than just formulas; it involves carrying out various processes. Mathematical processes involve not only following steps in the correct order, but also making decisions and repeating subprocesses. Mathematical processes are typically called *algorithms.* We start learning algorithms early in our mathematical education although we often do not use the name "algorithm" until later in our studies. Thus, we learn algorithms for the four basic arithmetic operations—addition, subtraction, multiplication, and division. These algorithms cover just whole numbers at first, then later include fractions, decimals, and negative numbers. Today, many of us rely on our calculators to carry out these algorithms for us.

Several techniques have developed in mathematics to simplify the way we communicate how to accomplish a calculation or carry out an algorithm. One such technique, the use of algebraic formulas, is so much a part of mathematics that it is easy to forget that calculations were once described in sentences. Do you recognize the famous formula that Euclid described in his *Elements* this way:

In right-angled triangles the square on the side subtending the right angle is equal to the squares on the sides containing the right angle.

(Source: James R. Newman (1956), *The World of Mathematics,* New York: Simon and Schuster, p. 190)

If we look at the "right-angled triangle" in Figure 1 and label its sides and angles, we recognize the Pythagorean theorem in Euclid's words. In this triangle, angle C is the right angle and we use lowercase letters to denote the lengths of the sides. Euclid's words correspond to our familiar formula

$$c^2 = a^2 + b^2.$$

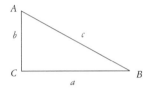

FIGURE 1 A "right-angled triangle."

For modern readers of mathematics, the formula is a more compact and clear way of conveying the Pythagorean theorem than Euclid's sentence.

A recent development in the mathematical sciences is the use of a standard language to write down algorithms. This standard language is called *pseudocode,* referring to its resemblance to some coding languages used for communicating with computers and calculators. Pseudocode does not demand the precision of punctuation and word usage that programming code requires, but it is more formal than everyday written language. Just as algebraic formulas are more compact and precise than ordinary language for communicating calculations, pseudocode is more compact and precise for communicating algorithms.

In pseudocode, the way that we communicate the order of steps is formalized into "control struc-

tures." The way that we describe the individual steps will be less formal.

The "sequence" Control Structure

The first control structure is very simple; it is called "sequence" and simply states that in the absence of any other instructions, we follow the steps in the order in which they are written. For example, if we want to write pseudocode to explain how to evaluate $3 + 7x$ for a specific value of x, we might write

```
Get the x-value
Multiply x by 7
Add 3 to the answer you just got
Store the result in a place called answer
The value in answer is the value of 3 + 7x
```

This example not only introduces us to pseudocode but also reminds us that for simple calculations, formulas are more compact than pseudocode.

The "if-then-else" Control Structure

The remaining control structures allow us to include decisions and repetitions in pseudocode. First, we will look at decisions. The control structure that we need is the "if-then-else." A decision is implied in the definition of the absolute-value function; our first step in finding $|x|$ is to determine if x is negative or nonnegative:

$$|x| = \begin{cases} x & \text{if } x \geq 0, \\ -x & \text{if } x < 0. \end{cases}$$

We can express that decision using pseudocode and the "if-then-else" control structure.

```
If x ≥ 0 then
        |x| should be set equal to x
else
        |x| should be set equal to −x
```

Since the phrase "should be set equal to" occurs frequently in pseudocode, several symbolic notations have been devised for it. We will use a left-pointing arrow. The left-pointing arrow will mean that the calculation on the right of the arrow will be made and the answer stored in the variable/location named at the left. Thus, we rewrite the pseudocode defining $|x|$:

```
If x ≥ 0 then
        |x| ← x
else
        |x| ← −x
```

We call the step "$|x| \leftarrow x$" the "then branch" of the above pseudocode and the step "$|x| \leftarrow -x$" the "else branch." Sometimes there are several individual steps in either the then or the else branch. There are two ways to indicate which steps are in a particular branch: First, we indent all the steps so that they start a few columns to the right of the column where the "if" and the "else" begin. Second, we agree on special words to end the branch. When there are both then and else branches, the word "else" signals us that the then branch has ended. To signal the end of the else branch, or a then branch when there is no else branch, we use the expression "end if." Let us incorporate this into our definition of the absolute value:

```
If x ≥ 0 then
        |x| ← x
else
        |x| ← −x
end if
```

Here are some examples of pseudocode for familiar nonmathematical decisions.

E X A M P L E 1 **What I Do When the Sun Shines Brightly**

```
If the sun is shining brightly then
        I put on sunscreen
        I wear a big straw hat
        I wear sunglasses
end if ▲
```

E X A M P L E 2 **How I Pack for a Vacation**

```
If my vacation is in the mountains then
        I pack my hiking boots
        I pack my compass
else    if my vacation is at the beach then
                I pack my bathing suit
                I pack my kite
        else
                I pack several books to read
                I pack my frisbee
        end if
end if
I pack my camera ▲
```

In Example 1 there is no "else" branch, because there are no steps to be taken if the sun is not shining brightly. In Example 2 the person packs a camera regardless of where the vacation happens to be, because the step "I pack my camera" is reached in sequence as the next step after the "end if." Also in Example 2, we have "nested ifs"—the else branch of the outer if is itself an if.

Documenting the Applicability of the Algorithm

We all understand that if we are sharing n sandwiches among k people, then the number of sandwiches that each person gets can be found using a division algorithm. Other algorithms, such as those for addition and subtraction, are not useful to us in solving our problem. An important but often overlooked part of a pseudocode description of an algorithm is including a clear description of what the algorithm achieves and when it is applicable. This description, called *external documentation,* gives us a brief description of what the algorithm does so we can decide whether the algorithm will help us solve the problem at hand.

In the pseudocode presentations in this text, we have four kinds of information listed before the actual pseudocode:

Input(s) and *Output(s)* introduce the notation and identify the variables in the algorithm;

Preconditions are facts that must be true about the inputs in order for the algorithm to work properly;

Postconditions are features that are supposed to be true about the output—showing that these features are indeed true requires proof, just as any theorem must be proven, not merely stated.

These four kinds of data are sometimes called external documentation.

Pseudocoded algorithms are also usually documented with *internal documentation,* comments explaining the steps as they occur. We will set off these comments by writing them in italic and enclosing them in brackets.

Example 3 shows how our pseudocode definition of absolute value looks with both kinds of documentation.

EXAMPLE 3 Fully Documented Pseudocode Definition of Absolute Value

Input: A real number, x
Output: The absolute value of x
Preconditions: None
Postconditions: The absolute value is never a negative number

```
If x ≥ 0 then
       |x| ← x {a nonnegative number is
       its own absolute value}
else
       |x| ← −x {the negative of a
       negative number is its absolute
       value}
end if ▲
```

Control Structures for Repetition— the "for"

There are circumstances when carrying out an algorithm requires the repetition of some portion of the algorithm. For example, in doing long division, we repeat these steps several times: guess at the next digit in the divisor, multiply, subtract, and verify whether our trial divisor was correct. (In this oversimplified description of long division, there are hidden repetitions; for example, those needed to do the multiplication and subtraction.)

Two types of control structures indicate repetition. One type is used when we know in advance the number of repetitions that will be required. The other type is used when we do not know this number in advance, but we do know some event or condition that will be a clear indication to us to stop.

Our first example will be one in which we know the number of repetitions needed. Suppose we wish to use a calculator to add k numbers. How many times will we press the + key? One method requires

$k - 1$ presses. Example 4 presents pseudocoded directions for this method.

EXAMPLE 4   Algorithm for Adding k Numbers

Input: An integer k, and a list of k numbers
Output: The sum of the k numbers
Preconditions: $k > 0$. The k numbers should be of a type the calculator can process. For example, not all calculators can process fractions or complex numbers.
Postconditions: The desired sum is now showing in the calculator

```
Preset the calculator to the type of
 numbers being added, if necessary
Clear the calculator, if necessary
Enter the first number on the list into the
 calculator
For i from 2 to k
        Add the ith number on the list to
        the number in the calculator
end for  ▲
```

In Example 4 we introduce the "for" control structure for repetition. A control structure indicating repetition is usually called a *loop*. Both the indentation and the "end for" are used to indicate which step(s) is (are) to be repeated. The indented steps between the "for" and the "end for" are called the *body of the loop.* In Example 4 the first time we perform the loop, $i = 2$, the initial value. The value of i is increased by 1 each time we repeat the loop, and the last repetition of the loop occurs when $i = k$. Example 5 shows us that in a "for" control structure, we can "count by 2s" as we form pairs of students from a line of k students. There are exactly $k/2$ repetitions of the for loop.

EXAMPLE 5   Pseudocode for Forming Pairs from a Line of k Students

Input: An integer k, a line of k students
Output: $k/2$ pairs of students
Preconditions: k should be an even, positive integer
Postconditions: None

```
For i from 2 to k by 2s
        Make a pair from the students in
        positions i and i − 1 in line
end for  ▲
```

In general, we can "count by" any number, even negative or fractional ones, as long as we set up the "for" so that the initial value, plus some finite number of additions of the increment value, eventually equals (or passes) the final value. The general format of a "for" is:

```
For i from "initial" to "final" by
 "increment."
```

Now we investigate loops for which we do not know the number of repetitions in advance.

A Control Structures for Repetition—the "while"

Sometimes we have no idea how many times we need to repeat a process before we have accomplished our purpose. For example, suppose you are snacking on your favorite brand of chips. Example 6 is pseudocode that might describe your snacking algorithm:

EXAMPLE 6   Snacker 1

Input: One person and one package of chips
Output: One nonhungry person
Preconditions: The person likes the chips; there are enough chips
Postconditions: There may or may not be chips left at the end

```
Eat an initial handful of chips
While you are still hungry
      Eat another handful of chips
end while  ▲
```

In Example 6 we introduced the "while" control structure. The "while" tells us to repeat the steps between the words "while" and "end while" as many times as necessary, provided the condition stated after the word "while" is true. Thus we repeat "Eat another handful of chips" over and over, ending our repetitions only when the condition "you are still hungry" becomes false. Example 7 is another snacking pseudocode algorithm, which looks very much like the one in Example 6 but is different in an important technicality.

E X A M P L E   7   Snacker 2

Input: One person and one package of chips
Output: One nonhungry person
Preconditions: The person likes the chips; there
            are enough chips
Postconditions: There may or may not be chips
              left at the end

```
While you are hungry
      Eat a handful of chips
End while  ▲
```

In Snacker 2 the person does not necessarily eat any chips. This is because of a convention concerning the "while." The test to see if the loop is "repeated" precedes the "repetition," even for the very first "repetition." In Snacker 2, our person could start out either hungry or not hungry. Since the first step in the algorithm is to test for hunger, if the person is not hungry, then the "while" condition is false from the outset, and thus the steps in the loop are never followed.

Another Control Structure for Repetition—the "repeat-until"

The "repeat-until" control structure is like the "while" in that the number of repetitions is controlled by the true/false value of some condition. But the convention we follow with the "repeat-until" is that there is always at least one "repetition" of the steps in the loop. Example 8 is our Snacker 1 algorithm, rewritten using the "repeat-until."

E X A M P L E   8   Snacker 3

Input: One person and one package of chips
Output: One nonhungry person
Preconditions: The person likes the chips; there
            are enough chips
Postconditions: There may or may not be chips
              left at the end

```
Repeat
      Eat a handful of chips
Until you are not hungry  ▲
```

Snacker 3 and Snacker 1 have exactly the same structure. Both tell the person to eat a handful of chips before thinking about whether or not hunger exists. In the "repeat-until" control structure, the test for further repetition can be thought of as taking place *after* each repetition, so that the condition following the word "until" is the condition that tells us to stop. By contrast, the condition following the word "while" is the one that tells us to start or continue repeating. Thus, the "while condition" in Snacker 1 is exactly the opposite of the "until condition" in Snacker 3.

As you can see, although the "repeat-until" is a useful control structure, it is not strictly necessary since we can express the same algorithm using a "while." In fact, we could use a "while" instead of a "for," but this is rarely done in practice. In place of a structure such as

```
For i from m to k by js
        Do some stuff
end for
```

we could write this pseudocode:

```
i ← m
While i ≤ k
        Do some stuff
        Add j to i {can also be written as
         i ← i + j}
end while
```

There are two things of note in these equivalent examples of pseudocode. First, the "while" will not "Do some stuff" if the initial value of the variable i, namely m, is greater than the final value, namely k.

Since the two loops are equivalent, we now know that when the initial value for i is greater than the final value, a "for" loop will result in no repetitions at all. Second, in the comment that "Add j to i" can be written as $i \leftarrow i + j$, we see an example of how pseudocode is typically used to indicate modifying the value of a variable. The pseudocode $i \leftarrow i + j$ means that we add j to i and then keep the result of the calculation as the new value of i.

The following reference shows that any arbitrary flow of control can be achieved, that is, adequately described, using just the control structures we discussed here: sequence, selection (if-then-else), and repetition: Bohm, C. and G. Jacopini (1966), "Flow diagrams, Turing machines, and languages with only two formation rules," *Communications of the Association for Computing Machinery,* **5,** no. 5 (May):366–371.

SOLUTIONS TO ODD-NUMBERED EXERCISES

SECTION 1.2

1. (a) $a_n = 3n \Rightarrow \{3, 6, 9, 12, 15, 18, \ldots\}$
    (b) $a_n = 5n + 2 \Rightarrow \{7, 12, 17, 22, 17, 32, \ldots\}$
    (c) $a_n = (1/2)^n \Rightarrow \{1/2, 1/4, 1/8, 1/16, 1/32, 1/64, \ldots\}$
    (d) $a_n = n^2 - 4 \Rightarrow \{-3, 0, 5, 12, 21, 32, \ldots\}$
    (e) $a_n = n^3 + n \Rightarrow \{2, 9, 30, 68, 130, 222, \ldots\}$
    (f) $a_n = (n^2 + 2)/(n^2 + 1) \Rightarrow \{3/2, 6/5, 11/10, 18/17, 27/26, 38/37, \ldots\}$
    (g) $a_n = 2^n/3^n \Rightarrow \{2/3, 4/9, 8/27, 16/81, 32/243, 64/729, \ldots\}$
    (h) $a_n = n^2/2^n \Rightarrow \{1/2, 4/4, 9/8, 16/16, 25/32, 36/64, \ldots\}$

3. Derive these graphs:

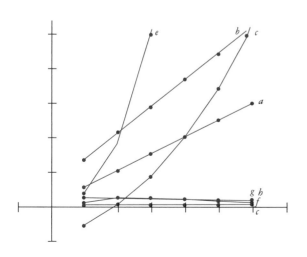

5. $a_n = \{2, 5, 8, 11, 14, 17, \ldots\} \Rightarrow \Delta a_n = 3 \Rightarrow a_n = 3n + b$
    $a_1 = 2 \Rightarrow b - 1 \Rightarrow a_n = 3n - 1$
    $a_7 = 20, a_8 = 23$

7. $a_n = \{2, 5, 10, 17, 26, 37, 50, 65, 82, \ldots\} \Rightarrow \Delta a_n = \{3, 5, 7, 9, 11, 13, 15, 17, \ldots\} \Rightarrow \Delta^2 a_n = \{2, 2, 2, 2, 2, 2, \ldots\} \Rightarrow a_n = An^2 + Bn + C \Rightarrow A = 1, B = 0, C = 1 \Rightarrow a_n = n^2 + 1$
    $a_{10} = 101, a_{11} = 122$

9. $a_n = \{2/5, 4/25, 8/125, \ldots\} \Rightarrow a_n = \{2^n/5^n\} = \{2/5\}^n$
    $a_4 = 16/625, a_5 = 32/3125$

11. $a_7 = \{1, 5, 15, 35, 70, 126, \ldots\}$

n	a_n	Δa_n	$\Delta^2 a_n$	$\Delta^3 a_n$	$\Delta^4 a_n$
1	1	4	6	4	1
2	5	10	10	5	1
3	15	20	15	6	1
4	35	35	21	7	
5	70	56	28		
6	126	84			
7	210				

Therefore, $a_7 = 210$

13. $a_n = 1/2(n)(n + 1) \Rightarrow a_{10000} = 50,005,000$

15. $a_n = 7n \Rightarrow \Delta a_n = 7$

17. $a_n = n^4 \Rightarrow \Delta a_n = 4n^3 + 6n^2 + 4n + 1$

19. (a) $s_{n+1} = s_n + 2$
  (b) 33
  (c) 680

21.

n	a_n	Δa_n	$\Delta^2 a_n$
1	2	2	2
2	4	4	4
3	8	8	8
4	16	16	
5	32		

23.

n	a_n	Δa_n	$\Delta^2 a_n$
1	-2	-2	6
2	-4	4	12
3	0	16	18
4	16	34	24
5	50	58	
6	108		

(d)

n	a_n	Δa_n	$\Delta^2 a_n$
1	1.3	1.2	-7.4
2	2.5	-6.2	4.2
3	-3.7	-2.0	-1.2
4	-5.7	-3.2	1.9
5	-8.9	-1.3	
6	-10.2		

(e)

n	a_n	Δa_n	$\Delta^2 a_n$
1	22	9	-9
2	31	0	0
3	31	0	-9
4	31	-9	9
5	22	0	9
6	22	9	
7	31		

SECTION 1.3

1. (a)

n	a_n	Δa_n	$\Delta^2 a_n$
1	2	3	-2
2	5	1	-3
3	6	-2	1
4	4	-1	0
5	3	-1	-4
6	2	-5	
7	-3		

(f)

n	a_n	Δa_n	$\Delta^2 a_n$
1	0	-1	2
2	-1	1	0
3	0	1	-2
4	1	-1	0
5	0	-1	2
6	-1	1	0
7	0	1	-2
8	1	-1	0
9	0	-1	
10	-1		

(b)

n	a_n	Δa_n	$\Delta^2 a_n$
1	-2	4	0
2	2	4	0
3	6	5	1
4	11	3	5
5	14	8	
6	22		

(c)

n	a_n	Δa_n	$\Delta^2 a_n$
1	1	0	1
2	1	1	0
3	2	1	1
4	3	2	1
5	5	3	2
6	8	5	3
7	13	8	
8	21		

3. (a)

n	a_n	Δa_n	$\Delta^2 a_n$	$\Delta^3 a_n$
1	3.4	0.2	0.8	0
2	3.6	1.0	0.8	0
3	4.6	1.8	0.8	0
4	6.4	2.6	0.8	0
5	9.0	3.4	0.8	
6	12.4	4.2		
7	16.6			

Therefore, the general term is defined by a polynomial of degree 2.

(b)

n	a_n	Δa_n	$\Delta^2 a_n$	$\Delta^3 a_n$
1	1	-2	2	0
2	-1	0	2	0
3	-1	2	2	0
4	1	4	2	
5	5	6		
6	11			

Therefore, the general term is defined by a polynomial of degree 2.

(c)

n	a_n	Δa_n	$\Delta^2 a_n$	$\Delta^3 a_n$	$\Delta^4 a_n$
1	−7.9	5.3	2.8	−0.6	0
2	−13.2	8.1	2.2	−0.6	0
3	−21.3	10.3	1.6	−0.6	0
4	−31.6	11.9	1.0	−0.6	
5	−43.5	12.9	0.4		
6	−56.4	13.3			
7	−69.7				

Therefore, the general term is defined by a polynomial of degree 3.

(d)

n	a_n	Δa_n	$\Delta^2 a_n$	$\Delta^3 a_n$	$\Delta^4 a_n$
1	−24.1	6.9	12	6	0
2	−17.2	18.9	18	6	0
3	1.7	36.9	24	6	0
4	38.6	60.9	30	6	
5	99.5	90.9	36		
6	190.4	126.9			
7	317.3				

Therefore, the general term is defined by a polynomial of degree 3.

5.

n	w_n	Δw_n	$\Delta^2 w_n$
1	10	−3	8
2	7	5	1
3	12	6	−3
4	18	3	−6
5	21	−3	
6	18		

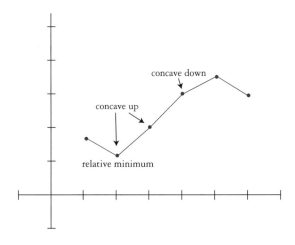

concave down

concave up

relative minimum

7. (a) $h = 17.89/\pi r^2$

(b) $a = 0.02\pi[4r^2 + 35.78/r]$

(c)

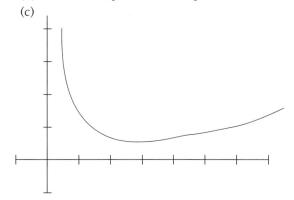

9. {20, 18, 15, 11, 6, 0, . . .}

11. {0, 5, 10, 5, 0, 5, . . .}

13. {0, 1, 3, 6, 10, 15, . . .}

SECTION 1.4

1. The graph is smooth; its slope is approximately 0.054.

3. The model is reasonable. The slope is approximately 0.008.

5. (a) Let a_n represent the amount of digoxin after n days. Then

$$a_{n+1} - a_n = ka_n,$$

$$a_0 = 0.5.$$

For the data given, an estimate of k is -0.31, giving

$$a_{n+1} = 0.69a_n,$$

$$a_0 = 0.5.$$

n	a_n
0	0.5
1	0.3450
2	0.2381
3	0.1643
4	0.1133
5	0.0782
6	0.0540
7	0.0372
8	0.0257
9	0.0177
10	0.0122
11	0.0084
12	0.0058
13	0.0040
14	0.0028
15	0.0019

(b) $a_{n+1} = 0.69a_n + 0.1$

$a_0 = 0.5$

(c)

n	a_n
1	0.4450
2	0.4071
3	0.3809
4	0.3628
5	0.3503
6	0.3417
7	0.3358
8	0.3317
9	0.3289
10	0.3269
11	0.3256
12	0.3246
13	0.3240
14	0.3236

7. Let c_n represent the amount of carbon remaining after n years. If k is the proportion that decays each year, then

$$c_{n+1} = c_n - kc_n.$$

In order to determine k, we must use the information that, after 5700 years, only one-half remains. (In the next section we learn how to use this information to compute directly how long it would take to decay to 1% of the

initial value.) To construct a numerical solution, one would have to experiment with different values of k to find the k that results in one-half the original amount after 5700 years. Using that k, then iterate until only 1% remains. This exercise shows the limitations of numerical solutions.

SECTION 1.5

1. $a_k = 3^k c$

$a_0 = 1 = 3^0 c \Rightarrow c = 1$

$a_k = 3^k$

3. $a_k = 5^k c$

$a_0 = 5^0 c = 10 \Rightarrow c = 10$

$a_k = 5^k \cdot 10$

5. $x_k = (3/4)^k c$

$x_0 = 64 = (3/4)^0 c \Rightarrow c = 64$

$x_k = (3/4)^k \cdot 64$

7. $x_k = 1^k c$

$x_0 = 200 = c$

$x_k = 200$

9. Let a_n represent the balance after n quarters. Then

$$a_k = (1.04)^k - 500,$$

$$a_4 = (1.04)^4 \cdot 500 = \$584.93,$$

$$a_{40} = (1.04)^{40} \cdot 500 = \$2400.51.$$

11. Let a_n represent the amount of food after n years. Then

$$a_k = (1.01)^k a_0,$$

$a_{10} = (1.01)^{10} a_0 = 1.1046\, a_0$, or about 10.5% increase.

13. Let c_n represent the concentration of drug after n hours. Then

$$c_{n+1} = c_n - 0.2c_n = 0.8c_n,$$

$$c_k = (0.8)^k c_0 = (0.8)^k \, 640.$$

Find k such that $c_k = 100$:

$$100 = (0.8)^k \, 640,$$

$$100/640 = (0.8)^k,$$

$$k = 8.32, \text{ or about } 8.32 \text{ hours.}$$

For 10%, $c_k = 0.10c_0$. Then

$$0.10c_0 = (0.8)^k c_0,$$

$$0.10 = (0.8)^k,$$

$$k = 10.32, \text{ or about } 10.32 \text{ hours.}$$

15. Substitute, obtain an identity.

SECTION 1.6

1. (a) $b/(1 - r) = 30/(1 - 0.5) = 60$ is the equilibrium value.
   $a_k = (0.5)^k c + 60$
   The solution converges to the equilibrium value, which is stable.

   (b) $b/(1 - r) = 40/(1 - 3) = -20$, the equilibrium value
   $a^k = (3)^k c - 20$
   The solution remains at -20 if -20 is the initial value. For initial values greater than -20, the solution grows without bound. For initial values less than -20, the solution becomes infinitely negative. Unstable.

   (c) Since $r = 1$, no equilibrium value exists.
   $a_k = c + 20k$ which becomes arbitrarily large.

   (d) $b/(1 - r) = -20/(1 - 2) = 20$, the equilibrium value
   $a_k = (2)^k c + 20$
   If you start at 20, you remain there. For starting values greater than 20, the solution grows without bound. For starting values less than

20, the solution becomes infinitely negative. Unstable.

   (e) $b/(1 - r) = -30/(1 - 0.8) = -150$, the equilibrium value
   $a_k = (0.8)^k c - 150$
   The solution converges to the equilibrium value for all starting values.

   (f) Since $r = 1$, no equilibrium value exists.
   $a_k = c - 20k$
   The solution becomes infinitely negative.

3. Model: Let a_n represent the amount remaining after n months. Then

$$a_{n+1} = 1.0075a_n - b,$$

$$a_0 = 120{,}000,$$

$$a_{240} = 0.$$

   (a) The solution is

$$a_k = (1.0075)^k c - b/-0.0075$$

$$= (1.0075)^k c + 133.33b,$$

$$a_0 = 120{,}000 = c + 1333.33b,$$

$$a_{240} = 0 = (1.0075)^{240} c + 133.33b,$$

$$c = -23{,}956.15,$$

$$b = 1079.67.$$

Monthly payments are $1079.67.

   (b) $a_{72} = (1.0075)^{72}(-23{,}956.15) + 133.33(1079.67)$
   $a_{72} = \$102{,}929.83$
   (Solution to parts c, d, and e, are not included here.)

5. (a) Amount needed 20 years from now:

$$\text{Model:} \quad a_{n+1} = 1.005a_n - 1000,$$

$$a_0 = x,$$

$$a_{96} = 0.$$

Solution: $a_k = (1.005)^k c + -1000/(1 - 1.005)$,

$$a_k = (1.005)^k c + 200{,}000,$$

$$a_0 = x = c + 200{,}000,$$

$$a_{96} = 0 = (1.005)^{96} c + 200{,}000,$$

$$c = -123{,}904.78,$$

$$x = 76{,}095.21.$$

You will need \$76,095.21, 20 years from now.

(b) Amount to invest: $a_{n+1} = 1.005a_n + b$,

$$a_0 = 0,$$

$$a_{240} = 76{,}095.21$$

Solution: $a_k = (1.005)^k c - 200b$

$$a_0 = 0 = c - 200b$$

$$a_{240} = 76{,}095.21 = (1.005)^{240}c - 200b$$

$$c = 32{,}938.73369$$

$b = 164.6936$, or approximately
\$165.70 each month.

1.

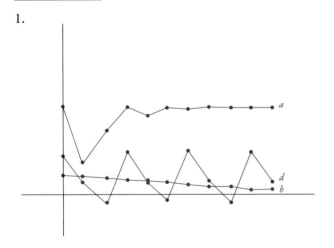

Note: c_n diverges.

3. (a) $a = 3 - \sqrt{3}, 3 + \sqrt{3}$
    (b) $b = 0, 5/2 + \sqrt{13}/2, 5/2, -\sqrt{13}/2$

5. Graphical iteration

7. (a) $r_{n+1} = r_n + c(r_n)(1000 - r_n)$
    (b) $r_{17} = 446, r_{18} = 569$, therefore, the 18th day.
        About 28 or 29 days.
    (c) About 14 days

9.

n	No rounding	With rounding
1	2	2
2	3.59	4
3	6.44	7
4	11.51	13
5	20.45	23
6	35.97	40
7	62.17	69
8	104.17	115
9	165.80	181
10	243.46	260
11	319.69	333
12	371.04	378
13	392.53	395
14	398.39	399

1. (a) $\{a_n\} = \{2, 2, 4, 4, 8, \ldots\}$
        $\{b_n\} = \{2, 0, 4, 0, 8, \ldots\}$
    (b) $\{c_n\} = \{0, 3, -3.6, 4.17, -4.496, \ldots\}$
        $\{d_n\} = \{5, -2, 1.9, -1.42, 2.369, \ldots\}$
    (c) $\{e_n\} = \{100, 62, 67, 55.82, 52.03, \ldots\}$
        $\{f_n\} = \{30, 90, 55.8, 60.3, 50.238, \ldots\}$
    (d) $\{g_n\} = \{0.15, -0.12, 0.9, -0.72, 5.4, \ldots\}$
        $\{h_n\} = \{0.13, 0.03, 0.78, 0.18, 4.68, \ldots\}$

3. Verify by substitution.

5. (a) $\{a_n\} = \{1, 2.375, 6.6090625, \ldots\}$
        $\{b_n\} = \{0.5, -1.2625, -1.601498, \ldots\}$
        $\{c_n\} = \{1.5, 0.05, -0.995, \ldots\}$
    (b) $\{d_n\} = \{0.5, 1, 2, \ldots\}$
        $\{e_n\} = \{0.5, 0.25, 0.25, \ldots\}$
        $\{f_n\} = \{1, 0, -0.5, \ldots\}$

7. (a) $a_{n+1} = 0.7a_n$,

$b_{n+1} = 0.6b_n + 0.3a_n$,

$c_{n+1} = 0.45c_n + 0.3b_n$

(b)

n	a	b	c
0	350	0	0
1	245	105	0
2	171.5	136.5	31.5
3	120.0	133.4	55.1
4	84.0	116.0	64.8
5	58.8	94.8	63.9

Total contamination $= a + b + c = 217.62$ ml

(c)

n	a	b	c
6	41.2	74.5	57.2
7	28.8	57.1	48.1
8	20.2	42.9	38.8
9	14.1	31.8	30.3
10	9.9	23.3	23.2

Total contamination $= a + b + c = 56.37$ ml

(d) Contaminant in water supply $=$

$$0.35 \sum_{n=0}^{\infty} c_n \approx 0.35 \sum_{n=0}^{50} c_n \approx 167 \text{ ml}$$

SECTION 1.9

1. (a) $\sum_{k=1}^{6} k^2 = 1^2 + 2^2 + 3^2 + 4^2 + 5^2 + 6^2 = 91$

(b) $\sum_{k=1}^{8} (k + 1) = \sum_{k=1}^{8} k + \sum_{k=1}^{8} 1 = (8)(9)/2 + 8 = 44$

(c) $\sum_{k=1}^{5} (-1)^k/k = -47/60$

(d) $\sum_{k=1}^{5} 2^k = \sum_{k=0}^{5} 2^k - 2^0 = (2^6 - 1)/(2 - 1) - 1 = 62$

(e) $\sum_{k=1}^{5} (1/2)^k = (1/2)^6 - 1/(1/2) - 1 - 1 = 31/32$

(f) $\sum_{k=1}^{5} (-1/2)^k = (-1/2)^6 - 1/-(1/2) - 1 - 1 = -11/32$

3. (a)

n	a_n	Δa_n
1	0	0
2	0	-3
3	-3	-3
4	-6	0
5	-6	3
6	-3	3
7	0	0
8	0	

(b)

n	a_n	Δa_n
1	3	1
2	4	-1
3	3	-3
4	0	-5
5	-5	-3
6	-8	-1
7	-9	1
8	-8	

5.

n	a_n	Δa_n	$\Delta^2 a_n$
1	2	2	2
2	4	4	2
3	8	6	2
4	14	8	2
5	22	10	2
6	32	12	2
7	44	14	2
8	58	16	2
9	74	18	2
10	92	20	
11	112		

7. No solution provided.

9. (a) $\sum_{k=1}^{\infty} 3/10^k = 3 \sum_{k=0}^{\infty} (1/10)^k - 3 = 3(10/9) - 3 = 1/3$

(b) $\sum_{k=1}^{\infty} (p/3)^{k-1} = \sum_{k=0}^{\infty} (p/3)^k = p(1/1 - 1/3) = 3p/2$

1. (a) $x^2 - 6x + 13, 1 \le x \le 6$

x	f	Δf
1	8	-3
2	5	-1
3	4	1
4	5	3
5	8	5
6	13	

x	f	Δf
2.8	4.04	-0.03
2.9	4.01	-0.01
3.0	4.00	0.01
3.1	4.01	0.03
3.2	4.04	

Relative minimum near $x = 3.0$

(b) $x^3 - 2x^2 - 4x + 5, 0 \le x \le 5$

x	f	Δf
0	5	-5
1	0	-3
2	-3	5
3	2	19
4	21	

x	f	Δf
1.9	-2.961	-0.039
2.0	-3.0	0.041
2.1	-2.959	

Relative minimum near $x = 2.0$

(c) $2^x - x^3, 0 \le x \le 9$

x	f	Δf
0	1	0
1	1	-5
2	-4	-15
3	-19	-29
4	-48	-49
5	-97	-55
6	-152	-63
7	-215	-41
8	-256	39
9	-217	

x	f	Δf
0.5	1.28921	0.01
0.6	1.29971	-0.018
0.7	1.28150	-0.052
0.8	1.22910	

Relative maximum near $x = 0.6$

x	f	Δf
8.0	-256	-1.067
8.1	-257.067	-0.234
8.2	-257.301	0.687
8.3	-256.614	

Relative minimum near $x = 8.2$

(d) $x^{5/2} - 4x^{3/2} - 3x^{1/2}, 0 \le x \le 4$

x	f	Δf
0	0	-6
1	-6	-3.89
2	-9.899	-0.50
3	-10.392	4.39
4	-6	

x	f	Δf
2.5	-10.672	-0.028
2.6	-10.706	0.009
2.7	-10.697	0.055
2.8	-10.642	

Relative minimum near $x = 2.6$

3. (a) $x^2 - 6x + 13$

x	f	Δf	$\Delta^2 f$
1	8	-3	2
2	5	-1	2
3	4	1	2
4	5	3	2
5	8	5	
6	13		

Appears to have no infection points

(b) $x^3 - 2x^2 - 4x + 5$

x	f	Δf	$\Delta^2 f$
0	5	-5	2
1	0	-3	8
2	-3	5	14
3	2	19	22
4	21	41	24
5	60	65	
6	125		

Appears to have no inflection points

(c) $2^x - x^3$

x	f	Δf	$\Delta^2 f$
0	1	0	-5
1	1	-5	-10
2	-4	-15	-14
3	-19	-29	-20
4	-48	-49	-6
5	-97	-55	-7
6	-152	-63	22
7	-215	-41	80
8	-256	39	
9	-217		

x	f	Δf	$\Delta^2 f$
6.1	-158.387	-6.424	-0.018
6.2	-164.811	-6.442	0
6.3	-171.253	-6.442	0.022
6.4	-177.695	-6.420	
6.5	-184.115		

Inflection point near $x = 6.2$

(d) $x^2 + 3 - (20/x)$

x	f	Δf	$\Delta^2 f$
1	-16	13	-4.666
2	-3	8.333	0.333
3	5.333	8.666	1.333
4	14	10	1.666
5	24	11.666	1.79
6	35.666	13.475	
7	49.1428		

x	f	Δf	$\Delta^2 f$
2.5	1.250	0.818	-0.003
2.6	2.068	0.815	-0.001
2.7	2.883	0.814	0.002
2.8	3.697	0.816	
2.9	4.513		

Inflection point near $x = 2.6$

5. (a)

n	left	right
0	0	1
1	0	0.5
2	0.25	0.5
3	0.25	0.375
4	0.3125	0.375
5	0.34375	0.375
6	0.34375	0.359375
7	0.34375	0.3515625

Root is near $x = 0.35$

(b)

n	left	right
0	1	2
1	1.5	2
2	1.5	1.75
3	1.5	1.625
4	1.5	1.5625
5	1.53125	1.5625
6	1.53125	1.546875
7	1.53125	1.5390625

Root is near $x = 1.53$

(c)

n	left	right
0	-3	0
1	-3	-1.5
2	-2.25	-1.5
3	-2.25	-1.875
4	-2.0625	-1.875
5	-1.96875	-1.875
6	-1.921875	-1.875
7	-1.8984375	-1.875
8	-1.8867187	-1.875

Root near $x = -1.88$

(d) There are no roots larger than 3 since $f(3) = 10$ and the function increases for greater x. Since $f(2) = -5$, we look for a root $2 < x < 3$.

n	left	right
0	2	3
1	2.5	3
2	2.5	2.75
3	2.5	2.625
4	2.5	2.5625
5	2.5	2.53125
6	2.515625	2.53125
7	2.515625	2.5234375

Root near $x = 2.52$

(e)

n	left	right
0	0	1.5708
1	0.785395	1.5709
2	0.785395	1.17809
3	0.785395	0.981743
4	0.785395	0.883569
5	0.834482	0.883569
6	0.859025	0.883569
7	0.859025	0.871297

Root near $x = 0.86$

(f)

n	left	right
0	0.2	2
1	1.1	2
2	1.1	1.55
3	1.1	1.325
4	1.1	1.2125
5	1.15625	1.2125
6	1.15625	1.184375
7	1.15625	1.1703

Root near $x = 1.16$

(Solutions not provided for Exercises 7 and 9.)

SOLUTIONS TO CHAPTER 1 EXERCISES

1. (a) $a_n = 7n - 5$
    (b) $b_n = 3n^2$
    (c) $c_n = 3n^2 - 7n - 5$
    (d) $d_n = 3n - 2$

3.

n	a_n	Δa_n	$\Delta^2 a_n$	$\Delta^3 a_n$
1	1	-4	0	6
2	-3	-4	6	6
3	-7	2	12	6
4	-5	14	18	6
5	9	32	24	
6	41	56		
7	97			

5. (a) $c = 2$
    (b) $a = -3, -5$
    (c) $b = 0, 3, -3$
    (d) $p = 0, 0.3$

7. (a) $a_n = \{1, -4, -23, -25\}$
        $b_n = \{2, 5, -7, -76\}$
    (b) $c_n = \{2, 4.5, 13, 35\}$
        $d_n = \{1, 8, 18, 42\}$

9. Forever

11. Five years: \$1,251.79; 20 years: \$2,455.45

SECTION 2.1

1. (a) scalar
   (b) vector
   (c) scalar
   (d) vector

3. (a) False
   (b) True

5. (a)

   (b)

$$0 = \overline{PA} + \overline{PB} + \overline{PC} + \overline{PD} + \overline{PE}$$
$$= (\overline{PO} + \overline{OA}) + \cdots + (\overline{PO} + \overline{OE})$$
$$= 5\overline{PO} + (\overline{OA} + \cdots + \overline{OE})$$
$$= 5\overline{PO} + 0 = 5\overline{PO},$$

   so P = 0.

7. (a) |**a** + **b**| = |**a**| − |**b**|, **a** + **b** has same direction as **a**.
   (b)
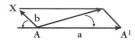
   (c) Switch **a** and **b**.
   (d)
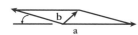

9. (a)
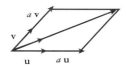

(b)
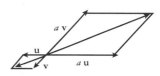

(c)

(d)

(e)

(f)

(g)

11. (a) D is two-thirds of the way from A to A' and from B to B'.
    (b) If AA' meets CC' at E, then by (a) $E = D$. So AA', BB', CC' all meet at D.
    (c) $\overline{OD} = (\overline{OA} + \overline{OB} + \overline{OC})/3$.

13. $[D, E]$ is parallel to $[A, C]$ and one-third as long.

SECTION 2.2

1. (a) Turn left $135°$ and move $2\sqrt{2} = 2.83$ ft.
   (b) Turn left $45°$ and move $2\sqrt{2} = 2.83$ ft.
   (c) Turn left $27°$ and move $2\sqrt{5} = 4.47$ ft.
   (d) Turn right $146°$ and move 7.21 ft.

(e) Turn 180° (either way) and move 10 ft.

3.

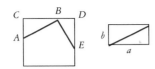

Angle *CBA* + angle *DBE* = angle *CBA* + angle *CAB* = 90°; hence angle *ABE* is a right angle and the inner quadrangle is a square. Then the area of the outer square equals the area of the inner square plus the area of two pairs of the triangle, i.e., $(a + b)^2 = 2ab + c^2$, from which the result follows.

5. (3/5, 4/5) and $(-3, -4)$

7. (a) The circle with center $(-2, 3)$ and radius 1.
    (b) The circle with center $(5, 10)$ and radius 4.
    (c) The point $(-1, -1)$.
    (d) The empty set.

9. $(\cos(\alpha + \beta) - 1)^2 + (\sin(\alpha + \beta) - 0)^2$
    $= (\cos \alpha - \cos \beta)^2 + (\sin \alpha + \sin \beta)^2,$

   $(\cos^2(\alpha + \beta) + \sin^2(\alpha + \beta)) - 2\cos(\alpha + \beta) + 1$
    $= (\cos^2 \alpha + \sin^2 \alpha) - 2\cos \alpha \cos \beta$
    $+ (\cos^2 \beta + \sin^2 \beta) + 2\sin \alpha \sin \beta$
    $= 2 - 2\cos \alpha \cos \beta + 2\sin \alpha \sin \beta$

11. $\tan(\theta + 180°) = \dfrac{\sin(\theta + 180°)}{\cos(\theta + 180°)}$

   $= \dfrac{-\sin \theta}{-\cos \theta} = \dfrac{\sin \theta}{\cos \theta} = \tan \theta,$

   $\tan(-\theta) = \dfrac{\sin(-\theta)}{\cos(-\theta)} = \dfrac{-\sin \theta}{\cos \theta} = -\tan \theta,$

   $\tan(\alpha + \beta) = \dfrac{\sin(\alpha + \beta)}{\cos(\alpha + \beta)} = \dfrac{\sin \alpha \cos \beta + \cos \alpha \sin \beta}{\cos \alpha \cos \beta - \sin \alpha \sin \beta}$

   $= \dfrac{\dfrac{\sin \alpha \cos \beta}{\cos \alpha \cos \beta} + \dfrac{\cos \alpha \sin \beta}{\cos \alpha \cos \beta}}{\dfrac{\cos \alpha \cos \beta}{\cos \alpha \cos \beta} - \dfrac{\sin \alpha \sin \beta}{\cos \alpha \cos \beta}}$

   $= \dfrac{\sin \alpha/\cos \alpha + \sin \beta/\cos \beta}{1 - (\sin \alpha/\cos \alpha)(\sin \beta/\cos \beta)} = \dfrac{\tan \alpha + \tan \beta}{1 - \tan \alpha \tan \beta},$

   $\tan(\alpha - \beta) = \dfrac{\tan \alpha + \tan(-\beta)}{1 - \tan \alpha \tan(-\beta)} = \dfrac{\tan \alpha - \tan \beta}{1 + \tan \alpha \tan \beta}$

13. (a) $-1.99\mathbf{i} + 3.79\mathbf{j}$
    (b) $-0.99\mathbf{i} + 1.79\mathbf{j}$
    (c) $-1.65\mathbf{i} + 8.18\mathbf{j}$
    (d) $6.46\mathbf{i} + 0.88\mathbf{j}$

15. $(-r, \theta)$: $(r, \theta + 180°)$, $(r, -\theta)$: $(r, 360° - \theta)$, $(-r, -\theta)$: $(r, 180° - \theta)$

17. Because

   $$\sin(\theta + 180°) = -\sin \theta,$$
   $$\cos(\theta + 180°) = -\cos \theta,$$
   $$\sin(-\theta) = \sin(360° - \theta), \text{ and}$$
   $$\cos(-\theta) = \cos(360° - \theta)$$

SECTION 2.3

1. (a)

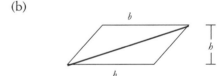

   (b)

   (c) Let h be the altitude from C to side AB. Then $h = b \sin \alpha$ and $Area ABC = (1/2)ch = (1/2)bc \sin \alpha.$
    (d) $(1/2)bc \sin \alpha = (1/2)ac \sin \beta = (1/2)ab \sin \gamma.$ Multiply through by $2/abc$ to get Eq. (2).

3. (a) Turn 45° and move 2.83 ft
    (b) Turn 135° and move 2.83 ft
    (c) Turn 60° and move 8.20 m
    (d) Turn 180° and move 8 m
    (e) Turn $-25.14°$ and move 3.62 cm

5. 54.93 mph

7. $120°$

9. $|\overline{OX}|^2 = |\overline{OX'}|^2 + |x_3\mathbf{k}|^2 = |x_1\mathbf{i}|^2 + |x_2\mathbf{j}|^2 + |x_3\mathbf{k}|^2 = x_1^2 + x_2^2 + x_3^2$, whence Eq. (11).

11. $-7/\sqrt{14}$

13. 0

15. (a)

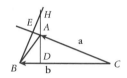

$$\overline{CH} = \overline{CA} + \overline{AH} = \overline{CA} + x\overline{DA}$$
$$= \overline{CA} + x(\overline{CA} - \overline{CD}),$$

$$\overline{CH} = \overline{CB} + \overline{BH} = \overline{CB} + y\overline{EB}$$
$$= \overline{CB} + y(\overline{CB} - \overline{CE}),$$

(b) $\mathbf{a} + x(\mathbf{a} - (\mathbf{a}\cdot\mathbf{b}/\mathbf{b}\cdot\mathbf{b})\mathbf{b}) = \mathbf{b} + y(\mathbf{b} - (\mathbf{b}\cdot\mathbf{b}/\mathbf{a}\cdot\mathbf{a})\mathbf{a})$

(c) $\overline{CH} = k\,[(\mathbf{b}\cdot\mathbf{b} - \mathbf{a}\cdot\mathbf{b})\mathbf{a} + (\mathbf{a}\cdot\mathbf{a} - \mathbf{a}\cdot\mathbf{b})\mathbf{b}]$, where $k = \mathbf{a}\cdot\mathbf{b}/[(\mathbf{a}\cdot\mathbf{a})(\mathbf{b}\cdot\mathbf{b}) - (\mathbf{a}\cdot\mathbf{b})^2]$

(d) $\overline{AB} = -\mathbf{a} + \mathbf{b}$. $\overline{AB}\cdot\overline{CH} = \cdots = 0$. Thus the altitude from C also passes through H.

17. (a) Given $\mathbf{u}, \mathbf{v}, \mathbf{w}$, if say $\mathbf{u}$ and $\mathbf{v}$ are parallel, then any plane parallel to $\mathbf{u}$ and $\mathbf{w}$ is parallel to all three vectors.

(b) $(p - s)\mathbf{u} + (q - t)\mathbf{v} = (x - r)\mathbf{w}$. If $x \neq r$, then $\mathbf{w}$ would be parallel to a plane parallel to $\mathbf{u}$ and $\mathbf{v}$. Hence $x = r$, $(p - s)\mathbf{u} + (q - t)\mathbf{v} = \mathbf{0}$, and by the result in Section 2.1, $p = s$ and $q = t$.

SECTION 2.4

1. At $(39.0, -7, 52.4)$, coordinates in miles.

3. (a) 15.9

(b) 24.46

5. Turn left $90°$ and move 4.4.

7. Yes, at $(8041/527, -40/527, -4312/527)$.

9. $4x_1 - x_2 - 3x_3 = 0$.

11. Only if P is the first or the fourth quadrant.

13. (a)–(d) The circle with center $(1/2, 0)$ and radius $1/2$.

15. $s = -(1/10)(x_1 + 4x_2 + 18)$, $t = 5 + x_2 - (3/10)(x_1 + 4x_2 + 18)$.
$10x_3 = 30 + 10s + 20t = \cdots = 4 - 7x_1 + 8x_2$.

17. Let A be $(2, 3, 4)$ and B be $(4, 2, 3)$. The curve is a circle with center C on $[A, B]$ and radius r satisfying $\sqrt{5^2 - r^2} + \sqrt{6^2 - r^2} = |AB| = \sqrt{6}$, from which $r = 5\sqrt{23}/2\sqrt{6}$, $|AC|/|AB| = 5/12 = t$, say. Then $\overline{OC} = (1 - t)\overline{OA} + t\overline{OB} = (1/12)(34\mathbf{i} + 31\mathbf{j} + 43\mathbf{k})$. Now $\overline{AB} = 2\mathbf{i} - \mathbf{j} - \mathbf{k}$, and by inspection (or easy calculation) the vectors $\mathbf{u} = \mathbf{i} + \mathbf{j} + \mathbf{k}$ and $\mathbf{v} = \mathbf{j} - \mathbf{k}$ are perpendicular to $\overline{AB}$, and therefore are parallel to the plane of our circle, and are perpendicular to each other. Then the curve is given by

$$\overline{OX} = \overline{OC} + r\cos\theta\mathbf{u}/|\mathbf{u}| + r\sin\theta\mathbf{v}/|\mathbf{v}|$$
$$= (2.83 + 2.83\cos\theta)\mathbf{i}$$
$$+ (2.58 + 2.83\sin\theta + 1.13\cos\theta)\mathbf{j}$$
$$+ (3.58 - 3.46\sin\theta + 2.83\cos\theta)\mathbf{k}$$

SECTION 2.5

1. $(67.5, -4.3)$

3. The ratio h is most sensitive to v_1 and v_2 at the upper end of their ranges and to v_3 at its lower end. If $v_2 = 7.00$ and $v_3 = 0.50$, a change of v_1 from 8.99 to 9.00 changes h from 17.43 to 17.44; so h is not very sensitive to v_1. If $v_1 = 9.00$ and $v_3 = 0.50$, a change of v_2 from 6.99 to 7.00 changes h from 17.42 to 17.44; so h is a little more sensitive to v_2 than to v_1. If $v_1 = 9.00$ and $v_2 = 7.00$, a change of v_3 from 0.51 to 0.50 changes h from 17.09 to 17.44; so h is much more sensitive to v_3.

5. Arbie would move around a $3k$-by-$3k$ square forever.

7. Arbie would bounce off the walls and then move around a $3k$-by-$3k$ square, tilted $45°$, forever. To escape, increase the standard length of a move from k to, say, $2k$.

9. Distance from (λ_1, ϕ_1) to $(\lambda_2, \phi_2) = 4000$ $\cos^{-1}(\cos \lambda_1 \cos\lambda_2 \cos(\theta_1 - \theta_2) + \sin \lambda_1 \sin\lambda_2)$.

11.

SECTION 2.6

1. (a) $-2x_1 + 3x_2 = 12$
   (b) $3x_1 + x_2 = -7$
   (c) $7x_1 - x_2 = 49$
   (d) $x_1 = -2$
   (e) $x_2 = -9$
   (f) $10x_1 + x_2 = 68$
   (g) $x_1 = 19$
   (h) $x_2 = -2$
   (i) $x_2 = 0$
   (j) $x_1 = 0$

3. (a)

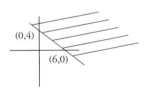

   (b)

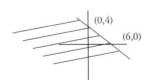

(c)

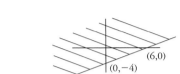

(d)

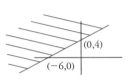

(e)

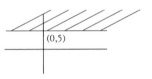

(f)

(g)

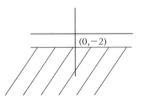

(h)

(i)

(j)

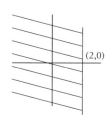

(2,0)

System	Solution	Corner point?
123	(0, 0, 0)	no
124	(0, 0, 2)	yes
125	(0, 0, 4)	yes
134	(0, 3, 0)	yes
135	(0, 6, 0)	yes
145	none	—
234	(1, 0, 0)	no
235	(3, 0, 0)	yes
245	(−3, 0, 0)	no
345	(−3, 0, 0)	no

5. (a)

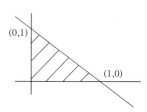

(0,1)

(1,0)

(b)

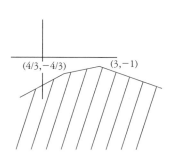

(4/3, −4/3)   (3, −1)

(c) The empty set, i.e., there are no feasible points. (This happens all the time in real-world applications!)

7. Buy 41.67 days production from Supergas and 122.22 days from Exxtrabig for the minimum cost of $1,138,888.

9. Denote the system consisting of the first three equations in the exercise by 123, the system consisting of the first two and the last equation by 125, etc.

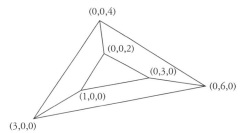

(0,0,4)

(0,0,2)

(0,3,0)

(0,6,0)

(1,0,0)

(3,0,0)

11. $k\mathbf{a} = \begin{bmatrix} ka_1 \\ ka_2 \end{bmatrix}$, $-\mathbf{a} = \begin{bmatrix} -a_1 \\ -a_2 \end{bmatrix}$, $\mathbf{a} + \mathbf{b} = \begin{bmatrix} a_1 + b_1 \\ a_2 + b_2 \end{bmatrix}$,

$|\mathbf{a}| = \sqrt{a_1^2 + a_2^2}$, $\mathbf{a} \cdot \mathbf{b} = a_1 b_1 + a_2 b_2$. If $\mathbf{a} = \begin{bmatrix} a_1 \\ a_2 \\ \vdots \\ a_n \end{bmatrix}$, $\mathbf{b} = \begin{bmatrix} b_1 \\ b_2 \\ \vdots \\ b_n \end{bmatrix}$, then

$k\mathbf{a} = \begin{bmatrix} ka_1 \\ ka_2 \\ \vdots \\ ka_n \end{bmatrix}$, $\mathbf{a} + \mathbf{b} = \begin{bmatrix} a_1 + b_1 \\ a_2 + b_2 \\ \vdots \\ a_n + b_n \end{bmatrix}$, $\mathbf{a} \cdot \mathbf{b}$

$= a_1 b_1 + a_2 b_2 + \cdots + a_n b_n$, etc.]

1. No matter what ABC is, triangle DEF is always equilateral. This theorem has been attributed to Napoleon.

3. (5/4, 11/4, 1).

SECTION 3.1

1. (a) $\begin{bmatrix} 7 \\ 2 \\ 1 \end{bmatrix}$

   (b) $[7, 4, 3, 1]$

   (c) $\begin{bmatrix} 4 \\ -1 \\ 6 \end{bmatrix}$

   (d) 6

   (e) 2

3. (a), (b) Both additions yield the vector $\begin{bmatrix} 1 \\ 5 \end{bmatrix}$

   (c) $\begin{bmatrix} -1 \\ 1 \end{bmatrix}$

   (d) $\begin{bmatrix} 2 \\ 8 \end{bmatrix}$

5. Greatest deviation is in diesel oil: The 1160 gallons produced are 460 in excess of 700.

7. $20x_1 + 4x_2 + 4x_3 = 500$
   $10x_1 + 14x_2 + 5x_3 = 850$
   $5x_1 + 5x_2 + 12x_3 = 1000$

9. (a) $\begin{bmatrix} 1/2 \\ 7/24 \\ 5/24 \end{bmatrix}$

   (b) $\begin{bmatrix} 7/16 \\ 11/32 \\ 7/32 \end{bmatrix}$

   (c) $\begin{bmatrix} 14/23 \\ 5/23 \\ 4/23 \end{bmatrix}$

11. (a) $\begin{bmatrix} 2/3 & 1/2 \\ 1/3 & 1/2 \end{bmatrix}$

    (b) 7/12

    (c) 11/18

13. (a) $\begin{bmatrix} .5 & .25 & 0 & 0 \\ .5 & .5 & .25 & 0 \\ 0 & .25 & .5 & 0 \\ 0 & 0 & .25 & 1 \end{bmatrix}$

    (b) $\begin{bmatrix} .375 \\ .5 \\ .125 \\ 0 \end{bmatrix}$

    (c) $\sim \begin{bmatrix} .31 \\ .47 \\ .19 \\ .03 \end{bmatrix}$

SECTION 3.2

1. $\begin{bmatrix} 5.8 & 7.5 & 8.8 \\ 7.7 & 6.0 & 8.8 \\ 7.8 & 7.2 & 8.0 \\ 5.5 & 5.3 & 6.2 \end{bmatrix}$

3. (a) $\begin{bmatrix} 12 & 3 & 15 & 12 \\ 0 & 9 & 12 & 27 \\ 6 & 9 & 3 & 18 \end{bmatrix}$

(b) $\begin{bmatrix} 4 & 6 & 2 & 8 \\ -4 & -6 & 8 & 2 \\ 6 & 10 & 0 & 8 \end{bmatrix}$

(c) $\begin{bmatrix} -6 & -9 & -3 & -12 \\ 6 & 9 & -12 & -3 \\ -9 & -15 & 0 & -12 \end{bmatrix}$

(d) $\begin{bmatrix} 6 & 4 & 6 & 8 \\ -2 & 0 & 8 & 10 \\ 5 & 8 & 1 & 10 \end{bmatrix}$

(e) $\begin{bmatrix} 14 & 11 & 13 & 20 \\ -6 & -3 & 20 & 21 \\ 13 & 21 & 2 & 24 \end{bmatrix}$

(f) $\begin{bmatrix} 8 & -3 & 13 & 4 \\ 4 & 15 & 4 & 25 \\ 0 & -1 & 3 & 10 \end{bmatrix}$

5. (a) $4\mathbf{I} + 2\mathbf{J}$
(b) $\mathbf{J} - \mathbf{A}$
(c) $4\mathbf{I} + 3\mathbf{J} - 2\mathbf{A}$

7. (a) -2
(b) -8
(c) 21
(d) 14

9. (a) 85

(b) $\begin{bmatrix} 3 \\ 47 \\ 43 \end{bmatrix}$

(c) Undefined
(d) Undefined
(e) 1865

11. (a) $\begin{aligned} x_1 + 3x_2 &= 4 \\ 2x_1 + 3x_2 &= -2 \end{aligned}$
(b) $\begin{aligned} 4x_1 - x_2 &= 2 \\ x_1 + 3x_2 &= 1 \end{aligned}$

(c) $\begin{aligned} 2x_1 + x_2 &= 2 \\ 5x_1 - x_2 + 4x_3 &= 3 \\ 5x_1 + 2x_2 + 3x_3 &= 4 \end{aligned}$

(d) $\begin{aligned} -x_1 &= 0 \\ 2x_3 &= 0 \\ x_2 &= 0 \end{aligned}$

13. (a) $\mathbf{Ax} = \mathbf{By} + \mathbf{c}$
(b) $\mathbf{Ax} - \mathbf{By} = \mathbf{c}$

17. (a)

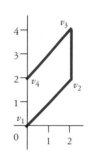

(b)

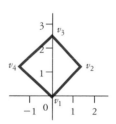

(c)

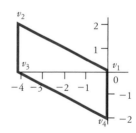

SECTION 3.3

1. $\mathbf{BA}$, a 3×5 matrix; $\mathbf{DC}$, a 4×2 matrix; $\mathbf{EC}$, a 4×2 matrix.

3. (a) Not possible

(b) $\begin{bmatrix} -2 & -3 & -4 & -5 \\ -2 & -4 & -6 & -8 \\ -1 & -1 & -1 & -1 \end{bmatrix}$

(c) $\begin{bmatrix} 16 & 14 & 20 \\ 32 & 28 & 40 \\ 41 & 35 & 47 \end{bmatrix}$

(d) $\begin{bmatrix} 16 & 31 & 46 & 61 \\ 7 & 12 & 17 & 22 \\ 10 & 19 & 28 & 37 \\ 11 & 19 & 27 & 35 \end{bmatrix}$

(e) $\begin{bmatrix} 13 & -7 & -6 \\ 1 & 2 & -3 \\ 7 & -3 & -4 \\ 2 & 1 & -3 \end{bmatrix}$

5. $\mathbf{AB} = \begin{bmatrix} 1 & 0 \\ 0 & 1 \end{bmatrix} = \mathbf{BA}$

7. (a) $\mathbf{BA} = \begin{bmatrix} 2.30 & 3.05 \\ 1.65 & 2.10 \end{bmatrix}$

(b) $\mathbf{CB} = \begin{bmatrix} 7000 & 12500 & 5500 \\ 14000 & 25000 & 11000 \end{bmatrix}$

9. $\mathbf{x}^{(k)} = \mathbf{A}^k \mathbf{x}$

11. $\mathbf{A}^2\mathbf{A}$ is the same as $\mathbf{AA}^2$; see (16) in Example 5.

15. (a)

(b)

(c)

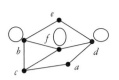

17. (a) $\begin{bmatrix} 2 & 0 & 2 & 0 \\ 0 & 2 & 0 & 2 \\ 2 & 0 & 2 & 0 \\ 0 & 2 & 0 & 2 \end{bmatrix}$

(b) $\begin{bmatrix} 3 & 1 & 3 & 1 & 2 \\ 1 & 3 & 1 & 3 & 2 \\ 3 & 1 & 3 & 1 & 2 \\ 1 & 3 & 1 & 3 & 2 \\ 2 & 2 & 2 & 2 & 4 \end{bmatrix}$

(c) $\begin{bmatrix} 1 & 0 & 1 & 0 & 0 \\ 0 & 2 & 0 & 1 & 0 \\ 1 & 0 & 2 & 0 & 1 \\ 0 & 1 & 0 & 2 & 0 \\ 0 & 0 & 1 & 0 & 1 \end{bmatrix}$

(d) $\mathbf{I}$

19. (a) $\begin{bmatrix} 6 & 0 & 0 \\ 0 & 2 & 0 \\ 0 & 0 & 15 \end{bmatrix}$

(b) The diagonal entries are $a_{11}b_{11}$, $a_{22}b_{22}$, and $a_{33}b_{33}$.

21. (a) $\begin{bmatrix} 1 & 0 & 0 \\ 0 & 1 & 0 \\ 0 & 2 & 1 \end{bmatrix}$

(b) $\begin{bmatrix} 1 & -4 & 2 \\ 0 & 1 & 0 \\ 0 & 0 & 1 \end{bmatrix}$

23. (a) 10^3

(b) 10^6

(c) 2000

(d) 500

(e) 2×10^3

(f) 10^4

25. (a) All columns in the matrix approach the vector
$$\begin{bmatrix} 14/23 \\ 5/23 \\ 4/23 \end{bmatrix}.$$

(b) All columns in the matrix approach $\begin{bmatrix} .4 \\ .2 \\ .4 \end{bmatrix}$.

(c) All columns in the matrix approach $\begin{bmatrix} 0 \\ 0 \\ 0 \\ 1 \end{bmatrix}$.

11. (a) $\begin{bmatrix} 1/2 \\ 0 \\ 1/2 \end{bmatrix}$

(b) $\begin{bmatrix} 1/3 \\ 1/3 \\ 1/3 \end{bmatrix}$

(c) $\begin{bmatrix} 1/3 \\ 1/3 \\ 1/3 \end{bmatrix}$

13. Add 1/2 first row to second row, add 1/4 first row and 1/2 second row to third row.

SECTION 3.5

3. (a) $x_1 + 2x_3 = 1, x_2 + 3x_3 = 0, x_1 + 4x_3 = 0$

(b) $\begin{bmatrix} 2 \\ 3/2 \\ -1/2 \end{bmatrix}$

5. (a) $\begin{bmatrix} 48 \\ 20 \\ -8 \end{bmatrix}$

(b) $\begin{bmatrix} 60 \\ -40 \\ -20 \end{bmatrix}$

(c) No solution

(d) $1/11 \begin{bmatrix} -20 \\ -240 \\ 460 \end{bmatrix}$

7. (a) $\begin{bmatrix} 0 & 2/7 & 1/7 & 0 \\ -1 & 16/7 & 1/7 & -1 \\ -2 & 15/7 & 4/7 & -1 \\ 0 & 3/7 & -2/7 & 0 \end{bmatrix}$

(b) $\begin{bmatrix} 1 & -1 & 0 & -1 \\ -5/4 & 7/4 & 1/4 & 6/4 \\ 2/4 & -2/4 & -2/4 & 0 \\ -1/4 & -1/4 & 1/4 & 2/4 \end{bmatrix}$

SECTION 3.4

1. (i) $(c) = (b) - 2(a)$
(ii) $(c) = (b) + 3/2(a)$

3. (a) $x_1 = 30, x_2 = 14, x_3 = -9$;
(b) $x_1 = 1/3, x_2 = -7/3, x_3 = 0$;
(c) many possible solutions;
(d) $x_1 = -21/11, x_2 = -76/11, x_3 = 164/11$.

5. (a) $\begin{bmatrix} 13/3 \\ 13/3 \\ -16/3 \\ -1/3 \end{bmatrix}$

(b) $\begin{bmatrix} 1 \\ -1 \\ 2 \\ -2 \end{bmatrix}$

7. (a) $x_1 = 30, x_2 = 14, x_3 = -9$;
(b) $x_1 = 1/3, x_2 = -7/3, x_3 = 0$;
(c) many possible solutions;
(d) $x_1 = -21/11, x_2 = -76/11, x_3 = 164/11$.

9. 75 tables, 50 chairs, 30 sofas

(c) $\begin{bmatrix} 13/6 & -2 & -1/2 & 11/6 \\ 7/6 & -1 & -1/2 & 5/6 \\ -8/3 & 3 & 1 & -7/3 \\ -2/3 & 1 & 0 & -1/3 \end{bmatrix}$

9. (a) $\begin{bmatrix} .001 & -.33 & .04 \\ -.005 & .63 & -.02 \\ .008 & -.28 & .004 \end{bmatrix}, \begin{bmatrix} 4.57 \\ 7.17 \\ 1.96 \end{bmatrix},$

(b) -1.32

(c) $-02k$

11. (a) $\begin{bmatrix} 1/2 & -1 & 1/2 \\ 1/2 & 0 & -1/2 \\ -3/2 & 2 & 1/2 \end{bmatrix} \begin{bmatrix} 100,000 \\ 50,000 \\ 100,000 \end{bmatrix}$

(b) $\begin{bmatrix} -100,000 \\ 0 \\ 200,000 \end{bmatrix}$

(c) $\begin{bmatrix} -5000k \\ -5000k \\ 15,000k \end{bmatrix}$

13. $\begin{bmatrix} 1 & 0 & 0 & 0 \\ 0 & 1 & 0 & 0 \\ -a & 0 & 1 & 0 \\ 0 & 0 & 0 & 1 \end{bmatrix}$

15. (a) $\begin{bmatrix} 1/2 & \sqrt{3}/2 \\ -\sqrt{3}/2 & 1/2 \end{bmatrix}$

(b) $\begin{bmatrix} \cos\theta & \sin\theta \\ -\sin\theta & \cos\theta \end{bmatrix}$

17. (a) (i) $\begin{bmatrix} 18 & 5 \\ 5 & 23 \end{bmatrix}$

(ii) $\begin{bmatrix} 23 & 14 \\ 10 & 9 \end{bmatrix}$

(iii) $\begin{bmatrix} 9 & 5 \\ 3 & 14 \end{bmatrix}$

(b) (i) PG
(ii) QZ
(iii) WU

SECTION 3.6

1. (a) 0
(b) -2
(c) -17

3. $x_1 = 15D - kG/150 - 5k$, $x_2 = 10G - 5D/150 - 5k$, $k = 30 \Leftrightarrow \det = 0$; nonuniqueness means no solution or multiple solutions.

5. (a) 0
(b) 0
(c) -4
(d) $\sim .33 \times 10^{-8}$

SECTION 3.7

1. (a) $-2, -3$
(b) $-1, 3$
(c) $-2, 2, 1$

3. (a) $\begin{bmatrix} 1349 \\ 645 \end{bmatrix}$

(b) $3 \cdot 7^n \begin{bmatrix} 1 \\ 1 \end{bmatrix}$

(c) $\begin{bmatrix} 2721 \\ 2017 \end{bmatrix}$

5. (a) (i) $\lambda_1 = 6, \lambda_2 = -1$
(ii) $\lambda_1 = 3$ (only one)
(iii) $\lambda_1 = 3, \lambda_2 = -3$
(iv) $\lambda_1 = 8, \lambda_2 = \lambda_3 = 2$
(v) $\lambda_1 = 3 + \sqrt{2}, \lambda_2 = 3 - \sqrt{2}, \lambda_3 = -1$
(vi) $\lambda_1 = 6, \lambda_2 = 4, \lambda_3 = \lambda_4 = 2$

(b) (i) $\begin{bmatrix} 2 \\ 5 \end{bmatrix}$

(ii) $\begin{bmatrix} 1 \\ 1 \end{bmatrix}$

(iii) $\begin{bmatrix} 5 \\ 1 \end{bmatrix}$ or $\begin{bmatrix} 1 \\ -1 \end{bmatrix}$

(iv) $\begin{bmatrix} 2 \\ 1 \\ 2 \end{bmatrix}$

(v) $\begin{bmatrix} 1 \\ \sqrt{2} \\ 1 \end{bmatrix}$

(vi) $\begin{bmatrix} 1 \\ 1 \\ -1 \\ 1 \end{bmatrix}$

7. (a) (i) $\lambda_1 = 1, \lambda_2 = .5$

(ii) $\lambda_1 = 1.2, \lambda_2 = 1$

(iii) $\lambda_1 = 1.24, \lambda_2 = .96$

(b) (i) $\mathbf{u} = \begin{bmatrix} 3 \\ 1 \end{bmatrix}$

(ii) $\mathbf{u} = \begin{bmatrix} 2 \\ 1 \end{bmatrix}$

(iii) $\mathbf{u} = \begin{bmatrix} 1 \\ \sqrt{2} \end{bmatrix}$

(c) (i) $\mathbf{v} = \begin{bmatrix} 1 \\ 2 \end{bmatrix}$

(ii) $\mathbf{v} = \begin{bmatrix} 2 \\ 3 \end{bmatrix}$

(iii) $\mathbf{v} = \begin{bmatrix} 1 \\ -\sqrt{2} \end{bmatrix}$

(d) (i) $\mathbf{x}^{(k)} = 2\mathbf{u} + 4(.5)^k\mathbf{v}$

(ii) $\mathbf{x}^{(k)} = 2.5(1.2)^k\mathbf{u} + 2.5\mathbf{v}$

(iii) $\mathbf{x}^{(k)} = 8.536(1.24)^k\mathbf{u} + 1.463(.96)^k\mathbf{v}$

9. (a) $\lambda_1 = 3, \lambda_2 = 1, \lambda_3 = 0, \mathbf{u} = \begin{bmatrix} 1 \\ -2 \\ 1 \end{bmatrix}, \mathbf{v} = \begin{bmatrix} 1 \\ 0 \\ -1 \end{bmatrix}, \mathbf{w} = \begin{bmatrix} 1 \\ 1 \\ 1 \end{bmatrix}$

(b) $\mathbf{p} = 10\mathbf{u} + 20\mathbf{v} + 20\mathbf{w}$

(c) $10 \cdot 3^{12} \begin{bmatrix} 1 \\ -2 \\ 1 \end{bmatrix}$

11. The first column of $\mathbf{A}^n$ equals $\mathbf{A}^n\mathbf{i}_1$. The long-term behavior principle implies that $\mathbf{A}^n\mathbf{i}_1$ "normally" approaches a multiple of the dominant eigenvector. This is an exception to the normal behavior caused by the fact that $\mathbf{i}_1$ is an eigenvector associated with a smaller eigenvalue.

SECTION 3.8

1. (a) $7/25, \sim74°$

(b) $5/\sqrt{50}, 45°$

(c) $2/\sqrt{18}, \sim62°$

(d) $4/\sqrt{42}, \sim52°$

3. (a) parallel

(b) orthogonal

(c) neither

(d) orthogonal

5. (a) $.08$

(b) $-.765$

7. (a) $\begin{bmatrix} 3/2 & 1/2 \\ 2 & 1 \end{bmatrix}$

(b) $\begin{bmatrix} -3 & 2 \\ 2 & -1 \end{bmatrix}$

9. (a) 0

(b) $-2/5$

(c) $-37/25$

(d) 0

(e) 0

11. (a) $\begin{bmatrix} 8/5 \\ -1/15 \end{bmatrix}$

(b) $\begin{bmatrix} -4/5 \\ 11/20 \end{bmatrix}$

(c) $\begin{bmatrix} -37/49 \\ 10/49 \\ 1/49 \end{bmatrix}$

(d) $\begin{bmatrix} 0 \\ 1 \\ 0 \end{bmatrix}$

13. $\hat{y} = 1.82x$, terrible fit

Notation: AP = addition principle, MP = multiplication principle, C = combination, P = permutation

SECTION 4.1

1. AP: $4 + 5 + 3 = 12$

3. MP: $4 \cdot 7 \cdot 3 \cdot 2 = 168$

5. MP: $26 \cdot 26 \cdot 10 \cdot 10 \cdot 10 = 676{,}000$. We get more possibilities either by replacing some numbers with letters or by using six, not five, symbols.

7. MP: $8 \cdot 8 \cdot 10 = 640$ exchanges, each of which can have $10 \cdot 10 \cdot 10 \cdot 10 = 1000$ numbers, giving (MP) $640{,}000$ possibilities.

9. Using generalized AP, the number of people with pets is $3020 + 705 - 345 = 3380$, so the number of people without pets is $5000 - 3380 = 1620$.

11. The minimum size of the town is $20{,}450$, assuming that all $13{,}700$ people with arthritis also have allergies. We cannot tell the maximum size of the town because we have no information about the number of people who have neither condition.

SECTION 4.2

1. Chocolate topped by fruit: MP: $4 \cdot 6 = 24$
    Fruit topped by chocolate: MP: $6 \cdot 4 = 24$
    AP: $24 + 24 + 48$

3. Letters left, numbers right: MP: $26 \cdot 26 \cdot 26 \cdot 10 \cdot 10 \cdot 10 = 17{,}576{,}000$
    Letters right, numbers left: MP: $10 \cdot 10 \cdot 10 \cdot 26 \cdot 26 \cdot 26 = 17{,}576{,}000$
    AP: $17{,}576{,}000 + 17{,}576{,}000 = 35{,}152{,}000$

5. First question: AP: $4 + 9 = 13$ possibilities
    Second question: AP: $7 + 3 = 10$ possibilities
    Questions 3–7: MP: $37 \cdot 36 \cdot 35 \cdot 34 \cdot 33 = 52{,}307{,}640$ possibilities
    Whole test: MP: $13 \cdot 10 \cdot 52{,}307{,}000 = 6{,}799{,}993{,}200$ possibilities

7. Full-time options:

    | Study 2, Work 2: | MP: $3 \cdot 4 = 12$ |
    | Work 1, Study 2, Work 1: | MP: $4 \cdot 3 \cdot 1 = 12$ |
    | Work 2, Study 2: | MP: $4 \cdot 3 + 12$ |
    | Part-time options: | MP: $5 \cdot 7 = 35$ |

    Total number of options: AP: $12 + 12 + 12 + 35 = 71$

9. Scheme 1 is same as Exercise 3: $35{,}152{,}000$
    Scheme 2:
    2 letters, 4 numbers: MP: $26 \cdot 26 \cdot 10 \cdot 10 \cdot 10 \cdot 10 = 6{,}760{,}000$
    2 numbers, 2 letters, 2 numbers: MP: $10 \cdot 10 \cdot 26 \cdot 26 \cdot 10 \cdot 10 = 6{,}760{,}000$
    4 numbers, 2 letters: MP: $10 \cdot 10 \cdot 10 \cdot 10 \cdot 26 \cdot 26 = 6{,}760{,}000$
    AP: $6{,}760{,}000 + 6{,}760{,}000 + 6{,}760{,}000 = 20{,}280{,}000$
    Scheme 1 is better by $14{,}872{,}000$ plates

SECTION 4.3

1. Fussy, order counts: P: $_{37}P_{10} = 1.26 \cdot 10^{15}$
    Nonfussy, order doesn't matter: C: $_{37}C_{10} = 348{,}330{,}136$

3. P: $_{20}P_{20} = 20! = 2.43 \cdot 10^{18}$. With thousands of books, a library without a system for shelving its books would be unable to locate the books it has.

5. P: $_5P_5 = 5! = 120$

7. For both cases, C to divide the shrubs into two sets: $_5C_2 (= {}_5C_3) = 10$. That is also the answer for a nongardener. A gardener, however, sees P of each set of plants into the locations: P and MP: (10, from the division) $\cdot {}_2P_2 \cdot {}_3P_3 = 10 \cdot 2 \cdot 6 = 120$.

9. The son is nonfussy; different orders of the *same* set of hats all count as the same display to him: C: $_8C_6 = 28$. The daughter counts different orders of the same set of hats as different displays: P: $_8P_6 = 20{,}160$.

11. Each passer can pass to any of the other four players:
MP: $4^7 = 16,384$

13. MP for each case (which three departments) and AP for total:

No H:	$16 \cdot 8 \cdot 17$	$= 2176$
No FL:	$20 \cdot 8 \cdot 17$	$= 2720$
No SS:	$20 \cdot 16 \cdot 17$	$= 5440$
No NS:	$20 \cdot 16 \cdot 8$	$= 2560$
	Total	$= 12,896$

15. If you can use exactly the same streets to go from location A to location B as for the reverse trip, then traveling between A and B in either direction would be essentially one item to learn, so we would count combinations. However, often there are one-way streets in a city, so the two routes could not be the reverses of each other. Further, the reverse of a left turn is a right turn, etc. Thus, it is likely that we need to count permutations.

17. Since $f(n) = n!/(n - k)!$, to check the initial condition, we substitute k for n, giving $f(k) = k!/(k - k)! = k!/0! = k!$, which checks. For $n > k$, we would have the difference equation $f(n) = (n/(n - k)) \cdot f(n - 1)$. We substitute as above for $f(n)$ and $(n - 1)!/(n - 1 - k)!$ for $f(n - 1)$, giving $(n \cdot (n - 1)!)/((n - k) \cdot (n - k - 1)!) = (n \cdot (n - 1)!)/((n - k) \cdot (n - 1 - k)!)$, which checks.

19. Since $f(k) = n!/(n - k)!$, to check the initial condition, we substitute 0 for k, getting $f(0) = n!/(n - 0)! = 1$. The difference equation is $f(k) = (n - k + 1) \cdot f(k - 1)$. We check by substituting as above for $f(k)$, and $n!/(n - (k - 1))!$ for $f(k - 1)$, giving $n!/(n - k)! = (n - k + 1) \cdot (n!/(n - (k - 1))!) = (n - k + 1) \cdot (n!/(n - k + 1)!)$, which is true.

21. Since $f(k) = n!/(n - k)!k!$, to check the initial condition, we substitute 0 for k, getting $f(0) = n!/(n - 0)!0! = 1$. For $n > k$, we have the difference equation $f(k) = ((n - k + 1)/k) \cdot f(k - 1)$. We substitute as above for $f(k)$ and $n!/(n - (k - 1))!(k - 1)!$ for $f(k - 1)$, giving $n!/(n - k)!k! = ((n - k + 1)/k) \cdot (n!/(n - (k - 1))!(k - 1)!) = ((n - k + 1)/k \cdot (n!/(n - k + 1)!(k - 1)!)$, which is true.

1. P: $_nP_n = n!$ and $m \le n!$ since $n < n!$, the square matrix can be formed.

3. AP: $7 + 3 + 5 = 15$

5. If the lists are pairwise disjoint, AP: $54 + 61 + 44 = 159$.

7. MP: $4^{1500} = 1.23 \cdot 10^{903}$

9. MP for each length, then AP. If we did not treat each enzyme as being able to recognize just one length, then we could not use the AP in this way.

$$4^4 + 4^6 = 256 + 4096 = 4352$$

11. MP: 5^k

13. AP then MP: Dems$(14 + 5 + 1) \cdot$ Reps$(17 + 4 + 3) = 20 \cdot 24 = 480$

15. P: $_9P_9 = 9! = 362,880$

17. C: $_{10}C_3 = 120$

19. P and MP:
    Choose goalie first:
    $_{17}C_1 \cdot _{16}C_{10} = 17 \cdot 8008 = 136,136$
    Choose goalie last:
    $_{17}C_{10} \cdot _7C_1 = 19,448 \cdot 7 = 136,136$, the same.

21. C and MP, then AP:
    First round: $6 \cdot _4C_2 = 6 \cdot 6 = 36$
    Elim. round: $8 + 4 + 2 + 1 = 15$
    Extra game: 1
    Total $= 52$

23. Two digits, 1 and 0, can be used as the second digit of an exchange (now, in the "old days" they couldn't), but they do not correspond to letters, so there are more possibilities with all digits.

25. Neither P nor C, since both rule out having the same flavor for both scoops.

27. P: $_{13}P_{13} = 13! = 6{,}227{,}020{,}800$ possible orders

29. Number of trios of drugs: $_{1500}C_3 = 651{,}375{,}500$

31. We know from Theorem 1 of Section 4.3 that $\binom{n}{0} = \binom{n}{n}$. So we just have to show that one of them equals 1:

$$\binom{n}{0} = \frac{n!}{(n-0)!0!} = 1.$$

33. We show that $\binom{n}{k} = \binom{n-1}{k} + \binom{n-1}{k-1}$, for $0 < k < n$, by directly substituting:

$$\frac{n!}{(n-k)!k!} = \frac{(n-1)!}{((n-1)-(k))!(k)!}$$

$$+ \frac{(n-1)!}{((n-1)-(k-1))!(k-1)!}$$

$$= \frac{(n-1)!}{(n-k-1)!k!} + \frac{(n-1)!}{(n-k)!(k-1)!}$$

$$= \frac{(n-1)!(n-k) + (n-1)!(k)}{(n-k)!k!}$$

$$= \frac{(n-1)!(n)}{(n-k)!k!},$$

which is equal to the original formula above.

35. When we expand $(x - y)^n$, we have n factors, each of which is the binomial $(x - y)$. Any term in the product has k factors that are x and $n - k$ factors that are y, where $0 \le k \le n$, so it looks like $x^k y^{n-k}$. In how many ways can such a term be created? That is the same as asking, "In how many ways can we choose to take k of the binomial factors and let them contribute an x to the term, while the other $n - k$ binomial factors contribute a y?" The number we desire is exactly $\binom{n}{k}$.

1. (a) No
   (b) No

3. Answers will vary.

5. Not deterministic

7. Only three assignments can be completed on time.

9. Four assignments can be completed on time.

11. The island and the three outer regions are nodes and the bridges are edges, giving the graph shown below.

By Euler's theorem, the desired walk cannot be accomplished.

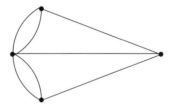

13. Answers will vary.

15. (a) It is not deterministic, because it doesn't tell exactly what node to start at nor at each step which edge to select. We could number the nodes and start at the lowest-numbered node of odd degree, or else at node 1; similarly, we could number the edges and pick an available edge with the smallest number.

(b) The fundamental idea is that Euler's theorem guarantees that at any stage in drawing the path, there has to be some way to finish it, so it cannot happen that every available edge leads to disconnecting the remaining edges.

(c) At each step we draw a new edge, we never undraw one, and there are only finitely many of them.

17. Answers will vary. Here is one:

Inputs: integers a, b
Output: greatest common divisor of a and b
Preconditions: 0 < a < b
Postconditions: output is greatest common divisor of a and b

```
repeat
        c ← b mod a
        a ← c
        b ← a
until (c = 0)
write ('greatest common divisor is', b)
```

SECTION 5.2

1. (a)

Cycle	Cost
BGMaMiB	443
BGMiMaB	402
BMiGMaB	369
BMiMaGB	443
BMaMiGB	402
BMaGMiB	369

(b) BMaMiGB, for a cost of 402

3. Add in order the edges BMa (50), BMi (73), MiG (114), and GMa (132), for the cycle MaGMiBMa, with cost 369. This is a lower-cost cycle than the one found

by applying the nearest-neighbor heuristic starting at Beloit.

5. Answers will vary.

7. (a) Each of the nondiagonal entries in the matrix is 1.

(b) 7 time units

(c) No. The problem becomes more interesting for general n, as you may enjoy exploring.

9. (a) The wallpaper roll looks like:

ABCDABC
BCDABCD
CDABCDA
DABCDAB
ABCDABC
⋮

You may find it helpful to make yourself some "wallpaper" like this, to help in thinking through the problem.

The wall, and the four different sheets of wallpaper, must look like:

ABCDABC	DABCDAB	CDABCDA	BCDABCD
BCDABCD	ABCDABC	DABCDAB	CDABCDA
CDABCDA	BCDABCD	ABCDABC	DABCDAB
DABCDAB	CDABCDA	BCDABCD	ABCDABC
ABCDABC	DABCDAB	CDABCDA	BCDABCD

We have $r = 4$ and $d = 3$, so $n = r/\text{GCD}(d, r) = 4$. We also have $l = 5$. The matrix of weights is

		Cut second			
		1	*2*	*3*	*4*
Cut first	1	4	2	1	0
	2	0	4	2	1
	3	1	0	4	2
	4	2	1	0	4

(b) The nearest-neighbor algorithm says to cut in the order 1-4-3-2-1, for a total waste of 0.

Spanning cycle	Cost
12341	8
12431	4
13241	4
13421	4
14231	4
14321	0

11. (a) For the currency with denominations 1, 3, and 4 and a payout of 6, the greedy algorithm gives a 4 and two 1s, but two 3s would do it.

 (b) By induction on the number of denominations

SECTION 5.3

1. GMiBMaE, for a cost of 413

3. (a) Answers will vary.
 (b) Not necessarily.
 (c) Answers will vary.

5. Answers will vary.

7. $\binom{v}{2}$, which is realized by a complete graph.

9. Answers will vary. If the two trees are not identical, since one is contained in the other, the larger one must have at least one more edge. But both trees have the same number of nodes, so by Lemma 3 they must have the same number of edges.

SECTION 5.4

1. We exhibit the game tree in compressed form, using bold to indicate the player's choice of move, italic to indicate a forced continuation of sowing, and L and W for lose and win:

$11110 \rightarrow 02110 \rightarrow 00220 \rightarrow 10201$ L
$11110 \rightarrow 10210 \rightarrow 10021 \rightarrow$
    $10021 \rightarrow 01021$ L
    $10021 \rightarrow 20002 \rightarrow 01102$ L
$11110 \rightarrow 11020 \rightarrow 21001 \rightarrow 02101$ L
$11110 \rightarrow 11101 \rightarrow$
    $11101 \rightarrow 02101 \rightarrow 00211$ L
    $11101 \rightarrow 10201 \rightarrow 10012$
        $10012 \rightarrow 01012$ L
        $10012 \rightarrow 10003 \rightarrow$
            $10003 \rightarrow 01003$ L
    $11101 \rightarrow 11011$ L

3. (a) From 8 or 9 sticks, leave 7; from 11 or 12, leave 10; from 14 or 15, leave 13. The other numbers are losing positions.

 (b) If possible, leave your opponent $3n + 1$ sticks.

 (c) If the opponent takes 1, you take 2; if the opponent takes 2, you take 1. This works because you always leave the opponent $3n + 1$ sticks, for some n that decreases with each turn, until $n = 0$, when the opponent must take the last stick.

5. (a) Positions of the form $3n + 1$
 (b) Positions of the form $3n$
 (c) Positions of the form $4n$

7. The safe positions are the same except that $(1, 1, 0)$ is not safe but now $(1, 0, 0)$ is. The strategy is the same as for normal-play Nim, except that when the strategy calls for you to leave $(1, 1, 0)$, leave either $(1, 1, 1)$ or $(1, 0, 0)$ instead.

9. First Player loses.

11. Second Player loses. This type of game is Nim with two rows (corresponding to the two directions), with a player allowed to take exactly one stick from only one of the two rows.

13. (a) $\binom{6}{0} + \binom{5}{1} + \binom{4}{2} + \binom{3}{3} = 1 + 5 + 6 + 1 = 13$

 (b) $\binom{8}{0} + \binom{7}{1} + \binom{6}{2} + \binom{5}{3} + \binom{4}{4} = 1 +$

$7 + 10 + 15 + 1 = 34$. Notice that this total is the sum of the number of playings for F_6 and for F_7.

15. (a) 3: a corner square and its neighbor (call this variant A), two of the four center squares (variant B), or one of the four center squares and its neighbor on an edge (variant C),

(b) Answers will vary. Exhaustive computer analysis (e.g., using the Gamesman's Toolkit) shows that the Second Player to win is the natural outcome of variants A and B but First Player to win is the natural outcome of variant C. If First Player begins by removing a central square, then Second Player can choose an edge and be sure that he or she plays variant B, which Second Player can win. If First Player begins by removing a corner, he or she plays variant A, which Second Player can win. Summary: The natural outcome is Second Player to win.

SECTION 5.5

1. **a & J, b & L, c & M, d & K**, after two rounds

3. **a & L** is one blocking pair; **d & J** is another.

5. **J & d, K & c, L & a**, and **M & b**, after four rounds. This is not the same matching as in Exercise 4.

7. (a) Using the blocking pair **a & L** gives the matching **a & L, b & M, c & J**, and **d & K**, which has the blocking pair **d & J**. Matching **d** with **J** gives **a & L, b & M, c & K**, and **d & J**, which is stable.

(b) Repeated searching for blocking pairs appears to be less efficient than the GS algorithm.

9. There are only three possible matchings: **A & B** and **C & D**; **A & C** and **B & D**; and **A & D** and **B & C**. For

the first, **A & C** is a blocking pair; for the second, **B & C**; for the third, **A & B**.

11. No

SECTION 5.6

1. Yes

3. (a) Since $76 = 64 + 8 + 4$, we need to multiply 3^{64}, 3^8, and 3^4.

(b) 19,487,171

5. Answers will vary.

7. (a) For $n = 6$, we have $n - 1 = 5 = 2^0 \cdot 5$, so $l = 0$ and $m = 5$. For $a = 3$, we have the sequence 3^5, which we take modulo 6 to get the pattern 3. We conclude that 6 is not prime.

(b) As in (a), with the sequence 4^5, which we take modulo 6 to get the pattern 4. Again, the pattern says that 6 cannot be prime.

9. Answers will vary.

11. (a) Yes

(b) No

13. Answers will vary.

15. (a) 4, 1, 2, 1, 1; 7, 3; 9; 2

(b) 9, 1; 7, 3; 4, 2, 2, 1, 1

(c) 3. We are guaranteed that first-fit will require no more than twice as many bins as needed, or 6, and that decreasing first-fit will require no more than $11 \cdot 3/9 + 4 = 7.67$.

17. The machine writes three 1s at and to the right of the start position, writes three 1s to the left of the start position, then halts.

19.

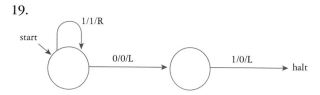

Notation: ST = separate tasks principle; NL = nested loops principle

1. $f(n, m) = nm$. When $m = n$, the formula becomes n^2.

3. The product $\mathbf{A} \times \mathbf{B}$ will be of the shape $n \times n$, which means that we will need to compute n^2 scalar products of m-element vectors. Each scalar product requires $2m - 1$ arithmetic operations, so the entire computation requires $n^2(2m - 1)$ arithmetic operations (using NL). For the product $\mathbf{B} \times \mathbf{A}$, the roles of n and m are reversed, so there are $m^2(2n - 1)$ arithmetic operations needed. Both expressions become $2n^3 - n^2$ when $n = m$, when the matrices are square (see Example 3).

5. The main problem in writing the formula is knowing how many proposals are accepted in each round. In the best case, every proposer proposes to a different person, and since each of them receives just one proposal, it is accepted and we are done with just n proposals being made. In the worst case, every proposer has the exact same preference list, and in each round only one proposer is accepted. In this case round 1 contains n proposals, round 2 contains $n - 1$ proposals, round 3 contains $n - 2$ proposals, and so on until round n, which contains just 1 proposal. Using ST we get the sum $n + (n - 1) + (n - 2) + \cdots + 1$, which is the sum of the first n positive integers and is equal to $n(n + 1)/2$. It is much harder to develop a formula for the nonextreme cases.

SECTION 6.2

1. For $n = 1$, there is one multiplication, so $w(1) = 1$. Comparing the number of multiplications in the n by n case with those in the $(n - 1)$ by $(n - 1)$ case, we see that in the larger case, there is one more row of n items and also an additional column of n items, but that one item is in both the new row and the new column. Thus (using ST) we get the recurrence equation $w(n) = w(n - 1) + 2n - 1$, $n > 1$.

3. In Exercise 1 of Section 1 we developed the formula n^2 for scalar multiplication of an n by n matrix. Substituting this formula into the initial condition in Exercise 1 of this section, we get $w(1) = 1^2 = 1$, which checks. Making the substitution into the recurrence equation $w(n) = w(n - 1) + 2n - 1$, we get $w(n) = n^2$ for the left-hand side and $w(n - 1) + 2n - 1 = (n - 1)^2 + 2n - 1 = n^2 - 2n + 1 + 2n - 1 = n^2$ on the right-hand side, so the formula satisfies.

5. If $n = 0$, then nothing is printed, so $w(0) = 0$. For $n > 0$, the module makes two calls to itself for a tower one ring shorter and prints one line, so (using ST) we get the recurrence equation $w(n) = 2w(n - 1) + 1$. Checking the proposed solution, we see that $w(0) = 2^0 - 1 = 1 - 1 = 0$, which checks. Making the substitution into the recurrence equation: On the left we get $w(n) = 2^n - 1$. On the right we get $2w(n - 1) + 1 = 2(2^{n-1} - 1) + 1 = 2^n - 2 + 1 = 2^n - 1$, which equals the left.

SECTION 6.3

In all cases, the bigger of the two functions is the less preferred.

1. $n = 25$; A is bigger on $(0, 25)$; B is bigger on $(25, +\infty)$.

3. $n = 34$; B is bigger on $(0, 34)$; A is bigger on $(34, +\infty)$.

5. $n = 7$; B is bigger on $(0, 7)$; A is bigger on $(7, +\infty)$.

7. $n \approx 11.5$; A is bigger on $(0, 11)$; B is bigger on $(12, +\infty)$.

9. $n \approx 6312$; A is bigger on $(0, 6312)$; B is bigger on $(6312, +\infty)$.

11. $n = 50, 70, 120$; B is bigger on $(0, 50)$ and $(70, 120)$; A is bigger on $(50, 70)$ and $(120, +\infty)$.

13. $n = 55, 70, 130$; A is bigger on $(0, 55)$ and $(70, 130)$; B is bigger on $(55, 70)$ and $(130, +\infty)$.

SECTION 6.4

1. (a) $4n^3$
   (b) $5n^2$
   (c) $3n^2 + 2n$
   (d) n^3
   (e) 7^n

3. (a) $O(n^3)$
   (b) $O(n!)$
   (c) $O(n^2)$
   (d) $O(n \log(n))$

SOLUTIONS TO CHAPTER 6 EXERCISES

1. $n \approx$ (is approximately equal to) 108; A is bigger on $(0, 108)$ and B is bigger on $(108, +\infty)$. A: $4 \log_2 n$ is $O(\log n)$ and B: $n/4$ is $O(n)$. Since $O(\log n)$ is a proper subset of $O(n)$, B is bigger on the unbounded interval.

3. $n \approx 8$; B is bigger on $(0, 8)$ and A is bigger on $(8, +\infty)$. A: $n!$ is $O(n!)$ and B: $720n^5$ is $O(n^5)$. Since $O(n^5)$ is a proper subset of $O(n!)$, A is bigger on the unbounded interval.

5. (Using NL) The loop on k has 2 arithmetic operations and is repeated $n - 1$ times for a total of $2(n - 1)$ operations each time the loop is reached. The loop on j contains the loop of k plus 1 additional operation and is repeated n times, for a total of $n(2(n - 1) + 1) = 2n^2 - n$ operations. The loop on i contains just the loop on j and repeats n times, for a grand total of $2n^3 - n^2$ operations. All formulas for squaring a matrix have the same order of magnitude, namely $O(n^3)$, and this pseudocoded algorithm would run slightly faster than the one in Section 6.1.

7. A solution to the recurrence equation $f(n) = nf(n - 1)$, $f(0) = 1$, is $f(n) = n!$.

9. (Using NL) In the loop on k, which repeats $n - 1$ times, there are 2 operations needed to find the "multiplier," $n + 1$ multiplications to form the temporary row, and $n + 1$ additions to do the addition, for a total (using ST) of $(n - 1)(2n + 4)$ operations. The loop in i contains just the loop on k, and repeats n times, so the entire method takes $n(n - 1)(2n + 4)$ operations. This expression has order of magnitude $O(n^3)$.

11. Gaussian elimination has order of magnitude $O(n^3)$, and finding just one determinant has $O(n!)$, so Gaussian elimination is preferable to Cramer's rule.

13. At about $n = 83$, the percentage error drops below 0.1%.

SECTION 7.1

1. (a) Either I am a college student or my parents like my hair style.
   (b) I play the guitar and I voted in the last Presidential election.
   (c) It is not true that I am a college student and that I voted in the last Presidential election.
   (d) I play the guitar, and either I voted in the last Presidential election or my parents dislike my hair style.
   (e) I do not play the guitar and I am a college student.

3. (a) $C \wedge {\sim}G$
   (b) $C \vee {\sim}H$
   (c) $C \wedge G \wedge {\sim}V$
   (d) $V \vee (C \wedge H)$
   (e) ${\sim}(C \wedge G)$

5. (a) Not equivalent
   (b) Equivalent

(c) Equivalent

7. (a) Tautology
   (b) Neither

9. (a) $A|A$
   (b) $((A|B) | (A|B))$
   (c) $(A|A) | (B|B)$
   (d) $(A | (B|B)) | ((A|A) | B)$

11. (a)

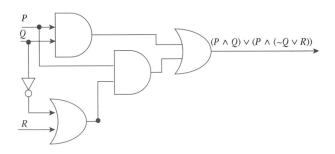

$(P \wedge Q) \vee (P \wedge (\sim Q \vee R))$

(b)

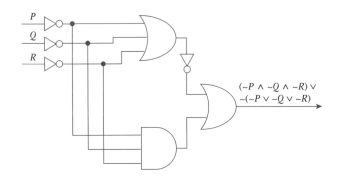

$(\sim P \wedge \sim Q \wedge \sim R) \vee \sim(\sim P \vee \sim Q \vee \sim R)$

13. (a) $((P \oplus Q) \wedge (Q \vee R)) | (Q \vee R)$
    (b) $((P \vee Q) \vee (Q \wedge R)) \downarrow R$

SECTION 7.2

1. (a) $pqr + pqr' + p'qr' + p'q'r'$

(b)

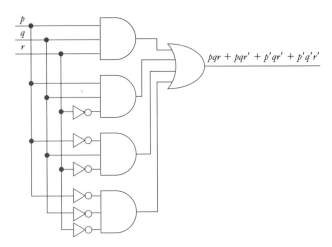

$pqr + pqr' + p'qr' + p'q'r'$

3. Suppose the number of input variables is n. Then each conjunction in the disjunctive normal form will require $n - 1$ (binary) *and* gates. If there are k lines on which the output is a 1, then there will be $k(n - 1)$ *and* gates, with $k - 1$ (binary) *or* gates between the conjunctions, for a total of $k(n - 1) + k - 1 = kn - 1$ (binary) gates.

5. (a)

W	E	S
1	1	0
1	0	0
0	1	1
0	0	0

The disjunctive normal form for S is $S = W'E$

(b)

$W'E$

In the answers to 7, 9, and 11, only the disjunctive normal form is given. The circuit diagrams are similar to Figure 10, with one and gate for each conjunction, and negations as needed.

7. $O_1 = I'_3I'_2I'_1I'_0 + I'_3I_2I'_1I'_0 + I'_3I_2I'_1I_0 + I'_3I_2I_1I'_0 + I_3I'_2I'_1I'_0 + I_3I'_2I'_1I_0$.

9. $O_4 = I'_3I'_2I_1I'_0 + I'_3I'_2I_1I_0 + I'_3I_2I'_1I_0 + I'_3I_2I_1I'_0 + I_3I'_2I_1I'_0 + I_3I'_2I'_1I'_0 + I_3I'_2I'_1I_0$.

11. $O_6 = I_3'I_2'I_1'I_0' + I_3'I_2'I_1'I_0 + I_3'I_2'I_1I_0' + I_3'I_2'I_1I_0 + I_3'I_2I_1'I_0' + I_3'I_2I_1I_0 + I_3I_2'I_1'I_0' + I_3I_2'I_1'I_0.$

In the answers to 13 and 15, the lines corresponding to impossible inputs have been set with 0 as the output. The circuit diagrams are closely related to those in the text, and thus are not given here.

13.

# of sticks on the board			# of sticks the device should remove			Is this board a possibility?
I_1	I_0	(dec)	O_1	O_0	(dec)	
1	1	(3)	1	0	(2)	yes
1	0	(2)	0	1	(1)	yes
0	1	(1)	0	0	(0)	no
0	0	(0)	0	0	(0)	no

So we have $O_1 = I_1I_0$ and $O_0 = I_1I_0'$.

15.

# of sticks on the board				# of sticks the device should remove			is this board a possibility?
I_2	I_1	I_0	(dec)	O_1	O_0	(dec)	
1	1	1	(7)	0	0	(0)	no
1	1	0	(6)	0	0	(0)	no
1	0	1	(5)	1	0	(2)	yes
1	0	0	(4)	0	1	(1)	yes
0	1	1	(3)	0	0	(0)	no
0	1	0	(2)	1	0	(2)	yes
0	0	1	(1)	0	1	(1)	yes
0	0	0	(0)	0	0	(0)	no

So we have $O_1 = I_2I_1'I_0 + I_2'I_1I_0'$ and $O_0 = I_2I_1'I_0' + I_2'I_1'I_0$.

SECTION 7.3

1. (a) Every counting number is a positive number. (False if we consider zero as a counting number.)

(b) Every prime is odd. (False, 2 is prime and is even.)

(c) There is a counting number which is negative. (False, counting numbers are all non-negative.)

(d) There is an integer which is negative. (True, -1 is a negative integer.)

3. (a) $D = \{\text{points on a circle of radius } r\}$; $D(x) = $ "x is a distance r from the center of the circle"; $(\forall x)D(x)$

(b) $D = \{\text{triangles}\}$; $P(x) = $ "$c^2 = a^2 + b^2$ is true on x"; $(\exists x)P(x)$

(c) $D = \{\text{square matrices}\}$; $I(x) = $ "x has an inverse"; $(\exists x) \sim I(x)$

(d) $D = \{\text{things which come to net}\}$; $F(x) = $ "x is a fish"; $(\forall x)F(x)$

(e) $D = \{\text{people ("bodies")}\}$; $B(x) = $ "x throws a brick"; $(\exists x)B(x)$

(f) $D = \{\text{things}\}$; $W(x) = $ "x loves a (any) wall"; $(\exists x) \sim W(x)$

(g) $D = \{\text{the parts of the world}\}$; $S(x) = $ "x is a stage"; $(\forall x)S(x)$

(h) $D = \{\text{deepest feelings}\}$; $S(x) = $ "x shows itself in silence"; $(\forall x)S(x)$

5. $D = \{\text{straight lines}\}$, $(\forall x)$ (x has a y-intercept). The negation of this is $(\exists x)$ (x does not have a y-intercept), which corresponds to "There is a straight line which does not have a y-intercept." To show the original false, find a counterexample.

7. $D = \{\text{straight lines}\}$, $(\exists x)$ (x intersects neither the x-axis or the y-axis). The negation of this is $(\forall x)$ (x intersects either the x-axis or the y-axis), which corresponds to "Every straight line intersects either the x-axis or the y-axis." To show the original false, we would need to construct a proof.

SOLUTIONS TO CHAPTER 7 EXERCISES

3. (a) Start with $\sim(\sim A \wedge B)$; use DeMorgan to get $A \vee \sim B$, which is not the same as $A \wedge \sim B$, so the expressions are not equivalent.

(b) Start with $\sim(\sim M \wedge N)$; use DeMorgan to get $M \vee \sim N$, which is the other expression, so they are equivalent.

(c) Start with $P \wedge (\sim Q \wedge R)$, and use associativity to get $(P \wedge \sim Q) \wedge R$, so they are equivalent.

I_2	I_1	I_0	O_7	O_6	O_5	O_4	O_3	O_2	O_1	O_0
1	1	1	1	0	0	0	0	0	0	0
1	1	0	0	1	0	0	0	0	0	0
1	0	1	0	0	1	0	0	0	0	0
1	0	0	0	0	0	1	0	0	0	0
0	1	1	0	0	0	0	1	0	0	0
0	1	0	0	0	0	0	0	1	0	0
0	0	1	0	0	0	0	0	0	1	0
0	0	0	0	0	0	0	0	0	0	1

5. In Circuit 1, the *exclusive-or* gates are set to 0 whenever the A and B inputs are identical and set to 1 if they are different. The *nor* gate is a 1 only under the circumstance that all the inputs are 0s, which would mean that the As and the Bs all match. Circuit 1 is the comparator. Circuit 2 is a one-bit adder and carries out these definitions of S, out (sum bit) and C, out (carry-out bit):

A, in	B, in	C, in	C, out	S, out
1	1	1	1	1
1	1	0	1	0
1	0	1	1	0
1	0	0	0	1
0	1	1	1	0
0	1	0	0	1
0	0	1	0	1
0	0	0	0	0

For Exercises 7, 9, 11, and 13, only the defining truth tables are given.

7.

H	D	A
1	1	0
1	0	1
0	1	0
0	0	0

So the disjunctive normal form is $A = HD'$.

9.

C	L	A	S
1	1	1	1
1	1	0	1
1	0	1	1
1	0	0	0
0	1	1	1
0	1	0	0
0	0	1	0
0	0	0	0

So the disjunctive normal form is $S = CLA + CLA' + CL'A + C'LA$.

11. We use subscripts to indicate the power of 2 corresponding to each input and the decimal value corresponding to each output:

13. The defining table for the demultiplexer is similar to that for the decoder in Exercise 11, except that now each of the old output lines is *anded* with the data input line. Since only one output line will be a 1, only one of the *ands* will duplicate the data input line, and the rest will be 0s.

15. (a) $D = \{\text{people}\}$; $M(x) = $ "x can be a master"; $\sim(\forall x)M(x)$

(b) $D = \{\text{royal roads}\}$; $G(x) = $ "x leads to geometry"; $\sim(\exists x)G(x)$

(c) $D = \{\text{people}\}$; $V(x) = $ "x is tired on the day of victory"; $\sim(\exists x)V(x)$

(d) $D = \{\text{appearances worn by Nature}\}$; $M(x) = $ "x is mean"; $(\forall x) \sim M(x)$

(e) $D = \{\text{people who have not been in the interior of a family}\}$; $D(x) = $ "x can say what the difficulties of any individual of that family might be set"; $\sim(\exists x)D(x)$

(f) $D = \{\text{flowers that do not wilt}\}$; $L(x) = $ "we don't like x"; $(\forall x)L(x)$

17. $D = \{\text{quadratic equations with real coefficients}\}$; $R(x) = $ "x has roots of $2 + 3i$ and 7." The original translates as $(\exists x)R(x)$. The original is false. Its negation is $\sim(\exists x)M(x)$ or $(\forall x) \sim M(x)$, which translates as "Among all quadratic equations, none has roots of $2 + 3i$ and 7." A proof would be needed to convince us of this.

19. $D = \{\text{systems of linear equations}\}$; $S(x) = $ "x can be solved." The original translates as $(\exists x) \sim S(x)$, and it is true.

21. $D = \{\text{matchings produced by the Gale–Shapley algorithm}\}$; $S(x) = $ "x is stable"; The original translates as $(\forall x)S(x)$, and it is true.

SECTION 8.1

1. (a) 16 feet
   (b) 144 feet
   (c) 256 feet

3. $P = 100,000 \ 2^{120/30} = 1,600,000$

5. 10 years

SECTION 8.2

1. $(26)(1000) + (26)(26)(100) = 93,600$

3. $(20)(16)(8)(17) + (20)(16)(8) + (20)(16)(17) + (20)(8)(17) + (16)(8)(17)$

5. (a) $724/11429 \sim .063$
   (b) $357/11429 \sim .031$

7. $\binom{5}{3} \bigg/ 2^5 = 5/16$

9.

Die 1 → Die 2 ↓	1	2	3	4	5	6
1	2	3	4	5	6	7
2	3	4	5	6	7	8
3	4	5	6	7	8	9
4	5	6	7	8	9	10
5	6	7	8	9	10	11
6	7	8	9	10	11	12

11. (a) .7

SECTION 8.3

1.

	Juniors	Sophomores	Total
Prefer Diet	17	21	38
Prefer Regular	23	24	47
Totals	40	45	85

Answers are 38/85, 23/85, 17/38

3. 8/9

5. (a) 1/2

(b) 26/51
(c) 25/51
(d) 1/2

7. (a) B
   (b) .008
   (c) .5735
   (d) .406

9. 1/3

11. 2/3; 5/6; 1

13. 80

SECTION 8.4

1. $(1/27)^3 \approx .00005$; $(5/27)(21/27)(5/27) \approx .0267$ (assuming vowels are A, E, I, O, U)

3. 1/4,000,000

5. Two \$10,000 machines. The probability that both will fail to detect a weapon is .0001.

7. $(714/8399)^4 \approx .00005$

9. Assume reliability of amplifier is .96
   (a) .835
   (b) .931
   (c) .903
   (d) .069
   (e) .925
   (f) .000028

11. Hint: What is $\Pr(A \cap B)$?

SECTION 8.5

1. $f(z) = \begin{cases} .13 & \text{if } z = 5,000 \\ .07 & \text{if } z = 8,000 \\ .8 & \text{if } z = 10,000 \\ 0 & \text{otherwise} \end{cases}$;

   $F(z) = \begin{cases} 0 & \text{if } z < 5,000 \\ .13 & \text{if } 5,000 \le z < 8,000 \\ .20 & \text{if } 8,000 \le z < 10,000 \\ 1 & \text{if } z \ge 10,000 \end{cases}$

3. ($500)(.92) + (−$700)(.08) = $404

5. 244/495; −7/9 dollars

7. Expected profit for high school is $9,200 but for the elementary school it is $9,625.

9. (a) 8.4
   (b) in front of elevator #3

11.

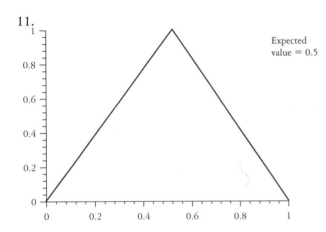

Expected value = 0.5

13.

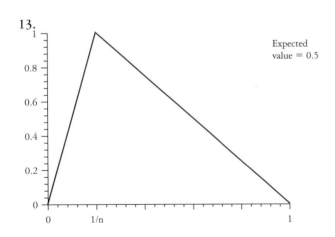

Expected value = 0.5

15.

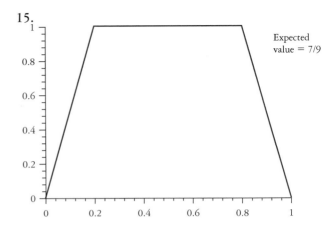

Expected value = 7/9

SECTION 8.6

1. If no gender bias, the probability of 13 or more women being convicted is .444;
   If no gender bias, the probability of 11 or more women being convicted is .868.

3. Assuming no bias, the probability that 339 or fewer Mexican-Americans would be chosen is

$$\sum_{k=0}^{339} \frac{\binom{.79P}{k}\binom{.21P}{870-k}}{\binom{P}{870}}$$

where P is the population of the county.

5. (Russian class) (a) $(1/15)^2$
   (b) $(14/15)^7$
   (c) 49.8
   (d) 10

7. Policy 1: assume real danger only when all devices sound.
   Policy 2: assume real danger only when at least one device sounds.
   Policy 3: assume real danger only when a majority of devices sound.

   The following tables show the probability of acting correctly if true danger exists and the probability of acting wrongly if there is no danger,

(a) Two devices

	True Danger	No Danger
Policy 1	.9025	.0025
Policy 2	.9975	.0975

(b) Three devices

	True Danger	No Danger
Policy 1	.857	.000125
Policy 2	.999875	.143
Policy 3	.99275	.00725

9. If danger exists, probability of acting correctly is

$$(.1)^5 + 5(.1)^4(.9) + 10(.1)^3(.9)^2(.9)^5 + 5(.9)^4(.1) + 10(.9)^3(.1)^2 = .99144$$

if there is no danger, probability of acting wrongly is

$$(.1)^5 + 5(.1)^4(.9) + 10(.1)^3(.9)^2 = .00856$$

SECTION 8.7

1. Expected utilities are: Senate (.536) and Secretary of State (.396)

3.

	Strategy 1	Strategy 2	Strategy 3
Excellent	.44	.35	.31
Good	.36	.525	.465
Fair	.16	.10	.22
No Job	.04	.025	.005
Expected Value of Salary	$30,200	$30,500	$30,300

5. Accept an Excellent or Good offer the first time you get one; otherwise move on to the next interview. Expected salary is $29,800

7. Accept only an Excellent or Good offer at interview 1, but accept an Excellent, Good, or Fair offer at Interview 2. Expected salary is $29,131.95.

9. Expected value of jury's judgment is $180,000.

SECTION 9.1

1. Three rotations: $0°, 120°, 240°$, and three reflections across lines from vertices to midpoints of opposite sides.

3. no

5. D_{2n} has n rotations of the form $k(360°/n)$, where $k = 0, \ldots, n - 1$. D_{2n} has n reflections. When n is odd, the axes of reflection are the lines from the vertices to the midpoints of the opposite sides. When n is even, half of the axes of reflection are obtained by joining opposite vertices; the other half, by joining midpoints of opposite sides.

7. $HD = DV$, but $H \neq V$.

9. R_0, R_{180}, H, V

	R_0	R_{180}	H	V
R_0	R_0	R_{180}	H	V
R_{180}	R_{180}	R_0	V	H
H	H	V	R_0	R_{180}
V	V	H	R_{180}	R_0

11. See answer for Exercise 9.

13. In each case the group is D_{12}.

15. No

17. C_5, C_{12}, C_{22}

19. For any fixed vectors $\mathbf{v}$ and $\mathbf{w}$, translation by the vector $\mathbf{v}$ takes a vector $\mathbf{u}$ to $\mathbf{u} + \mathbf{v}$, and translation by $\mathbf{w}$ takes $\mathbf{u} + \mathbf{v}$ to $(\mathbf{u} + \mathbf{v}) + \mathbf{w}$. Thus, the composition

takes **u** to $(\mathbf{u} + \mathbf{v}) + \mathbf{w} = \mathbf{u} + (\mathbf{v} + \mathbf{w})$, which is the same as translation of **u** by $\mathbf{v} + \mathbf{w}$.

21. Let T be an isometry, p, q, and r the three noncollinear points, and s any other point in the plane. Then the quadrilateral determined by $T(p)$, $T(q)$, $T(r)$, and $T(s)$ is congruent to the one formed by p, q, r, and s. Thus $T(s)$ is uniquely determined by $T(p)$, $T(q)$, and $T(r)$.

23. $p6$, $p3m1$, $p31m$
$p2$, $p3$, pmm

25. (a) V
(b) I
(c) II
(d) VI
(e) VII
(f) III

27. t^5r

33. Reading down the columns starting on the left, we have:

pgg, pmm, $p2$, $p1$, cmm, pmg, pg, pm, $p3$, $p4$, $p4m$, $p4g$, cm, $p6$, $p3m1$, $p31m$, $p6m$.

31. th

33. Use the method shown in Figure 22.

SECTION 9.2

1. does not contain the identity; closure fails

3. Suppose $\begin{bmatrix} a & b \\ c & d \end{bmatrix}$ is the inverse. Then,

$$\begin{bmatrix} 2 & 2 \\ 1 & 1 \end{bmatrix}\begin{bmatrix} a & b \\ c & d \end{bmatrix} = \begin{bmatrix} 1 & 0 \\ 0 & 1 \end{bmatrix}$$

so that $2a + 2c = 1$ and $a + c = 0$.

5. 1 and 0 are real numbers but 1 divided by 0 is not.

7. (i) $2a + 3b$;
(ii) $-2a + 2(-b + c)$;
(iii) $-3(a + 2b) + 2c = 0$

9. (a) $75 = 3 \bmod 24$, so we need add only 3 hours.
(b) $87 = 7 \bmod 40$, so we need add only 7 spaces.
(c) An odometer resets to 0 at 100,000 miles, so $97{,}000 + 12{,}000 = 9{,}000 \bmod 100000$.

11. aca^{-1}

13. $abc = e$ implies that $bc = a^{-1}$. Then $e = a^{-1}a = bca$.

15. Clearly, the set satisfies the associative property and contains the identity 1. Now suppose that a and b are in the set, so that $a^n = 1 = b^n$. Then $(ab)^n = a^n b^n = 1 \cdot 1 = 1$. Also, $(a^{-1})^n = (a^n)^{-1} = 1^{-1} = 1$, and similarly for b. Thus, the set satisfies all the requirements to be a group. If n can vary with x, the set is a group because $a^m = 1$ and $b^n = 1$ imply that $(ab)^{mn} = (a^m)^n(b^n)^m = 1^n \cdot 1^m = 1$.

17. no; no

19. For n positive, use induction. For n negative, observe that $e = (ab)^n(ab)^{-n} = (ab)^n a^{-n} b^{-n}$ (since $-n$ is positive) so that $(ab)^n = b^n a^n = a^n b^n$. For $n = 0$, observe that $(ab)^0 = e = ee = a^0 b^0$. $(ab)^n \neq a^n b^n$ for non-Abelian groups.

21. Use Exercise 20.

23. FR_{324}

25. $\{1, 3, 5, 9, 13, 15, 19, 23, 25, 27, 39, 45\}$

27. Use the observation that $(b^{-1}ab)^3 = (b^{-1}ab) \cdot (b^{-1}ab)(b^{-1}ab) = b^{-1}a^3b$. The same argument works for all integers.

29. Consider the row labeled with a. Let x be any element of the group. Then $a^{-1}x$ is the label for some column and x is the element in the table with row heading a and column heading $a^{-1}x$. Now suppose x appears in a row labeled with a twice. Say $x = ab$ and $x = ac$. Then cancellation gives $b = c$. But the columns are labeled with distinct elements. The argument for columns is analogous.

31. Observe that $(ab)^2 = abab$ and $a^2b^2 = aabb$ and use cancellation. It follows that the group has the form $\{e, a, a^2, b\}$. Now look at the row of the group table that

corresponds to multiplying by a. We obtain $ab = b$, so that $a = e$.

33. R; M

35. There is no integer s so that $4 \cdot s = 1 \bmod 16$.

SECTION 9.3

1. 5

3. Since $a_9 + a_8 + \cdots + a_{i+1} + a_i + \cdots + a_1 + a_0 = a_9 + a_8 + \cdots + a_i + a_{i+1} + \cdots + a_1 + a_0$, the check digit for the incorrect number is the same as the check digit for the correct number when the check digit is not involved in the transposition. If the check digit c is transposed and the transposition is undetected, then $c = (a_9 + a_8 + \cdots + a_1 + a_0) \bmod 9$ and $a_0 = (a_9 + a_8 + \cdots + a_1 + c) \bmod 9$. Subtraction gives $c - a_0 = a_0 - c \bmod 9$ or $2c = 2a_0 \bmod 9$. Multiplying both sides by 5 and using mod 9, we have $c = a_0$. But transposing equal digits does not result in an error.

5. 3

7. The transposition $\cdots ab \cdots \rightarrow \cdots ba \cdots$ is detected if and only if $(a10^{i+1} + b10^i) \bmod 7 \neq (b10^{i+1} + a10^i) \bmod 7$. That is, if and only if $(10a + b) \bmod 7 \neq (10b + a) \bmod 7$. This reduces to $9a \bmod 7 \neq 9b \bmod 7$ or $2a \bmod 7 \neq 2b \bmod 7$. Multiplying by 4 on each side, we have $8a \bmod 7 \neq 8b \bmod 7$ or $a \bmod 7 \neq b \bmod 7$. That is, $|a - b| \neq 7$.

9. 9

11. 3

13. 9

15. (a) 3
    (b) 4
    (c) 9

17. 0-669-09325-4

19. W

21. 52

23. It would not change it.

27. 2, 2, 3

29. (a) Both $d(\mathbf{u}, \mathbf{v})$ and $wt(\mathbf{u} - \mathbf{v})$ equal the number of positions in which $\mathbf{u}$ and $\mathbf{v}$ differ.

   (b) Both $d(\mathbf{u}, \mathbf{v})$ and $d(\mathbf{v}, \mathbf{u})$ are the number positions in which $\mathbf{u}$ and $\mathbf{v}$ differ.

   (c) $d(\mathbf{u}, \mathbf{v}) = 0$ if and only if $\mathbf{u}$ and $\mathbf{v}$ agree in every position.

   (d) If $\mathbf{u}$ and $\mathbf{v}$ disagree in the ith position and $\mathbf{u}$ and $\mathbf{w}$ agree, then $\mathbf{w}$ and $\mathbf{v}$ disagree in the ith position.

   (e) Use part (a).

31. Argue that $wt(\mathbf{v}) + wt(\mathbf{u} + \mathbf{v}) \geq wt(\mathbf{u})$.

33. 0111001; 0101110

35. UHWUHDW; ATTACK; assign 0 to A, 1 to B, and so on. Then the rule assigns x to $x + 3 \bmod 26$.

37. 0000, 1121, 0212, 1011, 1102, 1220, 2022, 2110, 2201

39. $3^4 = 81$; $3^6 = 729$

41. Since the minimum weight of any nonzero member of C is 4, we see by Theorem 2 that C will correct any single error or detect any *triple* error. (To verify this, use $t = 3/2$ in the last paragraph of the proof for Theorem 2.)

43. We cannot determine the intended code word, because the code words 11100 and 11001 both differ from 11101 in exactly one position.

45. The code can correct two or fewer errors or detect five or fewer errors. (That's right, five. In Theorem 2, $t = 2.5$.)

47. 011110011101000101110010; ABAACBAAED

SECTION 9.4

1. $\alpha^{-1} = \begin{bmatrix} 1 & 2 & 3 & 4 & 5 & 6 \\ 2 & 1 & 3 & 5 & 4 & 6 \end{bmatrix}$,

   $\alpha\beta = \begin{bmatrix} 1 & 2 & 3 & 4 & 5 & 6 \\ 6 & 2 & 1 & 5 & 3 & 4 \end{bmatrix}$,

$$\beta\alpha = \begin{bmatrix} 1 & 2 & 3 & 4 & 5 & 6 \\ 1 & 6 & 2 & 3 & 4 & 5 \end{bmatrix},$$

$$\alpha^{100} = \begin{bmatrix} 1 & 2 & 3 & 4 & 5 & 6 \\ 1 & 2 & 3 & 4 & 5 & 6 \end{bmatrix},$$

$$\beta^5 = \begin{bmatrix} 1 & 2 & 3 & 4 & 5 & 6 \\ 1 & 2 & 3 & 4 & 5 & 6 \end{bmatrix}$$

3. $\epsilon = R_0$, $\rho = R_{90}$, $\rho^2 = R_{180}$, $\rho^3 = R_{270}$, $\phi = H$, $\rho\phi = D'$, $\rho^2\phi = V$, $\rho^3\phi = D$

5. (a) $\{\epsilon, \rho^3\phi\}$
   (b) $\{\epsilon, \rho\phi\}$

7. $(n - 1)!$

9. $\alpha = \begin{bmatrix} 1 & 2 & 3 & 4 & 5 \\ 3 & 4 & 5 & 1 & 2 \end{bmatrix}$, $\beta = \begin{bmatrix} 1 & 2 & 3 & 4 & 5 \\ 5 & 3 & 2 & 4 & 1 \end{bmatrix}$

11. 2, 6, Q, 3, 7, 10, 9, 5, J, K, 4, A, 8

13. 24

15. 24

17. 60

19. 24; no

21. Suppose $\mathrm{orb}_G(i) \cap \mathrm{orb}_G(j) \neq \emptyset$, and let $k = \alpha(i)$ and $k = \beta(j)$ for some α and β in G. Let $t \in \mathrm{orb}_G(i)$, so that $t = \gamma(i)$ for some $\gamma \in G$. Then $t = \gamma(\alpha^{-1}(k)) = (\gamma\alpha^{-1})(k) = \gamma\alpha^{-1}(\beta(j)) = (\gamma\alpha^{-1}\beta)(j)$ so that $t \in \mathrm{orb}_G(j)$. This proves that $\mathrm{orb}_G(i) \subseteq \mathrm{orb}_G(j)$. An analogous argument gives $\mathrm{orb}_G(j) \subseteq \mathrm{orb}_G(i)$.

23. 231

25. 21

SOLUTIONS TO CHAPTER 9 EXERCISES

1. Since $a^2 = b^2 = (ab)^2 = e$, we have $aabb = abab$. Now cancel on the left and right.

3. Suppose that not all the code words end with 0. Let A be the set of code words that end with 0, and let B be the set of code words that end in 1. Let $\mathbf{v} \in B$. Then $\mathbf{v} + B \subseteq A$ and $\mathbf{v} + B$ and B have the same number of elements. Thus, $|B| \leq |A|$. Also, $\mathbf{v} + A \subseteq B$ and $\mathbf{v} + A$ and A have the same number of elements. Thus $|A| \leq |B|$.

5. 4!

7. Consider the sequence $a, a^2, a^3, \ldots$, and use the fact that the group is finite.

9. (a) IV
   (b) III
   (c) VII
   (d) IV

11. (a) 3
    (b) The dot product is not 0; the correct number cannot be determined; yes.
    (c) No; the dot product is 0.
    (d) Say the transposition is $\cdots ab \cdots \to \cdots ba \cdots$ with a weighted with $i + 1$ and b weighted with i. The transposition is undetected only if $a(i + 1) + bi$ and $b(i + 1) + ai$ end in the same digit. This simplifies to their difference, $|a - b|$, ending in 0. Since a and b are distinct and between 0 and 9, this can never happen.

13.

$\{R_0, H\}$; $\{R_0, D'\}$; $\{R_0, H\}$
$\{R_0\}$; $\{R_0\}$; $\{R_0\}$

15. D_{56}

RECOMMENDED READINGS

Aarts, Emile H. L. (1992), "Simulated annealing," *The UMAP Journal* 13:79–89.

Expository article on a topic in the Spotlights.

Abernethy, K. and Allen, J. T., Jr. (1992), "The Stable Marriage Problem," Chapter 17 in *Experiments in Computing: Laboratories for Introductory Computer Science in THINK Pascal,* pp. 265–280, Pacific Grove, CA: Brooks/Cole.

Excellent introduction to programming the GS algorithm in Pascal, with exercises on variations of the algorithm.

Astrachan, Owen (1992), "On finding a stable roommate, job, or spouse: A case study crossing the boundaries of computer science courses," *SIGCSE Bulletin* 24(1) (March):107–112.

Babai László (1992), "Combinatorial optimization is hard," *Focus* (Mathematical Association of America), 12:4 (Sept.): 3, 6, 18.

Balakrishnan, V. K. (1982), "The Chinese postman problem," UMAP Modules in Undergraduate Mathematics and Its Applications: Module 582. In *UMAP Modules: Tools for Teaching 1982,* 1–15, Lexington, MA: COMAP.

Treats the problem of optimizing a route when repeating some edges is unavoidable.

Bartholdi, J. J. and McCroan, K. L. (1990), "Scheduling interviews for a job fair," *Operations Research* 38:951–960.

Bartholdi, J. J., III, Platzman, L. K. Collins, R. L., and Warden, W. H. III, (1983), "A minimal technology routing system for Meals on Wheels," *Interfaces* 13:3 (June):1–8.

Details of the application of space-filling curves to meals on wheels.

Beck, Anatole (1969), "Games," Chapter 5 in *Excursions into Mathematics,* by Anatole Beck, Michael N. Bleicher, and Donald W. Crowe, pp. 315–387, New York: Worth Publishers.

Introduces the ideas of tree game and natural outcome, with numerous examples and exercises.

Berlekamp, E. R., Conway, J. H., and Guy, R. K. (1982), *Winning Ways for Your Mathematical Plays,* Vol. 1: *Games in General;* Vol. 2: *Games in Particular,* New York: Academic Press.

This is the premier book about the mathematical theory behind recreational games.

Biggs, N., Lloyd, E., and Wilson, R. (1976), *Graph Theory 1736–1936,* Oxford: Oxford University Press.

Reprints in English translation many early and important papers in graph theory, including Euler's 1736 article on the Königsberg bridge problem.

Bland, R. G. and Shallcross, D. F. (1989); "Large traveling salesman problems arising from experiments in

X-ray crystallography: A preliminary report on computation," *Operations Research Letters* **8**:125–128.

Bouton, Charles L. (1901), "Nim, a game with a complete mathematical theory," *Annals of Mathematics* 2nd ser. 3(1) (Oct.):35–39.

This paper analyzed Nim with multiple rows and gave the game its name.

Cao, M. and Ferris, M. C. (1991), "Genetic algorithms in optimization," *The UMAP Journal* 12:81–90.

Expository article on a topic in the Spotlights.

Charlesworth, Arthur (1979), Infinite loops in computer programs. *Mathematics Magazine* 52 (1979) 284–291.

Shows why it is impossible to have a general program to detect if other programs are in an infinite loop.

Chavey, Darrah (1992), *Drawing Pictures with One Line: Exploring Graph Theory,* Lexington, MA: COMAP.

Concentrates on Eulerian paths and cycles, giving examples from folk art, several algorithms, and real-world problems.

Chavey, D. P., Campbell, P. J., and Straffin, P. D. (1994), "Professors' commentary: The politics of course time slots," *The UMAP Journal* 15:123–136.

Cipra, Barry (1992), "Theoretical computer scientists develop transparent proof technique," *SIAM News* **25**:3 (May): 1, 25.

Clarkson, Mark (1995), "Moody's evolving help desk," *Byte* (Feb.):76–80.

Connors, Joseph M. (1981), "National resident matching program," *New England Journal of Medicine* **305**:525.

Dantzig, G. B., Fulkerson, D. R., and Johnson, D. M. (1954), "Solution of a large-scale traveling-salesman problem," *Journal of the Operations Research Society of America* 2:393–410.

Davenport, H. (1992), *The Higher Arithmetic: An Intro-*duction to the Theory of Numbers. 6th ed., pp. 172–176. Cambridge, England: Cambridge University Press.

Discusses Rabin's primality test, with an extended example.

Degrazia, Joseph (1948), *Math Is Fun,* New York: Emerson Books, Inc. 1981. Reprinted under the title *Math Tricks, Brain Twisters, and Puzzles,* New York: Bell Publishing Company.

Dewdney, A. K. (1993), *The (New) Turing Omnibus.* New York: W. H. Freeman.

Chapter 50, "Detecting primes" (pp. 335–338), discusses Rabin's probabilistic algorithm for primality. Chapter 31, "Turing machines" (pp. 207–215), investigates a Turing machine that does multiplication. Chapter 39, "Noncomputable functions" (pp. 265–268), discusses the Busy Beaver problem. Chapter 51, "Universal Turing machines" (pp. 339–344), describes a universal Turing machine.

Dowling, William F. (1989), There are no safe virus tests. *American Mathematical Monthly* 96 (1989) 835–836.

———. (1990), Computer viruses: Diagonalization and fixed points. *Notices of the American Mathematical Society* 37 (1990) 858–861.

These articles by Dowling are the sources for the Spotlight on unsafe virus programs.

Dudley, Underwood (1987), *A Budget of Trisections.* New York: Springer-Verlag. 1994. Rev. ed., under the title *The Trisectors.* Washington, DC: Mathematical Association of America.

Except for special angles, it is impossible in general to trisect an angle with straightedge and compasses. The author explains Euclidean and non-Euclidean geometric constructions, relates personal experiences with people ("trisectors") who erroneously claim they can trisect angles, and offers a wide variety of their alleged trisections.

Ernst, Michael D. (1995), "Playing Konane mathematically: A combinatorial game-theoretic analysis," *The UMAP Journal* 16(2):95–118.

Discusses elementary principles of combinatorial game theory in the context of analyzing a traditional Hawaiian form of checkers that resembles peg solitaire.

Eves, Howard (1969), *An Introduction to the History of Mathematics,* third ed., New York: Holt, Rinehart, Winston.

Gale, D. and Shapley, L. S. (1962), "College admissions and the stability of marriage," *American Mathematical Monthly* 69:9–15.

Proves that the GS algorithm is optimal for the proposers.

Garey, M. R. and Johnson, D. S. (1979), *Computers and Intractability: A Guide to the Theory of NP-Completeness,* Freeman.

Garfinkel, Robert S. (1977), "Minimizing wallpaper waste, Part 1: A class of traveling salesman problems," *Operations Research* 25:741–751.

Gersting, Judith L. (1993). *Mathematical Structures for Computer Science,* third ed., New York: W. H. Freeman.

Gleason, Andrew (1966), *Nim and Other Oriented-Graph Games.* 16 mm B&W film, 63 min. Mathematical Association of America. Distributed by Ward's Modern Learning Aids Division (P.O. Box 1712, Rochester, NY 14603, or P.O. Box 1749, Monterey, CA 93940).

This film has not been reissued by the Mathematical Association of America on videotape. It may still be available from the university film rental libraries at Idaho State University, Pennsylvania State University, and others.

Graettinger, J. S. and Peranson, E. (1981), "The matching program." *New England Journal of Medicine* 304:1163–1165.

Grundy, P. M. (1939), "Mathematics and games," *Eureka* 2:6–8.

Gusfield, D. and Irving, R. W. (1989), *The Stable Marriage Problem,* Cambridge, MA: MIT Press.

Thorough treatment at an advanced level of many kinds of matching problems.

Guy, Richard K. (1989), *Fair Game: How to Play Impartial Combinatorial Games,* Arlington, MA: COMAP.

Contains short chapters on several dozen games, together with exercises and student projects.

Guy, Richard K., ed. (1991), *Combinatorial Games,* Providence, RI: American Mathematical Society.

Collection of articles, most surprisingly easy to read, that give the state of knowledge about combinatorial games, plus a substantial bibliography.

Hales, A. W. and Jewett, R. I. (1963), "Regularity and positional games," *Transactions of the American Mathematical Society* 106:222–229.

Although the result was known earlier, this paper formally proves that in a game in which it does not hurt to have an extra piece on the board (e.g., Tic-Tac-Toe), Second Player cannot have a winning strategy (just let First Player "steal" Second Player's strategy).

Hobbs, Arthur M. (1991), "Traveling salesman problem," In *Applications of Discrete Mathematics,* edited by John G. Michaels and Kenneth H. Rosen, pp. 263–287, New York: McGraw-Hill.

Very readable chapter in a book of applications of discrete mathematics, with further information on TSP.

Hodges, Andrew (1983), *Alan Turing: The Enigma,* New York: Simon and Schuster.

This is the main biography of Turing.

Holland, John H. (1992), "Genetic algorithms," *Scientific American* (July): 66–72.

Johnson, D. E., Hilburn, J. L., and Julich, P. M. (1979), *Digital Circuits and Microcomputers,* Englewood Cliffs, NJ: Prentice-Hall.

Johnson, D. S., A. Demers, J. D. Ullman, M. R. Garey, and R. L. Graham (1974), Worst case performance bounds for simple one-dimensional packing algorithms. *SIAM Journal on Computing* 3 (1974) 299–325.

Research article that summarizes the state of knowledge about bin-packing.

Jones, John Dewey (1994), "Orderly currencies," *American Mathematical Monthly* 101:36–38.

Julstrom, Bryant A. (1993), Noncomputability and the Busy Beaver problem. UMAP Modules in Undergraduate Mathematics and Its Applications: Module 728. *The UMAP Journal* 14 (1): 39–74. 1994. Updated 2nd ed. In *UMAP Modules: Tools for Teaching 1993,* edited by Paul J. Campbell, 31–67. Lexington, MA: COMAP, 1994.

Extended treatment of Turing machines, including exercises and a sample exam (with solutions). Proves that the Busy Beaver and shift functions are not computable, relates their noncomputability to the undecidability of the halting problem, and describes other noncomputable functions based on Turing machines.

Kosko, B. and Isaka, S. (1993), "Fuzzy logic," *Scientific American* (July).

Koza, John R. (1992), *Genetic Programming: On the Programming of Computers by Means of Natural Selection,* Cambridge, MA: MIT Press.

Kruse, Robert L (1994), *Data Structures and Program Analysis,* third ed., Englewood Cliffs, NJ: Prentice Hall.

Lawler, E. L., Lenstra, J. K., Rinnooy Kan, A. H. G., and Shmoys, D. B. (1985), *The Traveling Salesman Problem,* Chichester, U.K.: Wiley.

Research volume that summarizes known results on TSP and its history.

Loeb, Daniel E. (1995), "How to win at Nim," UMAP Modules in Undergraduate Mathematics and Its Applications, Module 746. *The UMAP Journal* 16(4):367–388. Reprinted in *UMAP Modules: Tools for Teaching 1995,* edited by Paul J. Campbell. Lexington, MA: COMAP, 1996.

Mairson, Harry (1992), "The stable marriage problem," *Brandeis Review* (Summer):38–42.

Malkevitch, J., Meyer, R., and Meyer, W. (1994), "Street Networks, Chapter 1 in *For All Practical Purposes: Introduction to Contemporary Mathematics,* 3rd ed., edited by Solomon Garfunkel, 5–32, New York: W. H. Freeman.

Considers Eulerian cycles and generalizations, including further applications.

Malkevitch, J. Meyer, R., and Meyer, W. (1994), "Visiting vertices," Chapter 2 in *For All Practical Purposes: Introduction to Contemporary Mathematics,* 3rd ed., edited by Solomon Garfunkel, pp. 33–74, New York: W. H. Freeman.

Considers Hamiltonian cycles and the traveling salesperson problem.

Explains Kruskal's algorithm for a minimum cost spanning tree.

Maurer, S. B. and Ralston, A. (1991), *Discrete Algorithmic Mathematics,* Reading, MA: Addison-Wesley.

McVitie, D. G. and Wilson, L. B. (1971), "The stable marriage problem," *Communications of the ACM* 14:486–492.

Meyer, Rochelle (1990), *ExploreSorts* (an integrated package of Macintosh software and user's manual), Consortium for Mathematics and Its Applications (COMAP), Boston.

Moore, J. Michael (1968), "An *n* job, one machine sequencing algorithm for minimizing the number of late jobs," *Management Science* 15:102–109.

This is the original source for the proof of Hodgsons's rule.

National Resident Matching Program (1991), *Handbook for Students Participating through U.S. Medical Schools,* Evanston, IL: National Resident Matching Program.

O'Beirne, T. H. (1965), *Puzzles and Paradoxes,* New York: Oxford University Press. 1984. Reprinted under the title *Puzzles and Paradoxes: Fascinating Excursions in Recreational Mathematics,* New York: Dover.

Chapter 8, "Gamesman, Spare that Tree" (pp. 130–150), and Chapter 9, " 'Nim' You Say, and 'Nim' It Is" (pp. 151–167), treat tree games, including Nim, Wythoff's game, Grundy's game, and Welter's game.

Padberg, M., and Rinaldi, G. (1987), "Optimization of a 532-city symmetric traveling salesman problem by branch-and-cut," *Operations Research Letters* 6:1–8.

Quisquate, J.-J. and Desmedt, Y. G. (1991), "Chinese lotto as an exhaustive code-breaking machine," *Computer* 24:11 (Nov.):14–20.

How to break a code, by using randomization and millions of processors.

Rabin, Michael O. (1980), Probabilistic algorithm for testing primality. *Journal of Number Theory* 12 (1980) 128–138.

Research article in which Rabin describes his algorithm, proves the theorem about the proportion of witnesses to nonprimality, and discusses implementing the algorithm for large *n*.

Ross, K. A. and Wright, C. R. B. (1988), *Discrete Mathematics,* second ed., Englewood Cliffs, NJ: Prentice-Hall.

Roth, Alvin E. (1984), "The evolution of the labor market for medical interns and residents: A case study in game theory," *Journal of Political Economy* 92:991–1016.

Roth, Alvin E. (1990), "New physicians: A natural experiment in market organization," *Science* 250 (14 Dec.):1524–1528.

Shier, Douglas R. (1982), Testing for homogeneity using minimum spanning trees," *The UMAP Journal* 3:273–283.

Skiena, Steven (1990), *Implementing Discrete Mathematics:* *Combinatorics and Graph Theory with Mathematica,* Reading, MA: Addison-Wesley.

Documents the Mathematica add-on package Combinatorica.

Sprague, R. (1936), "Über mathematische Kampfspiele." *Tohoku Mathematical Journal* 41:438–444.

Tanenbaum, Andrew S. (1989), *Structured Computer Organization,* third ed., Englewood Cliffs, NJ: Prentice-Hall. The material on circuits and the construction of a microarchetecture show how one can put together circuits that are not much more complex than those in this chapter and that create a programmable computer.

Vajda, Steven (1992), *Mathematical Games and How to Play Them,* Chichester, England: Ellis Horwood.

Chapter 1 (pp. 1–20) covers solitaires and Chapter 2 (pp. 21–85) treats two-player tree games, including subtraction games like Nim.

Watson, Andrew (1991), "Graph theory shows how to dig a better mine," *New Scientist* 129 (16 March):21.

Whitemore, Hugh (1987), *Breaking the Code.*

Williams, K. J., Werth, V. P. and Wolff, J. A. (1981), "An analysis of the resident match." *New England Journal of Medicine* 304:1165–1166.

Wolfe, David (1994), Gamesman's Toolkit. Computer program in C with source. Extended by Michael Ernst to handle Ancient Konane and Modern Konane. Available by sending email to wolfe@cs.berkeley.edu or to mernst@lcs.mit.edu.

Analyzes values of positions for the games Domineering, Toads and Frogs, and Ancient and Modern Konane.

Wolfe, David (1995), "Undergraduate research opportunities in combinatorial games." *The UMAP Journal* 16(1):71–77.

A gentle introduction to the ideas of combinatorial game theory, using Domineering as a sample game. Suggests where to find other games to explore.

Woolsey, R. E. D. (1992), "The fifth column: Survival scheduling with Hodgson's rule, or see how those salesmen love one another," *Interfaces* 22:5 (Sept.–Oct.):81–84.

Describes a highly amusing true instance of an application of Hodgson's rule.

Woolsey, R. E. D. and Swanson, H. S. (1975), *Operations Research for Immediate Application: A Quick and Dirty Manual,* New York: Harper & Row.

Gives Hodgson's rule and an example, together with many other algorithms important to management science and operations research.

INDEX